Modern Physics

Modern Physics intertwines active learning pedagogy with the material typically covered in an introductory survey, from the basics of relativity and quantum mechanics through recent developments in particle physics and cosmology. The flexible approach taken by the authors allows instructors to easily incorporate as much or as little active learning into their teaching as they choose. Chapters are enhanced by "Discovery" and "Active Reading" exercises to guide students through key ideas before or during class, while "ConcepTests" check student understanding and stimulate classroom discussions. Each chapter also includes extensive assessment material, with a range of basic comprehension questions, drill and practice calculations, computer-based problems, and explorations of advanced applications. A test bank and interactive animations as well as other support for instructors and students are available online. Students are engaged by an accessible and lively writing style, thorough explanations, "Math Interludes" which account for varying levels of skill and experience, and advanced topics to further pique their interest in physics.

Gary N. Felder is Professor of Physics at Smith College, Massachusetts. His research has focused on inflationary cosmology, numerical simulations, and the development of computer and written tools for teaching physics concepts. In 2017 he received an SGA Faculty Teaching Award from Smith's Student Government Association in recognition of his excellent teaching and positive impact on the student body. His paper "The Effects of Personality Type on Engineering Student Performance and Attitudes" received the American Society for Engineering Education 2003 "Best Paper of the Year" Award.

Kenny M. Felder has taught physics and mathematics at Raleigh Charter High School, North Carolina, for more than twenty years. He received his Bachelor's degree in Physics with highest honors from the University of North Carolina at Chapel Hill. His book *Advanced Algebra II* pioneered much of the active learning pedagogy included in this textbook and was described by the Connexions project at Rice University as "a non-traditional approach to a very traditional subject" that makes "the world of second-year algebra come alive."

Gary Felder and Kenny Felder also co-authored *Mathematical Methods in Engineering and Physics* (2016).

Modern Physics

Gary N. Felder

Smith College, Massachusetts

Kenny M. Felder

Raleigh Charter High School, North Carolina

CAMBRIDGE
UNIVERSITY PRESS

University Printing House, Cambridge CB2 8BS, United Kingdom

One Liberty Plaza, 20th Floor, New York, NY 10006, USA

477 Williamstown Road, Port Melbourne, VIC 3207, Australia

314–321, 3rd Floor, Plot 3, Splendor Forum, Jasola District Centre, New Delhi – 110025, India

103 Penang Road, #05–06/07, Visioncrest Commercial, Singapore 238467

Cambridge University Press is part of the University of Cambridge.

It furthers the University's mission by disseminating knowledge in the pursuit of education, learning, and research at the highest international levels of excellence.

www.cambridge.org
Information on this title: www.cambridge.org/highereducation/isbn/9781108842891
DOI: 10.1017/9781108913270

First published 2023

Printed in the United Kingdom by TJ Books Limited, Padstow, Cornwall, 2023

A catalogue record for this publication is available from the British Library.

ISBN 978-1-108-84289-1 Hardback

Additional resources for this publication at www.cambridge.org/felder-modernphysics

Contents

Preface

Notes to the Instructor

This textbook provides an introductory course in modern physics. Students are assumed to have already taken introductory courses in classical mechanics and electromagnetism, so this course might be the third in the physics sequence. Mathematically, the minimum prerequisite is a semester of calculus including both derivatives and integrals; more advanced mathematical topics are introduced or reviewed in the book as needed.

There are a good number of textbooks for such courses already on the market. So why did we feel the need to write a new one?

- We wanted to highlight the conceptual breakthroughs represented by relativity and quantum mechanics: how these breakthroughs build on classical physics, how they radically depart from our day-to-day physical intuition, and how the equations reflect those radical changes in physical paradigms. We felt that the existing books on the market did not do this well. Many existing texts seem to be written for professors more than for students, and can leave students solving equations with no clear idea of what those equations are telling them.

- We wanted questions that make sure students understand the material at a basic level, and also questions that push them into deeper exploration of the subtleties. We wanted lots of problems that provide drill and practice on core mathematical skills, and also problems that ensure students understand and can replicate key derivations. Existing texts generally don't provide enough in any of these categories.

- We wanted to review (or in some cases introduce) math topics that students might not have already mastered – topics such as standing and traveling waves, complex exponential functions, discrete and continuous probability distributions, partial derivatives, spherical coordinates, and of course partial differential equations – as they are needed for quantum mechanics.

- We wanted to make use of digital functionality to offer a suite of online materials to support both instructors and students. Later in this Preface we detail the resources available online, but just in case you don't read that far, we want to grab your attention now and urge you to browse and see what is available at www.cambridge.org/felder-modernphysics. We also recommend telling your students to bookmark this site at the beginning of the course, as they may need to refer back to it often.

- Most importantly, we wanted the kinds of tools that are available in many introductory mechanics and electricity and magnetism physics books, but are generally not available for higher-level courses. If you don't use active learning techniques in the classroom, we hope you will still see in our book all the advantages we list above. But if you have used active learning or are interested in trying it, you will find tools in this book that make it easy to experiment without sacrificing content. Much of this Preface is devoted to explaining how we envision those tools being used (and how we use them ourselves).

For both of us, at different times and in different schools, our introductions to modern physics came as life-changing revelations. Fast-moving objects do *what?* Individual photons do *what?* You're saying this stuff *really happens?*

We want to share that stunned feeling with students. We want them to gather around a conference table with Einstein, Bohr, and Schrödinger, and try to craft a coherent theoretical model to explain the startling experimental evidence. We want them to become comfortable with the mathematical processes involved, and simultaneously to see how the underlying physics defies their intuition.

One of our working titles for this book was "Childlike Wonder and Differential Equations." We want students to see all that in their introduction to special relativity and quantum mechanics. In the later chapters we want them to see how those theories helped answer some of the most daunting scientific problems of the twentieth and twenty-first centuries. And we also want to share with them the questions that are still open in 2022 – questions that they themselves may help to answer some day.

Topics and Organization

Broadly speaking, this book comprises three parts: special relativity (Chapters 1 and 2), quantum mechanics (Chapters 3–6), and advanced topics (Chapters 7–14). You could cover the entire book in two semesters. But in most institutions this course is only given one semester, forcing you to pick which sections to cover.

If your course does not include relativity, skip Chapters 1 and 2. If you do cover relativity, you could stop at any point from Section 1.2 (for a one- to two-day introduction) through the end of Chapter 2 (for two to three weeks). Relativity comes up occasionally in later chapters, most often in problems involving the formula $E^2 = p^2c^2 + m^2c^4$, so if you don't get that far in relativity you should avoid assigning those problems.

The first three sections of Chapter 3 – the mathematics of interference, the classical Young double-slit experiment, and then the same experiment repeated one photon at a time – are our introduction to quantum mechanics. Students should be given the chance to convince themselves that many counterintuitive aspects of the theory are *required* in order to explain these experimental results. The rest of the chapter presents blackbody radiation, the photoelectric effect, Compton scattering, and other results that led early twentieth-century physicists to propose and accept the quantum hypothesis; you can cover as many or as few of these as you like, but they are not essential for later chapters.

Chapters 4, 5, and 6 give the theory and the mathematics of quantum mechanics, at this introductory level. This material forms the largest part of a typical modern physics course. There are parts that can be omitted without loss of continuity, however.

- You could skip Section 6.3 (Momentum Eigenstates) and/or Section 6.4 (Phase Velocity and Group Velocity), although you should cover the former if you plan to do the latter.
- Section 6.6 introduces the time-dependent Schrödinger equation, and uses it to derive the rule for the time evolution of energy eigenstates that was introduced earlier. This section is more mathematically advanced than most, and can be omitted.

You could skip Sections 4.1 (Atomic Spectra and the Bohr Model) and 4.2 (Matter Waves), but we believe they provide important motivation for Schrödinger's wave mechanics and we recommend against skipping them. Section 5.1 is a review of concepts about force and potential energy that your students should have learned in introductory physics. This can be a quick review, but unless you are very confident that your students have mastered this information, you skip this section at your peril (or more to the point at theirs).

After Chapter 6, the book showcases applications of quantum mechanics. These later chapters are mostly independent of each other. Occasionally one section depends on an earlier section in another chapter, and such dependencies are noted at the beginning of the appropriate section. The only major inter-chapter dependency is that Chapter 9 (Molecules) depends on Chapter 8 (Atoms), which in turn depends on Chapter 7 (The Hydrogen Atom).

It might be fun to pick one or two of those chapters and explore them in depth. (Gary recommends Chapter 14, Cosmology!) On the other hand, you may want to give your students a taste of many different topics. To facilitate that latter option, we've organized many of these chapters so that the first section or two can be presented as standalone overviews. Consider doing only the first section of Chapter 7 (The Hydrogen Atom), the first two sections of Chapter 10 (Statistical Mechanics), the first two sections of Chapter 11 (Solids), the first section or two of Chapter 12 (The Atomic Nucleus), the first two sections of Chapter 13 (Particle Physics), and/or the first section or two of Chapter 14 (Cosmology). (For Chapter 7, you could also do the introduction in the first section and then jump to the discussion of spin in Sections 7.5 and 7.6.)

Active Learning

As we observed above, we believe that our book offers many advantages over the existing textbooks on this subject. If you have no interest in active learning, we hope that you will judge for yourself whether our Explanation sections are clearer, and our Questions and Problems more thorough and probing, than other texts you are considering.

But we are unabashed proponents of active learning, because all the research says that it works: students learn better, understand in greater depth, and retain longer when active learning techniques are applied properly.[1]

---

1 See, for example, Freeman, S., Eddy, S. L., McDonough, M., et al., Active learning increases student performance in science, engineering, and mathematics, *Proceedings of the National Academy of Sciences of the United States of America*, **111**(23), 8410–15, 2014.

At the same time, we are very aware of the challenges. How do you incorporate active, challenging moments into your already packed class time without sacrificing content? And even if you want to, how do you come up with the right activities to get students constructively engaged? This book makes it easy to experiment with active learning in small ways that take very little extra time, in or out of class. If you find that these experiments are genuinely fast, easy, and (most importantly) effective in increasing student understanding and retention, you may feel emboldened to try a few more.

So here is a guide to the elements of our book that are designed for that purpose.

Discovery Exercises

A "Discovery Exercise" is a self-guided tutorial meant to take 5–20 minutes *before* a lecture. It can be assigned as homework due on the day of the lecture, done in class at the start of the topic, or done in class at the end of the previous topic to prime them for the next class. The students work through some of the math and physics of a topic on their own. If they are able to solve it all, they start the lecture well prepared. If they get stuck, they come into the lecture with a clearer sense of what the topic is, and of exactly what they are confused about, and they are ready to learn.

We should emphasize that we are not urging you to use all the exercises in this book. We ourselves don't use all of them when we teach! The book gives you a lot of exercises so that you can pick and choose.

Active Reading Exercises

The following box appears in our section on scattering. We've already shown students how to solve for the energy eigenstates of a particle incident on a barrier whose height is greater than the particle's energy. Now we consider a barrier whose energy is less than the particle's energy. Having seen the earlier case, they should easily be able to write down the solutions mathematically, but we want to lead them to understand that in this case those solutions show us that some part of the wave will be transmitted.

> ### Active Reading Exercise: A Potential Step with $E > U_0$
>
> 1. Write the energy eigenstate at $x \geq 0$ for a particle with $E > U_0$.
> 2. Your answer to Part 1 includes two pieces. What does each of the two pieces represent about our particle?
> 3. Neither of these two pieces blows up as $x \to \infty$, but one of them has to equal zero anyway. Can you guess which one, on physical grounds?

For motivated students reading the text before class, pausing to try these exercises before moving on will greatly improve their understanding and retention of the material. But many students will not do that (if they read the book at all), and their first introduction to the material will be in your lectures.

So we would love you to try, just with one or two of these, pausing your lecture at that point. Put a slide up with those questions. (We provide all the slides online.) Give the students one minute to think about these questions on their own, and then another minute to compare and discuss answers with their neighbors. Research shows that one or two such interruptions, lasting no more than 1–2 minutes each, vastly improve student retention of a 50-minute lecture.

What then? In some cases (such as the one above), we answer the Active Reading Exercise immediately in the text. In such cases the answer is vital to the next steps in the lecture. So you might pick up your lecture by polling the students about their responses and then discussing the right answer. In other cases, the answer to the Active Reading Exercise is online; you can choose to discuss it in class, or tell students that they can find it later.

Quick Checks and ConcepTests

Among the online resources that accompany the book is a bank of multiple-choice questions designed for use in class. One common active learning technique is to ask students to vote on such a question with clicker systems, with smartphone apps, or just by holding up pieces of paper. If almost all students get the answer right, you simply note the correct answer and move on. If many of them don't, you can ask them to discuss it in pairs and then come back and talk about it.

These questions are divided into two categories.

- A "Quick Check" is a simple factual question. ("Which of the following defines a reference frame as 'inertial'?") Quick Checks are intended to quickly make sure the class is following the basic material. They are only offered online, not in the printed book.
- A "ConcepTest," a term coined by Harvard physicist Eric Mazur, is a subtler test of understanding. A ConcepTest will often garner as many wrong answers as correct ones. But if you give students a minute to answer on their own, and then another minute to confer with their neighbors, scores should be significantly higher after the second minute – and a lot of learning takes place in those two minutes. ConcepTests can also be used as parts of homework sets. All of our ConcepTests are offered online, but some are also included in the book as "Conceptual Questions" (described below).

As with the Exercises, we provide a large number of multiple-choice questions so that you can pick and choose, not to urge you to use all of them!

Other Key Features of the Book

In addition to the Exercises and Questions designed to explicitly support active learning, our book includes several other features that can help students learn and retain the material.

Math Interludes

How much math have the students in a modern physics class been exposed to? How much have they comfortably mastered? The answers vary tremendously from school to school. It is likely that most students at this level are comfortable with basic trigonometry and introductory calculus. They are probably less familiar with multivariate functions (a traveling wave $y = A \sin(kx - \omega t)$, for instance), even if they have seen them. They may not have seen complex numbers since high school. They may not have worked with partial derivatives, spherical coordinates, or partial differential equations. If you assume that students have mastered all these topics, or will pick them up as they go, the result may be frustration and superficial understanding. So we devote sections to these topics at the points where they are needed for the continued development of the physics.

For math topics that are not prerequisite to your course, you might want to spend class time covering these sections. For math topics that are prerequisite to your course, consider assigning a few homework problems from these sections to alert students to topics they might need to review.

Questions

After the Explanation in each section you'll find an extensive list of Questions and Problems.

Our questions are divided into "Conceptual Questions and ConcepTests" and "For Deeper Discussion." Unlike Problems, these questions require little or no calculation. A student who understands the topic presented in a section should be able to answer most of the Conceptual Questions, perhaps after some work and discussion, so these questions are appropriate for regular homework assignments and/or in-class discussions. But that same student may still struggle with the For Deeper Discussion questions. This struggle can be valuable, especially if it leads to discussion and debate! These questions make great extra credit problems, topics for longer class discussions, or extra challenges for highly motivated students. We recommend bracketing these with a caveat like "I don't necessarily expect you to get the right answer, but I do expect you to show me some real thought."

Problems

The last part of each section is Problems, which in many ways look like the problems you would see in any other textbook. The first difference you may notice is that we have many more of them. (In our experience, professors almost always want more problems than they have been given.) The problems marked as "Explorations" push deeper into real-world applications or advanced concepts, and might be appropriate for extra credit or as group projects.

Many derivations that are traditionally done in Explanations are done entirely or in part in our Problems. Our goal is to provide enough scaffolding to guide the students successfully through the derivations. For the derivations you really want your students to learn, we strongly believe the students will understand and retain them better if they do them (or at least attempt them) on their own before they see them from you.

When a problem explores a particular topic such as single-slit diffraction or pair annihilation, we mark it with a boldface problem title.

Because departments differ on whether they require multivariate calculus as a prerequisite for modern physics, problems that require multiple integrals are marked so that professors can skip those if their students haven't learned that skill yet.

Some problems are also marked as dependent on previous problems. If in some section you see "15. [*This problem depends on Problem 14.*]" that means you can assign just Problem 14 or both Problems 14 and 15, but you shouldn't assign Problem 15 without Problem 14.

Computer Problems

Using computers to solve problems is a crucial skill for physics students, but over-reliance on computers can prevent students from learning crucial skills. Our goal is not to dictate the right balance, but to support you wherever you find it. Scattered throughout our problem sections are problems marked with an icon (⌨) indicating that they require use of a computer. (We don't mark problems that simply require a graphing calculator.) You could have the students do these problems with Wolfram Alpha®, Wolfram Mathematica, Matlab®, Excel, or any other program. If you don't want computation in your course, don't assign these problems.

Online Resources

The online resources at www.cambridge.org/felder-modernphysics that accompany our book are designed to make teaching this course as easy as possible, and taking the course as productive as possible.

Some of these resources involve the active learning tools discussed above.

Discovery Exercises

All Discovery Exercises are in the printed book, so you can assign them by saying "Do Exercise 1.1." But they are also available in printable form on the website, including blank space for student work, so you can hand them to students to work on in class.

Active Reading Exercises

The Active Reading Exercises are also available in two forms: in the printed book, and on the website as slides that you can easily flash up in the middle of class.

Solutions to *Some* Active Reading Exercises

If the solution to an Active Reading Exercise is essential in order to continue to the lecture, the solution is provided in the textbook immediately under the Exercise. The assumption is that you will present the Exercise, discuss it, and then make sure the solution is clear before proceeding further. So the solutions that are online, where students can refer to them later, are the ones that are *not* necessary for the rest of the lecture.

Quick Checks and ConcepTests

All our multiple-choice questions are available on PowerPoint and PDF slides, formatted for easy use as in-class polls. Some, but not all, of the ConcepTests are also included with the Conceptual Questions in the printed book.

We have also created other resources that will support more traditional courses which don't use active learning tools.

Homework Sets

We have emphasized that our textbook gives professors more problems, and a greater variety of problems, than others. But that strength can become a weakness, if putting together homework assignments becomes too time-consuming. For each section we've selected a small sample of Questions and Problems that cover the key ideas from the section. If you assign daily homeworks, you can simply use that set. If you assign weekly homeworks, you can paste together the sets from whichever sections you've covered.

Animations

At a number of points in the book we have a link to an online animation, usually interactive, that illustrates an idea better than a static picture can. For example, the book shows images of a wave packet approaching a barrier and being partially reflected and partially transmitted. The online animation allows the student to watch the wave packet hit the barrier, penetrate in, and then create reflected and transmitted waves. Then the student can adjust the barrier height and rerun to see what changes. These animations can be shown in class, used as hands-on activities, or assigned as part of the reading.

Supplemental Sections

The book also includes links to several sections on topics that supplement the coverage in the book. These topics include, for example, antimatter, interpretations of quantum mechanics, and open questions in cosmology. These can be assigned as extra credit, or simply made available to interested students.

One Final Thought

We leave you with a few words about the amazing enterprise that is science.

> *We are physical machinery – puppets that strut and talk and laugh and die as the hand of time pulls the strings beneath. But there is one elementary inescapable answer. We are that which asks the question.*
>
> – Sir Arthur Stanley Eddington

Acknowledgments and Figure Credits

On the front cover of this book you see two names, Gary N. Felder and Kenny M. Felder. But this book would not have been possible without the contributions of a lot more people.

Vince Higgs and Melissa Shivers, our editors at Cambridge University Press, have walked this path with us from the beginning. Both of them immediately saw that we were trying to do something different from most other Cambridge textbooks (or *"different to* other Cambridge textbooks,"* as they said with the English accents that never lost their charm to our American ears). Our unusual approach didn't scare them; they were supportive of it, even excited by it, and they worked with us every step of the way to make our initially vague vision into a tangible reality. They got our book to dozens of early reviewers, whose feedback made every part of the book better. They pointed out aspects of the book that were non-standard in ways we hadn't intended, and gave us advice on how to fix them. (We're still not sure how Melissa found every instance of "stacked heads" without reading the entire book cover to cover. But they're all gone now.) Melissa and Vince, we sincerely want you to know that the two of us had conversations behind your backs about how thoroughly supported we always felt by both of you, and how much we appreciated it.

Cambridge assigned two different people, at two very different stages of the book's development, to copy-edit and proofread the entire text. First John King, and later Susan Parkinson, read through every Explanation, every Exercise, every Question and every Problem. And in both cases, we were stunned by the quality of their feedback. Some of it was detail-oriented ("this diagram is subtly wrong"). Some of it required content expertise ("that isn't actually a magnetic effect"). They caught mistakes that we had missed after dozens of re-readings, and they suggested improvements that we would never have thought of. It was clear from beginning to end that they were not just doing what was required; they cared about making this book the best it could possibly be. At one point we commented to each other that whatever Cambridge pays these people, it isn't enough. Toward the end of the project, much of the leadership of the process came from Rachel Norridge at Cambridge. Rachel continually made sure we understood where things stood, and why, and did her best to accommodate our (sometimes unorthodox) requests about how to handle the proofing process. She made smooth a process that could otherwise have felt very rocky.

We also received invaluable advice from early readers that we found on our own. Listing some of the most important (in alphabetical order): Richard Felder, Jessica Jiang, Chitose Maruko, Travis Norsen, Hyo Jung (Catherine) Park, and Barbara Soloman all read many first drafts and pointed out places where we were unclear, places where we needed fewer words and more pictures, new ways in which we could explain things, and just plain mistakes. Every part of the book has been touched by their expertise and care.

Jessica Jiang and Chitose Maruko (yes, the same ones from the previous paragraph) – both Gary's students at Smith College, by the way – are responsible for all of the wonderful animations on our website. We would say something like "It would be nice to have something that illustrates different eigenfunctions of the infinite square well combining," and one of them would say "OK, I'll do that." (Or surprisingly often, "Oh yeah, I wrote something like that last week.") They designed them, coded them, debugged them, and gave them to us fully ready to post. Jessica and Chitose, you are both amazing.

Hugh Churchill, Jerome Fung, and Will Williams taught entire semester-long modern physics courses from our book, occasionally getting PDFs the day before they needed them. The feedback from them and their students improved the book in many ways. Using an unfinished book involves hassles and logistical problems for both teachers and students, and we appreciate the faith they showed in us and the help they gave us.

This book benefited from discussions with subject experts on many topics. We particularly want to acknowledge Andrew Berke for advising us on atomic and molecular chemistry, Nat Fortune for helping us understand crystal structure, and Doreen Weinberger for making our optics discussion clearer.

Our highest and fondest gratitude goes always to our wives, Rosemary McNaughton and Joyce Felder. It goes without saying that they put up with a lot, and put in a lot of extra work, for years. (The same is true of our children: Mary, Benjamin, Jack, Shannon, James, and Cecelia Felder.) But Rosemary and Joyce are also both expert teachers who listened and gave us feedback throughout the process. The book you are holding would be a much inferior product without their contributions.

Figure Credits

4.0 (https://creativecommons.org/licenses/by/4.0)]. Access for free at https://openstax.org/books/university-physics-volume-3/pages/1-introduction; **Figure 12.5** OpenStax [CC BY 4.0 (https://creativecommons.org/licenses/by/4.0)]. Access for free at https://openstax.org/books/chemistry-2e/pages/1-introduction; **Figure 12.7** This figure is a part of the Nobel Prize lecture of Robert Hofstadter. Copyright © The Nobel Foundation 1961; **Figure 12.9** J.J. Thomson, 1913. Public Domain; **Figure 13.6** Reprinted with permission from: Carl D. Anderson, Physical Review, 43, 491-494, 1933. Copyright 1933 by the American Physical Society; **Figure 13.7** © 1970-2022 CERN (License: CC-BY-4.0); **Figure 13.8** © The Regents of the University of California, Lawrence Berkeley National Laboratory; **Figure 14.3** Henrietta S. Leavitt and Edward C. Pickering, 1912. Public Domain; **Figure 14.10** Mario De Leo [CC BY-SA 4.0 (https://creativecommons.org/licenses/by-sa/4.0)]; **Figure 14.15** © ESA and the Planck Collaboration.

1

Relativity I: Time, Space, and Motion

Einstein's theories have become part of popular culture. The fact that time passes differently for different observers ("time dilation") is a staple of science fiction, from *Planet of the Apes* (1968) to *Interstellar* (2014). If someone can only name one physics equation, it's probably $E = mc^2$. You have probably heard people use the word "spacetime," assure you that "science tells us everything is relative," and mention that Einstein's work somehow helped Oppenheimer invent the atomic bomb.

But how many of these people actually understand what the theory says?

As so often happens, the real physics is more exciting than the popular version. More subtle. More infuriating. More *weird*. And it is this last characteristic, perhaps – relativity's stubborn refusal to conform to our physical intuition – that marks the definitive break between "classical" and "modern" physics.

1.1 Galilean Relativity

Galilean relativity (or "Galilean invariance") is one of the cornerstones of Newtonian physics. If your mechanics professor ever said "Let's look at this problem from another reference frame" then you were learning about Galilean relativity, even if you never heard the name.

Here is how Galileo himself expressed this principle in his *Dialogue Concerning the Two Chief World Systems* (1632):

> *Motion exists relatively to things that lack it; and among things which all share equally in any motion, it is as if it did not exist.*

Galileo explained this principle with a thought experiment:

> *The goods with which a ship is laden leaving Venice pass by Corfu, by Crete, by Cyprus and go to Aleppo. Venice, Corfu, Crete, etc. stand still and do not move with the ship; but as to the sacks, boxes, and bundles with which the boat is laden and with respect to the ship itself, the motion from Verflice to Syria is as nothing, and in no way alters their relation among themselves. This is so because it is common to all of them and all share equally in it. If, from the cargo in the ship, a sack were shifted from a chest one single inch, this alone would be more of a movement for it than the two-thousand-mile journey made by all of them together.*

> *It is obvious, then, that motion which is common to many moving things is idle and inconsequential to the relation of these movables among themselves, and that it is operative only in the relation that they have with other bodies lacking that motion.*

Einstein did not overthrow or revise this principle. On the contrary, he enthusiastically reaffirmed it. A review of this classical idea will give you an important first step toward Einstein's modern theory.

1.1.1   Discovery Exercise: Galilean Relativity

Spaceman Spiff is hurtling through the solar system at 3 kajillion miles per hour. His enemy on a nearby planet shoots a deadly missile straight toward him at 4 kajillion miles per hour. Spiff's rocket can withstand an impact of 4.5 kajillion mph without harm; anything above that will puncture the hull.

1. Suppose the enemy is ahead of Spiff, so the missile and the rocket crash head-on. Does the missile penetrate the hull?

   See Check Yourself #1 at www.cambridge.org/felder-modernphysics/checkyourself

2. Suppose the enemy is behind Spiff, so the missile catches up and rear-ends the rocket. Does it penetrate the hull?

3. Suppose the missile is coming in from the side, and hits perpendicular to the rocket's direction of travel. Does it penetrate the hull?

1.1.2   Explanation: Galilean Relativity

Imagine a small girl in the back of a blue car, heading straight down the road at 60 mph. Her father is behind the steering wheel in the front. Outside her left window, a green car going 65 mph passes her car. Outside the right window, a row of trees lines the side of the road. Across the highway she sees cars going at 60 mph in the opposite direction (see Figure 1.1).

What does all this look like from the girl's perspective? She sees her own car, and her father, perfectly at rest. The green car drifts by slowly – at only 5 mph, it takes a while to pass her. The trees are rushing backward at 60 mph, and the cars across the highway are zooming by at a breakneck 120 mph.

You could have figured all that out yourself. But you may be tempted to think about it this way:

> *The girl sees all those incorrect velocities because she's moving. If she knew she was moving forward at 60 mph, she could use that information to calculate the true velocities of all those objects.*

Our goal is to talk you out of the preceding paragraph and into the following one instead:

> *The girl's perspective is perfectly correct. Her description of events is just as valid as the first description we gave (from the perspective of the road). "The green car is*

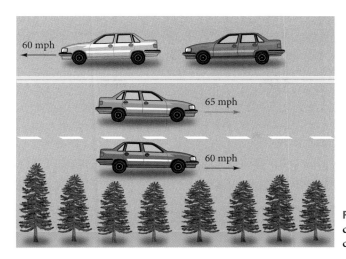

Figure 1.1 A blue car sees a green car passing it and red and yellow cars moving the other way.

going 65 mph from the road perspective" and "The green car is going 5 mph from the girl's perspective" are two equivalent ways of expressing the same fact, and neither one is more correct than the other.

Are you skeptical? Do you find yourself believing that, in some absolute sense, the road is "really" still and the car is "really" moving? Then consider this: the road is actually spinning around the Earth's center at 1000 mph. And the Earth itself is hurtling around the Sun at over 60,000 mph. How far do you back up to find that "really still" point? The answer is, don't try. Choose your non-moving perspective based on whatever makes the problem you're working easiest.

Example: Change of Perspective

An ocean liner is moving to the right (across the page) at 10 m/s. Elijah is standing in the back corner of a 10 m by 10 m cabin and throws a ball at 5 m/s directly toward the opposite corner. Does the ball hit the front wall, the side wall, or the corner?

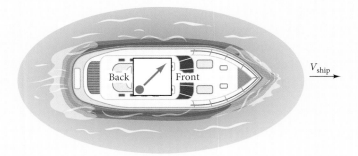

Answer: This is a standard vectors problem, right? The ball's velocity vector is the sum of the ship's velocity vector (10 m/s to the right) and the ball's thrown vector (5 m/s at 45°). Then you take into account the fact that the ship (and therefore the walls) are also moving to the right to see where the walls are when the ball hits . . .

That's how many beginning students would approach this problem. It's all perfectly correct, but it's also a lot of work. The more experienced student would work in the reference frame of the ship rather than the ocean. The cabin isn't moving. The ball moves in a straight line toward the opposite corner, so that's where it hits.

Formal Statement of the Principle

When Galileo first put forth his principle of relativity (p. 1), he took the viewpoint of the cargo in a ship on a 2000-mile voyage from Venice to Aleppo. Let's update his example.

Active Reading Exercise: Galileo's Spaceship

Imagine yourself on a spaceship far outside our solar system. It's a roomy ship with all the comforts of home, but it has no windows and no communication of any kind with the outside world. You hold your hand out in the center of the room with a ball in it – not on a table or the floor, just in mid-air – and then you open your hand and let go of the ball.

1. What does the ball do if the ship suddenly speeds up?
2. What does the ball do if the ship suddenly slows down?
3. What does the ball do if the ship makes a hard right turn?
4. What does the ball do if the ship stays the course, moving incredibly fast without change?

Jot down all four answers before you continue reading!

The answers to the first three questions are "It starts moving backward," "It starts moving forward," and "It starts moving left," respectively. But the answer to the fourth question is "nothing." An external observer would say that the ball is hurtling through space with the exact same velocity as the ship. From inside the ship, it looks to you like a non-moving ball.

So if you see the ball hovering, you can't tell if the ship is at rest or moving. If you really want to know which is happening, you might start running experiments. Throw the ball around the room and measure how it moves through the air and bounces off the walls. Bounce beams of light off mirrors and measure their reflection angles. It is all to no avail; you still have no idea if the ship is moving fast or slowly, forward or backward, or not moving at all.

We can summarize all that in one 12-word sentence.

The Principle of Galilean Relativity

The laws of physics are the same in any inertial reference frame.

(The word "inertial" in that sentence means "non-accelerating.")

In an *accelerating* reference frame, the laws of physics are quite different. Billiard balls start flying through the air of their own accord. Enough acceleration (for instance, about 10g, or 10 times

Earth-normal gravity, for over a minute) will kill you. Observers in all reference frames will agree that you are dead.

But the world inside a not-moving-at-all spaceship, and the world inside a zooming-steadily-at-half-a-billion-miles-per-hour spaceship, are completely indistinguishable. That's why you are perfectly within your rights to assert that the spaceship is *not* moving – even if you must then conclude that nearby stars are rushing by, backward.

To put it another way, acceleration is an intrinsic property that every object in the universe has at any given moment. But the velocity of an object can only be meaningfully defined in relation to a particular reference frame; other reference frames will record different velocities for the same object, and all are equally correct.

The Math of Galilean Relativity

You are standing by the side of the road as the little girl's blue car zooms by at speed u. You and the little girl are both measuring the positions of cars, trees, and other objects.

We will use primes to indicate measurements in the girl's reference frame. For instance, if x designates the position of a particular tree, both of the following might be true at once (Figure 1.2):

- $x = 30\,\text{m}$ (You see the tree 30 m ahead of you.)
- $x' = -10\,\text{m}$ (The girl sees the same tree 10 m behind her.)

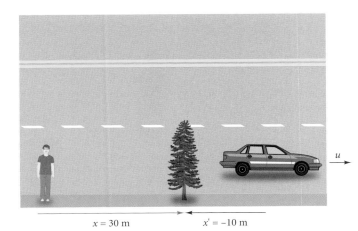

$x = 30\,\text{m}$ $x' = -10\,\text{m}$

Figure 1.2 A tree is 30 m ahead of you, but 10 m behind the girl in the car.

These primes do *not* indicate derivatives; throughout this chapter and Chapter 2, primed variables will designate the perspective of an alternative reference frame.

At the moment the girl's car passes you, you both set your stopwatches to zero. Hence, that particular place and time is $t = 0$ and $x = 0$ by your measurements. It is also $t' = 0$ and $x' = 0$ by hers.

Keep in mind that for you, $x = 0$ will always mean the spot where you are standing. The girl's x-value increases over time ($x_{\text{girl}} = ut$). For her, on the other hand, $x' = 0$ will always mean the spot where she is sitting. The x'-value she measures for your position is steadily receding ($x'_{\text{you}} = -ut'$).

More generally, here are the conversions from your system to hers; the variable v represents the velocity of any arbitrary object as measured by you, and v' the velocity of the same object as measured by the girl; accelerations are similarly a and a':

$$\left.\begin{aligned} t' &= t \\ a' &= a \\ v' &= v - u \\ x' &= x - ut \end{aligned}\right\} \tag{1.1}$$

For instance, in our earlier example ($u = 60$ mph), the green car zoomed by at $v = 65$ mph. The girl saw it creeping ahead at $v' = v - u = 5$ mph.

In 3D we might add two more equations to that list: $y' = y$ and $z' = z$. Because your relative motion is along the x axis, you will both agree entirely about the y and z coordinates. This important principle will remain unchanged as we graduate from Galileo to Einstein.

All those equations demonstrate an important property of classical physics: they may look confusing at first, but if you are willing to put in the time you will eventually conclude that they make perfect sense. Enjoy that feeling while you can, because in the next section we're going to undermine it.

1.1.3 Questions and Problems: Galilean Relativity

Conceptual Questions and ConcepTests

1. You are standing at rest and Mary is moving by you in the positive y direction. Ben is moving nearby in some way that we haven't specified (see Figure 1.3). Each of the following questions refers to a measurement of Ben taken some time after Mary has passed you, and asks you to compare that measurement in your frame (unprimed) with Mary's (primed). For each one, say whether the unprimed variable is greater, the primed variable is greater, they are equal, or there isn't enough information to tell. *Remember that all the variables in this problem refer to measurements of Ben, as measured by you (unprimed) or by Mary (primed).*

    (a) x and x'

    (b) y and y'

    (c) v_x and v'_x

    (d) v_y and v'_y

    (e) a_x and a'_x

    (f) a_y and a'_y

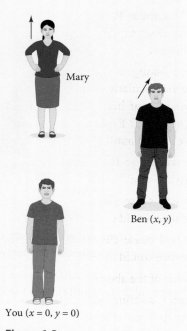

Mary

Ben (x, y)

You $(x = 0, y = 0)$

Figure 1.3

2. Two observers in different reference frames are measuring everything they can measure about the

same moving object. Based on Galilean relativity, list three quantities that these two observers must agree on, and three that they might disagree about. (Some of the answers might change in Einstein's theory, but don't worry about that for now.)

3. Explain the meaning of the equation $t' = t$ in Equations (1.1).

4. If you are driving at 35 mph when another car passes in the same direction as you, going at 50 mph, we say "the other car is going at 15 mph in your reference frame." What does that mean? Your answer should contain the number 15 and should explain what it means in this context. It should not include the phrase "reference frame."

5. A cart has a vertical spring with a ball on top of it. When you push the spring down and release it with the cart at rest, the ball goes straight up and down and falls back into the cart. For each of the scenarios below, will the ball fall (A) into the cart, (B) behind the cart, or (C) in front of the cart?

    (a) You give the cart a push and release the spring as it is moving at steady velocity along the floor. (Ignore friction.)

    (b) The same as Part (a), but this time assume the cart does experience significant friction as it slides.

6. You are biking at 20 mph when a frisbee moving perpendicularly to you at 10 mph slams into the side of your head. How much will the frisbee hurt your head? Explain your answer. (This might involve a bit of math.)

    A. It will cause the same injury that a 10 mph frisbee would cause if you were not riding.

    B. It will cause the same injury that a 20 mph frisbee would cause if you were not riding.

    C. It will cause the same injury that a 30 mph frisbee would cause if you were not riding.

    D. None of the above.

7. Superman is sitting calmly in outer space, not moving, when your spaceship passes by at constant velocity. Inside the spaceship you are still looking for an experiment that will tell you whether the ship is moving or not. You grab a pool ball and throw it up into the air. It rises, bounces off the ceiling, and comes back into your hand.

    (a) Describe the motion of the pool ball as it appears to you. What shape does its path trace out?

    (b) Superman is watching the whole thing with his X-ray vision. Describe the motion of the ball as it appears to him. What shape does its path trace out?

8. In Equations (1.1) we wrote conversions from your side-of-the-road reference frame to a little girl's zooming-by-in-a-car frame. Write equations to convert the other way. What's the only difference between the two sets of conversions?

9. Aristotle believed that there is a fundamental difference between the states "in motion" and "at rest." But Galileo said that this distinction depends on your reference frame: what one observer calls "at rest" another calls "in motion" and vice versa. How, then, to explain Aristotle's observation that moving objects tend to return to the state of rest? We see that all the time, don't we?

For Deeper Discussion

10. We said that every object in the universe has a well-defined acceleration at any given moment. If we want the velocity of a given object, we can just integrate its acceleration with respect to time. Given that the acceleration must be the same in every reference frame, how can the velocity come out differently?

11. Part of Galileo's motivation for proposing his relativity was to dispute a common objection to the Copernican solar system. On a galloping horse you feel the wind. If the Earth is hurtling rapidly through space, why aren't we buffeted by gale-force winds?

Problems

12. Your friend Al is driving at 60 mph when another car passes him at 70 mph. "In your reference frame," you explain to Al, "The other car is going at 10 mph." But Al doesn't believe you. Al somehow (annoyingly) always wins these debates, so you have

to hit him with an argument that is irrefutable at every step. "Let's suppose you and the other car both start together, at $t = 0$ hours and $x = 0$ miles," you say. Al has no objection so far.

(a) Find the positions of Al's car and the other car one hour after it passes him, in the reference frame of the road. Don't just answer with two numbers; justify those numbers beyond any possible doubt from Al.

(b) Now find the position of the other car *relative to Al* at that same moment, one hour after the other car passed him. Don't use any equations from this section; just use a common sense argument that Al can't possibly refute.

(c) "By the definition of velocity," you say, "if a car moves x miles in an hour then that car is moving (on average) x miles per hour." Al agrees with that. Using that definition and your result from Part (b), convince Al that the other car is going at 10 mph in his reference frame.

Postscript to Problem 12: You made a sound argument, but Al's going to come back to this one in the next section. Al somehow (annoyingly) always wins these debates.

13. A swallow is flying at exactly 25 mph directly northward. Let north be the y direction and east the x direction. Write the swallow's velocity from the perspective of . . .

(a) a tree that it flies over.

(b) another swallow, also flying 25 mph directly northward.

(c) a third swallow, flying 25 mph directly southward.

(d) a fourth swallow, flying 25 mph directly eastward.

14. Equations (1.1) give the Galilean conversions between two reference frames, one moving at speed u with respect to the other, for four quantities: time, position, velocity, and acceleration. Give the analogous conversions for mass, momentum, and kinetic energy.

15. María José runs by you at 20 m/s, moving along a straight line 30° above the positive x axis. At the moment she passes the origin (you), you both synchronize your stopwatches to time 0. At that same moment, a car goes by with velocity $\vec{v}$ in your frame. At some later time t you both measure its position, velocity, and acceleration. Write the conversions from your t, x, y, v_x, v_y, a_x, and a_y measurements to María José's corresponding primed coordinates. (Your answer will consist of seven equations, one for each variable.)

16. You are standing still in a field with some x and y axes drawn on it. (Of course you are standing at the origin.) James runs by at a constant velocity of $(3\,\text{m/s})\hat{i} + (4\,\text{m/s})\hat{j}$, and at that same moment Cecelia bikes by at 10 m/s in the positive y direction. She has her brakes on and is accelerating at $-(2\,\text{m/s}^2)\hat{j}$. At the moment they both pass you all of your stopwatches are set to 0.

(a) Two seconds after she passes you, what are Cecelia's position, velocity, and acceleration in your reference frame?

(b) At that same moment, what are Cecelia's position, velocity, and acceleration in James' reference frame?

17. Equations (1.1) give the conversions for measurements of any object from one reference frame to another, assuming the primed frame is moving at speed u in the x direction relative to the unprimed frame. Rewrite those conversions assuming the primed frame is moving at speed u at an angle ϕ counterclockwise from the x axis. Your answer should consist of seven equations, one for time and one each for the x and y components of position, velocity, and acceleration.

18. If you are in a train that is going at a fast (but constant!) speed around a small circle, you don't need to look out the window to know you're moving; you will perceive a pull toward the outside of the circle.

(a) Why does this *not* violate Galileo's principle that the laws of physics are the same in all inertial reference frames? (You can answer this using only words.)

(b) So why don't we feel the fact that we are in constant circular motion around the Earth's rotational axis? (This one will require some calculation.)

19. A piece of gum is stuck to the edge of a bicycle tire. The tire has a radius of $1/3$ m and makes a full rotation every $\pi/12$ seconds as the bicycle moves forward 8 m/s. At time $t=0$ the gum is at the highest point on the tire.

    (a) Write the function $x'(t)$ giving the horizontal position of the gum, relative to the center of the tire. Assume t is measured in seconds and give your answer in meters.

    (b) Write the function $x(t)$ giving the horizontal position of the gum relative to the road. Assume that $x(0)=0$. Assume t is measured in seconds and give your answer in meters.

20. Dudley Do-Right is floating safely downward toward the ground, parachute open, at a constant 10 m/s. His arch-enemy Snidely Whiplash, 500 m above him, drops a brick, which begins at rest but accelerates at 9.8 m/s² downward. How long does it take the brick to reach Dudley?

    (a) Solve this problem from the perspective of the ground. (The numbers given above were all from this perspective.)

    (b) Start over and solve the same problem from Dudley's perspective. You should find that although many of the individual measurements (positions and velocities) are completely different, the equation you end up solving for time is identical.

    (c) Why didn't we ask you to solve the problem from the brick's perspective?

21. Remember the little girl in the 60 mph car? Well, suddenly her father spots an emergency: a trailer 2 miles ahead is rolling along the road at only 20 mph, neither speeding up nor slowing down. Dad slams on the brakes! What acceleration must his brakes apply if his car is to avoid hitting the trailer? First you're going to solve this in the reference frame of the road.

    (a) Call the car's position $x=0$ at the moment Dad hits the brakes. Write an equation for the car's position as a function of time and another for the trailer's position. Set the two equal to represent the moment they collide. The result should be one equation with two unknowns: t and a.

    To finish solving for a you would need to solve that quadratic equation for t and then find an inequality involving a for when that solution for t has no real solutions, meaning they never collide. But you might prefer a different approach.

    (b) Start over and solve the problem in the reference frame of the trailer. If you do this correctly it's a much easier calculation. *Hint:* You do not need to use time at any point in the solution.

22. [*This problem depends on Problem 21.*] Finish doing the calculation from Problem 21 in the car's reference frame. Make sure you get the same answer you did in the trailer's reference frame.

23. In the Example on p. 3 we determined what would happen to a ball on a ship using the ship's reference frame. Now re-do that calculation in the ocean's reference frame. Take the x axis to point in the direction of the ship's motion.

    (a) What are the ball's x- and y- velocities in the ship's reference frame?

    (b) What are the ball's x- and y- velocities in the ocean's reference frame? (The rest of the problem will be in the ocean's reference frame.)

    (c) Write equations for the ball's positions $x_b(t)$ and $y_b(t)$. Take its initial position to be $(0,0)$.

    (d) Write equations for the position of the opposite corner as a function of time, $x_c(t)$ and $y_c(t)$.

    (e) Solve for the time at which $y_b = y_c$.

    (f) Find x_b and x_c at that time. If they are equal then the ball hits the corner. If not then it hits one of the walls. If you didn't get the same answer as we got in the Example, go back and find your mistake.

24. Two particles with masses m_a and m_b and initial velocities v_{a_0} and v_{b_0} collide and move off with velocities v_{a_f} and v_{b_f}. All the motion is in the x direction, and the velocities are given in the lab reference frame. Show that if momentum is conserved in the lab reference frame then it is conserved in any inertial reference frame moving with velocity u in the x direction relative to the lab.

25. Non-Inertial Reference Frames

Galilean relativity says that all inertial reference frames are equally valid. It's certainly possible to calculate things like position and velocity in a non-inertial reference frame, but the physics we are familiar with doesn't work. To see why that is, suppose you are in your ship floating at rest in space with nothing exerting a force on you. Asma is at rest next to you. At a time that you both call $t = 0$, she starts accelerating at a rate a_C in the x direction.

(a) Write your position, velocity, and acceleration in Asma's reference frame.

(b) Explain how you can tell from your answers that Newton's laws are not valid in Asma's reference frame.

1.2   Einstein's Postulates and Time Dilation

Newtonian physics can be exciting. You can model and predict the behavior of familiar systems like pendulums and ocean waves. You can notice possibly unfamiliar phenomena like the way a top wobbles as it slows down, and then explain them. It gives you the sense that the world around you follows a comprehensible set of rules.

Modern physics is exciting in a very different way, because its effects do *not* obey the laws that we see in our daily lives. In special relativity, time passes differently for different observers. Objects change length as they move. Different people can disagree about the order in which things happened, and they can all be right! (And that's the least confusing of the modern physics theories.)

So when you study modern physics, some of the fun is finding out that the universe is stranger and more complex than we normally think. You will need to set aside some of your lifelong intuitions and slowly start to build up new ones. We're going to start all that with Einstein's Special Theory of Relativity.

1.2.1   Discovery Exercise: Einstein's Postulates and Time Dilation

The speed of light, generally represented by the letter c, is approximately 3×10^8 m/s.

Spaceman Spiff is floating motionless in space when a spaceship zooms past at speed $c/3$ (one third the speed of light). At the instant the ship passes him Spiff turns on his flashlight, pointing the same direction that the ship is traveling. The beam leaves the flashlight at c, the speed of light, in Spiff's reference frame.

    1. After one second, the beam of light has traveled how far in front of Spiff?

    2. After one second, the spaceship has traveled how far in front of Spiff?

    3. So after one second, the beam of light has traveled how far in front of the spaceship?

      See Check Yourself #2 at www.cambridge.org/felder-modernphysics/checkyourself

    4. Use your previous responses to answer the question: how fast is the light beam traveling in the reference frame of the spaceship?

1.2.2   Explanation: Einstein's Postulates and Time Dilation

Every story of modern (twentieth-century) physics begins the same way: "nineteenth-century physics was working great, but there was a little problem . . ."

One high point of nineteenth-century physics was Maxwell's electromagnetic theory. Four equations elegantly encapsulated everything known about the electric and magnetic forces. Maxwell's 1861 paper "On Physical Lines of Force" established the connection between the fundamental constants that govern the electric and magnetic forces (ϵ_0 and μ_0) and the speed of light (c):

$$c = \frac{1}{\sqrt{\epsilon_0 \mu_0}} \approx 3 \times 10^8 \text{ m/s}. \tag{1.2}$$

You can derive Equation (1.2) directly from Maxwell's equations. The resulting value matches the speed of light, which had been accurately measured in 1848. This match between theory and experiment established the nature of light as a traveling wave of electric and magnetic fields. Nineteenth-century physics was working very well.

But Maxwell's theory caused a problem with Galilean relativity, which had at this point been accepted as a fundamental principle for centuries. Can you see the problem (before you continue reading)?

Well, if light travels at speed c in one reference frame, then Equations (1.1) say it must travel at speed $c - u$ in some other reference frame. But Maxwell's equations predict the same speed, $1/\sqrt{\epsilon_0 \mu_0}$, in any reference frame. What is the "correct" reference frame, the one that is actually described by Maxwell's equations? And what is going on in all other reference frames?

Nineteenth-century physicists answered this question for light by analogy to how they (and we) answer the same question for sound. Sound is an oscillating wave of air pressure – without air (or some other medium) sound cannot exist at all – and the speed of sound is roughly 343 m/s *relative to the air*. By analogy electromagnetic waves must be a disturbance in some kind of medium, moving at 3×10^8 m/s *relative to that medium*. Because light travels to us from the stars, its medium must fill all of space. They called this medium the "aether" (sometimes spelled "ether").

One nice thing about the aether hypothesis is that it makes a testable prediction.

Imagine honking a loud horn in the midst of a strong wind. The sound travels outward, moving at the same speed in every direction relative to the air. So relative to the ground you will measure the sound traveling unusually fast in the downwind direction, more slowly in a perpendicular direction, and slowest of all as it travels upwind.

Well, the Earth is moving through space (around the Sun) at about 30,000 m/s. In such a strong "aether wind," physicists reasoned, we should detect different light speeds in different directions. In 1887 Albert Michelson and Edward Morley tested the speed of light in different directions with a device more than sensitive enough to measure a 30,000 m/s variation. To their complete surprise, they found no detectable difference. Light traveling with the Earth's motion, light traveling opposite the Earth's motion, light traveling perpendicular to the Earth's motion – in all cases the light moved at exactly c.

Section 2.5 discusses that experiment in more depth, but our primary interest is its outcome: theorists struggled for years to reconcile the Michelson–Morley result with the existence of the aether. The struggle ended when Albert Einstein proposed that there is no aether and that Maxwell's equations, like all laws of physics, work identically in all inertial reference frames.

Einstein's Postulates

Einstein's 1905 paper "On the Electrodynamics of Moving Bodies" developed a new system – complete, self-consistent, and a radical departure from Newton – based on two postulates.

The Two Postulates of Special Relativity

1. The laws of physics are the same in any inertial reference frame.
2. The speed of light through a vacuum ($c \approx 3 \times 10^8$ m/s) is the same in any inertial reference frame.

The first postulate reaffirms Galilean relativity. The second completely flies in the face of Galilean relativity (Equations (1.1)), not to mention common sense. Einstein did not ignore this paradox; he explored it, promising that the latter postulate "is only apparently irreconcilable with the former." Following in his footsteps, we will explore these apparent contradictions as a way of developing what we now call his "special theory of relativity."

Just to give you a preview of where we're going, here are a few of the consequences we're going to derive from those two postulates:

- A clock ticks off time at a different rate when it's moving.
- An object in motion has a different length than it has at rest.
- Two observers can disagree about the order in which two events happened, and they can both be right!

Einstein's postulates also make complete sense of the Michelson–Morley result. Although this does not seem to have been one of the primary reasons he developed the theory, it is certainly one of the primary reasons why the theory was widely accepted.

The Paradox of the Second Postulate

Figure 1.4 illustrates a "thought experiment" that we are going to use many times over the next several sections to bring out ideas and equations of special relativity.

- There are two mountains labeled A and B. Their peaks (which we will treat as points) are separated by distance L. The two mountains are stationary with respect to each other, so they share an inertial reference frame.
- An airplane flies by at speed u. The airplane never slows down or speeds up, so it defines its own inertial reference frame.
- At the moment the plane passes by Mountain A, someone in the plane turns on a flashlight, sending a beam of light toward the ceiling of the plane. A passenger inside the plane would describe the beam's motion as vertical.
- At the moment when the plane passes Mountain B the light beam reaches the ceiling of the airplane, a height h above the spot where it originated.

It will aid our discussion to introduce a word here. In relativity an "event" refers to a particular place in space, at a particular moment in time. So "the plane travels from Mountain A to

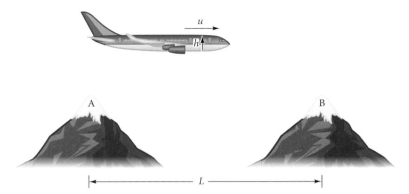

Figure 1.4 **The plane passing the mountains as seen in the mountain reference frame. In the relativity chapters we will generally draw stationary objects in red and moving ones in blue.**

Mountain B" is *not* an event in this scenario, since it extends over a span of distance and time. The interesting events are "light beam emitted from floor" (at Mountain A) and "light beam reaches ceiling"(at Mountain B). Observers in both reference frames witness these two events.

Active Reading Exercise: The Airplane and the Mountains, Part 1

Imagine our airplane-and-mountains scenario above with one change: instead of a flashlight shining straight up from the floor, a gun fires straight up from the floor. The bullet, like the light beam, travels at constant speed to the ceiling. (We don't want to think about what happens after that.) As with the light beam, assume the bullet reaches the ceiling exactly as the plane passes the second mountain. You can assume that the effect of gravity on the bullet is negligible during its flight.

1. Draw the path of the bullet twice: first in the plane reference frame, and then in the mountain reference frame. (They should not look the same!)
2. How far does the bullet travel in the reference frame of an observer in the airplane?
3. How far does the bullet travel in the reference frame of an observer on the mountains?
4. Based on purely Newtonian/Galilean assumptions, the two observers calculate different speeds. Which observer sees the bullet going faster? Briefly explain your answer.

Don't read on until you have written down answers to these questions. It's okay if you get some wrong, but commit to your best shot before reading our answers (below).

In the plane reference frame the bullet travels straight up a distance h.

In the mountain reference frame the bullet moves in a diagonal line, up h and over L, so it moves a total distance $\sqrt{h^2 + L^2}$. Because this perspective sees the bullet traverse a greater distance in the same amount of time, the bullet must be going faster in the mountain frame than in the airplane frame. (You can arrive at the same conclusion by thinking about velocity vectors. Both observers see the same vertical velocity, but the mountain observers also see a horizontal velocity and the airplane observers don't.)

One way or another, you should take the time to thoroughly convince yourself that the mountain observers *must* measure a higher bullet velocity than the airplane observers see. This conclusion is irrefutable, and it's also correct.

Now let's return to our original scenario, with a flashlight instead of a gun. We could walk through the same argument, step by step, to conclude that the light beam must be traveling faster in the mountain reference frame than in the airplane reference frame. But this time we have a problem, because that conclusion contradicts Einstein's second postulate.

Should we abandon relativity as a lost cause? Let's break down the argument.

- The light travels a longer distance in the mountain reference frame than it does in the plane reference frame.
- The light takes the same amount of time to travel in the mountain reference frame as it does in the plane reference frame.
- By the definition of speed as distance over time, the light is going faster in the mountain reference frame.

The last step is simply what we mean by the word "speed." The first step is basic geometry; the distance that you travel going up h and over L is longer than the distance h. So Einstein concluded that for his two postulates to work, the middle step must be wrong. *The two reference frames disagree about how much time passes between these two events.*

Active Reading Exercise: The Airplane and the Mountains, Part 2

Suppose the mountain clocks say one second elapsed between the two events (light-emitted and light-hits-ceiling). Does the plane clock say that it took more than one second, or less than one second?

Don't just write down an answer. Write down a sentence or two explaining how your answer allows the two frames to agree on the speed of light.

Since the plane reference frame sees the light move a shorter distance, and we want the ratio of distance over time to come out the same in both reference frames, the plane reference frame must also record a shorter time. This result, known as "time dilation," is generally summarized by the sentence "Moving clocks run slowly."

How slowly do they run? You already have everything you need to answer that question: the plane clock must run more slowly than the mountain clocks by a factor of $h/\sqrt{h^2 + L^2}$ to resolve the paradox. In Problem 31 you will turn that formula (which is specific to this scenario since it depends on L and h) into a general formula for time dilation (dependent only on the relative speeds of the two reference frames, u). Here's what you'll get:

Time Dilation

A clock is at rest in inertial reference frame R, and moving at speed u with respect to inertial reference frame R'. (For example, if the clock is on an airplane, R could be the plane frame and R' the ground frame.) Between the events "the clock measures t" and "the clock measures $t + \Delta t$," a time Δt passes in Frame R. (That should be obvious.) The time elapsed between those two events in Frame R' is

$$\Delta t' = \frac{1}{\sqrt{1 - (u/c)^2}} \Delta t \qquad\qquad \text{Time dilation: first formulation.} \qquad (1.3)$$

The factor $1/\sqrt{1 - (u/c)^2}$ occurs frequently in relativistic expressions and is identified by the Greek letter γ ("gamma"), so you can write this more concisely as

$$\Delta t' = \gamma \Delta t \qquad\qquad \text{Time dilation: second formulation.}$$

In words:

Moving clocks run slowly by a factor of γ.   Time dilation: Third formulation.

In the airplane-and-mountains scenario the plane clock is moving relative to the mountain reference frame (R' is the mountain reference frame in this case), so the mountain observers say the plane's clock is going slowly by a factor of γ. Notice that γ is always greater than or equal to 1 (can you see why from its definition?), so when a clock ticks off time Δt between two events, observers in every inertial reference frame other than the clock's measure a time gap longer than Δt.

Equation (1.3) and other important formulas from relativity are in Appendix B.

You can see an animated demonstration of airplanes flying by mountains at www.cambridge .org/felder-modernphysics/animations

When you watch that animation, pay particular attention to the relative speeds of the clocks in the mountain and airplane reference frames. As you go through this chapter, you may want to keep referring back to that animation, because it contains many of the relativistic phenomena that we will be exploring in later sections.

Example: Time Dilation

A space station needs to send relief medicine to a planet 3 light-days away. (A light-day is the distance light can travel in a day.) Their fastest ship can go at $0.9c$, but the medicine only lasts 2 days before expiring. Will the medicine reach the planet in time?

Answer: We can write the speed of the rocket as 0.9 light-days/day, so it takes the rocket $t = d/v = 3.333$ days to reach the planet. (As a sanity check, note that at the speed of light it would take 3 days, so at a slightly lower speed it must take slightly longer.)

The situation may look hopeless, but remember that the rocket clock is moving slowly by a factor $\gamma = 1/\sqrt{1 - 0.9^2} = 2.294$. So while the planet frame says 3.333 days pass, the rocket clock only ticks off 3.333 days/2.294 = 1.45 days. The medicine ages less than 1.5 days on the journey and the planet is saved.

This problem is an important reminder that "moving clocks run slowly" is not a statement about clocks. It's a statement about time, and everything that changes with the passage of time.

It's worth saying a few words about the units we used in the preceding time dilation example. You may have been tempted to convert everything to SI units: 3 light-days becomes 7.771×10^{13} m and so on. You can do that in Problem 24 and you should end up with the same final answer we did (the medicine ages by 1.45 days). But you'll find yourself doing a lot more work.

So we urge you to adopt what are sometimes called "relativistic units." If you're measuring time in seconds, measure distance in light-seconds. The speed of light therefore always becomes $c = 1$: for instance, 1 light-second/second. Other speeds expressed in terms of c are also easier to handle: for instance, $c/2$ becomes $1/2$ light-second/second. It may sometimes be more convenient to measure time in hours, days, or years, but as long as you measure distance in the corresponding measure (light-hours, light-days, etc.) you'll get the same simplification.

When You Can and Can't Use the Time Dilation Equation (Equation (1.3))

There is a subtlety hiding behind the definition of time dilation. Think of Δt as the time gap between two events such as "the clock strikes $t = 0$" and "the clock strikes $t = 1$," as measured in the clock's reference frame (which we'll call R). Then $\Delta t'$ gives the time gap between those same two events in some other frame R′. But *Equation (1.3) doesn't apply to all time gaps; it only applies if the two events occur at the same place in Frame R*. Two chimes on the same clock make a wonderful example, as does the medicine in the relief ship. But two events that don't occur in the same place according to either reference frame require a more general formula (Section 1.4).

So, you have two events that occur in the *same place* in one reference frame (R), and Equation (1.3) gives you their time gap in a different frame (R′). You can see from that equation that the new time gap is always *longer* than the old.

The inertial reference frame in which two events occur at the same place measures the shortest gap between those two events of any inertial reference frame.

That's just a more careful way of saying "moving clocks run slowly." For instance, in the medicine scenario above, the two interesting events are "ship leaves first planet" and "ship arrives at second planet." The ship itself is the reference frame in which these two events occur at the same place. We are therefore guaranteed that the planet frame, or any other inertial reference frame, will measure more than 1.45 days between those events.

If you look back at our original airplane-and-mountains scenario you might now object: why does Equation (1.3) apply to our flashlight beam, when the two events (beam-emitted and beam-reaches-ceiling) actually happened in different places, even in the plane reference frame? Hey, we're glad you asked! They were different places vertically, but in the airplane frame they happened at the same place horizontally. And the horizontal direction is the one that matters, because that is the direction of the relative velocities of the two reference frames. If the flashlight beam had been fired from the tail of the plane to the cockpit, Equation (1.3) would not apply.

The Twin Paradox

To finish this section we're going to describe (and then of course resolve) one more seeming paradox. This is one of the most commonly used thought experiments in relativity, and we're going to return to it repeatedly to illustrate different aspects of the theory.

Active Reading Exercise: The Twin Paradox

Emma goes on a long voyage in a near-light-speed rocket ship. She travels in a straight line to her destination, turns around (more or less instantaneously), and returns to Earth. Her twin, Asher, has been waiting for her in California the whole time.

Just before Emma steps out of her rocket, Asher thinks: "Because she has been moving the whole time, her clocks were running slowly. So she has aged less than I have." Meanwhile, Emma thinks: "From my point of view Asher was moving this whole time so his clock was running slowly. When I get home he will have aged less than I have."

Question: Who was right, and what was wrong with the other one's argument?

Don't keep reading before you think about this one a while. As the two twins stand side-by-side at the end, they *must* agree about which twin has aged more. Does this paradox prove that Einstein's theory is logically inconsistent?

Hint: Time dilation is a result of Einstein's two postulates. Look back at p. 12 and re-read them. The answer is there!

When the twins reunite, Emma is younger than Asher. You might imagine that 50 years have elapsed on Earth, but only a few years passed on the ship. (The exact numbers depend of course on Emma's speed and distance, which we never specified.) Asher is stooped and gray-haired, while Emma has hardly aged.

That answer makes perfect sense from Asher's viewpoint: his clock was running normally, and Emma's was running slowly, the entire time. But why can't Emma legitimately make the same argument the other way?

The answer lies in the word "inertial." Asher stayed in an inertial (non-accelerating) reference frame throughout the experiment.[1] Emma's rocket at some point slowed to a stop and then sped up again in the opposite direction. In a word, she accelerated. You cannot analyze this scenario from "Emma's inertial reference frame" because she doesn't have one.

---

1 You might object that the Earth accelerates as it rotates about its axis and revolves about the Sun. That's true, but those are tiny effects and can be ignored for our purposes here.

Relativity works perfectly well to describe accelerating objects, as long as the analysis is done from the reference frame of an inertial object. So Asher's statement that Emma's clock was slow is valid, while Emma's analysis isn't.

Remember that the statement "Emma accelerated" is *not* reference-frame-dependent (in either Galilean or Einsteinian relativity). It is an intrinsic fact about her motion. If she were sealed up inside her rocket with no view of the outside, she would still be able to determine when the ship was accelerating.

To mathematically analyze the situation, you have to recognize that the scenario involves three different inertial reference frames: the frame of Asher and the Earth, the frame of Emma departing, and the frame of Emma returning. We'll give you two different opportunities to do those calculations, first using the relativity of simultaneity (Section 1.3 Problem 25), and then with velocity transformations (Section 1.5 Problem 19). This problem is one of the best examples of how relativity, despite all its apparent paradoxes, ultimately emerges as an amazingly consistent system; all reference frames agree about Asher's (old) age and Emma's (young) age when the twins reunite.

So it's all logically consistent, but is it true? We don't have the technology to send people on near-light-speed journeys, but in 1971 Joseph Hafele and Richard Keating performed a different version of the twin experiment. They took several highly accurate atomic clocks on round-the-world airplane trips while leaving other clocks at rest on the ground. When they reunited the clocks they found that the moving ones had fallen behind the stationary ones by exactly the amount predicted by relativity (see Problem 26).

Coordinate Time and Proper Time

In the twin paradox, Asher remains in one inertial reference frame – that of the Earth – for the entire story. But Emma switches between two different inertial reference frames, one outgoing and one returning, and that's what makes the story complicated (and interesting).

In such situations we distinguish between "coordinate time" (the time as measured in some inertial reference frame) and "proper time" (the time measured by a particular person, or rocket, or – to use the easiest example – clock). The former is often designated as Δt and the latter as $\Delta \tau$.

Consider three different descriptions of the time between the two events "Emma leaves Earth" and "Emma returns to Earth."

- The time measured by Asher is both a coordinate time Δt (the time between those two events in the Earth reference frame) and a proper time $\Delta \tau$ (the amount that Asher's clock measured, and therefore the amount by which he aged), and the two are equal.

- The time measured between those same two events in the outgoing reference frame is a coordinate time Δt. It is not a proper time, because no single object in that reference frame experiences both events.

- The time Emma aged throughout the trip is her proper time $\Delta \tau$. It is not a coordinate time, because it involves more than one inertial reference frame.

It's important that Asher's reference frame is the *only* frame that can measure the time between those two events as both a coordinate and a proper time. To put it another way, his is the only inertial frame in which the same clock is at both events.

Active Reading Exercise: Coordinate Time, Proper Time

An airplane flies by Mountain A, and then later flies by Mountain B.

1. The clock on the airplane registers a certain time between these two events. Is that a proper time, a coordinate time, or both?

2. The perfectly synchronized clocks on the two mountains also register a certain time between these two events. Is that a proper time, a coordinate time, or both?

After you write down your best answers, you can see ours – along with a brief discussion of the particular importance of the airplane's reference frame in this scenario – at www.cambridge.org/felder-modernphysics/activereadingsolutions

Stepping Back

We started this section with theoretical reasons (Maxwell's equations) and experimental evidence (the Michelson–Morley experiment) for believing that the speed of light is the same in all inertial reference frames. We saw that this rule leads to seeming contradictions, and to resolve them we had to introduce time dilation.

So if different observers disagree about the flow of time, why haven't you noticed it before? Why is it that, after you take a long round-trip flight, your friends don't seem a bit older than you?

The answer lies in the equation $\gamma = 1/\sqrt{1 - u^2/c^2}$. At normal speeds, such as the speed of a bullet ($c/400{,}000$) or the Earth's rotation ($c/10{,}000$) or the fastest spacecraft to date ($c/4000$), you get $\gamma \approx 1$ and relativity becomes effectively irrelevant.

So our everyday world is essentially Newtonian. You can get a mechanical engineering degree from Georgia Tech without ever mentioning relativity. That's also why our physical intuition rebels when confronted with modern ideas. If we had evolved in a world of predators and prey moving at $c/4$, time dilation might seem as natural to us as Galileo's relative velocities.

1.2.3 Questions and Problems: Einstein's Postulates and Time Dilation

Conceptual Questions and ConcepTests

1. Define each of the following phrases in your own words:
   (a) Time dilation
   (b) Coordinate time
   (c) Proper time

2. The central idea of this section can be expressed in one sentence: "If you accept that the speed of light is the same in all reference frames, then you *must* accept time dilation as an inevitable consequence." Explain in your own words why.

3. A very long train zooms past a station at speed $c/2$. Observers on the train and ground measure the time it takes for the train to pass the station. Choose one of the following and explain your choice.

   A. The time elapsed on the train clocks, as the train passes the station, is *longer* than the time elapsed on the station clock.

   B. The time elapsed on the train clocks, as the train passes the station, is *shorter* than the time elapsed on the station clock.

   C. The two times are the same.

4. In this book the time dilation equation is $\Delta t' = \gamma \Delta t$. In another textbook you might see the time dilation equation written as $\Delta t = \gamma \Delta t'$. Those are *not* the same equation (since γ is not generally 1). Instead, the two different equations represent different definitions of Δt and $\Delta t'$. Clearly explain the two different definitions.

5. A rocket ship flying at near-light speeds goes from Space Station A to Space Station B (at rest relative to each other). During the trip Aisha gets up and walks forward to the front of the ship. Her trip takes a time Δt in the rocket frame but takes some other time $\Delta t'$ in the frame of the space stations. Explain why we *cannot* use Equation (1.3) to relate those two times.

6. The Hogwarts Express is moving at speed $0.87c$. At the front of the train Hermione stands up, and a second later (in the train frame) Ron stands up at the back of the train. Harry proudly announces that he's been studying relativity and, because $\gamma = 2$, Earth-frame clocks will show that Hermione stood up 2 seconds before Ron. Hermione gives him a withering look and tells him he doesn't understand relativity. What did Harry get wrong?

7. The solution to the famous "twin paradox" (p. 17) hinges on the time period when Emma reverses direction – thus accelerating – to come home. Now suppose that Asher stays home (as before) but Emma never slows down or speeds up; instead she travels at constant speed in a big circle, ending back at home anyway. What do they find when Emma returns home? (Choose one and briefly explain.)

   A. Emma is younger than Asher (just as she was in the original twin paradox).

   B. Asher is younger than Emma.

   C. They disagree about which one is younger.

   D. They are the same age.

8. Equation (1.3) implies a fundamental limitation on the possible values of u, the relative speed between two different moving objects. Say what that limitation is, and explain how it comes from the equation.

9. We said in the text that $\gamma \geq 1$. Under what circumstance would it equal 1 and in what limiting case would it approach ∞?

10. You are inside a spaceship in which you feel no acceleration. You sleep for your usual eight hours. For each of the questions below, briefly explain how you know your answer.

    (a) Is there any inertial reference frame in which your night's sleep lasted longer than eight hours?

    (b) Is there any inertial reference frame *other than the one you are in* in which your night's sleep lasted exactly eight hours?

    (c) Is there any inertial reference frame in which your night's sleep lasted less than eight hours?

11. Zhang Minh is at rest (in her inertial reference frame) while an hour ticks by on her watch. Call the start and end of that hour on her watch Events A and B. In any other inertial frame, more than an hour passed between those two events. Any other observer present at those two events will experience a proper time of an hour or less between them. How can you reconcile those two (correct) statements?

12. When Neil Armstrong flew to the Moon and back, why did he not seem to have aged less than the people who remained on Earth?

For Deeper Discussion

13. A rocket flies by a planet at speed $0.9c$. Isabella is in the rocket and throws a ball forward (in the direction of the rocket's motion) at 20 m/s (relative to the rocket, as measured in the rocket frame). Pick one of the following for the speed of the ball as measured in the planet frame, and explain your answer:

    A. $v_b < 20 \text{ m/s} + 0.9c$

    B. $v_b = 20 \text{ m/s} + 0.9c$

    C. $v_b > 20 \text{ m/s} + 0.9c$

14. Much of Einstein's original motivation for special relativity came from electromagnetic theory. Consider two parallel wires, each carrying an identical current I in the same direction.

The moving charges create magnetic fields, and react to each other's magnetic fields, causing the wires to attract each other. (Curl your right-hand fingers and thumb around until you're convinced that's true.) But now consider the same scene from a reference frame that is moving with the current. In that frame the electrons are (on average) not moving, so they exert no net magnetic field. Do the wires still move toward each other? Why or why not?

15. Can you describe a series of events from the reference frame of a light beam? Why or why not?

16. We derived the formula $\Delta t' = \gamma \Delta t$ in the context of two inertial reference frames. We went on to say that in the twin paradox we had to use Asher's time as $\Delta t'$ and Emma's as Δt because only Asher had an inertial reference frame. Does that mean R′ has to be inertial but R doesn't? Explain.

Problems

17. Figure 1.4 shows a plane passing two mountains. Suppose the plane is moving at $0.8c$ and the mountains are 6×10^8 m apart. (It's a thought experiment, OK?) The two mountains have synchronized their clocks. At the moment the plane passes Mountain A, the observers on the airplane and Mountain A both see that their clocks show $t = 0$.

    (a) What time will the clock on Mountain B show when the plane passes it?

    (b) What time will the clock on the plane show when the plane passes Mountain B?

18. The USS Enterprise heads out on impulse engines at a leisurely pace of $(9/10)c$ to explore the galaxy. After 5 years have elapsed on board the ship, it arrives back home. How much time has elapsed on Earth?

19. You travel in a rocket ship at half the speed of light from Earth to Alpha Centauri, 4.37 light-years from Earth. How much older are you when you arrive?

20. In the book *Ender's Game*, Earth sends the military champion Mazer Rackham on a round trip at near light speed so that he can still be young enough to lead Earth's army in a future war. When he returns 80 years have passed on Earth but he has aged only 8. How fast was his ship moving?

21. In the song "39" by the rock group Queen – a song that deserves to be much better known than it is, by the way – a traveler returns home to discover that "The Earth is old and gray . . . though I'm older but a year." The song is not specific about how much time has passed on Earth, but several generations have clearly passed, so let's call it an even 100 years. How fast must the traveler have been going to cause this 100-to-1 time dilation?

22. The center of the galaxy is about 30,000 light-years from Earth. How fast would you need to travel in a rocket ship to arrive at the center of the galaxy 10 years older than you started?

23. A very long train is moving past a station at speed $c/2$. Of course, from the train's point of view, the train station is moving at speed $c/2$ past a very long train. In the train reference frame it takes 10 s for the station to go past the train. In the station reference frame, how long does it take for the train to go past?

24. On p. 15 we did an example with a shipment of space medicine. Re-do this problem, beginning by converting all the units to meters and seconds. You should end up with the same answer as we did (the medicine ages 1.45 days). But you should also be convinced that relativistic units are a good time-saver!

25. **GPS**

    Global positioning satellites (GPS) work by sending out signals with their current location and time. When your device on Earth gets signals from multiple satellites it can tell by the time that they each report which signals took the longest to reach you. For this to work the satellites need very accurate clocks. They are moving at roughly 14,000 km/hr relative to the Earth, though, so they need to consider the effect of time dilation. How much time does a GPS satellite lose each day relative to a clock on Earth? (If the clocks didn't correct for this, GPS systems couldn't work.) (*Note*: You are only calculating the effect of special relativity on the GPS clocks. There is another, stronger effect that comes from Einstein's *general* theory of relativity. For GPS to work it needs to account for both effects.)

26. One of the most famous tests of relativity is the Hafele–Keating experiment, first performed in 1971 (and repeated since many times). Joseph Hafele and Richard Keating put several cesium-beam atomic clocks on commercial airliners that flew around the world, and then compared those clock times to other clocks that had remained at the United States Naval Observatory.

    (a) The planes took three days to circumnavigate the globe. Assuming the radius of the Earth is 4000 miles, how fast were the planes going?

    (b) How much of a time difference would you expect to see between the traveling clocks and the stationary clocks?

    The actual calculations were more complicated than this, for two reasons. First, it was necessary to take the Earth's rotation into account. Second, the higher-altitude clock experienced another time differential due to gravitational effects predicted by Einstein's *general* theory of relativity. But with all that properly taken into account, the final results were remarkably close to predicted values.

27. Particles called muons are constantly being created by high-energy collisions in the upper atmosphere. Suppose a muon covers 100,000 m from the upper atmosphere to a detector on Earth, traveling at 99.999% of the speed of light.

    (a) A muon at rest in a lab typically decays after 2.2×10^{-6} seconds. Show that, according to non-relativistic calculations, the muon will not reach the detector.

    (b) But in fact, most of the muons created at that height do make it to Earth. Show how relativistic calculations explain why.

    This experiment is a common test of relativity.

28. Solve Equation (1.3) for Δt. Based on the resulting equation, if we gave you $\Delta t'$, would Δt come out higher or lower? Explain why this never leads to the conclusion that "moving clocks run fast" from *any* reference frame.

29. It is a general property of modern physics that it reduces to classical physics in normal life. Billiard balls, spinning tops, skyscrapers, and rocket ships all obey Newton's laws for all practical purposes. As an example, suppose Hermione is dutifully applying Equation (1.3) for relativistic time dilation. Ron is carelessly applying the Galilean equivalent, $\Delta t' = \Delta t$.

    (a) Calculate the percent difference in their results for $u = 1000$ m/s, which is roughly the fastest airplane speed ever recorded.

    (b) Calculate the percent difference in their results for $u = 75,000$ m/s, the approximate speed of NASA's Juno spacecraft, the fastest macroscopic object made by people.

    (c) How fast would u have to be for Ron's and Hermione's calculations to differ by 0.1%?

    (d) The Large Hadron Collider (LHC) accelerates protons to $0.99999999\ c$. (That's eight "9"s.) Now what is Ron's error?

30. According to the aether model, light on the surface of the Earth should move more slowly in the direction of the Earth's motion than it should perpendicular to that motion. What percent difference should this cause in the speed of light? (It is a testament to the accuracy of Michelson's interferometer that he very reasonably expected to measure this difference!)

31. **Deriving the Time Dilation Formula**

    This problem refers to the airplane-and-mountains scenario illustrated on p. 13. There are two events in the problem. At the first event the plane passes Mountain A and the light is emitted. At the second event the plane passes Mountain B and the light hits the ceiling.

    Let Δt be the amount of time between those events in the plane reference frame (Reference Frame P). During that time the light moves a distance h at speed c and the plane doesn't move. (The mountains move backward at speed u, but that doesn't matter for this problem.)

    Let $\Delta t'$ be the amount of time between those two events in the mountain reference frame (Reference

Frame M). During that time the plane moves a distance L at speed u, and the light moves a distance $\sqrt{h^2 + L^2}$ at speed c.

(a) Using the fact that distance equals speed times time, and the fact that light travels at speed c in both frames, write an expression for the ratio $\Delta t' / \Delta t$. Your answer should only depend on L and h. Algebraically rewrite this expression so that it only depends on the ratio L/h.

(b) That ratio $\Delta t' / \Delta t$ is what we want, but we don't want a formula that depends on details of this setup, so we want to eliminate L and h. Write an equation for h in

terms of Δt and another equation for L in terms of $\Delta t'$ and u. (Once again, this is all about the fact that distance equals speed times time.)

(c) Plug your expressions for L and h from Part (b) into your equation from Part (a). The result should be an equation with $\Delta t' / \Delta t$ on one side and a combination of Δt, $\Delta t'$, u, and c on the other.

(d) Solve your equation for the ratio $\Delta t' / \Delta t$ so you only have u and c on the other side. If you didn't get the formula we gave for time dilation, go back and find your mistake.

1.3 Length Contraction and Simultaneity

We continue here the program that we began in Section 1.2, using thought experiments and seeming paradoxes to draw out the consequences of Einstein's two postulates. Also, milking our airplane-and-mountains scenario for all it's worth.

1.3.1 Explanation: Length Contraction and Simultaneity

Let's briefly review the thought experiment we developed in Section 1.2 (Figure 1.5):

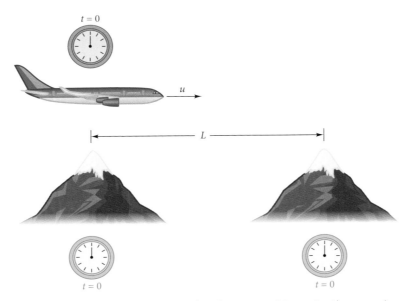

Figure 1.5 The first interesting event (airplane passes Mountain A), as seen in the mountain reference frame. All three clocks are synchronized to $t = 0$ at this instant.

- The two mountains are a distance L apart in their frame.
- The airplane flies by, moving to the right at speed u. The two events of interest are "airplane passes Mountain A" and "airplane passes Mountain B." (We don't need the flashlight any more.)
- The time between these two events, as registered on the mountain clocks (one clock on each mountain), is L/u. Remember that this is just the definition of speed.
- The time between these two events as registered on the airplane clock (just one clock) is *less* than L/u, demonstrating that "moving clocks run slowly."

That entire description is from the mountain reference frame. But another paradox arises when we attempt to describe the same scenario from the airplane reference frame.

- The airplane is not moving.
- The two events are "Mountain A passes the airplane, moving to the left at speed u" and "Mountain B passes the airplane at the same speed."
- The clock on the airplane is running at a perfectly normal speed of one second per second, because it is not moving. (The mountain clocks are off, but we'll return to them later.)
- So *how do we explain, in this reference frame, why the airplane clock records a time less than L/u between the two events?* Think about that one for a moment before reading on.

You may have guessed the answer, if only based on the name of this section: from the point of view of the airplane, the distance between the two mountains is not L! Same speed, different time, therefore different distance.

Active Reading Exercise: Length Contraction

In the mountain frame, the distance between Mountains A and B is L. In the airplane frame, is the distance between the mountains greater than L, or smaller than L?

The mountain frame says the plane is moving to the right, and the plane frame sees the mountains moving to the left, but they both agree about their relative speed u. Distance equals speed times time. Both frames agree on the speed, and the plane frame records a shorter time, so it must record a shorter distance. More specifically, since the airplane frame measures a shorter elapsed time by a factor of γ, it must also record a shorter distance by a factor of γ.

Thinking of the mountain range (extending from Mountain A to Mountain B) as an object, we can generalize this conclusion to "moving objects appear short." In the mountain frame the mountain range is not moving and it has length L, which we call its "proper length." In the plane frame the mountain range is moving with speed u and has length L/γ.

Length Contraction

A ruler is at rest in an inertial reference frame R, and moving at speed u along the direction of the ruler's length with respect to an inertial reference frame R'. In its rest frame R the ruler has length L. The ruler's length in R' is

$$L' = \frac{L}{\gamma} \qquad \text{Length contraction.} \qquad (1.4)$$

In words, *moving objects are shortened by a factor of γ*. Recall that $\gamma = 1/\sqrt{1 - (u/c)^2}$ and $\gamma \geq 1$.

The mountain frame and the airplane frame agree about how tall the mountain range is and how far it goes back into the page; length contraction only affects distances along the axis of the object's velocity. Because objects are contracted in one direction (the direction of motion) but not in the other directions, their shapes can distort significantly, as illustrated in the Example below, "Length Contraction."

When we used the airplane-and-mountains scenario to introduce time dilation, we used the plane frame as Frame R. Using that same scenario to introduce length contraction, we are now calling the *mountain* frame R. Did you notice we had switched? More to the point, can you see why?

The time dilation formula (Equation (1.3)) defines R as the frame in which the two events occur at the same place. The two events were plane-passes-A and plane-passes-B, so that makes R the plane frame. The length contraction formula (Equation (1.4)) starts from R as the frame in which the object is at rest. The "object" in question here is the mountain range, which is at rest in the mountain frame.

There's a subtlety to be aware of with length contraction. When you measure a ruler in its rest frame, it doesn't matter if you measure the positions of the two ends at different times. In any other frame, however, it's important to define the length of the ruler as the distance between the ends of the ruler at one fixed moment in time. In the airplane-and-mountains scenario the mountain frame will always measure Mountains A and B to be a distance L apart, while the plane frame will measure that *at any particular instant* the two mountains are a distance L/γ apart.

Example: Length Contraction

A rocket is approaching the Earth at $0.9c$. What shape does the Earth have according to the rocket frame?

Answer: Length contraction happens in the direction of travel, so the radius of the Earth perpendicular to the direction of motion is its proper length (roughly 4000 miles). But the radius of the Earth in the direction of motion is foreshortened:

$$r' = \frac{r}{\gamma} = (4000 \text{ miles}) \sqrt{1 - 0.9^2} \approx 1700 \text{ miles.}$$

The resulting shape is an ellipsoid.

The problem above asked what shape the Earth has according to the rocket frame. That turns out to be different from the question "What does the Earth look like to an observer in the rocket?" because for that question you also have to take into account the light travel times from different parts of the Earth to the rocket. See Problem 22.

Now, about Those Mountain Clocks

Our earlier discussion of length contraction didn't need mountain clocks at all: just mountains zooming by, and a clock on the airplane. But now let's get back to the two clocks that are at rest in the mountain frame.

Active Reading Exercise: The Plane Gazes Back

When the airplane crew look at the airplane clock, they see it running at a perfectly normal speed. When the mountain dwellers look at a mountain-based clock, they likewise see it running normally.

But when the mountain folk look at the clock on the airplane, they see it running too slowly. From their perspective, the airplane clock advances by less than one second when a second goes by.

Question: What do the airplane riders see when they look at the clock on a mountain? Do they see it running too slowly, too quickly, or at normal speed?

As always, jot down your answer, along with a brief justification, before you read ours.

It's natural – almost inevitable – to conclude that, if the mountain observers see the airplane clock running slowly, then the airplane observers must see the mountain clock running too quickly. But it's also wrong, and we can prove it's wrong in four words:

Moving clocks run slowly!

From the airplane point of view, the mountain clocks are moving. The symmetry of the situation – the principle that the laws of physics are the same in any inertial reference frame – *demands* that the airplane observers must see the mountain clocks running too slowly.

But common sense rebels. Two observers both say to each other "your clock is running slower than mine"? Surely there must be a real paradox hiding in there, an irreconcilable contradiction that demolishes the whole system!

So let's try to nail that down. We'll start by putting specific numbers to it. Let's say the plane is moving at $u = \left(\sqrt{3}/2\right) c$, so $\gamma = 2$. (Observers in the two reference frames will disagree about who is traveling, and in what direction, but they will agree about $|u|$ and therefore about γ.) And let's say the rest length between the two mountains is just long enough that the airplane will take one second to journey from Mountain A to Mountain B, as measured in the mountain frame. So here's what this experiment looks like in the mountain frame:

- As the airplane passes Mountain A, all three clocks (one on each mountain, and one on the airplane) are synchronized to read $t = 0$.

- When the airplane reaches Mountain B, the mountain clocks read $t = 1$.

- Therefore, when the airplane reaches Mountain B, the slow-running airplane clock reads $t = 1/2$.

All of these times are shown in Figure 1.6.

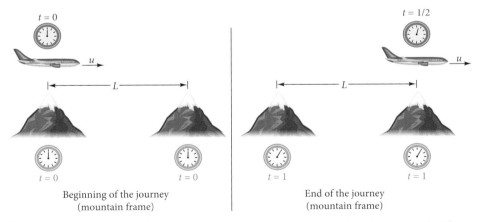

Beginning of the journey
(mountain frame)

End of the journey
(mountain frame)

Figure 1.6 In the mountain frame the two mountain clocks stay synchronized and tick off one second, while the plane clock ticks off half a second.

A statement such as "the airplane clock reads $t = 1/2$ when the airplane passes Mountain B" must be agreed on by all reference frames. An observer on Mountain B, and an observer on the airplane, might both have taken pictures to prove it.

So how does all that look from the airplane frame?

- The airplane clock reads $t = 0$ as Mountain A goes zooming by, and $t = 1/2$ as Mountain B goes zooming by. We've already established both of those facts in a frame-independent way. Therefore the trip took half a second. (We knew that anyway, since the mountain-to-mountain distance was shortened by a factor of 2.)

- The clock at Mountain A read $t = 0$ when Mountain A went zooming by. Later, the clock at Mountain B read $t = 1$ when Mountain B went zooming by. These facts are also frame-independent and indisputable.

- But because the mountain clocks are running slowly by a factor of 2, and because the entire trip took $1/2$ a second, the mountain clocks only advanced $1/4$ of a second during this trip.

Are you following these numbers? We've just concluded that the mountain clocks started at $t = 0$, advanced by $1/4$ of a second, and ended up reading $t = 1$ anyway. We set out to find a paradox; surely this qualifies!

And here's the way out. The clock at Mountain B read $t = 1$ at the end of a $1/4$-second journey. So at the beginning of the journey, just as the airplane was passing over Mountain A, the clock on Mountain B must have read $t = 3/4$.

"No way!" the mountain-based observers cry. "We synchronized the clocks on our two mountains!" But the airplane observers disagree (Figure 1.7).

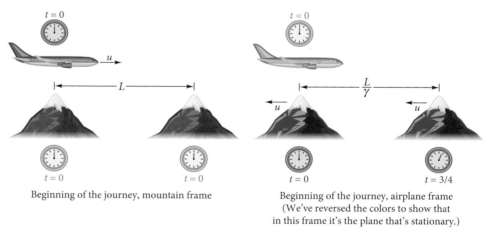

Figure 1.7 In the airplane frame, the two mountain clocks are out of sync, with the second mountain clock three-fourths of a second ahead of the first one.

Take a moment to digest that. At the beginning of the experiment, the mountain-based observers synchronized their two clocks. That is, there were two events – "Mountain A clock is set to zero" and "Mountain B clock is set to zero" – that the mountain observers declared to be simultaneous. From the airplane point of view these two events were *not* simultaneous. Therefore the two clocks were never properly synchronized.

Active Reading Exercise: The Other Clock

In the reference frame of the airplane, what did the clock on Mountain A read at the moment when the airplane reached Mountain B? Jot down your answer along with a brief explanation.

You can check your answer to that Active Reading Exercise against ours at www.cambridge.org/felder-modernphysics/activereadingsolutions

Once you have disagreement about simultaneity, it isn't a big step to conclude that two events that occur in sequence, one after another, can occur in the *opposite order* in another reference frame. Imagine that James is working on Mountain B and he sneezes at the instant his clock says 0.5 s. Which event happens first: "James sneezes on Mountain B" or "the airplane passes Mountain A"? Can you see the two different answers that the two different reference frames must arrive at?

The result that different frames disagree about which events are simultaneous with each other is called "the relativity of simultaneity." You can derive the following formula in Problem 18.

The Relativity of Simultaneity

The short version:

Events that are simultaneous in one reference frame are not simultaneous in other reference frames.

The long version:

Consider a series of clocks that are all synchronized in Frame R. That is, from the perspective of that reference frame, all these clocks strike $t = 0$ at the same moment.

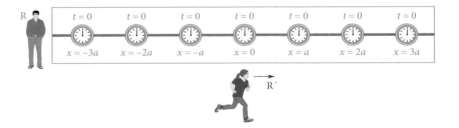

In Frame R, the $t = 0$ events on any two clocks are simultaneous.

Now consider Frame R′ moving at speed u with respect to Frame R. According to that frame, at that moment when the $x = 0$ clock in the R frame strikes $t = 0$, a clock at position $x = a$ in Frame R reads

$$t = \frac{ua}{c^2}. \tag{1.5}$$

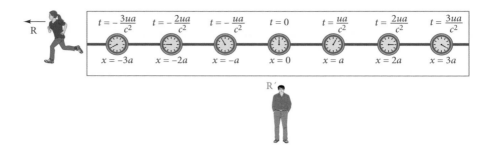

According to Frame R', all these very different readings on the R clocks happen at the same instant: $t' = 0$. (As usual we have recolored the figures so that in both cases the red image is the still reference frame, with respect to which the blue image is moving.)

The simultaneity relationship can be expressed in many ways. The wording we chose above makes no reference to t'-values (different times in the primed reference frame) or x' values (different positions as measured in the primed reference frame). We are looking at the (unprimed) t-values of clocks at different (unprimed) x-values, all at one moment that is simultaneous in the primed reference frame. You could call that instant $t' = 0$ at all points.

That wording has the virtue of easily mapping onto our airplane-and-mountains scenario. The initial synchronizing event is "airplane passes Mountain A," which both systems agree to designate as $x = t = 0$. Equation (1.5) is designed to answer the question: "What does the Mountain B clock read at that same instant, according to the airplane's reference frame?" The answer is ua/c^2, where u is the speed of the plane and a is the distance between the mountains, as measured in the mountain frame.

To calculate that, we'll measure distance in light-seconds and express the speed of light as 1 light-second per second. Recall that in our mountain-and-airplane example we chose $u = (\sqrt{3}/2)c$. We also said that the airplane takes one second to make the trip in the mountain frame, so $a = \sqrt{3}/2$ light-seconds. We conclude that the Mountain B clock reads:

$$t = \frac{ua}{c^2} = \frac{\left(\sqrt{3}/2\,\text{ls/s}\right)\left(\sqrt{3}/2\,\text{ls}\right)}{(1\,\text{ls/s})^2} = \frac{3}{4}\,\text{s}.$$

This is the same conclusion we reached previously without Equation (1.5).

Example: Simultaneity

Two alien ships are flying northward over the Earth at half the speed of light, one thousands of miles ahead of the other. At the exact same moment – in their reference frame! – one ship drops a bomb on Quito, Ecuador, and the other drops a bomb on New York City. If a Quito clock says exactly noon at the moment its bomb drops, what does a New York clock say as it is bombed?

Answer: Equation (1.5) says that at that instant in the ship frame the New York clock says $t_{NY} = ua/c^2 = a/(2c)$. We can make the calculation easier by expressing the speed of light as 1 light-second per second. The constant a is the distance between the two cities in the Earth frame, which is about 4600 km, or about 0.0153 light-seconds:

$$t_{NY} = \frac{0.0153\,\text{ls}}{2(1\,\text{ls/s})} = 0.008\text{ s}.$$

Make sure you follow what that calculation tells us! According to the ship reference frame, $t = 0$ on a Quito clock and $t = 0.008$ on a New York clock are simultaneous events. Because the bombs were dropped simultaneously in the ship frame, they were dropped 0.008 seconds apart in the Earth frame.

One of the lessons of this example is that to notice the effect of the relativity of simultaneity you need both very large speeds and very large distances.

Causality and Relativity

You drop a rock. Some time later, that rock hits the ground. Those two events have a "causal relationship," meaning that the first event led to the second. Any two observers would agree that if you had not dropped the rock, it would not have hit the ground. Therefore any two observers must agree that the first event happened before the second! Similarly, everyone will agree that our plane reaches Mountain A before it reaches Mountain B.

However, the two events "plane reaches Mountain A" and "James sneezes" in our example do *not* have a causal relationship. Two different observers can disagree about the order of these two events, and neither order leads to the impossible conclusion that an effect preceded its own cause.

What's the difference? A core principle of relativity is that no information can travel faster than the speed of light. Another way of expressing this principle is that no *causal influence* can travel faster than light. For instance, if the Sun were to explode, we wouldn't know it for eight minutes. We would feel no difference in heat, see no difference in light, detect no change in gravity. The event "the Sun exploded" could not cause any event of any kind on Earth before the eight minutes had elapsed.

So in relativity we may describe one pair of events as having a "timelike separation" and another pair of events as having a "spacelike separation." The speed of light forms the dividing line between these two types of relationships.

Timelike separation	Spacelike separation
$\Delta x < c\Delta t$	$\Delta x > c\Delta t$
Example: An event occurring on the Sun 20 minutes ago, and an event occurring on Earth right now.	*Example*: An event occurring on the Sun two minutes ago, and an event occurring on Earth right now.
The two events may or may not have a causal connection.	The two events cannot have a causal connection.
All observers will agree about the order in which the two events occurred.	Different observers will disagree about the order in which the two events occurred.
There will be one reference frame in which the two events happened at the same place.	There will be one reference frame in which the two events happened at the same time.

To complete this set of terms, we should say that two events separated by exactly enough distance and time for a light beam to go between them have a "lightlike separation." Different observers will disagree about the distance between them and the time between them, but everyone will agree that the distance equals the time multiplied by c.

1.3.2 Questions and Problems: Length Contraction and Simultaneity

Conceptual Questions and ConcepTests

1. The distance from Earth to Trantor is 20,000 light-years, and they are non-moving with respect to each other. A rocket makes the trip at $(4/5)c$, so a total of $25,000$ years elapses in the Earth/Trantor reference frame. However, less than $25,000$ years elapses on the rocket. Why?

    (a) Answer this question from the Earth/Trantor reference frame.

    (b) Now answer the same question from the rocket reference frame.

2. The twenty-second-century Intra-Terran Bullet Train plows straight through the center of the Earth from Hawaii to Botswana. When the train leaves, its clock is in synch with the Hawaii clock, but when it arrives in Botswana it is behind the Botswana clock (after accounting for time zones).

    (a) Explain this discrepancy from the Earth's reference frame.

    (b) Explain this same discrepancy from the train's reference frame.

3. A train follows a straight-line route from New York to Los Angeles at near-light speed. (Let's ignore all the reasons why that would be a Very Bad Idea.) The train and the ground each define a reference frame. Label each of the following measurements as "both reference frames must agree" or as "the two reference frames might disagree":

    (a) The relative speed of the train and the ground

    (b) The amount of time the journey takes

    (c) The setting of the New York clock as the train leaves New York

    (d) The setting of the train clock as the train leaves New York

    (e) The setting of the Los Angeles clock as the train leaves New York

    (f) Whether the gap between the two events ("train leaves N.Y." and "train reaches L.A.") is spacelike, timelike, or lightlike

4. Much science fiction – even fiction written with a careful eye toward modern physics – involves some sort of phone that allows people in different solar systems to have real-time conversations with each other. Explain why such technology is explicitly forbidden by special relativity. *Don't just say that nothing can go faster than light; explain why it would lead to a paradox if something could.*

5. On p. 28 we presented pictures of the beginning of our airplane's journey from the mountain perspective and the airplane perspective. Draw similar pictures of the *end* of the journey, when the airplane reaches Mountain B. Make sure to note the times on all three clocks in both reference frames. (Six times in all.)

6. Consider a variant of our airplane-and-mountains scenario in which the rest length of the airplane is L: the same as the distance between the mountains in their rest frame. If this airplane were lying down it would fit perfectly between the two mountains. But the airplane is zooming by, and moving lengths are contracted.

    (a) Does the airplane fit between the two mountains or not? Answer this question twice: once from the perspective of the mountain, and once from the airplane. Is this the sort of question that both observers must agree on, or the sort that they can disagree on?

    One way to frame this question is in terms of two events: "front of airplane reaches Mountain B" and "back of airplane reaches Mountain A."

    (b) Which of these events happens first if the airplane is shorter than the mountain gap (and therefore fits)?

(c) Which of these two events happens first if the airplane is longer than the mountain gap (and therefore does not fit)?

(d) Putting all your answers together, which of those two events happens first in each reference frame?

7. If a rocket is traveling at near-light speed, each atom in that rocket is foreshortened in the direction of travel. This alters the geometry of the atom, and in particular the average distance between the electrons and the nucleus. Why doesn't that fundamentally change the behavior of the atoms?

8. Suppose a plane is flying at near-light speed past a chain of mountains. First it passes Mountain A, then Mountain B, and so on. The mountains all have synchronized clocks according to the mountain frame. At the moment the plane passes Mountain B that mountain's clock says noon.

(a) According to the plane frame, does the Mountain C clock at that same moment show a time before, after, or exactly at noon?

(b) According to the plane frame, does the Mountain A clock at that same moment show a time before, after, or exactly at noon?

9. Suppose a chain of equally spaced planes is flying at near-light speed past a mountain. First plane A passes it, then plane B, and so on. The planes all have synchronized clocks according to the plane frame. At the moment plane B passes the mountain that plane's clock says noon.

(a) According to the mountain frame, does the plane C clock at that same moment show a time before, after, or exactly at noon?

(b) According to the mountain frame, does the plane A clock at that same moment show a time before, after, or exactly at noon?

10. Give an example of two events with a timelike separation, an example of two events with a spacelike separation, and an example of two events with a lightlike separation. Of course none of these should be examples that we have already given.

11. Rena and Ezra each own a meter stick (each with a rest length of 1 m, of course). Rena's has sharp blades attached at the two ends. Here is what Ezra sees one day: Rena's meter stick zooms by, and suddenly swings down and slices his meter stick. Because Rena's ruler is shortened, Ezra's is sliced perfectly at the 1/3 and 2/3 meter marks (Figure 1.8). He sends her a photograph of his stick cut into three equal parts and claims that this proves his perspective was correct; her meter stick must have been shorter than his. How does Rena explain this from her reference frame?

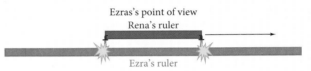

Figure 1.8

For Deeper Discussion

12. One of the most famous apparent paradoxes in relativity involves two farmers trying to fit a long beam into a short barn. For definiteness let's say the beam is twice as long as the barn (when they are both at rest). Farmer Ben has been studying relativity, so he decides that he will carry the beam into the barn so fast that in the barn frame the beam will be contracted by a factor of 3, so it will easily fit. He tells his brother Jack to stand at the barn door and slam the door behind him the moment the beam is entirely in the barn. Does the beam end up fitting in the barn? Whichever way you answer, explain it from both reference frames. *Note:* This becomes particularly interesting if you assume that both the beam and the barn are unbreakable and unbendable.

Problems

13. A football field is 100 yards long. Calculate the length according to each of the following reference frames. (Assume all flight is along the same direction as the long way across the field.)

(a) A human spacecraft flying overhead at $75,000$ m/s

(b) An alien spacecraft flying past the Earth at $c/10$

(c) Superman, flying past the Earth at $0.9c$

14. How fast would you have to travel to shrink all lengths by a factor of 5?

15. (a) Based on Equation (1.4), what is L' when $u = 0$? Briefly explain why this result was predictable without doing the math.

    (b) Based on Equation (1.4), what is $\lim\limits_{u \to c} L'$?

    (c) What happens if you try to plug in $u = 2c$ into Equation (1.4)? What does your answer tell you about the theory of relativity?

    (d) Explain how Equation (1.4) allows us to conclude that the proper length of an object is always as long or longer than its length viewed in a different reference frame.

16. The Example on p. 15 discussed a spaceship journey of 3 light-days at $0.9c$. The medicine on board, which could only last 2 days, made the journey intact and saved the planet because its "clock" was running slowly. Now consider the same scenario from the rocket perspective: the rocket is not moving, and the medicine is expiring at a normal pace.

    (a) How fast is the planet moving toward the rocket?

    (b) How far is the planet from the rocket, in the reference frame of the rocket?

    (c) How long does the planet take to reach the rocket?

    (d) In both the planet and the rocket reference frames, the medicine reaches the planet before expiring. But in one frame this result is due to time dilation, in the other to length contraction. Can you construct a scenario in which the two reference frames *disagree* about whether the medicine reaches the planet unexpired?

17. Section 1.2.2 Problem 27 discussed muons that are created in the upper atmosphere and make the $100,000$ m trip to the Earth's surface, traveling at 99.999% c. We explained that the muons survive the trip, despite having an average life expectancy of only 2.2×10^{-6} s, because their clocks are running slowly. Now consider the same scenario from the muon's perspective. The muons are stationary and their clocks running normally, while the Earth rushes toward them. Perform calculations to show why the surface of the Earth reaches the muons before they decay.

18. **The Relativity of Simultaneity**

In the airplane-and-mountains scenario the plane passes Mountain A with speed u at the moment when the plane clock and the clock on Mountain A both say 0. In this problem you will calculate what time the clock on Mountain B reads at that same instant, according to the plane frame. You can assume that the two mountain clocks are synchronized in the mountain frame. In other words, you are going to derive Equation (1.5).

    (a) In the mountain frame the two mountains are separated by a distance L. What time does the Mountain B clock read when the plane reaches it?

    (b) What time does the plane clock read when the plane reaches Mountain B? *Hint*: This is easiest to answer by continuing to analyze the motion in the mountain frame.

    (c) According to the plane frame, how much time elapsed on the Mountain B clock while the plane was flying between the two mountains?

    (d) Combining your answers to Parts (a) and (c), what time does the plane frame say the Mountain B clock read at the moment the plane passed Mountain A? Simplify your answer as much as possible. You should be able to verify Equation (1.5).

19. The Explanation beginning on p. 23 stepped through the airplane-and-mountains scenario for an airplane speed of $\left(\sqrt{3}/2\right) c$, which led to $\gamma = 2$. In this problem you will re-evaluate the entire scenario with $\gamma = 3$. Assume that, at the moment when the airplane flies past Mountain A, the airplane clock and the Mountain A clock are both set to $t = 0$.

    (a) At what speed must the airplane be flying if $\gamma = 3$?

    (b) Suppose the trip from Mountain A to Mountain B takes the airplane exactly one second, as measured in the mountain frame. What is the proper distance between the two mountains?

(c) What is the distance between the two mountains, as measured in the airplane frame?

(d) How much time elapses on the airplane clock between "airplane flies past Mountain A" and "airplane flies past Mountain B"?

(e) According to the reference frame of the airplane, how much time elapses on the mountain clock between those two events?

(f) In the mountain frame, the two mountain clocks are perfectly synchronized. That is, at the moment when the airplane passed Mountain A, the clock on Mountain B read $t = 0$. According to the *airplane* frame, what did the clock on Mountain B say when the plane passed Mountain A?

20. In the far future, humans colonize the galaxy and synchronize a set of clocks throughout it. A rocket flies past Earth moving at 0.9c at the moment that Earth's clock says 0.

(a) If the rocket is moving toward the center of the galaxy, 30,000 light-years from Earth, what time does the center-of-the-galaxy clock say at the moment the clock passes Earth, according to the rocket frame?

(b) When the rocket reaches the center of the galaxy, what time does the Earth clock read, according to the rocket frame?

(c) When the rocket reaches the center of the galaxy, what time does the Earth clock read, according to the Earth frame?

(d) Suppose that at the moment when the Earth clock says 0 another rocket flies by at 0.9c, flying away from the center of the galaxy. In that rocket's frame, what time does the center-of-the-galaxy clock say at that moment?

21. The star Alpha Centauri is about 4 light-years away from us. Suppose that in our mutual reference frame an Earth comedian named Rebo tells a joke and a week later an alien at Alpha Centauri laughs.

(a) How fast would a rocket have to be moving in order for its reference frame to conclude that the laugh happened at the same time as the joke?

(b) Is it possible that the joke caused the alien to laugh? How do you know?

22. **The Length You See**

In this problem you'll show in one very simple example that there is a difference between what size something has according to your reference frame and the size you actually see when you look at it. The difference occurs because of light travel times.

You are standing at the origin and an object of proper length L is moving toward you along the positive x axis with speed u (in the negative x direction). See Figure 1.9.

Figure 1.9

You are going to do all the calculations in this problem in your reference frame. When we talk about the front end of the object we mean the end closer to you, so the front end has a lower x-value than the back end at any given time.

(a) How long is the object according to your reference frame?

(b) Assuming the front of the object reaches you (the origin) at time $t = 0$, write a function $x_B(t)$ for the position of the back of the object at all times. (Remember that you are working entirely in your frame.)

(c) Remember that, except for objects at $x = 0$, you always see things in the past: you see them when the light reaches your eyes. If an event takes place at position x and time t, when do you see that event? (Assume $x > 0$.)

Our question here is: *how long does the object actually look to you?* We will call the answer L_{app}, the "apparent length" of the object. Remember that at $t = 0$ the front of the object reaches you. So you can find L_{app} by determining where you see the *back* of the object at that moment.

(d) Use your answers to Parts (b) and (c) to answer the question: at what time does the back of the object emit a light beam that reaches you at precisely $t = 0$?

(e) Find L_{app} by finding the location of the back of the object at the moment you calculated in Part (d). Simplify your answer as much as possible.

You should have found that the apparent length doesn't equal the proper length or the length of the object in your frame. You can do similar calculations in two and three dimensions and show that the shapes of objects appear to distort and rotate, not just because of length contraction but also because of light travel time (see Problem 24).

23. [*This problem depends on Problem 22.*]

(a) In Problem 22 you calculated the apparent length of an object moving toward you at speed u. Find the apparent length of the same object moving directly *away* from you at speed u.

(b) Describe what an object would look like to you at the moment its (proper) midpoint passed you.

24. [*This problem depends on Problem 22.*] Another object of proper length L is moving toward you at speed u, but this object is oriented perpendicular to its direction of motion. See Figure 1.10.

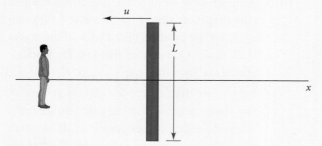

Figure 1.10

When the center of this object is a distance x from the origin (you), the top and bottom are farther away, so light has to travel farther to reach you from the ends.

(a) At a particular moment t, you see the light that was emitted from the center of the object at a particular position x_C. At that same moment you see the light emitted from the top and bottom of the object at what x-value?

(b) Sketch the object's apparent shape from your perspective at several different values of x_C.

25. **The Paradox of the Twins, Revisited**

On p. 17 we met Asher and Emma, two twins who ended up at very different ages. We finally have all the pieces we need to analyze this scenario from Emma's point of view. Just to be specific, let's suppose Emma travels away from Earth at $0.9c$ for precisely 10 years in Earth's (and Asher's) reference frame. When she reaches a space station that is stationary in the Earth reference frame she turns around (more or less instantaneously), and travels back at the same speed.

We begin by analyzing the entire scenario from Asher's reference frame.

(a) How much time did Emma's journey take? (This is the same question as "How much did Asher age during Emma's journey?")

(b) How far away is the space station where Emma turned around?

(c) How much did Emma's slow clock advance during the journey? (This is the same question as "How much did Emma age?")

This is the part where you want to say "Now let's analyze from Emma's perspective." But Emma doesn't have just one inertial reference frame; she has two, an outgoing frame and a returning frame.

(d) First, do one more calculation using Asher's frame. What times did Emma's clock and the space station clock show when Emma arrived there?

(e) How does Emma explain the fact that the slow space station clock shows a later time than her clock does?

(f) Just before she arrives at the space station, Emma is moving along with the outgoing reference frame. According to that frame, what time does Asher's clock show at the moment Emma arrives at the space station?

(g) When she arrives at the space station, Emma jumps over to the returning frame. According to that frame, what time does Asher's clock show at the moment Emma arrives at the space station?

(h) How much time ticks off on Emma's clock during the return journey?

(i) What time does Emma's clock read when she gets back to Earth?

(j) According to the returning frame, how much time ticks off on Asher's clock during Emma's return journey?

(k) Using values from the returning frame, add the time on Asher's clock when Emma arrived at the space station to the time that elapses on his clock during the return journey to find the time on his clock when Emma gets home.

(l) From Asher's point of view, Emma flew to the space station with a "body clock" that was ticking slowly the whole time, and then flew back to Earth with a body clock that was ticking slowly the whole time, so when she got back she was younger than Asher. From Emma's point of view, the whole time she was flying out to the space station her body clock was normal and Asher's was slow, and the whole time she flew back her body clock was normal and his was slow. Based on what you've calculated in this problem, how does Emma explain from her point of view that Asher ends up older than her?

That's a lot of calculations, but make sure you step back at the end and see the point of it all. The Earth clock (Asher) and the rocket clock (Emma) agree at the beginning of the journey that $t = 0$ (the two twins are the same age). They disagree about almost everything that happens during the trip, but at the end, when the twins are reunited, both twins agree again – as they must. Asher is the old twin. (Emma spends the next two weeks trying to teach him how to text with his thumbs, but he never really gets the hang of it.)

1.4 The Lorentz Transformations

We now take up in full generality the fundamental question of relativistic kinematics: given the location and time of an event in one reference frame, how do you calculate its location and time in a different frame?

1.4.1 Explanation: The Lorentz Transformations

The equations we have developed are important, but they cover special cases. The time dilation formula applies to the time gap between two events that occur at the same place in one reference frame. The length contraction formula requires one of the two reference frames to measure both ends of an object at the same time. All of our examples have been designed to fit those criteria.

The "Lorentz transformations" provide the general version of these rules: the generic transformation of spacetime coordinates from one inertial reference frame to another.

The Lorentz Transformations

Consider inertial Reference Frame R′ moving in the x direction with velocity u relative to Frame R. These two frames have synchronized their axes and clocks such that the place and time designated as $x = y = z = t = 0$ is also $x' = y' = z' = t' = 0$.

An event occurs at coordinates (x, y, z, t) in Frame R. The coordinates of this same event in Frame R′ are given by

$$t' = \gamma \left(t - \frac{ux}{c^2} \right)$$
$$x' = \gamma (x - ut)$$
$$y' = y$$
$$z' = z$$

(1.6)

Equations (1.6), along with many other important equations of relativity, are listed in Appendix B.

As you look through the following Example, take a moment to convince yourself that the equations we have previously learned for time dilation and length contraction cannot be directly used to answer this question.

Example: Lorentz Transformations

You are working at Space Station Alpha when you spot a stolen ship flying by at $c/2$. You radio ahead to Space Station Beta, 3×10^{10} m away. In your frame they receive the signal 100 s later. When and where do they receive the signal in the ship's frame?

Answer: As usual we will find it easier to work in relativistic units, so we will call the distance between the two stations 100 light-seconds.

Let R be your reference frame and R' be the ship's. Define $x = x' = t = t' = 0$ when the ship passes you. In your frame the event "signal reaches Space Station Beta" occurs at $x = 100$ light-seconds and $t = 100$ s. Now use the Lorentz transformations with $u = c/2$ to convert those to the ship's frame:

$$\gamma = \frac{1}{\sqrt{1 - (1/2)^2}} = \frac{2}{\sqrt{3}}$$

$$t' = \gamma \left(t - \frac{ux}{c^2} \right) = 58 \text{ s}$$

$$x' = \gamma(x - ut) = 58 \text{ light-seconds}$$

If you want SI units, it's easy to convert to $x' = 1.7 \times 10^{10}$ m. But in the (vastly superior) units we used, t' and x' ended up being exactly the same number. Can you see why that had to happen in this particular scenario?

To answer the question we posed at the end of the Example, t' in seconds equals x' in light-seconds because in the ship reference frame, just as in the station frame, the signal traveled at precisely the speed of light.

Tricky Issues in Measurement

Issues of measurement lie just under the surface of almost every example we have presented in this chapter. In our airplane-and-mountains scenario, the airplane passed Mountain A at a particular moment. An observer in the airplane, and an observer on Mountain A, both synchronized their clocks to $t = 0$ at that moment – implying that they could agree on what both of their clocks said. But they did not agree on whether the Mountain B clock also said $t = 0$ at that moment. Why couldn't they just look?

The answer – this may sound like a detail, but it's a central theoretical issue in relativity – is that when you look at far-away objects, you see them in the past. No information of any kind can ever travel faster than c. You see the Sun as it was eight minutes ago, and nearby stars as they were years ago. So two observers in different reference frames can synchronize their clocks only when they are at the same place.

This may lead you to question our space station scenario. We found the time when Space Station Beta received your signal, measured in the ship frame. But the ship's captain wasn't at Space Station Beta at that event and couldn't directly measure that time.

A common way to think about this issue in relativity is to imagine that each reference frame carries an infinite grid of clocks and rulers that are all stationary and (for the clocks) synchronized, *in that reference frame*. (This is illustrated in Figure 1.11 on p. 41.) If your rulers are each one light-second long and the end of each one is labeled with a number, starting at 0 at Space Station Alpha, then your ruler labeled 100 recorded an event "your message reached Space Station Beta" at the moment that the clock attached to that ruler showed 100 s. (Remember, these clocks are all synchronized in your frame, and we said that your clock was set to 0 when you sent the signal.)

Meanwhile, the ship is moving along with its own infinite grid of rulers, each of which appears short to you because of length contraction, and its infinite set of clocks that appear to you to be out of synch with each other and to all be running slowly. The event when your message reaches Space Station Beta occurs right next to the ship frame's 58 light-second ruler while its clock shows 58 s, thus defining the coordinates of the event in the ship's frame.

The issue of measurement divides facts in relativity into "frame-independent" and "frame-dependent." Everyone in any reference frame must agree about what happens at any particular event. (You can take a picture of what happens at an event and everyone will agree about what's in the picture.) For example, everyone agrees that, at the event "ship passes Station Alpha," the ship clock and the station clock both say 0. That is a "frame-independent" statement.

But anything that involves comparing two events cannot be directly measured. What time did Station Beta's clock say at the moment the ship passed Station Alpha? How much time did it take the ship to get from Station Alpha to Station Beta? How long is the ship? These are "frame-dependent" questions that can only be answered in the context of a particular reference frame.

The Spacetime Interval

One of the most important properties of any coordinate transformation is its "invariants": the quantities that remain the same through the transformation. For instance, if you rotate your x and y axes, the x and y coordinates of every point change, but for any two points the new and old systems agree about the distance $\sqrt{\Delta x^2 + \Delta y^2}$. So distance is invariant under rotation.

When we change reference frames in relativity we know that distances are not invariant, and neither are time intervals. However, in Problem 22 you will show that the following quantity is invariant under the Lorentz transformations:

$$s = \sqrt{c^2 \Delta t^2 - \Delta x^2 - \Delta y^2 - \Delta z^2} \qquad \text{The spacetime interval.} \qquad (1.7)$$

Take a moment to convince yourself that s is real and positive for two events with a timelike separation, imaginary for a spacelike separation, and zero for a lightlike separation.[2]

Coordinate Time, Proper Time, and Spacetime Interval

The spacetime interval is important because, mathematically, any invariant quantity can be a great aid in problem solving. (See for instance Problem 11.) But Equation (1.7) also suggests some important physical interpretations.

Consider two timelike-separated events A and B. There must exist some reference frame (which we'll call R) in which those events occur at the same place. Let Object O be at rest in Frame R and be present at both events, and let T be the time it measures between them.

- T is the coordinate time Δt between the two events as measured in Frame R. Any other inertial frame will measure a longer coordinate time between them.

- T is also the proper time $\Delta \tau$ experienced between the events by Object O. To be present at both events, any other observer would have to accelerate, and would experience a shorter proper time between them.

As an example, you can think (as usual) about the twin paradox. The two events, Emma-leaves-Earth and Emma-returns, have a timelike separation. The "special" reference frame between those two events is Asher's (aka Earth's), because in that frame the two events happen at the same place. (Again – this isn't obvious until you think about it – there is exactly one such frame for any pair of timelike-separated events.) In that one frame the proper time between those two events is also a coordinate time. What we have just said is that any other inertial reference frame will measure a *longer* coordinate time (moving clocks run slow), and any non-inertial observer who reaches both events will measure a *shorter* proper time (Emma ends up younger than Asher).

But what about the spacetime interval? In Asher's frame $\Delta x = 0$, so $s = c\Delta\tau$. That brings us to one important interpretation:

The spacetime interval between two timelike-separated events is the proper time between them, *as measured in the inertial reference frame in which the two events happen at the same place,* times c.

What does that mean to an observer who is not in that particular reference frame? Let's call such an observer "Emma" for the time being. She and Asher will experience different proper times between the two events, but they will agree about the spacetime interval. Emma, Asher, and everyone else will agree that $s = c\Delta\tau_{\text{Asher}}$, where "Asher" refers to an observer in the inertial frame in which the two events occur at the same place.

---

2 Some texts define the interval with opposite signs, as $s = \sqrt{-c^2\Delta t + \Delta x^2 + \Delta y^2 + \Delta z^2}$, which reverses our statements about timelike and spacelike separations.

Derivation of the Lorentz Transformations

We began this section with Equations (1.6), the general conversions of spacetime coordinates between reference frames. You may feel that we should have opened the chapter with those equations, and built up all of relativity from that starting point. But Einstein began with only two postulates: Galilean relativity, and the invariance of light speed. These two postulates led inexorably to the time dilation and length contraction formulas in earlier sections. Now let's see how we can build from there to derive the Lorentz transformations.

We will once again assume that Reference Frame R$'$ is moving in the positive x direction at speed u relative to Frame R, and we will use the image described above of each reference frame having an infinite array of rulers and clocks.[3] We'll put the distance labels at the junctions, next to the clocks. In either reference frame the spacetime coordinates of any event are simply the number written next to the clock and the time on the clock, at that event. We'll assume the origins of the two systems pass each other at time 0, so they both agree that there is an event with $x = x' = t = t' = 0$ (Figure 1.11).

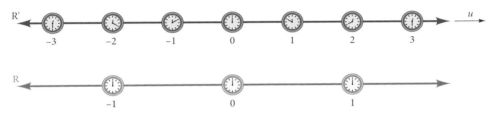

Figure 1.11 The two reference frames, as seen from the R reference frame at $t = 0$. We have drawn the R$'$ grid "above" the R grid, but the two actually lie on top of each other.

From the point of view of Frame R, all of the red clocks tick off time in unison. The blue clocks all start at different times at $t = 0$ (because of the relativity of simultaneity) and all tick time slowly (because of time dilation). Assume an event occurs at the red junction marked with some number x while the clock at that junction reads some time t. Our task is to figure out which blue junction is at that event (x') and what the clock at that blue junction says then (t').

Active Reading Exercise: The Position

Let x' be the blue junction at the event with R coordinates x and t. Each of the questions below is easiest to answer by considering how the blue (R$'$) rulers are behaving from the point of view of Frame R.

1. How far does the x' junction travel from $t = 0$ to the event at time t?

2. Knowing that, at time t, the x' junction reached position x, where was it at time 0?

3. So now that you know *where* it was at $t = 0$, how many rulers are between $x' = 0$ and that junction?

---

3 We don't lose any generality by assuming the motion is in the x direction. Given two reference frames you can always define the x axis to point along their relative velocity vector.

Try your best with those three questions, and then compare your answers with ours at: www.cambridge.org/felder-modernphysics/activereadingsolutions

You should end up with the x' part of the Lorentz transformations. You can pick up the t' part in Problem 21.

1.4.2 Questions and Problems: The Lorentz Transformations

Conceptual Questions and ConcepTests

1. Under what circumstances do we need to use the Lorentz transformations given in this section, instead of being able to use the time dilation and length contraction formulas given earlier in the chapter?

2. James and Cecelia both have watches on their wrists. Under what circumstances will they both agree that their watches are synchronized? (Read all the answers carefully, then choose one.)

   A. They can synchronize their watches if they are both in the same reference frame.

   B. They can momentarily synchronize their watches if they are in the same place.

   C. A and B. That is, they can only synchronize their watches if they are in the same reference frame AND in the same place.

   D. A or B. That is, they can synchronize their watches if they are in the same reference frame, OR if they are in the same place for an instant.

3. The two events $x = 0$ and $x = L$ at $t = 0$ represent two sides of an object of length L lying along the x axis at time 0 in some reference frame R. If R' is moving at velocity u along the x axis, then the Lorentz transformations tell us that these two events occur at $x' = 0$ and $x' = \gamma L$, which seems to say that Frame R' sees the object as *bigger* than L, its length in Frame R. (Recall that $\gamma > 1$.) But length contraction says that R' should see the object as smaller than L. What went wrong with our argument?

4. On p. 40 we offer the following interpretation of s: "The spacetime interval between two timelike-separated events is the proper time between them,

as measured in the inertial reference frame in which the two events happen at the same place, times c."

   (a) Write a similar sentence that begins with the phrase "The spacetime interval between two *spacelike*-separated events is . . ."

   (b) Write an example to illustrate your generalization, beginning with "For instance, suppose you are measuring the length of a couch."

   (c) What can you say about the spacetime interval between two events with a lightlike separation?

5. Use the invariance of the spacetime interval to argue that the proper time measured by an inertial clock between two events is the shortest possible time interval that any inertial frame can measure between those two events.

6. We said that the proper time between two events measured by an inertial clock that goes from one event to the other is the *shortest* time that any inertial reference frame will measure between those two events. We also said that the proper time experienced by that clock is the *longest* time experienced by any observer between the two events. How can you reconcile these two seemingly contradictory claims?

7. **The Three Ways of Defining a Time Interval**

   We now know three ways to describe the time interval between two timelike-separated events. One is the coordinate time as measured in an inertial frame. One is the proper time measured by an observer present at both events. And one is the spacetime interval (divided by c). For each pair of these, under what circumstances are they or are they not equal?

For Deeper Discussion

8. Assume Frame R has an infinite grid of clocks and rulers that cannot be moved from their positions (in the R frame of course). How can it be verified that the clocks are synchronized? Think about why your scheme wouldn't work for clocks in different reference frames from each other.

Problems

9. I'm lounging on my asteroid when you fly by at $(3/4)c$. Coincidentally, both our watches say exactly noon at that time. Twenty light-minutes past my asteroid, my friend Nalini is on another asteroid, with a watch that is perfectly synchronized to mine (in our mutual reference frame). At exactly 12:10 by her watch Nalini drops her phone. According to your frame, what were the time and position at which she dropped the phone?

10. You are working in the Space Toll Booth when the Galaxy's Longest Spaceship flies by (Figure 1.12).

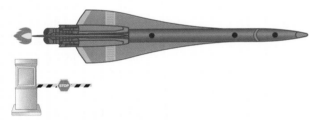

Figure 1.12

Its proper length is $1/10$ of a light-second. (At almost 20,000 miles long, this ship would make a fair-sized planet.) It is moving at $(2/3)c$. Your nose is buried in a book, so you don't notice the ship going by until the *back* of the ship is passing you. At this point you frantically radio the captain (in the front of the ship) to say "Hey, you! Come back and pay the toll!" Assume that the event of your sending the message is at time and position 0 in both frames.

(a) Where and when does the captain receive your signal, according to the ship frame? *Hint*: This requires almost no calculation.

(b) Where and when does the captain receive your signal, according to the toll booth frame? *Hint*: Be careful with your signs.

11. Planets A and B are 10 light-years apart. A very fast ship flies from one to the other in 35 years (as measured in the planets' frame).

(a) Calculate the ship's speed, find γ, and use time dilation to figure out the elapsed time in the ship's frame.

(b) What you just calculated is the ship's proper time for this voyage. But there is an easier way to calculate the same quantity, by using the spacetime interval. Solve the same problem that way (and make sure you get the same answer!).

12. A rocket with a 100 m particle accelerator is moving at $0.9c$ past a young prince. At the instant the back of the ship passes by the prince's asteroid, a particle is launched at $0.9c$, relative to the ship. Suppose that the event of the particle being fired has position and time 0 in both the ship and prince reference frames.

(a) At what moment does the particle reach the front of the ship, in the ship's reference frame?

(b) When and where does the particle reach the front of the ship, in the prince's reference frame?

(c) So, according to the prince, how fast is the particle traveling? (If you get exactly c, go back and re-do the problem, keeping more decimal places.)

(d) What would be the answer to that same question, the particle's speed in the prince's frame, according to Galilean relativity?

13. Equations (1.6) convert the spacetime coordinates of an event from Reference Frame R to R'. Write the equations to convert coordinates from R' to R.

14. Equations (1.6) convert the spacetime coordinates of an event from Reference Frame R to R' assuming they both agree on the origin $x = x' = t = t' = 0$. Rewrite the x and t equations if Frame R' decides to label the point $x = t = 0$ with coordinates $x' = X$ and $t' = T$. (Don't bother with the y and z equations.)

15. Equations (1.6) convert the spacetime coordinates of an event from Reference Frame R to another frame R' that is moving with respect to R in the

x direction. Write the equations to convert from Reference Frame R to a different reference frame R′ that is moving, relative to Frame R, with speed u in the positive y direction.

16. James and Cecelia head off from Earth in opposite directions. They both use $x = t = 0$ to represent the moment of takeoff. James travels at $(1/2)c$ in the $-x$ direction, while Cecelia travels at $(3/4)c$ in the $+x$ direction. Ten hours later Cecelia reaches the dwarf planet Pluto. (The speeds and time given in this problem are all relative to Earth, which we will treat as stationary.)

    (a) Give the time and position of Cecelia arriving at Pluto, in Cecelia's reference frame.

    (b) Give Cecelia's t and x coordinates in James' reference frame.

    (c) How fast was Cecelia traveling, according to James?

17. The formula for time dilation (see Appendix B) applies to the special case of two events that occur at the same place in one reference frame. Show how this equation can be derived as a special case of the Lorentz transformations.

18. The formula for length contraction (see Appendix B) applies to the special case of two events that occur at the same time in one reference frame. Show how this equation can be derived as a special case of the Lorentz transformations.

19. Use the Lorentz transformations to prove that the speed of light is the same in all reference frames. (This is admittedly circular logic, since we used that postulate to derive the Lorentz transformations, but it's a nice consistency check.)

20. Event A at position x_A happens at time t_A, and Event B at position x_B happens at a later time t_B. If a beam of light starting at Event A could reach x_B at a time earlier than t_B, then the two events have a timelike separation, and it is possible for Event A to have caused Event B.

    (a) Use the Lorentz transformations to show that if the separation between two events is time-like in one frame, it is timelike in all frames.

Hint: You can make the math a bit easier by assuming (with no loss of generality) that x_A and t_A are zero in both reference frames, and that $x_B > 0$.

    (b) Use the Lorentz transformations to show that if the separation between two events is timelike in one frame, then $t_B > t_A$ in all frames.

21. Deriving the $t′$ Transformation

In the Active Reading Exercise on p. 41 you derived the $x′$ formula in the Lorentz transformations. In this problem you will use that formula to derive the $t′$ transformation. As usual we will consider inertial Reference Frame R′ moving in the x direction with velocity u relative to Frame R, with their central clocks synchronized at the moment they pass, such that $x = t = 0$ is also $x′ = t′ = 0$.

    (a) If an event occurs at position x and time t in the unprimed frame, what is its position in the primed frame? (This is just asking you to write down the Lorentz transformation for position, which we already derived.)

    (b) Now turn it around: if an event occurs at position $x′$ and time $t′$ in the primed frame, what is its position in the unprimed frame? (Think about the velocity.)

    (c) Solve your two equations simultaneously to eliminate $x′$ and write $t′$ as a function of t and x. You'll know you're done when you have derived the $t′$ portion of Equations (1.6)!

22. The Spacetime Interval

The spacetime interval between Events 1 and 2 is defined as

$$s = \sqrt{c^2(t_2-t_1)^2 - (x_2-x_1)^2 - (y_2-y_1)^2 - (z_2-z_1)^2}.$$

Use the Lorentz transformations to show that the interval is the same in any two reference frames. *Hints:* It's sufficient, and algebraically easier, to show that s^2 is the same. You can also make the algebra a lot simpler (with no loss of generality) by taking the first event to be the origin: $t_1 = x_1 = y_1 = z_1 = 0$.

1.5 Velocity Transformations and the Doppler Effect

We conclude our survey of relativistic kinematics by considering how velocities and light frequencies transform at relativistic speeds.

1.5.1 Discovery Exercise: Velocity Transformations

A spaceship is traveling in the positive x direction at $(3/4)c$ relative to you.

1. On board the spaceship is the captain's chair, which in the frame of the spaceship is immobile. What is the speed of this chair relative to you?

2. The captain shines a flashlight forward (in the positive x direction). What is the speed of the light beam in the ship's reference frame? What is the speed of the light beam in your reference frame?

Now a crew member at the back of the ship launches a small missile toward the front of the ship. (Don't ask us why.) The speed of that missile, in the reference frame of the ship, is also $(3/4)c$.

Question: What is the speed of the missile in your reference frame?

3. What is the answer to that question according to Galilean relativity?

4. Why is that clearly not the right answer?

5. Without doing any calculations, you can see that the right answer is between what (lower bound) and what (upper bound)?

1.5.2 Explanation: Velocity Transformations and the Doppler Effect

If you compare the Galilean transformations (Equations (1.1)) to the Lorentz transformations (Equations (1.6)) you may well feel that something is missing. We talked about how velocities transform in Galileo's system, but we have not yet discussed the issue in Einstein's.

The Galilean equation $v' = v - u$ cannot be correct, as it would have some objects traveling faster than light in some reference frames. If Object A is going at $0.9c$ relative to Object B, which is going at $0.9c$ relative to Object C, and so on for 23 more iterations, Object A *still* must be going less than c relative to Object Z.

So the Galilean transformations don't work. Here are the equations that do.

The Lorentz Velocity Transformations

Consider inertial Reference Frame R′ moving in the x direction with velocity u relative to Frame R.

An object is moving with velocity $\vec{v} = \langle v_x, v_y, v_z \rangle$ in Frame R. The velocity of that object in Frame R′ is given by:

$$v'_x = (v_x - u) \frac{1}{1 - v_x u/c^2}$$

$$v'_y = \frac{v_y}{\gamma} \frac{1}{1 - v_x u/c^2}$$ (1.8)

$$v'_z = \frac{v_z}{\gamma} \frac{1}{1 - v_x u/c^2}$$

Equations (1.8), along with many other important equations of relativity, are listed in Appendix B.

You might be surprised that we have to transform v_y and v_z. If the relative velocity u is in the x direction, won't the two frames agree about the y and z coordinates of any event? Yes they will – but they won't agree about the *time* gap between two events, so they won't agree about dy/dt or dz/dt.

More generally, the velocity transformations follow directly from the transformations we have already seen for position and time. You can derive Equations (1.8) in Problems 16 and 17.

Example: Velocity Transformations

A particle accelerator slams two particles into each other. Particle A is traveling at $0.9c$ and Particle B at $0.95c$ in the lab frame. Find the effective speed of the collision – that is, find the speed of Particle B in the reference frame of Particle A – if . . .

1. the collision is head-on.
2. the particles collide at a right angle.

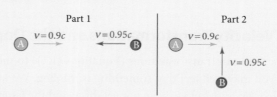

Answers: Note that we will treat the velocity of Particle A in the lab frame as defining the positive x direction.

1. In the first scenario Particle B is moving in the negative x direction.

$$v'_x = (v_x - u) \frac{1}{1 - v_x u/c^2} = (-0.95c - 0.9c) \frac{1}{1 - (-0.95c)(0.9c)/c^2} = -0.997c$$

$$|\vec{v}| = 0.997c$$

2. In the second scenario Particle B has velocity $\langle 0, 0.95c, 0 \rangle$ in the lab frame. Remember that γ is based on the relative velocities of the two reference frames: the frame of Particle A, and the lab frame.

$$\gamma = \frac{1}{\sqrt{1 - (0.9)^2}} = 2.294$$

$$v'_x = (v_x - u) \frac{1}{1 - v_x u/c^2} = (0 - 0.9c) \frac{1}{1 - 0} = -0.9c$$

$$v'_y = \frac{v_y}{\gamma} \frac{1}{1 - v_x u/c^2} = \frac{0.95c}{2.294} \frac{1}{1 - 0} = 0.414c$$

$$|\vec{v}| = \sqrt{(0.9c)^2 + (0.414c)^2} = 0.991c$$

There are a few ways in which you can reality-check those answers. It was perfectly predictable, with no math, that v'_x in the second scenario would be $-0.9c$: do you see why? It makes sense that both answers came out less than c but greater than the individual velocities of the particles. And it also makes sense that the first answer came out higher than the second one.

The Doppler Effect

Imagine a sound wave hitting your ear. The note you hear is based on the frequency of the wave: that is, how many cycles per second hit your ear.

It's not hard to imagine that if you are moving *toward* the incoming sound, more cycles per second will reach your ear. You will be running toward the peaks at the same time they are approaching you, so you will pass from one peak to the next faster than a stationary observer. The higher frequency will be perceived as a higher pitch. If you are moving *away* then fewer cycles per second will reach your ear, lowering the pitch.

For an animation that might make this effect clearer, see www.cambridge.org/felder-modernphysics/animations

The change in frequency that occurs when the source and/or receiver is moving is called the Doppler effect. It's the reason why a train whistle sounds higher as the train is approaching you, and lower as it speeds away.

You can substitute light waves for sound waves in the description above and quite correctly conclude that if you are moving toward a light source, the color is shifted toward the blue end of the spectrum; if you are moving away, the shift is toward the red.

However, although these two phenomena – Doppler shifting of sound waves, and Doppler shifting of light waves – can be described with similar words, they are actually quite different. The reason for this difference is that sound travels through a medium (usually air). If you are moving through the air toward the source of the sound, that's quite different from the source moving through the air toward you. (Both increase the frequency, but differently.)

With light, on the other hand, there is no medium. Einstein's first postulate makes a solemn promise: an observer moving toward a light source, and the light source moving toward the observer, are indistinguishable.

That difference in physics leads to a difference in the math. The equation for Doppler-shifted sound includes two different velocities: the velocity of the source of the sound (relative to the air), and the velocity of the receiver (relative to the air). The equation for Doppler-shifted light includes only u, the relative velocities of the source and the observer. This is the formula we present below.

The Doppler Shift for Light

Consider two objects moving with relative speed u. If one object emits light with frequency v, the frequency of that light in the reference frame of the other object is

$$v' = v\sqrt{\frac{1 \pm |u|/c}{1 \mp |u|/c}}. \tag{1.9}$$

Use the top signs ($+$ in the numerator and $-$ in the denominator) if the source and receiver are getting closer. Use the bottom signs if they are moving apart.

Equation (1.9) comes pretty quickly from the Lorentz transformations we've already seen. You can go through the derivation in Problem 18. (You can find definitions of wave properties like frequency and wavelength, and how they are related in Appendix F.)

Example: The Doppler Shift for Light

Astronomers frequently measure hydrogen clouds emitting radiation with frequency $v = 1.42\,\text{GHz}$. (That's 1.42 billion cycles per second, or equivalently 1.42 cycles per nanosecond.) Suppose they measure this radiation from a distant galaxy moving away from Earth at $0.1c$. What frequency of light do they detect?

Answer: Remember that when two objects are moving away from each other, we use the bottom signs in Equation (1.9):

$$v' = (1.42\,\text{GHz})\sqrt{\frac{1-0.1}{1+0.1}} = 1.28\,\text{GHz}.$$

As expected, we find that the frequency is lower than the original. (If that isn't obvious to you, hold onto the visual image of the star moving away rapidly as it emits those 1.42 waves each nanosecond. Can you see why *fewer* than 1.42 waves per nanosecond reach us?)

1.5.3 Questions and Problems: Velocity Transformations and the Doppler Effect

Conceptual Questions and ConcepTests

1. If the relative motion of two observers is in the x direction only, they will agree about the y coordinates of all objects. So why do they not agree about v_y for all objects?

2. A loudspeaker emits waves that travel out in circles. (They actually travel out in spheres but we'll just pay attention to the part that stays at ground level.) If the loudspeaker is at rest, the circles are all concentric (Figure 1.13).

The largest circles represent the earliest waves, and the smaller ones are waves emitted more recently. Redraw this figure for a loudspeaker moving to the right at constant speed. Each circle should be centered on the point where it was emitted, so the larger circles will be centered on points farther to the left than the smaller ones. (Assuming the loudspeaker is moving more slowly than the speed of sound, the circles will not overlap.) Explain how your drawing illustrates the Doppler effect on both the right and left sides of the loudspeaker.

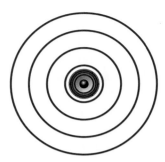

Figure 1.13

3. Look at Equations (1.8) on p. 46 and consider the signs of the results.

(a) Under what circumstances does v'_x come out positive?

(b) Under what circumstances does v'_y come out positive?

(c) Are these answers the same, or different, from the answers you would expect from Galilean relativity?

4. Two objects are approaching each other along the x axis, both moving at speed v in the lab frame. What is their relative approach speed as $v \to c$, according to . . .

(a) Galilean relativity?

(b) Einstein's relativity?

5. From 1915 to 1917 Vesto Slipher measured light coming from 25 galaxies (not including our own) and found that in 21 of them the light had lower frequencies than their gases usually emit. What did that tell him about the movements of those galaxies?

For Deeper Discussion

6. You are standing at rest on the ground. Object A is shining a light while moving toward you at half the speed of light. Object B is playing a sound while moving toward you at half the speed of sound. Which one experiences a greater Doppler shift (fractional shift in frequency)? Don't just look up the formulas; explain why one effect would be larger than the other.

Problems

7. Spaceman Spiff is illegally speeding at $0.4c$. The Space Police Car is chasing him at $0.9c$.

(a) How fast would the police car be approaching in Spiff's reference frame according to a Galilean calculation?

(b) How fast is the police car approaching in Spiff's reference frame, according to special relativity?

(c) The police car has a bright blue light, shining at 650 THz. What light frequency does Spiff see?

8. Particle A is moving in the positive x direction (to the right) at $0.99c$. Particle B is moving at $0.95c$, down-and-left, at an angle $30°$ below the x axis. What is their relative speed when they collide?

9. (a) How fast would you have to go to make a red light (400 THz) appear green (600 THz)?

(b) Would you be heading toward the light, or away from it?

(c) How fast would you have to travel in the opposite direction to make a green light appear red?

10. As of this writing, the most distant known object in the universe is the galaxy GN-Z11. Astronomers were able to measure the types of light typically emitted by hydrogen gas coming from this galaxy, but all of the light appeared at 1/12 of its usual frequency. How fast was GN-Z11 moving relative to us when it emitted that light?

11. The Andromeda galaxy is moving toward us at about a thousandth of the speed of light. Hydrogen gas in Andromeda emits radiation with frequency 1.420405 GHz. At what frequency do we receive that radiation?

12. Relativistic equations should always approach Newtonian physics as $c \to \infty$. Show that Equations (1.8) (p. 46) have this property.

13. Frame R′ is moving at speed u in the x direction relative to Frame R. (Isn't it always?) Show that if a photon is traveling at speed c in the x direction in Frame R, then it is also traveling at speed c in the x direction in Frame R′.

14. Show that if a photon is traveling at speed c in one inertial reference frame, it is traveling at speed c in all inertial reference frames. (This is conceptually similar to Problem 13, but working in three dimensions instead of one makes the algebra messier.)

15. Addition of Velocities

A spaceship flies by you with speed u_1 in the positive x direction. At the same moment the passengers of that ship see a second ship going by at speed u_2 in the positive x direction.

(a) How fast is the second spaceship moving according to your reference frame? This equation is sometimes called the formula for addition of velocities. It is not quite identical to the v_x part of Equations (1.8), but can be immediately derived from it.

(b) If the first spaceship is moving at $c/2$ in your frame and the second one at $c/2$ in the frame of the first ship, how fast is the second one moving in your frame?

16. Deriving the v_x Transformation

Inertial Reference Frame R$'$ is moving in the x direction with velocity u relative to Frame R. An object is moving along the x axis with speed v_x in the unprimed frame. The object starts at the place and time that has been mutually designated $t = t' = x = x' = 0$. At a later time t the object reaches the position $x = v_x t$ in the unprimed frame. Calculate the time t' and position x' of this event in the primed frame, and divide them to get $v'_x = x'/t'$. (You should not of course use Equations (1.8) in your calculations; rather, you should end up with one of them!)

17. Deriving the v_y Transformation

Inertial Reference Frame R$'$ is moving in the x direction with velocity u relative to Frame R. An object is moving with velocity $\vec{v} = v_x\hat{i} + v_y\hat{j} + v_z\hat{k}$ in the unprimed frame. The object starts at the place and time that has been mutually designated $t = t' = x = x' = y = y' = z = z' = 0$. At a later time t the object reaches the point $(v_x t, v_y t, v_z t)$ in the unprimed frame. Calculate the time t' and vertical position y' of this event in the primed frame, and divide them to get $v'_y = y'/t'$. (You should not of course use Equations (1.8) in your calculations; rather, you should end up with one of them!)

18. Deriving the Doppler Shift

The frequency of a wave is given by one over the period, $\nu = 1/T$ (note the difference between nu, the symbol for frequency, and italic vee, the symbol for velocity), and the period is the time from when one wave crest passes you to when the next one passes you. An object is moving past you in the x direction at speed u, emitting light waves in your direction. We're going to call the object's frame R and yours R$'$. The object emits a wave crest (the highest point of the wave) at the exact moment it passes you, and of course you receive that crest immediately.

(a) According to the object's frame, it emits the next wave crest at time $t = T$, the period of the wave. According to your frame, at what time t' was that second wave crest emitted?

(b) According to your frame, at what position was that second crest emitted?

(c) What time do you say elapsed between your receiving the first and second crests?

(d) You should now have a relationship between the periods T' and T. Using the fact that $\nu = 1/T$, rewrite this as a relationship between ν' and ν. When you check your answer against Equation (1.9) remember that in this scenario the object is moving *away* from you.

19. The Paradox of the Twins, Re-Revisited

On p. 17 we met Asher and Emma, two twins who ended up at very different ages. Let's suppose Emma travels away from Earth at $0.9c$ for precisely 10 years in Earth's (and Asher's) reference frame. When she reaches a space station that is stationary in the Earth reference frame she turns around (more or less instantaneously), and travels back at the same speed. (If you have already worked Problem 25 in Section 1.3 you will find the first part of this problem familiar, but from there on you will analyze the situation in a different way.)

We begin by analyzing the entire scenario from Asher's reference frame.

(a) How much time did Emma's journey take? (This is precisely the same question as "How much did Asher age during Emma's journey?")

(b) How far away is the space station where Emma turned around?

(c) How much did Emma's slow clock advance during the journey? (This is the question "How much did Emma age?")

This is the part where you want to say "Now let's analyze from Emma's perspective." But Emma doesn't have an inertial reference frame; she has two! We will analyze the scenario from *Emma's outgoing perspective:* a reference frame traveling away from Earth at 0.9*c*. (The process for doing it from the incoming reference frame is similar.)

(d) During the first part of the journey, Emma's rocket is stationary while the Earth rushes away from her and the space station rushes toward her, both at 0.9*c*. What is the distance between these two objects?

(e) How long does the space station take to reach her? (Don't use the results from Asher's frame; do your calculations in the outgoing frame.)

(f) How much time does Asher's clock tick off during that part of the trip?

(g) During the second part of the journey, the Earth and the space station are still moving backward at 0.9*c*. But Emma is now moving backward much faster. How fast?

(h) What is Emma's speed relative to Earth on her return journey? *Remember that you are calculating everything according to the outgoing frame.*

(i) How long does it take Emma to return to Earth?

(j) During that second time period, how much time does Asher's slow clock tick off? By how much time does Emma's even slower clock advance?

(k) In total, according to the outgoing reference frame, how much older is each twin when Emma returns to Earth?

Chapter Summary

Our survey of Einstein's special theory of relativity is divided between Chapters 1 and 2.

The equations for both chapters are collected in Appendix B. You may want to cross-reference that appendix as you read through this summary.

Our "Reference Frame Exploration" animation uses airplanes, mountains, and birds to demonstrate time dilation, length contraction, the relativity of simultaneity, and spacetime diagrams (Section 2.1): www.cambridge.org/felder-modernphysics/animations

Section 1.1 Galilean Relativity

Galilean relativity says: "The laws of physics are the same in any inertial reference frame."

- If you are in a rocket ship moving at a constant velocity of 60 m/s, or 1,000,000 m/s, the laws of motion inside your rocket ship will be indistinguishable from the laws inside a non-moving rocket ship. You can therefore do all your calculations in the "inertial reference frame" in which the rocket is not moving.

- If the rocket ship slows down, speeds up, or turns, then it no longer qualifies as an inertial reference frame. This is not an arbitrary definition: the acceleration of your reference frame changes the physical behavior of objects, even when both the objects and the measuring devices are all sealed inside the rocket.

- Equations (1.1) on p. 6 (not in Appendix B) are the *classical* transformations of position, time, and velocity from one reference frame to another.

Section 1.2 Einstein's Postulates and Time Dilation

The two postulates of special relativity are "the laws of physics are the same in any inertial reference frame" (as Galileo said), and "the speed of light is the same in any inertial reference frame."

These two postulates lead to "time dilation," expressed by the maxim "moving clocks run slowly" by a factor of γ (defined in Appendix B).

- The time dilation formula $\Delta t' = \gamma \Delta t$ applies to two events that occur at the same place in Reference Frame R (such as two chimes of a clock at rest in R).

- This formula applies as long as Frame R' is inertial, whether or not Frame R is inertial.

- The time gap between two events measured in any inertial reference frame (even if no observer is present at both) is called a "coordinate time." The time gap measured by any observer (even if the observer isn't inertial) is called a "proper time."

Section 1.3 Length Contraction and Simultaneity

Two more consequences of Einstein's postulates are "length contraction," which says that moving rulers are shortened by a factor of γ, and the fact that different reference frames disagree about which events occur simultaneously.

- The length contraction formula $\Delta x' = \Delta x/\gamma$ applies to the length of a ruler at rest in Frame R. Observers in Frame R' must measure both ends of the ruler at the same time t'.

- Length contraction happens only in the direction of motion; other directions are unaffected, so shapes are distorted in different reference frames.

- No inertial reference frame will see two events as simultaneous unless the events have a "spacelike separation" ($\Delta x > c\Delta t$). In that case no information of any kind can travel between the two events, so they cannot have any causal relationship. Two events with a "timelike separation" ($\Delta x < c\Delta t$) may or may not have a causal relationship.

Section 1.4 The Lorentz Transformations

The Lorentz transformations are the relativistic transformations of time and position from one inertial reference frame to another.

- The time dilation, length contraction, and simultaneity equations are all special cases of these transformations.

- The "spacetime interval" between two events (in Appendix B) is an invariant quantity, meaning all reference frames will agree on it.

Section 1.5 Velocity Transformations and the Doppler Effect

- The Lorentz velocity transformations are the relativistic transformations of velocity from one reference frame to another.

- The Doppler effect for light is the relativistic transformation of a light wave's frequency from one reference frame to another.

2

Relativity II: Dynamics

You've been through an entire chapter on special relativity and you haven't seen the equation $E = mc^2$ even once. How weird is that?

Chapter 1 focused on relativistic "kinematics": time, and space, and motion. If you stop your relativity study here you'll have a solid introduction to some of the core ideas of relativity and how they arise from Einstein's two postulates.

But if you continue through this chapter you will learn how to extend many Newtonian concepts to a relativistic framework. You'll see how to represent interactions in spacetime visually in Section 2.1, and how to extend our notions of scalars and vectors to 4D spacetime in Section 2.4. In Sections 2.2 and 2.3 you'll learn how the laws of conservation of mass, energy, and momentum are redefined in relativity, leading among other things to Einstein's famous relation between mass and energy. Finally, in Section 2.5 you'll revisit the Michelson–Morley experiment that we introduced in Section 1.2 and learn the details of this experiment, which played a key role in convincing the physics world to accept Einstein's radical ideas.

2.1   Spacetime Diagrams

The human brain evolved with a huge capacity for processing visual information,[1] so we often translate information into visual terms to leverage our natural strengths. Curves representing functions, bar graphs and pie charts displaying data, and even political cartoons are all clever techniques for harnessing our visual processing capacity to understand abstract information. (If we say "this function attains a local maximum where its derivative is zero" do you think about numerical differences and limits, or do you picture the top of a hill?)

In classical mechanics we use plots of position, velocity, and acceleration to develop our intuition for motion. In this section we will make plots of position vs. time in relativity. The goal, as with any visualization technique, is to give us a useful tool for understanding the theory.

---

1 Some evidence suggests that in blind people these parts of the brain can be repurposed for other types of processing, e.g. detecting subtle differences in sensation while reading braille.

2.1.1 Discovery Exercise: Spacetime Diagrams

 A "spacetime diagram" is a plot with time on the vertical axis and space on the horizontal axis. Notice that this is backward from how you are probably used to plotting $x(t)$! Figure 2.1 shows an example with two moving objects.

1. Do the two objects start at the same time, the same place, or both?

2. Which object is moving faster, the blue or the green? How can you tell?

3. Copy the sketch. (Don't worry about distinguishing the blue and green lines.) Sketch the path of a light beam on your diagram. Assume you are using relativistic units, measuring time in seconds and distance in light-seconds.

Figure 2.1 A spacetime diagram for two moving objects.

2.1.2 Explanation: Spacetime Diagrams

In classical mechanics you graph $x(t)$ with time on the horizontal axis and position on the vertical one. In relativity it's standard to make those plots the other way, with time on the vertical axis. Such a plot is called a "spacetime diagram." The path of an object on a spacetime diagram is called its "world line."

Active Reading Exercise: Spacetime Diagrams

Draw a spacetime diagram with world lines for the following three objects. Assume all three start at $x = 1$ at time 0.

1. An object at rest.
2. An object moving in the negative direction at a constant velocity.
3. An object moving in the positive direction and slowing down as it goes.

Make all three sketches before reading further!

None of this has anything to do with relativity so far, right? We're just getting used to plotting time on the vertical axis. The object at rest traces out a vertical line. The object moving in the negative x direction follows a line that goes up (increasing time) and left (decreasing position). The third world line goes up and right but curves upward as it goes, approaching the vertical line of an object at rest. You can see our three drawings at www.cambridge.org/felder-modernphysics/activereadingsolutions

Spacetime diagrams are drawn in relativistic units (such as seconds and light-seconds, or years and light-years). In such units $c = 1$, so light beams travel both left and right at a

$45°$ angle. The slope of a world line is $\Delta t/\Delta x$, the reciprocal of speed, so anything slower than light has a steeper line than $45°$.

The following Example shows how we can use these diagrams to visualize a complicated situation.

Example: Sending Messages

On p. 17 we met the twins Asher and Emma. As you may recall, Asher stays on Earth while Emma journeys out at a constant speed and then returns at a constant speed. For this example we will suppose she travels at $c/2$ and her destination is 5 light-days away, so her journey takes 10 days in each direction in Asher's reference frame. Every two days Asher sends his sister a radio message. How many of those messages does Emma receive on her way out, and how many on her way home?

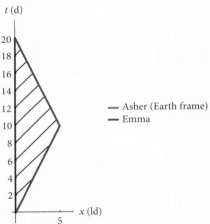

Answer: You can certainly solve that algebraically, but you can see it in an intuitive way on a spacetime diagram. Emma's speed is 1/2 light-day per day, so her world line has a slope of ± 2. (Remember that speed is one over slope.) The radio messages all have a slope of 1. The red and blue lines in this figure are Asher's and Emma's world lines, respectively.

You can simply count to get the answer. Emma gets 2 messages on her way out and 7 on her way home.

This diagram gives us another way to visualize Doppler shifts. Asher sends out messages (pulses) at regular intervals, but for Emma they get spread out when she is moving away and bunched up when she is moving toward Asher.

The discussion and drawing above show Asher's perspective. The x axis shows the entire (one-dimensional) world at the time that he would call $t = 0$. The t axis contains all the events that he would consider to be happening at $x = 0$: his own position throughout the story.

Of course, much of relativity involves seeing events from different perspectives. Let's use x' and t' to represent coordinates in Emma's outgoing reference frame. Below we invite you to take the first step toward that visualization, drawing the axes.

Active Reading Exercise: Axes on Spacetime Diagrams

1. Draw the axes of a spacetime diagram. The coordinates x and t are of course defined in some reference frame R.

Now consider a frame R′ moving at speed $c/2$ to the right, relative to R.

2. Draw the t' axis on your Frame R spacetime diagram: that is, all the points for which $x' = 0$.

3. Add the x' axis into the same drawing.

Will either axis be horizontal, or vertical? Will they be perpendicular to each other? None of these answers is obvious without some thought. Try to go back and forth between two modes of thinking: purely visual (think about the twin story above), and algebraic (the Lorentz transformations are in Appendix B).

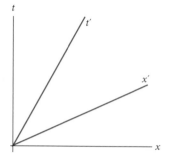

Figure 2.2 Spacetime diagram drawn in Frame R, showing the axes of Frame R′ that is moving to the right at speed c/2.

The first thing to notice is that, in this Example, R′ is Emma's inertial reference frame on the outgoing part of her trip, moving to the right at $c/2$. Emma's outgoing world line is the lower blue line in the diagram in the Example. In R′ she is at rest during that time period: that is, $x' = 0$ from the time she leaves to the time she reaches the outpost. So that is the first axis that the above Active Reading Exercise calls for, the t' axis for Emma's outgoing reference frame.

Algebraically you can set $x' = 0$ in the Lorentz transformations, using $c = 1$ and $u = 1/2$ (relativistic units). You arrive immediately at $t = 2x$, indicating that the t' axis is a line with a slope of 2: the same conclusion we reached above.

By a very similar argument, setting $t' = 0$ in the Lorentz transformations leads quickly to $t = (1/2)x$. That tells us that the x' axis is a line with slope $1/2$ (Figure 2.2).

This still isn't Emma's perspective since we are still using Asher's frame to define the horizontal and vertical axes of our diagram. See Question 17.

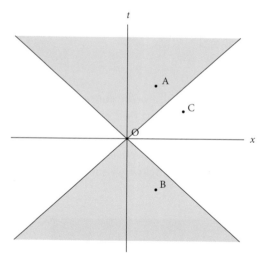

Figure 2.3 The past and future light cones of Event O. Events A and B each have timelike separations from O, while O and C are spacelike separated.

Light Cones and Causality

Remember that an "event" in relativity happens at one particular location in space, at one particular moment in time. So every point on a spacetime diagram represents an event. Figure 2.3 labels four events, with Event O representing $x = t = 0$. The diagonal lines show light beams coming both to, and from, Event O.

The two outgoing light beams (in the upper half of the figure) form the "future light cone" of Event O. The shaded region inside the future light cone is sometimes called Event O's "absolute future." Any point inside that region (such as Event A) has a timelike separation with Event O. Observers in all reference frames will agree that Event O came before Event A. It is possible that Event O had some influence on Event A.

Similarly, the incoming light beams form the "past light cone" that defines the "absolute past" of Event O. So Event B could potentially influence Event O. Events in the unshaded region, such as Event C, have a spacelike separation from Event O; they can neither influence it nor be influenced by it.

(The term "light cone" comes from 3D spacetime diagrams with x, y, and t. In such a diagram the future and past light cones literally have the shapes of cones. Of course our actual spacetime with x, y, z, and t is 4D, but we can't draw that very well.)

Remember that we're talking about events, not objects. Earth's North and South Poles can influence each other back and forth in all kinds of ways. But an event happening at one pole, and an event happening a nanosecond later at the other (in the Earth reference frame), are spacelike separated. *Every* frame will agree that neither of those events can have an effect on the other.

World Lines, the Spacetime Interval, and Proper Time

A proper time interval is the duration experienced by a single person, or clock, between two given events. The proper time is not just a function of the two events; it also depends on the particular path the person follows between them. For example, in the twin paradox Asher and Emma started and ended at the same two events (Emma leaves and Emma returns). But they followed different world lines between those two events, so they experienced different proper times (Asher aged more than Emma).

By contrast, if you want the spacetime interval between two events, you can just calculate $s = \sqrt{c^2 \Delta t^2 - \Delta x^2}$ between the starting and ending points; the path doesn't matter.

These two quantities are related in an important way, however.

Along any straight world line, the spacetime interval from start to end is c times the proper time along that world line.

We made the same point on p. 40, only then we talked about an object at rest in an inertial reference frame instead of one traveling along a straight world line. Can you see that the two are equivalent?

So what if the entire journey is *not* a straight line? In that case, just knowing the spacetime interval from beginning to end is not enough to calculate the proper time. But if the journey is a sequence of straight lines, you can run that calculation piecewise.

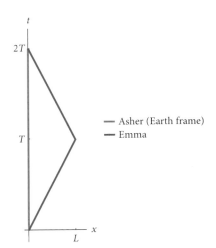

- For each linear leg of the journey, you can calculate the spacetime interval just from the endpoints of that leg. Then divide by c to get the proper time for that leg.
- Add up all the proper times to get the total proper time for the journey.

Which brings us back to Asher and Emma. Look at their world lines on Figure 2.4.

Asher's world line is one straight line. His proper time for the entire trip ($2T$), times c, is the spacetime interval ($\sqrt{c^2(2T)^2 - 0^2}$). It's easy enough to confirm that that works.

Figure 2.4 Emma (blue world line) travels to a space station a distance L away and returns. The red world line represents her twin Asher, who remains on Earth. The journey takes time $2T$ in Asher's frame.

Emma's world line is two straight lines. We can do similar calculations for each of them individually, and then add the two proper times to see how much she ages.

Active Reading Exercise: Proper Time and Spacetime Interval

Figure 2.4 shows the world lines of Emma (blue) and Asher (red). *The five items below should take less than 30 seconds of calculation each. If it gets messier than that, try a different approach.*

1. Write the spacetime interval for the first (outgoing) leg of Emma's trip. Your answer will be a function of the constants T and L.

2. Write Emma's proper time for that same leg of the journey.

3. Write Emma's proper time for the second (incoming) leg of the journey.

4. Write Emma's proper time for the entire journey.

5. Is the final result greater than, or less than, Asher's proper time $2T$?

The Active Reading Exercise above is considerably easier than the calculations you went through for the same scenario in Chapter 1. The relationship between proper time and the spacetime interval, properly understood, can save you a lot of algebra.

You should have ended up concluding that Emma's proper time is $2\sqrt{T^2 - (L/c)^2}$. You can see that this is less than $2T$, offering yet another proof that Emma ends up younger. We can generalize this result.

The longest possible proper time between any two timelike-separated events is a straight world line.

Look what happened with the Emma-and-Asher calculations and you can see why this works in general: any world line that goes out to the side and back adds extra Δx, which subtracts from the proper time of the journey.

2.1.3 Questions and Problems: Spacetime Diagrams

Conceptual Questions and ConcepTests

1. Sketch the world line for an object that moves in the positive direction at half the speed of light for one second, stops abruptly and stays at rest for two seconds, and then gradually (over the course of several seconds) speeds back at up to half the speed of light in the negative direction.

2. Object A traveled in an inertial reference frame, but Object B did not. How would this fact be reflected on their world lines?

3. A and B represent two distinct (non-identical) events, but the spacetime interval between them is zero.

    (a) What does that tell you about those two events, physically?

    (b) What does that tell you about those two events on a spacetime diagram?

    (c) Would it be possible for Event A to cause Event B? Briefly explain your reasoning.

4. We said that two objects are causally connected if either one could potentially have caused the other (even if, in fact, they didn't).

    (a) Is it possible for Events A and B to be causally disconnected, B and C to be causally disconnected, and A and C to be causally connected? If so, show an example on a spacetime diagram. If not, explain why not.

(b) Is it possible for Events A and B to be causally connected, B and C to be causally connected, and A and C to be causally disconnected? If so, show an example on a spacetime diagram. If not, explain why not.

5. Draw a spacetime diagram in Frame R and show what the x' and t' axes would look like on that diagram for Frame R', moving to the right at $(4/5)c$ . . .

(a) using Einsteinian relativity

(b) using Galilean relativity.

In both cases use relativistic units.

6. What shape would the world line of an object have if it were moving at constant speed in a circle around the origin? (Because it's moving in 2D the spacetime diagram would be 3D.)

7. With one spatial dimension, spacetime diagrams are 2D and the future light cone at any moment in time looks like two points. If you similarly looked at the future light cone of an event in three spatial dimensions, what shape would it have at a single moment in time?

8. Figure 2.3 shows (among other things) a gray shaded region representing the absolute future of Event O. Any world line starting at O that goes outside that region is physically impossible. Can there be a world line that stays entirely in that shaded region and is still physically impossible? If so, draw an example. If not, explain why not.

9. Rank all of the world lines in Figure 2.5 (left to right, red, orange, green, blue) from shortest to longest proper time. (It's possible that some may be equal.)

Figure 2.5

10. Consider two events with a timelike separation. If you wanted to travel between them in the *longest* possible proper time then you would travel a straight world line; that is, you would choose a reference frame in which the two events happened at the same place. (Imagine jumping on a train that leaves precisely at one event and arrives precisely at the other.) There is no *shortest* possible proper time you could experience between the two events, however. Why not?

11. On p. 55 we drew a spacetime diagram of the twin paradox in which Asher stays on Earth while Emma travels away and then returns. Below that we drew the x' and t' axes of Emma's outgoing reference frame on that diagram, but we were still drawing the diagram in Asher's frame, meaning his x and t axes were horizontal and vertical. Draw a spacetime diagram in Emma's outgoing frame (with that frame's x' and t' axes horizontal and vertical) and draw Asher's x and t axes on it. You don't need to include anything else on the diagram (unless you go on to do Problem 17).

12. **(a)** Draw a spacetime diagram and label two points: A at $x = t = 0$ and B at $x = 0, t = 1$.

(b) The points in which we are interested are all the points with the following property: the spacetime interval between this point and Point A is the same as the spacetime interval between A and B. (Point B itself is obviously one such point, but not the only one.) Draw the set of all such points. What shape does it describe?

13. Use a spacetime diagram to illustrate the relativity of simultaneity. That is, show how two events can occur in a different order in different inertial reference frames.

For Deeper Discussion

14. Mathematicians define a "distance function" d as any function between pairs of objects A and B in a set (e.g. points in a space) that satisfies the following three conditions:

(a) If $A = B$ then $d(A, B) = 0$. If $A \neq B$ then $d(A, B) > 0$.

(b) $d(A, B) = d(B, A)$.

(c) The "triangle inequality": $d(A, C) \leq d(A, B) + d(B, C)$.

Is the spacetime interval between two events with a timelike separation a valid mathematical distance function between two events? Why or why not?

Problems

15. Draw a spacetime diagram (first quadrant only) and label the following three points:

> A: $x = 1$ light-second, $t = 3$ seconds;
>
> B: $x = 2$ light-seconds, $t = 6$ seconds;
>
> C: $x = 4$ light-seconds, $t = 4$ seconds.

For each pair of events (A-B, B-C, and A-C):

(a) Identify the separation as timelike, spacelike, or lightlike.

(b) Calculate the spacetime interval.

(c) If the separation is timelike, calculate the proper time that would elapse if you traveled from one to the other along a straight-line path.

16. Draw a spacetime diagram (first quadrant only) and label the following three points:

> A: $x = 1$ light-second, $t = 3$ seconds;
>
> B: $x = 2$ light-seconds, $t = 6$ seconds;
>
> C: $x = 3$ light-seconds, $t = 8$ seconds.

Imagine traveling along a straight world line from A to B, and then from B to C.

(a) Calculate the proper time for the first leg of your journey, and then calculate the proper time for the second leg of your journey.

(b) If you add those two proper times do you get the proper time for your entire journey? Why or why not?

(c) Calculate the spacetime interval between Events A and B, and then calculate the spacetime interval between Events B and C.

(d) Calculate the spacetime interval between A and C. Is it the sum of the two spacetime intervals you have already calculated? If not, is it bigger or smaller than that sum?

17. On p. 55 we drew world lines for the following scenario: Asher stays put for 20 light-days; Emma travels to the right at $(1/2)c$ to a point 5 light-days away, and then returns home at the same speed.

The entire drawing was done from Asher's reference frame. Redraw the entire scenario from Emma's outgoing reference frame.

18. In Frame R, Event A happens at $x = 0$, $t = 1$ year.

(a) Draw a spacetime diagram (in Frame R) showing Event A.

(b) Frame R$'$ is moving at velocity $(3/5)c$ relative to R. Draw a spacetime diagram in Frame R$'$, with horizontal and vertical x' and t' axes. Show Event A on your diagram.

(c) Frame R$''$ is moving at velocity $-(4/5)c$ relative to R. Draw a spacetime diagram in Frame R$''$, with horizontal and vertical x'' and t'' axes. Show Event A on your diagram.

(d) In each of your diagrams A appears at a different location, but you should have found that all of these locations are on the hyperbola $t^2 - x^2 = 1$. Sketch that hyperbola on your diagrams to confirm this. Then explain why it makes sense that the locations would all lie on that curve.

19. Two events A and B have a spacetime interval $s = 5$ light-seconds, but you know nothing else about them.

(a) Explain how we know that the separation between these two events is timelike.

(b) If you traveled along a straight world line from A to B, how much time would your watch tick off?

(c) If you traveled along a different world line from A to B, would the resulting spacetime interval be greater than, equal to, or less than 5 light-seconds?

(d) Still on your alternative path, would the amount of time recorded on your watch be equal to, less than, or greater than, the amount you calculated in Part (b)?

20. Frame R$'$ Axes

(a) Draw an empty spacetime diagram to represent a reference frame R. You may want to use graph paper, and/or have a ruler handy.

(b) Draw in the world line for an object that starts at $x = t = 0$ and moves to the right at $(3/5)c$.

(c) The object whose world line you just drew represents an inertial reference frame that we will call R′. Explain why the line you drew is the $t′$ axis for that object's reference frame.

(d) Use the Lorentz transformations (Appendix B) with $c = 1$ and $u = (3/5)c$ to reach the same conclusion about where the $t′$ axis lies on the R spacetime diagram.

(e) Now use the Lorentz transformations to find the $x′$ axis and draw that into your plot as well.

21. [*This problem is a generalization of Problem 20. You don't need to do that one to do this, but if you find this one difficult, you may want to do Problem 20 first.*] Frame R′ is moving to the right at speed u with respect to Frame R.

(a) Draw an empty spacetime diagram (first quadrant only). Your x axis represents "the set of all events for which $t = 0$"; your t axis represents "the set of all events for which $x = 0$."

(b) The $t′$ axis represents "the set of all events for which $x′ = 0$." Using the Lorentz transformations, find the equation for this line. Express your answer as a function $t(x)$. (Your answer will involve the relative velocity u, but it will not contain any primed variables. It will not contain c either, because we are working in units in which $c = 1$.)

(c) The $x′$ axis represents "the set of all events for which $t′ = 0$." Find the equation for this line, also expressed as a function $t(x)$.

(d) On your spacetime diagrams, draw and label the $t′$ and $x′$ axes for $u = 1/10$, $u = 1/2$, and $u = 9/10$. Don't worry about pinpoint accuracy, but do worry about the slopes of these six lines relative to each other.

(e) In general, for any u such that $0 < u < 1$, what can you say about the slopes of the $t′$ and $x′$ axes?

(f) Do points on these lines have a spacelike relationship to the origin, or a timelike relationship to the origin? Briefly explain why these answers must be correct (in terms of causality, not in terms of the calculations you just did).

22. The spacetime diagram in Figure 2.6 is drawn in Frame R, and it also shows the axes for Frame R′, which is moving at velocity $(4/5)c$ relative to R. Distance is measured in light-years and time in years.

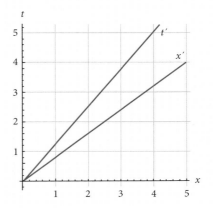

Figure 2.6

Make tick marks on the $t′$ axis showing years as measured in Frame R′. How far apart are those marks, as measured in the Frame R diagram?

23. **Time Dilation and Simultaneity**

You are standing on the planet Narn (Frame R). Your friend G'Kar flies to Centauri Prime at $(3/5)c$. In your frame (R), Centauri Prime is 3 light-years away and G'Kar's trip takes 5 years. As usual, we'll assume that your clock and G'Kar's clock were both set to zero at the start of the journey.

(a) What time does the clock on G'Kar's ship show when he reaches Centauri Prime?

(b) Draw a spacetime diagram in your frame that shows the world line of G'Kar's trip, the $x′$ axis, and the $t′$ axis. *Hint*: Two of the three are the same. Mark an "E" on your diagram at the event "G'Kar reaches Centauri Prime."

(c) Mark a line on your diagram showing all the events simultaneous with Event E, according to Frame R. Mark another line showing all the events simultaneous with Event E, according to R′. Make sure to clearly indicate which line is which.

(d) What is the time on the Narn clock (the t axis) on each of the two lines you drew?

(e) You're going to answer the following question twice, once for each reference frame. At the instant G'Kar arrives on Centauri Prime, is his clock ahead of or behind your clock on Narn, and by what factor? *The answers should be trivial based on what you've already done.*

   i. Answer according to your reference frame (R).

   ii. Answer according to G'Kar's reference frame (R').

24. Length Contraction

The length of an object in any frame is the distance from one end of the object to the other, both measured at the same time.

(a) Explain why the length of an object (in any frame) is thus equal to the magnitude of the spacetime interval between the two ends at the moments you measure them. (Why do we need the word "magnitude" in that sentence? Based on its definition, s cannot be negative – but it can be imaginary.)

(b) If a rod of length L in Frame R is at rest between points $x = 0$ and $x = L$, draw a spacetime diagram and shade in the (two-dimensional) region occupied by the rod.

(c) Frame R' is moving to the right at $c/2$ relative to Frame R. Mark two points on the spacetime diagram that you might use to measure the length of the rod in Frame R'.

(d) Using the marks you made in Part (c) and the fact that the measured length in R' is the spacetime interval between those two events, argue whether the length in R' is larger or smaller than L. *You should not need to use the Lorentz transformations or the phrase "length contraction."*

2.2 Momentum and Energy

We have been discussing kinematics, the description of motion. Now we turn to dynamics, the laws that determine how things will actually move in different situations. In special relativity the most important dynamic principles are the conservation of momentum and the conservation of energy.

2.2.1 Discovery Exercise: Momentum and Energy

Consider the following scenario: an object starts at rest, and is subject to a constant force in the positive x direction.

1. In Newtonian mechanics, as you know, $\vec{F} = m\vec{a} = m(d\vec{v}/dt)$. Based on that equation, draw a quick sketch of this object's speed as a function of time. Then answer the question: what is $\lim_{t \to \infty} v$?

2. In relativity, as you know, speed can never reach or exceed c. Draw a second graph that copies your first graph for $v \ll c$, but obeys this universal speed limit in the long term.

3. If you keep accelerating an object, its kinetic energy will increase without bound (just like in classical physics). With that in mind, explain why the classical equation for kinetic energy $K = (1/2)mv^2$ cannot be correct in relativity.

As a bonus, you might want to think about how you could modify the function $K(v)$ so it would work in this limit. You don't have to know the correct answer, but think about whether you can find some function that would behave correctly.

2.2.2   Explanation: Momentum and Energy

In classical mechanics there are two equivalent ways to express the laws of motion. One is Newton's laws: given certain forces, here's how a system will accelerate. The other is conservation laws: the total momentum and energy of an isolated system are conserved.[2]

There are situations where one approach is more useful than the other. A block sliding down a plane with friction is easy to analyze with Newton's laws, but conservation of momentum and energy aren't useful because you would have to include the momentum of the Earth (pulled upward by the block) and the energy of all the molecules of the plane and surrounding air that get energy from the frictional sliding. Conversely, a car collision surely obeys Newton's second law but the forces are complex and unknowable; conservation of momentum is the practical way to figure out what will happen.

Either Newton's laws or the conservation laws are *in principle* sufficient to predict how objects will move in any given situation. Starting from either set of laws you can derive the other. That's still true in relativity: you can start with a postulate about forces (the equivalent of Newton's second law), or one about conservation of momentum and energy. In relativity, however, the conservation laws are generally easier to use.

The Definitions of Momentum and Energy

Newtonian momentum is defined as $\vec{p} = m\vec{v}$. Newtonian kinetic energy is defined by $K = (1/2)mv^2$. From these you can quickly conclude that $K = p^2/(2m)$. (As usual, when we write p or v without the arrow, that refers to the magnitude.)

When you switch reference frames using the Galilean transformations, momentum and kinetic energy both change. However, you can show mathematically that if these quantities are conserved in one reference frame then they are conserved in all others. In Problem 16 you can show that Newtonian momentum and kinetic energy cannot be conserved in special relativity; if they are conserved in one reference frame, then the Lorentz transformations ensure that they are *not* conserved in other frames. So if special relativity is going to retain the conservation of energy and momentum, we need new definitions of these quantities.

At a minimum, those definitions need to satisfy two requirements:

- If momentum or energy is conserved in one inertial reference frame, then it should be conserved in all inertial reference frames.
- In the limit $v \ll c$ our definitions should reduce to their Newtonian values.

And there is a third requirement, more important than any theoretical construct: if we claim that these quantities are conserved, that claim should hold up under experiment. Collisions at

---

2 Also angular momentum, but we're not going to talk about angular momentum in relativity.

relativistic speeds are commonplace events in particle accelerators. If we make predictions based on conservation laws, and those predictions don't hold up in the lab, our proposed conservation laws are wrong.

Below are the relativistic definitions of momentum and energy. We will not "derive" these definitions. (You never really derive a definition.) But they do obey the three requirements above. In Section 2.4 we'll present further arguments for these equations as the only reasonable relativistic generalizations of classical momentum and energy.

Relativistic Momentum and Energy

A particle moving at velocity $\vec{v}$ (with speed v) has the following momentum and energy:

$$\vec{p} = \gamma m \vec{v} = \frac{m\vec{v}}{\sqrt{1 - (v/c)^2}} \tag{2.1}$$

$$E = \gamma mc^2 = \frac{mc^2}{\sqrt{1 - (v/c)^2}} \tag{2.2}$$

These definitions lead to the relation

$$E = \sqrt{p^2 c^2 + m^2 c^4}. \tag{2.3}$$

The momentum equation looks like a reasonable relativistic extrapolation of the classical definition, and clearly reduces to $\vec{p} = m\vec{v}$ for $v \ll c$, just as it should. The energy equation is not so obviously related to a classical limit. Also, there are many forms of energy in classical physics, so which ones are included in this formula? We will address these issues in this section and the next.

But first – with those definitions in place – we are ready to state the relativistic law of conservation of momentum and energy.

For an isolated system, relativistic momentum and energy are conserved.

The word "isolated" means no particles enter or leave the system and no external forces act on it. That law is really four equations because it says that the energy and all three momentum components are conserved. In 1D we only need two of those four.

Before we dive deeper into what the equations are telling us, let's look at an example of how to use that conservation law.

Example: An Elastic Collision

A ball of mass $3m$ moving at $(4/5)c$ collides elastically with a stationary ball of mass m.

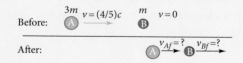

1. What are their velocities after the collision?

2. Using your answer to Part 1, show that momentum is conserved in the frame in which the mass $3m$ ball is initially at rest.

Answer:

1. We handle this exactly as we would in Newtonian physics, by setting the momentum and energy before the collision equal to the momentum and energy after the collision. Only now we use the relativistic formulas:

$$p_0 = \frac{3m(4/5)c}{\sqrt{1 - (4/5)^2}} = 4mc$$

$$p_f = \frac{3mv_{Af}}{\sqrt{1 - (v_{Af}/c)^2}} + \frac{mv_{Bf}}{\sqrt{1 - (v_{Bf}/c)^2}}$$

$$E_0 = \frac{3mc^2}{\sqrt{1 - (4/5)^2}} + mc^2 = 6mc^2$$

$$E_f = \frac{3mc^2}{\sqrt{1 - (v_{Af}/c)^2}} + \frac{mc^2}{\sqrt{1 - (v_{Bf}/c)^2}}$$

Setting $p_0 = p_f$ and $E_0 = E_f$ gives two equations with two unknowns. The algebra is a mess, but you can hand it to a computer and get $v_{Af} = (8/17)c$, $v_{Bf} = (12/13)c$. You can plug these back in to verify that the final momentum and energy come out to $4mc$ and $6mc^2$.

2. Having assumed conservation of energy and momentum to derive those results in the original frame, we now want to show that they are consistent with the conservation laws in the frame in which the $3m$ mass is at rest. In this frame $v_{A0} = 0$ and $v_{B0} = -(4/5)c$. To find the final velocities we use the velocity transformation rule with u (the velocity of the new frame relative to the old) equal to $(4/5)c$:

$$v_{Af} = \frac{(8/17)c - (4/5)c}{1 - (8/17)(4/5)} = -\frac{28}{53}c$$

$$v_{Bf} = \frac{(12/13)c - (4/5)c}{1 - (12/13)(4/5)} = \frac{8}{17}c$$

You can plug in the initial and final velocities to find $p_0 = p_f = -(4/3)mc$ and $E_0 = E_f = (14/3)mc^2$. So the momentum and energy are different in this frame, but they are still conserved.

In Section 2.3 we will discuss the relativistic definition of the term "elastic," which is not identical to its classical meaning. In the meantime, though, the way you analyze elastic collisions is the same in relativity as in Newtonian mechanics: set the momentum and energy before the collision equal to the momentum and energy after. In relativity you just have more complicated formulas for $\vec{p}$ and E.

What Are Momentum and Energy in Relativity?

Whenever you see a new equation – especially a complicated one – you should take a few minutes to look at it and see what the math is telling you. What do Equations (2.1) and (2.2) for momentum and energy mean?

Let's start with the easier of the two. Equation (2.1) says that momentum increases with both mass and velocity, just as we're used to. If $v \ll c$, that's pretty much the whole story: as always, the relativistic formula reduces to the classical at low speeds. That's why the two p vs. v curves in Figure 2.7 look alike on the left side.

But at high speeds that new factor of γ makes a big difference: *even though velocity is limited, momentum can grow without bound.* That's why the relativistic curve in Figure 2.7 approaches a vertical asymptote at $v = c$.

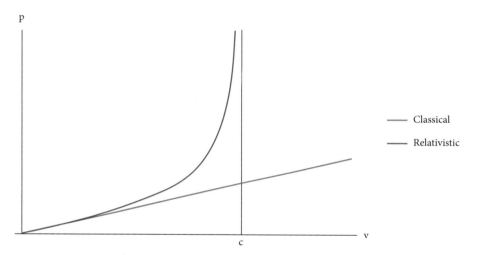

Figure 2.7 Momentum as a function of velocity.

That momentum asymptote makes it possible to formulate Newton's second law in relativity. It can't still be true that $\vec{F} = m\vec{a}$ because that would lead to speeds greater than c. But it can be (and is) still true that $\vec{F} = d\vec{p}/dt$. See Problem 18.

Equation (2.2) for energy has the same property of approaching infinity as $v \rightarrow c$. That tells us that adding energy to a system indefinitely will cause it to approach but never reach the speed of light (Figure 2.8). But what kind of energy does $E = \gamma mc^2$ represent? Is it kinetic energy? Potential energy? Thermal or chemical energy? Some new relativistic energy that has no classical analogue?

The answer is, yes to all of the above. *Equation (2.2) represents the total energy of an object.*

By using a Maclaurin expansion (Problem 17), you can break down that relativistic energy formula as follows:

$$E = \gamma mc^2 = mc^2 + \left(\frac{1}{2}mv^2 + \text{terms involving } \left(\frac{v}{c} \right) \text{ raised to increasing powers} \right). \quad (2.4)$$

The latter part – the part in parentheses – is the kinetic energy of the object, concisely defined as:

$$\text{Kinetic energy } = \gamma mc^2 - mc^2.$$

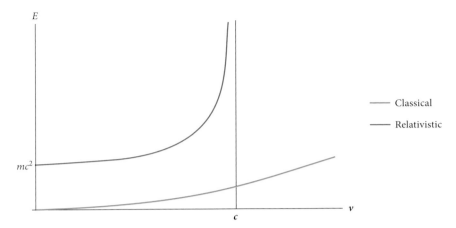

Figure 2.8 Energy as a function of velocity.

That is the energy that depends on the object's motion, just as Newtonian kinetic energy is. The relativistic kinetic energy reduces to the Newtonian formula for low velocities, but grows without bound as $v \to c$.

The mc^2 term, sometimes called the "rest energy" of the object, has no classical analogue. It indicates that every massive particle has enormous energy simply by virtue of having mass. But it goes far beyond that. We have already said that Equation (2.2) represents the total energy of an object. The rest energy therefore includes *every form of energy except kinetic.*

As an example, suppose you put a poker into the fire. As the metal heats up its internal energy increases. Part of that internal energy is in the form of kinetic energy of the molecules in the poker, but the v in Equation (2.4) refers to the net velocity of the entire object, and the poker isn't going anywhere. That means $v = 0$ in Equation (2.4) and the kinetic energy of the poker as a whole is still 0. So the increase in internal energy is reflected in the poker's rest energy, its mc^2.

As another example, consider stretching a spring from its relaxed position. That increases its potential energy by $(1/2)kx^2$. In relativity that's once again reflected in an increase in mc^2.

But how can mc^2 increase, when c is a constant? The only possible answer is that m, the mass of the poker or spring, increases. This leads to one of the most important and un-Newtonian ideas in relativity:

> *Whenever you increase the internal energy of an object – thermal energy, chemical energy, or any other form – that increases the object's mass.*

How much? Since $E = mc^2$ is the rest energy, $\Delta m = \Delta E/c^2$ gives the increase in mass.[3]

That may sound like a bookkeeping trick: we don't want to count those increases as kinetic, so we enter them in the "rest energy" ledger and awkwardly conclude that the mass has increased. But those changes really do measurably change an object's mass. We will discuss the idea of mass more carefully in Section 2.3.

In the meantime, there is one more question that you should always ask in relativity.

---

3 Einstein's 1905 paper "Does the Inertia of a Body Depend Upon its Energy Content?" didn't actually include the form $E = mc^2$ of this equation. He wrote "If a body gives off the energy L in the form of radiation, its mass diminishes by L/c^2."

Question: What happens to all this energy if you look at the same object from a different reference frame?

Answer: The kinetic energy is different. (That is of course also true classically.) But the rest energy is the same. You can think of Equation (2.4) as breaking the total energy down into "the part that depends on what reference frame you look from" and "the part that doesn't."

In fact, as we will discuss in the next section, the relativistic definition of mass is carefully crafted to ensure that it is frame-independent. As we have already stressed, calculations are much easier if you have a few invariant quantities you can count on.

Defining a System (or, "Does Potential Energy Count, or Not?")

Imagine that a brick is lying on the ground in front of you, and you pick it up and put it on a table. In an introductory mechanics class it's common to say that the brick has gained potential energy mgh. But its mass hasn't changed (nothing about the brick is different), and its velocity v hasn't changed (still zero), which means its γmc^2 is the same on the table as it was on the ground. Is this an exception to the rule that γmc^2 represents the total energy?

There are no exceptions to that rule, but you have to be careful to define the system you're looking at. The brick itself has the same energy it started with, and so does the Earth. But the *Earth–brick system* has gained energy, and therefore that system's mass has increased. (Even in a careful Newtonian analysis, that extra energy does not *really* reside in the brick, although we often ignore that subtlety in our calculations.)

So, does γmc^2 include potential energy, or not? The answer is that it includes all the potential energy inside an object. That includes molecular bonds and potential energy within atoms and nuclei. On a macroscopic level it includes the gravitational potential energy for the Earth–brick system or the spring potential energy for a stretched spring. It does not include potential energy that exists in relation to anything outside the object under consideration.

One final thought on our brick: when you moved the brick up to the table, the energy to lift it came from the stored chemical energy in your muscles. So the energy and mass of the Earth-and-everything-on-it system remained the same, as they must for an isolated system.

Non-Relativistic Particles, Ultra-Relativistic Particles, and Massless Particles

If an object is moving much slower than the speed of light, you can use Newtonian formulas for its momentum and kinetic energy. We call such objects "non-relativistic," even though they still obey the equations of relativity, because those equations only provide tiny corrections to the simpler Newtonian ones.

At the other extreme, an object moving very close to the speed of light is called "ultra-relativistic." In that case the equations simplify in a different way, because $E^2 = p^2c^2 + m^2c^4$ can be approximated as $E = pc$.

There's no exact definitions of the speeds at which you can call particles non-relativistic or ultra-relativistic, but as a rough guide you can generally use Newtonian formulas if an object's speed is less than a few percent of c, and you can use $E = pc$ if an object's speed is within a few percent of c. In between, for speeds like $v = c/2$, you need to use the full relativistic formulas. If you know an object's kinetic energy, you can figure out which category it's in without bothering to calculate its speed: a non-relativistic particle has $K \ll mc^2$ and an ultra-relativistic one has $K \gg mc^2$ (Problem 14).

The extreme case of an ultra-relativistic object is a massless particle. Equations (2.1) and (2.2) for momentum and energy suggest that a particle with zero mass would have zero momentum and zero energy. But light is made of massless particles called "photons" (see Chapter 3) that do carry momentum and energy. Looking more closely at those definitions, the only way you can have $m = 0$ and not get a zero result is if $v = c$. In that case both the momentum and energy formulas become indeterminate: $0/0$. So a massless particle must always travel at exactly $v = c$, and it has energy and momentum that are not given by Equations (2.1) and (2.2). (You'll see in Chapter 3 that the energy and momentum are actually related to the frequency of the light wave.)

The relationship between energy and momentum (Equation (2.3)) still holds, though, and with $m = 0$ it reduces exactly to $E = pc$. We will make use of that equation when we examine the nature of light more carefully in the chapters on quantum mechanics.

2.2.3 Questions and Problems: Momentum and Energy

Conceptual Questions and ConcepTests

1. Under what circumstances are the energy and momentum of a system conserved?

2. In Newtonian physics a constant force causes speed to increase linearly with time. Explain why this cannot be universally true in relativity.

3. A subatomic particle in a particle accelerator is accelerated from rest at time $t = 0$, up to *almost* the speed of light at time $t = T_f$. With an eye on Figure 2.8, describe how the energy of this particle grows from the beginning to the end of its journey . . .

   (a) classically

   (b) relativistically

   (c) According to which description does this acceleration require more energy?

   (d) Which description is correct?

4. Explain why a massless particle must travel at exactly c.

5. A "tachyon" is a hypothetical particle (much discussed in science fiction) that moves faster than light. Give some reasons why it's unlikely that such particles could exist.

6. Two objects are floating in space orbiting each other. You fly up in your rocket and pull them farther apart, and then leave. (Assume you leave them each with the same velocities they had before you arrived.) Does the mass of the two-object system . . .

A. decrease?

B. increase?

C. stay the same?

Briefly explain your answer.

7. Four protons deep within the Sun fuse together to form a helium nucleus, emitting radiation in the process. *Note:* You do not have to Google any trivia about protons, or helium nuclei, to answer this question. The fact that they fuse tells you everything you need.

   (a) What effect does fusing have on the potential energy of the four-particle system? (Choose one and explain.)

      A. The potential energy increases.

      B. The potential energy stays the same.

      C. The potential energy decreases.

   (b) How does the mass of the resulting helium nucleus compare to the sum of the masses of the four protons before the fusion? (Choose one and explain.)

      A. The helium nucleus has more mass than the four protons combined.

      B. The helium nucleus has the same mass as the four protons combined.

      C. The helium nucleus has less mass than the four protons combined.

8. Which of the following equations are valid for a massive particle? (Choose all that apply.)

A. $p = \gamma mv$

B. $E = \gamma mc^2$

C. $E^2 = p^2c^2 + m^2c^4$

D. $E = pc$

9. If you apply the equation $p = \gamma mv$ to a photon (a particle of light), which of the following statements about the equation is true? (Choose one.)

A. It tells you correctly how to find the momentum of a photon.

B. It tells you the momentum of a photon, but its answer is not valid.

C. It gives you 0/0, which doesn't really tell you anything.

Problems

10. In 2018 the (roughly) 600 kg Parker Solar Probe surpassed 150,000 miles per hour (67,000 m/s), becoming the fastest man-made object relative to the Sun. (That's what the news reported, anyway. We assume they're not counting particle accelerators.)

(a) Calculate the probe's momentum using both the classical and the relativistic formulas. Then calculate the percent error if you were to use the classical formula to estimate the relativistic momentum.

(b) Similarly, calculate the percent error if you were to use the classical formula to estimate the kinetic energy.

11. (a) How much energy would you have to add to an object to increase its mass by 1 mg? (That's *milligram*, not kilogram.)

(b) Burning a gallon of gasoline releases about 10^8 J. How much gasoline would you have to burn to produce the amount of energy you calculated in Part (a)?

12. The Large Hadron Collider can accelerate protons up to 6×10^{12} eV. How fast are those protons moving?

13. The fastest particles known on Earth are cosmic rays that strike our upper atmosphere. These include protons with energies up to 10^{20} eV.

(a) How much mass would an object at rest need, to have an equivalent amount of energy?

(b) A proton is accelerated until its energy is 10^{20} eV. If you try to calculate its velocity on most calculators, you may find that round-off error leads to the obviously incorrect conclusion that $v = c$. So instead of finding v, find γ for that proton.

(c) In the proton's frame, how long does it take to cross our galaxy (roughly 100,000 light-years across)?

14. A particle's kinetic energy K is defined as its total energy minus its rest energy. How fast would a particle have to move for its kinetic energy to equal its rest energy?

15. Particle A has mass $3m$ and is initially moving at $(1/2)c$ in the positive x direction. Particle B has mass m and is initially moving at $(1/2)c$ in the negative x direction. If they collide elastically, find their final velocities. *You might want to have a computer do the algebra for you.*

16. **Why $p = mv$ Can't Be Right**

In this problem you're going to demonstrate that the *Newtonian* formula for momentum is incompatible with *relativistic* transformations. So you'll do a purely Newtonian calculation to find the final velocity of this system. But then, when you switch to a different reference frame, use the Lorentz velocity transformations!

Our demonstration is a simple inelastic collision. Block A with mass m is moving at speed $(3/5)c$. It runs into stationary Block B, also with mass m. The two stick together and move off together (Figure 2.9).

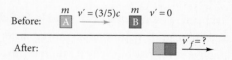

Figure 2.9

(a) Use conservation of momentum to find the final velocity of the blocks. *Do this part using only Newtonian formulas, exactly as you would have done in an introductory physics course.*

Our description of the initial situation was in Block B's initial reference frame (because we said Block B was initially stationary). We'll call that Frame R, and we'll define Frame R′ as Block A's initial frame. In Frame R′, Block A is initially at rest and Block B is initially moving with $v'_{B0} = -(3/5)c$ (Figure 2.10).

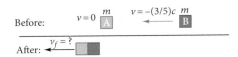

Figure 2.10

Frame R′ is moving at velocity $u = (3/5)c$ relative to Frame R.

(b) In Frame R′, what is the initial Newtonian momentum of the whole system?

(c) You previously found the velocity of the final-stuck-block system in Frame R. Use the Lorentz transformation of velocities (not conservation of momentum!) to convert that final velocity to Frame R′.

(d) What is the final Newtonian momentum of the whole system?

(e) Was Newtonian momentum conserved in Frame R′?

17. A particle moving with speed v has momentum $\vec{p} = \gamma m\vec{v}$ and energy $E = \gamma mc^2$. You're going to find low-speed approximations to these formulas using the binomial approximation, which (to second order) says

$$(1 + x)^n \approx 1 + nx + \frac{n(n-1)}{2}x^2 \text{ for } x \ll 1.$$

(This approximation can easily be derived as a Taylor series.) As x gets smaller the later terms become smaller much faster, so it becomes a good approximation to use just the first few terms.

(a) Write a binomial approximation for γ valid when $v \ll c$. (Using the formula above, your answer should be a fourth-order polynomial of v.)

(b) Use only the first term in that expansion to write an approximation for $\vec{p}$ at small speeds.

(c) What's the next-order correction to this approximation?

(d) Use the first two terms in your expansion for γ to write an approximation for E at small speeds.

(e) What is the next-order correction to this approximation for energy? Based on that next term, is the relativistic energy higher or lower than $mc^2 + (1/2)mv^2$?

18. You can define the force on a particle as $d\vec{p}/dt$. An object of mass m, initially at rest, experiences a constant force F_0 in the x direction. Find and plot $v(t)$.

19. **Deriving Kinetic Energy from the Work–Energy Theorem**

In classical physics you can derive the formula for kinetic energy from the work–energy theorem and the definition of momentum:

$$K = \int F\,dx = \int \frac{dp}{dt}\,dx = \int \frac{dx}{dt}\,dp = \int v\,d(mv)$$

$$= m\int v\,dv = \frac{1}{2}mv^2 + C.$$

You can find that $C = 0$ by setting $K = 0$ when $v = 0$ and conclude that $K = (1/2)mv^2$.

In relativity you can do the exact same process, just substituting the relativistic definition of momentum. You'll know you did it correctly if you get the right answer for K. *Hint:* When you set $K(0) = 0$ the constant of integration won't equal 0 this time.

20. **A Two-Dimensional Collision**

A particle of mass m moving with speed v_0 in the positive x direction collides elastically with an initially stationary particle of mass m. The two move off at equal and opposite angles θ, as shown in Figure 2.11.

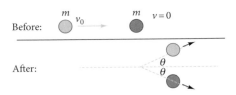

Figure 2.11

(a) Prove that in Newtonian physics $\theta = 45°$. (More generally, in Newtonian physics the angle between the two will always be 90° even when they aren't symmetric with respect to the original direction. You can test this with pool balls.)

(b) Prove that in the limit $v_0 \to c$ the angle θ approaches 0.

2.3 Mass and Energy (and the Speed of Light Squared)

The previous section and this one form one large topic, discussing the intermingled ideas of momentum, energy, and mass in special relativity. Section 2.2 by itself provided a mostly self-contained introduction to the topic. Continue to this section if you want a deeper understanding of the interaction between mass and energy, and in particular the fabled $E = mc^2$ formula.

2.3.1 Discovery Exercise: Mass

What does the word "mass" mean? If you ask introductory physics students that question – which we've done a lot, and it's a fun exercise – the most common answer is that mass is a measure of how many molecules (or atoms or fundamental particles) an object has. To push you beyond that (inadequate) answer, here's a more specific question.

Fact: A proton has roughly 2000 times the mass of an electron.

Question: What experiments could you do to test that fact, or to measure the mass ratio more precisely?

This is not a relativity-specific question, so feel free to give a purely classical answer. Also, don't feel constrained by practical considerations: assume you have an unlimited budget and unlimited technology. The goal is to articulate, in a measurable way, what that true fact means. (It does *not* mean that a proton is composed of 2000 bound electrons!) You can give a perfectly good answer to this question in one or two short sentences.

2.3.2 Explanation: Mass and Energy (and the Speed of Light Squared)

In Section 2.2 we presented definitions of momentum and energy and used them to analyze an elastic collision. Although the definitions were more complicated than their Newtonian counterparts, the procedure was essentially identical.

That example hides some important subtleties, however, that can be brought out by analyzing an inelastic collision.

Active Reading Exercise: An Inelastic Collision

Two balls of putty, each of mass m, are moving toward each other. Before they collide, each ball is moving with speed $(4/5)c$ in the lab frame. After they collide, they stick together in one lump.

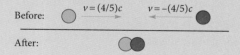

1. How will the lump move after the collision? (This is not a trick question; the answer should be obvious.)
2. Use Equation (2.2) to calculate the energy before and after the collision. Did the two come out equal?

The initial momenta are clearly equal and opposite, so the sum is zero. Conservation of momentum therefore dictates that the final lump must be at rest. (You probably could have guessed that anyway.)

Now let's write the total energy before and after the collision. Each ball of putty has mass m. We'll use M for the mass of the stationary lump you end up with.

$$E_0 = 2\frac{mc^2}{\sqrt{1 - (4/5)^2}} = \frac{10}{3}mc^2$$

$$E_f = \frac{Mc^2}{\sqrt{1 - (v_f/c)^2}} = Mc^2$$

You may have been taught that you can't use conservation of energy for an inelastic collision, but in relativity an isolated system always conserves both momentum and energy. So E_0 must equal E_f. That equation forces us to conclude that the final lump must have mass $M = (10/3)m$.

How can two balls, each with mass m, compose a lump whose mass is greater than $2m$? You may be able to figure out the answer to that question by thinking about inelastic collisions, and about our discussion of mass in Section 2.2. In the meantime we're going to carefully track the energy: first after the collision (one stationary lump), and then before the collision (two moving balls).

The Energy after the Collision

The original two balls have kinetic energy. When they collide and combine, where does that energy go?

At a microscopic level, the kinetic energy of the two balls becomes the internal kinetic and potential energy of molecules jiggling around in the resulting lump. (This happens in any such "sticky" inelastic collision. Some energy may also go into the surrounding air and surfaces – as sound, for instance – but we'll assume that initially it primarily goes into the balls themselves.) To put it another way, the final lump is *hotter* than the original two balls were.

A Newtonian analysis would say that the kinetic energy of the two balls has become thermal energy. This is why, in classical mechanics, we cannot use conservation of energy to analyze an inelastic collision.

But in relativity, as we described in Section 2.2, such internal changes are part of an object's rest energy. The difference $(10/3)mc^2 - 2mc^2$ is equal to the original kinetic energy of the two balls before the collision, which became thermal energy in the lump. If we wait a while, the lump will cool down to room temperature, releasing all that energy, and its mass will reduce to $2m$.

The Energy before the Collision

In Section 2.2 we discussed the importance of clearly defining the system you're looking at. If you consider each ball before the collision to be its own system then each ball has mass m, velocity v, and total energy γmc^2. That is how we originally concluded that the total energy of the system is $(10/3)mc^2$ before the collision.

But what if you look at the two balls before the collision as one big system? The velocity of that system is zero (because its total momentum is zero). That means *all* its energy is rest energy, $M_{\text{system}}c^2$. But the total energy has to be $(10/3)mc^2$ no matter how you look at it.

The conclusion is inescapable. Each ball individually has mass m, but the two-ball system has mass $(10/3)m$, *even before the collision!*

Just as molecular motion contributed to the mass of the post-collision lump, the ball motion contributes to the mass of the pre-collision system. The motion counts as *external* motion (which contributes kinetic energy) if you consider each ball as its own system, but as *internal* motion (which contributes mass) if you consider the entire two-ball system.

From this example we can make two generalizations about mass in relativity.

- *Mass is conserved.* The total mass of the system is $(10/3)m$ before the collision, and $(10/3)m$ after the collision. Mass is always conserved in Newtonian mechanics, and it is still always conserved in relativity.[4]

- *The mass of a composite system does not in general equal the sum of the masses of its parts.* This is of course quite different from classical dynamics! In our system (before or after), each ball has mass m but the total mass is increased to $(10/3)m$ by the energy of motion internal to the system.

One practical upshot of all this is that, when you analyze an inelastic collision in relativity, you generally include the final masses as unknowns. We demonstrate that in the following Example.

---

4 Some older textbooks say that mass and energy are no longer separately conserved in relativity. That comes from outdated definitions. Using modern definitions of mass and energy, each one is conserved for any isolated system.

Example: An Inelastic Collision

A particle of mass $2m$ moving at speed $(4/5)c$ hits a stationary particle of mass m and they stick together. Find the final mass and velocity of the combined particle.

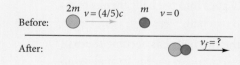

Answer: In Newtonian physics we would have one equation (conservation of momentum) and one unknown (final velocity). In relativity we have two equations (conservation of momentum and conservation of energy) and two unknowns (final mass and velocity):

$$\frac{2m(4/5)c}{\sqrt{1-(4/5)^2}} = \frac{Mv_f}{\sqrt{1-(v_f/c)^2}}$$

$$\frac{2mc^2}{\sqrt{1-(4/5)^2}} + mc^2 = \frac{Mc^2}{\sqrt{1-(v_f/c)^2}}$$

You can find v_f by dividing the first equation by the second, and then plug that in to get M, with the result $v_f = (8/13)c$ and $M = \sqrt{35/3}\, m$.

We said in Section 2.2 that counting all these non-kinetic forms of energy – the internal potential energy of a system, the internal motion of a system, and others – as forms of "mass" is not just a matter of bookkeeping. The examples above may help convince you that we *have* to broaden our idea of mass in order to conserve relativistic energy. But how does this new quantity relate to the classical idea of mass? And for that matter . . .

What *Is* Mass, Anyway?

What does the word "mass" mean in physics? Even in classical mechanics, what do we mean when we claim that a proton has roughly 2000 times the mass of an electron? There are two very different answers to this question.

- Mass is inertia, as expressed in the equation $F = ma$. If you put a proton and an electron near each other, they exert equal-and-opposite forces on each other. But the electron accelerates about 2000 times as much as the proton. So "large mass" means "difficult to accelerate."
- Mass determines how an object exerts, and reacts to, gravity. Jupiter pulls on objects much more strongly than Earth, and a brick and a feather feel different gravitational pulls toward the Earth.

So when we say that an object's internal energy contributes to its mass, we are making a specific, testable prediction.

> *Heating an object increases its mass, making it more difficult to accelerate, and causing it to exert and feel stronger gravitational forces.*

If that's true (which it is), how did people go for centuries without noticing? The answer to that question is always the same in relativity: "because c is so big." In this case $E = mc^2$ means a huge amount of energy corresponds to a very small mass. For example, a liter of water has 1 kg of mass. If you heat it by $10°$ Celsius you add about 42,000 J of energy to it. That may sound like a lot, but divide that energy by c^2 and you find you increased the mass of the water by less than a trillionth of a kilogram. Unless your bathroom scale is a lot more accurate than ours, you're not going to notice.

Why does increasing the temperature of an object make it harder to accelerate? Here's one way to think about it. Suppose you exert a certain force on an object for a certain amount of time, increasing its velocity from $v = 0$ to $v = 0.1c$. Now you exert the same force, for the same time, on an object that is already traveling at $v = 0.95c$. Clearly you cannot increase its velocity by $0.1c$ this time! So the same force causes less dv/dt, meaning the object's inertia has increased.

That line of thinking can lead you to conclude that "more velocity" means "more mass" – and indeed you will see relativity described that way in some very old textbooks. But that definition makes an object's mass different in different reference frames; it is more useful to define mass in an invariant way. We therefore define an object's mass as its inertia (force over acceleration) *measured in its rest frame.* All reference frames will agree on the answer to the question: "What is the Earth's inertia in its rest frame?"

As an example, consider the inelastic collision that began this section. In the reference frame of one of the two balls – that is, the reference frame in which it is not moving – its mass is $F/a = m$ for any applied force. The same is true of the other ball (although in a different reference frame). But in the reference frame of the two-ball system, which is the frame that the problem was presented in, both balls are moving, and are therefore harder to accelerate. That's why the combined system has a greater mass than the individual parts. A similar argument applies to the final lumped-together system, since in its rest frame it is full of molecules that are moving fast and are therefore hard to accelerate.

This explanation doesn't account for why internal *potential* energy also contributes to mass for a composite object (see Question 12), but it gives you some flavor of how internal energy and mass are related.

Converting Between Mass and Energy

People often describe the equation $E = mc^2$ by saying that you can convert mass to energy and vice versa. But what does that mean when we've said that mass and energy are each separately conserved? We can address this with the example of nuclear fusion.

Put two protons and two neutrons sufficiently close to each other and they are attracted by the strong nuclear force, which (at sufficiently close range) is much stronger than the electric force that repels them. They combine to form a helium nucleus. When objects that attract each other get closer, they lose potential energy, so the whole system has lower potential energy when it is bound than when it is unbound.

We have discussed other examples of potential energy changes, such as the stretching of a spring or the lifting of a brick. In those cases the potential energy was increased, so something external had to put energy into the system. When the potential energy decreases in nuclear fusion, where does that energy go? The answer is that it is emitted as radiation. This is the reaction that powers the Sun.

None of that requires relativity. A classical system can certainly lose potential energy and radiate out the difference. But in relativity, any change to internal energy is a change in mass. The nucleus is easier to accelerate than the four separate particles, and exerts less gravitational force. Those predictions are unique to relativity – and they are demonstrably, experimentally true.

The whole system – protons, neutrons, and emitted radation – has the same mass as the original four particles. But if we just focus on what's in front of us, first four particles and then a helium nucleus, that now has a lower mass and the difference (times c^2) has been emitted as radiation.

It's a common mistake to think that although nuclear processes like this involve changes in mass, chemical processes like burning something or dissolving salt in water don't. In fact both types of processes change a system's energy and thus its mass. The difference is that nuclear processes are governed by the strong nuclear force while chemical processes are governed by the electromagnetic force. The strong force is, as you might guess, much stronger. So when oxygen and hydrogen combine to make water, the changes in their potential and kinetic energies change their total mass by less than one part in a billion. In nuclear reactions, however, these changes in mass can be on the order of 1%. If you could convert 1% of the mass of a water drop to useable energy every hour, you could satisfy the power needs of a fair-sized city.

Creating and Destroying Particles

The equivalence of mass and energy leads to a type of interaction that does not occur in Newtonian physics: the creation and destruction of particles.

Example: Pair Creation

A photon hits a stationary electron. The photon annihilates and produces two new particles: a second electron and a "positron" (an antimatter particle with the same mass as an electron). Assume the three particles all move with the same velocity after the collision, but far enough apart from each other that we can neglect their potential energy.

1. What is their final velocity?
2. What was the initial energy of the photon?

Answer: The photon's energy and momentum are related by $E_{ph} = p_{ph}c$. The electron initially has zero momentum and energy $m_e c^2$. So setting initial momentum and energy equal to final momentum and energy gives

$$p_{ph} = \frac{3 m_e v_f}{\sqrt{1 - (v_f/c)^2}}$$

$$p_{ph}c + m_e c^2 = \frac{3 m_e c^2}{\sqrt{1 - (v_f/c)^2}}$$

The algebra on this one isn't so bad to do by hand, and the result is $v_f = (4/5)c$, $E_{ph} = 4 m_e c^2$.

This "pair creation" problem is similar to the collision problems we've already solved: we set momentum and energy before equal to momentum and energy afterward and solved for our two unknowns. But we want to draw your attention to a few novel features of this problem:

- The particles coming out are different from the ones coming in. (As we will discuss below, that qualifies the collision as inelastic.)
- A photon is a massless particle, so it obeys $E = pc$.
- Even though this isn't an elastic collision, we didn't treat the final masses as unknowns. That's because these are fundamental particles. A chair can have more or less mass depending on what its internal atoms are doing, but an electron isn't made of anything else so it always has the same mass.

Elastic and Inelastic Collisions

When particles collide, conservation of momentum and of energy aren't enough to determine what will happen. If you know that the collision of two particles is perfectly elastic then you can solve for the outcome. If you know it is perfectly inelastic (they stick together) you can also solve. For an intermediate case, or if you don't know whether the collision is elastic, then you need more information. And on a macroscopic level, "perfectly elastic" is always at best an approximation.

The preceding paragraph applies perfectly to both Newtonian and relativistic physics. The two systems differ, however, in the definition of "elastic."

In Newtonian physics an elastic collision is one in which mechanical energy (potential plus kinetic) is conserved. In an inelastic collision, some of that energy gets converted to thermal energy, sound, heat, or other forms. But in relativity, energy is always conserved for an isolated system; thermal and other forms of energy are accounted for in the total mass.

Nonetheless, there is still an important distinction in relativity: the particles that leave an elastic collision are the same ones that entered, with the same masses. No particles are created or destroyed and no objects change their masses. We implicitly used those assumptions when we solved the elastic collision problem on p. 64 by setting the masses of the final objects equal to their initial masses.

When the colliding objects are particles like electrons, protons, or photons, you can assume that they always have the same mass, so particle collisions are somewhat simpler to analyze than macroscopic ones.

Summary: Momentum, Energy, and Mass

The relativistic concepts of momentum, energy, and mass closely parallel their classical counterparts. Momentum is based on an object's mass and velocity, and is conserved in any isolated system. Energy is also conserved, but can be converted between different forms. Mass is reference-frame-independent, and quantifies how difficult it is to accelerate an object.

But different formulas lead to different results and some new concepts. In relativity both momentum and energy approach infinity in the limit $v \to c$. The energy mc^2 of a composite system at rest includes all the rest energies of its constituents, plus their internal kinetic and potential energies. When the whole system is moving, its mass is the same but its energy now also includes a new kinetic energy term.

The "external potential energy," such as the energy of a brick on top of a building, is not included in $E = \gamma mc^2$ for the brick. That energy belongs to the Earth–brick system, but not to either object individually. (If you ask very strictly "where is that energy, then?" the answer is "in the gravitational field." You can write a formula for the energy density of a gravitational or electric field and then integrate to find its total energy.)

We've summarized some of the key properties of these quantities in Table 2.1. In reading the table, keep in mind the following definitions:

- A quantity is "invariant" if it is the same in all reference frames. The number of atoms in a table is invariant; every observer will agree about it. The velocity of an object is not.

- A quantity is "conserved" if (for an isolated system) that quantity doesn't change during interactions.

Table 2.1 **Key properties of some system variables**

	Invariant	Conserved	Additive
Number of particles	Yes	No	Yes
Mass	Yes	Yes	No
Momentum	No	Yes	Yes
Energy	No	Yes	Yes

- A quantity is "additive"[5] if the whole equals the sum of its parts. The number of chairs in a room is additive; two 10-chair rooms between them have 20 chairs. Temperature is not; two $70°$ rooms between them are not at $140°$.

We want to highlight the lower right part of this table. The momentum and energy of a system are additive; you can just sum up the momenta and energy of the parts. The mass is not. But the total mass, momentum, and energy are all conserved in interactions.

---

5 In thermodynamics the term "extensive" is used in the way we are using "additive."

2.3.3 Questions and Problems: Mass and Energy (and the Speed of Light Squared)

Conceptual Questions and ConcepTests

1. **(a)** Under what circumstances is it correct to say that a composite object's mass equals the sum of the masses of its constituents?

   (b) Under what circumstances is it correct to say that a composite object's mass equals the sum of the masses of its constituents plus the sum of their individual kinetic and potential energies?

2. We said that in relativity you can use conservation of energy, even for an inelastic collision, but you have to treat the final mass as unknown. Why can't you just ignore both of those and solve for the final velocity using only conservation of momentum, as you would in Newtonian physics?

3. A nuclear bomb is often described as converting a certain amount of mass into energy. But we said above that mass is conserved. Explain.

4. In very old textbooks you will find the equation $m = \gamma m_0$. In words, the mass of an object increases as its speed increases. (You can see why they might say this, since a faster object is harder to accelerate than a slow object.) But today we define mass in terms of the speeds of the components of a system, *in the rest frame of that system*. What is the big advantage of this clumsier-sounding definition?

5. When two balls collide and stick together, their Newtonian mechanical energy ($K = (1/2)mv^2$) is not conserved, but their relativistic energy ($E = \gamma mc^2$) is. How can that be true when Newtonian physics is the low-speed limit of special relativity?

6. In the Example on p. 77 we showed that for a photon to collide with a stationary electron and produce two new particles each with the mass of an electron, it needed $4m_ec^2$ of energy. Since only two new particles were created, why couldn't the photon just have $2m_ec^2$ of energy?

7. P1 and P2 are fundamental particles of mass m, and L is a photon.

   (a) Can the system of P1 and P2 have a mass greater than $2m$?

   (b) Can the system of P1 and P2 have a mass less than $2m$?

   (c) Can the system of P1 and L have a mass other than m?

8. A photon strikes an object and gets absorbed.

   (a) Does the object's mass increase?

   (b) Does the total mass of the system (object plus photon) increase?

9. An atom transitions from a high-energy to a low-energy state and gives off two photons. These photons have identical energy and go in opposite directions. Analyze this scenario from the original rest frame of the atom.

   (a) Explain how we know that the atom is still at rest (in the same frame) at the end of this story.

   (b) After the photons are emitted, is the total mass of the atom higher, lower, or the same as before? Briefly explain how you know.

   (c) After the photons are emitted, is the total mass of the atom-and-both-photons system higher, lower, or the same as the original mass of the atom? Briefly explain how you know.

10. In our two-balls-colliding example, the final mass of the system *had to be* equal to the initial mass of the system. But in our particles-fusing-into-a-nucleus example, the final mass of the nucleus was less than the initial masses of the particles. What is the difference between these two examples that allows the total mass to change in one case but not the other?

For Deeper Discussion

11. A photon has zero mass, so the formula $PE = mgh$ for gravitational potential energy says that it has zero potential energy. Now suppose a positron and an electron annihilate to form two photons near Earth's surface. The photons then travel freely to the upper atmosphere where they annihilate to reproduce the electron–positron pair. However, now that they are high up, the particles have potential energy that you can harvest by having them fall and then turn a motor when they reach sea level. Have you just created energy from nothing? If so, rush out now and solve the world's energy problems! If not, find the flaw in this argument.

12. On p. 75 we made an argument, based on the inviolable speed of light, that an object's internal kinetic energy must contribute to its mass. But that particular argument does not apply to an object's internal *potential* energy. In this problem you're going to show the contradiction that would result if internal kinetic energy contributed to mass but internal potential energy didn't.

   Imagine a system consisting of a light box containing two heavy balls connected by a spring, oscillating back and forth.

   (a) If the mass of the total system depended on the kinetic energy of the balls but not their potential energy, graph what the system mass over time would look like.

   (b) Describe how by alternately lifting and dropping the system you could get an endless supply of energy.

   The conclusion of all of this is that if internal kinetic energy contributes to a system's mass (which we argued it must), then internal potential energy must as well.

13. We said the mass of an Earth–brick system differs from the sum of their individual masses by the amount of potential energy between them, divided by c^2. But the amount of potential energy is arbitrary; you can set the zero anywhere. And the inertial mass is not arbitrary; you can exert a force and see how much the system accelerates. So what determines the actual mass of the system?

Problems

14. Two balls of putty, each of mass m, move toward each other at $(3/5)c$ until they collide and stick together in one big lump (Figure 2.12).

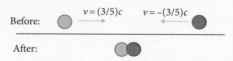

Figure 2.12

   (a) Explain how we can conclude, mathematically, that the final lump must be at rest.

   (b) Calculate the initial energy (total, not just kinetic) of one of the moving balls of putty.

   (c) Double your answer to Part (b) to find the total energy of the system before the collision.

(d) Write a formula for the total energy of the system after the collision. Use M to represent the mass of the lump.

(e) Use conservation of energy to find M, the mass of the lump, in terms of m, the original mass of each ball.

(f) What is the mass of the original, pre-collision two-ball system?

15. [*This problem depends on Problem 14.*] In this problem you will re-analyze the collision from Problem 14 from a reference frame moving to the right at $(3/5)c$ with respect to the original frame. In this reference frame, one of the two balls begins at rest. Remember that mass is an invariant quantity!

(a) Calculate the initial velocity of the other ball in this reference frame.

(b) Calculate the total energy of this system, pre-collision, by calculating the energy of each ball and then adding them.

(c) How fast is the final lump moving in this reference frame? (This should be a very quick and easy calculation.)

(d) Calculate the total energy of this system, post-collision, based on its mass (which you know) and its velocity (which you know).

(e) Was the energy of the system conserved in this reference frame?

(f) What is the mass of the original, pre-collision two-ball system in this reference frame?

16. A system comprises three objects, each of mass m. One object is at rest, one is moving away at $(1/2)c$ to the left, and one is moving away at $(1/2)c$ to the right.

(a) Calculate the total mass of this system.

(b) If we add a fourth ball of mass m, also at rest, how much does that add to the total mass?

17. A system consists of a ball of mass m at rest and a ball of mass m moving to the right at $(4/5)c$.

(a) Find the total energy of the system.

(b) Find the total momentum of the system.

18. [*This problem depends on Problem 17.*] Let R$'$ be the reference frame in which the two-ball system in Problem 17 is at rest ($V' = 0$). Find the velocities of the balls in frame R$'$. Explain why your answer makes sense.

19. When a neutron collides with a uranium 235 nucleus it can cause nuclear fission. There are many possible products of this reaction; one typical example is

$$n + {}^{235}\text{U} \rightarrow 3n + {}^{141}\text{Ba} + {}^{92}\text{Kr}.$$

The masses are $m_n = 1.00867$ amu (atomic mass units), $m_U = 235.0439$ amu, $m_{Ba} = 140.9144$ amu, $m_{Kr} = 91.9262$ amu. If a kilogram of ${}^{235}\text{U}$ were to completely undergo fission by this reaction (not a realistic assumption), how much energy would be produced, in joules (J)? *Hint*: You do not need to convert the atomic masses to kg to solve this.

20. The Sun releases about 4×10^{26} J of energy per second. This mostly comes from hydrogen fusion, in which four protons fuse into a helium nucleus, producing two positrons and two neutrinos. (There are many intermediate steps in this process involving neutrons, other nuclei, and other particles. In the Explanation (Section 2.3.2) we simplified this by saying we started with two neutrons and two protons, but this is more accurate.) The helium nucleus has a mass of 6.64648×10^{-27} kg. You can neglect the masses of the positrons and neutrinos. Since a single proton is the nucleus of a hydrogen atom, this is usually described as fusing hydrogen into helium. How many kg of hydrogen does the Sun fuse each second? (Don't just find the change in mass; find the mass of the original, pre-fusion hydrogen.)

21. A particle of mass m is accelerated from $v = (1/2)c$ to $v = (4/5)c$. (The particle is clearly not an isolated system.) Find the change in the particle's momentum, energy, and mass.

22. A 3 kg object is moving at $v = (1/2)c$ and a 2 kg object is moving at $v = -(3/4)c$.

(a) Find the total energy E of the system.

(b) Find the total mass M of the system of two particles. *Hint*: Use Equation (2.3).

(c) Does $E = Mc^2$? Explain why your answer makes sense.

23. A 1000 kg car and a 2000 kg car are moving toward each other at 30 m/s each (Figure 2.13), when they collide and stick together. (Miraculously nobody is hurt.)

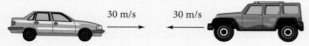

Figure 2.13

(a) How fast does the combined mass of metal move immediately after the collision? Explain why it's reasonable to use Newtonian calculations to answer this.

(b) What is the final kinetic energy of the car-heap? (Again, Newtonian calculations should be fine.)

(c) What was the combined kinetic energy of the two cars before the collision?

(d) Just after the collision, how much more than 3000 kg is the mass of the heap of car? *Hint*: You can do this without calculating any γ factors.

(e) Since we asked you to do a Newtonian calculation in Part (a) you presumably assumed the final mass was 3000 kg. Was that a reasonable approximation? Explain.

(f) Over time, after the collision, does the mass of the heap increase, decrease, or stay the same? Explain.

24. Object A has mass $3m$ and is initially moving at $(1/2)c$ in the positive x direction. Object B has mass m and is initially moving at $(1/2)c$ in the negative x direction. If they collide and stick together, find their final velocity.

25. A photon of energy E_{ph} is absorbed by an atom of mass m_0 at rest. Find the final velocity of the atom.

26. A moving proton collides with a proton at rest and produces a proton–antiproton pair. (An antiproton is another particle with the same mass as a proton.) The four particles (two original plus two new ones) move off together. Neglect the potential energy between the particles.

(a) What was the initial speed of the moving proton? (Once you set up the equations you might want to give them to a computer to solve. You can do them by hand but the algebra gets messy.)

(b) Could you make this happen with a smaller-energy initial proton by having all four final particles be at rest? Why or why not?

27. You are in a 10,000 kg rocket at rest in outer space. At the captain's command you cause 1 kg of matter to completely annihilate into radiation (photons) moving out of the rear of the rocket. How fast do you end up going?

28. A particle of mass m is moving at $(15/17)c$ when it strikes a stationary particle of mass m.

(a) What is the total energy of the system?

(b) If all of that energy could go into creating new particles of mass m, how many new ones could be created?

(c) Calculate the initial velocities in the center-of-mass frame of the system (meaning the one in which total momentum is zero).

(d) In the center-of-mass frame, what's the initial energy and how many new particles of mass m could theoretically be created?

(e) Which of your two conclusions about creating new particles is correct? Explain why the conclusion in the other reference frame was wrong, from the point of view of that reference frame.

29. **Pair Annihilation**

An electron moving at $v = (4/5)c$ in the positive x direction strikes a positron (which has the same mass as an electron) at rest. They annihilate each other, producing two photons. If one moves perpendicular to the original electron's motion, in what direction does the other go?

30. In Section 2.2 Problem 16 we presented a conservation-of-Newtonian-momentum problem in one reference frame, and showed that the results are not compatible with conservation of

momentum in a different reference frame. In this problem you're going to do the same calculation using relativistic momentum and energy and show that it *does* work. (You do not need to have done the previous problem to do this one.)

Our demonstration is a simple collision. Block A with mass m is moving at speed $(3/5)c$. It runs into stationary Block B, also with mass m. The two stick together and move off together (Figure 2.14).

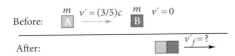

Before: m $v' = (3/5)c$ m $v' = 0$

After: $v'_f = ?$

Figure 2.14

(a) Use conservation of momentum and energy to find the final mass and velocity of the blocks.

Our description of the initial situation was in Block B's initial reference frame (because we said Block B was initially stationary). We'll call that Frame R, and we'll define Frame R' as Block A's initial frame.

In Frame R', Block A is initially at rest and Block B is initially moving with $v'_{B0} = -(3/5)c$ (Figure 2.15).

Before: $v = 0$ m A $v = -(3/5)c$ m B

After: $v_f = ?$

Figure 2.15

Frame R' is moving at velocity $u = (3/5)c$ relative to Frame R.

(b) In Frame R', what is the initial momentum of the whole system?

(c) You previously found the velocity of the final-stuck-block system in Frame R. Use the Lorentz transformation of velocities (not the conservation of momentum!) to convert that final velocity to Frame R'.

(d) What is the final momentum of the whole system in Frame R'?

(e) Was momentum conserved in Frame R'?

2.4 Four-Vectors

One of the key messages of Section 2.1 on spacetime diagrams was that in relativity it's useful to think of space and time together as a four-dimensional "spacetime." Space and time coordinates get mixed together in the Lorentz transformations. The ideas of special relativity are easiest to understand in terms of events, each of which has four spacetime coordinates. Spacetime diagrams allow you to see how events look in different reference frames.

In this section we'll formulate the laws of special relativity in terms of scalars and vectors in 4D spacetime. We'll use this idea of 4D vectors to justify the definitions we gave for momentum and energy in Section 2.2.

We're going to briefly introduce the properties of 4D vectors, but we will not delve into the full mathematical basis for those properties, which involves linear algebra, group theory, representation theory, and other topics beyond the scope of this book.

Nonetheless, this is a long and difficult section. You could skip it entirely: nothing else in the book depends on this material in any way, and you will still come out with a solid background in special relativity. You could also cover the first half, stopping before the introduction of four-momentum. Finally, you could very reasonably cover four-momentum but stop before the discussion of "Four-Momentum and Spacetime" at the end, which is arguably the most abstract material in the chapter.

2.4.1   Discovery Exercise: Coordinate Transformation

Figure 2.16 is not a spacetime diagram. In fact, nothing in this exercise directly involves relativity.

You have just laid down some x and y axes to map out a space. Unfortunately your friend has laid down different axes, x' and y', rotated from yours by an angle θ (less than 90°).

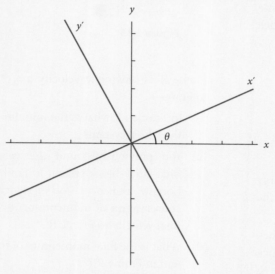

Figure 2.16 Two sets of axes rotated relative to each other.

You can convert any point from the unprimed coordinate system to the primed system by using the equations

$$\left. \begin{array}{l} x' = x\cos\theta + y\sin\theta \\ y' = -x\sin\theta + y\cos\theta \end{array} \right\} \tag{2.5}$$

1. Look at the point $x = 1$, $y = 0$ on the drawing. Answer by looking at the diagram: Is its x' coordinate positive or negative? Is its y' coordinate positive or negative?

2. Confirm that Equations (2.5) match your visual prediction from Part 1.

3. A bicycle rides smoothly across the page, and at a certain moment you measure everything about this bicycle in your unprimed coordinates. For each quantity specified below, indicate if you would use Equations (2.5) to *convert* your numbers to the primed coordinates, or if your numbers would be the *same* in the primed coordinates. For example, if we asked about the bicycle's position you would say that you would use Equations (2.5) to convert it, while if we asked about the height of the bicycle you would say that you and your friend agree about it; no conversion is needed.

   (a)   The bicycle's velocity

   (b)   The bicycle's speed

   See *Check Yourself #3* at www.cambridge.org/felder-modernphysics/checkyourself

   (c)   The bicycle's mass

   (d)   The bicycle's acceleration

2.4.2   Explanation: Four-Vectors

Figure 2.17 shows a town with two sets of axes. It also shows a vector $\vec{d}$ representing the path of a watermelon seed that Gaston spit from his house.

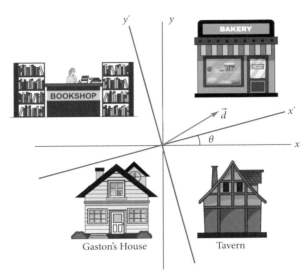

Figure 2.17 A single vector $\vec{d}$ can be described in the xy coordinate system, or in the $x'y'$ system.

The vector $\vec{d}$ is independent of the axes: everyone will agree that it is 12 feet long and it points in the direction from Gaston's house toward the bakery. But people using the two different sets of axes will disagree about the components of $\vec{d}$. If you know the components x and y you can find the primed components using

$$\left. \begin{array}{l} x' = x\cos\theta + y\sin\theta \\ y' = -x\sin\theta + y\cos\theta \end{array} \right\} \tag{2.6}$$

We need to emphasize three points about this rotated-axes system:

- If you express the components of $\vec{d}$ in the unprimed coordinate system, Equations (2.6) will convert those components to the primed system. In this example, $\vec{d}$ is a displacement, but those same equations would work to convert the components of velocity, momentum, electric field, or any other vector. In fact – and this is much of our point here – the word "vector" is often *defined*[6] as a set of components that transforms from one coordinate system to a different, rotated, coordinate system with Equations (2.6).

- On the other hand, if you measure a scalar quantity such as mass, temperature, or "number of watermelon seeds" in the unprimed coordinate system, those numbers will remain unchanged in the primed system. You can define "scalar" as a quantity that is invariant under rotation.

---

6 Mathematicians use more abstract ways to define "vector" and "scalar," but the definitions we are giving here correspond to how the terms are used in much of physics.

- Although the primed system and the unprimed system disagree about the x component of vector $\vec{d}$, and also about its y component, they agree about its magnitude:

$$\sqrt{x^2 + y^2} = \sqrt{x'^2 + y'^2}.$$

In other words, *the magnitude of any vector is a scalar*: an invariant quantity. For instance, velocity is a vector but speed is a scalar.

You probably learned that "vector" means a quantity that has a magnitude and a direction. Here we're giving another definition in terms of how it transforms under rotation. The definition we're giving is somewhat more precise and will be easier to generalize to relativity, but it is essentially equivalent to the magnitude-and-direction definition. In the example here, Equations (2.6) guarantee that both sets of observers will agree on the magnitude and direction of $\vec{d}$, and in particular they will agree on the physically meaningful statement that $\vec{d}$ points from Gaston's house toward the bakery.

While this definition of "vector" may be a case of expressing already-familiar ideas in unfamiliar language, this definition of "scalar" is more likely to be a new idea. Many introductory physics students end up with the impression that "scalar" is just a fancy word for "number," but it actually refers to an invariant quantity. For instance, the magnitude of a vector is a scalar; people using different axes will agree on it. But the x component of a vector is *not* a scalar, because it is not invariant under rotation.

None of that has anything to do with relativity, but this discussion of vectors and scalars in a familiar setting will help set up the different type of vectors we use in relativity.

Four-Vectors

In relativity we no longer treat space and time separately, but talk instead about events occurring in 4D spacetime. So the vector quantities we define have four components. To distinguish these from the vectors we use in Newtonian physics, we call them four-vectors.

In this section we will use bold capital letters for four-vectors ($\mathbf{V}$) and arrows to indicate three-vectors ($\vec{d}$). The components of four-vectors will be capital, but not bold (V_t).

Four-vectors can be plotted as arrows in spacetime diagrams just as three-vectors can be shown as arrows in space. As usual our drawings will show only one spatial dimension, not three! We will continue to use the term "four-vector" even when there are only two components (x and t).

The four-vector that we shall start with is the "position four-vector" $\mathbf{R} = (ct, x, y, z)$ (also sometimes called the "spacetime four-vector"). The c in the time component gives the components the same units as each other. In a spacetime diagram we always use units where $c = 1$, so that extra factor of c has no effect on our drawings.

Just as in three-space, we can represent an object's displacement in spacetime by subtracting two position four-vectors, which we can still do using the tip-to-tip method you may have learned for vector subtraction.

Active Reading Exercise: Four-Displacement

The figure shows the journey of an object from the event represented by four-vector $\mathbf{R}_1$ to the event represented by four-vector $\mathbf{R}_2$.

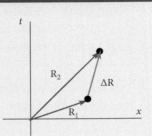

1. What does the horizontal component of $\Delta\mathbf{R}$ tell us about the journey?

2. What does the vertical component of $\Delta\mathbf{R}$ tell us about the journey?

3. What does the slope of $\Delta\mathbf{R}$ tell us about the journey?

You can find our answers at www.cambridge.org/felder-modernphysics/activereadingsolutions

You might reasonably have expected us to also ask about what the length of the vector represents. But that length, $\sqrt{c^2\Delta t^2 + \Delta x^2}$, doesn't mean much of anything. As we will see, the concept of the "magnitude" of a four-vector does have meaning, but that's not the right way to define it.

None of our answers to this Active Reading Exercise talked about the direction of the journey. Does the displacement four-vector tell us that? It does, but remember that we are working with only one spatial dimension here. So all there is to say about direction is that the journey was in the positive x direction. If we were also using y and z components then the ratios of the x, y, and z components would tell you about the spatial direction of the journey, just as they do for three-vectors.

Definition of Four-Vectors and Scalars

At the beginning of this Explanation section we defined the word "vector" (what we are now calling a "three-vector"), and the word "scalar," based on how they transform when the axes are rotated.

In special relativity the definitions of "four-vector" and "scalar" are based on how a quantity transforms from one inertial reference frame to another.

The rotational vector definitions given above make a great analogy, but they are not the same thing. For instance, energy is a scalar under rotation: an object's total energy is the same no matter how you tilt your head. But energy is *not* a scalar in the sense that we will use the term below, because an object's kinetic energy – and therefore its total energy – is different in different inertial reference frames.

Active Reading Exercise: Four-Vectors and Scalars

On p. 85 we made three specific points about rotating axes: a vector transforms according to Equations (2.6), a scalar remains invariant, and the magnitude of a vector (defined as $\sqrt{x^2 + y^2}$) is invariant (a scalar).

> Before reading further, can you figure out the three equivalent points about four-vectors and inertial reference frames? What equations are used to translate a four-vector such as (ct, x, y, z) to a different inertial reference frame? What formula, involving the components of a four-vector, remains invariant when you do so?

This sentence is your last chance to answer that Active Reading Exercise for yourself before you read the answers below!

Four-Vectors and Scalars

A four-vector consists of four components (V_t, V_x, V_y, V_z) that transform from one inertial reference frame to another (moving at relative speed u in the positive x direction) via the Lorentz transformations:

$$\left.\begin{aligned} V'_t &= \gamma \left(V_t - uV_x/c\right) \\ V'_x &= \gamma \left(V_x - uV_t/c\right) \\ V'_y &= V_y \\ V'_z &= V_z \end{aligned}\right\} \tag{2.7}$$

A scalar is a single quantity that is *invariant*, meaning it has the same value in any inertial reference frame.

The magnitude of a four-vector, defined as $\sqrt{V_t^2 - V_x^2 - V_y^2 - V_z^2}$, is a scalar.

You may have noticed that these transformations look a bit different from the Lorentz transformations in Appendix B. That's because the position four-vector is defined as (ct, x, y, z). If you replace V_t with ct and (V_x, V_y, V_z) with (x, y, z) these become the familiar Lorentz transformations. The definition of "magnitude" becomes the spacetime interval.

Active Reading Exercise: What is a Scalar?

We define a scalar as a quantity that is the same according to all inertial reference frames. A simple example is any counting number: how many there are of something. If I go to the store and buy a box of marbles, everyone will agree that the box has 20 marbles in it. Another scalar is an object's temperature.

List at least two quantities that are scalars (besides counting numbers and temperature) and at least two that aren't.

There are many possible answers to the preceding Active Reading Exercise, but here are some scalars that we have discussed already:

- *The spacetime interval between any two events.* It's worth noting that we're saying the same old things with different language. We previously said that this interval is "invariant" and now we're saying it's a "scalar": the meaning is identical. Also, we will now sometimes refer to "the magnitude of the displacement four-vector"; once again, those new words refer to the same old $s = \sqrt{c^2 \Delta t^2 - \Delta x^2 - \Delta y^2 - \Delta z^2}$.

- *An object's mass.* Recall that mass is defined as the inertial mass (force over acceleration) in the object's rest frame, so everyone agrees on that.

- *The proper time of a journey.* In the twin paradox the two twins aged different amounts. But everyone in every reference frame will agree about how much Asher aged, so his proper time from start to end is a scalar. Emma's proper time between the same two events is a different scalar.

Examples of non-scalars include an object's speed, its energy, its length, the coordinate time between two events, and the frequency of a light beam.

Four-Momentum: The Wrong Definition

An object moves from position (x_1, y_1, z_1) to position (x_2, y_2, z_2) at constant velocity in a time Δt. In Newtonian physics the object's velocity and momentum are

$$\vec{v} = \left(\frac{\Delta x}{\Delta t}, \frac{\Delta y}{\Delta t}, \frac{\Delta z}{\Delta t} \right) \qquad \text{Newtonian velocity}$$

$$\vec{p} = \left(m\frac{\Delta x}{\Delta t}, m\frac{\Delta y}{\Delta t}, m\frac{\Delta z}{\Delta t} \right) \qquad \text{Newtonian momentum}$$

Suppose we try to define relativistic momentum in the same way, starting with the position four-vector:

$$\mathbf{P} = \left(m\frac{c\Delta t}{\Delta t}, m\frac{\Delta x}{\Delta t}, m\frac{\Delta y}{\Delta t}, m\frac{\Delta z}{\Delta t} \right) \qquad \textit{Wrong!} \qquad (2.8)$$

Remember that a four-vector is defined by the fact that it transforms from one inertial reference frame to another according to the Lorentz transformations. Equation (2.8) definitely does *not* have that property. Do you see why?

The numerators Δt, Δx, and so on all convert between reference frames by the Lorentz transformations. When we multiply all four components by a scalar such as m, the same transformations still work. (In math lingo this is because the Lorentz transformations are "linear.") But Equation (2.8) divides the components by Δt, which is itself different in different frames. So these fractions will transform in a more complicated way between reference frames, meaning Equation (2.8) does not meet our definition of a four-vector.

In Section 2.2 we made another argument against Equation (2.8) as a definition of momentum. You showed in Problem 16 of that section that if $p = mv$ is conserved in one reference frame, it is not conserved in others. Four-vectors never have that problem: if a four-vector is conserved in one inertial reference frame then it's conserved in all of them. (You can show that in Problem 21 of this section.)

So – what operation could we do to the position four-vector that would lead to a plausible momentum four-vector? The answer is not obvious, but you do have all the pieces.

Four-Momentum: The Right Definition

The mistake we made in Equation (2.8) was dividing by *coordinate* time. We need to divide by a scalar, the *proper* time.

Make sure you remember the difference. The coordinate time Δt is the time between two events as measured in a particular reference frame. The proper time $\Delta \tau$ is the time between two events as measured by a particular observer (or object or clock).

So if we want to measure the momentum of a particular object, we can use the following recipe:

1. Choose a period of time during which the object is moving. We assume for the moment that the velocity during this period was constant. (We will revisit that assumption shortly.)

2. Calculate the object's displacement for this period – in your own reference frame, of course. This is a four-vector.

3. Divide that displacement by the object's proper time, the time elapsed *in its own reference frame*. Because $\Delta \tau$ is a scalar (all observers must agree on it), the result of this division is still a four-vector.

4. Multiply by the object's mass. That's another scalar, so we still have a four-vector:

$$\mathbf{P} = \left(m\frac{c\Delta t}{\Delta \tau}, m\frac{\Delta x}{\Delta \tau}, m\frac{\Delta y}{\Delta \tau}, m\frac{\Delta z}{\Delta \tau} \right) \quad \text{Correct, but we're not done.} \quad (2.9)$$

Will different observers agree on the result? No! For example, the object itself will always measure the last three components of Equation (2.9) to be zero, but other observers won't. But if you use Equation (2.9) to calculate four-momentum in one inertial frame, you can convert that to any other inertial frame using Equations (2.7). That qualifies this definition of four-momentum as a four-vector.

But it's an awkward definition. To calculate it you have to find the elapsed time and distance in your reference frame, and you have to find the elapsed time in the object's own reference frame. Fortunately we know how those are related.

Active Reading Exercise: Δt and $\Delta \tau$

We're describing a scenario in which, in your reference frame, an object moves at constant velocity for a time Δt, and its own clock ticks off $\Delta \tau$ in that time. Write an equation relating the two.

This Active Reading Exercise is just a good old-fashioned time dilation problem. You say the object's clock is ticking slowly so $\Delta t = \gamma \Delta \tau$. (Can you see why it's $\Delta \tau$ that has to be multiplied by γ?) We can plug this into Equation (2.9) to express an object's momentum in terms of quantities you can measure in your own frame:

$$\mathbf{P} = \left(\gamma mc, \gamma m\frac{\Delta x}{\Delta t}, \gamma m\frac{\Delta y}{\Delta t}, \gamma m\frac{\Delta z}{\Delta t} \right) \quad \text{Still correct, but still not quite done.}$$

Just one step to go! We said above that these equations are predicated on the idea that the velocity is constant. If the velocity is changing, we can treat it as a constant over a sufficiently small time interval. In the limit as $\Delta t \to 0$ this approximation becomes exact, and our fraction $\Delta x/\Delta t$ becomes a derivative dx/dt, or v_x. This bring us to our final form of this four-vector:

$$\mathbf{P} = \left(\gamma mc, \gamma mv_x, \gamma mv_y, \gamma mv_z\right) \quad \text{Four-momentum.} \tag{2.10}$$

Recall that we defined the position vector not as (t, x, y, z) but as (ct, x, y, z), thus giving all four components the same units. In light of the equations for momentum and energy in Appendix B, we can rewrite Equation (2.10) as:

$$\mathbf{P} = \left(\frac{E}{c}, p_x, p_y, p_z\right).$$

Just as the spacetime four-vector represents both position and time, the momentum four-vector represents both energy and momentum (with a factor of c thrown in to match the units). When we assert that this quantity is conserved, we express in one equation the conservation of both momentum and energy. (This four-vector is sometimes called the "energy–momentum four-vector.")

But is it conserved? We can't prove that it is from Einstein's original two postulates. The conservation of $\mathbf{P}$ is a dynamic postulate, equivalent to $F = ma$ in Newtonian mechanics. (Recall that Newtonian mechanics can also be written starting with conservation of momentum and energy as its postulates.) We assert that it works based on conformity with experimental results. What we have proven is that momentum thus defined is a four-vector, which means that if it's conserved in one inertial reference frame then it's conserved in all others.

Now, remember that every four-vector has a "magnitude" $\sqrt{V_t^2 - V_x^2 - V_y^2 - V_z^2}$ that is invariant. Is the magnitude of Equation (2.10) a brand-new invariant quantity? Actually it's an old one. You will calculate in Problem 19 that the magnitude of $\mathbf{P}$ is just a constant times m, the object's mass.

A Brief Summary

In addition to being unusually abstract, this is also one of the *longest* sections in this book, and it presents a lot of material. So you may find it useful to outline the ground we've covered so far.

1. We began by defining the words "vector" and "scalar" in two very different ways:

    (a) An ordinary vector (which we here call a "three-vector") is defined by the particular way (Equations (2.6)) in which the quantity transforms through a rotation. A scalar is invariant under rotation.

    (b) In special relativity, a four-vector is defined by the particular way in which the quantity transforms between inertial reference frames. A scalar is invariant under that transformation.

    (c) The relativistic definition of a scalar is more restrictive; scalars in relativity are the same in any reference frame and any orientation of the axes. So a relativistic scalar is also a scalar in the old sense, but the reverse is not always true. For instance, "speed" is a scalar by the first definition, but not by the second. The word scalar means "invariant under whatever transformation I'm talking about right now," so it must be interpreted in context.

2. A four-vector has the following properties:

    (a) It transforms from one inertial reference frame to another via the Lorentz transformations (Equations (2.7), which are slightly different from the transformations in Appendix B because of the factor of c included with the time).

    (b) Its magnitude $\sqrt{V_t^2 - V_x^2 - V_y^2 - V_z^2}$ is the same in all inertial reference frames: it is a scalar.

    (c) If a four-vector is conserved in any inertial reference frame, it is conserved in all inertial reference frames.

3. The "position four-vector" or "spacetime four-vector" $\mathbf{R} = (ct, x, y, z)$ represents the time and location of an event. Its magnitude is the spacetime interval between that event and the spacetime origin.

4. The "momentum four-vector" $\mathbf{P} = (\gamma mc, \gamma mv_x, \gamma mv_y, \gamma mv_z)$ represents the momentum and energy of an object, and is a conserved quantity. Its magnitude is proportional to the object's mass. (Remember that relativistic mass is not the same as classical mass, and includes many forms of internal energy, but it is an invariant quantity.)

Four-Momentum and Spacetime

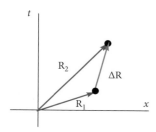

Figure 2.18
$\mathbf{\Delta R} = (c\Delta t, \Delta x)$.

If you stop reading right here, you have completed a solid introduction to special relativity – including the relativistic definitions of energy and momentum, and several motivations for those definitions. Below we present one final motivation, a geometric argument for the definition of $\mathbf{P}$. This argument will not be needed for any later material in the book; it presents relativistic dynamics in a way that is more abstract and visual, and that lays some groundwork for general relativity.

Just as a three-vector has a direction in space, a four-vector has a direction in 4D spacetime. For instance, consider the first four-vector we drew in this section, representing a journey at constant velocity (Figure 2.18).

The direction of the red arrow on the spacetime diagram tells us both the spatial direction and the speed of the journey.

Active Reading Exercise: The Direction of Four-Momentum

If you were to draw the object's four-momentum on the spacetime diagram in Figure 2.18, where would it point? Parallel to $\mathbf{\Delta R}$, above it, or below it?

Remember that we arrived at the formula for $\mathbf{P}$ by starting with the formula for $\mathbf{\Delta R}$. We then divided it by a scalar ($\Delta\tau$) and multiplied it by a different scalar (m). Multiplying a vector by a scalar does not change its direction, so the two four-vectors must point in the same direction.

An object's momentum four-vector points in the same direction as the world line of the object.

If the object is accelerating, then this means that $\mathbf{P}$ is tangent to the world line.

The conclusion that four-momentum is parallel to an object's world line comes from the particular way we defined **P**. Could we have defined it in some other way and obtained a different result? No.

To see why, consider an object at rest. Its world line is straight up, reflecting the fact that it has a Δt but no Δx (Figure 2.19). Its momentum four-vector is also straight up, reflecting the fact that $E \neq 0$ but $p = 0$. So once again we can pose the question, could someone come up with an alternative definition of four-momentum that would point in some other direction? If they did, it would have to point in either the $+x$ or $-x$ direction. Since it's arbitrary which way you call positive, there's nothing that should physically pick out one or the other.

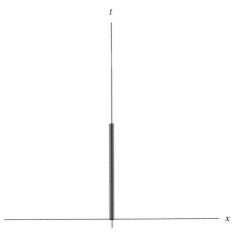

Figure 2.19 World line of an object at rest. By symmetry the four-momentum must point straight up.

Does that argument only apply to objects at rest? At any given moment every object is at rest in some inertial reference frame. If you conclude that $\Delta \mathbf{R}$ and **P** must be parallel in some inertial reference frame, then the definition of a four-vector guarantees that they will be parallel in all inertial reference frames. So symmetry demands that the four-momentum for an object must point along the object's world line.

Another way to frame this conclusion is that the direction of a vector – three-vector or four-vector – is physically meaningful. In the Example with which we started Section 2.4.2, all observers had to agree that Gaston spat toward the bakery, and the components they assigned to the vector $\vec{v}$ had to reflect that. Similarly, everyone has to agree that **P** must point parallel to an object's world line because there's nothing that would physically single out any other direction.

The upshot of this argument is that any physically meaningful definition of four-momentum would have to have the same direction in spacetime as the definition we gave. In Question 12 you'll argue why it also has to have the same magnitude as the way we defined it. Once you have described both the direction and the magnitude, you have uniquely specified the four-vector.

2.4.3 Questions and Problems: Four-Vectors

Conceptual Questions and ConcepTests

1. Suppose $\mathbf{V_A}$ and $\mathbf{V_B}$ are four-vectors and S is a scalar. Which of the following is a four-vector? (Choose all that apply.)

   A. $\mathbf{V_A} + \mathbf{V_B}$

   B. $S\mathbf{V_A}$

   C. $V_{Bt}\mathbf{V_A}$

   D. $d\mathbf{V_A}/dt$

2. Rank the magnitudes of the four-vectors in Figure 2.20 from largest to smallest. (They all have the same length on the diagram, but their magnitudes are not all equal.) Explain how you know.

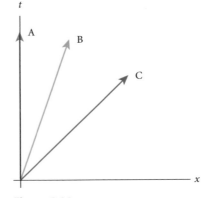

Figure 2.20

3. What does it tell you about a four-vector, drawn on a spacetime diagram, if its magnitude is imaginary? (Choose one.)

    A. The vector points downward – that is, in the direction of negative time.

    B. The vector has an angle below 45°, pointing in a spacelike direction.

    C. The vector has an angle above 45°, pointing in a timelike direction.

    D. This is mathematically impossible.

4. What's the four-momentum of a particle of mass m at rest?

5. Our usual spacetime diagrams show time and one spatial dimension. Now imagine a diagram that shows time and all three spatial dimensions. (You're going to have to imagine it, because we surely can't draw it.) Two objects have world lines that start at the same point. One of them ends up at (t_1, x_1, y_1, z_1) and the other ends up at (t_2, x_1, y_1, z_1): same spatial coordinates but different time. How are the journeys described by these world lines the same as each other, and how are they different?

6. An electron in a particle accelerator begins at rest and is gradually accelerated to $0.9c$.

    (a) Draw its world line. This can be purely qualitative, but the shape should be correct.

    (b) Draw in three plausible four-momentum vectors at the beginning, middle, and end of your drawing.

    (c) Based on your drawing, briefly describe what is happening to the *ratio* of the electron's momentum to its energy during this journey.

7. Explain how we know that $d\mathbf{R}/d\tau$ is a four-vector, and $d\mathbf{R}/dt$ is not.

8. Below is a set of statements about the scenario depicted in Figure 2.17 on p. 85: one vector represented with two different sets of axes. For each one, give the equivalent statement for a four-vector represented in two different inertial reference frames. (Assume for simplicity that the four-vector has no y or z components.) Here is an example to make sure it's clear what we're looking for.

    Example 3D statement: The primed and unprimed coordinate systems will use different horizontal and vertical components to describe the vector $\vec{d}$.

    Example response: Two different reference frames will use different time and space components to describe the four-vector $\mathbf{D}$.

    (a) The vector $\vec{d}$ could represent different things such as a displacement or a momentum.

    (b) The primed and unprimed coordinate systems will agree about the magnitude $|\vec{d}| = \sqrt{d_x^2 + d_y^2}$.

    (c) You convert the components of $\vec{d}$ from one coordinate system to the other by using Equations (2.6).

    (d) If $\vec{d}$ represents an object's displacement, then the primed and unprimed coordinate systems will agree about where the trip starts and ends (e.g. "the corner of Gaston's house" and "the spot where the watermelon seed landed").

    (e) If an object moves in a straight line and $\vec{d}$ represents the object's displacement, its momentum throughout that trip will point in the same direction as $\vec{d}$.

    (f) The direction of $\vec{d}$ is the same in all reference frames, although expressed in different coordinates.

9. Would $d\mathbf{P}/dm$ be a four-vector or not? Explain briefly how you know.

10. The Explanation (Section 2.4.2) looks at two different definitions of the word "scalar." A traditional scalar (or three-scalar) is invariant under rotation; a relativistic scalar (or four-scalar) is invariant under change of reference frame.

    (a) The mass of a particle is a scalar under both definitions. Name another quantity that is also a scalar under both definitions.

    (b) The energy of a particle is invariant under rotation, but is not invariant under reference frame transformation. Name another such quantity.

11. Why is it impossible for a particle to have four-momentum $(1 \text{ kg ls/s}, 2 \text{ kg ls/s}, 0, 0)$?

12. One way to justify the formula for the four-momentum **P** is to consider its direction and magnitude separately. The Explanation argued that the direction must point along the world line of the moving object. Once you have that, you can change its magnitude – that is, multiply every component by a scalar – and you will still have a valid four-vector, and it will still point in the right direction. So, to complete the argument, explain why multiplying our formula for **P** by any scalar (even a dimensionless one) would make it a bad candidate for a momentum four-vector (quite apart from whether or not it would match experimental evidence). *Hint*: Think about the requirements we listed for defining relativistic momentum in Section 2.2.

For Deeper Discussion

13. An object's momentum four-vector is tangent to its world line. If an object is moving with constant velocity there is always some reference frame in which its world line points straight up on a spacetime diagram, and others in which it points at an angle. Yet we claimed that each four-vector has a unique direction in spacetime. Discuss.

14. What does it mean to say that energy is the time component of four-momentum (up to a factor of c)? Don't just say something trivial like "the first component of a four-vector is the time component" or "if you draw four-momentum on a spacetime diagram it's the part that points along the t axis." What does it mean to say this represents "momentum in the time direction"?

Problems

15. Calculate the four-momentum of an electron moving at $c/2$ in the positive x direction.

16. Dmitri flies in a 10,000 kg rocket at constant speed $(3/5)c$ to a planet 5 light-years away in the positive x direction.

   (a) Calculate his displacement four-vector $(c\Delta t, \Delta x, \Delta y, \Delta z)$ for the journey.

   (b) Calculate his proper time $\Delta\tau$ for the journey.

   (c) Calculate his four-momentum by using Equation (2.9). (You have everything you need to just write this down.)

   (d) Calculate his momentum and energy by using Equations (2.1) and (2.2).

17. Object A is on your left, hurtling toward you at $(3/5)c$. Object B is on your right, hurtling toward you at $(4/5)c$ (see Figure 2.21). Both objects have mass m. In this problem you will describe Object B according to the reference frame of Object A. You will do this twice: first without four-vectors, and then with.

Figure 2.21

   (a) Using the velocity transformation formula in Appendix B, calculate the velocity of Object B according to the reference frame of Object A. (It isn't $(7/5)c$. That would be bad.)

   (b) Using the equations for mass and momentum in Appendix B, and based on the velocity you just calculated, write the equations for the energy and momentum of Object B according to the reference frame of Object A.

   (c) Now let's start over. Write the four-momentum of Object B in your reference frame.

   (d) Transform that four-momentum to the reference frame of Object A, using Equations (2.7).

   (e) Make sure your answers match. (Remember that the first component of the four-momentum is not E, but E/c.)

18. A particle has four-momentum (2 kg ls/s, 1 kg ls/s, 0, 0) in Frame R.

   (a) What is the particle's mass?

   (b) What is the particle's velocity?

   (c) What are the components of the four-momentum in a frame moving at speed $c/2$ in the positive x direction with respect to R?

19. Using the definition given in Equation (2.10), calculate the magnitude of the four-momentum and simplify your answer as much as possible. *Hint*: Remember to use the definition of γ.

20. If you call an object's position four-vector **R** then you can define a velocity four-vector as **V** = $d\mathbf{R}/d\tau$.

    (a) How fast would an object have to move in the x direction for V_x to be greater than c?

    (b) Why doesn't this violate the rule that nothing can move faster than light?

21. Frame R′ is moving at speed u in the positive x direction relative to Frame R.

    (a) Write the equations to convert E and p_x from Frame R to Frame R′. (Be careful with the factors of c.)

    (b) Particles A and B collide, and in Frame R their momentum and energy are conserved: $E_{A0} + E_{B0} = E_{Af} + E_{Bf}$, and $P_{xA0} + P_{xB0} = P_{xAf} + P_{xBf}$. Using just the transformation equations you found in Part (a), show that momentum and energy are also conserved in Frame R′.

    (c) If momentum is conserved in Frame R but energy is not, will momentum be conserved in Frame R′?

    Although you did this calculation for the momentum four-vector, the same result would apply to any conserved four-vector. If it's conserved in any inertial reference frame it's conserved in all of them.

22. The magnitudes of three-vectors obey the triangle inequality $|\vec{a} + \vec{b}| \leq |\vec{a}| + |\vec{b}|$. Does this same relationship hold for four-vectors? If so, prove it. If not, give a counterexample. *Hint*: You may find it helpful to draw some four-vectors on a spacetime diagram.

23. The dot product of two four-vectors **A** and **B** is defined as $A_t B_t - A_x B_x - A_y B_y - A_z B_z$. Prove that this dot product is a scalar.

24. In this problem you will show that the magnitude of a three-vector is preserved under rotations and the magnitude of a four-vector is preserved under the Lorentz transformations.

    (a) Using Equations (2.6), show that $x'^2 + y'^2 = x^2 + y^2$.

    (b) Using Equations (2.7), show that $V_t'^2 - V_x'^2 = V_t^2 - V_x^2$. (Including the y and z components wouldn't matter since you'd be adding the same thing to both sides.)

25. You can use the four-momentum to derive the velocity transformation equations. To keep things simple we'll stick to one spatial dimension, so let's consider a particle of mass m moving in the x direction with speed v. *Warning*: There will be several different "γ"s in this problem. To keep them straight you can either use subscripts or don't bother using γ and just write out its definition each time it comes up.

    (a) What is the particle's four-momentum?

    (b) Transform that four-momentum into a Frame R′ moving at speed u in the positive x direction relative to R.

    (c) From Part (b) you should have an expression for P_t' that depends on v and u. Remember what that represents: it's E'/c, where E' is the energy of the particle in the primed reference frame. So set that equal to $mc/\sqrt{1 - (v'/c)^2}$ and solve for v', the particle's velocity in R′.

26. You can add four-vectors graphically in the same way as you do for three-vectors, using the tip-to-tail method. As an example, consider a 2 kg particle moving at $c/2$ in the positive x direction, colliding and sticking to a 1 kg object at rest.

    (a) Draw the initial four-momenta of the two particles.

    (b) Add them graphically tip-to-tail to find the four-momentum of the final, combined particle.

    (c) Estimate the slope of the final four-momentum to find the final speed.

27. **Four-Force**

    You can define a four-vector force as **F** = $d\mathbf{P}/d\tau$.

    (a) Explain how we know this is a four-vector.

    (b) How are the spatial components of **F** related to the three-force, which is defined as $\vec{F} = d\vec{p}/dt$?

2.5 More about the Michelson–Morley Experiment

In 1887 Albert Michelson and Edward Morley conducted one of the most famous experiments in the history of physics, which we now call the "Michelson–Morley experiment." This bit of 130-year-old science, conducted in the basement of a dormitory on the eastern edge of Cleveland, is worth our careful study for two reasons. First, Michelson invented a remarkable tool capable of measuring small changes in waves traveling at high speeds: a tool that is still used today. Second, Michelson and Morley used that tool to look for small difference in the speeds of different beams of light – and, much to their surprise, they could not find any such differences. That legendary failure was one of the most important steps on the road to modern physics.

This section relies on the idea of phase shifts leading to constructive and destructive interference. That idea is reviewed in Section 3.1. If you have not previously studied wave interference we recommend going through that section before this one, even if you are not yet studying quantum mechanics.

2.5.1 Discovery Exercise: The Michelson–Morley Experiment

Ze and Maria are both capable of swimming with speed v in still water. A stream of width L is flowing with a steady current u (less than v) to the right.

Maria aims her body directly across the stream, swims to the other side, and then swims back. Note that her swimming velocity v is directed across the stream; she is also being carried downstream at speed u, but she doesn't care.

Ze, on the other hand, swims downstream a distance L, and then swims back upstream.

Both of their paths are shown in Figure 2.22. All speeds in this problem are far below the speed of light, so ignore relativistic effects.

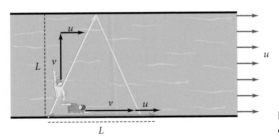

Figure 2.22 Two swimmers moving along different paths.

1. How long does Maria take to travel across the stream and back? *Hint*: The current is irrelevant for this question.

2. Ze's speed as he travels downstream is $v + u$. How long does it take him to travel distance L?

3. What is Ze's speed upstream, and how long does it take for him to return to his starting point?

4. Find the ratio of Ze's total time to Maria's. Simplify as much as possible.

This example introduces the basic idea of an interferometer. (It would be a more accurate analogy if Maria swam directly across instead of drifting downstream (Problem 8), but this conveys the basic idea with slightly simpler math.)

2.5.2   Explanation: The Michelson–Morley Experiment

Imagine that outer space is not empty. Just as the space around you right now is filled with a substance called "air" that you generally don't notice, imagine the entire universe is filled with a substance called "the luminiferous aether" (or just "aether," or sometimes "ether"). Just as sound is actually vibrations in the air, light (in this scenario) is actually vibrations in the aether. The aether provides the ultimate non-moving reference frame, the still point against which all motion should properly be measured. The Earth is zooming through this still aether at over 30,000 m/s (or 67,000 mph), so we are constantly in the midst of a strong "aether wind."

We should pause to remind you that nothing in the preceding paragraph is true. But the aether theory held sway for hundreds of years, and is seen in the writings of Boyle, Huygens, Newton, Cauchy, Stokes, and Maxwell, among others. Since light is a wave, they assumed there must be something waving – some physical object or substance, oscillating in harmonic motion – and that's what Michelson and Morley set out to measure. As you follow their reasoning you need to hold onto what you have learned from Galileo, but suspend everything you have learned from Einstein.

So, how do you measure an aether wind? If you are riding on a bicycle you feel the wind hitting you, but the aether in this model passes right through us and we don't feel it. There is another way, though. The speed of sound is 343 m/s *relative to the air that carries it*. If you are moving relative to the air, you will measure different speeds of sound in different directions. By analogy, the speed of light should be *c* relative to the aether. Our Earth-bound laboratories, moving through the aether in a particular direction, should measure light moving at different speeds in different directions.

Active Reading Exercise: The Speed of Sound

Imagine that you are on an airplane traveling at 100 m/s through still air. The airplane emits a sound that travels 300 m/s in every direction, relative to the air. Using Galilean relativity (or common sense), and ignoring Einsteinean relativity (which is not relevant at these speeds anyway), answer the following questions from your perspective on the airplane.

1. How fast do you measure the sound moving forward – that is, moving in the same direction the airplane is traveling?
2. How fast do you measure the sound moving in the opposite direction?

You can answer the exercise above using the equations of Galilean relativity, or you can think about it in this way: the sound is moving at 300 m/s *relative to the air*. In one second the airplane travels 100 m and the sound travels 300 m in the same direction, so an airplane scientist should measure the sound traveling at 200 m/s in the forward direction. By similar logic the backward sound wave should travel at 400 m/s. A sideways sound wave would travel somewhere in between those two speeds. (Can you see why it's not exactly 300 m/s?)

In 1887 Albert Michelson and Edward Morley set out to measure the variations in the speed of light as measured by the Earth as it moves through the aether. The basic idea is similar to the sound-on-the-airplane scenario we just discussed. The hard part is, how do you measure something that travels hundreds of thousands of miles every second? Michelson's ingenious answer was a device we now call a "Michelson interferometer."

The Michelson Interferometer

The Michelson interferometer is represented in Figure 2.23. Light from a single source reaches a half-silvered mirror and splits into two beams traveling at right angles to each other. The beams reach mirrors that send them *back* to the splitter, where they travel together to an eyepiece.

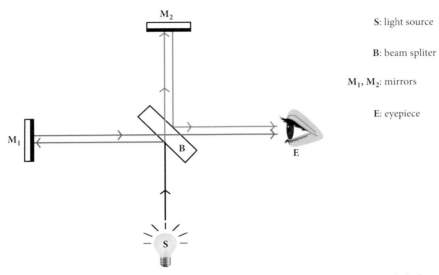

S: light source

B: beam spliter

M_1, M_2: mirrors

E: eyepiece

Figure 2.23 Schematic representation of the Michelson interferometer. A single light source S sends beams to a single eyepiece E along two different paths. The two paths are shown here in blue and red so we can see which is which, but remember that the actual experiment initiated only one beam (white) which split into two paths.

We're going to ask whether or not the two beams arrive in phase – and therefore constructively interfere – at the eyepiece. So imagine following one crest of the wave. At the splitter, that one crest splits into two different crests that follow different paths to the eyepiece.

- If the two paths are identical in length, *and* the two beams travel identical speeds, then the two crests arrive at the eyepiece together. Since all crests do likewise, the two beams are in phase, and the interference is constructive.

- If the two paths are different lengths then the two crests take different amounts of time to complete the journey. The two beams may therefore arrive out of phase.

- If the two paths are identical lengths but are traveled at *different speeds,* then once again the two crests arrive at different times, and the two beams may be out of phase.

The cases where the beams arrive out of phase end up looking like Figure 2.24.

Now suppose the Earth is traveling along the horizontal axis from M_1 to **B**. The blue beam travels first against, and then with, the Earth's motion, while the red beam travels always perpendicular to that motion. (We're only considering the paths from the splitter to the mirrors and back again. The paths before and after that are identical for both rays so they don't affect the interference pattern.)

The aether model predicts that for part of the trip the blue beam is going faster, and for part of the trip the red beam is going faster. But these two effects do not fully cancel out, and there is an overall difference in the times the two paths take (see Problem 8). So even if the two path

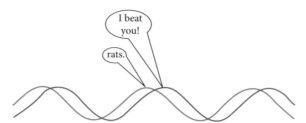

Figure 2.24 In the last leg of the journey through the interferometer, the two beams move along the same path at the same speed. But one of them may arrive at that last leg ahead of the other, either because it traveled a shorter distance to get there or because it traveled faster on the way there.

lengths are identical, the beams will arrive out of phase. (Once again remember that we are using "blue beam" and "red beam" to identify the two paths shown in the figure, but they are really the same color light.)

Do you see the brilliance of this scheme? When you turn on a light at **S** you see the result at **E** – the entire result, from both paths – instantaneously (for all practical purposes). But what you see at point **E**, constructive or destructive interference, depends on whether two paths took different times by even a small fraction of a second. That time difference, far too small to measure directly, becomes visible on a wall.

Michelson's device is so powerful that essentially the same design was used 130 years later to detect gravitational waves by measuring a spatial distortion of about 10^{-19} m: roughly 10,000 times smaller than a proton (Problem 9).

Rotating the Interferometer

We said above that the two beams may arrive at the eyepiece out of phase for two reasons: either the path lengths are different, or the path speeds are different. Michelson and Morley were looking for differences in speed, but they could not guarantee that their path lengths were identical to the needed level of accuracy. (The wavelength of visible light is less than 10^{-6} m.) So they found a clever trick for factoring out length-related differences and isolating the difference due to speed.

After measuring the type of interference they observed, they rotated the entire apparatus. A uniform rotation does not change the geometry of the experiment; whatever phase change was caused by differences in path length is still exactly the same as it was. But a phase change caused by differences in speed, due to orientation with respect to the aether wind, will change.

So the experiment was not to see whether the two beams constructively or destructively interfered on the wall, since such interference might have multiple causes. Rather, they looked to see how the interference pattern *changed* as they rotated the device. Such a change would only result from light traveling at different speeds in different directions, because of the aether wind.

Results

Michelson and Morley went into this experiment, as scientists often do, basically knowing what they were going to find. They were completely surprised when they didn't find it.

Their experiment found no evidence at all that the two beams had traveled at different speeds. Michelson wrote in a letter that the result was "decidedly negative." Even allowing for measurement errors, his experiment seemed to show that "the relative velocity [of the Earth and the aether] is less than one sixth of the Earth's velocity."

There was no real doubt, then as now, about how fast the Earth is moving through the solar system. But the Michelson–Morley experiment seemed to show that this motion had no effect on the speed of light beams in the reference frame of the Earth. We accept and expect that to be the case now, but it caused deep theoretical problems that engaged the top minds of the day. Michelson wrote in his 1887 paper:

> Stokes has given a theory of aberration which assumes the ether at the Earth's surface to be at rest with regard to the latter, and only requires in addition that the relative velocity have a potential; but Lorentz shows that these conditions are incompatible. Lorentz then proposes a modification which combines some ideas of Stokes and Fresnel, and assumes the existence of a potential, together with Fresnel's coefficient. If now it were legitimate to conclude from the present work that the ether is at rest with regard to the Earth's surface, according to Lorentz there could not be a velocity potential, and his own theory also fails.

Einstein rendered all those ideas obsolete in a single stroke. People often think of his radical idea as "space and time are reference-frame-dependent" or perhaps "mass is a form of energy." But you can boil his theory down to "there is no aether." Light, unlike sound, can travel with no medium of any kind.[7]

Einstein believed that Maxwell's equations were fundamental laws of physics and should be valid in all inertial reference frames, and that led him to follow the consequences of eliminating the aether as a preferred reference frame for light. Whether Einstein was influenced by the Michelson–Morley result is debatable. He does seem to have heard of it, but he said on occasion that it was not a significant influence in his thinking. Regardless, he noted in later years that it was an important reason why people accepted his theory.

Some Experimental Details (or, "Science is Hard")

We want you to understand why nineteenth-century physicists believed that they were moving through an "aether wind," and why that wind would lead to different light speeds in different directions. We want you to understand how small changes in nearly instantaneous motion can be measured by looking for the interference caused by a phase shift. Most of all, we want you to understand how the results of this historic experiment justify the axioms of special relativity.

That's the theory. What about the experiment? There is an enormous gap between the simple schematic of Figure 2.23 and an actual lab setup. Let's get our hands dirtier and look at some of the issues Michelson and Morley had to consider.

---

7 Some authors have shown that the Michelson–Morley experiment doesn't preclude the existence of aether. If you use the Lorentz transformations rather than the Galilean ones to convert between inertial reference frames, you can still define one particular reference frame as "the aether frame" and say that light consists of waves of a medium that's at rest in that frame. But the physics will appear the same in all inertial reference frames, so there is no way to determine which reference frame is the aether frame.

We'll start with a big one: the eyepiece is not an infinitesimal point, but a finite area on which the light shines. At some places on this area the two beams are out of phase by an integer number of wavelengths, so they constructively interfere and create a bright light. At other places the two beams are out of phase by a half-integer number of wavelengths, leading to dark patches. You therefore see an "interference pattern," alternating between light and dark. (If you don't quite see why that happens, hold that thought; we'll explain it carefully in Section 3.2.) You'll see an interference pattern no matter how the light beams traveled, so in and of itself it doesn't tell you anything. But for reasons we explained above, rotating the device should have caused the interference fringes on the eyepiece to move. It didn't.

Here are a few more issues – presented not because of any theoretical importance, but to offer a glimpse at the rigors of experimentation.

- The interferometer was extremely sensitive to vibrations even when still. And remember, the entire apparatus had to be rotated without disturbing anything! To minimize distortion Michelson and Morley did their work in a closed room in the basement of a heavy stone dormitory, and built their device on top of a large block of sandstone floating in a pool of mercury. When they gave the block a push it rotated smoothly, so they could watch the interference pattern change with the angle to the aether wind (or not change, as it turned out).

- But working in a basement gave them very little space. This created a different problem, because the phase difference is easier to detect if the path lengths are longer. So they put four mirrors in each place where our diagram shows only one, bouncing the light back and forth between the mirrors many times. By this mechanism they achieved a complete path length of 11 m.

- They used yellow light when aligning the mirrors, because that light has a consistent frequency. But they ran the actual experiment with white light because it was easier to detect changes in the resulting colored fringe pattern by eye.

- Finally, when their experiment failed to produce any measurable change in the interference pattern as the device rotated, they began looking for explanations. One of their hypotheses is particularly interesting, and we quote here Michelson's and Morley's original paper:

> In what precedes, only the orbital motion of the Earth is considered. If this is combined with the motion of the solar system, concerning which but little is known with certainty, the result would have to be modified; and it is just possible that the resultant velocity at the time of the observations was small though the chances are much against it. The experiment will therefore be repeated at intervals of three months, and thus all uncertainty will be avoided.

So there's the final piece of the experiment: six months later the entire experiment was repeated. As we would now expect, the results were unchanged.

This last point testifies to a very general rule in the development of science: when an experiment doesn't fit your theory, you fix the experiment or tweak the theory. Only when all else fails do you abandon or rewrite fundamental principles.

2.5.2 Questions and Problems: The Michelson–Morley Experiment

Conceptual Questions and ConcepTests

1. You are on a train going northward (at less than 65 mph). Outside the train is a highway, with four cars each going at 65 mph and heading north, south, east, and west.

   (a) Rank-order the speed of the four cars as measured in your reference frame from fastest to slowest, indicating if any are equal.

   (b) Briefly (but specifically) explain what this question has to do with Michelson and Morley's expectations about the speed of light.

2. The Discovery Exercise (on p. 97) compared Maria, who traveled a distance $2L$ in the y direction with a constant y-velocity of v, with Ze, who traveled a distance of L at speed $v + u$ and then the same distance at speed $v - u$. Your conclusion was that, even though Ze was faster than Maria for part of the trip and slower for part of the trip, the two swimmers did not take the same amount of time. Explain what this result has to do with Michelson's expectations for his experiment.

3. If light is a vibration in the aether just as sound is a vibration in the air, how do we know that the aether must exist all through space – throughout the galaxy, and even in between galaxies – instead of just in the Earth's atmosphere?

4. Michelson and Morley used multiple mirrors to make their effective path length longer. Why does a longer path length help?

5. Why did the Michelson–Morley experiment have to be repeated six months later? (Choose the best answer.)

   A. Good experiments should always be repeated to guard against error.

   B. Six months later the Earth would be pointing a different way, effectively rotating the apparatus.

   C. They wanted to account for possible changes in the aether over time.

   D. They wanted to account for possible motion of the solar system through space.

   E. Michelson needed more grant funding.

6. What result did Michelson find so surprising? Why is that result perfectly predictable in the framework of special relativity?

For Deeper Discussion

7. One possible explanation for Michelson's null result is that, as the Earth moves through the aether, it drags the surrounding bubble of aether with it. So we don't experience an aether wind for the same reason we don't experience an atmospheric wind blowing always against the Earth's motion. How would you test such a theory? What problems can you see with it?

Problems

8. **Calculating the Phase Shift**

   In this problem you will analyze a Michelson interferometer under the assumption that the speed of light is c relative to the aether: a pre-Einstein analysis. We're going to use a slightly simpler geometry than the actual experiment, but it won't change anything important.

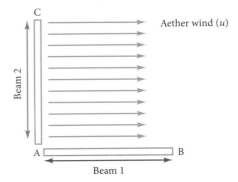

Figure 2.25

The setup is shown in Figure 2.25. Two light beams are emitted from Point A. The aether wind is flowing to the right (from A to B) at speed u. Beam 1 goes downstream to Point B and then upstream back to A. Beam 2 goes perpendicular to the aether wind, to Point C and back. The distances AB and AC are each equal to L. You (the observer) are at Point A the whole time.

(a) When Beam 1 goes from A to B, how fast is it moving relative to you?

(b) How fast is Beam 1 moving relative to you on the way back?

(c) What's the total time taken by Beam 1 on its journey?

(d) How fast is Beam 2 moving relative to you?[8] *Hint*: It's not c. On the return trip the speed will be the same, but downward instead of upward.

(e) What's the total time taken by Beam 2 on its journey?

(f) Now let's put some numbers to it. Assuming $c = 3 \times 10^8$ m/s, $u = 3 \times 10^4$ m/s, and $L = 11$ m, calculate the time difference between the two paths.

(g) Finally, assume the oscillation period of the light is 5.1×10^{14} Hz. Calculate the phase difference between the two paths.

(h) The actual Michelson interferometer shown in Figure 2.23 has a slightly more complicated geometry than that presented in this problem, but it's exactly equivalent. Assume that in Figure 2.23 the aether wind is moving horizontally, in the direction from $\mathbf{M_1}$ to the eyepiece. Which points in that diagram correspond to Points A, B, and C in Figure 2.25, and why is it okay for us to ignore the parts of the path in Figure 2.23 that aren't included in the L formed by those three points?

9. **LIGO**

The Laser Interferometer Gravitational-Wave Observatory (LIGO) was designed to detect gravitational waves predicted by Einstein's *general* theory of relativity. The theory predicts that certain cosmic events will cause ripples of spatial distortion. LIGO made global news in 2015 when it detected the gravitational waves generated by a pair of colliding black holes 1.3 billion light-years away, and since then many more detections have been made.

The fundamental structure of LIGO is closely modeled on Michelson's work (Figure 2.26). A laser travels along two perpendicular arms, bounces off mirrors, and merges back. (Just like Michelson's device, LIGO bounces the light back and forth many times to elongate its path length. But the scale is quite different: LIGO bounces 400 times along arms 4 km long!) Of course we know that the *speed* of the light along the two paths is the same; LIGO is looking for changes in the *distance*. A gravitational wave will elongate one arm and shrink the other arm slightly, changing the interference of the two beams.

Figure 2.26 *Source: Caltech/MIT/LIGO Lab © 2008.*

The laser beam that LIGO uses has a frequency of 2.8×10^{14} Hz.

(a) How much would one arm have to grow relative to the other to cause the phase difference to shift by 10% of a period? (Remember that each arm has an effective length 400 times its actual length, and if one grows a bit its effective length grows by 400 times that much.)

(b) In the first gravity wave ever detected, the interferometer arms grew and shrank by approximately 2×10^{-17} m. What phase shift would that cause in the interference pattern? You should find that LIGO's measuring devices are capable of detecting shifts a whole lot smaller than Michelson and Morley could have detected by eye.

---

8 When Michelson analyzed his original interferometer experiment he treated the perpendicular speed as c. Lorentz pointed out the error and Michelson corrected it in a subsequent paper.

Chapter Summary

Our survey of Einstein's special theory of relativity is divided between Chapters 1 and 2.

The equations for both chapters are collected in Appendix B. You may want to cross-reference that appendix as you read through this summary.

Section 2.1 Spacetime Diagrams

A "spacetime diagram" is an important tool for visualizing relativistic events. Unlike traditional position vs. time graphs, these put time on the vertical axis and position on the horizontal.

- These diagrams are traditionally drawn using units where $c = 1$, which makes the world line of a light beam a line at $\pm 45°$. The light beams coming into and out of a point on the diagram form the "past light cone" and "future light cone" of that event.

- Events inside the past or future light cone of some Event E have a "timelike separation" from E. Events on E's light cone have a "lightlike separation" from E. Events on or inside E's light cone can either affect E or be affected by it.

- Events outside E's light cone have a "spacelike separation" from E and cannot have any causal relationship with it.

Section 2.2 Momentum and Energy

The relativistic definitions of momentum and energy (Appendix B) are conserved for any isolated system in all inertial reference frames.

- The relativistic formulas for $\vec{p}$ and E reduce to their Newtonian values for $v \ll c$. Both approach ∞ in the limit $v \to c^-$.

- Relativistic energy can be broken into the "rest energy" mc^2, and the kinetic energy $K = E - mc^2$. The rest energy includes all forms of internal energy such as thermal energy or the potential energy of a compressed spring. Therefore any increase in such forms of energy is an increase in an object's mass.

- You can simplify calculations if you know that objects are moving much slower than c or very close to c.

  - For a "non-relativistic" particle ($v \ll c$, $K \ll mc^2$), you can use Newtonian formulas for momentum and energy.

  - For an "ultra-relativistic" particle ($v \approx c$, $K \gg mc^2$), you can use the approximation $E \approx pc$. Massless particles are always ultra-relativistic, with $v = c$ and $E = pc$.

Section 2.3 Mass and Energy (and the Speed of Light Squared)

- Just as in classical mechanics, mass determines a body's inertia: the same force will cause greater acceleration in an object with less mass.

- Just as in classical mechanics, mass is conserved over time and is the same in all reference frames (invariant).

- *Unlike* in classical mechanics, mass is not an "additive quantity": the mass of System A plus the mass of System B is not necessarily the mass of the total system A+B. This is because what counts as potential and kinetic energy in the individual systems may count as mass energy in the combined system.

- All collisions conserve energy. An "elastic collision" ends with all the same particles with which it began, with no changes to their individual masses. If particles are created or destroyed or change their masses, the collision is inelastic.

Section 2.4 Four-Vectors

A "four-vector" has a time component and three spatial components. The position four-vector (or "spacetime four-vector") is defined as (ct, x, y, z).

- A four-vector is defined by the fact that its coordinates convert from one inertial reference frame to another via the Lorentz transformations (adjusted by an extra factor of c to make all the units the same).

- "Scalar" in this context means a quantity that is invariant, i.e. it has the same value in all inertial reference frames.

- The magnitude of a four-vector, defined as $\sqrt{V_t^2 - V_x^2 - V_y^2 - V_z^2}$, is a scalar.

- The "momentum four-vector" gives the energy and momentum of an object: $\mathbf{P} = (E/c, p_x, p_y, p_z)$.

Section 2.5 More about the Michelson–Morley Experiment

Nineteenth-century physicists believed that the universe was permeated with a substance called the "aether," and that the nature of light was vibrations in the aether. Michelson and Morley set out to measure the "aether wind" caused by the motion of the Earth.

- If the aether theory were correct, the speed of light would be faster in the direction of the wind, and slower in the opposite direction.

- They sent one light beam on a path perpendicular to the Earth's motion and another parallel to it, and measured the interference of the two beams when brought back together. Then they rotated the apparatus 90° to measure how the interference pattern changed.

- Their experiment (from their point of view) failed; no difference in light speeds in different directions was detected. Einstein later described this result as one of the main reasons why his theories were accepted.

3

The Quantum Revolution I: From Light Waves to Photons

According to the theory of relativity, Newton's laws only work for objects traveling much slower than the speed of light. This does not mean that we need one set of laws for fast objects and a different set for slow objects; the equations of relativity work at all speeds. But Einstein's equations and Newton's laws make essentially the same predictions as each other for slow objects, and diverge significantly for fast ones. Early twentieth-century physicists were able to measure objects moving close to the speed of light, and such objects followed Einstein's laws – not Newton's.

Around that same time, measurements of very small objects revealed another domain where classical predictions stopped working. We now view Newtonian physics as an approximation that works for objects of ordinary speed *and size*. Just as fast objects require relativity, small objects such as atoms require quantum mechanics.

Whereas special relativity more or less sprang fully formed from the head of Einstein in 1905,[1] quantum mechanics was developed over several decades by many physicists. That succession of experiments and theories is a fascinating story, but it is not the main topic of this book. We are therefore going to present the development, not in historical order, but in the way that we believe most clearly illustrates the ideas. Appendix A summarizes the results in chronological order to give you a better sense of how quantum physics emerged historically.

We are going to start with the "Young double-slit experiment," which we think demonstrates in the clearest way possible why we now believe that a particle can exist in more than one place at once, how the very fact of measurement changes a system, and other core concepts from quantum mechanics. In order to explain that, though, we need to cover some background first. Section 3.1 discusses "interference," an important property of waves. This well just form a review of your earlier physics courses, but we encourage you not to skip it as it provides vital background for the section that follows it. Section 3.2 explains how the Young double-slit experiment was used in the nineteenth century to establish the wave nature of light. Section 3.3 shows how this same experiment, with a twentieth-century twist, demonstrates that light also has properties of a particle. By the time you're done with these two sections, you won't just know what it means to say "light has properties of both particles and waves" (which may not sound like the most amazing revelation right now). You will see why this dual nature forced physicists to rethink their most fundamental assumptions about the nature of reality.

---

1 That is of course an exaggeration; other people had developed some of the key ideas before that.

3.1 Math Interlude: Interference

We're going to introduce the central concepts of quantum mechanics through the Young double-slit experiment, which had a huge impact on both nineteenth- and twentieth-century physics. Before we can properly introduce the experiment, however, we need to discuss the math and physics of "interference."

3.1.1 Explanation: Interference

Alice and Bob are holding the two ends of a long rope tightly enough that it is more or less horizontal between them. We're going to describe three simple experiments they can do. (You can try these at home, too!)

- *Alice gives the rope one quick shake.*
  The result is a bump that starts next to her and travels along the rope to Bob.

- *Alice and Bob give the rope quick shakes at the same time.*
  The result, shown in Figure 3.1, is that bumps form next to both of them heading toward each other. When the two bumps meet in the middle – this is the part we're interested in – they form one big bump. Then the bumps pass through each other, and each bump moves on as if the other one hadn't been there at all.

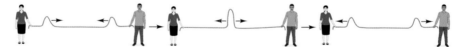

Figure 3.1 Alice and Bob each shake one end of a rope.

Active Reading Exercise: The Third "Alice's Rope" Scenario

Did you follow what happened when Alice's and Bob's bumps collided? Let's find out by changing the story just a bit.

- *Alice does her same signature shake. But this time Bob gives the rope a downward shake, causing an upside-down bump to move toward Alice.*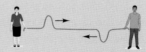

  1. Draw and/or describe what the rope will look like when the two bumps meet in the middle.
  2. Draw and/or describe what the rope will look like a second or two after that.

Don't read further until you have written down both guesses!

Here are the answers. When Alice's up-bump meets Bob's down-bump they will cancel, and the rope will be flat. A moment later the two bumps will re-form, the up-bump continuing to the right and the down-bump to the left.

The point of all this rope-shaking is to illustrate "interference," the interaction of two waves. Interference is not hard to understand physically, but it's important to also see the underlying mathematical principle.

Alice and Bob's Rope, Mathematically

We're going to assume that the rope moves only up and down. (There is no motion into, or out of, the page.) Even with that assumption, a full mathematical treatment would require a "multivariate function" $y(x, t)$ specifying the height as a function of both horizontal position and time. We will make much use of such functions over the next few chapters.

But for the moment we're going to sidestep the time variable. *At any specific moment,* the shape of the rope can be described by a single-variable function $y(x)$. The independent variable x is a position on the rope – say, the horizontal distance from Alice toward Bob. The dependent variable y is the height of the rope. In its relaxed position, the rope is described by the function $y(x) = 0$ everywhere along its domain.

Suppose only Alice shook the rope. At some instant a few seconds later, the rope would be in some shape (such as the bump in all three examples above) that we will call $y_A(x)$. At that same moment the shape of the rope if only Bob had shaken it is some other function that we will call $y_B(x)$. Our concern here is what happens when they both shake the rope. The answer, in most cases, is that their two functions just add. This is called the "principle of linear superposition."

The Principle of Linear Superposition

A wave system is said to follow "linear superposition" if the wave produced by two sources is the sum of the waves produced by each source.

For instance, if at some instant Alice has given the rope the shape $y_A(x)$ while Bob has given it $y_B(x)$, then the actual position of the rope at that instant will be given by:

$$y_{rope}(x) = y_A(x) + y_B(x).$$

When Alice and Bob both gave the rope up-shakes, their positive y-values added to give higher y-values. This phenomenon (bump plus bump equals higher bump) is called *constructive interference.*

When Bob gave the rope a down-shake, his negative y-values canceled Alice's positive y-values. This phenomenon (bump plus anti-bump equals flat rope) is called *destructive interference.*

You may be tempted to think of linear superposition as "obvious," but it is not a given. If Alice and Bob both give giant shakes, the rope will *not* bend all the way up to $y_A(x) + y_B(x)$; after all, there's only so far a rope can stretch! So linear superposition is not a universal law of nature, but a property that characterizes certain systems (such as small waves on a rope) and not others (such as large waves on a rope).

What systems do obey this principle? We can give you a few important examples: electromagnetic waves in a vacuum obey it, and mechanical systems like ropes and water obey it to a very good approximation provided the amplitudes of the waves aren't too big. If you've studied differential equations we can give you the more general answer, which is that a system obeys this principle if the differential equation(s) describing it are "linear" and "homogeneous." (If $y_A(x)$ and $y_B(x)$ are both solutions to such an equation – that is to say, valid states of the system – then $y_A(x) + y_B(x)$ is also a solution.) The equations for electromagnetic waves in vacuum are linear and homogeneous, while the equations for ropes and water waves are well approximated by linear homogeneous equations for small amplitudes. If you haven't studied differential equations, the examples are all we need for now.

Superposition of Sine Waves

We've been looking at traveling bumps, but many physical phenomena are represented by extended sinusoidal shapes. When such waves add, the result can be very complicated. (Try plotting $\sin(5x) + \sin(6x) + \sin(7x)$; the result looks almost random.) But when different sines with the *same frequency* add, the results may remind you of Alice's rope.

Consider $y_1(x) = \sin x$ and $y_2(x) = \sin x$. (These could once again represent waves on a rope at some moment in time, produced by two different sources.) What does $y_1(x) + y_2(x)$ look like? Of course it is $2 \sin x$ but think about this visually. The two waves reach their positive values together (resulting in higher peaks) and their negative values together (resulting in lower troughs). This is constructive interference, illustrated on the left side of Figure 3.2.

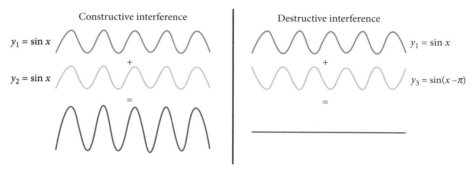

Figure 3.2 Two sine waves can add constructively or destructively.

Now consider $y_3(x) = \sin(x - \pi)$. What does $y_1(x) + y_3(x)$ look like? The answer is less clear algebraically, but you can easily see it on the right side of Figure 3.2. We now have a classic case of destructive interference: the peaks of y_1 cancel the troughs of y_3 and vice versa, so $y_1(x) + y_3(x) = 0$.

But it's important to see how we got there. We didn't change the amplitude or frequency of our sine wave, or multiply it by -1. We just started it at a different point in its cycle, and that made all the difference. We say that $y_1(x)$ and $y_2(x)$ are "in phase," while $y_1(x)$ and $y_3(x)$ are "out of phase."

Superposition of Sine Waves

When two sine waves of the same frequency meet, the resulting interference depends on their relative phases. If they are "in phase" (they peak together), the result is constructive interference. If they are "out of phase" by precisely half a cycle, the result is destructive interference.

When two waves aren't perfectly in phase or out of phase, they can partially enhance their amplitudes in some places and partially cancel in others. See Problem 16.

The math discussed above becomes physically important when two waves coming from different places meet.

Active Reading Exercise: What Does the Microphone Hear?

Consider two speakers emitting sinusoidal sound waves that are identical in every way: same amplitude, same frequency, and perfectly in phase (meaning the speakers peak at the same time). A microphone is placed nearby.

What will the microphone pick up if . . .

1. the microphone's distance to the two speakers is the same?
2. the microphone is exactly one wavelength closer to one speaker than the other?
3. the microphone is exactly half a wavelength closer to one speaker than the other?

Write down your answers to all three questions before reading further.

The answers to the first two questions are the same: the waves arrive at the microphone in phase, so their constructive interference results in a loud noise.

In the third case the waves arrive perfectly out of phase, so their destructive interference results in silence. Such a spot (depicted on the right in Figure 3.3) is called a "node."

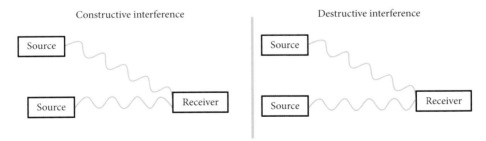

Figure 3.3 A snapshot of the waves going from the sources to the receiver at one instant. On the left a trough reaches the receiver simultaneously from both waves. On the right, a trough of the top wave and a crest of the bottom wave reach the receiver.

On Alice's rope, individual bumps from each side passed each other, momentarily interfered (constructively or destructively), and then moved on. Our speakers pumping out continuous sine waves represent a more typical and important situation. At certain points the two waves *always* constructively interfere because the peaks from the two sources always arrive together and the troughs from the two sources always arrive together. So at those points there will be a large oscillation. At other points the two waves always interfere destructively. A peak from one always arrives with a trough from the other, so there is no oscillation at those points.

For example, the receiver in the right image of Figure 3.3 is getting a trough from the top wave and a peak from the bottom one. A moment later it will get a peak from the top and a trough from the bottom. The two will always cancel and the receiver will not register any signal.

In practice, when you play two speakers, you don't notice alternating loud and soft spots as you walk across the room. You hear a jumble of waves of different frequencies and phases, some coming directly from the speakers and some bouncing off other surfaces. When scientists experiment with interference, as we will discuss in the next section, they have to carefully control all these variables.

Example: Two Channels

The figure shows two narrow water-filled channels: a straight path from Point A to Point B, and a bent path that goes up to Point C and then back down. At Point A you are tapping the water repeatedly, creating waves that travel along both channels with wavelength 0.1 meters. Describe what the water does at Point B. (Assume there are no reflected waves from any of the surfaces. Also assume that the ripples go around the corner at Point C without changing significantly.)

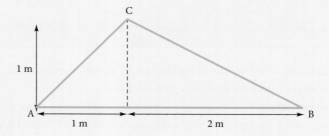

Answer: The waves traveling along the two channels start out perfectly in phase. The wave traveling along the straight channel goes a distance $D_1 = 3$ m. The wave traveling along the bent channel goes a distance

$$D_2 = \sqrt{(1\,\text{m})^2 + (1\,\text{m})^2} + \sqrt{(1\,\text{m})^2 + (2\,\text{m})^2} \approx 3.65\,\text{m}.$$

The difference between the two paths is almost exactly six and a half wavelengths. If it were an integer the two waves would be perfectly in phase and they would interfere constructively. But it's halfway between two integers, so the two waves arrive almost perfectly out of phase and interfere destructively; when one is going up the other is going down.

The water at Point B remains still.

Waves in Time, Waves in Space

Alice's rope changes in both space and time. Above we represented it as $y(x)$, a function of space only, by taking a snapshot at one instant. We can also do the opposite: focus on one x-value and watch the rope move up and down. The height of that point would be $y(t)$, a function of time only.

So we're going to end this section by briefly reviewing the terminology that is used to describe sinusoidal oscillations in time, and the significantly different terminology that is used to describe sinusoidal oscillations in space. These terms and their relationships are summarized in Appendix F. In Section 6.1 we will look at the more general equations that describe a system oscillating in both space *and* time.

We'll start with a sinusoidal oscillation in time, plotted in Figure 3.4. This could represent the position of a mass on a spring, the height of a cork bobbing up and down in water, or the electric field at a point as a light wave passes by.

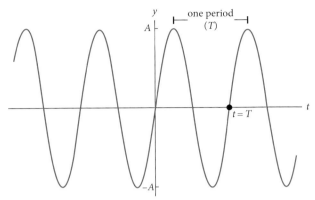

Figure 3.4 **A sine wave function of time.**

- The "amplitude" (A) tells you how much the function changes from its middle value to its maximum (or minimum) value. On the graph this is the vertical distance from the midline to a peak.
- The "period" (T) tells you the time one oscillation takes. On the graph this is the horizontal distance from one peak to the next. (But remember that this horizontal *distance* on the graph represents a period of *time* in the oscillation.)
- The "frequency" (ν) tells you how many full cycles occur in one unit of time. On the graph this is the number of oscillations per unit on the horizontal axis.
- The "angular frequency" (ω) counts the number of *radians* in one unit of time. Because there are 2π rad in one cycle (by definition), this is 2π times the frequency.

It's important to understand how period, frequency, and angular frequency all represent the same information in different ways. To illustrate this point we will make one statement about a sine wave – one fact – in three equivalent ways:

- $T = (1/30)$ s: The wave goes through one full cycle in 1/30 of a second.
- $\nu = 30$ Hz: The wave cycles up and down 30 times each second. (The unit "hertz" (Hz) means cycles per second.)
- $\omega = 2\pi(30)$ rad/s: The wave goes through 60π rad each second.

Figure 3.5 makes the same point graphically, showing how "high frequency" and "low period" express the same property in different words.

High frequency, low period
Low frequency, high period

Figure 3.5 **Two waves.**

Of these three interchangeable variables, angular frequency is the one that actually shows up when you write the function. For instance, $\sin(4t)$ has angular frequency $\omega = 4$. From that you can calculate $\nu = \omega/(2\pi) = 2/\pi$ and $T = 1/\nu = \pi/2$.

Now imagine a photograph of an ocean wave, or of a wave on a string. Or imagine the electric field along the length of a laser at one moment in time. These are functions that oscillate sinusoidally as a function of *space*. The math describing a sine wave is the same whether you call its independent variable t or x, but the letters and terminology we typically use are different.

For a sine wave function of x,

- we call the horizontal distance from one peak to the next the "wavelength" (λ) instead of the "period."
- we call the number of oscillations per unit distance the "spatial frequency" (f) instead of just "frequency."
- we call the number of *radians* per unit distance the "wave number" (k) instead of the "angular frequency."

The vertical distance from the midline to the peak is still called the "amplitude."

All of these terms and how they are related to each other are summarized in Appendix F. You might want to keep a copy of that appendix handy as you work problems in the next few chapters.

3.1.2 Questions and Problems: Interference

Conceptual Questions and ConcepTests

1. For each term below, give a brief description of what it represents.

    (a) Linear superposition

    (b) Constructive interference

    (c) Destructive interference

    (d) Node

2. Electromagnetic waves in a vacuum perfectly obey the principle of linear superposition. Electromagnetic waves in a medium such as air or water do not. Briefly explain what that means. (We're not asking *why* superposition applies here but not there – just a clear explanation of *what* we have just said is true here but not there.)

3. Alice sends her favorite up-bump down the rope toward Bob, while Bob sends a down-bump toward Alice. However, her bump has twice the amplitude of his. Assuming the rope still obeys the principle of linear superposition, what does it look like when the two bumps meet?

4. Alice sends her favorite up-bump down the rope toward Bob. At the same time, Bob sends toward Alice a *sideways* bump – that is, it points out of the page. Assuming the rope still obeys the principle of linear superposition, what does it look like when the two bumps meet?

5. Two sources emit sine waves. You have found a perfect node: a place where these two waves cancel perfectly, so the net effect at your spot is nothing at all, no matter how long you keep watching. Which of the following can you reasonably conclude? (Choose all that apply.)

    A. The two sources must be emitting at the same frequency.

    B. The two sources must be emitting at the same amplitude.

    C. The two sources must be emitting at the same phase.

6. Crowds at sporting events often simulate a wave as follows. A few fans stand up. As they sit down, the

people to their right stand up. As that group sits down, the people to their right stand up, and so on. The visual effect, from a distance, is of a wave rippling through the crowd. Now, suppose a fan at the far left of the stadium starts a wave that propagates to the right, while a fan at the far right starts a wave that propagates to the left. When the two waves meet in the middle, will they obey the law of linear superposition exactly, approximately, or not at all? Explain.

7. Figure 3.6 shows two speakers emitting sound waves of the same wavelength λ and the same initial phase. Let Point A be the point halfway between the two speakers, and let Point B be the point opposite Point A on the far wall. Is there any combination of D_{sp}, L, and λ for which the two waves would interfere destructively at Point B? Why or why not?

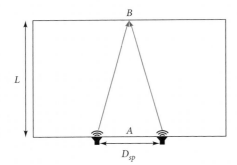

Figure 3.6

8. Immediately after Alice and Bob both give up-shakes, the rope has two separate bumps. When the two bumps reach the exact same x-position, it has one big bump. But what about *right before* that moment, when the two bumps have just started to merge? Draw what the rope might look like at such a moment. *Hint*: It may help to use a graphing application to look at a possible answer – but of course you have to figure out what would be a good function to graph!

9. Here is a common way to tune a guitar (really). You play two strings that are supposed to play exactly the same note. If they are slightly off, even if your ear can't hear the difference in pitch, you hear a "wah-wah-wah" sound of the note getting louder and then quieter. (Musicians describe listening for the "beats.") Why?

10. All waves carry energy. For example, when you drop a rock in a pond the resulting ripples spread out the energy of the impact across the surface of the pond. When you drop two rocks near each other there will be places where the two ripples interfere destructively and the water doesn't move. What happened to the energy carried by the two waves?

11. Imagine a sinusoidal wave traveling across the ocean at speed v. For each of the following quantities give a brief description of what it represents in that context, and units that could be used to measure it. Do not just give an equation; say what the quantity represents physically.

    (a) Amplitude A

    (b) Wave number k

    (c) Frequency ν

    (d) Wavelength λ

    (e) Period T

12. A string is waving up and down and at each moment it has a sine wave shape. What would it tell you about that oscillating string if it has a large wave number and a small angular frequency?

13. Source A emits light of only one frequency, radiating outward in a sphere. Source B emits light of that same frequency, also radiating out from itself in a sphere. Light from both sources hits the wall (see Figure 3.7). Will you see constructive or destructive interference on the back wall? (Choose one.)

Figure 3.7

    A. Only constructive interference (bright light)

    B. Only destructive interference (dim or no light)

C. Constructive interference in some places, destructive in others (some regions of bright light, some of dim)

D. The answer depends on the placement of the sources and their relative phases.

14. How would your answer to Question 13 change if...

   (a) the frequency of source A were twice the frequency of source B?

   (b) the frequency of source A were π times the frequency of source B?

Problems

15. For this problem, all graphs should be drawn on the domain $0 \leq x \leq 4\pi$.

   (a) Graph $f_1(x) = 3 \sin x$.

   (b) Graph $f_1(x) + f_1(x)$. Does this demonstrate constructive or destructive interference?

   (c) Graph $f_2(x) = 3 \sin(x - \pi)$.

   (d) Graph $f_1(x) + f_2(x)$. Does this demonstrate constructive or destructive interference?

   (e) Graph $f_3(x) = 3 \sin(x - 2\pi)$.

   (f) Graph $f_1(x) + f_3(x)$. Does this demonstrate constructive or destructive interference?

16. [*This problem depends on Problem 15.*] In Problem 15 you investigated $3 \sin x + 3 \sin(x - \phi)$ for $\phi = 0$, $\phi = \pi$, and $\phi = 2\pi$. Now, using a graphing application, investigate a variety of ϕ-values in between. What happens to the superposition as the waves become increasingly out of phase with each other?

17. Let $y_1(x) = \sin(2x)$.

   (a) Graph $y_1(x)$ on the domain $0 \leq x \leq 3\pi$.

   (b) The function $y_2(x) = \sin(2x + \phi_2)$ is in phase with $y_1(x)$. That is, they reach their peaks at the same x-values, so $y_1(x) + y_2(x)$ yields constructive interference.

   i. Graph $y_2(x)$ on the domain $0 \leq x \leq 3\pi$.

   ii. Find a non-zero value that could be ϕ_2.

   (c) The function $y_3(x) = \sin(2x + \phi_3)$ is completely out of phase with $y_1(x)$. That is, it reaches its peaks at the x-values where

y_1 reaches troughs, so $y_1(x) + y_3(x)$ yields destructive interference.

   i. Graph $y_3(x)$ on the domain $0 \leq x \leq 3\pi$.

   ii. Find a value that could be ϕ_3.

18. The functions $\sin x$ and $\sin(x + \phi)$ are in phase with each other for infinitely many different ϕ-values. We can identify them all by writing "$\phi = 2n\pi$ where n is any integer." Use similar notation to identify all the ϕ-values that make these two functions perfectly *out* of phase with each other.

Problems 19–21 refer to the following function, which could describe a rope shortly after Alice shakes it:

$$Sample\ bump : y(x) = \begin{cases} 0 & 0 \leq x \leq 2\pi \\ \sin x & 2\pi < x < 3\pi \\ 0 & 3\pi \leq x \leq 6\pi \end{cases}$$

$$(3.1)$$

19. Draw a graph of the function represented by Equation (3.1).

20. Equation (3.1) represents a flat line from 0 to 6π with an "up-bump" extending from 2π to 3π.

   (a) Write a function that represents a flat line from 0 to 6π with an "up-bump" extending from 3π to 4π. This might be Alice's shake a second after Equation (3.1). Be careful to make sure your bump still goes up, not down!

   (b) Write a function that represents the rope a second later, when the up-bump extends from 4π to 5π.

21. Equation (3.1) represents a flat line from 0 to 6π with an "up-bump" extending from 2π to 3π.

   (a) Write a function that represents a flat line from 0 to 6π with a "down-bump" extending from 2π to 3π. (This might be from Bob's shake.)

   (b) Add the function you just wrote to the original function from Equation (3.1). What kind of interference does the math show?

22. Alice is at position $x = 0$ and Bob is at position $x = 10$.

   (a) Soon after they shake the rope, Alice's shake has produced the shape $y_{A1}(x) = e^{-(x-3)^2}$ and

Bob's has produced $y_{B1}(x) = e^{-(x-7)^2}$. Draw, on one set of axes, those individual functions and the resulting shape of the rope (which is just $y_{A1}(x) + y_{B1}(x)$).

(b) A second later, Alice's shake has produced the shape $y_{A2}(x) = e^{-(x-4)^2}$ and Bob's has produced $y_{B1}(x) = e^{-(x-6)^2}$. Draw, on one set of axes, those individual functions and the resulting shape of the rope.

(c) Write functions $y_{A3}(x)$ and $y_{B3}(x)$ that describe the rope a second after Part (b). Draw, on one set of axes, those individual functions and the resulting shape of the rope.

(d) Write functions $y_{A4}(x)$ and $y_{B4}(x)$ that describe the rope another second later. Draw, on one set of axes, those individual functions and the resulting shape of the rope.

Problems 23–25 refer to the setup in the Example "Two Channels" on p. 112.

23. If the waves had wavelength 0.05 m instead of 0.1 m, would the interference at Point B be constructive or destructive? Explain.

24. Assume Points A and C remain where they are. How far to the right would you have to move Point B to see perfectly constructive interference at that point?

25. Assume Points A and C remain where they are, but Point B moves a bit up (so the path from A to B is still a straight line, but no longer horizontal). How far up would Point B move to make the interference at that point perfectly constructive?

Problems 26–28 refer to the setup shown in Figure 3.8. Two speakers are set up a distance D_{sp} apart from each other on one wall of a room. The opposite wall is a distance L away. The speakers emit sound waves in phase with each other with wavelength λ.

26. Assume $D_{sp} = 1.0$ m, $L = 5.0$ m, and $\lambda = 1.0$ m. Find at least one spot on the far wall where the two waves interference constructively, and at least one where they interfere destructively.

27. Let Point P be a spot on the far wall directly across from one of the two speakers.

(a) Take $D_{sp} = 1$ m and $L = 5$ m and find a value of λ that would cause constructive interference at Point P.

(b) For those same values of D_{sp} and L find a value of λ that would cause destructive interference at Point P.

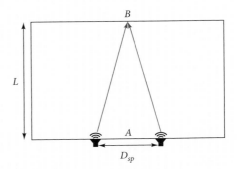

Figure 3.8

28. Suppose we gave you the values of D_{sp} and L. What wavelengths λ would make the point on the far wall *directly across* from one of the two speakers a node (a point of perfect destructive interference)? Note that there are infinitely many such wavelengths, so your general answer will end with the phrase "for any positive integer n."

29. For the function $y = (1/3)\sin(x/3)$,

(a) identify the amplitude, wavelength, and wave number.

(b) graph two full wavelengths.

30. The sine wave function $f(t)$ has amplitude 5 and frequency 2.

(a) Plot $f(t)$. Plot enough of a range in time to clearly show the behavior of the function.

(b) What are the angular frequency and period of f?

31. One ordinary differential equation that leads to sinusoidal waves is $d^2y/dx^2 = -9y$.

(a) Show that the function $y = 2\sin(3x)$ is one solution to this equation.

(b) Show that the function $y = 4\cos(3x)$ is another solution to this equation.

(c) Show that the function $y = 2\sin(3x) + 4\cos(3x)$ is another solution to this equation.

(d) Show that the principle of linear superposition holds for a system governed by this equation.

That is, show that if $y_1(x)$ and $y_2(x)$ are any valid solutions, then $y(x) = y_1(x) + y_2(x)$ is also a valid solution.

3.2   The Young Double-Slit Experiment

A thorough understanding of the Young double-slit experiment can take you deeply into quantum mechanics. This section explains the nineteenth-century version of the experiment, which illustrates some important but still classical properties of light. Section 3.3 will then explain how a twentieth-century variation of the same experiment illustrates many key ideas of quantum mechanics.

This section relies on your understanding of interference as explained in Section 3.1.

3.2.1   Discovery Exercise: The Young Double-Slit Experiment

Figure 3.9 looks straight down into a box filled with a shallow layer of water. The water at the bottom of the figure is being repeatedly struck by a small paddle, creating circular ripples that spread outward. These ripples pass through the narrow slits in Wall A. Whenever a wave passes through a narrow opening it spreads out circularly from there, so the slits in Wall A create two new circular ripples. The subject of this experiment is the impact of *those* ripples on Wall B.

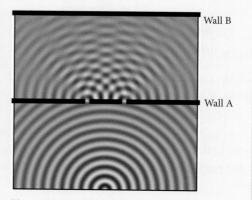

Figure 3.9 A double-slit experiment with water.

The two slits are equidistant from the paddle, which means the waves emanating from the two slits are perfectly in phase with each other.

Figure 3.10 shows the same box with two points marked on the back wall. It also shows the paths taken by the wave as it travels to those two points.

1. Point P_1 is in the middle of Wall B, and is therefore equidistant from the two slits. The solid lines in Figure 3.10 show the paths taken by the wave as it goes from the paddle, through each slit, and to Point P_1. Will the waves coming from the two slits reach P_1

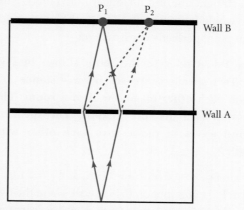

Figure 3.10 Path lengths in the double-slit experiment.

"in phase" (leading to constructive interference) or "out of phase" (leading to destructive interference)?

2. The dashed lines in Figure 3.10 represent the paths to P_2 from each slit. Which of these two dashed paths is longer, the left or the right?

3. Suppose the longer path is longer by precisely half a wavelength. (That is certainly true at *some* point to the right of P_1.) Will the two waves reach this point in phase or out of phase?

   See Check Yourself #4 at www.cambridge.org/felder-modernphysics/checkyourself

4. Will there be another "in phase" spot to the right of P_2? Why or why not?

Before reading on, think about what it would look like if you graphed the amplitude of the wave as a function of position along Wall B. You can check your answer against the following Explanation.

3.2.2   Explanation: The Young Double-Slit Experiment

Isaac Newton argued that a light beam is a collection of particles, like a gust of wind. A candle emits these particles in all directions and when those particles hit our eyes we experience vision. Others such as Robert Hooke and Christiaan Huygens believed that light is a wave, just like sound consists of waves of pressure in the air. Both sides produced theoretical arguments and experimental evidence to back up their points of view. In 1802, Thomas Young's double-slit experiment seemed to settle the question by conclusively showing that light is a wave.[2]

We're going to start with a simpler "single-slit" experiment and then move on to Young's double-slit version.

The Single-Slit Experiment

Consider two boxes without tops so we can look straight down into them. Each box has a wall in the middle that we'll call Wall A, with a small slit cut down the middle of the wall. The back wall of the box is Wall B.

In the front of the first box (left side of Figure 3.11) is a machine gun shooting bullets at Wall A. The machine gun rattles around so not every bullet comes from exactly the same place or follows exactly the same path. Most of the bullets hit Wall A and stop, but a few go through the slit, come out with a small range of different angles, and hit Wall B. The goal of this experiment is to see which regions on Wall B receive many bullet holes, which receive few, and which receive none at all.

The second box has an inch of water covering the bottom (right side of Figure 3.11). In place of the machine gun we have a small paddle rhythmically tapping the water to produce waves. The waves emanate in circles from their point of origin, gradually reducing in amplitude as they spread out. The slit in Wall A extends both above and below the surface of the water so that the waves pass through it. Whenever a wave passes through a narrow opening it spreads out circularly from there, so the effect is to create a new set of circular ripples. Once again we look for the resulting pattern on Wall B, but instead of bullet hole density we measure water wave height.

---

2 Spoiler: in the next section we'll see that this conclusion is not so clear.

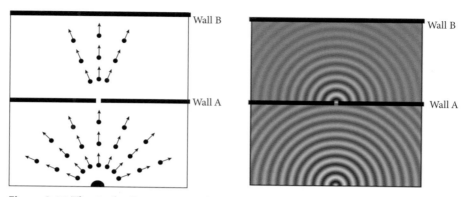

Figure 3.11 The single-slit experiment for particles and for waves.

Active Reading Exercise: The Single-Slit Scenario

Draw two graphs. One graph represents bullet-hole density along Wall B in the first box, and the other graph represents wave height along Wall B in the second box. These graphs will be qualitative (we haven't given you any numbers to work with), but pay particular attention to how the two shapes are similar and how they differ.

Do not continue reading until you have drawn the graphs!

Here are the answers.

Most of the bullets hit more or less directly behind the slit, but a few hit off to one side or the other. As you move along Wall B from the center toward the edges you find fewer and fewer bullets hitting.

The water wave spreads out in a circular pattern from the slit, but reduces in amplitude the farther it travels. Hence, the middle of Wall B (for which the journey was shortest) receives the highest-amplitude waves, and the waves reduce in amplitude as you move to the left or the right.

In short, the two graphs look essentially the same, although for different reasons (Figure 3.12).

Figure 3.12 The results of the single-slit experiment.

Now imagine that instead of a gun or a water wave we shine a beam of light at Wall A, and look for the pattern of brightness on Wall B. Whether light is a particle (that behaves like bullets) or a wave (that behaves like water ripples), the result will be the same: a bright spot behind the slit that diminishes as we move off to either side. So if our goal is to determine if light is a particle or a wave, this version of the single-slit experiment won't help.

The Double-Slit Experiment

Young's brilliant innovation was doing this experiment with two slits in Wall A.

With bullets the result is simple. Lots of bullets hit behind each slit, and the number gets smaller as you move away from the slits. This experiment is depicted on the left side of

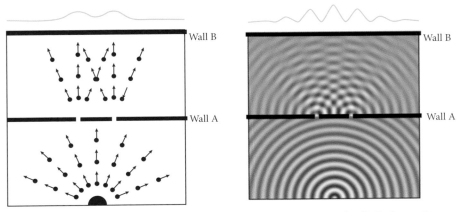

Figure 3.13 The double-slit experiment. The graph above the particles (left) shows the density of impacts along Wall B. The graph above the waves (right) shows the intensity, or amplitude, of the wave along Wall B. In the overlap region of the two waves, interference changes what the pattern looks like. The gray spokes are "nodal lines" where the two waves interfere destructively and the water doesn't move.

Figure 3.13. Above the schematic is a graph showing the resulting bullet-hole density along Wall B.

But with water ripples, things get more complicated. Each slit now acts as a source of a circular wave. These two waves have identical frequencies, and are initially in phase with each other (assuming the slits are equidistant from the original source). At one spot on Wall B the two waves travel identical distances, and therefore end up in phase. At nearby spots the two distances differ by half a wavelength, so the waves are out of phase. A bit farther along, the two distances differ by a full wavelength so the waves are in phase again. (If this is not clear, Figure 3.10 on p. 118 may help.) The result along Wall B is alternating bands of constructive and destructive interference. The right side of Figure 3.13 shows this experiment and the resulting graph of wave amplitude along Wall B.

Young ran the double-slit experiment with light and his results conclusively showed that light is a wave. His 1804 paper on these experiments neglected to include any photography (which would be invented 22 years later), so we've included a photograph from a more recent reproduction of his experiment (Figure 3.14).

Make sure you are clear on what Young saw, and what it proved. If you're confused about the geometry, please look through the Discovery Exercise that begins on p. 118. In particular, make sure you understand why some points on Wall B remain dark. Such points are receiving light waves

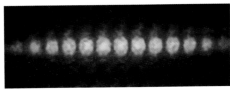

Figure 3.14 A photograph of the back wall in a Young double-slit experiment. The dark bands are nodes where waves from the two slits interfere destructively. *Source: Jordgette © 2010.*

from both slits, but those light waves are out of phase, and therefore cancel each other out. That kind of cancellation has no counterpart for particles: "some bullets" plus "some bullets" always results in "more bullets," not fewer.

Example: Double Slit

Consider a double-slit experiment with the two slits 1 mm apart and Wall B 5 m behind Wall A. You shine green light (wavelength 5×10^{-7} m) on the slits. Will a point on Wall B 2.5 mm from the center show constructive or destructive interference?

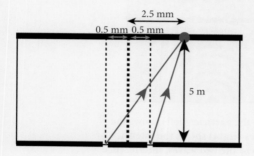

Answer: The figure (not to scale!) shows the paths taken by the light going through each slit.

The light from the left slit travels a distance $\sqrt{(5\,\text{m})^2 + (0.0025\,\text{m} + 0.0005\,\text{m})^2}$ while the light from the right slit travels $\sqrt{(5\,\text{m})^2 + (0.0025\,\text{m} - 0.0005\,\text{m})^2}$. All we care about is the difference between these distances, which is 5×10^{-7} m. Since the two paths differ by one wavelength of the light, there will be constructive interference at that spot.

The distance to the back wall is often much larger than the distance between the slits, which is in turn much larger than the wavelength of the wave you are measuring. In that case there is a simple approximation you can use for the locations of nodes and antinodes on the back wall (see Problem 15).

Diffraction

We began this section by discussing a single-slit experiment, and then we moved up to two slits. What if you use more than two? A set of many closely spaced slits – often called a "diffraction grating" – produces bands similar to the two-slit interference pattern we discussed above, but the bright spots are narrower and brighter, so the effect is easier to measure. Such an effect was dramatically demonstrated in 1913 by William Bragg and Lawrence Bragg, who realized that the layers of a crystal could act as microscopic slits and produce diffraction in X-rays (see Problem 16).[3] In 1915 they won the Nobel Prize for their work, thus making them the only father–son pair to jointly win the Nobel Prize, and making Lawrence Bragg the youngest physics Nobel Laureate at 25 years old.

Figure 3.15 Single-slit diffraction produces alternating bright and dark bands with intensity falling off rapidly as you move away from the center. *Source: Jordgette © 2010.*

Thus far, our discussion of interference has treated each slit as a point source for a new wave. That's a good model if the slits are much narrower than one wavelength of the light going through them. When the width of a slit becomes comparable to the wavelength, however, you can get interference between the wave fronts that pass through different parts of the *same* slit. The result is the familiar interference pattern, with alternating light and dark bands, but all coming from a single slit (see Problem 17). The intensity of the light bands falls off rapidly as you move away from the center of the back wall (directly behind the slit), as shown in Figure 3.15.

---

3 The phenomenon of X-rays showing bright and dark spots when they reflect off crystals was already known by then, but the Braggs explained why it happened.

When light goes through a wide slit and spreads out with bright and dark bands, that's called "diffraction." When it goes through two narrow slits and creates bright and dark bands that's called "interference." When it goes through many slits that's "diffraction" again. So what's the difference? We'll let Richard Feynman answer that one for us.

> *No one has ever been able to define the difference between interference and diffraction satisfactorily. It is just a question of usage, and there is no specific, important physical difference between them. The best we can do, roughly speaking, is to say that when there are only a few sources, say two, interfering, then the result is usually called interference, but if there is a large number of them, it seems that the word diffraction is more often used.*

In the case of a single slit the pattern appears because of infinitely many waves passing through different parts of the slit, so that falls under "diffraction."

You can find many of the equations describing diffraction and interference in Appendix E. In further studies of optics and/or electromagnetism you may become much more familiar with the mathematics of interference and diffraction. But for our purposes here, focus on the following facts.

- A wave passing through a very narrow slit spreads out from there in all directions (Huygens' principle). The result is indistinguishable from particles spreading out.

- So if you want to distinguish waves from particles you need to set up some kind of interference, because destructive interference is a property of waves that has no counterpart with particles.

- One way to set up that interference is to use two (or more) very narrow slits, so the waves emanating from these slits interfere with each other.

- A different way to set up that interference is to make one slit that is wide compared to the wavelength of the light, so the waves emanating from different parts of that one slit interfere with each other.

- Either way you do this experiment, if your source was a wave, you will see the "interference pattern" of light and dark bands. Because we see such a pattern with light, we know that light exhibits wave behavior.

3.2.3 Questions and Problems: The Young Double-Slit Experiment

Conceptual Questions and ConcepTests

1. If light were a stream of particles – not a wave at all – what would Young have seen on the back wall of his experiment?

2. You are doing slit experiments with very narrow slits (much narrower than the wavelength λ). Explain in your own words why such a single-slit experiment can't tell you whether light is a particle or a wave, but a double-slit experiment can. Would a triple slit experiment show a difference between the two?

3. If you do a double-slit experiment with light you will see alternating bright and dark spots. Is the following true or false: if you cover up one of the slits, the dark spots will become brighter? Explain.

4. The interference pattern in the Young double-slit experiment is alternating bands of light and dark. But if you look more closely you will see that the

middle bright band is the brightest; as you move away in either direction, the bright bands grow dimmer. Why?

5. If you shine two flashlights onto a wall, you will not see an interference pattern.

   (a) How would you have to change your two flashlights in order to get an interference pattern? Think about the frequencies, amplitudes, and phases of the waves they would have to emit.

   (b) Young chose to make his two different light sources slits, through which shone light from a single source. How did that help him achieve the conditions you listed in Part (a)?

6. Sound, like light, is a wave. In some ways playing a sound from two speakers is like doing the double-slit experiment, but you don't hear the sound getting alternately softer and louder as you walk along the back wall of your room. Why not? (There are a number of valid answers to this question.)

7. It's possible to do the double-slit experiment with ordinary sunlight. When you do, the bright spots in the interference pattern have their colors spread out in a rainbow rather than just being spots of white, as shown in Figure 3.16.

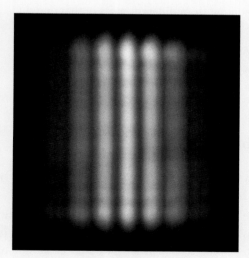

Figure 3.16 *Source: Berdnikov © 2010.*

Explain why that happens. *Hint*: The one place where this doesn't happen is the central spot.

For Deeper Discussion

8. Suppose you do a double-slit experiment with light and you see alternating bright and dark spots. Now you are going to modify the experiment in a few ways. For each one of the following, specify whether it will cause the bright spots to get closer together, get farther apart, or remain the same distance apart.

   (a) You increase the wavelength of the light.

   (b) You increase the intensity (amplitude) of the light.

   (c) You move the slits farther apart.

   (d) You move the back wall where you are doing the measurements farther away from the wall with the slits. *Hint*: The simplest point to consider here is a point on the back wall that is directly across from one of the two slits.

Problems

9. A source emits sine waves of wavelength 3 cm that travel through two different channels as shown in Figure 3.17. The first one travels a distance 2 m before reaching Point P and the second one travels a distance $(2\,\text{m}) + x$ before reaching Point P.

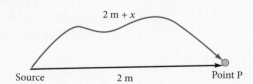

Figure 3.17

   (a) List three different values of x that would lead the two waves to meet and interfere constructively at Point P.

   (b) List three different values of x that would lead the two waves to meet and interfere destructively at Point P.

10. Double-Slit Interference

Figure 3.18 shows a double-slit experiment, *not drawn to scale*. The two slits are 4 cm apart, and the distance from Wall A to Wall B is 1 m. Two identical waves with wavelength 0.2 mm (not cm!) leave the two slits perfectly in phase.

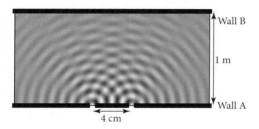

Figure 3.18

(a) Point P_1 is on Wall B precisely between the two slits. Calculate the distance from the left slit to P_1.

(b) Divide your answer to Part (a) by the wavelength of the waves, to find how many complete wavelengths there are along the path from the left slit to P_1. *Important:* If you get exactly 5000, you need to re-do Part (a) with less rounding.

(c) Similarly, find how many wavelengths are on the path from the right slit to P_1.

(d) Will the two waves interfere constructively or destructively at this point?

(e) Point P_2 is 0.25 cm to the right of P_1. Calculate the distance from the left slit to P_2, measured in wavelengths. Calculate the distance from the right slit to P_2, measured in wavelengths. Will the two waves interfere constructively or destructively at this point?

(f) Answer the same questions for Point P_3, 0.25 cm to the right of P_2.

(g) Sketch a graph of the resulting amplitude along Wall B for -1 cm $\leq x \leq 1$ cm, where x is the distance from P_1. You don't need numbers on the vertical axis but you should have them on the horizontal (x) axis. This should not require any calculations that you haven't already done.

Problems 11–15 refer to a double-slit experiment in which the slits are separated by a distance $2d$, the slits are a distance L from the back wall, and the waves have wavelength λ (Figure 3.19). Note that to make the math simpler we called the slit separation $2d$, so for example if you are told $d = 1$ cm that means the slits are 2 cm apart. (The geometry you need here involves right triangles. If you're stuck, it may help to work through Problem 10 first.)

11. $d = 1$ cm, $L = 5$ m, $\lambda = 2$ mm (not cm!). Is the interference at a point 25 cm from the center of Wall B constructive or destructive?

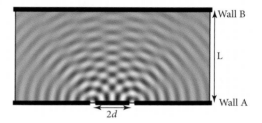

Figure 3.19

12. $d = 1$ cm, $L = 50$ cm, $\lambda = 1/2$ cm. Calculate the distance between the center point on Wall B and the nearest point of destructive interference.

13. $d = 1$ cm, $L = 1$ m, $\lambda = 1$ cm. Figure 3.20 shows the amplitude of the wave as a function of position on Wall B. Copy this sketch and label the horizontal axis with numbers (including units).

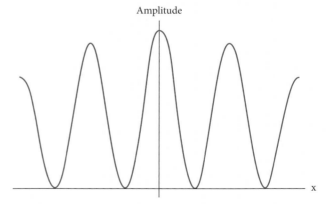

Figure 3.20

14. Leave d, L, and λ as letters. Find the distance from the center of Wall B to the first node (point of

destructive interference). *Warning*: The algebra is a bit messy. As you go, keep checking that the units in your equations match.

15. [*This problem depends on Problem 14.*] In Problem 14 you derived an expression for the distance from the center of the back wall to the first node.

    (a) Assuming $L \gg d \gg \lambda$, simplify this expression. Your answer will be one term with each of the letters L, d, and λ appearing in either the numerator or denominator.

    (b) Does your answer to Part (a) tell us the distance from a bright spot to the next bright spot ("peak to peak") or from a bright spot to a dark spot ("peak to trough")? How do you know?

    (c) Young wrote that the separation between the slits in his original experiment was "about one thirtieth of an inch." Assuming he was using light with a wavelength of 600 nm, and the distance to his back wall was 4 m, use your approximation from Part (a) to estimate the distance he would have seen between the fringes.

16. **Bragg's Law**

    In 1913 William and Lawrence Bragg explained why X-rays shining on a crystal produce a diffraction pattern. The setup is shown schematically in Figure 3.21. Two parallel beams of X-rays are incident on a crystal, which is modeled as parallel layers of atoms a distance d apart. The angle of incidence θ

is also the angle of reflection. The two beams begin perfectly in phase. However, one beam is reflected off the top layer of the crystal; the other beam makes it past that layer and is reflected off the second layer, and therefore travels a greater distance.

    (a) Explain why the difference between the two distances is twice the blue segment in the drawing.

    (b) Calculate that difference – twice the length of the blue segment in the drawing – as a function of d and θ.

    (c) What wavelengths λ would end up perfectly in phase after reflecting in this way? (Your answer will of course be a function of d and θ, but in order to list infinitely many wavelengths you will also need an arbitrary integer n.)

    (d) Suppose you shine an X-ray with wavelength 1 nm (which is 10^{-9} m) onto a crystal. The smallest angle at which you see a bright reflected spot is when you shine it at the angle $\theta = 10°$. What is the spacing between the layers of the crystal? What is the next-smallest angle at which you would see reflection?

17. **Single-Slit Diffraction**

    A wave of wavelength 2.0 mm passes through a single slit with width 4.5 mm. Assume the wave reaches each part of the slit with the same initial phase. The wave that goes through the slit then strikes a back wall 1 m away. We'll refer to the

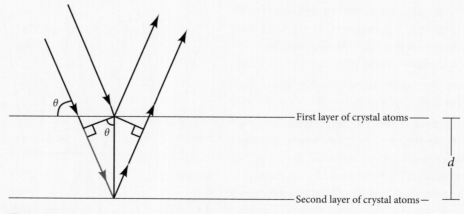

Figure 3.21

spot on the back wall directly behind the slit as the "center" of the wall.

(a) First consider two waves, one coming from the far right edge of the slit and the other coming from the exact center of the slit. At a point 0.5 m to the right of the center of the back wall, do those two waves interfere constructively or destructively?

(b) Now consider two other waves originating 1 mm to the left of the right edge of the slit, and 1 mm to the left of the slit's center. At that same point on the back wall do those two waves interfere constructively or destructively?

(c) What will that point on the back wall look like? Explain how you know, based on your previous answers.

3.3 One Photon at a Time

Young's interference fringes conclusively demonstrate the wave nature of light. But a variation on the same experiment makes an equally convincing case for the *particle* nature of light. Reconciling these two results brings out many of the core ideas of quantum mechanics.

3.3.1 Explanation: One Photon at a Time

The pattern produced in a double-slit experiment is one difference between particles and waves, but here's a simpler difference: particles are discrete. A given part of Wall B might be hit by 1000 bullets, or 100 bullets, or 1 bullet, but no region will ever be hit by half a bullet. A water wave, on the other hand, can get arbitrarily small. You can halve its height, and then halve it again, indefinitely.

As far as Young knew, light passed this second test of wave-hood as well. In 1802 there was no reason to suppose that you couldn't keep making a light beam dimmer and dimmer, allowing the brightness (the wave amplitude) to approach zero without any qualitative change in the phenomena.

But a modern lab can run *that* experiment far beyond the capability of nineteenth-century tools, and the result is not what Young would have expected. We can turn our light source so low that Wall B lights up one point at a time. A somewhat brighter source produces several dots at a time. Make the source even brighter and we get so many simultaneous dots that they merge into a blur – that is, we see what we expect to see when light shines on a wall. But we know the dots are still there.

Now, our goal is still to determine the nature of light. The light that Young described, a wave, would always produce a hazy blur. An individual dot implies the opposite: an indivisible packet of light that traveled from the source to one point on the wall. The word "photon" was first coined by Gilbert N. Lewis in 1926 for this particle of light.

This chapter presents some of the developments that led the physics community to accept the idea that light comes in indivisible packets, but we're not taking these developments in order. Planck proposed the idea in 1900 (Section 3.4), and Einstein pushed the theory further in 1905 (Section 3.5). It was not until 1909 that G. I. Taylor first saw discrete spots on the back wall of a double-slit experiment. (Taylor used smoked-glass screens to attenuate the light beams.) But we are starting with Taylor's experiment because we believe it offers the clearest demonstration of the seemingly contradictory nature of light.

The Double-Slit Experiment, One Particle at a Time

Do you see the deep problems caused by the two different results we presented above?

Remember that Young saw bright bands (lots of light) and dark bands (no light). Why is a dark band dark? It's not because light from the left-hand slit and light from the right-hand slit are avoiding that particular region. The patch is being hit by light waves from both slits, out of phase with each other, causing destructive interference. The explanation relies on the fact that light waves spread out through space, traveling through both slits at once.

On the other hand, a very dim light source creates only one dot at a time on the back wall. That result clearly points to a particle that occupies only one place at any given time.

A theory is needed that coherently explains both of these results. So imagine performing the double-slit experiment one photon at a time.

Active Reading Exercise: The One-at-a-Time Double-Slit

Imagine the double-slit experiment with a one-photon-at-a-time light source. You fire the first photon and a dot appears at one point on Wall B. You fire a second photon, and Wall B now has two dots. Then a third and a fourth. Eventually you will have so many dots that you can plot their density on Wall B.

Question: Will you see a "particle-like" pattern, with high density behind each slit and lower as you move away? Or will you see the alternating bands of high and low density that we associate with a wave?

Formulate your own answer to that question. Write it down, along with a few sentences explaining your reasoning. Don't be afraid to be wrong! But don't just guess haphazardly either; think it through.

Do not read further until you have written down your best guess.

Now that you have hypothesized, here is the experimental answer: you get alternating light and dark areas! With just a few dots on Wall B, there may appear to be no pattern. But as you keep watching you will see bands that receive a lot of dots, alternating with bands that never happen to receive any dots at all. Wait long enough and you will see the interference pattern that Young saw.

You can see an online animation of the one-at-a-time double slit experiment at the link below. The initial pattern on the back wall appears random. But as the dots accumulate, it becomes apparent that some regions are hit frequently, while others are never hit at all. It is impossible to predict where the next dot will hit, but after many dots the pattern is evident. www.cambridge.org/felder-modernphysics/animations

This is the most important moment in the chapter. If you think deeply enough about this one result, it can take you pretty far down the road toward the full-blown theory of quantum mechanics. Make sure you understand the experimental facts, and then try to formulate any theory consistent with those facts.

We have framed the problem as most physicists did, and do, frame it: "Is light a particle or a wave?" But there's no obvious reason that everything in the universe has to fall into one of those two categories, so it might be better to start with this question:

Is a photon something that exists at only one point in space at any given time, or is it spread out across many places at one time?

That question, it seems, demands a simple binary answer. But you get into trouble either way you go. If a photon is "smeared out" across a large expanse of space, how can we explain why we got individual dots rather than a hazy blur? If a photon exists at only one place at a time, how can we explain the interference pattern? (See Figure 3.22.)

Figure 3.22 Three versions of what the back wall of a double-slit experiment would look like with a dim source. (Left) If light were made of classical particles you would get lots of particles behind the slits and fewer off to the sides, but with no interference fringes. (Middle) If light were a classical wave you would get bands of light and dark, but no matter how dim you made the source you would not see individual dots. (Right) In the actual experiment you see individual dots that build up an interference pattern.

The next few chapters will explain these results with the most arbitrary-sounding, counterintuitive theory you have ever encountered. You may find yourself repeatedly wondering "Who came up with this nonsense, and why did they actually believe it?" Whenever you have that thought, think back to the light that travels through both slits in Wall A as an interfering wave and then lands on Wall B as an indivisible particle. That won't make you think "This all makes sense now," but it may help you arrive at the conclusion that "Quantum mechanics seems to be the best and simplest explanation of the experimental results."

The "Orthodox" Interpretation of Quantum Mechanics

The interference pattern on the back wall of a double-slit experiment has a clear *mathematical* implication: a wave is spreading out from the two slits and then interfering with itself. But beneath those calculations is a troubling *physical* question: what is waving? The classical answer is that an electric field and a magnetic field are both oscillating, and a high amplitude of the wave corresponds to high strengths of these two fields. But that answer would never lead to an individual spot at one point on the back wall. So we have an issue of "interpretation," meaning a problem in associating a physical meaning with the mathematical equations.

In 1926, Max Born proposed that the wave represents *probabilities*. A photon is a particle that hits the back wall in a particular spot, but associated with that particle is a wave – called its "wavefunction" – that determines the probabilities of measuring that photon at various positions. A high amplitude of the wave means "the photon is likely to be found here"; a zero amplitude means "it's definitely not here."

The theory of quantum mechanics, which we will present in Chapter 5, allows us to describe how the wavefunction behaves and to use that to predict the probabilities of finding the particle in different places. That theory evolved to explain results such as the one we presented above.

While all physicists agree that quantum mechanics makes successful predictions, they do not in general agree about what the math implies about what is actually happening. The "orthodox interpretation" says that the particle exists as a wave, spread out throughout space, until the moment you measure it.[4] At that moment the wave "collapses" into a state of being in one particular place. In this interpretation the act of measurement itself – regardless of any details about how the measurement is made – fundamentally changes the state of the particle.

Regardless of what they fundamentally believe, a strong majority of working physicists use the orthodox interpretation as a convenient way of describing quantum mechanics and making predictions, so we are going to mostly use that language in this book.

An orthodox description of the one-at-a-time double-slit goes like this. The wavefunction passes through both slits simultaneously, sending waves from each slit to the back half of the box. Those waves arrive at Wall B, constructively interfering to create regions of high probability in some places, and destructively interfering to create regions of low (or zero) probability in others. When the photon interacts with the measuring device at Wall B, that interaction causes the photon to collapse to having one particular position (Figure 3.23).

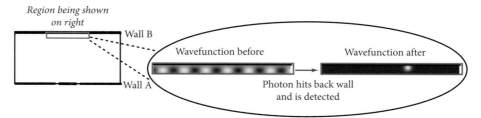

Figure 3.23 A photon's wavefunction along a section of the back wall in a double-slit experiment, shown just before and after the photon hits the back wall. Light areas show regions of high amplitude and dark areas show low amplitude. (Left) Just before it hits the wall the photon has a wavefunction with alternating bands of high and low amplitude as a result of interference from the two slits. (Right) Just after it hits the wall the wavefunction has high amplitude in one small spot (where the photon was detected) and is essentially zero everywhere else.

The position to which the photon collapses is most likely to be in a region where the wavefunction was highest just prior to the measurement. After you fire many photons, the regions where the wavefunction amplitude is high record many hits and the regions where it is low record very few, so you end up with an interference pattern.

Active Reading Exercise: The Orthodox Interpretation

Re-read the description above of the orthodox interpretation of quantum mechanics. Make sure you see how this model explains the alternating bands in the double-slit experiment.

Then make a list of things that don't fully make sense. You may want to break your list into two categories: *questions* about the orthodox interpretation, and *objections* you might raise. (If it helps, think of yourself as Einstein, who wrote long letters full of vehement objections to Bohr.) Think about the mathematics of waves, and the mathematics of probability. Think about particles and measurements. Think about alternative explanations.

---

4 This interpretation is often associated with the Danish physicist Niels Bohr and is therefore sometimes called the "Copenhagen interpretation." There is some dispute over how much this actually reflects Bohr's views, and there is variation in what people mean by the phrase "Copenhagen interpretation," so we'll stick with the less controversial term "orthodox interpretation."

> *Those who are not shocked when they first come across quantum theory cannot possibly have understood it.*
>
> – Niels Bohr

Now that you've done that, we'll give you a partial list we made. Some have answers that are simple and satisfying; others, not so much.

- *Remind me again why we can't just say that the photon went through one slit or the other?*

  Because then there would be no way to explain the dark bands in the interference pattern.

  You need to have the photon pass through both slits of Wall A simultaneously to explain the interference pattern, but you need to have it be at only one place on Wall B to explain the fact that you see just one dot show up there.

- *What happens to the wavefunction at Wall B?*

  It suddenly goes from being a big complicated wave, with high probabilities in some places and low in others, to being very high right at one spot (where you measured it) and zero everywhere else. This is called the "collapse of the wavefunction," and it is caused by a measurement.

- *The wavefunction is sometimes negative and sometimes positive, but probability can never be negative.*

  Good catch if you noticed that! The simple answer is that the probability is proportional to the *square* of the wavefunction. But the wavefunction is actually complex-valued, so a more precise statement is that the probability is proportional to the square of its modulus.

  Note that we cannot get around this problem by saying that the wavefunction is always positive, because then there would be no destructive interference. The wavefunction is positive in some places and negative in others, but the probabilities are of course never negative.

- *What happens when we fire the second photon?*

  Its wavefunction propagates exactly like the first one, so it reaches Wall B with the same probability distribution. However, when its wavefunction collapses at Wall B, it may end up someplace different, according to the laws of probability. We say that the propagation of the wavefunction is "deterministic" (it will always happen the same way under the same conditions), but its collapse is "random" (it can be predicted only probabilistically).

- *What exactly counts as a "measurement"? Is Wall B still the site of a measurement if no one looks at it?*

  The answer to the second question is yes. The experimental results discussed in the following Active Reading Exercise suggest that a measurement can collapse the wavefunction even if nobody looks at it. The more general question of what constitutes a measurement, however, has no clear answer in the orthodox interpretation. Most physicists just do the calculations without worrying about this issue. (So far this has worked better than you might suppose.) Others consider this a fatal flaw in the orthodox interpretation.

The issue of measurement is fundamental and subtle here, so let's compare the orthodox quantum mechanical description to a classical view.

A classical particle has a definite position at every given moment. When you want to know that position you perform a measurement. That measurement may not change the system in any significant way, other than letting you know the position the particle already had.

In orthodox quantum mechanics, by contrast, a particle literally doesn't have a well-defined position, only a wavefunction that associates different probabilities with different positions, or a "superposition" of different possible positions. The act of measurement fundamentally changes the state of the system, collapsing that superposition into one specific position. In the double-slit experiment, Wall B is the measuring device, and the individual spot that appears on the wall is evidence that the photon now has one well-defined position.

The central role of measurement applies to all particles (not just particles of light) and all properties (not just position). For instance, when you measure the momentum of an electron you force it to have a definite momentum.

That brings us to our next question/answer:

- *Waves of probability that collapse into randomly chosen states because someone measured them? Isn't there some better explanation?*

    There are *different* explanations. We discuss some of them in the online section "Interpretations of Quantum Mechanics." www.cambridge.org/felder-modernphysics/onlinesections

    All interpretations agree about the experimental results and the math, but not all of them agree about the fundamental nature of reality. The orthodox interpretation is the most commonly accepted and used. Other interpretations avoid some of its peculiarities, but introduce their own. None of them brings you home to the good old Newtonian world. Sorry about that.

Variations on the Double-Slit Experiment

To help clarify these results, we want to discuss three variations of this experiment.

Active Reading Exercise: Double-Slit Variations

Read the first scenario below and *write down* your best prediction of the pattern you'll see on the back wall. Then check it against our online answer. Then do the same for the second scenario, and then the third.

1. Run the double-slit experiment one photon at a time, but put detectors at both slits so you can measure a photon going through the left slit, the right slit, or both simultaneously.

2. The same as above, but with no detector in the right slit. (If the detector goes off, a photon went through the left slit. If something appears on Wall B but the detector didn't go off, the photon passed through the right slit.)

3. Repeat the original experiment (no detectors) but emit electrons – still one at a time – instead of light.

You can see the results of all three experiments at: www.cambridge.org/felder-modernphysics/ activereadingsolutions

We close with one final question/answer:

- *If the world really works in this really weird way, why haven't I ever noticed?*

    It's very difficult to tell that water is made of individual atoms. An atom is so small, and there are so many of them in a thimbleful of water, that it requires tremendous technology to determine that it isn't continuously liquid. In exactly the same way, it's very difficult to tell that a beam of light is made of individual photons. In everyday experience, light just acts like a wave.

3.3.2 Questions: One Photon at a Time

Conceptual Questions and ConcepTests

1. In this question you're going to step through the one-photon-at-a-time double-slit experiment as described by the orthodox interpretation. Assume the back wall is a photographic plate, so every time light strikes any point on that wall it makes a permanent white mark that you can see from then on.

   (a) You begin by firing a single photon – a particle of light – from your source. A wave propagates outward, eventually going through both slits. Briefly explain what that wave represents.

   (b) The waves from the two slits hit the back wall. They constructively interfere in some spots, destructively in others. Briefly explain why. (*Hint*: Your explanation will be the same one Young would have used. It may involve a picture.)

   (c) A classical analysis would have described the regions of constructive interference as having strong electric and magnetic fields, and the regions of destructive interference as having weak fields (or none). How does the orthodox interpretation describe the regions of constructive and destructive interference? (*Hint*: Your answer will not involve electric or magnetic fields.)

   (d) At this point, the wavefunction "collapses." Briefly explain what that means.

   (e) What do you see on the back wall when the collapse happens?

   (f) Finally, you shoot out a second photon. Once again a wave goes out, goes through both slits, and interferes with itself constructively in some places and destructively in others. Once again the wavefunction collapses and you see something on the back wall. Assuming everything about the source and the experimental setup is identical both times, what part of this process might come out differently for the two photons?

2. In your own words, explain why you can't explain the single-photon-at-at-time double-slit results with just a particle view of light, or with just a wave view.

3. A particle is moving along the x axis. Its wavefunction is shown in Figure 3.24. If you measure the position of this particle, where are you most likely to find it: near Point A, near Point B, or near Point C? Where are you least likely to find it?

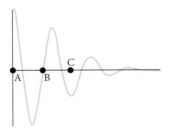

Figure 3.24

4. Sketch a possible wavefunction for a particle that has equal probability of being very near $x = 0$ or $x = 2$, and zero probability of being anywhere else.

5. In what ways are photons like and unlike Newtonian particles?

6. One of the most problematic aspects of the orthodox interpretation is the "collapse of the wavefunction." Answer each of the following parts briefly.

    (a) When does the wavefunction collapse? (Don't say "at Wall B"; we want a more general answer than that.)

    (b) What is the difference between the wavefunction before a collapse and the wavefunction after a collapse?

    (c) Suppose a particle has a wavefunction that sharply peaks in two places. When you measure the position of the particle will you find it in one place, two places, or more than two places?

7. If you measure a photon as it goes through the slits you always find it in one slit or the other. The orthodox interpretation says that if you don't measure it, though, then it is simultaneously in both. Suppose a friend says to you that this is just your ignorance; the photon is in one slit or the other but you don't know which because you haven't measured it. How would you answer your friend?

8. If you do the double-slit experiment one photon at a time you see some dark bands on the back wall where no photon ever hits. If you then cover up one of the two slits those bands become brighter. How is this explained in the orthodox interpretation? *Hint*: A good answer to this question will include the word "collapse" (and an explanation of its meaning in this context).

9. Explain briefly, in your own words, the following statement from the Explanation (Section 3.3.1): "The propagation of a wavefunction is deterministic, but its collapse is random."

10. We've been making a big deal about the fact that light has both a wave nature and a particle nature, but the same thing is true of water or ropes. In both cases the wave (which can interfere constructively or destructively) is made of particles (molecules). Explain why this model – a wave made up of particles – does *not* offer a good model of light in the double-slit experiment.

For Deeper Discussion

11. Electrons and atoms show the same interference patterns as those shown by light. But we previously said that things like bullets don't. Why not? You may not know the answer but try to speculate why this might be.

12. Suppose you do a double-slit experiment with a weak detector at one slit that only flashes one in ten times that a photon passes through that slit. What kind of pattern would you see on the back wall?

13. Many physicists object to the orthodox interpretation. One of the most common objections is that the theory describes every "measurement" as an event that fundamentally alters the state of a particle, but leaves the term "measurement" entirely undefined. Several alternatives have been proposed, following the same math and therefore making the same experimental predictions but interpreting the math quite differently. Research one of these. (The best known are the "many worlds" (Everett) and "pilot-wave" (Bohm) theories.) Answer the following three questions.

    (a) In the orthodox interpretation, the relationship between the "wave" and "particle" natures of light is that the former collapses into the latter when a measurement is performed. What is the relationship between those two states in your chosen alternative interpretation?

    (b) In the one-at-a-time double-slit experiment, a wave propagates simultaneously through the two slits on Wall A. How does this interpretation explain the fact that a measurement at Wall B finds the photon at only one place?

    (c) What are some conceptual difficulties raised by this interpretation?

3.4 Blackbody Radiation and the Ultraviolet Catastrophe

The first quantum crack in the classical wall came in 1900 when Max Planck proposed a solution to a theoretical problem called the "ultraviolet catastrophe." In this section we're going to explain this problem and Planck's solution.

Planck did not view his proposal as a radical departure from classical physics, and to this day physicists debate how Planck did interpret the math he presented. What is clear is that classical physics leads to an incorrect result when calculating the radiation in an enclosed space, and that we can solve this problem by assuming that the amplitude of a light wave cannot take any value, but is confined to a discrete set of values.

3.4.1 Discovery Exercise: Blackbody Radiation and the Ultraviolet Catastrophe

An enclosed cavity is filled with electromagnetic radiation, constantly being emitted and absorbed by the walls. The "spectrum" of that radiation tells how much of the energy is in the frequency range of blue light, how much in red, infrared, and so on.

Classical physics predicts that the spectrum in an enclosed cavity is $8\pi k_B T \nu^2 / c^3$, where k_B and c are fundamental constants and T is the temperature. This function is called the "Rayleigh–Jeans spectrum." In 1900 Max Planck proposed a radical hypothesis – quantized energy levels – that led to a different formula, $\left(8\pi h\nu^3 / c^3\right) / \left(e^{h\nu/(k_B T)} - 1\right)$. (Note the introduction of a new universal constant h.) These formulas can be written as:

$$S(\nu) = A\nu^2 \qquad \text{Rayleigh–Jeans (classical) spectrum} \qquad (3.2)$$

$$S(\nu) = \frac{B}{e^{C\nu} - 1}\nu^3 \qquad \text{Planck's spectrum} \qquad (3.3)$$

At 300 K (a typical room temperature), the constants are $A = 3.86 \times 10^{-45}$ J/(m^{3}Hz^3), $B = 6.17 \times 10^{-58}$ J/(m^{3}Hz^4), and $C = 1.6 \times 10^{-13}$ s.

1. Plot Equations (3.2) and (3.3) on the same graph, using the domain $0 \leq \nu \leq 2 \times 10^{12}$ s^{-1} and range $0 \leq S \leq 2 \times 10^{-20}$ J s/m^3. You should see that they track each other very well, but start to diverge as the frequencies get higher.

2. Plot Equations (3.2) and (3.3) on a second graph, using the domain $0 \leq \nu \leq 6 \times 10^{13}$ s^{-1} and range $0 \leq S \leq 3 \times 10^{-19}$ J s/m^3. For these higher frequencies you should see a dramatic difference.

The questions below are not asking for calculations; you can answer them quickly by looking at the graphs you just made.

3. Based on Planck's model, in roughly what frequency range would you expect to find the most radiation?

4. The energy density – the total energy in the cavity, divided by its volume – is obtained by integrating the spectrum function over all frequencies (0 to ∞). Explain why Planck's model might give a reasonable value for total energy, and the classical model cannot.

3.4.2 Explanation: Blackbody Radiation and the Ultraviolet Catastrophe

Every physical object is constantly radiating energy into its environment, and absorbing energy from its environment. The amount and type of that radiation depends on the object's temperature. Put a hot object next to a cold object and they will exchange energy unevenly, so energy will generally flow from the hot to the cold. Put two objects of the same temperature next to each other and they will exchange equal amounts of energy, a condition known as "thermal equilibrium."

Figure 3.25 Inside an enclosed cavity in equilibrium, the walls emit radiation at the same rate they absorb it, and the interior is filled with radiation of various wavelengths and amplitudes.

So, imagine a set of walls surrounding an empty cavity. If this system is in thermal equilibrium, then every day looks pretty much like the day before it. But that doesn't mean that nothing is going on! On the contrary, the walls are constantly emitting and absorbing radiation, so the space between them is filled with energy (Figure 3.25).

This contrived-sounding scenario is actually quite important because many objects, ranging from stars to a human body, emit radiation that is approximately the same as what we find inside a cavity in equilibrium. Such objects are called "blackbodies" and this energy is called "blackbody radiation."

Chapter 10 will discuss blackbody radiation in much more detail: what it is, why it's important, and how to predict its properties. Here we're going to leave some of those results in the categories of "you can work this out in the problems" or "we'll get back to this in Chapter 10." But we will show you how a classical analysis of blackbody radiation led to predictions that were not only measurably wrong, but theoretically impossible. And we will see how Planck resolved the problem by suggesting that – at least in this particular case – the allowable energy levels of a light wave are quantized.

Three Variables and Two Equations

The energy inside our cavity is in the form of electromagnetic waves. As you know, each such wave is characterized by two numbers: a frequency and an amplitude. The frequency of a wave determines what kind of light it is (X-ray, blue light, red light, etc.). The amplitude determines its energy.

The question that concerns us is the "spectrum" in the cavity: how much of the total energy is in the blue frequencies, in the green frequencies, in the microwave frequencies, and so on. We will look at three different but related variables that we call E_w, S, and ρ. In the examples below we consider the frequencies $\nu = 550\,\text{THz}$, which is green light, and $\nu = 650\,\text{THz}$, which is blue light.

- The function $E_w(\nu)$ represents the energy of each wave of frequency ν; E_w has units of energy (such as J or eV).

  If in a particular cavity we found $E_w(550\,\text{THz}) > E_w(650\,\text{THz})$, that would mean that in that cavity each green wave had more energy than each blue wave. Note that we are comparing two waves of different given *frequencies*, and asserting that one has a higher *amplitude* than the other.

- From $E_w(\nu)$ we will derive $S(\nu)$, the spectrum function itself. S has units of energy, per unit frequency, per unit volume, or equivalently energy density per unit frequency.

  If in a particular cavity we found $S(550\,\text{THz}) < S(650\,\text{THz})$, that would mean that in that cavity there was less energy in green light than in blue light. This might be true even if $E_w(550) > E_w(650)$: each green wave would have more energy than each blue wave, but there would be so many more blue waves that their energy would add up to more.

- From $S(\nu)$ we will derive ρ, the total energy density – meaning energy per unit volume – in the cavity. In equilibrium the energy is evenly distributed throughout the cavity, so ρ is constant and the total energy is ρ times the volume.

How are these variables related? Here's how you get from the first to the second:

$$S(\nu) = \frac{8\pi}{c^3} \nu^2 E_w(\nu). \tag{3.4}$$

That equation is certainly not obvious! You will derive it in Problem 39. Here it is enough to note that $S(\nu)$ grows much faster than $E_w(\nu)$ because there are more waves (each with its own energy E_w) at any given high frequency than there are at a low frequency.

You integrate $S(\nu)$ to find the energy density in a given range. For instance, $\int_{520\,\text{THz}}^{610\,\text{THz}} S(\nu)d\nu$ gives the density of energy in the green frequency range. The integral from 0 to ∞ gives the *total* energy density in our cavity:

$$\rho = \int_0^\infty S(\nu)\,d\nu. \tag{3.5}$$

(You can use Equations (3.4) and (3.5) to confirm that ρ has units of energy per unit volume. Remember that c has units of m/s, and both ν and $d\nu$ have units of s^{-1}.)

With those two equations we can calculate the spectrum of radiation, and the total energy density, once we have the function $E_w(\nu)$.

Example: Relating the Three Variables

Suppose the energy of one wave could be modeled by the function $E_w = a/[\nu(1 + b\nu^2)^3]$ for some positive constants a and b. Calculate the total energy density of the radiation.

Important note: Later we will see the (incorrect) classical E_w function, and then the (correct) quantum mechanical E_w function. This is neither of those; it is a made-up example to demonstrate how $E_w(\nu)$ leads to $S(\nu)$ and thence to ρ.

Answer: From Equation (3.4) we get the spectrum function

$$S(\nu) = \frac{8\pi}{c^3} \nu^2 E_w(\nu) = \frac{8\pi}{c^3} \nu^2 \frac{a}{\nu\left(1 + b\nu^2\right)^3} = \frac{8\pi a}{c^3} \frac{\nu}{\left(1 + b\nu^2\right)^3}.$$

Integrate across all frequencies to get the total energy density:

$$\rho = \int_0^\infty S(\nu)d\nu = \frac{8\pi a}{c^3} \int_0^\infty \frac{\nu}{\left(1 + b\nu^2\right)^3}\,d\nu = \frac{2\pi a}{bc^3}.$$

Now use the equation for E_w that we gave at the start of the Example to figure out the units of the constants a and b, and use them to check that our answer for ρ has units of energy density.

The Ultraviolet Catastrophe

The key takeaway from this section so far is that you can calculate everything once you know $E_w(v)$, the energy of each individual wave of frequency v.

We're going to outline below how nineteenth-century physicists calculated E_w, but first let's jump to the result: they concluded that E_w should be uniform across all frequencies. The energy of each wave should depend only on the temperature of the emitting body, based on "Boltzmann's constant," $k_B \approx 10^{-23}$ J/K:

$$\text{Classical prediction: } E_w = k_B T \qquad (3.6)$$

Active Reading Exercise: The Ultraviolet Catastrophe

In the preceding Example we used a made-up $E_w(v)$ function to predict the spectrum $S(v)$ and the energy density ρ inside a cavity. Now you use the classically predicted $E_w(v)$ (Equation (3.6)) to calculate the same variables. (This calculation is easier than the one we did in the Example.) Write down your answers before you look at ours. Can you see why this result was called a "catastrophe"?

From Equation (3.4), $S(v) = 8\pi k_B T v^2/c^3$. This classically predicted spectrum is called the "Rayleigh–Jeans spectrum." Because it is a constant times v^2, its integral from 0 to ∞ diverges. In other words, classical theory predicted that a cavity in equilibrium should have infinite energy density! The underlying theories were sound and well tested, the logic was irrefutable, and the final conclusion was completely absurd. Because the blow-up occurs at high frequencies, this prediction was called the "ultraviolet catastrophe."

You probably won't be surprised to hear that experimental results did not find infinite amounts of energy hiding inside every cavity. Figure 3.26 shows the Rayleigh–Jeans spectrum and the empirically measured one, both at $T = 2.725$ K. Notice that the area under the curve,

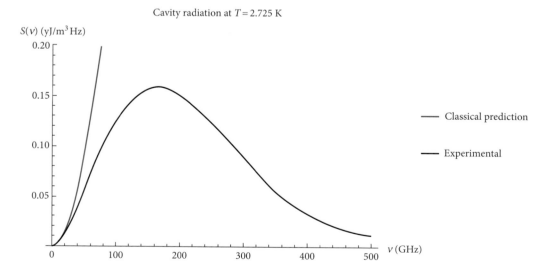

Figure 3.26 The vertical axis is the spectrum of radiation in a cavity (energy density per unit frequency). Integrating this gives the energy density. The energy here is measured in yJ (yoctojoules; 1 yJ = 10^{-24} J) and the frequency in GHz (gigahertz; 1 GHz = 10^9 Hz).

which gives the total energy density in the cavity, converges for the experimental result but clearly blows up for the classical prediction.

But we would also like you to notice something else: the classical prediction tracks the measured result quite well at the far left side of the graph. So the challenge faced by late nineteenth-century physicists was to change the theory in a way that did not significantly alter the predicted energies at low frequencies, but forced them down toward zero as the frequencies got higher.

"Where Did $E_w = k_B T$ Come From?" Part 1: A Probability Distribution

Before we see how Planck found an alternative to the catastrophic Equation (3.6), we have to see where $E_w = k_B T$ came from in the first place.

To do that, we have to mention a detail that we have hitherto ignored: the energy of each wave in our cavity is constantly changing. (The waves are caused by electrons buzzing around in the conducting walls, so no millisecond looks like the millisecond before it.) The E_w we're looking for is not precisely "the energy" of a wave of frequency v, but the *average* energy over time of such a wave.

In order to predict that average we're going to invoke an important rule – presented, for the moment, without proof.

The Boltzmann Distribution

Inside a cavity in thermal equilibrium, the probability that a given wave has a given energy E is proportional to $e^{-E/(k_B T)}$:

$$P(E) \propto e^{-E/(k_B T)}. \tag{3.7}$$

This probability distribution is called the "Boltzmann distribution." In Chapter 10 we will see where this distribution comes from and we'll see that it applies to much more than cavity radiation. Here we need to understand what it means in this context, and see where it leads.

Planck could have attempted to resolve the ultraviolet catastrophe by modifying Equation (3.7), but that would have disrupted basic tenets of statistical mechanics. Instead he focused on the next step: how do you get from Equation (3.7) (the probability of any particular energy) to $E_w(v)$ (the average energy of a wave over time)?

The answer depends on what energies you consider.

Example: Calculating the Average Energy with Only Two Options

Suppose a wave in our cavity could only have two possible energies: $E = 0$ and $E = 2\,k_B T$. If you measured its energy many times, what would be the average of all your measurements?

Like our previous example, this "only two possible energies" scenario is entirely hypothetical. But if you can follow the probability work in this example, then you can replicate many of Planck's calculations.

Answer: You might suppose that the average would be $k_B T$ (the average of the two possible measurements), but that doesn't take into account the probabilities. Equation (3.7) says that lower energy levels are more likely than higher energy levels, so you will measure 0 far more often than $2k_B T$. The average must still be between 0 and $2k_B T$, but it should come out considerably lower than $k_B T$.

Now let's make that quantitative. Equation (3.7) gives a proportionality relationship. We can express that relationship as an equation by introducing a constant of proportionality, which we will call C. So here are the probabilities of our only two possible measurements:

$$P(0) = Ce^{-0/(k_B T)} \qquad = C$$
$$P(2\,k_B T) = Ce^{-(2\,k_B T)/(k_B T)} = Ce^{-2}$$

Now, imagine that you measure the energy of this wave exactly $C + Ce^{-2}$ times. Based on the above probabilities, you would expect C of your measurements to get an energy of 0, and Ce^{-2} of your measurements to get an energy of $2\,k_B T$. (You might object that this is only possible if $C + Ce^{-2}$ happens to be an integer – you can't make half a measurement – but that's irrelevant for the following calculation.) The average energy you measure is the total of all your measured energies, divided by the number of measurements:

$$E_w = \frac{(0)(C) + (Ce^{-2})(2k_B T)}{C + Ce^{-2}} \approx 0.24k_B T.$$

This calculation is called finding the "expectation value" based on a probability distribution.

We did not just determine that the average energy of a wave is $0.24k_B T$. It isn't, because we just made up this example. Rather, we showed that this *would* be the average energy of a completely hypothetical wave for which 0 and $2\,k_B T$ are the only possible energies.

We encourage you to make sure you understand the calculation in this example. If you do, then the rest of the section should make sense. If you don't, there is no way for you to understand what Planck did to fix the ultraviolet catastrophe.

A key point from the example, which Planck understood, is that the average energy you calculate depends on what energies you allow in your calculation.

"Where Did $E_w = k_B T$ Come From?" Part 2: The Allowable Energies

If you followed the preceding Example, then you understand how to find E_w from Equation (3.7), given a finite set of allowable energy levels. Now suppose we allowed more energy levels: not only 0 and $2k_B T$ but also $4k_B T$, $6k_B T$, and $8k_B T$. Then our calculation would look like this:

$$E_w = \frac{(0)(C) + (Ce^{-2})(2k_B T) + (Ce^{-4})(4k_B T) + (Ce^{-6})(6k_B T) + (Ce^{-8})(8k_B T)}{C + Ce^{-2} + Ce^{-4} + Ce^{-6} + Ce^{-8}} \approx 0.31k_B T.$$

$$(3.8)$$

You can keep going from there – add in $10k_B T$, $12k_B T$, and as many more levels as you like – but the average energy will stay at $0.31\ k_B T$. Those high energies have probabilities too low to make any noticeable difference in the average.

What *does* make a difference is the way we "chunk" our energy. Equation (3.8) used discrete energy levels with $\Delta E = 2k_B T$. You can repeat the calculation with different step sizes and see how your answers change. Table 3.1 shows some that we got. (You can do calculations similar to these in Problems 17–25.)

Table 3.1 Average energy of a given wavelength as a function of the "chunking" of discrete allowed energy levels

ΔE	Allowed energy levels	Calculated E_w
$5k_B T$	$0, 5k_B T, 10k_B T, \ldots$	$0.034k_B T$
$2k_B T$	$0, 2k_B T, 4k_B T, \ldots$	$0.31k_B T$
$k_B T$	$0, k_B T, 2k_B T, \ldots$	$0.58k_B T$
$\frac{1}{2}k_B T$	$0, \frac{1}{2}k_B T, k_B T, \ldots$	$0.77k_B T$
$\frac{1}{10}k_B T$	$0, \frac{1}{10}k_B T, \frac{2}{10}k_B T, \ldots$	$0.95k_B T$

Every row in Table 3.1 represents the same calculation (expectation value), done on the same probability formula (the Boltzmann distribution), with an infinite set of evenly spaced energy levels. But with each successive row the allowed energy levels are more tightly packed. You can clearly see two trends:

- As ΔE approaches 0 (moving down the rows), E_w rises toward $k_B T$.
- As ΔE approaches ∞ (moving up the rows), E_w drops toward 0.

Both trends make sense based on $P = Ce^{-E/(k_B T)}$. Notice that $E = 0$ is the most likely outcome of any measurement, and the probabilities get precipitously lower as the energies rise. With that in mind, think about the two limits above.

- Consider first a small ΔE. There are many different energy levels in the low-energy range, with high probabilities. There are also plenty of high energy levels, but their probabilities diminish quickly.

  To make that at least roughly quantitative, let's say that any energy level above $E = 5k_B T$ will not contribute significantly to the expectation value. (Note that $E = 5k_B T$ has less than 1% the probability of $E = 0$.) With many different waves whose energies range from $E = 0$ to $E = 5k_B T$, but clustered heavily toward the bottom of that range, the average will clearly fall well above 0 and well below $2.5k_B T$. You can show in Problems 23 and 24 that, in the limit as $\Delta E \to 0$, the average energy approaches $E = k_B T$ – and while that exact number is certainly not obvious, we hope we've convinced you that it's plausible.

- And what about the other limit? When ΔE is very high, you generally don't have enough energy to get to any energy above zero. Picture climbing up a staircase with 20-foot steps: you're going to stay on the bottom.

  For instance, suppose $\Delta E = 10k_B T$. So the $E = 0$ energy level has probability C, and the next available level is $E = 10k_B T$ with probability Ce^{-10}. You will get over 20,000 zero-energy measurements for every one of those. So the average over many electrons will be almost exactly zero, because virtually every measurement will give $E = 0$. You can examine this in Problem 22.

To nineteenth-century physicists, the next step was obvious. Energy can be any real positive number, so the correct E_w is the limit of this process as $\Delta E \to 0$. In other words, the correct calculation is the integral in Problem 24 that leads to $E_w = k_B T$. Equation (3.6). Catastrophe.

So Planck said, what if we *don't* take that limit? What if, instead of an integral, we use discrete sums? As you have seen, exactly how much difference that makes depends on the ΔE we choose.

Active Reading Exercise: Tweaking ΔE

The traditional calculation of E_w is an integral, the limit of the Table 3.1 as $\Delta E \to 0$. Planck decided to use sums instead.

1. Suppose Planck's goal was to achieve a very low value for E_w. Should he choose a high ΔE, or a low one? How low could he get E_w to go?

2. Suppose Planck's goal was to achieve a very *high* value for E_w. Now what kind of ΔE should he choose? How high could he make E_w?

3. What did we say (earlier) that Planck's goal actually was?

Here are our answers.

1. By choosing very high values for ΔE, Planck could get E_w to approach zero.

2. By choosing very low values for ΔE, Planck could get E_w to approach the classical $E_w = k_B T$.

3. Remember Figure 3.26? Planck's goal was to closely match the classical prediction for low frequencies, but drive E_w down toward zero for high frequencies.

We hope it is now clear that Planck wanted a low ΔE for low frequencies, and a high ΔE for high frequencies. He accomplished this in the simplest way possible: he made ΔE proportional to the frequency, introducing a constant of proportionality that we now call "Planck's constant":

$$\Delta E = h\nu. \tag{3.9}$$

Equation (3.9) disrupted the classical chain of reasoning at a key point. Instead of integrating over a continuous stream of energy values, Planck *summed* over a discrete set of values. This led to a different $E_w(\nu)$ function, which in turn led to a different $S(\nu)$ function. You can go through those calculations with Planck in Problem 38.

What We've Done So Far

We're almost done, but before we take the last step it's worth reviewing the chain of reasoning that got us this far.

1. It all starts with Equation (3.7), the probability that a given wave will have energy E. We did not prove or justify this equation in any way (we will in Chapter 10), but its pedigree is solid enough that Planck did not challenge it.

2. From that probability distribution you can calculate $E_w(\nu)$, the average energy a given wave will have over any reasonably long time period. *This is the step where Planck did his work.* If you assume (as everyone had previously assumed) that the energy of a wave can take any value, you end up with catastrophe. By postulating that the energy could only take discrete values, and carefully tuning the discretization, Planck obtained the final result he needed.

3. Once you have an average $E_w(\nu)$ you can calculate the spectrum function $S(\nu)$ and the total energy density ρ. These are the experimental data that Planck needed to match.

Here are the results.

	Classical	**Planck**
Probability of energy E	$P(E) \propto e^{-E/(k_B T)}$	
	(see Chapter 10)	Same as classical
$E_w(\nu)$	$k_B T$	$\dfrac{h\nu}{e^{h\nu/(k_B T)} - 1}$
	(see Problem 24)	(see Problem 38)
$S(\nu)$	$\dfrac{8\pi \nu^2 k_B T}{c^3}$	$\dfrac{8\pi h\nu^3}{c^3 \left(e^{h\nu/(k_B T)} - 1\right)}$
	(from Equation (3.4))	(from Equation (3.4))
ρ	diverges	converges
	(from Equation (3.5))	(from Equation (3.5))

Planck's Blackbody Spectrum

If you were asked to summarize this entire section (and indeed much of this chapter) while standing on one foot, you would do pretty well to just quote Equation (3.9). Understanding that equation is a huge step on the road to understanding quantum mechanics.

As we said at the outset, every wave in our cavity is characterized by a frequency (that determines what kind of light it is) and an amplitude (that determines its energy). Classically these two variables are independent of each other, and the energy can be any positive real number. In fact, in Section 3.3 we cited this as one of the key properties of a wave: you can multiply its energy by 1/2, or 2/3, or anything else you like, and just have a smaller wave.

Planck said you can't. Certain energy levels are allowed, and the levels in between them are not. Such a restriction was not implied by any rules, physical or mathematical, known at the time. But Planck determined that this restriction could bring the theory into line with the data.

Planck could have said that the energy of any wave must be 1 μJ, or 2 μJ, or 3 μJ, but no fraction. That would be a simple form of quantization. But what makes his solution work is the way he tied the quantization of energy (or amplitude) to the frequency:

- An infrared wave with frequency 15 THz can have energy 10^{-20} J, or 2×10^{-20} J, or any other integer multiple of 10^{-20} J, but cannot have any energy in between those values.
- Ultraviolet light with frequency 1500 THz can only have 10^{-18} J, 2×10^{-18} J, and so on. Its allowed energies are spaced 100 times as far apart as those of the infrared wave.

In the preceding discussion we have shown how this rule left the low-frequency predictions for average energy per wave alone, but drove the high-frequency predictions down to zero, in accordance with experimental results. In Problem 38 you can use Planck's hypothesis to calculate $E_w(\nu)$, and from that $S(\nu)$, with the following result.

Planck's Blackbody Spectrum

The spectrum of radiation inside a cavity in equilibrium is

$$S(\nu) = \frac{8\pi h\nu^3}{c^3} \frac{1}{e^{h\nu/(k_B T)} - 1}. \tag{3.10}$$

Integrating this spectrum between any two frequencies tells you the energy density of radiation in that frequency range. Integrating it from $\nu = 0$ to ∞ tells you the total energy density in the cavity.

By carefully fine-tuning the value of h ("Planck's constant") Planck was able to bring his predicted results into very close alignment with experimental measurements. You can see this vividly in Figure 3.27, showing measurements of the "cosmic microwave background," the most perfect blackbody radiation known in the universe. (See Chapter 14 for much more information about the cosmic microwave background.)

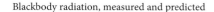

Blackbody radiation, measured and predicted

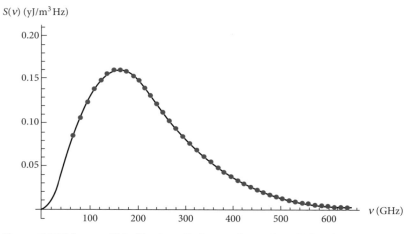

Figure 3.27 Measured blackbody radiation vs. the predicted Planck spectrum. The error bars on the data are too small to be visible.

Example: A Closed-Off Room

A closed-off room has been sitting empty for a while at $T = 300$ K (a typical room temperature).

1. Find the total energy density of the radiation in the room.
2. How much of that energy comes from waves with frequencies between 10^{13} Hz and 10^{14} Hz (a range in the infrared)?

Answer: Both answers come from integrating the Planck spectrum. The analytical result is a messy combination of strange-sounding functions (try it on a computer), so we used a computer to numerically approximate the integrals.

1. The energy density is the integral of the Planck spectrum over all frequencies:

$$\rho = \int_0^\infty \frac{8\pi h\nu^3}{c^3} \frac{d\nu}{e^{h\nu/(k_B T)} - 1}$$

$$= \int_0^\infty \frac{6.17 \times 10^{-58} \, \text{J}\,\text{s}^4/\text{m}^3}{e^{(1.6\times 10^{-13}\,\text{s})\nu} - 1} \nu^3 d\nu = 6.1 \times 10^{-6} \, \text{J}/\text{m}^3.$$

2. To find the portion of this energy density that comes from the given frequency range we integrate only over those frequencies:

$$\rho = \int_{10^{13}\,\text{Hz}}^{10^{14}\,\text{Hz}} \frac{8\pi h\nu^3}{c^3} \frac{d\nu}{e^{h\nu/(k_B T)} - 1}$$

$$= \int_{10^{13}\,\text{Hz}}^{10^{14}\,\text{Hz}} \frac{6.17 \times 10^{-58} \, \text{J}\,\text{s}^4/\text{m}^3}{e^{(1.6\times 10^{-13}\,\text{s})\nu} - 1} \nu^3 d\nu = 5.4 \times 10^{-6} \, \text{J}/\text{m}^3.$$

About 88% of the thermal radiation at room temperature is in that range of infrared frequencies.

Stepping Back

Our buildup to the rule $\Delta E = h\nu$ was not perfectly modeled on Planck's own process, but it did follow his thinking in one important respect: we presented quantization as a mathematical trick. That's pretty much how Planck thought about it in 1900. He wasn't redefining the very nature of energy (and ultimately of matter too). He just noticed that he could fit the theory to the data by cleverly *not* taking a limit that everyone else took.

Five years later Einstein showed that quantized light could also explain the photoelectric effect (Section 3.5). Four years after that, the double-slit experiment was "turned down" to reveal individual spots on the wall. With our knowledge of those and other experiments we can appreciate the importance of Planck's contribution far more than he could at the time. It is a fundamental property of light to come in discrete packets, each with energy $h\nu$. A cavity can have one blue photon, or two, or three, but never two and a half.

Planck's constant h is incredibly small, about 7×10^{-34} J s. That's why Young observed light to be continuous; only with twentieth-century technology could we create and measure light so dim that we could see photons of energy $h\nu$ individually. Even without being able to observe individual photons, however, Planck was able to show that discrete packets of light would resolve the ultraviolet catastrophe. That realization was the first step in the quantum revolution.

3.4.3 Questions and Problems: Blackbody Radiation and the Ultraviolet Catastrophe

Conceptual Questions and ConcepTests

1. For each quantity give a brief description of what it represents, and units that could be used to measure it.

(a) $E_w(\nu)$

(b) $S(\nu)$

(c) ρ

(d) $e^{-E/(k_B T)}$

2. We've discussed two blackbody spectra, the classically predicted (and incorrect) Rayleigh–Jeans spectrum and the (experimentally accurate) Planck spectrum. For each of the following, say whether it applies to (A) only the classical spectrum, (B) only the Planck spectrum, or (C) both. (All of them will apply to at least one of the spectra.)

   (a) It goes to zero in the limit $\nu \to 0$.

   (b) It goes to zero in the limit $\nu \to \infty$.

   (c) It has a maximum value.

   (d) It predicts that higher frequencies always carry more energy than lower ones.

   (e) It has units of energy per volume per frequency.

3. Suppose you evaluated the integral in Equation (3.5) for a given cavity with a given energy distribution, and you ended up with $\rho = 0.00001 \, \text{J/m}^3$. Explain briefly what that number tells you about the cavity.

4. Without doing any calculations, and without reference to any experimental results, how can you know that the classical result for the spectrum (Equation (3.2)) cannot be right?

5. Describe in one sentence what Planck hypothesized that led to the solution to the ultraviolet catastrophe.

6. Suppose the spectrum inside an enclosed space looks like the one in Figure 3.28. If you peer into that space, what will the light look like? (You may find it helpful to look up the spectrum in Appendix D.)

7. If $S(550 \, \text{THz}) > S(100 \, \text{THz})$ in a particular cavity, that means this cavity has more energy in and around the 550 THz frequency than the 100 THz frequency. But it does *not* mean there is more total energy in the green part of the spectrum than in the infrared. Why not? (You may find it helpful to look at Appendix D. It may also help to consider this question: if we told you that a, e, i, o, and u are the five most prevalent letters in a book, would that necessarily mean that the vowels outnumber the consonants overall?)

8. Equation (3.7) says that the probability of a wave in a cavity in equilibrium having a particular energy E is independent of the frequency of the wave. So how did Planck calculate that the average energy per wave was different for different frequencies? At what point in the calculation did frequency come in?

9. For each hypothetical quantization scheme below, say whether it would avoid the ultraviolet catastrophe and briefly say why or why not. (Of course Planck's is the only one that not only avoids catastrophe but also correctly matches observed data.) In each case assume h is a constant with appropriate units for that scheme.

   (a) $\Delta E = 1 \, \text{J}$

   (b) $\Delta E = k_B T$

   (c) $\Delta E = h\nu^2$

   (d) $\Delta E = h/\nu$

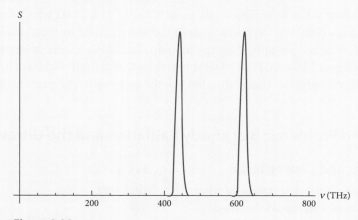

Figure 3.28

Problems

10. Calculate the average energy of a 60 THz wave in a cavity at 300 K. *Hint*: It's not $k_B T$.

11. An LED light emits 5 watts of red light. Estimate the number of photons per second emitted by the bulb.

12. If you integrate Planck's blackbody spectrum (Equation (3.10)) with respect to frequency ν, what units does the result have? (Show your work.) Explain why this makes sense, given the physical meaning of the spectrum $S(\nu)$. *You do not need to evaluate the integral for this problem.*

13. The frequency $\nu = 300$ THz is in the infrared range, and the frequency $\nu = 650$ THz is in the blue range. The energy carried by an electromagnetic wave is proportional to the square of its amplitude.

   (a) Consider a cavity with a 300 THz wave with amplitude A, a 650 THz wave with amplitude $2A$, and no other waves. What percentage of the total energy in this cavity is carried in the blue frequency?

   (b) Consider a cavity with a 300 THz wave with amplitude A, *two* 650 THz waves with amplitude $2A$, and no other waves. What percentage of the total energy in this cavity is carried in the blue frequency?

14. Imagine that you do a study in which you measure the heights of adult women. Your findings are summarized in Figure 3.29.

   (a) Calculate the average height of the women in your study.

   (b) Let $F(h)$ be the frequency at each height. For instance, $F(59) = 11$. Write a formula for the calculation you did in Part (a). Your formula should have the function $F(h)$ in it but should not have any specific numbers. In other words, if we gave you a different set of measurements

for, say, adult men's heights, you should be able to plug those numbers into your formula to get the average value for that. *Hint*: Your answer will involve a sum, which you should write with the summation symbol $\sum$.

15. From the energy per wave $E_w(\nu)$ you can calculate the spectrum function $S(\nu)$ and the total energy density ρ. In the Explanation (Section 3.4.2) we saw how the classically predicted $E_w(\nu)$ led to catastrophe, while Planck's $E_w(\nu)$ led to a perfect match with experiment. But we also did an Example on p. 137 with a completely fictitious $E_w(\nu)$ function, just to show how the equations work. Now you're going to do the same. In a hypothetical universe in which $E_w = ae^{-b\nu}$ for positive constants a and b, calculate the energy density ρ. Is the resulting answer reasonable, or catastrophic?

16. Statistical mechanics tells us that the probability of a wave having energy E is proportional to $e^{-E/(k_B T)}$, where $k_B \approx 1.4 \times 10^{-23}$ J/K and T is the temperature. What is that formula telling us? Let's take $T = 300$ K (a normal room temperature).

   (a) Calculate the relative probabilities of energies 0 and 10^{-22} J.

   (b) Repeat Part (a) for $E = 10^{-21}$ J. Keep multiplying E by 10 until the relativity probability is 100 times smaller than it is for $E = 0$.

   (c) In the Explanation (Section 3.4.2) we argued that it is reasonable to neglect very high energies when calculating the expected value E_w. If you neglect energies whose relative probability is lower than 1% of what it is for $E = 0$, how high do you have to go? Your answer should be somewhere between the last two values in your answer to Part (b).

Equation (3.7) gives the relative probability that a wave will have a particular energy E. From that probability you can calculate the average energy E_w that the wave will have over time. This is the

Height (inches)	59	60	61	62	63	64	65	66	67	68	69	70	71
Number of women	11	21	31	32	53	50	54	75	49	32	22	13	11

Figure 3.29

most important calculation in this section, because this is the part of the overall calculation that Planck modified.

On p. 139 we performed this calculation in a simple fictional universe with only two allowed energies. In Problems 17–25 you're going to do similar calculations for different scenarios.

17. Consider a universe in which the energy can be any non-negative integer multiple of $k_B T$.

   (a) The larger the E-value, the smaller the probability. At what E-value does the probability drop to less than 1% of the probability of $E = 0$? (Your answer will have T in it.)

   (b) Calculate the expected energy E_w. You may ignore all energy values higher than the one you found in Part (a). The calculation will be similar to what we did in Equation (3.8).

18. Consider a universe with only three allowable energies: $3k_B T$, $4k_B T$, and $5k_B T$.

   (a) The average of these three values is obviously $4k_B T$. Before you do any calculations, explain why the expected energy value E_w is *lower* than $4k_B T$.

   (b) Now calculate the expected energy value.

19. Consider a universe in which the energy can be any non-negative integer multiple of $3k_B T$.

   (a) First calculate the expected energy E_w if the only allowable energy values are 0, $3k_B T$, $6k_B T$, and $9k_B T$ (nothing higher).

   (b) Now add $12k_B T$, $15k_B T$, $18k_B T$, and $21k_B T$ to the list (eight values in all) and recalculate E_w.

   (c) You should have found that the two answers are almost identical. Why didn't adding those four extra allowable energy values affect the expected outcome?

20. Consider a room at $T = 300$ K.

   (a) Calculate E_w for a 1 THz wave using only the first two possible energies: 0 and $h\nu$. Express your answer as a multiple of $k_B T$.

   (b) Re-do your calculation using the first three energies: 0, $h\nu$, and $2h\nu$. How much difference does adding that extra energy level make?

   (c) Repeat Parts (a) and (b) for a 30 THz wave.

   (d) Why did adding the third energy make a big difference for one of the waves and essentially no difference for the other one? (If you didn't find that then go back and find what you did wrong.)

21. Calculate E_w if the energy can be any non-negative integer multiple of $(1/5)k_B T$. *Hint*: There are infinitely many possible energy levels here. However, as you may have shown in Problem 17 or 19, past a certain point they can be ignored.

22. Consider $\Delta E = 10k_B T$.

   (a) Calculate the expected value based on only three possibilities: $E = 0$, $E = 10k_B T$, and $E = 20k_B T$.

   (b) Look at the calculation you just did: not just the final answer, but the contribution of each of the three energy levels. Use this to explain why E_w will approach zero for any sufficiently large ΔE.

23. Classically, the energy of a wave can be any non-negative real value. We can treat that situation like Problems 19–21 except that instead of ΔE being $3k_B T$ or $k_B T$ or $k_B T/5$, it approaches zero. Use a computer to estimate E_w for successively smaller values of ΔE to find $\lim\limits_{\Delta E \to 0} E_w$.

24. **Deriving the Classical Blackbody Spectrum**

   Classically, E can be any non-negative real value. In Problem 23 you found the classical E_w by taking the limit as $\Delta E \to 0$. In that limit, the sum becomes an integral that you can evaluate analytically:

   Probability of finding $E_1 \leq E \leq E_2$

   $$= \int_{E_1}^{E_2} Ce^{-E/(k_B T)}\, dE.$$

   (a) Explain briefly why $\int_0^\infty Ce^{-E/(k_B T)}dE$ must be 1.

   (b) Use the condition in Part (a) to find C. (This process is called "normalizing" the probability distribution. Note that C is *not* a constant of integration; you are taking a definite integral.)

(c) To find the average energy you once again integrate over all probabilities, but this time multiplied by the energy: $E_w = \int_0^\infty E(Ce^{-E/(k_B T)})dE$. (That isn't obvious, but you should be able to see that this is the continuous version of the sum we used for discrete energy levels.) Using the value of C you found, evaluate this integral to find the classical prediction for E_w.

25. Fixing the Classical Blackbody Spectrum

In Problem 24 you found that E_w, the average energy of one wave, can be calculated as the following integral:

$$E_w = \frac{1}{k_B T} \int_0^\infty E e^{-E/(k_B T)} dE.$$

Then you evaluated that integral analytically and arrived at the catastrophic classical result. (If you didn't do Problem 24, you can take our word for this result.)

As you know, Planck solved the catastrophe by replacing that integral with discrete sums. In this problem you will look at those sums visually, as Riemann sums. You will need two identical printouts of the graph of the integrand. You can print them from any graphing program, using $k_B = 1.4 \times 10^{-23}$ J/K and $T = 300$ K, with the x axis in $[0, 2.5 \times 10^{-25}$ J] and $[0, 0.4$ J]. Alternatively, you can print the image that we have supplied at www.cambridge.org/felder-modernphysics/figures Throughout this problem you can use either left-hand or right-hand Riemann sums.

(a) Draw a Riemann sum using $\Delta E = 10^{-21}$ J. You don't need to numerically compute the result, but answer this question based on your drawing: Is the Riemann sum much less than, very close to, or much greater than, the actual area under the curve?

(b) Draw a second Riemann sum using $\Delta E = 2.5 \times 10^{-20}$ J, and use it to answer the same question.

(c) Based on your Riemann sums, explain why switching from an integral to a sum enabled Planck to match his theory to the experimental data. Use Planck's formula $\Delta E = h\nu$, and

remember what he needed to achieve for both high and low frequencies.

26. Consider a cavity at $T = 300$ K. *Note:* If you evaluate the integrals below using software that can integrate either analytically or numerically, you should integrate numerically. (Yeah, we figured that out the hard way.)

(a) Calculate the energy density in the range $0 \leq \nu \leq 10^8$ Hz twice, once using the classical blackbody spectrum and then again using the Planck spectrum.

(b) Repeat Part (a) for the range 10^8 Hz $\leq \nu \leq 10^{13}$ Hz.

(c) Repeat Part (a) for the range 10^{13} Hz $\leq \nu \leq 10^{15}$ Hz.

(d) When are the Planck numbers very close to the classical values, and when are they very different?

(e) Find the total energy density in the cavity according to the Planck spectrum.

(f) Why couldn't we ask you for the total energy density in the cavity according to the classical spectrum?

27. The core of the Sun can be considered a cavity at about 2 million kelvins.

(a) Find the total energy density of radiation in the center of the Sun.

(b) What fraction of that energy density in the Sun is in the visible range? (See Appendix D.)

(c) The surface of the Sun is at about 6000 K. Repeat your calculations to find what fraction of the radiation near the surface of the Sun is in the visible range.

28. A room with all the lights out and doors and windows sealed is in thermal equilbrium at $T = 300$ K.

(a) What fraction of the energy density in radiation in the room is in the visible range?

(b) What fraction of the energy density is in the infrared?

(c) Why can't you see the walls glowing in a closed room with no lights?

29. Sketch the spectrum $S(\nu)$ of a cavity at temperature 20 K. You should show numbers, with units, on both the horizontal and vertical axes. *Hint*: The peak occurs at about $\nu = 2.82\, k_B T/h$.

30. The probability of a wave at temperature T having energy E is proportional to $e^{-E/(k_B T)}$. With $T = 300$ K you can quickly confirm that the probability drops to effectively zero by the time you reach an energy of 10^{-19} J (as you found if you did Problem 16), so we will consider only energies between 0 and that level.

    (a) Radio waves have a very low frequency. A typical radio wave might have $\nu = 3 \times 10^8$ Hz. List the first three allowable non-zero energy levels for a radio wave in Planck's system.

    (b) A typical infrared ray might have a frequency of 3×10^{13} Hz. List the first three allowable non-zero energy levels for that infrared ray in Planck's system.

    (c) How many allowable energy levels of radio waves of $\nu = 3 \times 10^8$ Hz are there between 0 and 10^{-19} J?

    (d) How many allowable energy levels of infrared waves of $\nu = 3 \times 10^{13}$ Hz are there between 0 and 10^{-19} J?

Figure 3.27 shows Planck's blackbody spectrum. This is not only the spectrum of cavity radiation; if you multiply it by a constant it also approximately describes the spectrum of thermal radiation emitted by most objects. You can confirm from Equation (3.10) that $S(\nu) \to 0$ as $\nu \to 0$ and as $\nu \to \infty$, and that S reaches a maximum somewhere in the middle. Finding the local maximum by setting $S'(\nu) = 0$ is not so easy – you get an equation that you cannot solve algebraically – but "Wien's law" places that maximum at $\nu_{max} \approx 2.82 k_B T/h$. Problems 31–34 use Wien's law. Some of them will require you to look up the electromagnetic spectrum in Appendix D.

31. Human hands are typically at about 293 K (or 86 °F). What is the peak frequency of thermal radiation emitted by human hands? In what band of light (ultraviolet, visible, infrared, etc.) is this frequency?

32. The surface of the Sun is at about 5800 K. What is the peak frequency of thermal radiation emitted by the Sun? In what band of light (ultraviolet, visible, infrared, etc.) is this frequency?

33. Incandescent light bulbs (which is to say almost all light bulbs throughout the twentieth century) emit light by thermal radiation. The filament of the light bulb is heated until it glows. In order to emit a noticeable amount of visible radiation the peak frequency has to be near the visible range. It is typically around 1.7×10^{14} Hz.

    (a) Based on that fact, estimate the temperature of the filament of an incandescent light bulb. Express your answer in kelvins and also in degrees Celsius or Fahrenheit (whichever you prefer).

    (b) The efficiency of a light bulb is the fraction of its light energy in the visible range. Estimate the efficiency of an incandescent light bulb. You may conclude that there's a good reason why we have been switching to other forms of lighting.

34. For several hundred thousand years after the Big Bang, the universe was filled with a dense plasma (ionized gas) emitting thermal radiation. Today that radiation still fills all of space and is called the "cosmic microwave background," or CMB. Figure 3.27 shows the spectrum of that radiation as measured by the COBE (COsmic Background Explorer) satellite. Using that figure, estimate the temperature of the CMB. (Asking the temperature of thermal radiation is the same as asking the temperature that a cavity would have in equilibrium if it were filled with that radiation.)

35. Take the limit $h \to 0$ of Planck's blackbody spectrum. Explain why your result makes sense.

36. **Stefan's Law**

    The energy density in a cavity at equilibrium is given by the integral of Equation (3.10) (p. 144) with respect to ν from $\nu = 0$ to ∞.

    (a) Do a u-substitution $u = h\nu/(k_B T)$ and simplify the resulting integral as much as possible.

(b) When you bring all the constants to the front, you should have no letters left inside your integral except the variable of integration. Call the value of that integral I and write the resulting expression for the energy density. Your answer should be in the form $\rho = QT^p$, where Q and p are constants. (The constant Q will include the integral I.) What is that power p?

(c) If you double the temperature of a cavity, by what factor does its energy density increase?

(d) Evaluate the constant Q. That will involve evaluating the integral I on a computer or calculator and looking up values of the other constants.

The law you just derived is called "Stefan's law." Here you're expressing it as energy density, but it is more commonly given in terms of the power emitted by a radiating blackbody. You'll learn how to derive it in that form in Chapter 10.

37. The Wavelength Spectrum

The energy density in a cavity at equilibrium is given by the integral of Equation (3.10) with respect to ν from $\nu = 0$ to ∞. The wavelength of a beam of light is related to the frequency by $\lambda = c/\nu$. Do a u-substitution to rewrite the energy density as an integral over wavelength instead of frequency.

38. Exploration: Deriving the Planck Blackbody Spectrum

In this problem you are going to derive the average energy $E_w(\nu)$ and blackbody spectrum $S(\nu)$, based on Planck's hypothesis: a single wave of frequency ν must have energy $nh\nu$, where n can be 0, 1, 2, etc. but nothing in between.

(a) If you measure a particular wave of frequency ν at a particular moment, the probability that you will find energy E is proportional to $e^{-E/(k_B T)}$. Using C for the constant of proportionality, we can say that the measurement $E = nh\nu$ has probability $Ce^{-nh\nu/(k_B T)}$. Explain briefly why the following equation *must* be true:

$$\sum_{n=0}^{\infty} Ce^{-nh\nu/(k_B T)} = 1.$$

(b) Use the equation from Part (a) to solve for the constant C. (This is called "normalizing" the probability distribution.) You will need to know that the sum of an infinite geometric series with first term a_1 and common ratio $r < 1$ is $a_1/(1 - r)$.

(c) To find the average you evaluate the sum of each possible energy times its probability:

$$E_w = \sum_{n=0}^{\infty} (nh\nu)Ce^{-nh\nu/(k_B T)}$$

(Take a moment to convince yourself that this is the same calculation as we did in the text, although we never quite wrote it in this way.) Plug in the value of C that you found in Part (b) and then evaluate the sum.

(d) Plug the formula you found for E_w into Equation (3.4) to find the Planck spectrum.

39. Exploration: The Spectrum in a 3D Cavity

[*This problem requires multiple integrals and spherical coordinates, topics from multivariate calculus.*] Consider a 3D cavity that goes from 0 to L in all three directions. We'll assume for simplicity that the walls are conducting, so $\mathcal{E} = 0$ on the walls. (We're using $\mathcal{E}$ for the amplitude of the electric field to avoid confusion with the energy.) The standing waves of the electric field in such a cavity are of the following form:

$$\mathcal{E} = A \sin(px) \sin(ky) \sin(qz) \quad \text{where}$$

$$\nu = \frac{c}{2\pi}\sqrt{p^2 + k^2 + q^2}.$$

(a) Given the condition $\mathcal{E} = 0$ at $x = L$, what are the possible values of p? Your formula will have an integer a in it that can go from 1 to ∞.

(b) Write similar formulas for the possible values of k and q, using integers b and d.

(c) Express the frequency ν of the wave as a function of a, b, and d.

The total energy in the cavity is the sum of the energy of each wave. We can approximate that triple sum (over a, b, and d) as an integral. There is an extra factor of two in front because each combination of a, b, and d actually corresponds to two waves with two different polarizations:

$$E_{total} = 2 \sum_{a=1}^{\infty} \sum_{b=1}^{\infty} \sum_{d=1}^{\infty} E_w$$

$$\approx 2 \int_0^{\infty} \int_0^{\infty} \int_0^{\infty} E_w(a, b, d) \, da \, db \, dd.$$

We are using E_w to mean the average energy of a given wave of frequency ν. As you calculated, ν is a function of a, b, and d.

You can view that integral as a Cartesian integral in a, b, d space and then convert to spherical coordinates: $r = \sqrt{a^2 + b^2 + d^2}$, $\tan \theta = \sqrt{a^2 + b^2}/d$, $\tan \phi = b/a$.

(d) Rewrite the integral for E_{total} in spherical coordinates. Don't forget to include the Jacobian $r^2 \sin \theta$. When you write the limits of integration remember that the original integral didn't go over all of space, but only over positive values of a, b, and d.

(e) The variables θ and ϕ relate to the direction in which a wave is pointing inside the cavity. By symmetry, E_w shouldn't depend on that direction, so it is independent of θ and ϕ. Given that, explicitly evaluate the θ and ϕ integrals. Your final result should be a formula for E_{total} as a single integral over the variable r.

(f) Write an equation relating r and ν. Use a u-substitution to rewrite E_{total} as an integral over ν.

(g) Finally, to get an expression for the energy density ρ, divide E_{total} by L^3. You can check your answer against Equation (3.4).

In this problem we used a cubical cavity, but with some more work you can show that this applies to any shape. The one approximation we made was approximating our sum with an integral. That approximation becomes more accurate in the limit of large L, so Equation (3.4) is valid in large cavities.

3.5   The Photoelectric Effect

The "photoelectric effect" revealed problems with the wave model of light long before the one-photon-at-a-time double-slit experiment. Einstein's solution to these problems involved discrete packets of light, a key step toward the development of quantum mechanics.

3.5.1   Discovery Exercise: The Photoelectric Effect

This Discovery Exercise is part of a <u>classical</u> analysis of the photoelectric effect. As we shall see, the quantum mechanical understanding is quite different.

A laser beam with intensity 10 W/m^2 is shining on a plate coated with potassium. (Remember that one watt (W) means one joule per second.) An electron bound to a potassium atom requires an energy of roughly 2.3 eV ($3.7 \times 10^{-19} \text{ J}$) to be ejected from the atom.

1. Assume the atom has a circular cross-section with radius 10^{-10} m. How much power is striking the atom? Give your answer in watts.

   See Check Yourself #5 at www.cambridge.org/felder-modernphysics/checkyourself

2. Assume that all of the laser energy that strikes the atom is absorbed by the electron. How long will it take the electron to absorb enough energy to get ejected?

3.5.2 Explanation: The Photoelectric Effect

Figure 3.30 shows two metal plates in a vacuum. Plate B is large and is bent around Plate A. A battery (or any other voltage source) sets up a potential difference such that $V_A > V_B$.

Remember that electrons, being negatively charged, tend to move from low potential to high potential. If we closed this circuit, the electrons on Plate B would rush willy-nilly to Plate A. But with no connecting wire the electrons on Plate B are stuck, gazing longingly over the gap. On the other side, of course, the electrons on Plate A are certainly not going to jump to Plate B – crossing a gap with no wire, and simultaneously fighting *against* the electric force – unless someone gives them a really hard kick.

At this point, we give them a kick.

The "kick" is supplied by shining a beam of ultraviolet light onto Plate A. Such a beam can transfer energy to electrons in the metal, and an electron that absorbs enough energy will escape from the metal entirely. The energy left after that escape becomes the electron's kinetic energy as it moves through the vacuum. And if *that* energy is high enough, the electron can actually fight its way past the electric field, slamming into Plate B and through the exit wire.[5] An ammeter on that wire measures the current and thus determines how many electrons are leaving Plate A with sufficient energy to reach Plate B.

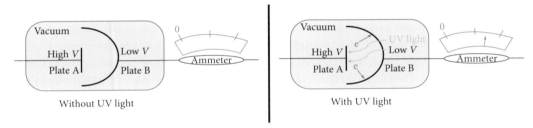

Figure 3.30 (Left) Electrons on Plate B are pulled toward Plate A by the electric field, but cannot cross the gap. (Right) UV light liberates some electrons from Plate A with enough energy to cross the gap *against* **the field and reach Plate B. The excess of electrons at Plate B causes a current in the ammeter.**

Once you have that setup, you can start to tinker with it. What happens if you change the intensity of the ultraviolet light? What happens if you change its frequency? What happens if you change the voltage difference between the two plates? Classical electromagnetic theory and a simple model of the atom give you the tools to answer these questions – and to get every one of them wrong. So this experiment provides one of the clearest illustrations of the differences between the classical and quantum views.

Follow the Energy

For an electron to make it to our ammeter it must absorb energy from the beam of light, use some of that energy in escaping from Plate A, and still have enough kinetic energy to overcome the potential difference and reach Plate B.

The required energy for an electron to initially escape Plate A depends on how tightly bound the electron is to its atom, and on how close it is to the surface (because it will lose additional

---

5 The wire connected to Plate A provides an endless supply of new electrons to replace the missing ones.

energy to collisions as it travels through the plate). We will use the letter w for that escape energy, but with the understanding that it is different for different electrons.

The second step, on the other hand, is always the same: the required kinetic energy for a charge e to cross a voltage gap V is Ve. Putting it together,

$$w + Ve \quad \text{Total energy required for an electron to escape Plate A and reach Plate B.}$$

Electrons leave Plate A with different kinetic energies. Only the ones with $K \geq Ve$ make it to Plate B. (We're using K for kinetic energy.) If the *maximum* kinetic energy of any escaped electron is below Ve then the current stops completely. So by increasing the potential gap until we find the "stopping potential" V_0, the potential difference that stops all current, we can calculate the maximum kinetic energy with which any electron leaves Plate A:

$$K_{max} = V_0 e \quad \text{The largest kinetic energy of any electrons leaving Plate A.}$$

A Classical Analysis

Classically, light is a wave of oscillating electric and magnetic fields. The energy of that wave is a function of its intensity, which is to say, of the amplitude (not the frequency) of that wave. And that energy is distributed through space: if a small area dA receives a small dose of energy dE, then half that area receives half the energy.

That means that each atom in Plate A receives a tiny fraction of the light beam's energy. As the electrons in those atoms absorb that energy they get closer to the energy required to escape. If the light beam is dim, there should be a measurable time lag before the first electron escapes Plate A. (You calculated this time lag for one specific example in the Discovery Exercise, on p. 152.) After that the current should build up to a steady value that continues for as long as the light is shining.

Active Reading Exercise: Photoelectric Predictions in the Classical Model

Imagine setting up the photoelectric effect and measuring the current of electrons flowing out of Plate B. Based on the classical model of light in the paragraphs above . . .

1. How would an increase in the intensity (or amplitude) of the light – holding the frequency constant – affect the time lag, the final current, and the maximum kinetic energy of released electrons?

2. How would an increase in the frequency of the light – holding the amplitude constant – affect the time lag, the final current, and the maximum kinetic energy of released electrons?

Your answers can be brief, but don't read further until you have written them down!

Hopefully your answers to the questions above looked something like this:

1. Higher intensity light conveys more energy. That means the time lag should be shorter, the final current should be larger, and K_{max} should be larger.

2. Changing the frequency of the light should have no effect, since the energy of the light is only a function of the amplitude.

Make sure you see clearly how the classical model leads to those predictions, so you can understand why the *failure* of those predictions helped lead to the quantum revolution!

The Actual Results

Some of the classical predictions are correct. Most notably, increasing the intensity of the light does increase the resulting current. However, there are three results that contradict the classical model. Recall as you read this that K_{max} refers to the highest kinetic energy that any electron has after leaving Plate A.

Classical prediction	Experimental result
K_{max} should increase at higher light intensities.	K_{max} is independent of the light intensity.
The frequency of the light should have no effect on K_{max}. Red, blue, and ultraviolet light with the same amplitude should all impart the same energy to an electron. (You can imagine classical theorists speculating that something interesting might happen if the light hits a particular resonant frequency of the atom. But the experimental effect of frequency turns out to be much simpler than that.)	K_{max} is a function of frequency: electrons liberated by high-frequency light (such as ultraviolet light) have more kinetic energy than electrons liberated by low-frequency light (such as visible light). Furthermore, there is a "cutoff frequency." Light below this frequency liberates no electrons, regardless of the intensity of the light or the potential difference between the plates.
A dim light should take a while to impart enough energy to liberate an electron, so there should be a measurable time lag – as much as several seconds in some cases – before any current is detected.	No time lag is observed, no matter how low the intensity of the radiation. Low-intensity light does produce a smaller current, but it still produces it instantly (to within our ability to measure, which is below 10^{-9} s).

Before you read further, we urge you to consider how the photon model of light that we have already discussed in this chapter makes sense of all these results.

Einstein's Explanation of the Photoelectric Effect in Terms of Photons

Faced with experimental results that stubbornly refuse to conform to theoretical predictions, a scientist's instinct is – and should be – to tweak. Try the experiment a bit differently. Adjust the theory around the edges. Only in desperation does a scientist overturn fundamental elements of a theory that has proven successful in many arenas.

So you can consider Einstein's 1905 paper "On a Heuristic Viewpoint Concerning the Production and Transformation of Light" to be an act of desperation. In this paper, in direct response to the problems caused by the photoelectric effect, Einstein proposed a radical new view of light as coming in discrete packets. The energy of each packet depends only on the frequency ($E = h\nu$). The intensity of the beam is a measure of the density of photons.[6]

---

6 The word "photon" didn't come around until later, and was applied retroactively to Einstein's proposal and to Planck's earlier one.

It is still true, in this model, that the total power carried in a light beam is proportional to its intensity. That fact had been well established by experiment, *including* the photoelectric effect: remember that the measured current is proportional to the intensity of the light! So the classical and quantum models must agree that a red light and a blue light, both of the same intensity, carry the same power. But in the photon model the blue beam comprises fewer photons, each of which carries a higher individual energy, than the red beam.

In the classical model, an electron gradually soaks up energy from the wave until it has enough to escape. In the photon model either a particular photon hits an electron and immediately imparts its energy $h\nu$, or it misses the electron and imparts no energy at all. If the energy $h\nu$ is larger than $w + Ve$ the electron will reach Plate B. Otherwise it will either fail to escape Plate A or will escape but get pulled back by the electric field.

Recall that w, the energy required to escape Plate A, is different for different electrons. There is a minimum possible value of w, for the electrons nearest the surface and most lightly bound to their atoms. That value, w_0, is called the "work function" of the material on Plate A. The electrons that are easiest to knock free require a total energy $w_0 + Ve$ to reach Plate B.

With that in mind, consider how Einstein's photon hypothesis explains the experimental results.

- If $h\nu$ is lower than $w_0 + Ve$ then no electrons cross the gap to Plate B, no matter how many photons you throw at them (how high the intensity of the light). This explains the "cutoff frequency" that is seen in experiment but is impossible to explain classically. Put another way, K_{max} depends on frequency (the energy per photon), not on intensity (the density of photons).

- Instead of the light energy being evenly distributed across the metal surface as in the classical model, some electrons get hit by a photon and instantly receive energy $h\nu$, while others aren't hit and receive zero energy. So there is no time lag; as soon as photons strike Plate A, electrons are liberated.

Both the classical and the photon models agree that higher intensity leads to higher current, but in the photon model that's because a higher intensity beam has more photons per second, so more electrons get knocked out.

Example: The Photoelectric Effect

A light source with frequency 4.5×10^{15} Hz hits Plate A. The "work function" of Plate A – the *smallest* amount of energy required to liberate an electron – is 7×10^{-19} J. Calculate the "stopping potential" V_0, and describe the behavior of the system below and above this potential.

Answer: One photon of this light carries energy $h\nu = 3.0 \times 10^{-18}$ J. The easiest-to-liberate electrons will therefore fly off Plate A with $3.0 \times 10^{-18} - 7 \times 10^{-19} = 2.3 \times 10^{-18}$ J. (That's our K_{max}.) The stopping potential occurs when $V_0 e$ exceeds that energy, so V_0 is K_{max} divided by the electron charge e, or around 14 V.

So if the potential difference between Plate A and Plate B exceeds 14 V, no current will flow. If the potential difference is less than that, current will flow. Decreasing the potential difference, or increasing the intensity of the light, will cause an increase in the current.

Stepping Back

The disparity between theory and experiment in the photoelectric effect was a huge problem for classical theory. After our discussions of the double-slit experiment and blackbody radiation, you may wonder why the photon solution wasn't obvious to contemporary physicists. So let's take a moment to put this into some historical context.

The double-slit experiment was still viewed at this time as the great proof of the *wave* nature of light. The technology to perform this experiment with one photon (or one electron) at a time was several years off.

Planck's solution to the ultraviolet catastrophe had been proposed in 1900. However, his quantization hypothesis was viewed as a specific property of light waves inside cavities. Einstein's 1905 proposal generalized that idea to a fundamental property of all electromagnetic radiation. Planck himself was reluctant to accept such a radical shift, and wrote in 1914 that Einstein had "missed the target . . . in his hypothesis of light quanta." Only seven years after those words were written, the existence of photons had become so indisputable that Einstein won the 1921 Nobel Prize for his explanation of the photoelectric effect.

Planck's blackbody paper proposed quantization in the emission of light waves; Einstein's photoelectric effect proposed quantization in their absorption. Both ideas relied on the linear relationship $E = h\nu$ and could be tested in the lab, resulting in experimental measurements of h. The fact that both processes produced almost identical values for this new fundamental constant was one of the most compelling early demonstrations of the quantum hypothesis.

3.5.3 Questions and Problems: The Photoelectric Effect

Conceptual Questions and ConcepTests

1. For each quantity below, give its units and a brief description of what it represents in the photoelectric effect. We have answered the first one for you, just to make sure you understand what we're asking.

   Example: V. *Answer*: The potential difference between Plate A and Plate B: $V = V_A - V_B$. It has units of potential, e.g. volts.

   (a) w

   (b) w_0

   (c) e

   (d) K_{max}

   (e) V_0

   (f) ν

2. Which emits more photons per second, a 5 watt red LED bulb or a 5 watt blue LED bulb?

   A. Red

   B. Blue

   C. The same number is emitted.

3. Explain why, even if you shine a light beam of perfectly consistent frequency and amplitude, not all liberated electrons will enter the vacuum with energy K_{max}.

4. A photon hits an atom on Plate A, increasing the energy of one electron in that atom. Explain briefly what will happen to this electron if . . .

   (a) The energy of the photon is less than the value of w for that electron ($h\nu < w$).

   (b) The energy $h\nu$ is greater than w, but less than $w + Ve$.

   (c) The energy $h\nu$ is greater than $w + Ve$.

5. Explain briefly how the photon view of light explains each of the following results in the photoelectric effect.

   (a) The maximum kinetic energy of an electron liberated from Plate A depends on the frequency of the incoming light, but not on its intensity.

(b) If the light frequency is high enough to induce a current, doubling the light intensity doubles the current.

6. Explain briefly why a classical view of light predicts a time lag between the beam of light striking Plate A and the current flowing through the ammeter. Then explain briefly why the photon view predicts no time lag.

7. Consider a red beam of light ($\nu \approx 474\,\text{THz}$) and a blue beam of light ($\nu \approx 638\,\text{THz}$) of the same intensity.

    (a) Briefly describe the two sine waves in a classical description of these two beams. How are they alike, and how are they different?

    (b) Briefly describe the photon collections in a quantum description of these two beams. How are they alike, and how are they different?

8. In the wave model of light, the intensity of radiation from a point source falls off as $1/r^2$, where r is the distance from the source. Would that still be true in the photon model? Why or why not?

9. Make a plot of K_{max} as a function of the frequency of the light beam for a photoelectric experiment. We're not giving you all the details you would need to put numbers on the plot, but it should have the correct shape and it should correctly show whether the plot intersects the origin and/or either of the two axes.

10. Figure 3.31 shows the current in a photoelectic experiment as a function of the potential difference between the plates. It's roughly (but not perfectly)

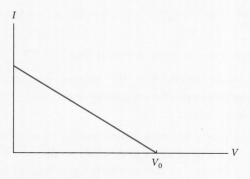

Figure 3.31

linear. Copy this sketch and then draw in a second plot showing the current for the same experiment but with light of twice the intensity. Be sure to clearly label which plot is which.

For Deeper Discussion

11. In many cases you can recover classical physics from quantum physics by taking the limit $h \to 0$.

    (a) Using the photon hypothesis, write an equation relating K_{max}, ν, and w_0.

    (b) In the limit $h \to 0$, does this formula reproduce what you would expect classically? If so, explain why. If not, explain why the procedure doesn't work in this case.

Problems

12. Remember that different electrons in a plate have different liberation energies. Let w_0 be the *smallest* liberation energy for any electron in Plate A, i.e. the work function.

    (a) Explain briefly why K_{max}, the *highest* kinetic energy of any liberated electron, is $h\nu - w_0$.

    (b) Explain briefly why V_0, the potential difference at which all current stops, is K_{max}/e.

    (c) In 1915 Robert Millikan carefully measured the relationship between V_0 and ν. (Millikan himself was firmly convinced that light was a wave, and was disappointed to discover that Einstein's predictions held up very well in the lab.) Based on your equations, calculate the slope of the $V_0(\nu)$ line that Millikan should have found.

13. You are performing the photoelectric effect experiment.

    (a) You turn up the voltage difference between Plate A and Plate B, and the current through your ammeter decreases. Eventually you find that at a potential difference of $V_0 = 1\,\text{V}$ the current stops entirely. Calculate K_{max}, the maximum kinetic energy of the liberated electrons.

    (b) Suppose the work function of the metal is $3.7 \times 10^{-19}\,\text{J}$. Calculate the frequency of the incoming light.

14. You perform a photoelectric experiment using light of frequency 1.5×10^{15} Hz. You find that the stopping potential – the potential at which the measured current drops to zero – is 3.86 V. What is the work function of the metal in Plate A?

15. Consider a beam of light with intensity 10 W/m^2 at frequency $\nu = 10^{10}$ Hz, shining on a plate filled with spherical atoms of radius 2.5×10^{-10} m. Inside the plate is an electron that requires 5×10^{-19} J of energy to escape its atom and leave the plate.

   (a) Classically, the power of the light is evenly distributed through space, and can be absorbed gradually by the atom. At what rate (in watts) does one atom absorb energy? Assuming all that energy goes to the electron, how long will it take for the electron to absorb enough to escape?

   (b) Quantum mechanically, how much energy does each photon have? Is that enough to liberate that same electron? If so, how much kinetic energy will that electron have after its release?

16. You perform a photoelectric experiment using light of frequency 2×10^{15} Hz and you measure a stopping potential of 3.3 V. What stopping potential will you measure if you repeat the experiment with light of frequency 4×10^{15} Hz? *Hint*: The answer is not 6.6 V.

17. Imagine a light bulb that emits 60 W of power through a surface area of 0.02 m^2. A normal light bulb emits many different frequencies of light, of course, but let's average all that out and say that this particular bulb emits light at $\nu = 6 \times 10^{14}$ Hz.

   (a) What is the energy of one photon of light at this frequency?

   (b) How many photons per second are emitted by this light bulb?

   (c) How many photons per second per square meter are coming through the surface of this light bulb?

   (d) Nineteenth-century physicists concluded that light was a wave, and did not come in discrete particles. Based on your answers above, explain why this view – which we now believe to be incorrect – worked so well in their experiments.

18. Suppose a laser beam of intensity 100 W/m^2 and frequency 10^{15} Hz is shining on a plate of area 0.1 m^2 in a photoelectric experiment. The actual current observed at Plate B will depend on many details about Plate A and the potential difference, but the maximum possible current would occur in the limit where every photon that hits Plate A causes an electron to escape and reach Plate B. Calculate that maximum possible current. Give your answer in amperes.

19. Lead has work function 4.25 eV.

   (a) What is the minimum frequency of light required to liberate electrons from a sample of lead?

   (b) Sketch the stopping potential vs. frequency for lead.

   (c) If you shine 2×10^{15} Hz light on a sample of lead, what will be the maximum kinetic energy of the liberated electrons?

20. You perform a photoelectric experiment using light of frequency 1.2×10^{15} Hz. For each of the following materials, look up the work function and then calculate the stopping potential, at which you will see the photoelectric current stop. *Note*: For one of these, the answer is that you will never see a current regardless of V.

   (a) Potassium

   (b) Cobalt

   (c) Thorium

21. In 1916 Robert Millikan published a paper in which he set out to test "the bold, not to say the reckless, hypothesis of an electro-magnetic light corpuscle of energy $h\nu$." He measured the photoelectric effect and found that the slope of the stopping potential vs. light frequency was between 4.124×10^{-15} and 4.131×10^{-15} V/Hz. Find the range of possible values of h that this corresponds to. Were his data compatible with our current best measurements of h?

3.6   Further Photon Phenomena

We've described three experiments that demonstrate the existence of photons: the one-photon-at-a-time double slit, blackbody radiation, and the photoelectric effect. In this section we will talk about the properties of photons and explore a few more of their consequences.

3.6.1   Explanation: Further Photon Phenomena

Here are some properties of photons that we have discussed.

*   When you look for a photon, you find it in one particular place.
*   Each photon has a well-defined energy, $h\nu$.
*   A room might contain 0, 1, 2, or 10^{29} photons, but never 2.4.
*   Photons can be created (e.g. as blackbody radiation) or absorbed (e.g. in the photoelectric effect). The object creating or absorbing them loses or gains the energy carried by the photon.
*   A typical beam of light is a very large collection of very low-energy photons. Hence we experience light as continuous rather than discrete, just as we experience water (a very large collection of very small molecules) as continuous.

Here are a few more properties that we haven't mentioned in this chapter, although you may recall the first two from relativity.

*   Photons move at the speed of light. (Of course – they are light.)
*   Photons have zero mass (see Problem 33).
*   Photons have momentum $p = h\nu/c$.

That formula for momentum follows immediately from the fact that photons have zero mass. In relativity the energy of a particle obeys $E^2 = p^2c^2 + m^2c^4$. So for a massless particle $E = pc$, which immediately gives $p = E/c = h\nu/c$.

With all of those particle-like properties, you might be tempted to think of light as a classical collection of particles. But that can't explain the double-slit experiment: light exhibits behaviors such as interference and diffraction that classical particles never do. The answer to the age-old "wave vs. particle" debate is that both models are required for a full explanation of light. The equation $E = h\nu$ provides a crucial link between the two models, since it expresses the energy of a single photon in terms of the frequency of a wave.[7]

In the double-slit experiment the light seems to act as a wave as it passes through the wall with the slits, and then as a collection of particles when it hits the back-wall detectors. We can generalize that observation as a rough guide: as it propagates through space light tends to act like a wave, but when it interacts with matter those interactions are generally discrete.

The rest of this section describes another three important phenomena in which light interacts in particle-like ways with matter. Over the course of the next few chapters we'll explore the mathematical framework that is used to describe this remarkable hybrid.

---

7   The relationship $E = pc$ holds in both pictures. We just explained why it's true in the particle picture. In Maxwell's theory you can calculate the energy density and momentum density of an electromagnetic wave and derive the same relationship between them.

Compton Scattering

Section 3.5 discussed how light liberates bound electrons from solids. Now consider a light beam of frequency ν striking a free electron. In classical electromagnetic theory the oscillating electromagnetic field would cause the electron to oscillate. That oscillation would in turn cause the electron to radiate in all directions. All of these oscillations – the original wave, the electron's motion, and the newly emitted wave – would occur at the same frequency ν. The net result would be that some of the energy of the incoming wave would get transferred to a new wave of the same frequency going outward in every direction.

In the quantum mechanical picture, a photon of definite energy and momentum collides with the electron.

Active Reading Exercise: A Photon–Electron Collision

An unbound electron is initially at rest when a photon of frequency ν collides with it. As in most collisions, the electron and photon move off in different directions afterward. Would the frequency of the photon after the collision be lower, the same, or higher than it was before the collision? Even if you're not sure, write down an answer and a justification for it before reading on.

We said the electron is initially at rest. After the collision the electron has picked up some kinetic energy, so the photon must have lost energy. That means (from $E = h\nu$) the photon after the collision must have a lower frequency, or equivalently a higher wavelength (Figure 3.32).

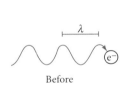

Before

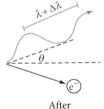

After

Figure 3.32 Compton shift.

You probably recognize this story: classical electromagnetic theory makes a prediction, the photon model makes a different prediction, and then experiments show that the photon model wins. Well, you're right. In 1923 Arthur Compton shone X-rays on a block of carbon. (That isn't hitting totally free electrons with photons, but it's pretty close. We'll come back to that in a minute.) Compton found that much of the outgoing light had a lower frequency than the incoming light, with exactly the frequency shift that the photon model predicted.

You can reproduce those calculations in Problem 34. They're the same sorts of calculations you used to analyze collisions in your introductory mechanics course: energy before equals energy after, and each component of momentum before equals that component of momentum after. You end up with one too many unknowns to solve, so your final energies and momenta come out as functions of the unknown scattering angle. The one thing that makes this a bit more complicated than those introductory mechanics problems is that the photon moves at the speed of light and the electron typically moves pretty close to it, so you have to use the relativistic formulas for energy and momentum.

Here we'll jump to the answer (which is usually written in terms of wavelength rather than frequency).

The Compton Shift

If a photon of wavelength λ scatters off a free electron at rest and comes out at an angle θ relative to its initial path, the wavelength of the photon will increase by

$$\Delta\lambda = \frac{h}{m_e c}(1 - \cos\theta). \qquad (3.11)$$

The combination $h/(m_e c)$ is called the "Compton wavelength" of the electron.

We have generally classified our photons by frequency ν, but we're following common practice by expressing Equation (3.11) in terms of the photon's wavelength λ. The relations between frequency, wavelength, and wave speed are summarized in Appendix F; the key relation you need in this section is $\lambda\nu = c$. You'll express the Compton shift in terms of frequency in Problem 35.

Example: Compton Scattering

Compton did his experiments with incoming light of wavelength 0.0709 nm.

1. What percentage shift did he measure in this wavelength at 135°?

    Answer: First we find the change in wavelength:

    $$\Delta\lambda = \frac{h}{m_e c}(1 - \cos\theta)$$

    $$= \frac{6.63 \times 10^{-34}\,\text{kg m}^2/\text{s}}{\left(9.11 \times 10^{-31}\,\text{kg}\right)\left(3.00 \times 10^8\,\text{m/s}\right)}\left(1 - \cos 135°\right) = 4.14 \times 10^{-12}\,\text{m}.$$

    The percentage shift is the change divided by the original (times 100 to make it a percentage):

    $$100 \times \frac{\Delta\lambda}{\lambda_0} = 100 \times \frac{4.14 \times 10^{-12}\,\text{m}}{7.09 \times 10^{-11}\,\text{m}} = 5.8\%.$$

2. What energy loss does this represent for each photon?

    Answer:

    $$\text{Incoming: } \nu = \frac{c}{\lambda} = 4.23 \times 10^{18}\,\text{Hz}, \qquad E = h\nu = 2.80 \times 10^{-15}\,\text{J}.$$

    $$\text{Outgoing: } \nu = \frac{c}{\lambda} = 4.00 \times 10^{18}\,\text{Hz}, \qquad E = h\nu = 2.65 \times 10^{-15}\,\text{J}.$$

    Each photon loses approximately 1.5×10^{-16} J or 1000 eV.

As we mentioned earlier, the electrons in a carbon block are not free (unbound), nor are they at rest. Compton's calculations work because the typical binding and kinetic energies of the outer electrons are much smaller than the energy of an X-ray photon.

However, many photons collide with much more strongly bound electrons. Such an electron is not knocked free, so the entire carbon atom recoils from the collision. Equation (3.11) still applies in this case, but instead of the mass of the electron you must use the mass of the entire atom, so the resulting shift in wavelength is typically too small to observe (see Problem 36).

You can see this in Compton's data in Figure 3.33. At each angle you see one peak with unchanged wavelength and one peak with a changed wavelength. The changed wavelength represents Compton scattering. The scattering off tightly bound electrons that doesn't significantly change the photon frequency is called "Rayleigh scattering." The distinction between the two types of scattering vanishes in the limit $\theta \to 0$, where the Compton shift approaches zero.

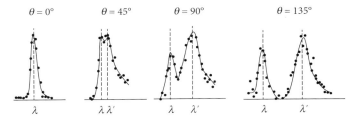

Figure 3.33 Different photon wavelengths are observed at various scattering angles. At $\theta = 0°$ essentially all the photons are observed at their original wavelength λ. At any other angle some are observed at λ and others at a different wavelength λ'. Source: Compton 1923.

For very low-energy radiation (e.g. visible or infrared) nearly all electrons in an atom are effectively tightly bound and there is only Rayleigh scattering. For high-energy gamma rays, all electrons are effectively free and there is only Compton scattering.

Bremsstrahlung Radiation

Imagine a high-energy electron passing near an atom at rest. The atom is made of charged particles, so the electron and the atom interact and the path of the electron is bent. Of course a bent path means acceleration. In a classical picture the electron would emit radiation constantly as it accelerated. In quantum mechanics it is still true that an accelerating charged particle emits radiation, but the radiation is quantized. So we would say that the interaction of the electron and the atom causes the electron to emit a photon. (It is possible but less likely that it can emit more than one photon.)

The final energies of the photon, the electron, and the atom must equal the initial energy of the electron. Because the atom is so massive, however, its recoil carries off negligible energy and we can say that the energy of the photon is equal to the energy lost by the electron (Figure 3.34).

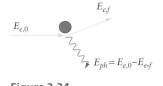

Figure 3.34 Bremsstrahlung radiation.

With all that in mind, consider a beam of electrons being accelerated to high velocity and then striking atoms in a target. Each electron will collide repeatedly with atoms until it loses most of its initial energy. We will assume that the electrons all come in with the same energy, but this does not mean that all the collisions are identical. Most electrons lose only a small percentage of their energy and therefore emit low-frequency photons. Some lose more, so you measure some higher frequencies.

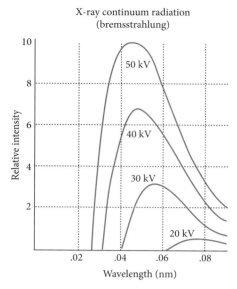

X-ray continuum radiation
(bremsstrahlung)

Figure 3.35 Spectra for light emitted by tungsten bombarded with electrons. Each curve represents a different kinetic energy of the incident electrons. *Source: Springer Nature © 2017.*

The most energetic photon that can possibly be released in this experiment results from an electron losing all its kinetic energy in a single collision.

Figure 3.35 shows spectra for an experiment in which tungsten was bombarded with electrons. Each curve on the plot represents a different electric potential through which the electrons were accelerated, and thus a different kinetic energy with which they struck the target. For each energy beam there is a maximum frequency ν_{max} (seen in the diagram as a minimum wavelength λ_{min}) of the emitted light.

Radiation emitted as an electron (or other charged particle) loses energy to collisions inside matter is called "bremsstrahlung" radiation after the German word for "braking." You can think of the emission of bremsstrahlung radiation as the opposite of the photoelectric effect. In the photoelectric effect light shines on an electron and gives it enough energy to leave a block of matter. With bremsstrahlung radiation, electrons entering a block of matter slow down or stop, turning their kinetic energy into light.

Pair Production and Pair Annihilation

To introduce our last interaction of photons with matter, we need to first introduce the idea of antimatter. We'll just say a few words about it here, but we have an online section that discusses antimatter at greater length: www.cambridge.org/felder-modernphysics/onlinesections

Every type of charged particle has a corresponding "antiparticle" with the same mass and opposite charge. For example, the antiparticle of an electron is a positively charged particle called a "positron." When a charged particle and its antiparticle collide they annihilate, producing radiation.

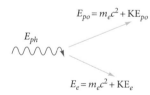

Figure 3.36 Pair production.

The discovery of antimatter[8] allowed the resolution of a mismatch between the predicted and observed rates at which X-rays lose energy as they pass through matter. Taking into account the photoelectric effect as well as Rayleigh and Compton scattering, theoretical calculations predicted too low a rate of energy loss. (Rayleigh scattering doesn't reduce the total energy in X-rays, but it reduces the energy of the direct beam that you measure after the scattering has taken place because the scattered photons are emitted in all directions.) This mismatch was corrected when people added one more process to the calculation.

Here's the new process.

1. We said above that an electron and a positron can collide and annihilate, producing radiation. The reverse of that process is "pair production": a photon spontaneously turns into an electron–positron pair. The initial energy of the photon is converted into a combination of the rest energies and kinetic energies of the new particles (Figure 3.36).

---

8 Antimatter was predicted theoretically by Paul Dirac in 1928 and observed by Carl Anderson in 1932.

Even in the limit where the new particles have zero kinetic energy, the photon must have an energy of at least $E_{min} = 2m_e c^2$ to produce the particles.

2. The newly produced positron inside a block of matter loses energy in collisions, as described in the preceding bremsstrahlung discussion.

3. Eventually the positron loses enough kinetic energy to become electrically bound to an electron. The resulting bound pair, called "positronium," lasts about 10^{-10} s before . . .

4. . . . the electron and positron annihilate each other and produce new radiation. Because of the positron collisions described in item 2 above, this radiation generally has less energy than the original photon that produced the pair.

In the end, the system has no positron and the same number of electrons it started with, but the energy of the initial photon has been scattered among various outgoing photons and the kinetic energy of some scattered particles.

You can show from conservation of momentum that an electron–positron pair can't annihilate into a single photon (see Question 9), so pair annihilation always produces at least two photons. Pair production can turn a single photon into an electron–positron pair provided there's a nearby atom to absorb the excess momentum. (The atom absorbs a negligible amount of the energy.)

Putting it All Together: A Light Beam Passing through Matter

When a beam of light passes through matter it loses energy. Visible light can pass through glass or water with relatively little loss, but loses its energy almost immediately going through opaque substances like wood or metal. X-rays can pass through wood or skin with little loss, but barely penetrate into metal or bone, and actually don't pass through glass very efficiently. The rate at which a light beam loses energy inside matter depends in complicated ways on the nature of the light and of the matter.

We have discussed three mechanisms that cause a light beam to lose energy as it passes through matter: the photoelectric effect, scattering (Rayleigh and Compton), and pair production. Generally speaking the photoelectric effect dominates for low photon energies, scattering for medium energies, and pair production for the highest-energy photons. The scattering itself is mostly Rayleigh at low photon energies and Compton at higher energies. Where all those cutoffs occur is different for different materials because it depends on the detailed structure of the atoms. Remember, however, that pair production is impossible for $h\nu < 2m_e c^2$ regardless of the material. (We haven't mentioned the absorption of light by atoms and molecules at particular resonant frequencies; we will talk about that in Chapter 4.)

3.6.2 Questions and Problems: Further Photon Phenomena

Conceptual Questions and ConcepTests

1. For each quantity give a brief description of what it represents in this section, and units that could be used to measure it.

(a) $\Delta\lambda$ (Compton scattering)

(b) λ_{min} (bremsstrahlung)

(c) E_{min} (pair production)

2. Explain in your own words why Compton-scattered light comes out with a longer wavelength than the incident light.

3. Why did Compton see two peaks in his wavelength graph for outgoing photons at each non-zero angle? Why was there only one peak at $\theta = 0$?

4. What type of radiation (radio, microwave, infrared, etc.) do you need to use to see the photoelectric effect? Based on that, explain without any calculations why in Compton's experiments with X-rays it was reasonable for him to treat the most loosely bound electrons as free and initially stationary.

5. A free electron cannot absorb a photon and conserve both energy and momentum. So why can a free electron undergo Compton scattering with a photon?

6. When an electron beam strikes a solid target, why is there a minimum wavelength of the emitted radiation?

7. Figure 3.35 shows spectra of bremsstrahlung radiation. Why do all the spectra approach zero in the limit $\lambda \rightarrow \infty$?

8. What observed feature of pair production supports the photon model of light (as opposed to the classical model)?

9. Explain why an electron–positron pair in a vacuum can't annihilate and produce a single photon without violating conservation of momentum. *Hint*: Work in the center-of-mass frame of the initial pair of particles.

10. Before the discovery of pair production, people noted that X-ray streams lost energy moving through matter at a faster rate than could be explained by other processes. Why wasn't this discrepancy observed for visible light passing through transparent matter?

11. Based on what you've learned throughout this chapter, list at least three independent pieces of evidence suggesting that light comes in discrete particles.

12. Based on what you've learned throughout this chapter, describe at least three independent ways in which you could measure the value of h.

13. How hot would you have to make a cavity before radiation escaping that cavity would have a significant chance of liberating electrons from the surface of an aluminum sample?

For Deeper Discussion

14. In a classical view of light, an incoming beam that scatters off an electron will come out at different angles, but always with the same wavelength. In the photon picture, we said this would violate conservation of energy because the electron gains energy so the photon must lose some. Why does the classical scenario not violate conservation of energy?

15. We said that pair production only happens in the presence of atoms, which must absorb the excess momentum in the process. In the reverse case of pair annihilation, however, momentum is conserved because two (or more) photons are produced. Why does pair annihilation happen that way but the reverse process – pair production by two or more photons together – almost never does?

16. A famous director has asked you to be the science consultant for his new science fiction movie. The movie takes place in an alternative universe whose laws are the same as the laws of our universe except that Planck's constant, instead of being on the order of 10^{-34} J s, is 1 J s. Briefly describe what such a universe would be like.

17. We said that a photon can only pair-produce an electron and a positron if it has enough energy to produce their combined mass energy. But in some other reference frame that photon would have been Doppler shifted and would have had a much lower energy. So how can an observer in that reference frame account for the pair production? *Hint*: Remember that a photon by itself in empty space can't pair-produce.

Problems

18. (a) Find the energy and momentum of a 20 nm photon.

    (b) With almost no extra work, find the energy and momentum of a 10 nm photon.

19. Compton used incoming light of 0.0709 nm for his experiments. Calculate the percentage change in that wavelength for Compton-scattered light measured at $0°$, $45°$, and $90°$.

20. If you do a Compton scattering experiment and measure a wavelength shift of 10^{-11} m at $90°$, what shift would you measure at $180°$?

21. You shine light of a single wavelength on a target and measure scattered light perpendicular to the incoming beam at two different wavelengths. The higher wavelength you measure is at 0.0612 nm. In that same experiment you measure the light coming out at $150°$ from the original and you once again measure it at two different wavelengths. What are they?

22. The third panel of Figure 3.33 on p. 163 shows two peaks for light scattered at $90°$. The lower peak is at $\lambda = 0.0709$ nm and the upper one is at 0.0731 nm. For the photons that emerged with the higher of those two wavelengths, find the kinetic energy and angle at which the electrons were scattered.

23. Suppose a gamma ray photon scatters off a free electron and comes straight back in the direction it came from. Find the energy of the backscattered photon. *Hint*: The answer is essentially the same for any incoming gamma ray, regardless of its energy.

24. Plot the change in wavelength of a photon scattered off a free electron as a function of scattering angle, from $0°$ to $180°$.

25. A photon with wavelength 0.12 nm scatters off an electron, which then collides with an atom and produces a new photon. Find the minimum possible wavelength of the new photon. *Hint*: It's not 0.12 nm.

26. A beam of electrons, each with kinetic energy 100 keV, is fired into a solid target. Find the minimum wavelength of the emitted radiation. What type of radiation does that wavelength correspond to?

27. A beam of electrons strikes a target and emits radiation with minimum wavelength 0.1 nm. What was the kinetic energy of the electrons before they hit the target?

28. A cloud of electrons starts with negligible kinetic energy and is accelerated through a potential difference before striking a target. How large a potential difference would you need in order for them to emit X-rays when they hit the target?

29. Suppose a 30 keV electron strikes a carbon atom and emits a 15 keV photon. The largest energy that the carbon atom could have would occur if the electron and photon both came out at $180°$ from the electron's original path. Use conservation of momentum and energy to find the carbon atom's final kinetic energy in this case. Express your answer as a fraction of the photon's energy. (You can safely treat the electron and the carbon atom non-relativistically.) Explain why we don't take the atom's recoil energy into account in discussing bremsstrahlung radiation. *Hint*: Be careful when you do your algebra to find the solution where the electron scatters backward, not the one where it scatters forward.

30. What is the minimum energy of a photon that can undergo pair production? Express your answer in eV, and also say what band of radiation it's in.

31. A photon with wavelength 10^{-13} m produces an electron and a positron. Assuming they fly off in opposite directions with equal kinetic energies, find their final speed. (Remember that if the speed is comparable to the speed of light you have to use the relativistic expression for energy.)

32. Rank the following in order from lowest to highest wavelength.

    A. The highest-wavelength (lowest-energy) photon that can photoliberate an electron from nickel

    B. The highest-wavelength (lowest-energy) photon that can pair-produce an electron and positron

    C. The lowest-wavelength (highest-energy) photon that can be produced from Compton scattering at $180°$ (meaning it comes straight back in the direction it came from)

In Problems 33–36 you will fill in missing steps in the Explanation (Section 3.6.1).

33. **The Mass of a Photon**

    (a) Write the total energy of a relativistic particle of mass m and speed v. Feel free to use Appendix B as a resource.

    (b) What happens to that total energy in the limit $v \to c$?

    (c) A photon has speed c and energy $h\nu$, which seems to contradict your answer to Part (b). What must be true of the photon's mass to resolve that contradiction? Explain how your answer solves the problem.

34. **The Compton Shift**

    In Compton scattering an incoming photon with momentum p_0 and energy E_0 strikes an electron at rest. The photon leaves at an angle θ with momentum and energy p_f and E_f and the electron at angle ϕ with momentum and energy p_e and E_e (Figure 3.37). The initial total energy is $m_e c^2 + E_0$ and the final energy is $E_f + E_e$.

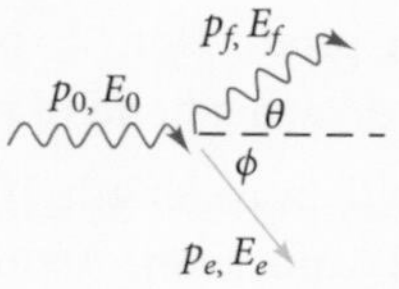

    Figure 3.37

    (a) Write an equation setting initial energy equal to final energy. Replace the final electron energy E_e with the relativistic expression $\sqrt{p_e^2 c^2 + m_e^2 c^4}$ and solve for p_e^2.

    (b) Write equations setting the initial momentum equal to the final momentum in the x and y directions. (Take the x direction to be the direction in which the photon is initially moving.) Solve one equation for $p_e \sin \phi$ and the other for $p_e \cos \phi$ and then square both equations and add them to get an equation for p_e^2.

    (c) You should now have two equations for p_e^2. Set them equal. You should have E_0 and E_f on one side and p_0 and p_f on the other.

    (d) For a photon, $E = hc/\lambda$ and $p = h/\lambda$. Rewrite your equation from Part (c) in terms of λ_0 and λ_f and solve for the difference $\Delta\lambda = \lambda_f - \lambda_0$. Your final solution should depend only on h, m_e, c, and θ.

35. Equation (3.11) gives the change in wavelength of a photon scattering off a free electron. Find the final frequency of the photon as a function of the initial frequency. (In the case of wavelengths, Equation (3.11) shows that final wavelength is just the initial wavelength plus a shift that is independent of the wavelength. The final frequency will depend on the initial one in a somewhat more complicated way. This is why the Compton shift is expressed in terms of wavelength.)

36. **Rayleigh Scattering**

    In Rayleigh scattering the incoming light effectively scatters off an entire atom instead of a single electron.

    (a) How many times smaller is the Compton shift for light scattered off a carbon atom (Rayleigh scattering) than light scattered off a single electron (Compton scattering)?

    (b) For incoming light of 0.0709 nm, calculate the percentage change in the wavelength of Rayleigh scattered light at $135°$.

37. **Exploration: Rube Goldberg Plays with Photons**

    A photon hits a free electron in a lead target and scatters off at a $45°$ angle. The electron recoils into an atom and releases a new photon. That photon liberates an electron from the surface of the lead (work function 4 eV) and sends it flying across an evacuated tube to a carbon target, where it hits an atom and releases yet another photon. That photon decays into an electron–positron pair. Find the minimum possible frequency of the initial photon.

Chapter Summary

This chapter has presented a variety of experiments that point to the same conclusion: light, although it displays many of the properties of a classical wave, comes in discrete packets of energy that cannot be subdivided. The clearest demonstration of this phenomenon (although not the first historically) is the one-particle-at-a-time Young double-slit experiment.

The most important equations for this chapter are in Appendix C (Quantum Mechanics Equations), Appendix D (The Electromagnetic Spectrum), Appendix E (Interference and Diffraction), and Appendix F (Properties of Waves). A timeline of major developments in quantum mechanics is given in Appendix A.

Section 3.1 Math Interlude Interference

When two waves are in phase as they reach a certain point in space, their amplitudes add ("constructive interference"). If they reach that point 180° out of phase with each other, their amplitudes subtract ("destructive interference").

- Two waves with identical frequencies, starting in phase and ending at the same point, will constructively interfere if their path lengths are the same or differ by an integer number of wavelengths. They will destructively interfere if their path lengths differ by a half-integer number of wavelengths.

- The words and letters used to characterize waves in space and time are summarized in Appendix F.

Section 3.2 The Young Double-Slit Experiment

In the Young double-slit experiment, light from a single source passes through two small slits in one wall and is measured on a second wall behind it.

- Thomas Young's 1804 paper described seeing a pattern of alternating constructive and destructive interference bands on the back wall, conclusively demonstrating the wave nature of light.

- The double-slit experiment and other experiments in interference and diffraction are described, and their equations summarized, in Appendix E.

Section 3.3 One Photon at a Time

It is possible today to create a beam of light so dim that it appears on a wall, not as a fuzzy halo or as alternating bands, but as one spot of light in one location at a time. This demonstrates the existence of particles of light, called photons.

If you run such a light source through a double-slit apparatus like the one used by Young, those individual spots of light gradually build up an interference pattern. In locations of destructive interference, no photons ever hit.

- The light seems to be in only one place at once (as evidenced by the individual spot on the wall), but seems to be interfering with itself (as evidenced by the regions of destructive interference).

- The most common modern explanation is that the light spreads out as a wave of probabilities – a "quantum mechanical wavefunction" – that obeys the traditional mathematics of waves and interference. But when the location of the particle is measured (by the back wall), the wavefunction "collapses" into one guaranteed position (the spot).

Section 3.4 Blackbody Radiation and the Ultraviolet Catastrophe

Historically, the first suggestion that light is quantized came in response to the "ultraviolet catastrophe," a classical calculation that says that the energy density in an enclosed cavity should be infinite. Max Planck showed in 1900 that the prediction could be brought in line with experimental results (which showed finite energy) if the light in the cavity can only come in discrete packets.

- Planck's theory matched the data by making the assumption that each light wave of frequency v must have energy nhv with integer n. He seems to have viewed this as a mathematical trick, but in hindsight we see it as evidence that the light is made of indivisible photons, each with energy hv.

Today we speak of the nature of light as "wave/particle duality." Consider the relationship of frequency to energy in the two models.

- In the classical (Maxwell) description, the energy of a light wave is a function of its amplitude. You can change the energy without changing the frequency, or vice versa.

- In the particle description, the energy of a photon is proportional to its frequency. But the energy of a beam of light depends on both the frequency and the number of photons. So you can still change frequency and energy independently, subject to the restriction $E = nhv$.

Using that quantization hypothesis and the "Boltzmann distribution," Planck was able to derive the correct spectrum of light in a cavity, meaning the amount of energy in each range of frequencies. (The Boltzmann distribution is presented in this section without justification; we will derive and explain it in Chapter 10.)

Section 3.5 The Photoelectric Effect

The photoelectric effect involves shining light onto a metal to excite electrons sufficiently for them to escape the surface of the metal. Einstein showed how the results of such experiments can be predicted by assuming that each excited electron absorbs a single photon of energy hv.

- If the frequency of light is too low then no electrons get excited, regardless of the amplitude.

- When you turn the light on, electrons begin flowing immediately (no time lag).

- Neither of those results can be explained by the classical wave theory of light.

Section 3.6 Further Photon Phenomena

- In Compton scattering, a photon is incident upon an almost free electron. The photon loses energy (increases its wavelength) precisely as quantum calculations predict.

- In bremsstrahlung radiation, a beam of electrons strikes atoms in a target. An electron can give up some of its energy in a collision, creating a photon. The wavelengths of the emitted light correctly match quantum calculations.

- In pair production, a photon turns into an electron and a "positron" (the antimatter equivalent of an electron, with a positive charge). The positron loses energy in collisions before annihilating both itself and an electron, causing a measurable energy loss.

4

The Quantum Revolution II: Matter and Wavefunctions

Chapter 3 presented a series of experiments that demanded serious changes to classical physics. These experiments pointed to particle-like behavior – quantization – of electromagnetic radiation.

We will begin this chapter with another such experiment, the spectra radiated and absorbed by hydrogen atoms. Like blackbody radiation, the photoelectric effect, and other phenomena, the hydrogen atom pointed toward quantized energy levels. And like the other problems, this one was solved by an ad hoc hypothesis that fitted the data but left larger questions unanswered. That hypothesis by Niels Bohr forms the "Bohr model" of the atom, a halfway point between classical and quantum models.

The first hints of a broader theory came when Louis de Broglie proposed another ad hoc hypothesis: "If waves act like particles, then maybe particles act like waves too. Why not?" The idea of matter waves eventually became a fully developed theory of *wavefunctions* that describe the state of a particle.

So we will spend most of this chapter exploring the idea of a wavefunction that describes a particle's position, momentum, and other properties. In Chapter 5 we will take up the math that allows us to calculate the wavefunctions of particles in various situations. And in Chapter 7 we will return to the hydrogen atom, and see how wavefunctions both explain and modify Bohr's original picture.

4.1 Atomic Spectra and the Bohr Model

Early twentieth-century models of the atom faced two major difficulties. In this section we will explain what those difficulties were, and how a model of the atom proposed by Niels Bohr resolved them. We'll end by describing some problems that Bohr's model didn't solve, and that will point us toward the actual theory of quantum mechanics.

4.1.1 Explanation: Atomic Spectra and the Bohr Model

In 1897, J. J. Thomson showed that atoms contain electrons, negatively charged particles much smaller and lighter than the atoms themselves. Since atoms are generally neutral, there had to be something to cancel that negative charge, and to account for most of the mass of the atoms. In 1904 Thomson proposed that the electrons in an atom were embedded in a uniform spherical distribution of positive charge. This description is called the "plum pudding" model of the atom because the electrons are distributed in the atom like raisins in a plum pudding.[1]

That model was conclusively disproven in 1911 when Hans Geiger and Ernest Marsden, under the direction of Ernest Rutherford, fired high-energy alpha particles (doubly ionized helium atoms) at a thin gold foil. We will discuss Rutherford's experiment in more detail in Chapter 12. Here we want to focus on the most important result of the experiment.

Electrons are too light to significantly deflect alpha particles, so any observed deflection resulted from interactions with the positive charges in the atom. The thinly spread out positive charge of Thomson's plum pudding model would have produced very small deflections, and for the most part that is what Rutherford observed. But some alpha particles recoiled at angles greater than 90 degrees, which would have been essentially impossible in Thomson's model. Rutherford wrote that "It was as incredible as if you fired a 15-inch shell at a piece of tissue paper and it came back and hit you."

Rutherford concluded that instead of being embedded in a diffuse medium of positive charge, the electrons must orbit about a tiny ball of positive charge that he called the "nucleus" (Figure 4.1). An alpha particle that happens to strike almost exactly next to this small, dense nucleus experiences a large force and can recoil at a large angle.

In Rutherford's experiment only about 1 alpha particle in 10,000 scattered at angles greater than $90°$, but in the plum pudding model the odds of an alpha particle doing so would have been roughly 1 in 10^{3000}! Rutherford's calculations accurately reproduced the angular distribution of scattered alpha particles, confirming his nuclear model.

But Rutherford's model faced two important problems. We can explain the first problem very quickly, but the second one takes more buildup.

The first problem is stability. As you know, any orbiting object has a centripetal acceleration. Electromagnetic theory states that accelerating charges emit radiation and lose energy, so an orbiting electron should rapidly spiral into the nucleus. In fact, Maxwell's equations predict that a Rutherford-style hydrogen atom should last about 10^{-12} s. This conflicts with the observation, widely accepted in the scientific community, that "matter exists at all."

The second problem is atomic spectra. We'll explain what those are and then circle back to the problem they posed for Rutherford's atoms.

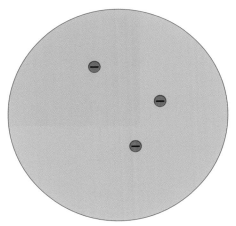

Thomson's "plum pudding" atom

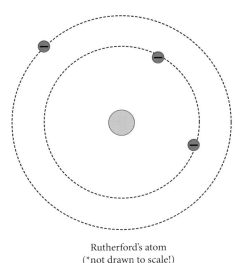

Rutherford's atom
(*not drawn to scale!)

Figure 4.1 Models of the atom.

---

1 Yes, plum puddings are made with raisins, not plums. It's a British thing.

Atomic Spectra

If you run a large electrical current through a tube filled with gas, the current gives energy to the atoms of that gas. The atoms then release that energy as electromagnetic radiation. If you put that radiation through a prism you find that it is not a uniform rainbow of colors. There are thin lines of very particular frequencies, and nothing is emitted in any of the frequencies in between. These colored lines are called "emission lines." Each element emits a unique pattern known as its "emission spectrum."

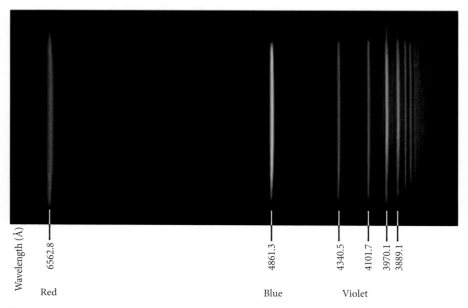

Figure 4.2 The Balmer series of emission lines from hydrogen. *Source: © Schawlow.*

Figure 4.2 shows the emission lines of hydrogen in the visible range.[2] This particular spectrum is called the "Balmer series" after Johann Balmer, who found an empirical formula for the frequencies of the lines. In 1888 Johannes Rydberg wrote a general formula that included Balmer's lines and predicted many others beyond the visible range.

In reading the formulas in the box below, keep a couple of definitions in mind. The "frequency" ν of a wave is one over the oscillation period. The "spatial frequency" f is one over the wavelength. We'll discuss these terms and their relationships in Section 4.2.

The Rydberg Formula for the Emission Lines of Hydrogen

The spatial frequencies of the emission lines of hydrogen are given by

$$f = \frac{1}{\lambda} = R_H \left(\frac{1}{n_1^2} - \frac{1}{n_2^2} \right). \tag{4.1}$$

Here n_1 can be any positive integer and n_2 can be any integer larger than n_1.

---

2 All but the first four are technically considered "ultraviolet," but several beyond that are close enough to the border that they can actually be seen by the human eye.

$R_H = 1.10 \times 10^7 \text{ m}^{-1}$ is the "Rydberg constant" (or sometimes the "Rydberg constant for hydrogen").

The frequency (defined as one over the oscillation period) of each line is $\nu = cf$.

The Balmer series consists of the lines with $n_1 = 2$.

In 1908 Friedrich Paschen measured the hydrogen spectrum in the infrared and found precisely the frequencies predicted by Rydberg's formula with $n_1 = 3$. Between 1906 and 1914 Theodore Lyman measured the $n_1 = 1$ series in the ultraviolet. These three series are shown in Figure 4.3.

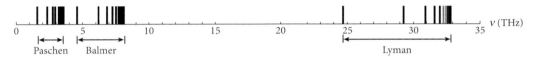

Figure 4.3 **Emission lines from hydrogen.**

There is another way in which you can see spectral lines. If you shine white light through a tube filled with hydrogen gas you find that most of the light makes it through the tube undisturbed, but certain colors are missing from the spectrum on the far side. The missing colors, called "absorption lines," represent some but not all of the emission lines. For example hydrogen at room temperature absorbs all the colors in the Lyman series, but nothing from the other series.

Do you remember that we promised to describe two problems with Rutherford's atom? We're ready now: the first problem is its stability (as discussed above), and the second problem is the spectral lines. Thomson's plum pudding atoms would emit light at one characteristic frequency (see Problem 22). Rutherford's electrons would radiate during their brief death spirals. But neither model offers any explanation of why an atom would emit or absorb these particular frequencies and no others.

The Bohr Model

In 1913 Niels Bohr (who was working in Rutherford's lab at the time) developed a model that solved both of the problems we described above. Bohr's model described electrons orbiting a nucleus in circular orbits – much like planets orbiting a star, but captured by the Coulomb force rather than by gravity. (Circular orbits make calculations easier, but you can reproduce the essential features of Bohr's model if you include elliptical orbits as well.) But Bohr also postulated that electron orbits, unlike planetary orbits, are quantized. Specifically:

- An electron can only be in an orbit with angular momentum equal to $nh/(2\pi)$ for positive integers n.
- Although an electron cannot exist in any orbit in between these, it can jump discontinuously between allowed orbits. Such a jump will emit or absorb a photon whose energy is precisely the difference between the energies of those two orbits.

A classical electron in orbit would spiral in, passing through all intermediate orbits and emitting radiation all the way. Bohr simply postulated that, contrary to Maxwell's equations, electrons in these discrete allowed orbits don't radiate. That takes care of the stability problem.

But the greatest triumph of the Bohr model was reproducing the Rydberg formula for spectral lines. To see how this comes about, we have to consider some implications of Bohr's hypothesis.

Active Reading Exercise: Quantization in the Bohr Model

Bohr's model explicitly says that the angular momentum of the electron orbits is quantized. That assumption implies that some other properties of the orbits must be quantized as well. Name two other properties and explain how quantization of angular momentum (for circular orbits) implies that they must also be quantized.

Don't read further until you have written down at least two responses.

The short answer is that virtually everything about the orbits must be quantized because once you have specified the angular momentum of a circular orbit you know pretty much everything about it: the radius, the velocity, the kinetic and potential energies, and so on. Just about the only things you *can't* calculate are the plane of rotation and the phase (which gives the position of the electron in its orbit at some specified moment).

Before we turn that insight into math, we need a tiny detour into notation. Because Planck's constant appears in the combination $h/(2\pi)$ in Bohr's hypothesis, that combination shows up throughout the calculations. In fact it shows up so much throughout quantum mechanics that it gets its own symbol, an h with a little line through it:

$$\hbar \equiv \frac{h}{2\pi} \quad \text{The reduced Planck constant (pronounced "h-bar").}$$

Okay, back to physics. You can show in Problem 23 that Bohr's rule for angular momentum leads to the following formulas for the energy and radius of the allowed orbits:

$$E = -\frac{m_e e^4}{2(4\pi \epsilon_0)^2 \hbar^2} \frac{1}{n^2} = -\frac{1 \text{ Ry}}{n^2} \tag{4.2}$$

$$r = \frac{4\pi \epsilon_0 \hbar^2}{e^2 m_e} n^2 = a_0 n^2 \tag{4.3}$$

The energy unit[3] "rydberg" (Ry) is roughly equal to 13.6 eV. The distance a_0, called the "Bohr radius," is about half an angstrom, or 5×10^{-11} m. So the model predicts that the lowest-energy state of hydrogen has angular momentum equal to $\hbar$, an energy of roughly -13.6 eV, and a radius of about half an angstrom. (In quantum mechanics the lowest possible energy level of a system is called its "ground state.")

Pause and see if you can figure out why the energy came out negative. (It's not enough to say that it's because the formula for potential energy has a minus sign in it. The question is, what does that tell you physically about the atom?)

Here's the answer: it's the potential energy that makes it negative. You define the potential energy of any system to be zero anywhere you choose, and then measure it everywhere else relative to that point. For Coulomb's law (just like for Newton's law of gravity) it's traditional to set the potential energy to 0 when the electron is infinitely far away from the nucleus.

---

3 Don't confuse the energy unit "rydberg" with Rydberg's constant, R_H, which has units of one over distance, not energy. Rydberg was an important guy.

Because the potential energy increases as you move away from the nucleus (just as it does as you move upward on Earth), that makes the potential energy negative at all finite distances. Figure 4.4 shows the potential energy "well" represented by the nucleus, drawing the electron inward.

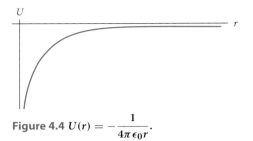

Figure 4.4 $U(r) = -\dfrac{1}{4\pi\epsilon_0 r}$.

The kinetic energy is of course positive by definition. If the total energy $E = U + K$ is positive, the electron has enough energy to leave the atom completely, so for any orbiting electron E will be negative. The $n = 1$ state has the lowest (most negative) energy, and energy levels for higher n-values grow less negative.

The Bohr Model and Atomic Spectra

The Bohr model says that an electron cannot gradually lose energy; it has to decay discretely from one level to another. The emission lines represent the energy differences between levels. You can show in Problem 25 that an electron dropping from level n_2 to level n_1 will emit photons with the frequencies predicted by the Rydberg formula. The Lyman series comes from electrons dropping from any of the higher levels into the ground state, the Balmer series comes from electrons dropping from higher levels into $n = 2$, and so on.

That also explains why the absorption spectrum matches the Lyman series but not the other ones. At room temperature essentially all hydrogen atoms are in the ground state. So if you shine light through that hydrogen, the atoms will absorb all the photons that have the right energy for transitions between $n = 1$ and higher n, which is the Lyman series.

It's worth noting that when Bohr formulated his model in 1913 only the Balmer, Paschen, and parts of the Lyman series had been measured. The subsequent completion of the Lyman series in 1914 and measurements of the Brackett ($n_1 = 4$) series in 1922 and Pfund ($n_1 = 5$) series in 1924 were all consistent with Rydberg's mathematical formula and Bohr's physical model.

Other Hydrogen-Like Atoms

The Bohr model can be applied to other "hydrogen-like" atoms, meaning atoms with any number of protons or neutrons but just one electron. This includes ions of other elements where all but one electron has been removed (e.g. singly ionized helium or doubly ionized lithium). It also includes isotopes of hydrogen and of these other ions, meaning atoms with more or fewer neutrons than usual in the nucleus.

In the Bohr model, the only difference between hydrogen and these other atoms is that the Coulomb attraction is stronger when there are more protons. For a hydrogen-like atom with Z protons, the Bohr model predicts that the nth orbit has energy $E = (-1\,\text{Ry})Z^2/n^2$. For instance, the spectral lines of ionized helium ($Z = 2$) should have energies exactly 4 times greater than the hydrogen lines.

Actual measurements of ionized helium spectral lines showed energies 4.0016 times greater than those of hydrogen, suggesting that the model was close but not perfect. Shortly after publishing his model Bohr amended it to include the motion of the nucleus. By analogy, think about the Earth orbiting the Sun. We typically imagine the Sun as stationary and the Earth moving, which is an excellent approximation because the Sun is so much more massive than the Earth. But in reality they both orbit about their mutual center of mass (which is deep inside the Sun).

It was known from classical mechanics that the exact orbit of a two-body system can be found by treating it as a single orbiting object with a "reduced mass" $\mu = m_1 m_2/(m_1 + m_2)$. If $m_2 \gg m_1$ then this approximately equals m_1, so for an atom it's a good approximation to just use the electron mass. Close, but not perfect. When Bohr corrected his model with the reduced mass of the electron/nucleus system he perfectly reproduced the measurements (see Problem 24), including the ratio of helium to hydrogen spectral line frequencies.

The Franck–Hertz Experiment

In 1914 James Franck and Gustav Hertz performed an experiment to test the quantization of atomic energy levels. They used a potential to accelerate electrons through a tube filled with mercury gas. At the other end of the tube the electrons were slowed slightly by a potential in the opposite direction and then collected on a plate (the anode in Figure 4.5). An ammeter leading from that plate measured the rate at which electrons reached it.

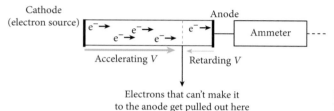

Figure 4.5 Schematic of the Franck–Hertz experiment.

The significance of the small retarding potential at the end is that if an electron were to collide with a mercury atom on the way and lose most of its energy then it wouldn't make it past the retarding potential and wouldn't be measured in the ammeter. The point was to measure under what circumstances this would occur.

As the accelerating potential was increased the electrons flowed across the tube faster and the current increased. As that potential passed 4.9 V, however, the current suddenly dropped dramatically (Figure 4.6). What's so important about 4.9 V? An electron accelerated through a 4.9 V potential acquires 4.9 eV of energy, which happens to correspond to the difference between two energy levels of mercury (as measured in emission spectra). So when the electrons had less

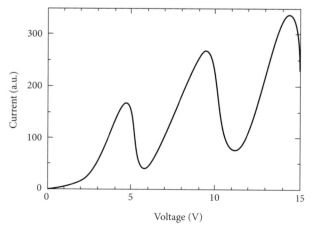

Figure 4.6 Measured current as a function of accelerating potential in the Franck–Hertz experiment.
Source: Ahellwig 2006.

energy than 4.9 eV they couldn't excite the mercury atoms, but as soon as they had that much energy they could collide with a mercury atom and give it all their energy, thus failing to make it to the ammeter.

Those excited mercury atoms should then drop to the ground state and emit radiation. Sure enough, subsequent experiments showed that when the voltage was 4.9 V there was ultraviolet light being emitted at the far right of the tube, where the electrons were colliding with the mercury atoms.

When the potential was raised even higher the current began to rise again. Electrons were still colliding with mercury atoms and losing all their energy, and that was detectable from UV light emission. But now that UV light was coming from closer to the middle of the tube, at the point where the electrons reached kinetic energies of 4.9 eV. From there they could get accelerated again and contribute to the current.

When the potential was raised to 9.8 V the current dropped again. This time there was UV emission from two spots, one in the middle of the tube and another at the far right. This suggests that each electron would accelerate to 4.9 eV, give all that energy to a mercury atom, accelerate to 4.9 eV *again*, and then lose all its energy in a second collision.

The conclusion of this experiment is that an electron can give exactly 4.9 eV of energy to a mercury atom, but it can't give the atom any smaller amount. The Franck–Hertz experiment provided dramatic independent evidence of the quantized energy levels of electron orbits.

Einstein reportedly said of these results "It's so lovely it makes you cry," and Franck and Hertz got the 1925 Nobel Prize in physics for their work. Today spectral lines and Franck–Hertz experiments are used to independently verify the energy levels of atoms.

The Correspondence Principle

The fundamental nature of the universe may be counterintuitive and relativistic and quantized, but our daily experience is decidedly Newtonian. Grains of sand, billiard balls, and planetary systems all seem to function according to $F = ma$.

In relativity we can explain the Newtonian nature of day-to-day life by looking at $\gamma = 1/\sqrt{1 - v^2/c^2}$. It is mathematically apparent that for $v \ll c$ this formula reduces to 1, which turns Einstein's equations into Newton's.

As Bohr proposed his own radical new paradigm, he knew that he also had to account for all the cases where classical physics worked fine. He thus suggested a "correspondence principle": in circumstances where classical physics is known to work experimentally, quantum predictions must reduce to the same predictions as classical theory. For example, Planck's quantization hypothesis about light is consistent with known observations because normal light beams consist of so many photons that the quantization is undetectable.

Bohr identified the classical limit with large values of n. All of the quantization rules, including Planck's and Bohr's, suggest that some quantity that in classical physics is continuous comes in discrete levels labeled by some integer n. Such integer labels are now called "quantum numbers." Bohr's proposal was that, for large quantum numbers, quantum and classical predictions should agree. In Problem 26 you can show that a hydrogen atom decaying from a very high energy level toward lower levels emits approximately the same spectrum of radiation that a classical electron would emit as it spiraled in toward the nucleus. As n gets close to 1 that approximation stops being valid.

Limitations of the Bohr Model

The Bohr model solved the stability problem of the Rutherford atom and explained the emission and absorption lines of hydrogen and other hydrogen-like atoms. Its discrete energy levels were verified in the Franck–Hertz experiment. Nonetheless, the model has a number of limitations that were known at the time it came out.

- The model correctly predicts the energies of photons released in atomic transitions, but it makes no predictions about how long an atom will remain in an excited state or what lower states it is most likely to go into when it decays. These values are measurable via the intensities of different spectral lines, but the Bohr model has no explanation for them.

- Careful measurements of spectra show that each line is divided into multiple lines with nearly identical wavelengths. That suggests that each energy level in the Bohr model actually represents several possible, very similar energies that an electron might have. The Bohr model can't account for this splitting of the spectral lines.

- The Bohr model doesn't account for interactions between electrons, so it only works for single-electron atoms. It can be extended approximately to some atoms in which all but one of the electrons are tightly bound to the nucleus and a single electron is more loosely bound, but for most atoms it simply doesn't work.

- Finally, the Bohr model, like so many other quantum hypotheses around that time, was an ad hoc solution to a particular problem rather than a full theory of how electrons (and other particles) behave.

That last point applies to many of the early hypotheses that make up what we now call "old quantum theory." Planck's hypothesis solved the ultraviolet catastrophe, Einstein's extension of it explained the photoelectric effect, Bohr explained atomic spectra, and de Broglie's model (which we'll explain in Section 4.2) predicted electron diffraction, but none of these models gave an overall description of how to apply these new rules to all particles in all situations. In 1926 Erwin Schrödinger published a theory that encompassed all of these previous hypotheses in one overarching framework, which we now call "quantum mechanics." We will begin to explore that theory later in this chapter.

4.1.2 Questions and Problems: The Bohr Model

Conceptual Questions and ConcepTests

1. Why would you expect large angle recoils to occur in the Rutherford model of the atom and not in the plum pudding model?

2. Why can't n equal zero in the Bohr model?

3. In 1908, Walter Ritz formulated the "Ritz combination principle", which says that many emission lines of any given element will have frequencies that are the sum of two other emission lines for that same element. How does the Bohr model explain this principle?

4. At room temperature the absorption spectrum of hydrogen matches the Lyman series. Why does extremely hot hydrogen also absorb radiation at the colors in the Balmer series?

5. If you shine light through a tube of hydrogen you find that the light on the other side is missing the colors of the Lyman spectrum. If you put detectors all around the tube, however, you find that the hydrogen is emitting light in all directions. At what frequencies would you expect that light to be emitted? Just at the Lyman frequencies? At

all the emission frequencies of hydrogen? At all frequencies? Explain.

6. Which of the following are true of the spectral lines of hydrogen? (Choose all that apply.)

A. There are infinitely many emission lines.

B. There is a highest frequency at which hydrogen can emit.

C. There is a lowest frequency at which hydrogen can emit.

D. All of the frequencies of absorption lines are also frequencies of emission lines.

7. (a) Two comets are both passing by the Sun. The total energy of Comet A (meaning kinetic plus potential, with the zero of potential at infinity as usual) is positive. The total energy of Comet B is negative. Briefly describe the difference in the long-term behaviors of these two comets.

(b) The energy of an allowed orbit in the Bohr model is negative. Briefly describe what this fact implies for the long-term behavior of an electron in such an orbit.

8. Careful measurements show that each series in the hydrogen spectrum also includes a faint continuous spectrum of emitted radiation at frequencies even higher than the highest spectral line in the series. Explain why that occurs.

9. Why do heavier elements like neon not follow Rydberg's formula for emission lines?

10. Suppose you perform a Franck–Hertz experiment on a gas of the element maduponium and you measure the output current as a function of voltage, as shown in Figure 4.7.

Which of the following would you conclude? (Choose one and explain.)

A. Maduponium has a 2 eV transition (meaning it has two electron states separated by 2 eV).

B. Maduponium has a 4 eV transition.

C. Madupomium has both a 2 eV transition and a 4 eV transition.

D. You cannot conclude that maduponium has either of these transitions.

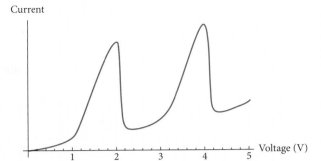

Figure 4.7

For Deeper Discussion

11. One result of Rutherford's experiments was that the number of recoils observed at very large scattering angles was proportional to t, the thickness of the foil. Explain why this is the dependence you would expect in the Rutherford model of the atom. Then suggest at least one good reason why the t dependence might be different in the Thomson model.

12. In the Rutherford model the atom should be unstable and the electrons should all collapse into the nucleus. The result would be something like the plum pudding atom, only much smaller. If atoms were nucleus-sized Thomson-like plum puddings, would you expect to see the large-angle recoils that Rutherford saw? Why or why not?

13. Bohr only considered circular orbits. How would his model change if you also allowed elliptical orbits with the same quantization rule?

14. In Chapter 3 we said that the spectrum of radiation in a cavity in equilibrium is independent of what the walls are made of. In this section we said that each element when heated emits a characteristic set of spectral lines. So if you have two cavities in equilibrium, one with walls of lead and the other with walls of tin, why isn't the lead one filled with extra radiation at the frequencies of lead spectral lines, and the tin one with extra radiation at frequencies of tin spectral lines?

Problems

15. What is the limiting spatial frequency of the Balmer series? (That's the spatial frequency that the Balmer lines asymptotically approach.)

16. The "Pfund series" is the series of spectral lines emitted when an electron drops from an energy level with $n > 5$ down to the $n = 5$ level. Calculate the largest two wavelengths in the Pfund series for hydrogen.

17. In Chapter 10 we will prove the following result (but right now take our word for it): a significant number of atoms will be in states above the ground state when the energy gap from the ground state to the first excited state is comparable to or less than $k_B T$. In this formula k_B is a fundamental constant ("Boltzmann's constant") and T is the temperature in kelvins. Based on this fact, estimate the temperature at which you would expect to see the Balmer series in the absorption spectrum for hydrogen. (This is observed when measuring light passing through the outer atmosphere of some stars.)

18. A "muon" is an elementary particle very similar to an electron but with roughly 200 times as much mass. What would be the longest wavelength of the "Lyman series" (for atoms falling into $n = 1$) for an atom consisting of a muon orbiting a proton? What type of radiation (infrared, visible, ...) would that series fall in?

19. Tritium is an isotope of hydrogen in which the nucleus consists of one proton and two neutrons. By what percentage would the frequencies of the spectral lines of tritium differ from those of ordinary hydrogen? (You can assume protons and neutrons have equal mass, roughly 2000 times the mass of an electron.)

20. The Bohr model tells you the energy of emitted photons, and then from $E = h\nu$ you can calculate their frequencies and from $\lambda = c/\nu$ their wavelengths. In glass, however, the speed of light is about 2/3 of its speed in vacuum. Calculate the highest wavelength in the Balmer series (electrons dropping down to $n = 2$) in vacuum and in glass.

21. Suppose you did a Franck–Hertz experiment with a tube of hydrogen and increased the voltage slowly from $1\,\text{V}$ to $11\,\text{V}$. Sketch the resulting current as a function of voltage that you would expect. *Hint*: Franck and Hertz did their experiment with mercury; an experiment with hydrogen would show qualitatively similar results, but with different numbers.

In Problems 22–26 you will fill in missing steps in the Explanation (Section 4.1.1).

22. We have said that the Thomson model cannot account for experimentally measured emission spectra. What kind of emission does that model predict? Consider a plum pudding hydrogen atom consisting of a single electron embedded in a uniform, spherical distribution with a positive charge equal to that of a proton. Use e as the magnitude of both the electron and the proton charge, m_e as the electron mass, R as the radius of the atom, and r as the electron's distance to the center of the atom.

    (a) Find the force exerted by the "pudding" on the electron as a function of the electron's radius r. *Hint*: Remember that from Gauss' law we know that the positive charges at radii greater than r will exert no force on the electron.

    (b) You should have found a force law of the form $F = -kr$, which is the force law for a simple harmonic oscillator. Using the value you found for k, what is the frequency of oscillations of the electron about the center of the atom?

    If the medium surrounding the electrons exerts a large enough damping force, then the electron will stay stationary and not emit. If not, the electron will oscillate and emit at the frequency you just found. In neither case will it emit the measured spectrum of hydrogen.

23. Bohr's model gave an explicit quantization rule for the angular momentum of an electron orbit. But because he also postulated circular orbits, we can use that rule to show how other electron properties must be quantized. Consider an electron in a classical, circular orbit around a proton at radius r. Treat the proton as stationary. Use e as both the electron charge and the proton charge, and m_e as the mass of the electron.

    (a) Setting the Coulomb force equal to the centripetal force needed to hold the electron in orbit, find the speed v of the electron.

    (b) Find the angular momentum L.

(c) Setting the angular momentum you found equal to $n\hbar$ as per Bohr's postulate, find r, v, and kinetic energy in terms of $\hbar$.

(d) The potential energy of a particle experiencing a Coulomb force is $-q_1 q_2/(4\pi\epsilon_0 r)$. Using the values you found for radius and kinetic energy, express the total energy (kinetic plus potential) in terms of $\hbar$.

If you didn't get the answers we gave in the Explanation (Section 4.1.1), go back and find your mistake.

24. [*This problem depends on Problem 23.*] Consider a singly ionized helium atom, which consists of one electron orbiting a nucleus of two protons and two neutrons.

   (a) Explain why the Bohr model predicts that the emission lines of this helium ion should occur at four times the frequencies of the hydrogen lines. (You could re-do all the calculations you did for Problem 23, but it's easier to look back through those calculations and see how they will change.)

   (b) Find the reduced mass of a hydrogen atom and the reduced mass of a helium atom. Use m_e for the mass of an electron, and m_p for the mass of a proton or a neutron.

   (c) Using those reduced masses instead of m_e, recalculate the ratio of the helium lines to the hydrogen lines. Estimate the numerical ratio using $m_p = 2000 m_e$.

25. In this problem you will use the energy levels of the hydrogen atom to calculate the Rydberg constant.

   (a) Use Equation (4.2) to find the energy lost by a hydrogen atom as it goes from level n_2 to n_1 in the Bohr model. Express your answer in terms of m_e, $\hbar$, and other constants.

   (b) If a photon of that same energy is emitted, what would its spatial frequency be? *Hint:* You may find it easiest to first find its frequency ν, then find its wavelength using $\lambda = c/\nu$, and after that find its spatial frequency.

   (c) Comparing your answer to the Rydberg formula, Equation (4.1), write an equation for the value of the Rydberg constant R_H.

(The equation you just derived is actually only valid in the limit of an infinitely heavy nucleus, and is sometimes written R_∞. See Problem 24 for how to correct it for a nucleus of finite mass.)

26. **The Correspondence Principle**

   (a) An orbit in the Bohr model is always a perfect circle. By setting the Coulomb force of a hydrogen atom equal to the centripetal force of an electron in a circular orbit, find the velocity and then the period of such an orbit. Using Equation (4.3) for the radius of the nth-level orbit in the Bohr model, express the period as a function of the electron charge e, the electron mass m_e, the energy level n, and fundamental constants.

   (b) A classical electron in a circular orbit would emit radiation with frequency equal to one over the period. Calculate the expected frequency emitted by an electron in the nth Bohr orbit of the hydrogen atom, if it were able to emit classical radiation.

   (c) Calculate the frequency of the photon emitted when an electron in the Bohr model drops from the nth level to the $(n-1)$th level. Your answer should once again be expressed in terms of e, m_e, n, and fundamental constants.

   (d) Find the ratio of the Bohr prediction you found in Part (c) to the classical prediction you found in Part (b) and take the limit of that ratio as $n \to \infty$.

   (e) Explain how your results show that a hydrogen atom with very large n would emit a classical spectrum. What assumption do you have to make about its decay in order to reach this conclusion?

27. **The Wilson–Sommerfeld Quantization Rule**

   In 1915–16 William Wilson and Arnold Sommerfeld independently proposed a general quantization rule for oscillatory systems. The Wilson–Sommerfeld rule provides an historical middle step between the ad hoc proposals of Planck and Bohr and the general system of Schrödinger.

   Consider a system described by some coordinate q that varies periodically in time. Any such coordinate has an associated "conjugate momentum"[4] p_q. The most common coordinates are position and angular position, for which the conjugate momenta are ordinary momentum and angular momentum, respectively. The Wilson–Sommerfeld rule says

$$\oint p_q dq = nh.$$

Here $\oint$ means an integral over one full period of the motion.

(a) Picture an electron in a circular orbit around a nucleus. Its angular position ϕ is a periodic function of time. Show that the application of the Wilson–Sommerfeld rule to the electron's angular momentum reproduces Bohr's quantization rule.

(b) For a simple harmonic oscillator, $x = A\cos(\omega t + \delta)$. What are the allowed energies of a simple harmonic oscillator according to the Wilson–Sommerfeld rule? (The simplest way to evaluate the integral is to write p and dx in terms of t.)

(c) An electromagnetic wave in a cavity is a simple harmonic oscillator. Show that your answer can be rewritten as Planck's $E = nh\nu$.

---

4 If you're familiar with Lagrangian mechanics, the conjugate momentum is $\partial L/\partial q$. If you're not, don't worry about it.

4.2 Matter Waves

The one-photon-at-a-time double-slit experiment, Planck's proposed quantization of blackbody radiation, Einstein's explanation of the photoelectric effect... What all of these have in common is that they ascribe quantization (a property classically associated with particles as separate entities) to light (classically considered a wave). The Bohr model of the atom seems to be a different topic: it still involves quantization, but it deals with an electron, classically considered a particle.

Where we're going with all this, eventually, is quantum mechanics. The system developed by Schrödinger and others doesn't divide the world into "particles" and "waves," but instead analyzes mathematical wavefunctions and uses them to predict the results of experiments.

One of the key stepping-stones that leads from all the individual results we have seen to the quantum theory that we're building toward is Louis de Broglie's 1924 proposal that objects classically considered as particles have some of the properties of waves.

4.2.1 Explanation: Matter Waves

By the early 1920s it was clear that light, which had classically been considered a wave, also displays behavior associated with particles. Its frequency and wavelength (properties of a wave) are related to the energy and momentum of a single photon (an indivisible particle) by the equations $E = h\nu$ and $p = h/\lambda$.

In 1924 Louis de Broglie[5] submitted a PhD thesis in which he proposed that matter that had classically been considered as particles, such as electrons and protons, might also display wave behavior. He postulated that their wavelengths were related to the momenta of the particles in the same way that they are for photons.

---

5 His last name is pronounced similarly to "broil."

The de Broglie Relation

Every particle has an associated "matter wave" with wavelength $\lambda = h/p$.

The idea of matter waves has the appeal of symmetry: if a wave (like light) sometimes acts like a particle, then maybe a particle (like an electron) sometimes acts like a wave. There is also a theoretical appeal: based on de Broglie's relation, you can show that the only possible orbits in which an electron can have a standing wave around a nucleus are the ones in the Bohr model (Problem 26).

All that said, what does it mean to "associate" a wave with a particle?

Active Reading Exercise: Matter Waves

Based on what you've learned so far about photons, how would you design an experiment to test de Broglie's hypothesis of matter waves? Try to write down at least one answer before reading on.

The answer, very generally, is that experiments that show the wave-like properties of photons should work for electrons as well. For instance, you might send a beam of electrons through a double slit. In 1961 Claus Jönsson did just that. Figure 4.8 shows the intensity of the electron beam on the back wall of his apparatus.

Double-slit interference has been observed for many particles including electrons, protons, atoms, and "Buckyballs" (molecules made of 60 carbon atoms each). Matter particles have also been shown to display single-slit and many-slit diffraction patterns. But these experiments took place many years after de Broglie's work; the first confirma-

Figure 4.8 Intensity on the back wall of Claus Jönsson's 1961 double-slit experiment with electrons. *Source: Springer Nature © 1961.*

tion of de Broglie's hypothesis came much sooner. In 1926, long before the electron double slit experiment was performed, Clinton Davisson and Lester Germer demonstrated diffraction with electrons. Shortly afterward the same feat was accomplished independently by G. P. Thomson.[6]

---

6 The 1937 Nobel Prize in physics was shared by Davisson and Thomson, and it's something of a mystery why Germer was left out. The year Davisson and Thomson won, the Nobel prize committee received four nominations for electron diffraction, all of which recommended sharing the prize between Davisson and Germer.

Example: The Davisson–Germer Experiment

Davisson and Germer performed the first clear test of de Broglie's hypothesis of matter waves. They accelerated a beam of electrons through a 54 V potential and then bounced them off a nickel crystal. The electrons scattered off the atoms on the surface of the crystal, which acted as sources of new beams, like the slits in a diffraction grating.

Detector

The nickel atoms were spaced about $w = 0.215$ nm apart. Davisson and Germer measured the intensity of the electron beam at different angles and found a strong peak at $\phi = 50°$.

1. The lowest non-zero angle at which a maximum occurs from a diffraction grating with slit spacing w is given by $w \sin\phi = \lambda$ (see Problem 25). Use this formula to estimate the wavelength of the electrons.

2. Does it match the predicted de Broglie wavelength?

Answer:

1. $\lambda_{\text{measured}} = (0.215 \times 10^{-9}\text{ m}) \sin(50°) = 1.65 \times 10^{-10}\text{ m}$

2. The electrons acquire energy $54\text{ eV} = 8.65 \times 10^{-18}$ J, so their speed is $v = \sqrt{2E/m_e} = 4.36 \times 10^6$ m/s. (This is barely over 1% the speed of light so the use of non-relativistic formulas is reasonable.)

$$\lambda_{\text{predicted}} = \frac{h}{p} = \frac{h}{m_e v} = \frac{6.63 \times 10^{-34}\text{ m}^2\text{ kg/s}}{(9.11 \times 10^{-31}\text{ kg})(4.36 \times 10^6\text{ m/s})} = 1.67 \times 10^{-10}\text{ m}.$$

The two values agree to within the measurement error of the experiment.

In 1965 Richard Feynman proposed the idea of doing a double-slit experiment with electrons one particle at a time. Think back to Chapter 3 and try to predict what you would see if you did that.

The answer that Feynman predicted was exactly what we described for photons in Chapter 3. Each electron should produce a single dot on the back wall. As the dots build up over time they should form a double-slit interference pattern, with lots of electrons hitting around the "bright bands" and few hitting around the "dark bands." This combination of showing up one at a time and building up an interference pattern shows that each electron is passing through both slits as a wave, but being measured on the back wall as a particle.

This experiment was first carried out in 1974 by Merli, Missiroli, and Pozzi. In a 2002 survey of readers of *Physics World*, the one-electron-at-a-time double-slit experiment was chosen as the all-time "most beautiful experiment in physics."

You can see a beautiful video of an actual running of the one-electron-at-a-time experiment at www.cambridge.org/felder-modernphysics/animations

The Other de Broglie Relation

Sometimes people use "de Broglie relations" in the plural: the $\lambda = h/p$ that we cited, and also the equation $E = h\nu$ relating the total relativistic energy of a particle to its frequency. (Note that the first involves a wavelength in space, the second a frequency in time.) For photons these two equations are equivalent. A wave that travels at the speed of light obeys $\nu\lambda = c$, and a massless particle obeys $E = pc$. With those two in hand you can easily derive either of these two equations from the other.

But matter waves don't travel at speed c, and massive particles don't obey $E = pc$, so for matter the two equations are two independent postulates. We (and many others) focus on the wavelength because that is a measurable quantity, as in the preceding double-slit example.

Electron Microscopes

Because of diffraction a microscope spreads out each point in the original image to a small disk in the magnified view. Features that are too close to each other end up with overlapping images, effectively limiting the resolution. The smallest feature that a microscope can resolve is roughly comparable to the wavelength of the light used to make the image, so ordinary microscopes can resolve features down to a few hundred nanometers. That's enough to see many types of cells, for example. To see smaller objects you need shorter wavelengths.

In 1931 Ernst Ruska and Max Knoll developed the first "electron microscope," which uses a beam of electrons instead of light. The beam is sent through a thin sample, spread out using an electromagnet, and captured on a fluorescent screen (similar to an old television screen). The resolution of an electron microscope is still diffraction-limited, but because electron beams have much smaller wavelengths than visible light they can resolve much finer details. See Problems 22 and 23.

Ruska's and Knoll's original model was a "transmission electron microscope" (TEM). Those are still used today, and other types now include a "scanning electron microscope" (SEM), which bounces electrons off the surface of the sample, and a "scanning tunneling microscope" (STM), which records the flow of electrons onto the sample from a very narrow tip that moves across the sample's surface.

Stepping Back: What is a Matter Wave?

While the term "matter waves" is generally used when talking about de Broglie's work,[7] in quantum mechanics that wave is now called a "wavefunction." Here are some facts on which pretty much everyone agrees.

- Associated with any particle is a mathematical wave, which we call that particle's "wavefunction." This wavefunction displays measurable wave phenomena such as interference and diffraction.

- If we measure something about the state of the particle, such as its position or momentum, the wavefunction can tell us the probabilities of finding different outcomes. It *cannot* in general give us definite predictions of what answers we'll measure.

- Schrödinger's equation tells us how the wavefunction evolves in between such measurements.

Calculating a particle's wavefunction, and then the particle's properties based on that wavefunction, gives accurate probabilities for a wide variety of experiments. As a tool for making predictions, quantum mechanics works.

But as a description of the nature of reality, quantum mechanics raises questions that seem impossible to answer experimentally. What is waving in a wavefunction? What does it mean to say that this mathematical function is "associated" with a particle? Does a particle really *not*

---

[7] This is despite the fact that he himself used the term "phase wave."

have a specific position before we measure it, or do we just not *know* its position? Answers to those questions are generally referred to as "interpretations" of quantum mechanics. Here is a brief summary of three of them.

The *orthodox interpretation* (sometimes called the "Copenhagen interpretation") entirely abandons the idea of a particle as a tiny little billiard ball with a particular position and momentum at any given moment. There is only the wavefunction. When you fire an electron at a double-slit apparatus the wave spreads out through space, passes through both slits simultaneously, and interferes with itself. When you measure its position on the back wall, that wave collapses. So you might carelessly say that when you measured the electron you "found out its position," but it would be more accurate to say that your measurement *forced the electron to have a position;* before your measurement, it wasn't anywhere specific at all.

Critics point out that the orthodox interpretation postulates that "measurement" fundamentally alters the wavefunction, but offers no satisfactory definition of what constitutes a "measurement." The two following interpretations, although different in important ways, share the feature that neither one ascribes special significance to the act of measurement.

In the *pilot-wave interpretation* (sometimes called "Bohmian mechanics") the particle and the wave are two distinct and real entities that interact with each other. The particle is much like a tiny billiard ball, with a definite position and momentum at any given moment, although we can never know exactly what that position and momentum are. The wavefunction pushes the particle around in ways that cause it to deviate from a classical trajectory. When we fire many "identical" particles at a double slit, they end up in different places because they started with different positions and/or momenta, even though everything we could know about them at the start was identical.

In the *many-worlds interpretation,* every quantum event that could happen in more than one way (e.g. an electron that could pass through either of two slits) causes the universe to split. You can think of this as the orthodox interpretation without the collapse: the wavefunction goes along many paths at once, so it strikes the wall at many places at once, so you measure it at all those places simultaneously. That is, one version of "you" (in one universe) measures a spot in this location, and another version of "you" measures it in a different location, and so on.

All of these interpretations lead to the same experimental predictions. You can calculate a particle's wavefunction and use that to calculate probabilities of different measurements, and experimentally those predictions work regardless of what reality you believe is behind them. For that reason, many physicists believe that these questions of interpretation are unresolvable and therefore unscientific. This attitude is often referred to in the physics community as "Shut up and calculate."[8]

We, your authors, find the issue of interpretation to be both fascinating and important, but we are not going to resolve these debates in this book. We have been using, and will continue to use, the language of the orthodox interpretation, because that is the language commonly used in physics. We explore the questions of interpretation in more detail in the online section "Interpretations of Quantum Mechanics." www.cambridge.org/felder-modernphysics/onlinesections

---

8 The idea that questions of interpretation are unscientific is also sometimes called the "Copenhagen interpretation," which is why we mostly avoid that term. This view is probably closer to Bohr's view than the orthodox interpretation. He once said "It is wrong to think that the task of physics is to find out how nature *is*. Physics concerns what we can say about nature."

4.2.2 Questions and Problems: Matter Waves

Conceptual Questions and ConcepTests

1. If you fire bullets at a double slit, why do you not see an interference pattern on the wall behind them? *Hint*: It is not because bullets are particles rather than waves. We now believe that all particles have associated wavefunctions.

2. An electron and a proton have the same kinetic energy. Which one of the following is true about them? Explain how you know.

   A. The electron has a larger de Broglie wavelength.

   B. The proton has a larger de Broglie wavelength.

   C. The two have equal de Broglie wavelengths.

3. Compare an electron's de Broglie wavelength to its Compton wavelength: is one always bigger than the other, or does it depend on the velocity? Briefly explain your answer.

4. Why is it harder to do a double-slit experiment with protons than it is with electrons?

5. In Chapter 3 we described several interactions of electrons with photons: the photoelectric effect, Compton scattering, bremsstrahlung radiation, and pair production and annihilation. Do any of those provide evidence that electrons have associated matter waves? Why or why not?

6. Some particular particle has a de Broglie wavelength of 10^{-10} m. What does that mean about that particle? What observable effects are we saying happen on that scale?

7. What happens to the de Broglie wavelength of an electron in the limit $v \to c$?

8. One reason to focus on $p = h/\lambda$ instead of $E = h\nu$ for matter waves is that "energy" is ambiguous in a way that momentum isn't. Is that kinetic energy? Kinetic plus potential? Total relativistic energy? Yet we used that formula for photons. Why was this ambiguity not a problem in that case?

9. In many cases we can view classical physics as the $h \to 0$ limit of quantum physics. What happens to the de Broglie wavelength in that limit and what does that imply about the behavior of particles?

For Deeper Discussion

10. We've now said that light and electrons each have essentially the same wave-like and particle-like properties. So why does classical physics view light as a wave and electrons as particles?

11. Do water waves and sound waves exhibit particle-like properties in the same way that light waves do? If so, why don't we ever observe them? If not, why not?

Problems

You can solve most of the problems in this section non-relativistically, but check the speeds and energies. If a particle's speed is much less than c, or (equivalently) its kinetic energy is much less than mc^2, non-relativistic calculations are fine. If the speeds and energies are comparable to or greater than these limits, re-do the calculations with relativistic equations.

12. Calculate the wavelength of an electron moving at 500 m/s.

13. (a) How fast would an electron have to move in order to have a de Broglie wavelength of 10^{-10} m, roughly the width of an atom?

    (b) If an electron were initially at rest, through what potential would you have to accelerate it to get to the speed you found in Part (a)?

14. In thermal equilibrium the average kinetic energy of a free particle is $(3/2)k_B T$, where k_B is Boltzmann's constant and T is the temperature in kelvins. Estimate the de Broglie wavelength of an electron in thermal equilibrium at room temperature. Do the same for a neutron.

15. Estimate the de Broglie wavelength of a baseball thrown by a pitcher.

16. What kinetic energy would an electron need in order to have a de Broglie wavelength of 10^{-8} m? What energy would a photon with that same wavelength have?

17. You do a double-slit experiment with a back wall 2 m behind the slits. In order to see a noticeable effect you want the light and dark bands

to be separated by 1 mm. (You can find the formulas you need in Appendix E.)

(a) If you fire photons at the slits with energy 100 eV, how close together would you have to put the slits?

(b) If you fire electrons at the slits with energy 100 eV, how close together would you have to put the slits?

18. You accelerate a beam of electrons through a 100 V potential, pass the electrons through a double slit, and measure them on a wall 2 m behind the slits.

(a) How close would you have to put the slits in order for the first dark band to be 0.01 mm from the center of the back wall? Use the small-angle approximation for a double slit in Appendix E.

(b) Was the small-angle approximation valid in this case? How do you know?

19. You accelerate a beam of electrons through a 1 MV potential, pass them through a double slit, and measure them on a wall 2 m behind the slits. How close would you have to put the slits in order for the first dark band to be 1 mm from the center of the back wall? *Hint*: This looks almost identical to Problem 18, but it's harder to solve.

20. Suppose you pass a stream of bullets with mass 1 g and speed 500 m/s through a double slit with slit separation 1 cm. Roughly how far away would you have to put the back wall to see a 1 cm separation between the central maximum and the first dark band?

21. The Davisson–Germer experiment described in the Example on p. 185 could be viewed as a test of the de Broglie relation. Equivalently, if you assume the de Broglie relation is correct then it could be viewed as a way of measuring Planck's constant. In reality their data were only accurate enough to say that the peak seemed to occur somewhere between 50° and 55°. What range of values for h do those results imply, and are they consistent with our current best measurements of h?

22. Suppose an electron microscope uses a beam of electrons accelerated through a 1 V potential.

(a) Estimate the resolving power of that microscope.

(b) Would it be able to image a 200 nm cell? A 4 nm cell membrane? A 0.1 nm hydrogen atom?

23. A particle accelerator can be viewed as a large electron microscope, and it is often used to probe nuclear structure. Through what potential difference would you need to accelerate an electron beam to resolve an atomic nucleus of size 10^{-14} m?

24. **1D Particle in a Box**

If an electron is confined to move along the x axis between points $x = 0$ and $x = L$ then its wavefunction will form a standing wave that goes to zero at those two endpoints.

(a) What is the longest wavelength such a standing wave could have? *Hint*: It's bigger than L and smaller than ∞. If you're stuck try drawing a picture of the longest wave you can draw that goes to zero at both ends of the domain.

(b) What is the minimum magnitude of momentum that the electron could have?

(c) Assuming the electron is non-relativistic ($v \ll c$), use your answer to Part (b) to figure out the minimum kinetic energy it could have.

25. **Exploration: The Davisson–Germer Experiment**

Consider the Davisson–Germer experiment described in the Example on p. 185. We're going to focus on two adjacent silver atoms, which we'll just call the left atom and the right atom.

(a) Suppose the detector is at a height h above the atoms and a distance x to the right of the midpoint between the two atoms. Write a formula for the difference in path lengths from the left and right atoms to the detector. Your answer should depend only on h, x, and w.

(b) In practice the atomic spacing w is *much* smaller than the distances x and h. Find the linear approximation to your answer to Part (a) for small w.

(c) Using the fact that $\sin \phi \approx x/\sqrt{x^2 + h^2}$, rewrite your answer so that it depends only on w and ϕ.

(d) The first non-zero maximum occurs when the path length difference equals λ. Using that fact, verify that your formula reproduces the one we used in the Example.

26. de Broglie's Interpretation of the Bohr Model
Louis de Broglie interpreted the electron in an atom as a standing wave wrapped around the nucleus.

Using the de Broglie relation between wavelength and momentum, show that the only possible circular orbits for such standing waves are the ones in the Bohr model.

4.3   Wavefunctions and Position Probabilities

In the Young double-slit experiment we saw that a photon spreads out as a wave, interferes with itself, and then appears as a particle at a particular point when measured. In Section 4.2 we saw that particles such as electrons display the same behavior.

These pieces begin to come together when we mathematically analyze a wave to determine the probabilities of different measurements.

4.3.1   Discovery Exercise: Wavefunctions and Position Probabilities

A meter stick lies on the ground in front of you. You drop a pin that is guaranteed to land somewhere on that meter stick. Any point on the meter stick is exactly as likely as any other point.

1. Suppose the meter stick is marked with lines every centimeter: 1, 2, 3, ..., 100. What is the probability that the closest line to your pin is the line marked 37? (This question is as trivial as it sounds.)

2. Now suppose the meter stick has lines every half-centimeter: 0.5, 1, 1.5, 2, ..., 100. Now what is the probability that the closest line to your pin is the one marked 37?

3. Forget about the marks now. What are the odds that the pin lands *exactly* 37 cm from the end of the rod? (Assume the pin is infinitely small; it's a thought experiment.)

4.3.2   Explanation: Wavefunctions and Position Probabilities

The one-particle-at-a-time double-slit experiment (whether it's done with photons or matter particles) points to three important conclusions:

- A particle moves through space as a wave that exists simultaneously throughout an extended region.

- Knowing the exact function that describes this wave (called the particle's "wavefunction") does not in general allow you to predict the outcomes of measurements of the particle. Instead it allows you to predict the *probabilities* of different measurements.

- When you measure a particle's position, you are most likely to find it in places where the wavefunction's amplitude is highest.

The wavefunction is usually represented by the Greek letter ψ. Since we are only considering a particle moving in one dimension, we write it as $\psi(x)$. The wavefunction is complex, so its value at any given point could be 1, $-\pi$, $3 + i$, or any other complex number. In this chapter, however, we will mostly work with real-valued wavefunctions.

A Discrete Universe

Consider a very simple universe that has only three points: $x = 1$, $x = 2$, and $x = 3$. This simple universe has one particle, and any measurement of that particle will find it on exactly one of those points. So ψ is just three numbers, one at each point: for instance,

$$\psi(x) = \begin{cases} 1/3 & x = 1 \\ 2/3 & x = 2 \\ -2/3 & x = 3 \end{cases}$$

Here's the rule for a discrete universe: $|\psi|^2$ at each position is the probability of finding the particle there. (We write $|\psi|^2$ instead of just ψ^2 because ψ can be complex; for real wavefunctions the two are interchangeable.) So the probability of finding the particle at $x = 1$ is $1/9$, the probability of finding it at $x = 2$ is $4/9$, and the probability of finding it at $x = 3$ is also $4/9$.

Active Reading Exercise: Three Probabilities

In our simple three-point universe, why would it be impossible for the particle to have the wavefunction $\psi(1) = \psi(2) = \psi(3) = 1/2$?

The problem in the exercise above is that those ψ-values say the probability at each point is $1/4$. Adding all of those up, the probability of finding the particle *somewhere* would only be $3/4$, but that total probability must always equal 1.

That may sound like a detail, but it is a fundamental principle called "normalization," to which we will return many times.

Normalization

A distribution in which all the probabilities add up to 1 is said to be "normalized." Any valid probability distribution must be normalized.

We will see later that when you calculate a wavefunction it generally has an arbitrary constant in front of it. You choose the value of that constant to normalize the wavefunction.

A Continuous Universe

We now move up to a bigger universe: still one-dimensional, and still with only one particle, but now comprising the entire number line. The wavefunction is therefore some function $\psi(x)$ that is defined for all real x-values. Let's take as an example one wavelength of a sine wave (Figure 4.9).

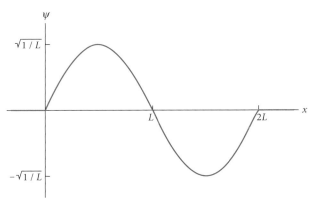

Figure 4.9 The wavefunction
$\psi(x) = \sqrt{1/L}\,\sin(\pi x/L)$ **from** $x = 0$ **to**
$x = 2L$, **and** $\psi(x) = 0$ **everywhere else.**

At any given point, ψ has some value: for instance, $\psi(L/2) = \sqrt{1/L}$. By analogy to what we did before, you might suppose that the square of this value $(1/L)$ gives the probability of finding the particle at $x = L/2$.

But that isn't quite true because $x = L/2$ is an infinitesimal point, one of infinitely many such points in the range $0 < x < 2L$. This, by the way, was the point of the Discovery Exercise on p. 191: as you increase the number of possible measurements, you decrease the probability of any given measurement. In the limit of a continuum of possible values, the probability of finding the particle at $x = L/2$ or any other specific value is zero.

So ψ for a continuous distribution does not give a probability, but a "probability density." The rule for interpreting a probability density is the following.

The integral of $|\psi|^2$ between any two points gives the probability of finding the particle between those two points.

For example, $\int_0^L |\psi|^2 dx$ gives the probability of finding the particle somewhere between $x = 0$ and $x = L$.

If you haven't worked with probability densities before, we urge you to play with the idea to get used to it. For starters, convince yourself that the following facts all come *mathematically* from basic properties of integrals, and also *physically* make sense for position probabilities.

- In any region where ψ is uniformly zero, the probability of finding the particle is zero.
- In areas where ψ is high, the probabilities tend to be high.
- The probability of finding the particle at any exact point is zero (as we said earlier), since the integral from any number to itself is always zero.
- The probabilities sum in a natural way. That is, the probability of finding a particle between $x = 10$ and $x = 13$, and the probability of finding it between $x = 13$ and $x = 20$, add up to the probability of finding it between $x = 10$ and $x = 20$.
- While ψ can be negative or even complex, the calculated probability is always real and positive. In Figure 4.9, the right half of the wave has exactly the same probability as the left half.

We still have a normalization condition, similar to the one we had in the discrete case. The integral $\int_{-\infty}^{\infty} |\psi|^2 dx$ gives the probability of finding the particle anywhere, and must equal 1.

(You can do a quick integral to convince yourself that $\sqrt{1/L}$ is the correct amplitude to normalize the function in Figure 4.9.)

Probabilities are always unitless, so the fact that $\int |\psi|^2 dx$ gives a probability means ψ must have units of $1/\sqrt{x}$. (Remember that dx has units of position, just like x.) You can use this to check your answers when you calculate wavefunctions. We'll see in Chapter 7 that the units of ψ are different when we go from 1D to 3D, though.

Example: A Wavefunction

A particle has the wavefunction $\psi(x) = A\sin(\pi x/L)$ from $x = 0$ to $x = L$ and $\psi = 0$ everywhere else.

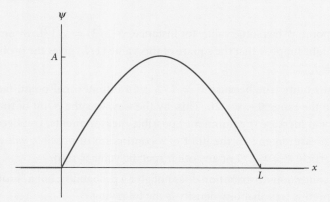

1. Calculate A.
2. What is the probability of finding the particle between $x = 0$ and $x = L/4$?

Answer: You can evaluate the integrals below by using trigonometric identities, by using a computer, or by looking up the integral of $\sin^2 x$. Here we just present the results.

1. You find the constant at the front of a wavefunction by normalizing:

$$\int_{-\infty}^{\infty} |\psi|^2 dx = \int_0^L A^2 \sin^2\left(\frac{\pi}{L}x\right) dx = A^2 \frac{L}{2}.$$

    Setting this equal to 1 gives $A = \sqrt{2/L}$.

2. Using our calculated value of A we can integrate to find the probability. But first let's look at the graph and make a rough prediction. The domain $0 < x < L/4$ covers $1/4$ of the region where $\psi \neq 0$, but ψ is smaller in this quarter than it is on average in the rest of the region, so the probability should come out less than $1/4$:

$$P\left(0 < x < \frac{L}{4}\right) = \int_0^{L/4} |\psi|^2 dx = \int_0^{L/4} \frac{2}{L} \sin^2\left(\frac{\pi}{L}x\right) dx = \frac{\pi - 2}{4\pi} \approx 0.1.$$

    As predicted, this is less than $1/4$. As another reality check, note that because L and dx both have units of length, our integral was unitless, as a probability should be.

In the preceding Example, ψ was given with an arbitrary constant A. We found the value of that constant by normalizing. This is a very common pattern.

On the other hand, what if we had asked the same question starting with the wavefunction $\psi(x) = Ax^2$? Normalization would require $\int_{-\infty}^{\infty} A^2 x^4 dx$ to equal 1, but no constant A will make that happen. We say that the wavefunction x^2 is not "normalizable." In practice this means that no actual particle could have that wavefunction across the entire real number line. (On the other hand, $\psi = Ax^2$ might serve as a valuable approximation to some actual wavefunction over a limited domain.)

As a quick reality check, any function that does not approach zero as $x \to \pm\infty$ is certainly not going to be normalizable. But that check doesn't work the other way; if a function *does* go to zero at both ends, its integral may still diverge.

Repeated Measurements

What happens if you measure a particle's position twice?

Active Reading Exercise: Repeated Measurements

A particle has the wavefunction we showed in the Example on p. 194. You measure its position and find it very near $x = 3L/4$. If you then immediately measure its position again, is the probability of finding it near that same spot the second time higher than, lower than, or the same as it was before the first measurement?

If you take the second measurement right away you are guaranteed to again find the particle near $x = 3L/4$. That's because when you took the first measurement you collapsed the particle's wavefunction, so that it changed from being spread out from $x = 0$ to $x = L$ into a narrow spike near wherever you found it (Figure 4.10). (Remember that this wavefunction collapse is how the "orthodox interpretation" of quantum mechanics explains the individual spots on the back wall of a double-slit experiment: the back wall itself is a measuring device that forces the position to a narrowly defined value.)

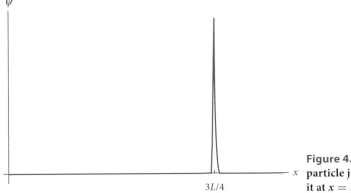

Figure 4.10 The wavefunction of a particle just after a measurement finds it at $x = 3L/4$.

Afterward the wavefunction will change in ways we'll describe in Chapters 5 and 6, so at later times the position probabilities will once again be spread out in some more complicated fashion.

Expectation Values

Imagine preparing a million particles with identical wavefunctions, and then measuring their positions one by one. You would not get the same result every time: the collapse of the wavefunction from "here is a range of probable positions" to "you found it right here" is a random process. But if you took the *average* of those million measurements you would find the average position represented by that wavefunction.

In statistics the average you expect to get from measuring the same thing many times is called the "expectation value" of that measurement. So in this case we are talking about the expectation value of position, which is often written $\langle x \rangle$.

To see how to calculate $\langle x \rangle$, let's return for a moment to our discrete world.

Active Reading Exercise: Discrete Expectation Value

Earlier we considered a particle that could be at $x = 1$ with probability 1/9, or at $x = 2$ with probability 4/9, or at $x = 3$ with probability 4/9. Suppose you took nine such particles, measured each of their positions, and then averaged the results. What would you expect to find?

Before reading further, do that calculation. Then see if you can generalize it to a formula.

In the preceding Example you would expect to see $x = 1$ once, $x = 2$ four times, and $x = 3$ four times. The average position would therefore be

$$\frac{1 + 2 + 2 + 2 + 3 + 3 + 3 + 3}{9} = \frac{21}{9}. \tag{4.4}$$

If instead of 9 measurements you had taken 18, that would double the numerator *and* the denominator, and the result would come out the same. So over many measurements you would expect to see an average result of roughly 2.3. Of course, no individual measurement would yield that value! But a large and equal number of 2s and 3s, with the occasional 1 mixed in, would give that average.

We can rearrange that calculation as

$$1 \times \frac{1}{9} + 2 \times \frac{4}{9} + 3 \times \frac{4}{9}.$$

That's the same calculation as Equation (4.4) (make sure you see that!) but in a form that lends itself to a general formula (in Problem 23 you can derive this equation):

$$\langle x \rangle = \sum_x x P(x). \tag{4.5}$$

Equation (4.5) applies to any discrete normalized probability distribution $P(x)$; if we are talking about a quantum mechanical position in particular, $P = |\psi|^2$. In a continuous world the formula looks similar, but you replace the sum with an integral over all the possible positions:

$$\langle x \rangle = \int_{-\infty}^{\infty} x |\psi|^2 dx.$$

Example: Expectation Value

A particle has the wavefunction $\psi(x) = (2x/L^{3/2})e^{-x/L}$ for $x > 0$ and 0 for $x \leq 0$. Find $\langle x \rangle$.

Answer: Just use the formula for expectation value. You can use integration by parts or do the integral on a computer:

$$\langle x \rangle = \int_{-\infty}^{\infty} x|\psi|^2 dx = \frac{4}{L^3} \int_0^{\infty} x^3 e^{-2x/L} dx = \frac{3}{2}L.$$

It's comforting to see that the expectation value of position does come out with units of position. (Can you see how we could tell from the equation for ψ that L had to have units of position?)

As a sanity check, we can plot the probability distribution $|\psi|^2$ for this wavefunction. It certainly looks plausible that on average the position measurements will come out somewhere between L and $2L$.

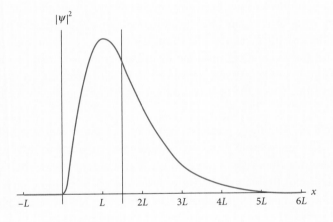

Summary: Discrete and Continuous Probabilities

In this section we've discussed probabilities for discrete and continuous variables. When you use a wavefunction to find position probabilities, that's a continuous distribution. The discrete case was just a warm-up example to illustrate some of the ideas. There are other quantities that actually do have discrete probabilities, though. On a macroscopic level we describe the roll of a die by saying that each outcome has a 1/6 chance of coming up. On a microscopic level we know that in some circumstances a particle's energy comes in discrete levels (e.g. in a hydrogen atom), and you'll soon see that we can use quantum mechanics to calculate probabilities for those discrete energies.

Here is a summary of the basic properties of discrete and continuous probabilities. We're using x to stand for whatever the quantity is you're measuring (position, energy, outcome of a die roll, ...). For the case of position probabilities you can substitute $|\psi(x)|^2$ for $P(x)$ in the right-hand column.

	Discrete probabilities	Continuous probability distribution
Interpretation:	$P(x)$ gives the probability of a particular measurement.	$P(x)$ gives the probability density at a particular x-value.
Example:	$P(70)$ is the probability of a given measurement finding $x = 70$.	$\int_{70}^{72} P(x)dx$ is the probability of a given measurement finding x between 70 and 72.
Normalization:	$\sum_x P(x) = 1$	$\int_{-\infty}^{\infty} P(x)dx = 1$
Expectation value:	$\langle x \rangle = \sum_x xP(x)$	$\langle x \rangle = \int_{-\infty}^{\infty} xP(x)dx$

Multiparticle Wavefunctions

Discrete or continuous, what we have done in this section – and will continue to do for the next several chapters – is to associate one wavefunction with one individual particle. We're going to end this section with a few paragraphs about the wavefunctions of multiparticle systems. This is an important and fascinating topic, but we are not going to return to it in this book. (We will model some multiparticle systems, such as an atom with many electrons, but we will not write their wavefunctions.) So if the next two paragraphs don't completely make sense at this point, don't panic.

For a system of (say) three particles, you might very reasonably assume that each particle gets its own wavefunction: $\psi_1(\vec{r}_1)$, $\psi_2(\vec{r}_2)$, and $\psi_3(\vec{r}_3)$, where $\vec{r}_1$ is the position of the first particle (generally a three-vector) and so on. But it doesn't work that way. Instead, there is one wavefunction for the state of the entire system: $\psi(\vec{r}_1, \vec{r}_2, \vec{r}_3)$. The squared amplitude of that wavefunction gives a probability density for finding the system near a particular state, where the state of the system includes the positions of all three particles.

Such wavefunctions raise mathematical difficulties: we generally cannot solve for them explicitly, and must use numerical approximation techniques. They also raise tricky issues of interpretation. Many physicists speak of a wavefunction as a physical quantity that has some value everywhere in space, somewhat like an electric or magnetic field. But that viewpoint doesn't work for the wavefunction of an N-particle system, which in general is a function of $3N$ independent variables! Such a wavefunction doesn't have a value at each *place*; it has a value at each *configuration* of everything that makes up the universe. This presents a significant challenge for the attempt to make sense of the physical universe described by the equations of quantum mechanics.

4.3.3 Questions and Problems: Wavefunctions and Position Probabilities

Conceptual Questions and ConcepTests

1. Give the units for $\psi(x)$, and explain briefly how you know, in . . .

(a) a 1D universe in which x can only take on a discrete set of values

(b) a 1D universe in which x can take any real value

2. What does it mean to say a wavefunction is "normalized"?

3. Why is it impossible for a particle to have the wavefunction $\psi(x) = Ax$ for some constant A?

4. Is it possible for a particle to have a wavefunction that does *not* approach 0 in the limit $x \to \infty$? If so, give an example. If not, say why not.

5. A particle has the following wavefunction: $\psi(x) = A\sin(\pi x/L)$ from $x=0$ to $x=L$. Compare the probability of finding the particle in the range $0 < x < L/2$ to the probability of finding it in the range $L/2 < x < L$. Choose one of the following. You do not need to do any calculations to answer this question, but you should briefly explain your reasoning.

   A. It's more likely to be found in $0 < x < L/2$.

   B. It's more likely to be found in $L/2 < x < L$.

   C. The two are equally likely.

6. Figure 4.11 shows the wavefunction $\psi(x)$ for a particle.

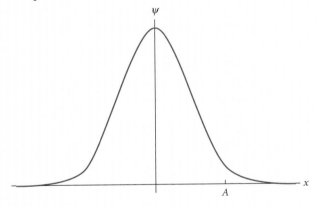

Figure 4.11

   (a) Estimate the probability that the particle is between $x = 0$ and $x = A$.

   (b) Estimate the probability that the particle is *exactly* at the position $x = A$.

7. In the Example on p. 194 we normalized a wavefunction and found $A^2(L/2)$ had to equal 1, so we set $A = \sqrt{2/L}$. Could we have instead chosen $A = -\sqrt{2/L}$? Would it have changed any of our predictions? Why or why not?

8. Equation (4.4) calculated the expectation value for a particular particle's position as 21/9. What does that mean, given that any measurement of the particle must find $x = 1$, $x = 2$, or $x = 3$, so 21/9 is not an option?

9. The calculation $\int_a^b |\psi(x)|^2 dx$ gives the probability of measuring a particle with ... (choose one, and explain your answer)

   A. $a < x < b$.

   B. $a \le x \le b$.

   C. Both of those, because they are both the same.

   D. Neither of those. (If you choose this, say what it does calculate instead!)

For Deeper Discussion

10. We tried to be careful, when writing this chapter, to avoid saying "the probability of the particle being at this position," and to say instead "the probability of *finding* the particle at this position." Why is the latter wording more correct?

Problems

11. If a particle can exist at only three locations and it is equally likely to be found at any of them, what is the value of the wavefunction at each of those positions? For simplicity you can assume that ψ is real and positive everywhere.

12. A particle can exist at only three locations and its wavefunction at those three locations is given by $\psi(1) = 2C$, $\psi(2) = 3C$, $\psi(3) = 5C$, where C is a constant.

   (a) What is the value of C, assuming it is real and positive?

   (b) What is the probability of finding the particle at position 2?

13. In a discrete universe with only three possible positions, $\psi(1) = i/2$ and $\psi(2) = -1/3$.

   (a) Give a real value that could be $\psi(3)$.

   (b) Give a non-real value that could be $\psi(3)$.

   (c) What is the probability of a given measurement finding a particle at $x = 3$?

14. For each of the following functions, if it could possibly be the wavefunction of a particle, find the constant A. If the function could not possibly be the wavefunction of a particle, explain why. (Assume k is a constant.)

    (a) $\psi_1(x) = Ax$

    (b) $\psi_2(x) = A/x$ for $x > 0$

    (c) $\psi_3(x) = A/x$ for $x > 1$

    (d) $\psi_4(x) = A\sin(kx)$

    (e) $\psi_5(x) = A\sin(kx)$ from $x = 0$ to $x = \pi/k$ and 0 everywhere else

15. For each of the following wavefunctions, find a value of A that normalizes it, or show that it is not normalizable. (Assume k is a positive constant.)

    (a) $\psi_1(x) = Ae^{-kx}$

    (b) $\psi_2(x) = \begin{cases} Ae^{kx} & x \le 0 \\ Ae^{-kx} & x > 0 \end{cases}$

16. A particle has the following wavefunction: $\psi = Ax(x - L)$ from $x = 0$ to $x = L$, and $\psi = 0$ everywhere else.

    (a) What is the value of A?

    (b) What is the probability of finding the particle in the region $0 < x < L/2$? (You can calculate this, but you can also figure it out with no calculations by making a graph.)

    (c) What is the probability of finding the particle in the region $0 < x < L/4$? (This one you have to calculate, although a graph can be a great reality check.)

17. A particle has the following wavefunction: $\psi = A\cos(\pi x/(2L))$ from $x = -L$ to $x = L$, and $\psi = 0$ everywhere else. What is the value of A? (Your answer will have L in it.)

18. A particle has the following wavefunction: $\psi = Ax(x - L)(x - 2L)$ from $x = 0$ to $x = L$ and $\psi = 0$ everywhere else.

    (a) What is the value of A?

    (b) What is the probability of finding the particle in the region $0 < x < L/2$?

    (c) What is the probability of finding the particle in the region $L < x < 2L$?

    (d) What is $\langle x \rangle$?

19. A particle has the following wavefunction: $\psi = Ae^{ikx}$ on the domain $0 \le x \le \pi/k$, and $\psi = 0$ everywhere else. Calculate A to normalize this wavefunction.

20. A particle has the following wavefunction: $\psi = Ax(x - L)(x + iL)$ from $x = 0$ to $x = L$ and $\psi = 0$ everywhere else. Note: This problem involves several integrals, all of which are straightforward but algebraically tedious. You might want to have a computer do them for you.

    (a) What is the value of A?

    (b) What is the probability of finding the particle in the region $0 < x < L/3$?

    (c) What is $\langle x \rangle$?

21. Consider a quantity whose probability distribution can be modeled by $P(x) = k/x^3$ for $x \ge 1$, where k is a positive constant. Assume it's 0 for $x < 1$. (This couldn't actually be a squared wavefunction but it could be some other probability distribution.)

    (a) Choose the constant k to properly normalize this distribution.

    (b) What is the probability of this quantity being in $1 \le x \le 2$?

    (c) What is the probability of finding $x > 5$?

    (d) If you took many uncorrelated samples of this quantity, and then averaged all your results, what would you get?

    (e) Show that the probability distribution $P = k/x^3$ cannot be properly normalized on the domain $x > 0$.

22. Many quantities can be modeled by the "Gaussian distribution"[9] $P(x) = ke^{-x^2}$, where k is a positive constant. (That's not a wavefunction, it's a probability distribution.) When working with this function it is helpful to know the following two facts:

$$\int_{-\infty}^{\infty} e^{-x^2}\,dx = \sqrt{\pi} \quad \text{and} \quad \int_{0}^{1} e^{-x^2}\,dx \approx 0.747.$$

---

9 This curve is also sometimes called the "normal distribution," which has nothing to do with being "normalized," which is really annoying.

Assume that some quantity has been found to follow the Gaussian distribution across the entire real axis.

(a) Choose the constant k to normalize this distribution.

(b) What is the probability of this quantity being between 0 and 1? *Hint*: It isn't 0.747.

(c) What is the probability of this quantity being between -1 and 1?

(d) What is the probability of this quantity being greater than 1?
Hint: From the above, you have enough information available to provide an answer to this question without the use of a calculator.

23. **Discrete Formula for Expectation Value**

Consider a particle that be found at three positions, x_1, x_2, and x_3, with probabilities P_1, P_2, and P_3.

(a) If you prepare a million particles in this state and measure all of their positions, how many will you find at position x_1?

(b) To calculate the average you need to add up all your results. There will be a million of them and they will all be either x_1, x_2, or x_3, so the result will be x_1 times the number of times you got that result, plus x_2 times the number of times you got that, plus x_3 times the number of times you got that. Then you have to divide by the number of measurements (a million in this case). Calculate that average. (Don't start with Equation (4.5); derive it!)

4.4 The Heisenberg Uncertainty Principle

The Heisenberg uncertainty principle places strict limits on our knowledge of a particle.

4.4.1 Explanation: The Heisenberg Uncertainty Principle

In 1927, Werner Heisenberg proposed that – to a certain, quantifiable extent – measurable properties of a particle must remain uncertain. But how can you quantify uncertainty?

Let's start with a qualitative look. Consider the two wavefunctions shown in Figure 4.12. (The positions are measured in angstroms, a common unit in atomic physics: $1\,\text{Å} \equiv 10^{-10}$ m.)

- Wavefunction A is a narrow spike. A measurement of that particle's position finds it at or very close to $x = 2\,\text{Å}$, so the particle's position uncertainty Δx is close to zero.
- Wavefunction B is more spread out. If you prepared many particles like this and measured their positions, most of the results would fall in the range $x = (2 \pm 1)\,\text{Å}$, so we estimate the uncertainty as $\Delta x \approx 1\,\text{Å}$.

Based on examples like these, we can offer an informal definition.

Position Uncertainty: First (Informal) Definition

Consider a large number of particles, all prepared with identical wavefunctions $\psi(x)$. Position measurements of all these particles yield a range of results. We will use the symbol $\langle x \rangle$ for the *average* of all those measurements. (This is the expectation value of x, as described in Section 4.3.)

Suppose that roughly 2/3 of those measurements fall between $\langle x \rangle - \Delta x$ and $\langle x \rangle + \Delta x$. Then we call Δx the "position uncertainty" associated with that wavefunction.

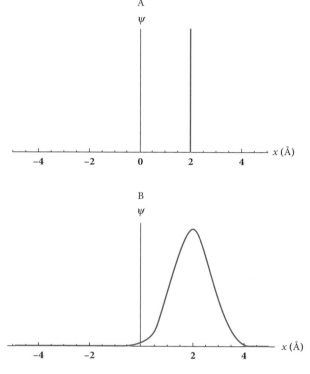

Figure 4.12 **Two wavefunctions.**

We'll see below that similar definitions can be applied to other quantities, like momentum. At the end of this section we will offer a more rigorous definition of uncertainty, based on the statistical idea of "standard deviation."

Active Reading Exercise: Position Uncertainty

Suppose you find that a wavefunction has $\langle x \rangle = 20$ Å, with $\Delta x = 3$ Å. What do you conclude?

If you measure the particle's position, there's about a 67% chance that you'll find it between $x = 17$ and $x = 23$ Å.

Momentum Uncertainty

Momentum, like position, is specified probabilistically by a particle's wavefunction, and we define Δp in the same way as we define Δx: prepare a large number of particles with the same wavefunction, measure their momenta, and see how far the individual measurements stray from the average $\langle p \rangle$.

In Chapter 6 we'll show you how to calculate momentum probabilities. For now, we need you to take our word for the following fact:

The more narrowly spiked a wavefunction is, the greater the uncertainty of its momentum measurement.

Wavefunction A in Figure 4.12 with $\Delta x \approx 0$ has a nearly infinite Δp. Wavefunction B has a larger Δx and a smaller Δp.

Active Reading Exercise: Slightly Modified Uncertainty

Draw a quick sketch of a wavefunction that is very similar to Wavefunction B in Figure 4.12, but with a slightly decreased position uncertainty Δx. How do you think your change affected the momentum uncertainty Δp?

Make your sketch and write your answer before going on. We'll give our answer after we present the uncertainty principle.

The Uncertainty Principle

We have seen a narrow wavefunction with a small Δx and a large Δp, and we have seen a wide wavefunction with a larger Δx and a smaller Δp. More importantly, we have seen a general rule that narrowly defined position distributions lead to widely defined momentum distributions.

The general relationship between position and momentum uncertainties can be derived from the mathematics of Fourier transforms. We're not going through that derivation, but below we present the all-important result.

The Heisenberg Uncertainty Principle

$$\Delta x \Delta p \geq \frac{\hbar}{2}. \tag{4.6}$$

The $\geq$ in this formula says that you can make both uncertainties as large as you want, but you can't simultaneously make them both arbitrarily small.

A "Gaussian" wavefunction $\psi(x) = Ae^{-k(x-x_0)^2}$ achieves Heisenberg's minimum allowed $\Delta x \Delta p = \hbar/2$ (see Problem 23). Any other wavefunction will have a greater value of $\Delta x \Delta p$, although many simple wavefunctions have $\Delta x \Delta p$ within an order of magnitude of this lower limit.

Example: Uncertainty

Wavefunction B in Figure 4.12 is a Gaussian function. Measurements of its momentum will give a spread of values centered on $p = 0$. If you prepare many particles with that wavefunction and then measure their momenta, what range of values will you get from most of your measurements?

Answer: We said above that we could see from the graph that $\Delta x \approx 1$ Å. Since this is a Gaussian we know the uncertainty is at its lower limit:

$$\Delta p = \frac{\hbar}{2\Delta x} = \frac{1.05 \times 10^{-34} \text{ kg m}^2/\text{s}}{2 \times 10^{-10} \text{ m}} = 5.3 \times 10^{-25} \text{ kg m/s}.$$

That tells us that most of the momentum measurements will fall within $\pm 5 \times 10^{-25}$ kg m/s.

We've been talking so far about one dimension only. In three dimensions, each direction has its own uncertainty relation: $\Delta y \Delta p_y \geq \hbar/2$ and so on.

The uncertainty principle is one of the most radical conclusions of quantum mechanics. It puts a fundamental limit on how precisely you can measure a particle's classical properties such as position and momentum, no matter how advanced your technology. In the orthodox interpretation, it does more than that: it says that the particle itself never *has* a precisely defined position and momentum. You can perhaps see why many (although not all) physicists have given up entirely on the idea of an objective reality, or on any statement stronger than "these equations successfully predict the results of experiments."

Having presented the uncertainty principle, we can now answer the Active Reading Exercise on p. 203. If you narrow the spread of a wavefunction you reduce Δx. Because the wavefunction in question is a Gaussian it was already at minimum uncertainty, so that would have to increase Δp.

The Time–Energy Uncertainty Principle

When people talk about the "uncertainty principle" without specifying it further, they generally mean Equation (4.6), which relates the uncertainties in position and momentum. There are many such relations in quantum mechanics, however. For example, an orbiting object has an uncertainty principle relating the uncertainties in its angular position and angular momentum.

Here we want to discuss one other such relation, which is qualitatively different from the others.

The Time–Energy Uncertainty Principle

$$\Delta t \Delta E \geq \frac{\hbar}{2}. \tag{4.7}$$

Can you see why this is so different, even though it looks a lot like the position–momentum uncertainty principle? ΔE is straightforward enough: prepare a bunch of particles with identical wavefunctions, measure their energies, and see how much the results differ from the average. But t is not a property of a particle.

Question: "What is that electron's time?"

Answer: "What are you talking about?"

Roughly speaking, Δt is the amount of time that it takes for any property of a system to change appreciably. If the system is oscillating, Δt is roughly equal to the oscillation period. If some property of the system is decaying (or growing) exponentially, Δt is roughly the time it takes for it to reduce its value in half (or double it).

More precisely, Δt is the time that it takes for any property of a system to change by an amount equal to the uncertainty of that property.

What if a particle's position is oscillating and its momentum is growing exponentially? Which property do you use? The answer is that they must all obey Equation (4.7), so whichever one is changing the fastest sets the minimum uncertainty in energy.

One consequence of the time–energy uncertainty principle is that states of definite energy can't change in any way; they are called "stationary states." We'll discuss some of the implications of that fact in Chapter 5.

Example: Line Width

An electron in the $n = 2$ state of hydrogen takes on average 1.6 nanoseconds to decay to the $n = 1$ state.[11] If you prepare many such atoms and measure the energies of the emitted photons, roughly what spread of energies will they have?

Answer: We can find the average expected energy from the Bohr model:

$$E_{photon} = (13.6 \text{ eV}) \left(\frac{1}{1^2} - \frac{1}{2^2} \right) = 10.2 \text{ eV}.$$

The Bohr model would predict that all the photons would be emitted with exactly the same energy. But the fact that this excited state spontaneously changes on a timescale of 1.6 ns means it can't have an exactly defined energy, which in turn means the photons will be emitted with a range of energies:

$$\Delta E \approx \frac{\hbar}{2\Delta t} = 2.1 \times 10^{-7} \text{ eV}.$$

So the spread of photon energies will be $\pm 2 \times 10^{-7}$ eV centered on approximately 10.2 eV.

This effect shows up in the emission spectrum as a broadening of the emission line. The "line width" of an emission line is a measure of the lifetime of an excited state of an atom.

---

11 You'll learn in Chapter 7 that there are different possible $n = 2$ states in hydrogen, and this is the one called $2p$. Don't worry about that for now.

The effect we just described in this example is called the "natural line width" of a transition. There are other effects that can broaden the measured line width, but it can never be smaller than the minimum value required by the time–energy uncertainty principle.

Here's a small, helpful tip. Now that you know the uncertainty principle, that makes it easy to remember the units of $\hbar$ (or of h, which are the same). It's position times momentum, or equivalently energy times time. From either of those you can quickly conclude, for example, that in the SI system $\hbar$ has units of kg m^2/s.

Standard Deviation

The average of a set of numbers is one way of thinking about a "typical" value for those numbers. If the average height of a second-grade class of students is four feet tall, then you would tend to think of the students far above that as tall and the ones far below that as short.

Table 4.1 Mr. Kotter's grades

Student	Grade (G)
Arnold	70
Freddie	80
Vinnie	60
Julie	90

But knowing the average doesn't tell you how spread out the numbers are. You could have a class where all the kids are between $3'11''$ and $4'1''$, or you might have a class where they range from two to five feet. Those two classes have the same average but different "standard deviations." We'll illustrate the idea through an example.

Mr. Kotter has finished his gradebook for the year. Looking at the grades in Table 4.1, he calculates that the *average* grade of his students is $\langle G \rangle = 75$.

Now Mr. Kotter wants to measure how spread out the scores are around that average.

Active Reading Exercise: Deviations from the Average

All of these questions refer to the gradebook above.

1. For each student calculate ΔG, the student's score minus the class average.
2. Find the average value of all the ΔG scores.
3. Why is the average of ΔG not a good measure of how spread out the class scores are?

For Arnold, $\Delta G = -5$, for Freddie it's 5, for Vinnie it's -15, and for Julie it's 15. If Mr. Kotter averages all those ΔG-values, he will get zero. You should be able to convince yourself that this will *always* be true for any set of numbers (see Problem 18). The problem is that some of the ΔG-values are negative and some are positive, but what you really care about is just how far they are from the average.

The solution is to *square* all the ΔG-values. You average all the resulting (non-negative) numbers, and then take the square root:

$$\sigma_G = \sqrt{\frac{\sum_n (\Delta G)^2}{n}}. \tag{4.8}$$

Make sure you see what Equation (4.8) accomplishes. Going back to our second-grade example, if the average is four feet and the standard deviation is 1 inch, then almost all the kids are very close to four feet. But a class with that same average and a standard deviation of one foot would have a much wider range of heights. In the former case, you can predict the height of a randomly chosen student with far more *certainty* than in the latter.

And that word brings us back to Heisenberg.

Quantum Uncertainty: Second (Stricter) Definition

Consider a large number of particles, all prepared with identical wavefunctions $\psi(x)$. A quantity Q is measured for all these particles. The uncertainty ΔQ is defined as the standard deviation of those measurements (sometimes written σ_Q).

In Problem 17 you will write the formula for calculating position uncertainty from $\psi(x)$. We will not use that formula much in this book, but we do recommend holding onto three facts about this important quantity:

- A large standard deviation means that individual measurements tend, on average, to be very far from the mean (whether above it or below it). A small standard deviation means that individual measurements tend to cluster around the mean.

- The position uncertainty Δx of a wavefunction is the standard deviation you would get if you prepared many particles with that same wavefunction and measured all their positions. A low uncertainty means that you can predict the outcome of a measurement with confidence. You can similarly define uncertainty for other measurable quantities.

- For a Gaussian distribution, roughly 2/3 of the measurements fall within one standard deviation of the mean.

4.4.2 Questions and Problems: The Heisenberg Uncertainty Principle

Conceptual Questions and ConcepTests

1. Explain the uncertainty relationship between position and momentum (in one dimension) in your own words.

2. Two different batches of particles have been prepared. All of them have wavefunctions centered at $x = 10$ m. That is, if you measure the positions of many of these particles, the average result will be $\langle x \rangle = 10$ m.

   (a) The first batch of particles has uncertainty $\Delta x = 1/4$ m. In what range of x-values will you find most of the particles?

   (b) The second batch of particles has uncertainty $\Delta x = 14$ m. Where will you find most of the particles?

3. Which of the following is the best estimate of the uncertainty of a die roll? (Choose one.) Explain your answer.

   A. 0

   B. 1.5

   C. 6

   D. ∞

4. Can an object's position uncertainty ever be negative? Explain.

5. You perform a very accurate measurement of a particle's position.

   (a) What effect does your measurement have on Δx for this particle? (Choose one and explain.)

   A. Δx increases.

   B. Δx stays the same.

   C. Δx decreases.

   (b) What effect does your measurement have on Δp for this particle? (Choose one and explain.)

   A. Δp increases.

   B. Δp stays the same.

   C. Δp decreases.

6. It is possible to measure a particle's position with very high accuracy and then measure its momentum with an accuracy better than $\hbar/2$ divided by the accuracy of your position measurement. Why doesn't that violate the uncertainty principle?

7. We said that the line width of an emission line is related to the uncertainty in the energy of the excited state. Why couldn't it also arise from uncertainty in the energy of the ground state? *Hint*: Think about how long the ground state lasts and what that tells us about its energy.

8. The correspondence principle (roughly) says that, as we approach the sorts of things Newton might have measured, we approach Newtonian behavior. Why does the uncertainty principle vanish in that limit?

9. In the limit in which a measurement of a particle's position approaches zero uncertainty, what happens to our knowledge of the particle's *velocity*? (Choose one comment and explain.)

   A. As $\Delta x \to 0$, that causes Δv to approach infinity. This means that we may sometimes have to say "This particle may be going faster than the speed of light." (If you choose this, explain why this is allowed by relativity.)

   B. Because v must always be between $-c$ and c, the uncertainty Δv can approach c at the most. This puts fundamental limits on the accuracy with which we can measure position.

   C. Because v must always be between $-c$ and c, the uncertainty Δv can approach c at the most. But this does *not* limit the accuracy with which we can measure position. (If you choose this, explain why it does not violate the uncertainty principle.)

For Deeper Discussion

10. Does the uncertainty principle say that we can't *know* exactly what a particle's position and momentum are, or that a particle can't *have* a definite position and momentum? Explain.

11. How could you experimentally verify the uncertainty principle?

Problems

12. Inside the nucleus a proton is confined in a region roughly 10^{-15} m wide. Estimate the uncertainty in its momentum.

13. A Gaussian wavefunction is $\psi(x) = Ae^{-k(x-x_0)^2}$. Assume A and k are positive constants, and x_0 is a real constant (positive or negative).

    (a) What x-value maximizes this function? *Hint*: The easiest way to answer this involves thinking about exponential functions, rather than taking derivatives.

    (b) What are the limits of this function as $x \to \pm\infty$?

    (c) How would you choose the constant A? (Don't do the math here; just briefly describe the procedure in words.)

    (d) What are the units of the constant k?

    (e) Draw two Gaussian wavefunctions on the same graph paper. One of them is $\psi_1(x) = A_1 e^{-k_1 x^2}$, and the other is $\psi_2(x) = A_2 e^{-2k_1 x^2}$. These can be qualitative; what we're really interested in is the effect of changing k_1 to $2k_1$.

    (f) Which curve has the higher Δx? Which has the higher Δp?

14. A carbon-14 nucleus decays into a nitrogen-14 nucleus, emitting in the process an electron and a neutrino. People often ask whether the electron was inside the nucleus all along, or whether it was created in the transformation. The answer is that the electron could not have been hiding inside the nucleus, because that would have violated the uncertainty principle. Demonstrate this fact, making use of the facts that a carbon-14 nucleus has a radius of 2.5×10^{-15} m and that the electron is typically emitted with about 50 keV of energy.

15. Suppose a particle is confined to move in one dimension in the region $0 < x < L$ and is equally likely to be anywhere in that region.

    (a) Estimate the particle's momentum uncertainty. (You can assume that Δp is close to the minimum allowed by Heisenberg.)

    (b) Assuming the momentum is equally likely to have any value from $-\Delta p$ to Δp, what is the average magnitude of the momentum?

    (c) What is the associated kinetic energy?

    (d) Does your answer to Part (c) tell you the average kinetic energy that a particle trapped in this region must have? The minimum kinetic energy? The maximum? Something else? How do you know?

16. Consider the following experiment: you roll a single, fair, six-sided die. You write down the resulting number. Then you repeat this experiment many times.

    (a) What is the expectation value $\langle x \rangle$ for this experiment?

    (b) What is the standard deviation σ_x for this experiment? *Hint*: 1.5 is close, but not exact.

(c) Now consider a different experiment. You roll 20 dice, average the result, and write down that number. Then you roll another 20 dice, average them, and write down the number, . . . and so on, many times. Will the expectation value of those numbers be higher, lower, or the same as the original (one-at-a-time) experiment? What about the standard deviation?

17. Equation (4.8) gives a formula for the standard deviation (or uncertainty) of a discrete quantity G.

   (a) Bearing in mind what you learned about discrete and continuous probability distributions in Section 4.3, write a formula for calculating Δx from $\psi(x)$.

   (b) Suppose a particle is equally likely to be found anywhere from $-L$ to L and has zero probability of being found anywhere else. Calculate Δx.

18. Suppose you have a list of N numbers $c_1, c_2, \ldots, c_N$. If $\Delta c_i = c_i - \langle c \rangle$, show that the average value of Δc_i must be zero.

19. **Size and Energy of the Hydrogen Atom**

   You can use the uncertainty principle to understand why the electron doesn't crash into the nucleus, and to estimate the energy and size of the ground state of hydrogen.

   (a) Classically, the lowest possible energy state for the electron would be at rest, right on top of the nucleus. Why isn't this possible in quantum mechanics? (You don't need to give a long answer to this.)

   (b) Assume the electron is somewhere within a sphere of radius $2R$ around the nucleus, where R is an as-yet-undetermined distance. Taking $r = R$ as a reasonable average value, what is the potential energy of the atom in this state?

   (c) The state we just described has a position uncertainty $\Delta x \approx R$. What is the minimum possible momentum uncertainty Δp_x?

   (d) The magnitude of the x momentum is on average about half of Δp_x. Explain briefly why that is, and then use it to calculate the average magnitude of p_x.

   (e) Assuming p_y and p_z are the same as p_x, find the kinetic energy associated with the average value of $|\vec{p}|$ you just found. Your answer should be in terms of R, the unknown radius with which we started.

   (f) Find the value of R that minimizes total energy (kinetic plus potential) and find the associated energy. Compare your answer to the ground state energy of the Bohr model.

20. **Uncertainty and Single-Slit Diffraction**

   A single electron is approaching a wall with a narrow slit at some speed v_y. Before it reaches the wall it has a wide range in x and an almost perfectly determined x-velocity $v_x = 0$. When the electron reaches the wall it happens to pass through the slit, suddenly reducing its range of x-positions to the width w of the slit (Figure 4.13). Over what range of angles will the electron emerge?

   Use the de Broglie relation to express your answer in terms of w and λ (the wavelength of the electron's wavefunction). For simplicity you can assume that the y-velocity is large enough that $|\vec{v}| \approx v_y$.

21. One of the $n = 3$ states of hydrogen is called $3p$ and it lasts about 0.164 ns before decaying to the ground state.

   (a) Estimate the spread of energies in the emitted photon.

   (b) Estimate the width of the associated emission line on a plot of frequencies.

22. The "W boson" is one of the fundamental particles in the standard model of particle physics. It can be produced in a lab, but it spontaneously decays into other particles after about 3×10^{-25} s. That means the W boson doesn't have a well-defined mass. You can find reported values of the W boson mass to very high accuracy, but these are averages over many measurements.

   (a) Estimate the uncertainty in the rest energy of a W boson.

   (b) Estimate the uncertainty in a W boson's mass.

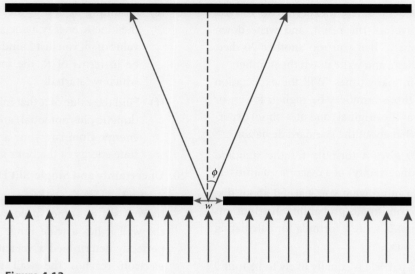

Figure 4.13

(c) The measured uncertainty in the W boson mass is about 4×10^{27} kg. Is that result consistent with the time–energy uncertainty principle?

(d) Explain what the phrase "uncertainty in the W boson mass" means.

(e) We believe all electrons in the universe have the same mass. What does that imply about the stability of electrons?

23. Exploration: Uncertainty of a Gaussian

A Gaussian function is any function of the form $Ae^{-k(x-x_0)^2}$ (assuming $k > 0$). For simplicity, we'll consider $\psi(x) = Ae^{-kx^2}$. (Setting $x_0 = 0$ means the Gaussian is centered on 0.)

(a) For any function centered on $x = 0$ the position uncertainty is given by

$$\Delta x = \sqrt{\int_{-\infty}^{\infty} x^2 |\psi(x)|^2 \, dx}.$$

Calculate Δx. *Hint 1*: You can use integration by parts with $u = x$. *Hint 2*: If you think about the normalization you can calculate this without bothering to calculate A.

(b) The probability distribution for momentum for this wavefunction is $|\hat{\psi}(p)|^2 = B^2 e^{-p^2/(2k\hbar^2)}$. (This is analogous to the probability distribution $|\psi(x)|^2$ for position. Chapter 6 will explain how to calculate $\hat{\psi}(p)$ from $\psi(x)$.) Calculate the momentum uncertainty.

(c) Calculate the product $\Delta x \Delta p$.

Chapter Summary

This chapter describes further evidence of quantization beyond that in Chapter 3, and shows how those early experiments and theories led to the core ideas of quantum mechanics.

The most important equations for this chapter are in Appendix C (Quantum Mechanics Equations), Appendix E (Interference and Diffraction), and Appendix F (Properties of Waves). A timeline of major developments in quantum mechanics is given in Appendix A.

Section 4.1 Atomic Spectra and the Bohr Model

Niels Bohr explained the stability of atoms and their emission and absorption spectra by postulating that electrons can only orbit at certain, quantized, levels of angular momentum and energy.

- Rutherford's gold-foil experiment suggested that an atom is a collection of small negative charges (electrons) orbiting a dense, positively charged nucleus. But this model does not explain why the orbiting electrons don't radiate away their energy and decay into the nuclei, and it does not explain why each element emits and absorbs light at certain discrete frequencies only.

- Bohr proposed that an electron can only be in a circular orbit with angular momentum $n\hbar$ for an integer n. That also implies quantization of radius, energy, and other properties.

- In the Bohr model an electron can only absorb or emit light by jumping discontinuously between allowed orbits. The resulting photon frequencies match observed spectra.

- The Franck–Hertz experiment demonstrated that atoms can only absorb energy from collisions at the same quantized levels at which they absorb photons, further supporting the Bohr model.

- The integer n in the Bohr model is a "quantum number," a physical property of a system that can only take on discrete values. Bohr proposed the "correspondence principle," which says that quantum behavior must approach classical predictions in the limit of large quantum numbers, much like relativistic behavior must approach classical predictions for low speeds.

Section 4.2 Matter Waves

Louis de Broglie proposed that every particle has an associated "matter wave" with wavelength $\lambda = h/p$.

- The model also says that the matter wave has frequency $\nu = E/h$, matching Planck's hypothesis for light.

- The Davisson–Germer experiment in 1926 provided the first clear test of de Broglie's hypothesis. Much later, in 1961, the Young double-slit experiment was replicated with electrons, again confirming de Broglie's ideas.

Section 4.3 Wavefunctions and Position Probabilities

Following de Broglie's proposal, we associate with every particle a "wavefunction," often denoted $\psi(x)$.

- The value of ψ at each point is in general a complex number.
- The function $|\psi(x)|^2$ is a "probability density" whose integral over any region gives the probability of finding the particle in that region.
- The wavefunction must be "normalized," meaning the integral of $|\psi(x)|^2$ over all space must equal 1.
- The "expectation value" of a measurement is the average value you would find if you repeated that measurement many times on identically prepared systems.

- The table on p. 198 collects all the equations you need for this section. In addition to presenting the rules for calculating position probabilities and expectation values for wavefunctions (in the "continuous" column), it shows the corresponding formulas for discrete probability distributions, which are useful for many other quantum mechanical properties.

Section 4.4 The Heisenberg Uncertainty Principle

A particle's wavefunction predicts *probabilities*, not definite values, for the outcomes of measurements. The "uncertainty" of a value tells you how spread out the possible outcomes of measurements will be. The "Heisenberg uncertainty principle" says that the uncertainty of position times the uncertainty of momentum for a particle must be at least $\hbar/2$.

- The uncertainty of a property is defined as the "standard deviation" of its probability distribution. In a series of many measurements of identical particles, how far on average will each measurement be from the expectation value? For example, if you take N measurements of particles with identical wavefunctions,

$$\Delta x = \sqrt{\frac{1}{N} \sum_N (\langle x \rangle - x_N)^2}.$$

- The product $\Delta x \Delta p$ can be arbitrarily large; it just can't be smaller than $\hbar/2$. So a narrowly spiked wavefunction (small Δx) must have large momentum uncertainty.
- The "energy–time uncertainty principle" $\Delta E \Delta t \geq \hbar/2$ says that the smaller the uncertainty in a particle's energy, the longer it takes the particle's properties to change. States of definite energy are therefore "stationary states," meaning they don't change.

5

The Schrödinger Equation

The one-particle-at-a-time double-slit and many other early twentieth-century experiments convince us that a photon or electron is associated with a "wavefunction." This function follows the normal mathematics of waves (including constructive and destructive interference), and probabilistically guides the position and other properties of the particle.

How do we find the wavefunction associated with a given particle? How will that wavefunction evolve over time? How can we use the wavefunction to predict the probabilities of different measurements? The answers are all given by the theory of quantum mechanics.

The first few sections of this chapter will introduce the basic model and use it to calculate the behavior of particles in different situations. The rest of this chapter and the next will apply that model to a variety of simple cases.

Everything in this chapter and the one after it will be about particles moving in one dimension. We'll extend the theory to three dimensions when we take up the hydrogen atom in Chapter 7.

5.1 Force and Potential Energy

Many physical phenomena can be explained in two different but equivalent ways: as forces, or as fields of potential energy. In quantum mechanics we work almost exclusively with potential energy, so it is worth some time reviewing this concept.

5.1.1 Explanation: Force and Potential Energy

A typical problem in Newtonian mechanics starts by specifying the forces acting on an object. "Near the surface of the Earth, gravity exerts a constant downward force $F = -mg$" and "An ideal spring exerts a restorative force $F = -kx$" are common examples. You plug those forces into Newton's second law $F = ma$ and solve the resulting equation to predict the motion.

A perfectly equivalent formulation starts with *potential energy* instead of forces. An object near Earth's surface has a potential energy that grows linearly as you move up: $U = mgh$. A spring has its lowest potential energy at its relaxed length, and higher potential energy when it is compressed or stretched: $U = (1/2)kx^2$. We can use the potential energy to predict motion by noting that all systems tend to accelerate toward regions of low potential energy.

This is not one of those "weird modern physics" things. Even in classical mechanics you can often simplify your calculations immensely by working with potential energy (a scalar) rather than force (a vector). If you have studied circuits you may recall that batteries, resistors, inductors, and capacitors are mathematically described in terms of potential energy functions.[1] These circuit elements do of course exert forces on electrons, but computing those forces would be a hopelessly impracticable approach to circuit analysis.

We mention all this here because in quantum mechanics, just as in circuit theory, we pretty much never use the word "force." A problem starts by specifying the potential energy function. For instance, instead of saying "an electron is attracted by a force whose magnitude is $|\vec{F}| = kQq/r^2$ and whose direction points toward the origin," we express the same fact as "an electron is in a potential energy field $U = -kQq/r$."

Potential Energy Graphs

A typical problem in classical mechanics might involve analyzing the potential energy field shown in Figure 5.1.

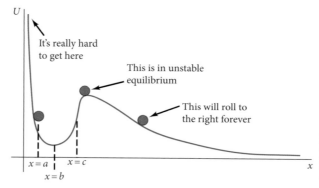

Figure 5.1 A sample graph of potential energy as a function of position. The descriptions in the figure are for objects starting at rest at the indicated positions.

It is vital to understand that Figure 5.1 represents an entirely one-dimensional situation. The potential energy is high at some x-values and low at others, and an object placed in this field will move left or right accordingly. We may speak metaphorically of an object moving "up" the graph (gaining potential energy), but nothing is moving up or down – only left and right.

Figure 5.1 shows three objects that begin at rest at different points in this potential field.

- The middle object is in "unstable equilibrium." It feels simultaneous pulls to the left and right, and these forces balance each other perfectly. So if it is left alone it will stay still forever, but if it is displaced, it will begin rolling.

- The right-hand object feels a force to the right. As it moves to the right it will gradually convert more and more of its initial potential energy to kinetic energy. Thus it will continue moving to the right, faster and faster, forever.

The left-hand object is the most interesting.

---

1 In circuits it's actually more common to talk about the "potential" than the potential energy. The electric potential is just the potential energy per unit charge.

Active Reading Exercise: Motion in a Potential Well

The left-hand object in Figure 5.1 is released at rest at the position marked $x = a$. How will it move initially? How will it move in the long run? As you describe its motion, describe its mixture of kinetic and potential energy at different times.

You can find our answer to that exercise at www.cambridge.org/felder-modernphysics/activereadingsolutions

Such a potential energy field is often described in terms of a ball rolling up and down hills, and that makes a great metaphor as long as you remember that it's only a metaphor. In reality the potential field could be caused by gravitational, electric, and other forces, and physical hills need not be involved.

You may also remember that potential energy as we have discussed it here applies only to "conservative forces": forces that conserve mechanical energy. You can associate potential energy with the force of gravity, the electrostatic force, the force of a spring, and many others. You cannot associate potential energy with forces such as friction or air resistance. This limitation will not bother us here because all forces are conservative at the microscopic level.

Potential Energy Mathematically

An object at $x = a$ in Figure 5.1 experiences a force to the right; an object at $x = b$ experiences zero force; an object at $x = c$ experiences a force to the left. Those three statements all come from the central idea that force points from high potential to low. We can express that idea as a mathematical relationship:

$$F = -\frac{dU}{dx}. \tag{5.1}$$

That is, on a potential energy curve with a positive slope, the force pushes in the negative direction. If you think of force as the fundamental property, then Equation (5.1) can be taken as the definition of potential energy: $U = -\int F(x)dx$.

But a curious thing happens when you define a quantity as an integral. Such a quantity is always arbitrary up to an additive constant. That is, the equations $U = mgy$ and $U = mgy + c$ lead to the same force law $F = -mg$, and therefore to the same physical behavior. You can reach the same conclusion graphically; if you move the entire curve in Figure 5.1 up or down, the behavior of the objects doesn't change.

We typically express this arbitrary choice by saying that you can pick any spot in the universe to be the place where $U = 0$. If you throw a ball off the top of a building you may choose the top of the building to be where $U = 0$, or you may choose the ground as $U = 0$; your calculations will look different, but your predictions for the ball's behavior will come out exactly the same.

5.1.2 Questions and Problems: Force and Potential Energy

Conceptual Questions and ConcepTests

1. An object is moving along the x axis, its potential energy at any point given by the constant function $U(x) = -3$ J. Describe the object's motion over time. *Hint*: If you said "it won't go anywhere," think harder.

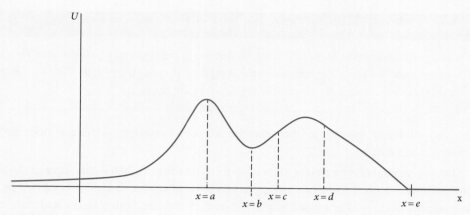

Figure 5.2

2. Figure 5.2 shows a potential field $U(x)$. The potential energy approaches zero asymptotically on the left, and is exactly zero for all $x \geq e$. For each part below, briefly describe the motion of the object with the given initial conditions. Your description should not just say what it will do immediately but what it will do over time.

   (a) It starts at rest at $x = a$.

   (b) It starts at rest at $x = b$.

   (c) It starts at rest at $x = c$.

   (d) It starts at rest at $x = d$.

   (e) It starts at $x = c$ moving rapidly to the right.

   (f) It starts at $x = d$ moving rapidly to the right.

3. Figure 5.3 shows a potential field $U(x)$. The potential energy approaches zero asymptotically on the left and right.

   Describe the long-term motion of a particle starting at rest at each of the four positions labeled on the plot (a, b, c, d).

4. The "strong force" holds together the nucleus of an atom. Here is a rough description of the force between two neutrons. If you put them closer together than $1/2$ fm, they repel each other (1 fm $= 10^{-15}$ m). Between $1/2$ fm and 2 fm, they attract each other. At distances beyond 2 fm, they exert no force.

   (a) Draw a qualitative sketch of the force exerted on Neutron B by Neutron A as a function of the distance between them.

   (b) Draw a qualitative sketch of the potential energy of Neutron B as a function of its distance to Neutron A.

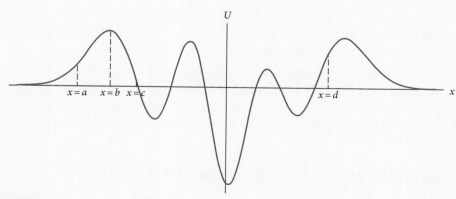

Figure 5.3

5. Explain why $x = a$ in Figure 5.2 is called an "unstable equilibrium" and $x = b$ is called a "stable equilibrium."

6. For each of the descriptions below, sketch a corresponding potential energy function.

   (a) An object is pushed in the positive x direction with a constant force.

   (b) An object is pushed in the positive x direction, but the force grows weaker as it moves to the right (positive x).

   (c) An object is pushed away from $x = 2$ on both sides.

7. If we tell you that "At $x = 10$ this object's potential energy is negative," what can you conclude about the behavior of the object? (Choose one.)

   A. The object will tend to move toward $x = 10$.

   B. The object will tend to move away from $x = 10$.

   C. The object cannot physically be at $x = 10$.

   D. You won't tell me that, because a negative potential energy is impossible.

   E. A negative potential energy is possible, but it doesn't physically tell you anything.

For Deeper Discussion

8. A block sliding along the floor with initial velocity v_0 eventually comes to rest. We typically say that the initial energy of the system was $(1/2)mv_0^2$, the final energy is zero, and the energy was lost due to the non-conservative force of friction. But at the level of atoms and molecules, energy is never lost. Where did it go?

Problems

9. The Sun creates a gravitational field drawing all objects toward itself. The potential energy of an object a distance r from the Sun is given by $U(r) = - GM_{Sun}m_{object}/r$, where G is a positive constant and r is the distance from the center of the Sun. (In this problem you will treat r as the one-and-only dimension. That's not valid in general, but it works, for instance, if the object starts at rest.)

   (a) Draw a quick qualitative sketch of $U(r)$. Include only positive values of r.

   (b) Describe the motion of an object that starts at rest in this potential field.

   (c) Describe the motion of an object that starts moving away from the Sun in this potential field. *Hint*: A good answer should start with "It depends on how fast it's moving..."

   (d) Write a function $F(r)$ for the force of the Sun on an object.

   (e) As $r \to 0$ this function predicts that the force, and the kinetic energy, approach infinity. That does not actually happen. Why not?

10. An object is moving along the x axis, its potential energy at any point being given by the function $U(x) = kx$, where k is a positive constant.

    (a) If U is measured in joules and x in meters, what are the units of the constant k?

    (b) Write an equation for the force $F(x)$ experienced by this object.

    (c) Where is the force negative? Where is it positive?

    (d) Draw the $U(x)$ curve for this object and briefly explain how this curve can be used to reach the same conclusions as those you reached based on the force law.

    (e) If the item begins at rest at a positive x-value, describe qualitatively its behavior in both the short term and the long term.

    (f) What physical system might this equation describe?

11. [*This problem depends on Problem 10.*]

    (a) Use Newton's second law $F = ma$ to turn your force law in Problem 10 into a second-order differential equation for $x(t)$.

    (b) Solve your differential equation to find the function $x(t)$. Your solution should introduce two new constants.

    (c) Find the particular solution that corresponds to the initial conditions $x(0) = x_0$, $x'(0) = 0$.

12. An object moves in the potential field $U(x) = ax^2 e^{bx}$, where a and b are positive constants.

(a) Find the x coordinates of all equilibrium points, and identify each such point as stable or unstable.

(b) Describe the long-term behavior of an object that starts at rest in this potential field at ...

　　i. $x = -3/b$

　　ii. $x = -2/b$

　　iii. $x = 0.1/b$

　　iv. $x = 3/b$

(c) In Chapter 6 we will look at particles that "come in from the left": that is, the initial condition is a high negative position and a positive velocity. Describe the behavior of such a particle in this potential field.

13. **The Simple Harmonic Oscillator**

An object is moving along the x axis, its potential energy at any point given by the function $U(x) = (1/2)kx^2$, where k is a positive constant.

(a) If U is measured in joules and x in meters, what are the units of the constant k?

(b) Write an equation for the force $F(x)$ experienced by this object.

(c) Where is the force negative? Where is it positive?

(d) At what x-value can the object stay at rest at equilibrium?

(e) Draw the $U(x)$ curve for this object and briefly explain how this curve can be used to reach the same conclusions as those you reached based on the force law.

(f) If the item begins at rest at some positive x-value, describe qualitatively its behavior in both the short term and the long term.

(g) What physical system might this equation describe?

14. **The Simple Harmonic Oscillator Solution**

[*This problem depends on Problem 13.*]

(a) Use Newton's second law $F = ma$ to turn your force law in Problem 13 into a second-order differential equation for $x(t)$.

(b) Show that the following function is a valid solution to your differential equation:

$$x(t) = A \sin\left(\sqrt{\frac{k}{m}}\, t\right) + B \cos\left(\sqrt{\frac{k}{m}}\, t\right).$$

(c) What are the units of the arbitrary constants A and B?

(d) Find the particular solution that corresponds to the initial conditions $x(0) = x_0$, $x'(0) = 0$.

5.2　Energy Eigenstates and the Time-Independent Schrödinger Equation

Question: In a classical (Newtonian) mechanics problem, you are given the forces that act on an object. How do you then find the position of the object?

Answer: You can't. Even if your knowledge of the forces is complete, the position and the velocity at any given time could be literally anything.

Why? Because you also need to know the initial conditions. You can plug the forces into a very important equation – Newton's second law – to calculate the "acceleration" of the object. From that, and from the initial conditions (usually the position and velocity at some specific time), you start making predictions.

We know, we're telling you things you already know. But that bit of classical review has important quantum parallels.

Question: In a quantum mechanics problem, you are given the potential energy field around a particle. (Remember that this is just an alternative way of specifying the forces.) How do you then find the wavefunction of the particle?

Answer: You guessed it – you can't! The wavefunction at any given time could be almost anything.

So what do you do? You plug the potential energy function into a very important equation – Schrödinger's equation – to calculate the "energy eigenstates" of the particle. From those, and from the initial condition (the wavefunction at some specific time), you start making predictions.

So our study of quantum mechanics begins with energy eigenstates: what they are, how to calculate them, and what to do once you know them.

5.2.1 Explanation: Energy Eigenstates and the Time-Independent Schrödinger Equation

At any given moment, the state of a particle is represented by its wavefunction.[2] The wavefunction generally can't tell you a particle's position, momentum, or energy, but it can tell you all the possible values you could get if you measured one of those quantities, and how likely you are to find each possible value. In Section 4.3 we talked about how, from a given wavefunction, you can calculate the probabilities for measurements of position. Now we are ready to tackle the same question for measurements of energy.

Energy Eigenstates

The key to finding probabilities for energy measurements is a special set of functions called "energy eigenstates."

Definition: Energy Eigenstate

If a particle's wavefunction ψ is an "energy eigenstate" (also called "energy eigenfunction"), then the particle has a definite energy. That is, there is a 100% chance of finding that particular energy, and zero chance of finding any other.

That particular energy is called the "eigenvalue" of that "eigenstate."

For instance, suppose a particle's current wavefunction is $\psi_1(x)$, which you know to be an energy eigenstate with eigenvalue 10 J. Then you know for certain that any measurement of the particle's energy will find the answer 10 J.

On the other hand, suppose the particle's current wavefunction is $\psi_2(x)$, which is *not* an energy eigenstate. Then you can predict the probabilities of various energy measurements (once we've taught you how to do that), but you can't predict the outcome with certainty.

Remember that every wavefunction is, mathematically, just a function. So what mathematical function has this wonderful property of being an "eigenstate of energy"? The answer is: it

---

[2] . . . plus another variable called "spin" that we'll discuss in Chapter 7.

depends on the potential energy field that the particle is in. For instance, in Chapter 7 we'll begin with the potential energy function that describes an electron in a hydrogen atom, and proceed mathematically to find the energy eigenstates of such an electron. We will see how this reproduces some of the predictions of the Bohr model, as well as making several new ones.

In the rest of this section we'll explain how to find energy probabilities once you know the energy eigenstates of a system, and then explain how to find those energy eigenstates. In the rest of this chapter we'll apply those techniques to a variety of simple systems.

Energy Probabilities

Active Reading Exercise: Energy Eigenstates

After analyzing a particular system, you have determined that two of its energy eigenstates and eigenvalues are

$$\psi_1(x) = Ae^{-kx^2} \text{ with associated eigenvalue } E_0$$
$$\psi_2(x) = Be^{-2kx^2} \text{ with associated eigenvalue } 4E_0$$

1. If the system is known to be in the state $\psi(x) = Ae^{-kx^2}$ when you measure its energy, what energy will you find?

2. If the system is known to be in the state $\psi(x) = (1/2)Ae^{-kx^2} + (\sqrt{3}/2)Be^{-2kx^2}$, what two possible energies might you find, and with what probabilities? (We haven't told you how to answer this part, but make your best guess before reading on. You may be able to guess at least part of the answer.)

In the first case the system is in an energy eigenstate and you are guaranteed to measure the associated energy eigenvalue, $E = E_0$. In the second case the wavefunction is a mixture of two energy eigenstates, so you might guess that you would get one of the two associated eigenvalues. You might also guess that, because the coefficient in front of ψ_2, is larger, you would be more likely to measure its associated energy eigenvalue. You would be right on both counts.

Specifically, the probability for each energy is the square of the coefficient in front of it.

Energy Probabilities

Suppose a system has energy eigenstates $\psi_1, \psi_2, \ldots,$ with associated eigenvalues E_1, $E_2, \ldots$ If the system's wavefunction is

$$\psi(x) = c_1\psi_1 + c_2\psi_2 + \cdots,$$

then the probability of measuring energy E_n is $|c_n|^2$.

In the preceding Active Reading Exercise, this means you would have a 1/4 chance of measuring E_0 and a 3/4 chance of measuring $4E_0$. Note that the squared coefficients in front of the energy eigenstates must add up to 1 in order for the sum to be normalized.

So now you know how to predict energy measurements when the wavefunction is an energy eigenstate (you're guaranteed to measure the associated eigenvalue) or when it's a sum of energy eigenstates (you could get any of the associated eigenvalues with probabilities given by $|c_n|^2$). What if the wavefunction isn't a sum of energy eigenstates, though? The answer is a remarkable mathematical fact that we will state without proof:

Once you have found the energy eigenstates of your system, you can (in principle) always rewrite your wavefunction $\psi(x)$ as a linear combination of those eigenstates.

That means, at least in principle, that the rule we gave above is all you need to find energy probabilities. The mathematical techniques for doing this in most cases are beyond the scope of this book, but we will show some simple examples.

That leaves one big question: given a potential energy function, how do you find the energy eigenstates? You use the Schrödinger equation.

The Time-Independent Schrödinger Equation

There are actually two Schrödinger equations. The "time-dependent Schrödinger equation" tells you how a given wavefunction will evolve over time; the "time-independent Schrödinger equation" tells you the energy eigenstates for a given system. Those two issues are more related than they sound. In Section 6.6 we'll see that the time-independent Schrödinger equation can be derived from the time-dependent one, and that (as a consequence) you calculate the time evolution of a wavefunction by first calculating its energy eigenstates.

Right now we're interested in the time-independent version. (It's common to use "Schrödinger equation" for either one and we will do so when it is clear from the context which one we mean.)

The Time-Independent Schrödinger Equation

If a particle with mass m is in a potential energy field $U(x)$, its energy eigenstates are the solutions to the time-independent Schrödinger equation,

$$-\frac{\hbar^2}{2m}\frac{d^2\psi}{dx^2} + U(x)\psi(x) = E\psi(x). \tag{5.2}$$

The solution to this equation is not one function $\psi(x)$ but many different functions, each corresponding to a different value of the constant E. Each such solution is an eigenfunction, representing the state of the system with definite energy E (its eigenvalue).

Solving Equation (5.2) can require advanced mathematical techniques, or computer-aided numerical approximations, for most potential energy functions. In this book we will work with some simple potential energy functions for which we can find the energy eigenstates by hand, and discuss the solutions and basic properties of a few more complicated ones. We will start with an example where we will present and test a solution, rather than finding one, simply as an illustration of the equation.

Example: The Schrödinger Equation

Question: A particle has potential energy $U(x) = (1/2)kx^2$ for some constant k. Show that $\psi(x) = Ae^{-\sqrt{mk}\, x^2/(2\hbar)}$ is an energy eigenstate and find its associated eigenvalue.

Answer: Calculate $\psi''(x)$. Then plug $\psi(x)$, $\psi''(x)$, and $U(x)$ into the Schrödinger equation:

$$-\frac{\hbar^2}{2m}\left(-\frac{\sqrt{mk}}{\hbar} + \frac{mk}{\hbar^2}x^2\right)Ae^{-\sqrt{mk}\, x^2/(2\hbar)} + \frac{k}{2}x^2 Ae^{-\sqrt{mk}\, x^2/(2\hbar)} = EAe^{-\sqrt{mk}\, x^2/(2\hbar)}.$$

That's a giant mess, but you can divide $Ae^{-\sqrt{mk}\, x^2/(2\hbar)}$ out of every term. Doing that, and distributing the factor $-\hbar^2/(2m)$ into the parentheses, turns it into something much more tractable:

$$\frac{\hbar\sqrt{k}}{2\sqrt{m}} - \frac{k}{2}x^2 + \frac{k}{2}x^2 = E.$$

This $\psi(x)$ is an energy eigenstate because it reduced the Schrödinger equation to "E equals a constant."

Conclusion: If the system is in the state $\psi(x) = Ae^{-\sqrt{mk}\, x^2/(2\hbar)}$ then an energy measurement is guaranteed to find $E = \hbar\sqrt{k}/(2\sqrt{m})$.

That example worked because E came out a constant. If E had ended up being a function of x, that would be invalid: that is, it would indicate that the $\psi(x)$ we were testing was not an eigenstate. You'll see an example like that in Problem 16.

Of course, testing an eigenstate is easier than finding one! We will spend much of this chapter (and later chapters), starting with Section 5.3, solving the Schrödinger equation for various potential energy functions to find their eigenstates.

Stepping Back: Energy Eigenstates and Other Wavefunctions

It is a common student misconception to think that the Schrödinger equation tells you what the wavefunction can be. In fact the wavefunction of any particle can be anything, as long as it satisfies three conditions.[3]

- The wavefunction must be everywhere continuous.
- The wavefunction must be everywhere differentiable.
- The wavefunction must be properly normalized.

Any function that obeys those restrictions is, in principle, a valid wavefunction that a particle might have. What the time-independent Schrödinger equation tells you is *which of those possible wavefunctions are energy eigenstates*. Finding those energy eigenstates turns out to be the most important step you can take to analyze a quantum mechanical system and predict its behavior.

---

3 We could technically leave out the first condition, since "differentiable" implies "continuous." But as a practical matter you often get one equation by requiring that the wavefunction be continuous and another by requiring that its derivative be continuous.

Why? What's so important about energy eigenstates? Here are some answers to that question.

- The first answer is the one we stressed above: when a particle is in an energy eigenstate, you know exactly what that particle's energy is. And when a particle is not in an energy eigenstate, you can express its wavefunction in terms of those eigenstates to find the probabilities of all possible energy measurements.

- When you know the energy eigenstates, you call the one with the lowest possible energy the "ground state." Many systems spend most of their time in their ground states. For example, most of the atoms that make up the objects around you are in their ground states. For that reason, much of chemistry is the study of how atoms interact when they are in their ground states.

- When a system transitions from one energy eigenstate to another it often emits or absorbs a photon. The energy of that photon is the difference between those two energy levels. So understanding the energy eigenstates allows us to understand and predict spectra. (Think, for instance, about the transition spectra in the Bohr model, Section 4.1.)

- Wavefunctions evolve over time. If a device emits an electron, then the electron will have a wavefunction sharply peaked around the exit of the device. A short while later that wavefunction will be peaked around some point farther away from the device.

  But if a particle is in an energy eigenstate, it will remain in that state as long as the potential energy field itself doesn't change. We say that energy eigenstates are the "stationary states" of a system. (This fact is implied by the time–energy uncertainty principle.)

Our last two points may seem to contradict each other. If a system in an energy eigenstate doesn't evolve, then how could a system like an atom transition from one energy eigenstate to another? The answer is that in practice all systems are interacting with the fluctuating electromagnetic fields around them. So the states that we calculate as energy eigenstates of an atom (or any other system in the real world) aren't perfect energy eigenstates and can therefore evolve over time (Problem 20).

Note also that when you measure the energy of a particle, you collapse its wavefunction into an energy eigenstate! That is, you force it to have one specific value of energy. Because the particle's new state is a "stationary" state, subsequent measurements will measure that same energy (provided the system is isolated).

5.2.2 Questions and Problems: Energy Eigenstates and the Time-Independent Schrödinger Equation

Conceptual Questions and ConcepTests

1. A thousand particles have been prepared in separate boxes, all with the same wavefunction. How would you experimentally determine whether that wavefunction is, or is not, an energy eigenstate?

2. The phrase "angular momentum eigenstate" has not been used at all in this section, but we're betting you can guess what it means. What can you say about measurements of a particle whose wavefunction is an angular momentum eigenstate? (Somewhere in your answer should appear the phrase "the eigenvalue of that eigenstate.")

3. Is it possible for a particle to be in a state in which any energy from 0 to ∞ is just as likely as any other energy? Why or why not?

4. If two eigenstates ψ_1 and ψ_2 are both associated with the *same* eigenvalue, they are said to be "degenerate." What can you say about a particle whose state is $A\psi_1 + B\psi_2$?

5. Can an energy eigenvalue be negative? Why or why not?

6. Under some circumstances, a particular eigenvalue E is associated with a mathematical function $\psi(x)$ that successfully solves the Schrödinger equation, but is not normalizable. What would you conclude about the probability of the particle having that particular energy?

7. If you start by knowing all the forces acting on a classical object, Newton's second law tells you the acceleration of that object. Write a sentence that describes Schrödinger's equation in similar terms: "If you start by knowing . . . the time-independent Schrödinger equation tells you . . ."

8. You have been given the potential energy function $U(x)$ around an object, and you have found a function $\psi(x)$ that satisfies Schrödinger's equation for that $U(x)$. You have also determined that $\int_{-\infty}^{\infty} |\psi(x)|^2 dx$ is 9. What do you conclude? (Choose one.)

    A. You can just divide your $\psi(x)$ function by 3. The resulting function will still solve Schrödinger's equation, and it will be properly normalized, so it represents an energy eigenstate.

    B. You need to start over again with Schrödinger's equation and find a completely different wavefunction; this one is not valid.

    C. Your $\psi(x)$ function is fine the way it is.

9. We said that knowing the energy eigenstates of a system allows you to predict the spectrum emitted by transitions between those states. But we also said that energy eigenstates are stationary, so how can such transitions occur?

For Deeper Discussion

10. A "position eigenstate" means a function $\psi(x)$ that is associated with one definite value of position (its eigenvalue). What kind of function does that? (This may well be a function you've never heard of

and therefore cannot name, but if you think hard you can describe it pretty accurately – or perhaps draw it!)

Problems

11. A particle has energy eigenstate $A_1\psi_1(x)$ with eigenvalue E_1, eigenstate $A_2\psi_2(x)$ with eigenvalue E_2, and so on. The A_n constants are properly chosen to normalize each eigenstate.

    (a) If the particle is in the state $A_4\psi_4(x)$, and you measure its energy, what will you find?

    (b) Suppose the particle is in the following state:

$$\sqrt{\frac{1}{10}}\, A_1\psi_1(x) + \sqrt{\frac{3}{10}}\, A_2\psi_2(x) + \sqrt{\frac{3}{5}}\, A_4\psi_4(x).$$

        i. What is the probability of measuring the energy to be E_2?

        ii. What is the probability of measuring the energy to be E_3?

    (c) Write down the equation that represents the normalization condition for correctly choosing the constant A_5.

    (d) Given the following equation, write down the equation that represents the normalization condition for correctly choosing the constants c_n:

$$\psi(x) = c_1 A_1\psi_1(x) + c_2 A_2\psi_2(x) + c_3 A_3\psi_3(x) \\ + c_4 A_4\psi_4(x) + c_5 A_5\psi_5(x).$$

    (e) Explain why the particle cannot be in the state $A_2\psi_2(x) + A_3\psi_3(x)$.

12. Suppose the following are all energy eigenstates and associated energy eigenvalues for some system. The coefficients are chosen to properly normalize each wavefunction: for instance, $\int_{-\infty}^{\infty} Ae^{-k_1 x^2} dx = 1$.

$$\psi_1(x) = Ae^{-k_1 x^2} \text{ with associated eigenvalue } E_0$$

$$\psi_2(x) = Be^{-k_2 x^4} \text{ with associated eigenvalue } 4E_0$$

$$\psi_3(x) = Ce^{-k_3 x^6} \text{ with associated eigenvalue } 9E_0$$

In each part below we'll give you the system's wavefunction, and you say what possible outcomes an energy measurement might have and how likely each outcome is.

(a) $\psi(x) = Be^{-k_2x^4}$

(b) $\psi(x) = (1/5)Ae^{-k_1x^2} + (\sqrt{24}/5)Be^{-k_2x^4}$

(c) $\psi(x) = (1/5)Ae^{-k_1x^2} + (2/5)Be^{-k_2x^4} + c_3 \times Ce^{-k_3x^6}$. *Hint:* You can answer this part even though we didn't tell you the value of c_3. Your answer should not have c_3 in it.

13. Consider three properly normalized energy eigenstates for some system: $\psi_1(x)$ has eigenvalue E_1, $\psi_2(x)$ has eigenvalue E_2, and $\psi_3(x)$ has eigenvalue E_3. Consider two combinations of those eigenstates:

$$\psi_a(x) = A(\psi_1(x) + 2\psi_2(x))$$

$$\psi_b(x) = B(\psi_1(x) + \psi_2(x) + \psi_3(x))$$

(a) Find A and give the probability of measuring $E = E_1$ for $\psi_a(x)$.

(b) Find B and give the probability of measuring $E = E_1$ for $\psi_b(x)$.

(c) If $\psi_3(x)$ happens to be the same function as $\psi_2(x)$, then the definitions we gave for ψ_a and ψ_b are identical, but you found different values for the normalization constants and, more importantly, different probabilities for $E = E_1$. Which one is correct, and what's wrong with the other calculation?

14. A particle moving in 1D experiences a constant force $F = k$.

(a) Write the associated potential energy function $U(x)$.

(b) Write (but do not solve) the time-independent Schrödinger equation for this particle.

15. A "simple harmonic oscillator" is any system that obeys the law $F = -kx$ for a positive constant k. A mass on a spring is an example.

(a) Write the potential energy function $U(x)$ for a simple harmonic oscillator.

(b) Write Schrödinger's equation for a simple harmonic oscillator. (Don't try to solve it.)

16. In the Example on p. 222 we considered a particle with potential energy $U = (1/2)kx^2$ and showed that $\psi(x) = Ae^{-\sqrt{km}x^2/(2\hbar)}$ is an energy eigenstate. Prove that $\psi(x) = A\sin\left(-\sqrt{km}x^2/(2\hbar)\right)$ is *not* an energy eigenstate by plugging it into the time-independent Schrödinger equation and showing that it doesn't reduce to E being a constant. (Don't worry about the fact that this is not normalizable.)

17. In the Example on p. 222 we showed that, for the simple harmonic oscillator potential $U(x) = (1/2)kx^2$, one eigenfunction is $\psi(x) = Ae^{-\sqrt{mk}\,x^2/(2\hbar)}$ with eigenvalue $E = \hbar\sqrt{k}/(2\sqrt{m})$. (There's nothing that follows that you can't do by hand, but you might want to use a computer to avoid some tedious algebra.)

(a) Show that $\psi(x) = Bxe^{-\sqrt{km}\,x^2/(2\hbar)}$ is also an energy eigenstate and find its (different) associated energy.

(b) Write an equation involving an integral that you would use to find the constant B. Do not solve it.

(c) For the following wavefunction, what are the possible energies of the particle, and what are their probabilities? (Assume the constants A and B have been chosen correctly to normalize their respective wavefunctions.)

$$\psi(x) = \sqrt{\frac{2}{7}}\, Ae^{-\sqrt{mk}\,x^2/(2\hbar)}$$
$$+ \sqrt{\frac{5}{7}}\, Bxe^{-\sqrt{km}\,x^2/(2\hbar)}$$

(d) Show that $\psi(x) = Cx^2e^{-\sqrt{km}\,x^2/(2\hbar)}$ is not an energy eigenstate.

18. In some region a particle feels a potential energy $U(x) = (\hbar^2 k/m)(\cot(kx))/x$, where k is a constant and m is the particle's mass. For each of the following, say whether it is an energy eigenstate for that particle, and if so give its associated energy eigenvalue.

(a) $\psi(x) = Ax(k - x)$

(b) $\psi(x) = B\sin(kx)$

(c) $\psi(x) = Cx\sin(kx)$

19. In some region a particle feels a potential energy $U(x) = U_0 + \dfrac{\hbar^2}{m(x^2 - Lx)}$, where L is a constant and m is the particle's mass. For each of the following, say whether it is an energy eigenstate for that particle, and if so give its associated energy eigenvalue.

(a) $\psi(x) = Ax(L - x)$

(b) $\psi(x) = B\sin(kx)$

(c) $\psi(x) = Cx\sin(kx)$

20. A hydrogen atom has an excited state called "2p." A hydrogen atom in the 2p state will stay there on average for about 1.6×10^{-9} s before decaying to the ground state.

(a) Use the time–energy uncertainty principle to estimate the energy uncertainty of the 2p state.

(b) How can we know that the ground state must have zero uncertainty?

(c) If the average photon energy emitted in the decay of a 2p hydrogen atom is 10.2 eV, find the range of possible wavelengths emitted by this decay.

5.3 The Infinite Square Well

Section 5.2 described the basic process we use to analyze many quantum mechanical systems: write the potential energy function $U(x)$, and then solve the Schrödinger equation with that potential energy to find the energy eigenstates.

In this section we'll apply this process to the simplest quantum mechanical system, a particle confined to a finite one-dimensional region and free to move with no forces within that region. This simple example will illustrate the general process of finding energy eigenstates and show some features that appear in many systems. Perhaps most importantly, we'll see how Schrödinger's equation leads mathematically to the conclusion that Planck, Einstein, and others had already reached based on experiments: energy levels are quantized.

5.3.1 Discovery Exercise: The Simple Harmonic Oscillator Equation

The following differential equation is an example of a "simple harmonic oscillator" equation:

$$\frac{d^2y}{dx^2} = -y. \tag{5.3}$$

You can read Equation (5.3) as "The second derivative of the mystery function is the same as the original function, but multiplied by -1."

1. Show that the function $y = 0$ is a solution to Equation (5.3).

2. Show that the function $y = e^{-x}$ is *not* a solution to Equation (5.3).

3. Show that $y = \sin x$ is a solution to Equation (5.3).

4. Find another solution. (We're asking for any function that satisfies Equation (5.3) *other than* $y = 0$ or $y = \sin x$. But don't just write something down that looks good: test it and make sure it works!)

5. Finally, find a non-zero solution to the following differential equation:

$$\frac{d^2y}{dx^2} = -9y.$$

Equations such as Equation (5.3) can be used to model a *classical* simple harmonic oscillator. We will see in Section 5.4 that the equations for a *quantum* harmonic oscillator are more complicated. But in this section we will see how equations such as Equation (5.3) apply quantum mechanically to a simpler scenario, the "infinite square well."

5.3.2 Explanation: The Infinite Square Well

Imagine a gas molecule that can move freely throughout a room, but can't get outside it. Imagine a proton bound tightly inside a nucleus, but able to move freely within that space. Imagine an electron moving in a short length of wire. All of these are examples of what is sometimes called a "particle in a box," a system that can move with virtually no forces within a certain region but cannot move outside it. The lack of forces and the impossibility of escape are always both approximations, but in the cases listed above and many others, they are pretty good approximations. Because this occurs frequently, and because it is the simplest quantum system to study, we will use it as our first example of how to find the energy eigenstates of a system.

First we need to write the potential energy function.

The Potential Energy Function

> **Active Reading Exercise: A 1D Particle in a Box**
>
> A particle can move with no forces in the region $0 < x < L$, but cannot move anywhere outside that region.
>
> 1. What must be true of the potential energy in that region, based on the fact that the particle feels no forces?
> 2. What must be true of the potential energy at the boundaries in order to prevent the particle from leaving that region?
> 3. Assuming the particle is purely classical and it begins at $x = L/2$ moving to the right, describe its behavior over time.

Our answers to the first two questions are below, and you can find our answer to the third one at www.cambridge.org/felder-modernphysics/activereadingsolutions

Because the particle experiences no forces within this region, the potential energy is constant there, and for simplicity we can set it to 0. To trap the particle in this region we need the potential energy to rise steeply at the edges. In the limit where we assume it is physically impossible for the particle to escape, the potential energy everywhere outside the region would be infinitely high.

The potential energy function is shown in Figure 5.4. Such a particle is often called a "particle in a box," and the potential field an "infinite square well."

In the regions $x < 0$ and $x > L$, where $U = \infty$, the only possible solution to Schrödinger's equation is $\psi(x) = 0$. (Look at Schrödinger's equation and see if you can convince yourself that, in the limit where $U \to \infty$, ψ must approach 0.) Since $|\psi|^2$ is the position probability

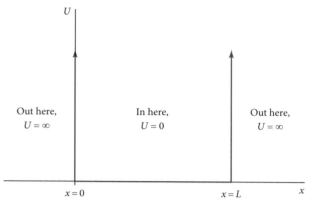

Figure 5.4 Potential energy $U(x)$ for an "infinite square well."

distribution, $\psi = 0$ means "the particle can't be here." Thus, in the region outside the well, the math leads us to a conclusion that we knew anyway.

In $0 < x < L$, where $U = 0$, we will write Schrödinger's equation and then rearrange it just a bit:

$$-\frac{\hbar^2}{2m}\frac{d^2\psi}{dx^2} + 0 = E\psi \quad \rightarrow \quad \frac{d^2\psi}{dx^2} = -\frac{2mE}{\hbar^2}\psi. \tag{5.4}$$

You may recognize Equation (5.4) as one of the most common differential equations: "The second derivative of $\psi(x)$ is $\psi(x)$ itself, multiplied by a negative constant." In Problem 16 you can show that the following solution works for any values of the constants A and B:

$$\psi(x) = A\sin\left(\frac{\sqrt{2mE}}{\hbar}x\right) + B\cos\left(\frac{\sqrt{2mE}}{\hbar}x\right) \quad (0 < x < L). \tag{5.5}$$

This solves Schrödinger's equation inside the well, but we're not done yet.

Continuity and the Boundary Conditions

In Section 5.2 we said that the wavefunction of a particle can be *almost* anything, regardless of the potential energy function. On the other side of that "almost" are three conditions: the function must be continuous, it must be differentiable, and it must be normalized.

Our infinite-square-well wavefunctions are going to violate one of these conditions because they will not be differentiable at the boundaries. This only happens because of the *infinite* potential change at those points, which could never happen in a real situation.

However, our wavefunctions must still be continuous.[4] What does that mean, mathematically? The function we described above is continuous for $x < 0$ and $x > L$ (since it is always zero), and it is also continuous for $0 < x < L$ (sines and cosines). But ψ must also be continuous where these functions meet, and that fact imposes boundary conditions: we must make Equation (5.5) go to zero at both $x = 0$ and $x = L$.

The first boundary is simple. Plugging $\psi(0) = 0$ into Equation (5.5) leads immediately to $B = 0$, so inside the well our solution simplifies to $\psi(x) = A\sin\left(\sqrt{2mE}\,x/\hbar\right)$.

---

4 See, for example, Griffiths, David, *Quantum Mechanics*, Cambridge University Press, 2017, for an explanation of why in this limit you lose differentiability but must still maintain continuity.

The second condition requires more thought. If $\psi(L) = 0$, then

$$A \sin\left(\frac{\sqrt{2mE}}{\hbar}L\right) = 0. \tag{5.6}$$

One obvious solution is $A = 0$: all the probabilities are zero so the particle can't be anywhere. That can't be right. But if $A \neq 0$, what do we have left to play with? We don't get to pick m (the mass of the particle), $\hbar$ (a universal constant), or L (the width of the well). So we can only satisfy this boundary condition by limiting the values of E.

Active Reading Exercise: Allowed Energies in an Infinite Square Well

What values of the constant E satisfy Equation (5.6)? This is a purely mathematical question, and it has infinitely many correct answers. If you're stuck, start by writing down one possible non-zero value of E that works. Then see if you can write a general formula for all of them.

The sine function gives you a zero at $\pm\pi$, $\pm 2\pi$, etc.:

$$\frac{\sqrt{2mE}}{\hbar}L = n\pi \quad \rightarrow \quad E = \frac{\hbar^2\pi^2}{2mL^2}n^2, \quad \text{where } n \text{ is any positive integer.} \tag{5.7}$$

Do you see how Equation (5.7) comes directly from Equation (5.6)? Just as importantly, do you see what Equation (5.7) is telling us? The energy of our particle can be exactly $\hbar^2\pi^2/(2mL^2)$. It can also be four times that value, or nine times, and so on. But no energies in between are allowed.

Energy quantization – which was an ad hoc hypothesis in the original formulations of Planck, Einstein, and Bohr – now emerges from the math of Schrödinger's theory.

Graphically, E controls the wavelength of the sine wave, and it has to be just long enough that it goes from $\psi = 0$ at $x = 0$ to $\psi = 0$ at $x = L$ (Figure 5.5).

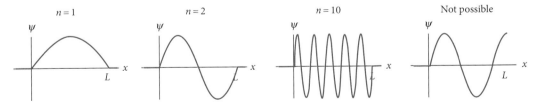

Figure 5.5 All four of these solve the Schrödinger equation inside the well, but the one on the right violates the condition of continuity at the right-hand edge.

Putting it All Together

The constants m, L, and $\hbar$ are fixed by the nature of the system. We've now determined B and the possible values of E. Plugging these in, we get $\psi(x) = A\sin(n\pi x/L)$. You can show in Problem 17 that normalization requires $A = \sqrt{2/L}$. Putting it all together,

$$\psi(x) = \begin{cases} 0 & x \le 0 \\ \sqrt{\dfrac{2}{L}} \sin\left(\dfrac{n\pi}{L}x\right) & 0 < x < L \\ 0 & x \ge L \end{cases}$$ Energy eigenstates of the infinite square well.

$$\tag{5.8}$$

Here n is any positive integer, and the energy is $E = \dfrac{\hbar^2 \pi^2}{2mL^2} n^2$.

Stepping Back 1: What Did We Do?

Solving the infinite square well is a model for how to solve many quantum mechanical systems, so it's worth reviewing what we did. We started with the assumption that a particle of known mass m was confined to a 1D region of width L and we set out to find its energy eigenstates.

1. We wrote down a potential energy function to describe the forces on our particle, and we wrote down Schrödinger's equation with that potential energy function. In this case it was defined as 0 inside the well and ∞ outside it.

2. We found the general solution to Schrödinger's equation in each region. Outside the well it was $\psi = 0$. Inside it had two arbitrary constants A and B, as well as the constant E from the Schrödinger equation.

3. We applied the condition of continuity at $x = 0$ to solve for B.

4. We applied the condition of continuity at $x = L$. This did not tell us the value of A, but instead told us the possible values for E.

5. Finally, we normalized to find A.

One point we want to highlight in this process is that quantization of energy did not come from solving the differential equation. Rather, quantization occurred when we imposed the boundary conditions. Energy quantization emerges in this same way for all "bound states": states for which the space allowed to the particle is restricted by the energy.

The result of this process was an infinite set of energy eigenstates and associated eigenvalues, each marked by a different positive integer n. An integer that distinguishes different eigenstates like that is called a "quantum number."

Stepping Back 2: What Did We Find?

One of the most common misconceptions among new quantum mechanics students is that solving the time-independent Schrödinger equation tells you what the wavefunction is. That's one reason we keep repeating that the wavefunction can be anything that obeys the three conditions: continuous, differentiable (except with an infinite potential jump like this one), and normalized.

So Equation (5.8) doesn't tell us what the wavefunction of a particle in our square well will be, or must be, or can be. It tells us its energy eigenstates. Keep that in mind as you answer the following Active Reading Exercise.

> ### Active Reading Exercise: What Did We Find?
>
> Equation (5.8) tells us the energy eigenstates of a particle in our infinite square well. Based on that:
>
> 1. If a particle is in the following state, what is its energy?
>
> $$\psi_A(x) = \sqrt{\frac{2}{L}} \sin\left(\frac{4\pi}{L}x\right)$$
>
> 2. If a particle is in the following state, what energies might a measurement find, and with what probabilities?
>
> $$\psi_B(x) = \sqrt{\frac{1}{8}}\left[\sqrt{\frac{2}{L}}\sin\left(\frac{4\pi}{L}x\right)\right] + \sqrt{\frac{7}{8}}\left[\sqrt{\frac{2}{L}}\sin\left(\frac{5\pi}{L}x\right)\right] \qquad (5.9)$$

This exercise is here to make sure you remember the meaning of those states we have found. If $\psi(x) = \psi_A(x)$, then the particle is in the $n = 4$ energy eigenstate, with a guaranteed energy of $16\hbar^2\pi^2/(2mL^2)$. If $\psi(x) = \psi_B(x)$, there is a 1/8 probability of measuring that same energy, and a 7/8 probability of measuring $25\hbar^2\pi^2/(2mL^2)$.

5.3.3 Questions and Problems: The Infinite Square Well

Conceptual Questions and ConcepTests

1. We said that continuity required $\psi(L) = 0$ and that led us to conclude that $\sqrt{2mE}\,L/\hbar$ had to equal $n\pi$. Each of the questions below has a different answer: a different reason why we didn't include certain values of n.

   (a) Why didn't we include non-integer values of n among our solutions?

   (b) Why didn't we include $n = 0$?

   (c) Why didn't we include negative (integer) values of n?

2. Equation (5.5) has both a sine and a cosine, but the next equation in the text has only a sine. Where did the cosine go?

3. Consider the infinite square well (Figure 5.4) with one change: on the domain $0 < x < L$, instead of $U = 0$, the potential energy is $U = U_0$ (a constant). How would that change the *behavior* (not the wavefunction) of a particle in this field? *Hint*: You will approach this question mathematically in Problem 12, but you can describe the effect on behavior right now with no math at all.

4. You are experimenting with 100 different particles. Each one is in its own infinite square well. Fifty of these particles are in the same state as each other, and it is a single energy eigenstate. The other 50 particles are in the same state as each other, and it is a superposition of different energy eigenstates. How would you experimentally determine which batch is which?

5. You have 1000 identical boxes, each of which contains a particle in an infinite square well. The first 500 boxes are labeled A and the second 500 B. At $t = 0$ you open all of the A boxes and measure the particles' positions. At $t = 1$ (one hour later) you do the same thing to all the B boxes. In each of the cases below, will you find the same distribution of positions in the A group and the B group? How do you know?

(a) Initially all the particles were in the ground state.

(b) Initially all the particles were in the state $(1/\sqrt{2})(\psi_1+\psi_2)$, where ψ_1 is the ground state and ψ_2 is the first excited state.

6. We'll use ψ_n to represent the nth energy eigenstate of a particle in an infinite square well. For example, $\psi_2 = (\sqrt{2/L})\sin(2\pi x/L)$ inside the well. Which of the following is a possible wavefunction for the particle? (Choose all that apply.)

   A. ψ_1

   B. ψ_2

   C. $\psi_1 + \psi_2$

   D. $(1/\sqrt{2})(\psi_1 + \psi_2)$

   E. $(1/\sqrt{2})(\psi_1 - \psi_2)$

   F. $Ax(L-x)$ (Choose this option if there is any value of A for which this is possible.)

7. We'll use ψ_n to represent the nth energy eigenstate of a particle in an infinite square well. For example, $\psi_2 = \sqrt{2/L}\sin(2\pi x/L)$ inside the well. List the following wavefunctions from lowest to highest in terms of the probability that you would find the particle in a small region near the middle of the well. (Some may be equal in the ranking.)

   A. ψ_1

   B. ψ_2

   C. $(1/\sqrt{2})(\psi_1 + \psi_2)$

For Deeper Discussion

8. The time-independent Schrödinger equation is a "linear, homogeneous differential equation" which implies – among other things – that the sum of any two solutions is still a solution. But for the infinite square well, $\sin(2\pi x/L)$ and $\sin(3\pi x/L)$ are both eigenstates of energy – solutions to the time-independent Schrödinger equation – and $\sin(2\pi x/L) + \sin(3\pi x/L)$ is not. That sounds like a contradiction: why isn't it?

9. The energy eigenstates of the infinite square well violate the condition of differentiability at the edges of the well. We said that was allowed because of the unphysical infinite jump in potential energy at those boundaries, but we also said that an infinite jump is the limit of physically allowed systems in which the potential energy jump at the boundaries approaches being infinitely steep. How can a non-differentiable solution be the limit of differentiable solutions?

Problems

10. Equations (5.7) and (5.8) represent the solution to Schrödinger's equation in an infinite square well of length L. Consider an infinite square well of length 1.

    (a) Write down the first three non-zero $\psi(x)$ functions on the domain $0 < x < 1$ and their associated energies.

    (b) Draw those three functions.

    (c) Can you draw the next function without looking at the equation?

    (d) Write the function $\psi(x)$ with $n = 1.5$, and sketch it. This function is a valid solution to Schrödinger's equation with $U(x) = 0$. Explain, based on your drawing, why it is *not* a valid state of our particle in a box.

11. Equations (5.7) and (5.8) represent the solution to Schrödinger's equation in an infinite square well of length L. Write down the solution for an infinite square well of length $2L$. *Hint*: If this takes you more than 30 seconds, you're working too hard.

12. Consider the infinite square well (Figure 5.4) with one change: on the domain $0 < x < L$, instead of $U = 0$, the potential energy is $U = U_0$ (a constant). (You examined this case qualitatively in Question 3; now you're going to do the math.)

    (a) Write down Schrödinger's equation for a particle in this field on the domain $0 < x < L$.

    (b) Rearrange your solution and write down the general solution, including two new constants A and B. (We haven't given you any "technique" for doing this, but the equation is similar enough to the one we solved that you should be able to work something out. Don't stop until you have mathematically confirmed that your $\psi(x)$ function works in your differential equation!)

(c) The boundary condition $\psi(0) = 0$ immediately tells you the value of one of your arbitrary constants. Apply that and write down a simpler $\psi(x)$ function than you had before.

(d) Now apply the boundary condition $\psi(L) = 0$ to limit the possible values of the constant E. Remember that U_0 is a constant!

(e) How is your solution like, and unlike, the solution we found for $U = 0$? Briefly explain the significance of your result.

13. A particle in an infinite square well is known to be in the following state:

$$\psi(x) = \sqrt{\frac{1}{2L}}\sin(\pi x/L) + \sqrt{\frac{1}{L}}\sin(2\pi x/L)$$

$$+ \sqrt{\frac{1}{2L}}\sin(3\pi x/L).$$

(a) Rewrite this as a set of coefficients times eigenfunctions, similar to Equation (7.13). Remember that each eigenfunction includes a factor of $\sqrt{2/L}$, so for example the first coefficient will *not* be $\sqrt{1/2L}$. (Use the notation ψ_n to indicate the nth energy eigenstate.)

(b) If you measure the energy of this particle, what are all the possible energies you might find, and with what probabilities? *Hint*: If your probabilities don't come out unitless, that means you did something wrong.

(c) If you then make a second measurement of the same particle, what energy will you find?

14. Equation (5.8) gives the wavefunction of a particle in our one-dimensional infinite square well, and the energy associated with that wavefunction.

(a) In general, the energy of a particle is a combination of kinetic and potential energy. Explain why the energy in Equation (5.8) represents only kinetic energy.

(b) What is the wavelength λ of the sine wave given in Equation (5.8)?

(c) The relationship between the kinetic energy E and the momentum p of a particle is $E = p^2/(2m)$. (If you're not familiar with that

form, you can get there very quickly from the classical $E = (1/2)mv^2$ and $p = mv$.) Use that equation to write the momentum of a particle with the energy given in Equation (5.7).

(d) You have now found the momentum and the wavelength of our particle in a box. Show that they obey the de Broglie relation (Section 4.2).

15. Find the energy eigenstates and eigenvalues for an infinite square well that goes from $x = -L$ to $x = L$.

In Problems 16 and 17 you will fill in steps left out of the solution in the Explanation (Section 5.3.2).

16. Show that Equation (5.5) is a valid solution of Equation (5.4) for any values of the constants A and B.

17. Normalize $\psi(x) = A\sin(n\pi x/L)$ in the region $0 < x < L$ to find A.

18. **Exploration: Perturbation Theory**

Consider an infinite square well with a bump in the middle: $U(x) = 0$ for $0 < x < 0.4L$, $U(x) = U_0$ for $0.4L < x < 0.6L$, $U(x) = 0$ for $0.6L < x < L$, and $U(x) = \infty$ everywhere else.

(a) Sketch this potential energy function.

Solving the Schrödinger equation for this potential energy is messy, but if U_0 is small enough then this system is very close to the regular infinite square well, and physicists use a technique called "perturbation theory" to approximate its ground state energy. The technique assumes that the ground state wavefunction of this system is the same as it is for the infinite square well. (It isn't, but it's a good approximation as long as U_0 is small.)

(a) The kinetic energy of the ground state is the same as the kinetic energy of the ground state of the simple well in the Explanation (since we're assuming the wavefunction is the same). What is that kinetic energy?

(b) What's the probability of finding the particle in the region $0.4L < x < 0.6L$?

(c) If you measure the potential energy you'll find either U_0 with the probability you calculated in Part (b) or 0 otherwise. What's the average value of potential energy that you'll find?

(d) Adding the kinetic and potential energies, what's the approximate ground state energy of this system?

(e) The energy you just found is the exact *expectation value* of the total energy for this wavefunction with this potential energy. Is that value higher or lower than the true ground state energy of this system? How do you know? (You can answer this question with no calculations.)

(f) We said this approximation only works when U_0 is small. Small compared to what?

5.4 Other Bound States

A "bound state" is one whose energy is less than the potential energy at both plus and minus infinity. For instance, a particle in a simple harmonic oscillator potential ($U \propto x^2$) is always bound because $\lim\limits_{x \to \pm\infty} U(x)$ is infinite. On the other hand, an electron attracted to a nucleus may be either bound or unbound, depending on how much energy it has.

Section 5.3 looked at the eigenstates of the infinite square well, which are all bound states. This section will discuss what aspects of that system do and don't apply to all bound states. Chapter 6 will look at unbound states.

5.4.1 Discovery Exercise: The Finite Square Well

🔍 A "finite square well" can be defined by a potential $U = 0$ for $0 < x < L$ and $U = U_0$ everywhere outside that region (Figure 5.6). A particle in this potential field is "bound" if $0 < E < U_0$.

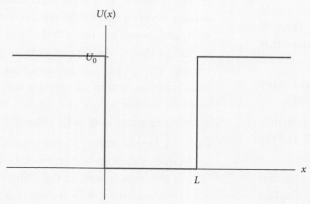

Figure 5.6 A finite square well.

In the region $x > L$, Schrödinger's equation becomes

$$\frac{d^2\psi}{dx^2} = \frac{2m(U_0 - E)}{\hbar^2}\psi. \tag{5.10}$$

We will express our solutions in terms of $\sqrt{U_0 - E}$ (a real quantity), not $\sqrt{E - U_0}$ (an imaginary quantity).

1. Two of the following functions are valid solutions of Equation (5.10) and the other two are not. Which two are? (Don't just guess: try them!)

   A. $\psi(x) = e^{\left(\sqrt{2m(U_0-E)}/\hbar\right)x}$

   B. $\psi(x) = e^{-\left(\sqrt{2m(U_0-E)}/\hbar\right)x}$

   C. $\psi(x) = \sin\left(\dfrac{\sqrt{2m(U_0-E)}}{\hbar}x\right)$

   D. $\psi(x) = \cos\left(\dfrac{\sqrt{2m(U_0-E)}}{\hbar}x\right)$

2. Of the two valid solutions you found, one of them is not a possible wavefunction for $x > L$. Which one, and why?

5.4.2 Explanation: Other Bound States

In Section 5.3 we found that a particle in an infinite square well can only have discrete energies, and we solved for those allowed energies and their associated energy eigenstates. In this section we will discuss two other systems that have "bound states": that is, states with $E < \lim\limits_{x\to\pm\infty} U(x)$. The potential functions are different and the solutions are therefore different, but the processes have much in common with the infinite square well – including the fact that the boundary conditions lead to quantized energy levels.

The Simple Harmonic Oscillator

You probably discussed masses on springs in your introductory physics class. That time was well spent, not because physicists spend so much of their time putting literal masses on literal springs, but because many other systems are governed by the same equations. Such systems are called "simple harmonic oscillators." Imagine a molecule alternately stretching and compressing, or a nucleus in a crystal oscillating about its equilibrium position.

A classical analysis of the mass-on-spring system starts with Hooke's law, $F = -kx$, which becomes a differential equation (often called the "simple harmonic oscillator equation") that is easy to solve (Problem 9):

$$F = -kx \quad\rightarrow\quad \frac{d^2x}{dt^2} = -\frac{k}{m}x \quad\rightarrow\quad x = A\sin\left(\sqrt{\frac{k}{m}}\,t\right) + B\cos\left(\sqrt{\frac{k}{m}}\,t\right). \tag{5.11}$$

As always, where we discuss forces in classical mechanics, we discuss potential energy functions in quantum mechanics. For $F = -kx$ the corresponding potential energy is $U = -\int F\,dx = (1/2)kx^2$. Equation (5.11) shows that any classical system in such a potential energy field will oscillate with angular frequency $\omega = \sqrt{k/m}$. Using that relationship, we can rewrite the potential energy in terms of ω as $U = (1/2)m\omega^2x^2$ (Figure 5.7).

Turning that argument around, we can *define* a simple harmonic oscillator as a system with potential energy function $U = (1/2)m\omega^2x^2$. Put that potential energy into Schrödinger's equation to find the energy eigenstates:

$$-\frac{\hbar^2}{2m}\frac{d^2\psi}{dx^2} + \frac{1}{2}m\omega^2x^2\psi = E\psi. \tag{5.12}$$

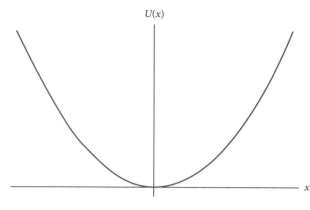

Figure 5.7 Potential energy function $U = \frac{1}{2}m\omega^2 x^2$ for a simple harmonic oscillator.

Equation (5.12) is not impossible to solve, but it's not trivial. You can see the method worked out in any advanced quantum mechanics book (and you'll do an important part of it in Problem 20), but the techniques involved are beyond the scope of this book. So we'll use a computer to find numerical solutions.

With or without a computer, the result depends on the constant E that you put in for the energy. Out of the other constants in the problem, the only combination that gives units of energy is $\hbar\omega$. So we typed Equation (5.12) into a computer with $E = \hbar\omega$, and it came back[5] with the graph in Figure 5.8.

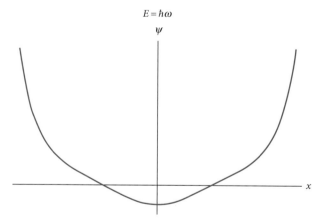

Figure 5.8 The solution $\psi(x)$ to the time-independent Schrödinger equation for the simple harmonic oscillator with $E = \hbar\omega$.

And there we run into a problem.

Active Reading Exercise: A Simple Harmonic Oscillator Solution

1. Why is the wavefunction in Figure 5.8 not physically possible?
2. What does that tell us about the energy eigenstates of the simple harmonic oscillator?

---

5 The general solution to a second-order differential equation has two arbitrary constants, so we set the value of ψ at two x-values to make this plot. Those choices won't have any effect on our conclusions.

It's not possible because the wavefunction approaches ∞ as $|x| \rightarrow \infty$, so this wavefunction can't be normalized. We haven't proven this – we've just drawn a suggestive plot – but you can prove that no solution to Equation (5.12) with $E = \hbar\omega$ can be normalized. That implies that a particle in a simple harmonic oscillator potential simply cannot have that energy. Only certain energies are possible.

To see what those are, we can try other values of energy. You can do this in Problem 11 and show that in the range $0 < E < \hbar\omega$ the only normalizable wavefunction is for $E = \hbar\omega/2$. When you fully solve for the eigenstates analytically you find that the allowed energies are $E = (n + 1/2)\hbar\omega$ for non-negative integers n.

The first few energy eigenstates are shown in Figure 5.9. These graphs show a pattern that we first saw in the infinite square well and that holds more generally for bound states: the wavefunction approaches zero at $\pm\infty$ (as it must), and in between it oscillates up and down a finite number of times. The higher the energy, the more oscillations it has.

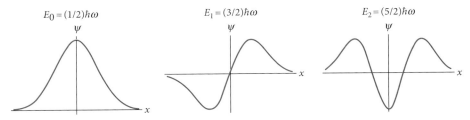

Figure 5.9 The first three energy eigenstates of the simple harmonic oscillator.

As we said earlier, the simple harmonic oscillator potential function can model a wide variety of important systems. When you encounter such systems later in this book – which you will, quite a few times – hold onto the following facts:

- There is an energetic ground state, $E = (1/2)\hbar\omega$, that is the lowest possible energy of the system. (In such a system, $E = 0$ is not an option.)

- All other energy levels are evenly spaced above that one. So a drop from any vibrational state to the next lower state will emit a photon with energy $\hbar\omega$.

You'll find this information summarized in Appendix C. You will also find the general formula for the eigenstates; that's useful reference information for the future, but we won't be using those functions in this book.

The Finite Square Well

Section 5.3 analyzed an infinite square well; a particle in it cannot escape from a certain region *no matter how much energy it has*. In reality a trapped particle can escape from any region if you give it enough energy,[6] so a more accurate model is a finite square well, e.g. $U = 0$ for $0 < x < L$ and U_0 everywhere outside that region (Figure 5.10).

---

6 except from a black hole.

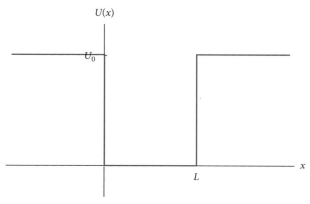

Figure 5.10 Potential energy function for a finite square well.

The Schrödinger equation and its solution inside the well are the same as they were for the infinite square well (it's the same potential):

$$\frac{d^2\psi}{dx^2} = -\frac{2mE}{\hbar^2}\psi \quad \rightarrow \quad \psi(x) = A\sin\left(\frac{\sqrt{2mE}}{\hbar}x\right) + B\cos\left(\frac{\sqrt{2mE}}{\hbar}x\right) \quad \text{for } 0 < x < L.$$

$$(5.13)$$

Outside the well, Schrödinger's equation looks almost exactly the same, but with E replaced by $E - U_0$:

$$\frac{d^2\psi}{dx^2} = -\frac{2m(E - U_0)}{\hbar^2}\psi \quad \text{for } x \leq 0, x \geq L. \qquad (5.14)$$

Pause on Equation (5.14) for a moment. Can you see the solution, simply by comparing it to Equation (5.13)?

Your first instinct might be to simply replace E in the first solution with $E - U_0$ for a new solution. That's just the replacement of one constant with another, right? Yes, absolutely – but note that the answer would now involve $\sqrt{E - U_0}$. If $E > U_0$ that's no problem, and we get our sines and cosines. If $E < U_0$ it's still no problem, if we don't mind complex solutions.

But we can find a real-valued solution for the $E < U_0$ case. Equation (5.14) becomes "The second derivative of this function is a *positive* constant times the function itself." Sines and cosines don't do that, but exponential functions do, so we conclude that

$$\psi(x) = Ce^{x\sqrt{2m(U_0-E)}/\hbar} + De^{-x\sqrt{2m(U_0-E)}/\hbar} \quad \text{for } x \geq L.$$

As we did with the simple harmonic oscillator, we have to throw out any solution that blows up. On the right side, e^x blows up as $x \to \infty$, but e^{-x} does not. So in the region $x \geq L$ we have to set $C = 0$. The reverse is true in the left region.

We have now arrived at the real-valued solution for a particle in a finite square well with $E < U_0$:

$$\psi(x) = \begin{cases} Fe^{x\sqrt{2m(U_0-E)}/\hbar} & x \leq 0 \\ A\sin\left(\frac{\sqrt{2mE}}{\hbar}x\right) + B\cos\left(\frac{\sqrt{2mE}}{\hbar}x\right) & 0 < x < L \\ De^{-x\sqrt{2m(U_0-E)}/\hbar} & x \geq L \end{cases} \qquad (5.15)$$

Those equations may look intimidating, but they paint a fairly simple picture. Inside the well the wavefunction oscillates sinusoidally. Outside the well (where the total energy is *less* than the potential energy), the wavefunction decays exponentially (Figure 5.11). Classically it would be impossible for a particle with $E < U_0$ to be found outside the well. Quantum mechanically it's not impossible, but the probability dies off rapidly as you move into the classically forbidden region.

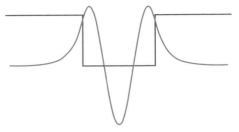

Figure 5.11 An energy eigenstate (red) of a finite square well (potential energy in blue).

Active Reading Exercise: The Finite Square Well

1. What physical requirements can we use to determine the unknown constants in Equation (5.15)?

2. Use those requirements to write down four equations involving the constants A, B, D, F, and E (the energy). Some of your equations should have L in them, but none of them should have the variable x.

The first answer – what physical requirements give us the constants – is that the wavefunction must be continuous and differentiable. (Full credit if you also included normalization. That gives us a fifth equation we'll be able to use to get the last of the constants.) You can see our equations at www.cambridge.org/felder-modernphysics/activereadingsolutions

You can solve three of these boundary conditions to eliminate three of the arbitrary constants, leaving just one constant A in front of all the solutions. The remaining equations can only be solved for certain discrete values of E. So, just as in the infinite square well, energy quantization arises from the boundary conditions.

You can do the calculations to rewrite Equation (5.15) with only one arbitrary constant in Problem 13. Unfortunately, the equation that then gives you the allowed values of E can't be solved analytically. You can solve it numerically to find the allowed energies of a finite square well in Problem 14.

Once you have the allowed energy eigenstates and eigenvalues, you can find the last arbitrary constant A by normalizing them.

Stepping Back: Bound States

Simple harmonic oscillators and finite square wells come up in many applications, but our goal is not to explore either of those in detail. We are using those two systems to illustrate the following features, which apply to all bound states:

1. Solving Schrödinger's equation gives you a general solution with arbitrary constants in it.

2. You can solve for the arbitrary constants using one or both of the following boundary conditions:

   (a) Require that ψ approach zero as $x \to \pm\infty$. (You might have a function that meets this criterion and is still unnormalizable, but most of the time this is sufficient.)

(b) For piecewise potential energy functions, impose continuity and differentiability at each boundary between regions.

You will always end up using all but one of these boundary conditions to solve for all but one of the arbitrary constants.

3. Solving the last boundary condition doesn't give you the last arbitrary constant, but instead restricts the energy to a discrete set of values.

4. You can then list the possible energy eigenstates and eigenvalues, labeled by an integer quantum number. (Sometimes you can't find analytical solutions, but you can still find numerical solutions.)

5. Finally, the last arbitrary constant is determined by normalizing the wavefunction. (This is separate from imposing *normalizability* in an earlier step.)

Before Schrödinger developed quantum mechanics, physicists like Planck and Bohr had introduced quantization of energy as an ad hoc hypothesis for certain systems. In quantum mechanics that quantization is not a postulate but a result that comes from solving for the energy eigenstates. It doesn't come from Schrödinger's equation, though, but rather from the boundary conditions that you apply after you solve Schrödinger's equation.

In Chapter 6 we'll see that unbound states do not have quantized energies.

5.4.3 Questions and Problems: Other Bound States

Conceptual Questions and ConcepTests

1. When does the potential field in Figure 5.12 represent a bound quantum mechanical system? (Choose one.)

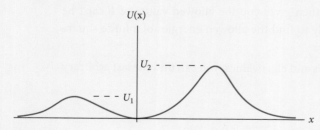

Figure 5.12

A. Never.

B. Only when $E < U_2$ and the system is between the two peaks.

C. Only when $E < U_1$ and the system is between the two peaks.

D. Always.

2. Define "bound state" in your own words.

3. Define "quantum number" in your own words.

4. Can the ground state energy of a system ever be lower than the minimum value of its potential energy function? Why or why not?

5. Figure 5.13 shows the fourth and tenth energy eigenstates of the simple harmonic oscillator. Which is which, and how can you tell? (We are *not* asking you to graph them on a computer: just look at the drawings and think about what they represent.)

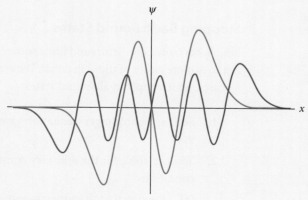

Figure 5.13

6. When we analyze the simple harmonic oscillator classically, one perfectly valid solution is $x(t) = 0$; the object sits motionless at the stable equilibrium point. Why does quantum mechanics rule out such a solution?

7. How will the shape of the bound energy eigenstates of the finite square well change as E increases? (This is asking about the limit as $E \to U_0$ from below.)

8. A particle is in a finite square well, in an energy eigenstate with $E < U_0$. If you measure the particle's position, is it possible to find it in a region where $U = U_0$? Discuss.

Problems

9. Fill in the missing steps in Equation (5.11). (This equation is purely classical.)

   (a) First, show how the force law leads to the differential equation.

   (b) Then, show that the given solution solves that differential equation.

10. The ground state vibrational energy of a hydrogen molecule is roughly $0.27\,\text{eV}$. Treating the molecule as a simple harmonic oscillator (an excellent approximation), what are its next two energy eigenvalues?

11. Figure 5.8 shows a function that solves Schrödinger's equation for the simple harmonic oscillator with $E = \hbar\omega$, but is not a valid wavefunction. In this problem you'll experiment to find an energy that does lead to a valid wavefunction.

   (a) To simplify your calculations, you can set $\hbar = \omega = m = 1$. (That is, effectively, choosing an unusual but useful set of units.) Write the Schrödinger equation for the simple harmonic oscillator using those values.

   (b) Have a computer solve the equation you wrote in Part (a) using the conditions $\psi(1) = \psi(-1) = 1$ for $E = 0.1$, $E = 0.3$, $E = 0.5$, $E = 0.7$, and $E = 0.9$. Plot the solution for each one in the domain $-3 < x < 3$ and describe its behavior.

   (c) The only energy eigenvalue in this range of energies is $E = 0.5$. What was wrong with all the other curves you drew?

12. This problem isn't about any particular bound state, but it illustrates a key mathematical idea.

   (a) Find the general solution to the differential equation $df/dx = 2kxf$. (You may be able to do this by just thinking about it, or you can use separation of variables.)

   (b) For some values of k the function $f(x)$ you found would be a possible wavefunction, and for others it wouldn't. For which values of k would $\psi(x) = f(x)$ be a possible wavefunction and why?

13. Equation (5.15) gives the energy eigenstates of the finite square well.

   (a) Use the requirements of continuity and differentiability to write four equations relating the different functions in this solution. Your equations should contain L but not x.

   (b) Solve three of your four equations to express F, B, and D in terms of A.

14. [*This problem depends on Problem 13.*] In Problem 13 you wrote four equations and solved three of them to express all of the other arbitrary constants in terms of A.

   (a) Plug those into the last equation to get something with only A. You should find that A cancels out, leaving an equation where the only unknown constant is E. Bring everything in that equation to one side, so that you can express it in the form $f(E) = 0$.

   (b) Simplify your equation as much as possible, using $\hbar = m = L = U_0 = 1$. That should leave E as the only letter in your equation.

   (c) Plot the function $f(E)$ over a large enough domain to see how many roots it has and to estimate their values.

   (d) What do your results in Part (c) tell you about this finite square well?

   (e) Repeat Parts (b) and (c) for $\hbar = m = L = 1$, $U_0 = 30$. How are the results different?

15. On pp. 239–240. we listed the steps involved in solving Schrödinger's equation for a bound state. Identify each of those steps in the solution for the infinite square well. This includes deciding which of those steps don't apply to the infinite square well and why.

16. A particle has the potential energy function $U(x) = U_0$ for $x \leq 0$, $U(x) = 0$ for $0 < x < L$, and $U(x) = 2U_0$ for $x \geq L$ (where $U_0 > 0$).

    (a) Write Schrödinger's equation in each of these three regions.

    (b) Write the general solution to Schrödinger's equation in each region, assuming $0 < E < U_0$. Your answer should have a total of six arbitrary constants. Use only real formulas, so make sure your square roots are all of positive quantities.

    (c) Which two of those arbitrary constants must equal 0 and why?

    (d) Figure 5.11 shows an energy eigenstate for a symmetric finite square well. How do the eigenstates of this problem's *asymmetric* well differ from that one, and what does that difference imply about the probability distribution for position?

17. **A Half Finite Well**

    A particle has the potential energy function $U(x) = \infty$ for $x \leq 0$, $U(x) = 0$ for $0 < x < L$, and $U(x) = U_0$ for $x \geq L$.

    (a) What is $\psi(x)$ in the region $x \leq 0$?

    (b) Write the general solution to the Schrödinger equation using real functions in each of the two remaining regions, assuming $0 < E < U_0$. Your answer should have a total of four arbitrary constants.

    (c) Two of the arbitrary constants from Part (b) must be zero. Which ones, and why?

    (d) Write an equation relating the two remaining arbitrary constants, and explain how you know it is true. Use it to rewrite the wavefunction in all of space with only one arbitrary constant.

    (e) Write an equation that the energy eigenvalue E must satisfy. Your equation should include m, L, U_0, and $\hbar$, but not any of the arbitrary constants. (You do not need to solve this equation.)

18. **Exploration: An Absolute Value Potential**

    You're going to solve for the bound states of a particle with the potential energy function $U(x) = k|x|$, where k is a positive constant.

    (a) Does this system also have unbound states? How do you know?

    (b) Write Schrödinger's equation twice, once for $x < 0$ and once for $x > 0$.

    The solutions to those equations involve "Airy functions," written Ai and Bi, and are

    $$
    \left.
    \begin{aligned}
    \psi(x) &= A\,Ai\left(-q(E + kx)\right) \\
    &\quad + B\,Bi\left(-q(E + kx)\right) \quad x < 0 \\
    \psi(x) &= C\,Ai\left(-q(E - kx)\right) \\
    &\quad + D\,Bi\left(-q(E - kx)\right) \quad x > 0
    \end{aligned}
    \right\} \quad (5.16)
    $$

    where $q = (2m)^{1/3}/(\hbar k)^{2/3}$. Note that q is positive. Don't worry if you've never heard of Airy functions before. All the information about them that you need to solve this problem is at www.cambridge.org/felder-modernphysics/figures

    (c) Without writing any more equations you can argue that two of the four arbitrary constants in Equations (5.16) must be zero. Which two, and why? Rewrite $\psi(x)$ (for both regions) with those constants set to zero.

    (d) Write an equation that must be satisfied for $\psi(x)$ to be continuous at $x = 0$. Use that equation to write a relationship between the two remaining arbitrary constants, simplifying as much as you can. Using that relationship, rewrite $\psi(x)$ for both regions with only one arbitrary constant.

    (e) Write an equation that must be satisfied for $\psi(x)$ to be differentiable at $x = 0$. Since there is no simple derivative of Airy functions, you can just write things like $Ai'(\ldots)$ in your equation, but don't forget that you still need to use the chain rule when you differentiate. Why can't you use this equation to solve for the last arbitrary constant?

    (f) Instead, use the equation from Part (e) to write an expression for possible energies of the system. Your answer should include α'_n, meaning the nth zero of $Ai'(x)$.

(e) Find the ground state energy of an electron in an absolute value potential with $k = 10\ \text{eV/nm}$.

19. [*This problem depends on Problem 18.*] The procedure you followed in Problem 18 only gave half the allowed energies.

   (a) Identify the point in the problem where you could have done the calculation differently to find a different set of allowed values.

   (b) The energy of the first excited state comes from this new set; find that energy for the electron from Part (e) of Problem 18.

20. **Exploration: Building Energy Eigenstates Inductively**

   A simple harmonic oscillator has potential energy $U = (1/2)m\omega^2 x^2$. In this problem you'll work out a method that lets you use one energy eigenstate to find infinitely many more. Suppose ψ_n is the nth energy eigenstate with energy eigenvalue E_n, and set $\psi_{n+1}(x) = -\hbar\psi_n'(x) + m\omega x\psi_n(x)$. You're going to prove that ψ_{n+1} is also an energy eigenstate and find its energy eigenvalue.

   (a) Calculate $\psi_{n+1}'(x)$ and $\psi_{n+1}''(x)$ using the definition we gave for ψ_{n+1}. Your answer will contain derivatives of ψ_n up to $\psi_n'''(x)$.

   (b) Write the Schrödinger equation satisfied by ψ_n and solve it for $\psi_n''(x)$.

(c) Take the derivative of the equation you derived in Part (b) to find a formula for $\psi_n'''(x)$ in terms of $\psi_n(x)$ and $\psi_n'(x)$.

(d) Now you're ready to show that ψ_{n+1} is an energy eigenstate. Calculate $-(\hbar^2/(2m))$ $\psi_{n+1}''(x) + U(x)\psi_{n+1}(x)$. Use everything you've calculated so far to show that this equals a constant times $\psi_{n+1}(x)$.

(e) What is the energy eigenvalue of ψ_{n+1}? (Your answer should include E_n, the energy eigenvalue of ψ_n.)

(f) The ground state of the simple harmonic oscillator is

$$\psi_0(x) = \left(\frac{m\omega}{\pi\hbar}\right)^{1/4} e^{-x^2 m\omega/(2\hbar)}.$$

Its energy eigenvalue is $E_0 = (1/2)\hbar\omega$. Use the method you just derived to find ψ_1 and E_1. (Don't worry about normalizing it.)

What you've shown is that, starting from the ground state, you can derive an infinite sequence of energy eigenstates. It can also be proven that this sequence includes *all* the simple harmonic oscillator energy eigenstates. The mathematical process "multiply a state by $m\omega x$, and then subtract $\hbar$ times the derivative of the state" is called a "ladder operator": it allows you to start with one eigenstate and generate all the higher ones, like climbing the rungs of a ladder.

5.5 Math Interlude: Complex Numbers

Many students at this level of physics have been exposed to complex numbers lightly, and a long time ago. They probably remember that $i^2 = -1$, and they may have seen Feynman's favorite equation, $e^{i\pi} + 1 = 0$, presented with the dazzle of a magic trick. But they may be less familiar with the complex plane, with modulus and phase, and with the oscillatory nature of the complex exponential function.

As we explore the time evolution of quantum mechanical wavefunctions, the energy eigenstates of a free particle, and more advanced topics, complex numbers – and the behavior of complex exponentials in particular – will be essential background.

The first half of this section reviews basic definitions and properties. The second half specifically focuses on the complex exponential function. Even if you don't need the review in the first half, you might find the second half useful.

5.5.1   Discovery Exercise: The Complex Exponential Function

Equation (5.17) is the simplest example of the "simple harmonic oscillator (or SHO) equation,"

$$\frac{d^2y}{dx^2} = -y. \tag{5.17}$$

1. Show that $y = A\cos x + B\sin x$ is a solution to Equation (5.17).
2. Show that $y = Ce^{ix}$ is also a solution to Equation (5.17).

This differential equation, with its two very different-looking solutions, suggests a fundamental connection between complex exponential functions and real-valued sines and cosines.

5.5.2   Explanation: Complex Numbers

The quantities we measure in physics (length, time, energy, momentum, force, etc.) are real. We ask questions about those quantities using real numbers, and we expect answers in terms of real numbers. But in many areas of physics, including circuit analysis, particle physics, and quantum mechanics, the calculations leading from those real questions to those real answers often involve complex numbers.

A Brief Introduction to Imaginary and Complex Numbers

We begin with a definition: $i \equiv \sqrt{-1}$, or (equivalently) $i^2 = -1$. Based on that definition and basic rules of arithmetic, you can convince yourself of facts such as $(-i)^2 = -1$, $(3i)^2 = -9$, and $(1+i)^2 = 2i$.

- A "real number" is any number that does *not* involve i, such as -5 or $\sqrt{2}$ or $\pi/6$.
- An "imaginary number" is a real number times i.
- A "complex number" is a real number plus an imaginary number:

$$z = a + bi \qquad \text{where } a \text{ and } b \text{ are } real\ numbers.$$

In this expression, a is referred to as the number's "real part" (sometimes written $Re(z)$), and b as its "imaginary part" ($Im(z)$).

All the functions we use with real numbers – square roots, exponentials, trigonometric functions, etc. – can be extended to complex numbers. So expressions such as 2^{-i} and $\sin(3+2i)$ can always, with some work, be written in the form $a + bi$ with real a and b.

Example: Real and Imaginary Parts

Find the real and imaginary parts of the number $z = \dfrac{2-i}{3+4i}$.

Solution: You cannot find the real and imaginary parts until you rewrite the number in the form $a + bi$. The trick in this case is to multiply the top and bottom by $3 - 4i$:

$$\frac{2-i}{3+4i}\frac{3-4i}{3-4i} = \frac{6-11i+4i^2}{9-(16i^2)} = \frac{2-11i}{25} = \frac{2}{25} - \frac{11}{25}i.$$

We conclude that $Re(z) = 2/25$ and $Im(z) = -11/25$.

The preceding Example used an important trick: we rewrote the fraction by multiplying the numerator and denominator by the "complex conjugate" of the denominator.

Complex Conjugate

The "complex conjugate" of the number $z = a + bi$ is $z^* = a - bi$. The real part stays the same, and the imaginary part changes sign.

In most cases, you can find the complex conjugate of a number by replacing every occurrence of i with $-i$. For instance, the complex conjugate of $i\sin(ix)$ is $-i\sin(-ix)$.

Modulus, Phase, and the Complex Plane

A number line is a representation of all real numbers. It includes positive and negative numbers, integers and fractions, rational and irrational numbers. But i is nowhere to be found. It is not possible to represent all complex numbers on a one-dimensional axis.

We can, however, represent all complex numbers in two dimensions (Figure 5.14). Each point on the "complex plane" represents one complex number by using the horizontal axis for its real part and the vertical axis for its imaginary part.

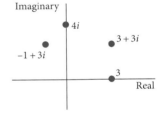

Figure 5.14 The complex plane.

Note that on the traditional (real-valued) xy plane, each point represents two numbers. For instance, when we graph a function, each point on the graph represents the number that was put into the function (x) and the number that came out (y). But on the complex plane, every point represents one number.

When we describe a complex number by giving its real and imaginary parts, we are specifying its location on the complex plane in Cartesian coordinates: the real part is the x coordinate, the imaginary part the y coordinate. We can instead specify that location in polar coordinates, which give the same information about the complex number in a different form.

- The point's distance to the origin is called the number's "modulus." The modulus of a number is designated by absolute value symbols, $|z|$. (We will consistently use the word "modulus" – the plural is "moduli" by the way – but in other texts you will see "magnitude" or "absolute value" meaning exactly the same thing.)

- The point's angle off the positive real axis is called the number's "phase." There is no universal symbol but we will generally use the notation ϕ_z for the phase of z.

Note that all four numbers that we use to describe a complex number – its real part a, its imaginary part b, its phase ϕ_z, and its modulus $|z|$ – are necessarily real numbers! The first

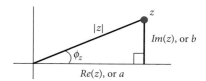

Figure 5.15 Phase (ϕ_z), modulus ($|z|$), and real and imaginary parts ($Re(z)$, $Im(z)$) of a complex number z.

two can be any real numbers. The phase can also be any real number, but any value outside $0 \leq \phi < 2\pi$ is redundant; $\phi = \pi$ and $\phi = 3\pi$ represent the same phase. The modulus, by definition, can never be negative. Figure 5.15 shows the relationships between these four variables:

$$a = |z| \cos \phi_z, \quad b = |z| \sin \phi_z, \quad a^2 + b^2 = |z|^2, \quad \tan \phi_z = \frac{b}{a}.$$

As we explore complex-valued wavefunctions, we will make much use of the following facts.

Important Facts about the Modulus

When you multiply any complex number by its complex conjugate, you get the square of its modulus:

$$zz^* = |z|^2.$$

When you multiply two numbers, their moduli multiply:

$$|z_1 z_2| = |z_1| \, |z_2|.$$

When you add two numbers, their moduli do *not* add:

$$|z_1 + z_2| \neq |z_1| + |z_2|.$$

Below we prove the first rule, just to give you the idea. (You can prove the other two in Problem 17.)

$$zz^* = (a + bi)(a - bi) = a^2 - (bi)^2 = a^2 + b^2 = |z|^2.$$

The Complex Exponential Function

Just about any soup you concoct by stirring together real numbers and "i"s can be expressed in the form $a + bi$, where a and b are real numbers. We showed above how to rewrite the number $(2 - i)/(3 + 4i)$ in that form, which required a less-than-obvious trick involving complex conjugates. Here are a few more examples:

- i^7 can be rewritten as $-i$ (so $a = 0$ and $b = -1$). You can confirm this by multiplying i by itself seven times.
- $\sqrt{i}$ can be rewritten as $(1/\sqrt{2}) + (1/\sqrt{2})i$. You will confirm this in Problem 12.
- 2^{3i-4}, $\ln(i - 2)$, $\sin(3i + \pi/6), \ldots$ All of these can be rewritten in $a + bi$ form.

Of particular interest to us are exponents. How can we rewrite the function e^{ix} in the form $a + bi$? The Discovery Exercise (p. 244) suggests an answer by pointing to a connection between complex exponential functions and real-valued sines and cosines.

> ### Euler's Formula for Complex Exponentials
>
> $$e^{ix} = \cos x + i \sin x \qquad (5.18)$$
>
> If x is real then Equation (5.18) breaks a complex exponential function into its real part ($\cos x$) and its imaginary part ($\sin x$). (If x is not real then Equation (5.18) is still true, but $\cos x$ no longer represents the real part!)

As an example, $e^{i\pi/6}$ can be written as $\sqrt{3}/2 + (1/2)i$.

You can show how the simple harmonic oscillator equation can be used to prove Euler's formula in Problem 22, and prove the same rule using Taylor series in Problem 23. In other problems you will prove the following rules, all of which will be important to us.

> ### Important Corollaries to Euler's Formula
>
> - For any real numbers ρ and α, the number $z = \rho e^{i\alpha}$ has modulus $|z| = \rho$ and phase $\phi = \alpha$ (Problem 18).
> - In particular, the number $e^{i\alpha}$, where α is real, has modulus 1. So multiplying any complex number by $e^{i\alpha}$ changes the number, but does not change its modulus.
> - $e^{-ix} = \cos x - i \sin x$ (Problem 21).

5.5.3 Questions and Problems: Complex Numbers

Conceptual Questions and ConcepTests

1. Figure 5.16 shows two numbers z_1 and z_2 on the complex plane.

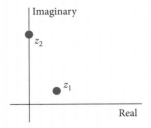

Figure 5.16

(a) Estimate ϕ_{z_1}, and give ϕ_{z_2} exactly.

(b) Which is bigger, $|z_1|$ or $|z_2|$?

(c) Estimate $|z_1/z_2|$.

2. Draw three different complex planes, and use them to show:

(a) the set of all numbers whose imaginary parts are greater than -2

(b) the set of all numbers whose moduli are greater than 2

(c) the set of all numbers whose phases are in $\pi/2 < \phi < 2\pi/3$.

3. Draw a complex plane.

(a) Show the number e^{i0}, and label it z_1.

(b) Show the number $e^{i\pi/2}$, and label it z_2.

(c) Show the number $e^{i\pi}$, and label it z_3.

(d) Show the number $e^{3i\pi/2}$, and label it z_4.

(e) A point moves around the complex plane, its position at any time t given by the complex number $z = e^{it}$, where $t \geq 0$. Describe its trajectory.

4. Does the function $z(t) = e^{it}$ represent (choose one and briefly explain your choice)

A. an increasing function?

B. a decreasing function?

C. a periodic function?

D. a constant function?

E. none of the above?

5. Which of the following represents all the points on the complex plane of the form $z = 5e^{i\alpha}$, where α can be any real number? (Choose one.)

A. A line

B. A half-line

C. A circle

D. A wedge

E. None of the above (If you choose this, specify the correct shape.)

6. Let $z_1(t) = e^{3it}$ and $z_2(t) = e^{(3i+2)t}$. Which of the following is true of the moduli $|z_1(t)|$ and $|z_2(t)|$? (Choose one and briefly explain your choice.)

A. $|z_1(t)| > |z_2(t)|$

B. $|z_1(t)| < |z_2(t)|$

C. $|z_1(t)| = |z_2(t)|$

D. It depends on the time t.

7. If z is a complex number and you multiply it by e^{2i} does that... (Choose one.)

A. increase the modulus?

B. decrease the modulus?

C. not change the modulus?

D. The answer depends on the original number z.

For Deeper Discussion

8. Imaginary numbers begin with this idea: "You can't actually take the square root of a negative number, but what if you could? Let's define i as $\sqrt{-1}$." Here's a similar-sounding idea: "You can't actually divide by zero, but what if you could? Let's define j as $1/0$." Show that we cannot introduce such a number, even imaginatively, into a coherent mathematical system.

Problems

9. Show that $1/i = -i$.

10. Find i^5.

11. Write $\sqrt{-8}$ in the form $a + bi$ with a and b real.

12. Show that $\sqrt{i} = (1/\sqrt{2}) + (1/\sqrt{2})i$. *Hint*: How would you show that $\sqrt{49} = 7$, if asked to do so?

13. In this problem you will find the modulus of the number $z = (4 + 7i)/(2 - 3i)$ in two different ways.

(a) Rewrite z in the form $a + bi$. *Hint*: Begin by multiplying the numerator and the denominator by the complex conjugate of the denominator.

(b) Find $|z|$ using the formula $|z|^2 = a^2 + b^2$.

Here's a different approach to the same question.

(c) Find z^* by replacing every occurrence of i with $-i$.

(d) Multiply zz^* and simplify your answer.

(e) Find $|z|$ using the formula $|z|^2 = zz^*$. If it doesn't match your answer to Part (b), something's gone wrong somewhere!

14. This problem gives you a bit of practice with complex exponentials and modulus, but it also sets up a fact that will turn out to be central to our study of the time evolution of wavefunctions. (You'll have to trust us on that for the moment.) Assume that t is a real number.

(a) Find the modulus $|e^{2it}|$.

(b) Find the modulus $|e^{3it}|$.

(c) Find the modulus $|e^{2it} + e^{3it}|$.

(d) If t represents time, which of your answers change over time, and which are constants?

15. Let $z_1 = 4 + 3i$ and $z_2 = 1 + i$.

(a) Multiply these two numbers to find the product $p = z_1 z_2$. Express your answer in $a + bi$ form.

(b) Find the moduli of all three numbers: $|z_1|$, $|z_2|$, and $|p|$. What is the relationship between these three numbers?

(c) Find the phases of all three numbers: ϕ_{z_1}, ϕ_{z_2}, and ϕ_p. Express your answers in decimal form. What is the relationship between these three numbers?

16. (a) Suppose you want to find the phase of the number $z_1 = 3 + 6i$. Use a drawing to show why $\phi_{z_1} = \tan^{-1} 2$.

    (b) Now suppose you want to find the phase of the number $z_2 = -3 + 6i$. Show that using the formula $\phi_{z_2} = \tan^{-1}(-2)$ gives an answer in the fourth quadrant $(-\pi/2 < \phi < 0)$. How do you know that's the wrong phase?

    (c) So what is the phase of z_2?

17. In this problem you will look at what happens to the modulus when you put complex numbers together.

    (a) Prove that, for any two complex numbers, $|z_1 z_2| = |z_1|\,|z_2|$.

    (b) Prove that, in general, $|z_1 + z_2|$ is *not* $|z_1|+|z_2|$. *Hint*: The easiest way to prove that something is not always true is to find any counterexample.

    (c) Can you find any case for which $|z_1 + z_2|$ *is* $|z_1| + |z_2|$?

18. Prove that, for any real numbers ρ and ϕ, the number $z = \rho e^{i\phi}$ has modulus $|z| = \rho$ and phase ϕ.

19. What happens to the modulus of a number when you . . .

    (a) multiply the number by 3?

    (b) multiply the number by -3?

    (c) multiply the number by $3i$?

    (d) multiply the number by e^{3it}, where t is a real number?

20. Let $f(t) = e^{(2+3i)t}$. Plot $|f|^2$ as a function of time.

21. Prove that $e^{-ix} = \cos x - i \sin x$. *Hint*: Remember that the cosine is an even function (meaning $\cos(-x) = \cos x$) and the sine is an odd function $(\sin(-x) = -\sin x)$.

22. The Discovery Exercise on p. 244 points to one way of proving Euler's formula. The function $A \cos x + B \sin x$ is the general solution to Equation (5.17); any other solution must be equivalent. Because e^{ix} is also a solution, there must be some constants A and B for which

$$e^{ix} = A \cos x + B \sin x. \qquad (5.19)$$

    (a) Plug $x = 0$ into both sides of Equation (5.19) to find one of the constants.

    (b) Take the derivative of both sides of Equation (5.19). Then plug $x = 0$ into both sides of the resulting equation to complete the proof of Euler's formula.

23. The following three Maclaurin expansions (Taylor series around $x = 0$) are well worth knowing:

$$e^x = \sum_{n=0}^{\infty} \frac{x^n}{n!}$$

$$= 1 + x + \frac{x^2}{2!} + \frac{x^3}{3!} + \frac{x^4}{4!} + \cdots$$

$$\sin x = \sum_{n=0}^{\infty} (-1)^n \frac{x^{2n+1}}{(2n+1)!}$$

$$= x - \frac{x^3}{3!} + \frac{x^5}{5!} - \frac{x^7}{7!} + \cdots$$

$$\cos x = \sum_{n=0}^{\infty} (-1)^n \frac{x^{2n}}{(2n)!}$$

$$= 1 - \frac{x^2}{2!} + \frac{x^4}{4!} - \frac{x^6}{6!} + \cdots .$$

Use these series forms to prove Euler's formula, $e^{ix} = \cos x + i \sin x$. *Hint*: If you show that it works for all terms up to the fifth order, we'll believe you from there.

5.6   Time Evolution of a Wavefunction

Let's circle back to the analogy that we introduced at the beginning of Section 5.2 between classical and quantum mechanics.

- In classical mechanics the fundamental question is: "Given an object with some initial position and velocity, and given a set of forces acting on that object, how will its position and velocity change over time?" Answering this fundamental question requires two skills: using Newton's second law to find the acceleration, and using the acceleration to predict the time evolution of the object.

- The equivalent question in quantum mechanics reads: "Given a particle with some initial wavefunction, and given its potential energy function, how will its wavefunction change over time?"

It turns out that predicting how a wavefunction will evolve depends on knowing the energy eigenstates of the system (even if the wavefunction isn't an energy eigenstate). So in quantum mechanics you also need two skills: using Schrödinger's equation to find the energy eigenstates, and using the energy eigenstates to predict the time evolution of the wavefunction.

With those two skills in mind, you can see where we are in this chapter and where we need to go. In Section 5.2 (and most of the rest of this chapter) the time-independent Schrödinger equation is used to find the energy eigenstates associated with a given potential function. In this section we will present a general rule that you can use to predict how any wavefunction will evolve over time, based on knowing its energy eigenstates. In Chapter 6 we will present the *time-dependent* Schrödinger equation, the fundamental axiom of quantum mechanics, and show how it leads to that time-evolution rule.

This section relies on the math from Section 5.5. Make sure you are comfortable with complex numbers, modulus, complex conjugates, complex exponentials, and Euler's formula, or you'll miss all the fun.

5.6.1   Discovery Exercise: Time Evolution of a Wavefunction

In each question below we'll give you a complex number and ask you to calculate its modulus squared. Assume x is real, and simplify your answers as much as possible. You should be able to write each answer in terms of all real quantities (no i).

1. $z_1 = e^{ix}$, so $|z_1|^2 =$
2. $z_2 = e^{2ix}$ so $|z_2|^2 =$
3. $z_3 = z_1 + z_2 = e^{ix} + e^{2ix}$ so $|z_1 + z_2|^2 =$

     Hint: The answer is *not* the sum of the previous two answers, $|z_1|^2 + |z_2|^2$.

5.6.2   Explanation: Time Evolution of a Wavefunction

You have a particle with a known wavefunction in a known potential energy field. How will that wavefunction evolve over time?

This section will build up a process for answering that question through three scenarios of increasing complexity. First, though, we need to introduce some notation.

Some Notation

It's common in quantum mechanics to write a wavefunction at one particular time as $\psi(x)$, and to write the wavefunction at all times as $\Psi(x, t)$. (Those symbols are the lowercase and uppercase versions of the Greek letter psi.) So if we say a particle's wavefunction started out as "$\psi(x)$ equals some function" and ask you what it will be at later times, the answer will be in the form "$\Psi(x,t) = \ldots$" Saying "the initial wavefunction is $\psi(x) = Ae^{-kx^2}$," for example, is equivalent to saying "$\Psi(x,0) = Ae^{-kx^2}$."

Throughout this section we will consider a system with energy eigenstates $\psi_1, \psi_2, \ldots$, and corresponding energy eigenvalues $E_1, E_2, \ldots$, so any time we write ψ with a numerical subscript that will mean an energy eigenstate.

Scenario 1: An Energy Eigenstate

If a particle is in an energy eigenstate its evolution is very simple.

Time Evolution of an Energy Eigenstate

Suppose at time $t = 0$ a particle is in energy eigenstate $\psi_n(x)$ with energy eigenvalue E_n. That is,

$$\Psi(x,0) = \psi_n(x).$$

Assuming the potential energy field doesn't change, the particle's wavefunction at any later time t will be given by

$$\Psi(x,t) = \psi_n(x)e^{-iE_n t/\hbar}. \tag{5.20}$$

You can relate Equation (5.20) to some of the early quantum ideas in Chapters 3 and 4. Euler's formula tells us that $e^{-iEt/\hbar}$ oscillates with angular frequency $\omega = E/\hbar$ and thus frequency $\nu = E/(2\pi\hbar) = E/h$. So the quantum mechanical time-evolution rule reproduces the relation $E = h\nu$ that Planck and Einstein proposed for photons and that de Broglie then extended to other particles.

In Chapter 6 we will show you how to derive this rule from the basic postulate of quantum mechanics, but here we want to explore some of its consequences.

Active Reading Exercise: Evolution of an Energy Eigenstate

Consider a particle in an infinite square well, whose initial wavefunction is given by the following energy eigenstate:

$$\Psi(x,0) = \psi_3(x) = \sqrt{\frac{2}{L}}\sin\left(\frac{3\pi}{L}x\right), \qquad E_3 = \frac{9\hbar^2\pi^2}{2mL^2}.$$

1. Write $|\Psi(x,0)|^2$, which determines position probabilities at $t = 0$.
2. Write $\Psi(x,\pi/3)$, the wavefunction at $t = \pi/3$, based on Equation (5.20).
3. Write $|\Psi(x,\pi/3)|^2$, which determines position probabilities at $t = \pi/3$. Simplify your answer.
4. How did the position probabilities change between these two times?

You can see our answers to that exercise at www.cambridge.org/felder-modernphysics/activereadingsolutions

The bottom line in those results is that the position probabilities did *not* change between these two times. This results mathematically from the fact (Section 5.5) that $|e^{ix}|^2 = 1$ for any real x.

How general is that result? It has nothing to do with the particular value $t = \pi/3$. It has nothing to do with the particular eigenstate $\psi_3(x)$, or even with the infinite square well. But on the other hand, it certainly does not mean that "For any wavefunction at any time, the position probabilities never change." (Think for a moment about the universe that would imply.) Here instead is the generalization:

> *A particle in an energy eigenstate will not change its position probabilities over time.*

The wavefunction is changing – that's important, as we shall see! – but the position probabilities are not. Our goal right now is that you can see how that generalization follows from Equation (5.20). But you should also see that it restates the claim we made in Section 5.2 that energy eigenstates are "stationary": if a particle is in one of those states, nothing measurable about it changes over time (see Question 1). You may also see how it connects to the time–energy uncertainty principle.

Scenario 2: A Sum of Energy Eigenstates

Now let's consider a particle whose wavefunction is the sum of two energy eigenstates: $\psi(x) = A\psi_m(x) + B\psi_n(x)$. Based on what we said above about how energy eigenstates evolve, take a moment and try to guess what will happen to this state over time.

The answer is that each individual eigenstate will be multiplied by $e^{-iEt/\hbar}$, just as in Equation (5.20). But the E in that equation is the eigenvalue *of that particular eigenstate*, so the individual pieces are not multiplied by the same exponential function!

$$\Psi(x,0) = A\psi_m(x) + B\psi_n(x) \quad \rightarrow \quad \Psi(x,t) = A\psi_m(x)e^{-iE_m t/\hbar} + B\psi_n(x)e^{-iE_n t/\hbar}.$$

Active Reading Exercise: Evolution of a Sum of Two Eigenstates

As an example of our two-eigenstate rule, suppose a particle starts out at $t = 0$ as an equal combination of two real-valued energy eigenstates ψ_1 and ψ_2:

$$\psi(x) = A[\psi_1(x) + \psi_2(x)]. \tag{5.21}$$

Will the probability position function for $\Psi(x,t)$ change over time? Don't just say yes or no: calculate the position probability distribution at a time $t > 0$ and see whether our result for the single eigenstate is still valid or not.

You can see us working through all the calculations at www.cambridge.org/felder-modernphysics/activereadingsolutions

On that page we find the position probability distribution for this wavefunction at some later time t:

$$|\Psi(x,t)|^2 = A^2 \left[\psi_1^2 + \psi_2^2 + 2\psi_1\psi_2 \cos\left(\frac{E_2 - E_1}{\hbar} t\right) \right]$$

(assumes $\Psi(x,0) = A(\psi_1 + \psi_2)$ and ψ_1 and ψ_2 are real). (5.22)

If we had multiplied both terms by the same complex exponential, the position probabilities would not have changed. But multiplying one term by $e^{-iE_1t/\hbar}$ and the other term by $e^{-iE_2t/\hbar}$ changed everything. Equation (5.22) shows that for this case, unlike the energy eigenstate, the position probability itself is a function of time.

The rule "Equation (5.20) applies to each energy eigenstate separately" gives you everything you need. That rule implies that if a particle is in an energy eigenstate, its position probabilities will not change. And that same rule also implies that if a particle is in a superposition of eigenstates, its position probabilities *will* change over time.

Example: Time Evolution of a Particle in an Infinite Square Well

A particle in an infinite square well starts with the wavefunction

$$\Psi(x,0) = \frac{1}{\sqrt{L}}\left[\sin\left(\frac{\pi}{L}x\right) + \sin\left(\frac{2\pi}{L}x\right)\right] \qquad (0 < x < L).$$

Make a series of sketches showing the probability distribution over time.

Answer: Looking back at Section 5.3 we recognize this as the sum of the ground state and first excited state of the infinite square well. The ground state has energy $E_0 = \hbar^2\pi^2/(2mL^2)$ and the first excited state has energy $E_1 = 4E_0$. So the wavefunction at all later times is

$$\Psi(x,t) = \frac{1}{\sqrt{L}}\left[\sin\left(\frac{\pi}{L}x\right)e^{-iE_0t/\hbar} + \sin\left(\frac{2\pi}{L}x\right)e^{-4iE_0t/\hbar}\right].$$

You can calculate the probability distribution directly from this formula. But because this is the equal sum of two real-valued eigenstates, we already did the work when we calculated Equation (5.22):

$$|\Psi(x,t)|^2 = \frac{1}{L}\left[\sin^2\left(\frac{\pi}{L}x\right) + \sin^2\left(\frac{2\pi}{L}x\right) + 2\sin\left(\frac{\pi}{L}x\right)\sin\left(\frac{2\pi}{L}x\right)\cos\left(\frac{3E_0}{\hbar}t\right)\right].$$

Since the only time dependence is in the cosine, the whole function will return to where it started after a time $T = 2\pi\hbar/(3E_0)$, so the sketches below show a sequence of representative times between 0 and that time.

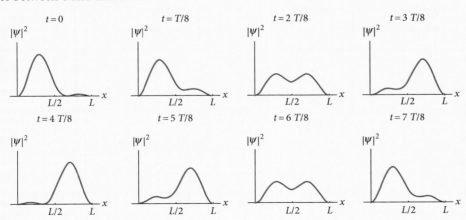

Initially the probability is almost (but not quite) entirely on the left side of the well. Over time it moves to the right, and then bounces back and moves left again. The probabilities will bounce back and forth like that forever. (Compare that to the classical motion of a particle in a well with no friction.)

The link below will take you to an animation of the infinite-square-well wavefunction. You can set the initial condition to any combination of the first five eigenstates. To replicate the preceding Example, set the first two eigenstates equal to each other, and the other three to zero.
www.cambridge.org/felder-modernphysics/animations

Scenario 3: Any Wavefunction Whatsoever

We've talked about the time evolution of an energy eigenstate, and a wavefunction that is a linear combination of energy eigenstates. What about wavefunctions that aren't sums of energy eigenstates?

If you remember Section 5.2 this question may seem familiar, and the answer is the same: *there is no such thing as a wavefunction that isn't a linear combination of energy eigenstates.* Every continuous, differentiable, normalizable function can be written as a sum of energy eigenstates. So no matter what your initial wavefunction $\psi(x)$ is, you can write it in the form $\psi(x) = A\psi_1(x) + B\psi_2(x) + \cdots$. Its time evolution is then $\Psi(x, t) = A\psi_1(x)e^{-iE_1 t/\hbar} + B\psi_2(x)e^{-iE_2 t/\hbar} + \cdots$.

Energy Probabilities

We've given you the rule for how Ψ evolves over time, and we've talked about the implications of that rule for position probabilities. (Short version: for a single energy eigenstate, position probabilities don't change, and for anything else they generally do.) But what about energy probabilities?

> ### Active Reading Exercise: Energy Probabilities
>
> In Section 5.2 we said that if a particle is in the state $\psi(x) = A\psi_1(x) + B\psi_2(x)$ and you measure its energy you will have probability $|A|^2$ of getting E_1 and $|B|^2$ of getting E_2. If you let the particle evolve forward until some later time t and then measure its energy, what is the probability you'll get the answer E_1?

At a later time the wavefunction is $\Psi(x, t) = A\psi_1(x)e^{-iE_1 t/\hbar} + B\psi_2(x)e^{-iE_2 t/\hbar}$. A very short calculation (Problem 10) shows that the probabilities of measuring E_1 and E_2 are the same at later times as they are initially.

It turns out that energy is unique in that way. In general:

- If a system is in an energy eigenstate then the probability distributions for all the system's measurable quantities (position, momentum, energy, angular momentum, . . .) are constant.

- If a system is not in an energy eigenstate then the probabilities for energy are still constant, but the probability distributions for all other measurable quantities will generally change.

This result is analogous in some ways to conservation of energy for a classical particle experiencing conservative forces. Its position, momentum, kinetic energy, and other properties might change over time, but its total energy does not.

Time Evolution and Measurement

The rule that we've presented here describes how a system's wavefunction evolves over time in a smooth, predictable way.

In the orthodox interpretation of quantum mechanics, however, *measurement* is a fundamentally different process that causes the wavefunction to collapse in a way not described by Schrödinger's equation.

In other interpretations, measurement is just a complicated interaction between the system and the measuring device, both of which follow Schrödinger's equation at all times. But the results are still only predictable as probabilities.

5.6.3 Questions and Problems: Time Evolution of a Wavefunction

Conceptual Questions and ConcepTests

1. Since particles in energy eigenstates don't measurably change, why does it matter that their wavefunctions evolve?

2. The Example on p. 253 showed an evolving probability distribution for a particle in an infinite square well. Name at least one way in which the time evolution shown there is similar to that of a classical particle in a well (with no friction), and at least one way in which it is different. (Don't just say "classical particles don't have probability distributions." Say something about the time evolution.)

3. Suppose a particle in an infinite square well (from $x = 0$ to $x = L$) is not in an energy eigenstate. Answer each of the questions below with "Yes," "No," or "Not enough information." Briefly explain each of your answers.

   (a) Will $\langle x \rangle$, the expectation value of its position, change over time?

   (b) If you average the value of $\langle x \rangle$ over a long time, will it on average be $L/2$?

   (c) If you wait a long time will $\langle x \rangle$ eventually settle down to $L/2$ and stay there?

4. Suppose you have two particles each in the state $\psi(x) = Ae^{-k(x-x_0)^2}$ (a bump centered on $x = x_0$). One of the particles has zero potential energy every-

where (no force) and the other has a simple harmonic oscillator potential energy function. A short while later, will the two particles' wavefunctions be the same or different? Explain how your answer results from the time-evolution rule presented in this section.

5. A particle in a simple harmonic oscillator is known at time $t = 0$ to have one of two energy levels: either E_1 (with probability P_1) or E_2 (with probability P_2). Which of the following best describes this particle a few seconds later?

   A. It still definitely has one of those two energy levels, with those two probabilities.

   B. It still definitely has one of those two energy levels, but the probabilities are different.

   C. It might have energy levels other than those two.

6. If two different energy eigenstates of a system have the same energy eigenvalues as each other, they are said to be "degenerate." If the system of a state is currently the superposition of two degenerate eigenstates, will its position probabilities change over time? Explain.

7. If we tell you the wavefunction of a particle in a finite square well, would you be able in principle to write it as a linear combination of the energy

eigenstates of a simple harmonic oscillator? How do you know? (Why you would want to is a question even we don't have an answer for.)

For Deeper Discussion

8. We said "If a system is not in an energy eigenstate then the probabilities for energy are still constant, but the probability distributions for all other measurables will generally change." You can view that as a quantum mechanical version of conservation of energy. Why isn't the momentum of an arbitrary wavefunction similarly conserved?

Problems

9. One of the most important results in this section is: "If a system is in an eigenstate of energy, then although the system's wavefunction changes over time, its position probability distribution function does not change." The text demonstrates this result for a specific case; now, prove it in general. (This should require a bit of math, but not a whole lot.)

10. Consider a particle with the initial wavefunction $\psi(x) = A\psi_1(x) + B\psi_2(x)$. Write $\Psi(x,t)$ and show that the probability of measuring $E = E_1$ is the same at all later times as it is initially.

11. The eigenstates and eigenvalues of a particle in an infinite square well are given in Appendix C. (For this problem you can assume all wavefunctions equal 0 outside $0 < x < L$ so you don't have to keep writing it.)

   Consider first a particle whose initial state is $\Psi(x,0) = \psi_2(x)$.

   (a) Write $\Psi(x,t)$ for this particle. (Your answer should not contain the letters ψ or E.)

   (b) Write the position probability distribution function $|\Psi(x,t)|^2$ of this particle. Simplify as much as possible.

   (c) What will you find if you measure the energy of this particle at time t?

   Now consider a particle whose initial state is $\Psi(x,0) = (1/2)\psi_2(x) + \left(\sqrt{3}/2\right)\psi_5(x)$.

   (a) Write $\Psi(x,t)$ for this particle.

   (b) Write the position probability distribution function $|\Psi(x,t)|^2$ of this particle. Simplify

as much as possible. *Hint*: You *cannot* use Equation (5.22).

   (c) At what time T will $|\Psi(x,t)|^2$ return to exactly what it was at $t = 0$?

   (d) What will you find if you measure the energy of this particle at a time t (not necessarily equal to T)?

12. [*This problem depends on Problem 11.*] Recall that $\langle x \rangle$, the expectation value of position, is given by $\int_{-\infty}^{\infty} x|\psi(x)|^2 dx$.

   (a) Calculate the expectation value of position, $\langle x(t) \rangle$, for the first particle in Problem 11.

   (b) Calculate the expectation value of position, $\langle x(t) \rangle$, for a particle in energy eigenstate ψ_5 of this infinite square well. *Hint*: You can do this with no calculation.

   (c) Calculate the expectation value of position, $\langle x(t) \rangle$, for the second particle in Problem 11.

   (d) Could you have calculated your third answer from your first two? If so, how? If not, why not?

13. [*This problem refers to one of the wavefunctions described in Problem 11. If you did that problem you can use your calculations from there as a starting point, but if you didn't then you can just do this problem entirely on a computer without doing that one first.*] Plot $|\Psi(x,t)|^2$ for a particle in an infinite square well with initial wavefunction $\Psi(x,0) = (1/2)\psi_2(x) + \left(\sqrt{3}/2\right)\psi_5(x)$. Make plots at several different times from $t=0$ until it returns to its original state. How are your plots similar to and different from the ones we made in the Example on p. 253?

14. The eigenstates and eigenvalues of a simple harmonic oscillator are given in Appendix C. For this problem we'll set $m = \omega = \hbar = 1$ in those formulas. (That isn't a physical change, just a choice of units.)

   (a) Suppose the wavefunction is initially ψ_0. Sketch $|\Psi(x,0)|^2$ and describe how that sketch would change if you plugged in a different value for t.

Now suppose the wavefunction is initially $\psi = (1/\sqrt{2})(\psi_0 + \psi_1)$.

(b) What is $\Psi(x,t)$? Your answer should have nothing in it but x, t, and numbers.

(c) Calculate $|\Psi(x,t)|^2$.

(d) At what time T will $|\Psi(x,t)|^2$ return to exactly what it was at $t = 0$?

(e) Plot $|\Psi(x,t)|^2$ at several times between $t = 0$ and $t = T$ (the period you found in Part (d)). Describe its behavior.

15. Suppose a particle starts out as the *difference* of two real-valued energy eigenstates: $\psi(x) = A(\psi_1(x) - \psi_2(x))$. Find $|\Psi(x,t)|^2$ and simplify as much as possible. Your answer should look similar to, but not identical to, Equation (5.22) for the *sum* of two energy eigenstates. *Hint*: Use Euler's formula while simplifying.

16. A classical simple harmonic oscillator will return to the state it started in after a time $2\pi/\omega$. Prove that the same is true for the position probabilities of a quantum simple harmonic oscillator that starts out in an equal superposition of two eigenstates, $\Psi(x,0) = A(\psi_m + \psi_n)$.

17. A classical simple harmonic oscillator will return to the state it started in after a time $2\pi/\omega$. Prove that the same is true for the position probabilities of a quantum simple harmonic oscillator, no matter what the initial wavefunction is. (If you're stuck you might want to try Problem 16 first, in which you solve the same problem for one specific case.)

Chapter Summary

Chapters 5 and 6 together provide an overview of quantum mechanical wavefunctions in one dimension: how to find them, how they evolve in time, and most importantly how to interpret them. These two chapters therefore provide the basic mathematical framework for analyzing the quantum world.

The equations for both chapters are collected in Appendix C. You may want to cross-reference that appendix as you read through this summary.

Section 5.1 Force and Potential Energy

Rather than specifying the forces on a particle and predicting behavior from $F = ma$, you can specify the potential energy field around a particle and predict behavior by saying that the particle will move toward lower energy.

- In classical mechanics the two formulations are equivalent as long as all the forces are conservative.

- In quantum mechanics the forces are always conservative, and the energy formulation is always used rather than forces.

Section 5.2 Energy Eigenstates and the Time-Independent Schrödinger Equation

A wavefunction that has a definite value of energy is called an "energy eigenstate" (or "energy eigenfunction"). The value of energy associated with an energy eigenstate is called its "energy eigenvalue."

You can find the energy eigenstates for a particle in a given potential energy field by solving the "time-independent Schrödinger equation" for that potential energy function.

- If a particle's wavefunction is not an energy eigenstate, you can write that function as a linear combination of energy eigenfunctions. The square of the coefficient in front of each energy eigenstate in that sum tells you the probability of measuring the corresponding energy.

- We caution you *not* to make the common beginner's mistake of believing that the solutions to the time-independent Schrödinger equation are "the possible wavefunctions of the particle." The solution tells you the energy eigenstates. But when you rewrite the particle's wavefunction as a combination of energy eigenstates you can make predictions for measurements of the particle's energy, and you can also predict how the wavefunction will evolve over time (Section 5.6).

Section 5.3 The Infinite Square Well

An "infinite square well" is a system with a potential energy of 0 within a finite region and ∞ outside that region. This section uses that system as a simple example to illustrate the process of solving Schrödinger's equation.

1. Plug the potential energy function into Schrödinger's equation.
2. Find the general solution, which will involve arbitrary constants.
3. Apply the boundary conditions: the wavefunction must be everywhere continuous, and in general it must also be everywhere differentiable. (This second condition cannot be met at an infinite potential jump.) This step will often lead to quantized energy levels, as well as solving for all but one of the arbitrary constants.
4. Normalize the wavefunction. This will solve for the final arbitrary constant, giving you the eigenfunctions and eigenvalues of the system.

Section 5.4 Other Bound States

The process for finding energy eigenstates and eigenvalues of other systems is similar to the one for the infinite square well. For a bound particle (one whose energy is less than the potential energy at $\pm\infty$) the result is always quantized energy levels.

- A simple harmonic oscillator is one with potential energy $U = (1/2)kx^2 = (1/2)m\omega^2 x^2$. A full solution of Schrödinger's equation with this potential is beyond the scope of this book, but you can trace out some of the steps. You end up with regularly spaced energy values.

- For a particle in a finite square well, whose energy is less than the energy of the potential jump, you arrive at the surprising conclusion that the particle's wavefunction is *not* zero in the classically forbidden region, although it decays rapidly toward zero.

- The energy eigenstates and eigenvalues for all three bound systems that we have discussed are given in Appendix C.

Section 5.5 Math Interlude Complex Numbers

- Euler's formula decomposes a complex exponential into its real and imaginary parts:

$$e^{ix} = \cos x + i \sin x.$$

- The complex exponential function $z = \rho e^{i\omega t}$ has a constant modulus of ρ, and a phase that oscillates with period $2\pi/\omega$.

Section 5.6 Time Evolution of a Wavefunction

If a particle's wavefunction begins in an energy eigenstate $\psi_n(x)$ with eigenvalue $E_n(x)$, that wavefunction will evolve over time as

$$\Psi(x, t) = \psi_n(x)e^{-iE_n t/\hbar}.$$

- Multiplying a wavefunction by $e^{-iE_n t/\hbar}$ changes the phase but not the modulus. So if a particle is in an eigenstate of energy, its measurable properties will not change over time. For this reason we describe energy eigenstates as "stationary states" of a system.

- If a particle's wavefunction is not an energy eigenstate, you can write its wavefunction as a linear combination of energy eigenstates. The wavefunction evolves by multiplying each eigenstate by $e^{-iE_n t/\hbar}$ with the appropriate value for E_n.

- The result for wavefunctions that aren't energy eigenstates is that the probability distribution for energy does not change, but the probability distributions for other quantities such as position and momentum generally do change. We will explore the resulting behavior more carefully in Chapter 6.

6

Unbound States

This chapter and Chapter 5 together form an introduction to the Schrödinger equation in one dimension.

In Chapter 5 you studied "bound" particles, meaning ones that don't have enough energy to escape toward infinity in either direction. In this chapter you'll learn about "unbound" particles, which do have that much energy. Boundary conditions led our earlier systems to discrete energy levels, so the unbound particles in this chapter do not have quantized energy levels.

Along the way you'll see a lot of math. Some of it may be review (standing and traveling waves, partial derivatives); some is likely to be new (Fourier transforms, partial differential equations). All of it will be presented in the context of quantum systems. We will see how an infinite number of continuous complex-valued wavefunctions can form a "wave packet" that in some ways resembles a classical particle – and how that packet sometimes acts in distinctly non-classical ways.

We will end the chapter with a look at the "time-dependent Schrödinger equation" and see how this one equation leads to several of the results that we presented as axiomatic in Chapter 5.

6.1 Math Interlude: Standing Waves, Traveling Waves, and Partial Derivatives

We assume that you are quite familiar with the function $y = \sin x$ and its variations. But "standing waves" and "traveling waves," oscillating in both time and space, may be less familiar. The equations that describe those waves are described in this section and summarized in Appendix F.

6.1.1 Discovery Exercise: A Traveling Wave

🔍 Consider the function $y(x, t) = 3 \sin(2x + \pi t)$.

1. At the instant $t = 0$ this represents a wave $y(x)$ spread out along the x axis. What is the wavelength of that wave?

2. At the position $x = 0$ this represents an oscillation $y(t)$ going up and down over time. What is the period of that wave?

6.1.2 Explanation: Standing Waves, Traveling Waves, and Partial Derivatives

Section 3.1 discussed sinusoidal functions with t (time) as the independent variable, and sinusoidal functions with x (position) as the independent variable. These two kinds of waves are mathematically the same, but we generally use different letters and different words to discuss them. For instance, $\sin(\omega t)$ has a "period" (time between crests) of $2\pi/\omega$, and $\sin(kx)$ has a "wavelength" (distance between crests) of $2\pi/k$.

In this section we will consider two different kinds of waves, called "standing waves" and "traveling waves," that vary in both time *and* space. Such a wave has both a period and a wavelength, and they don't mean the same thing.

The information in this section is summarized in Appendix F.

Standing Waves

Figure 6.1 shows a sine wave function of x. Its amplitude is decreasing over time; that is, every point is moving closer to the x axis. (Note that the zeros remain fixed.)

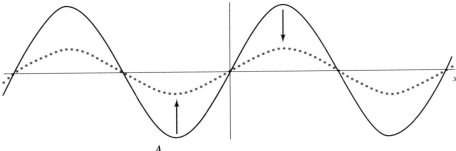

Figure 6.1 $A \sin(kx)$ becomes $\dfrac{A}{3}\sin(kx)$.

Now imagine that the amplitude continues to reduce until you are left with a horizontal line. Then the sine wave opens on the other side until it becomes a perfect mirror image of its original form, and then it starts to shrink again. When it returns to its original shape, the entire cycle starts over again.

This kind of "standing wave" might describe a vibrating guitar string. It might also describe the thermal radiation inside a cavity that we discussed in Section 3.4.

Active Reading Exercise: Standing Wave Wavelength and Period

A wave initially described by the function $y = 10\sin(4x)$ oscillates as described above: it shrinks until it reaches the x axis, grows until it reaches its original size on the other side of the x axis, and then oscillates back over and over.

Warning: We're about to ask you three questions. One of the three is a trick question!

1. What is the highest y-value ever reached at any place or time?

2. The "wavelength" is the distance along the x axis from one crest to the next. What is the wavelength of this wave?

> 3. The "period" is the time it takes for a peak to go all the way down to a valley and then back up to the peak. What is the period of this oscillation?
>
> Don't read further until you have written down all three guesses. Then, if you want to take one more adventurous step, see if you can write down the mathematical formula that describes this vibrating string as a function of both space and time. (You will have to introduce a new constant whose value you don't yet know.)

The answers to the first two questions above are that the highest value of y is 10 and the wavelength is $\lambda = \pi/2$. (If finding the wavelength of a sine wave isn't second nature, you might want to keep Appendix F handy.) The answer to the third question is: we haven't told you the period! A full oscillation could take three seconds, or 1.5 nanoseconds, or 10 years.

And here's our point. If you graph a sinusoidal function, the words "wavelength" and "period" refer to the same feature of the graph, distance between peaks. We use "wavelength" if it's a function of space and "period" if it's a function of time. Now, however, we are talking about a wave that varies in space *and* time. Its wavelength is a distance, its period is a time, and knowing one of them tells you nothing about the other.

We could make the same point by discussing "spatial frequency" (space) and "frequency" (time), or by discussing "wave number" (space) and "angular frequency" (time). In order to fully describe a standing wave you need to specify its maximum amplitude, *and* its spatial frequency (or wavelength), *and* its temporal frequency (or period).

Which brings us to the last question we asked above. Here is the equation for a standing wave:

$$y = A \sin(kx) \sin(\omega t) = A \sin\left(\frac{2\pi}{\lambda}x\right) \sin\left(\frac{2\pi}{T}t\right). \tag{6.1}$$

There is no universal way to draw or visualize a "multivariate function" such as Equation (6.1), but it is often helpful to consider the independent variables one at a time.

Active Reading Exercise: The Standing Wave Equation

Equation (6.1) represents a "standing wave." Suppose it gives the height of a vibrating string.

1. Imagine taking a picture of this vibrating string at $t = T/4$. The resulting snapshot is a sine wave in x. What are its amplitude and wavelength?
2. Now imagine watching the y-value at $x = \lambda/8$ moving up and down. The motion of that point is a sine wave in t. What are its amplitude and period?
3. Now, can you put all that together to see a mental video of the standing wave?

You can see our answers to the first two questions, along with a video of the resulting motion, at www.cambridge.org/felder-modernphysics/activereadingsolutions

Traveling Waves

Figure 6.2 shows a sine wave moving a fixed distance ϕ to the right. The amplitude and frequency are unchanged. Such a transformation is called a "phase shift."

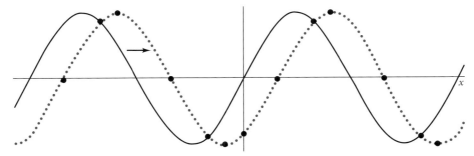

Figure 6.2 $A \sin(kx)$ becomes $A \sin[k(x - \phi)]$.

Now imagine that the wave moves further and further to the right at speed v. This kind of "traveling wave" might describe Alice's rope if she keeps shaking her end up and down.[1] It might also describe a sound wave or a beam of light.

Active Reading Exercise: Traveling Wave

Suppose a wave initially equal to $y = 5 \sin(\pi x)$ moves to the right at speed 6.

1. If you take a snapshot of this wave – that is, look at the function $y(x)$ at a particular t-value – you see a sine wave. What are its amplitude and wavelength?

2. If you stand at one particular x-value and watch the y-value over time, you see a sinusoidal oscillation that could be described by a $y(t)$ function. What are its amplitude and period? (Unlike the Exercise on p. 262, this time we have given you enough information to determine the period.)

3. What function $y(x, t)$ describes this traveling wave?

Don't read further until you have written down all three answers!

The first answer is the easiest: you see $y = 5 \sin(\pi x)$ moved to the right by $\Delta x = vt$. This motion doesn't change the wave's amplitude or wavelength, so they are $A = 5$ and $\lambda = 2$ at all times.

To answer the second, consider that the wave looks exactly like it did at the start ($t = 0$) after it has moved to the right by exactly one wavelength, or $\Delta x = 2$. At a speed of 6, that takes a time $\Delta t = 1/3$. Make sure you follow this point: snapshots of this wave taken at $t = 0$ and $t = 1/3$ and $t = 2/3$ will all look identical, so every point on the wave is oscillating sinusoidally with amplitude $A = 5$ and period $T = 1/3$ (Figure 6.3).

In this example, the function is $y(x, t) = 5 \sin(\pi x - 6\pi t)$. In the box "Traveling Waves" on the next page, we present the more general function for a traveling wave, and the relations between its parts. Make sure you see that the equations in this box describe relationships we have already explained.

---

1 Remember Alice? There's a section about Alice in Chapter 3.

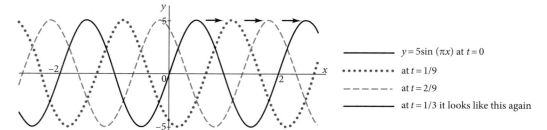

Figure 6.3 A traveling wave seen at several different times.

Traveling Waves

(This information can be found in Appendix F.)

The general form for a traveling sine wave is

$$y(x, t) = A \sin(kx - \omega t). \tag{6.2}$$

The wave travels to the right if the constants k and ω have the same sign as each other. The speed of the wave is the ratio of angular frequency to wave number: $v = |\omega/k|$.

Here are a few important points about traveling waves and Equation (6.2).

- Since $\omega = 2\pi v$ and $k = 2\pi/\lambda$, the relation $v = \omega/k$ can also be written as $v = \lambda v$. (For light waves, that becomes $c = \lambda v$, which we used in Chapter 3.)

- In quantum mechanics we generally express our traveling waves as complex exponentials. Recalling what you learned in Section 5.5, see if you can convince yourself that both the real and the imaginary parts of the function $Ae^{i(kx-\omega t)}$ are traveling waves (Problem 24).

- A standing wave can be described as the sum of two traveling waves moving in opposite directions (Problem 25).

Last, but Not Least, Some Calculus

Equations (6.1) and (6.2) are examples of "multivariate functions": one dependent variable with multiple independent variables. What does a derivative mean in that context?

Remember what a derivative means in an ordinary (one-variable) context. The quantity dy/dx ("the derivative of y with respect to x") tells you how fast y will change in response to a change in x. If $dy/dx = 3$, then we might roughly say that adding 0.01 to x will cause y to increase by 0.03.

With a multivariate function you ask how much the dependent variable will change in response to a change in one of the independent variables, *holding all others constant*. In practice this means that you actually treat all but one of the independent variables as constants when you evaluate the derivative. The result is called a "partial derivative" and it is written with a ∂ symbol instead of the d in a regular derivative.

In our case we have y depending on x and t. So there are two partial derivatives, and they have very different meanings.

- The quantity $\partial y/\partial x$ (read "the partial derivative of y with respect to x") asks how y is changing as you move along the x axis at one fixed moment in time. In a word, it tells you the *slope* of the curve.
- The quantity $\partial y/\partial t$ asks how y is changing over time at a particular (constant) x-value. It tells you the *vertical velocity* of the curve.

Both of these quantities are in general functions of both x and t. That is, you can look at one particular point on the curve, at one particular moment, and ask for its slope and/or vertical velocity. At a different point or a different moment the answers might be different.

Example: Partial Derivatives

Consider a traveling wave described by the equation

$$y(x,t) = \frac{1}{2}\sin(\pi x - 6\pi t).$$

At the point $x = 1/3$, $t = 0$, evaluate y and its two partial derivatives and briefly describe what each means.

Answer: Plug those two numbers in: $y(1/3,0) = \sqrt{3}/4$. This is the height of the function at that particular x-value and moment.

Take the derivative with respect to x, treating t as a constant. After you take the derivative, plug in the values for x and t (do you see why you can't plug in the values *before* taking the derivative?):

$$\frac{\partial y}{\partial x} = \frac{\pi}{2}\cos(\pi x - 6\pi t) \quad \rightarrow \quad \frac{\partial y}{\partial x}\left(\frac{1}{3},0\right) = \frac{\pi}{4}.$$

If you took a snapshot of the function at time $t = 0$, the curve at $x = 1/3$ would have a slope of $\pi/4$ in that snapshot.

Now take the derivative with respect to t, treating x as a constant, and then plug in values:

$$\frac{\partial y}{\partial t} = -3\pi\cos(\pi x - 6\pi t) \quad \rightarrow \quad \frac{\partial y}{\partial t}\left(\frac{1}{3},0\right) = -\frac{3}{2}\pi.$$

If you were an ant sitting at the point $x = 1/3$ and riding up and down as the function oscillates, then at the moment $t = 0$ you would be moving *down* at a speed of $3\pi/2$.

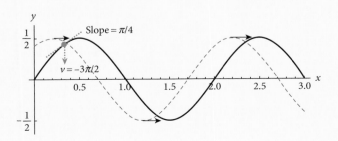

If you have previously taken a course on multivariate calculus, that was a brief review of a topic you have covered in far more depth. You're ahead of the game! If you've never taken multivariate calculus, give this important concept some time to sink in. Evaluating partial derivatives is easy (just treat one variable as a constant), but following what they mean is harder. You'll be ahead of the game when you do take that calculus course!

In either case, you can't skip this one. Traveling waves, standing waves, and partial derivatives are going to appear throughout our discussion of quantum mechanics.

6.1.3 Questions and Problems: Standing Waves, Traveling Waves, and Partial Derivatives

Conceptual Questions and ConcepTests

1. Describe (using a combination of words and a picture) a standing wave with a very long wavelength and a very short period.

2. Describe (using a combination of words and a picture) a traveling wave (not a standing wave) with a very long wavelength and a very short period. *Hint*: Think about what this implies about the velocity v of the wave.

3. Standing Waves A and B have the same wavelength, but Wave A has a higher frequency than Wave B. If you look at them, what looks different about them?

4. A real-world example of a standing wave is a plucked guitar string. Give one or two others.

5. A real-world example of a traveling wave is an electromagnetic wave (e.g. light) going from the Sun to the Earth. Give one or two others (that aren't electromagnetic waves).

6. Explain why the equation $\lambda = vT$ must hold for a traveling wave, using words and maybe a picture or two.

7. What is the period of the following function: $f(t) = 2\sin(4t) + 3\sin(2t)$? (Choose one and explain how you know.)

   A. 2

   B. 4

   C. $\pi/2$

   D. π

   E. This function doesn't have a well-defined period.

8. Equation (6.2) describes a sine wave traveling to the right at speed v. Write the equation for a sine wave traveling to the left at speed v.

9. We've written an equation for a sine wave traveling to the right, but waves don't always have the shape of a sine wave. Write a generic formula for a function with shape $y = f(x)$ moving to the right with speed v.

10. A function $f(x, t)$ has the following two properties:

    • If you look at the entire function at any particular moment in time, you see a sine wave in x.

    • If you follow the function at any particular x-value, you see it oscillating sinusoidally in t.

    Choose one:

    A. This function is neither a standing wave nor a traveling wave.

    B. This function might be a standing wave, but it is not a traveling wave.

    C. This function might be a traveling wave, but it is not a standing wave.

    D. This function might be a traveling wave, or it might be a standing wave.

    E. This function could be both a traveling wave and a standing wave, because they are two mathematical forms expressing the same motion.

11. Suppose $y(x, t)$ represents the height of each point on a vibrating string over time.

(a) The derivative $\partial y/\partial x$ represents the *slope* of a curve at a given point. In one word, what does the second derivative $\partial^2 y/\partial x^2$ represent?

(b) The derivative $\partial y/\partial t$ represents the *vertical velocity* of a curve at a given point. In two words, what does the second derivative $\partial^2 y/\partial t^2$ represent?

12. A curve is changing over time, so its height is given as a function $y(x,t)$. The partial derivative $\partial^2 y/\partial x^2$ is, in general (choose one):

A. a constant

B. a function of x but not t

C. a function of t but not x

D. a function of both x and t.

13. Suppose you measure temperature $u(x,t)$ along a rod and you find that at the point $x=2$, $t=1$ the partial derivative $\partial u/\partial x$ equals 10. You could interpret that by saying that if you walked from $x=2$ to $x=2.01$ the temperature would increase by approximately 0.1 (the derivative times the distance). Why is this only approximate? *There are at least two valid answers to this question.*

For Deeper Discussion

14. The function $A\sin(kx-\omega t)$ represents a sine wave traveling to the right at constant speed. Write a function for a sine wave traveling to the right, for $-\infty < t < \infty$, but always going faster and faster.

(a) The most common student answer to this question is $A\sin(kx-\omega t^2)$. Why isn't that completely correct?

(b) What would be a better answer?

Problems

15. Consider the following function: $y(x,t) = 5\sin(\pi x/3)\sin(2\pi t)$.

(a) Imagine that you are watching $x=3/2$, ignoring all other x-values. Write the function $y(t)$ that describes how the y-value at that x-value evolves over time. Then briefly describe in words what you see.

(b) Now imagine you switch your gaze to $x=3/4$. Briefly describe in words how the function

at this x-value is like $x=3/2$, and how it is different.

(c) What does the motion at $x=0$ look like?

(d) Finally, consider $x=-3/2$. How is this point like, and not like, the motion in Part (a)?

(e) You've been looking at the time evolution at fixed x-values; now do the opposite. Draw graphs of $y(x)$ at times $t=0, t=1/8, t=1/4$. Then briefly describe how $y(x)$ will continue to evolve after those three drawings. (You will have to think about what range of x- and y-values you should plot to see the behavior clearly.)

(f) What are the wavelength, period, wave number, and frequency of this wave?

16. Consider the following function: $y(x,t) = 5\sin(\pi x/3 - 2\pi t/3)$.

(a) Imagine that you are watching $x=0$, ignoring all other x-values. Write the function $y(t)$ that describes how the y-value at that x-value evolves over time. Then briefly describe in words what you see.

(b) Now imagine you switch your gaze to $x=3/4$. Briefly describe in words in what ways the function at this x-value is like the function at $x=0$, and in what ways it is different.

(c) You've been looking at the time evolution at fixed x-values; now do the opposite. Draw graphs of $y(x)$ at times $t=0, t=3/8, t=3/4$. Then briefly describe how $y(x)$ will continue to evolve after those three drawings. (You will have to think about what range of x and y values you should plot to see the behavior clearly.)

(d) What are the wavelength, period, and speed of this wave?

17. $f(x,t) = x^2\sin(2t)$. Calculate each of the following at the point $x=2$, $t=\pi/4$:

(a) $\partial f/\partial x$

(b) $\partial f/\partial t$

(c) $\partial^2 f/\partial x^2$

(d) $\partial^2 f/\partial t^2$

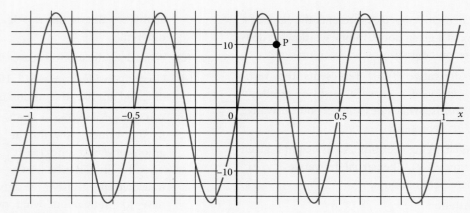

Figure 6.4

Problems 18–20 refer to Figure 6.4. This sine wave extends vertically from $y = -15$ to $y = 15$, and a full wavelength extends between $x = 0$ and $x = 0.5$. Some problems also refer to the point labeled P.

18. Consider Figure 6.4.

    (a) Identify the amplitude, wavelength, and wave number.

    (b) Why couldn't we ask you about the angular frequency?

    (c) Write the equation $y(x)$.

19. Imagine that Figure 6.4 represents a snapshot of a standing wave. The wave is shown at its maximum amplitude. It oscillates with period 10.

    (a) Identify the wavelength, the wave number, and the frequency.

    (b) At the point labeled P, is $\partial y/\partial x$ positive, negative, or zero? Explain briefly how you know (without doing any calculations).

    (c) At the point labeled P, is $\partial y/\partial t$ positive, negative, or zero? Explain briefly how you know (without doing any calculations).

    (d) Write the equation $y(x, t)$ that represents this standing wave.

20. Imagine that Figure 6.4 represents a snapshot of a traveling wave, moving to the right at speed 10.

    (a) Identify the wavelength, the wave number, the period, and the frequency.

    (b) At the point labeled P, is $\partial y/\partial x$ positive, negative, or zero? Explain briefly how you know (without doing any calculations).

    (c) At the point labeled P, is $\partial y/\partial t$ positive, negative, or zero? Explain briefly how you know (without doing any calculations).

    (d) Write the equation $y(x, t)$ that represents this traveling wave.

    (e) Write the equation $y(x, t)$ for an identical wave traveling to the *left* at speed 10.

21. An A note (technically A4, to distinguish it from other A notes) has a frequency of 440 Hz and travels through the air at 343 m/s. Calculate the wavelength of its sound wave.

22. Orange light with a wavelength of 600 nm travels through the vacuum at speed c. Calculate its frequency in Hz.

23. Consider the function $y(x, t) = 10 \sin(2\pi x)e^{-t}$ for $t \geq 0$.

    (a) Draw graphs of $y(x, t)$ at $t = 0$, $t = 1$, $t = 2$, and $t = 3$.

    (b) Describe in words how this function will evolve as $t \to \infty$.

24. **Traveling Waves Expressed as Complex Exponentials**

    Consider the function $y(x, t) = Ae^{i(kx-\omega t)}$, where A, k, and ω are positive real numbers.

    (a) Break this function down into its real and imaginary parts. (Review Section 5.5 if necessary.)

(b) The real part of this function is a traveling wave. What are its amplitude, angular velocity, and wave number? Is it traveling to the right or to the left? How fast?

(c) The imaginary part of this function is also a traveling wave. Are any of the properties we asked about in Part (b) different for the real and imaginary parts of this wave? If so, which ones and how are they different?

(d) Write the function – in complex exponential form – that represents an identical wave traveling in the opposite direction.

25. Building a Standing Wave from Traveling Waves

(a) Write a generic equation for a traveling sine wave $y_R(x, t)$ with amplitude A, wave number k, and angular frequency ω moving to the right.

(b) Write a generic equation for a wave $y_L(x, t)$ identical to y_R except that it is moving to the left.

(c) Use trigonometric identities to simplify the sum $y_R + y_L$ and show that it is a standing wave.

(d) Where are the nodes of that standing wave?

26. Consider a sinusoidal standing wave $y(x, t)$ on the domain $0 \leq x \leq 10$ that is subject to the boundary conditions that $y = 0$ at $x = 0$ and $x = 10$. (This might, for instance, be a rope that is tied down at both ends.)

(a) The highest wavelength such a wave could possibly have is 20. Draw what that would look like.

(b) What are the next three (lower) allowable wavelengths?

(c) Now express *all* possible wavelengths.

(d) Now express all possible wave numbers.

27. Exploration: The Wave Equation

A differential equation that involves partial derivatives is called a "partial differential equation" (or PDE). Section 6.6 will introduce the "time-dependent Schrödinger equation," the PDE at the heart of all of quantum mechanics. In this problem we will consider a different important PDE, called the "wave equation":

$$\frac{\partial^2 y}{\partial x^2} = \frac{1}{v^2} \frac{\partial^2 y}{\partial t^2} \qquad \text{where } v \text{ is a constant.}$$

(a) Show that the following traveling wave, $y = A \sin(kx - kvt)$, is a valid solution to this equation for any constants k and A.

(b) Show that the following standing wave, $y = A \sin(kx) \cos(kvt)$, is a valid solution to this equation for any constants A and k.

You have seen that ordinary differential equations can come with "initial conditions," sometimes called "boundary conditions." For a *partial* differential equation those are two different things: initial conditions in time, and boundary conditions in space. For instance, imagine a guitar string vibrating according to the wave equation. The boundary conditions say that the string is tacked down at $x = 0$ and $x = L$. The initial condition is the shape of the string before it is set free to vibrate, which could be almost any shape.

(c) Consider the boundary condition $y(0, t) = 0$. Show that the first (traveling wave) solution cannot meet this boundary condition unless y is uniformly zero at all times. Then show that the second (standing-wave) solution meets this boundary condition.

For the rest of this problem, therefore, we will focus on the standing wave.

(d) Find all values of k for which that solution meets the other boundary condition, $y(L, t) = 0$, for non-zero v and A.

(e) Show that the principle of linear superposition holds for any system governed by this equation. That is, show that, if $y_1(x, t)$ and $y_2(x, t)$ are any valid solutions, then $y(x, t) = y_1(x, t) + y_2(x, t)$ is also a valid solution.

(f) Write a solution $y(x, t)$ that satisfies this PDE, the boundary conditions (y goes to zero at $x = 0$ and $x = L$), and the initial condition $y(x, 0) = a \sin(2\pi x/L) + b \sin(7\pi x/L)$. *Hint*: This requires no new calculations. Just find a linear superposition of the solutions you've found so far that matches this.

6.2   Free Particles and Fourier Transforms

We're going to demonstrate the general properties of unbound states through the simplest example, a particle with no forces on it. Along the way we will introduce an important mathematical technique, the "Fourier transform."

6.2.1   Discovery Exercise: Free Particles and Fourier Transforms

A particle with no forces on it can be described by the potential energy function $U(x) = 0$, for which the time-independent Schrödinger equation is $-(\hbar^2/(2m))\psi''(x) = E\psi(x)$.

1. Write the general solution to this Schrödinger equation using sines and cosines.
    See Check Yourself #6 at www.cambridge.org/felder-modernphysics/checkyourself

2. What problem can you see with these functions being the energy eigenstates for this system?

3. Rewrite the general solution to this Schrödinger equation using complex exponentials.

4. Does the complex exponential form have the same problem as the one you identified in Part 2? Why or why not?

6.2.2   Explanation: Free Particles and Fourier Transforms

A particle that experiences no forces is called a "free particle." This is the system classically described by Newton's first law, which says that such an object will keep moving at constant velocity forever. Just as we used a particle in an infinite square well as our prototype of a bound state, we will use the free particle to introduce properties that turn out to apply to all unbound states.

Energy Eigenstates of a Free Particle

Saying the force on a particle is zero is equivalent to saying its potential energy is constant, and we lose no generality by defining that constant as $U(x) = 0$. So the time-independent Schrödinger equation for a free particle is $-(\hbar^2/(2m))\psi''(x) = E\psi(x)$.

Active Reading Exercise: A Free Particle

If you haven't done it yet, do Parts 1 and 2 of Discovery Exercise 6.2.1.

The Schrödinger equation for a free particle is the same as the one we wrote for a particle in a square well (infinite or finite), and it has the same general solution:

$$\psi(x) = A \sin\left(\frac{\sqrt{2mE}}{\hbar}x\right) + B\cos\left(\frac{\sqrt{2mE}}{\hbar}x\right) \qquad \text{Free-particle energy eigenstates, first version.}$$

(6.3)

Do you remember how this worked for bound states? The next steps, throwing out all unnormalizable solutions and matching boundary conditions, gave us quantized energy levels and specific values for some of the arbitrary constants. The final step, actually normalizing the wavefunction, nailed down the last arbitrary constant.

None of that applies to our free particle. There are no boundary conditions. Equation (6.3) is continuous and differentiable no matter what A, B, and E are. Most problematically, there are no possible values of A and B that normalize that $\psi(x)$ function. (Take a moment to convince yourself of that.)

All that leads us to two conclusions, which turn out to apply to all unbound states:

- The allowed energies for unbound states are not quantized.
- Because unbound energy eigenstates can't be normalized, a system can never be in an unbound state with definite energy.

We're going to talk about the time evolution of these eigenstates, and then come back to the normalization problem. But before we do any of that, it will be helpful to express the energy eigenstates in a different form. Take a moment to convince yourself that the following function is a valid solution to the free-particle Schrödinger equation $-(\hbar^2/(2m))\psi''(x) = E\psi(x)$. (That is, take its second derivative and see what you get.)

$$\psi(x) = Ae^{i\sqrt{2mE}\,x/\hbar} + Be^{-i\sqrt{2mE}\,x/\hbar} \qquad \text{Free-particle energy eigenstates, second version.}$$

This is not a new solution, but a different way – a more instructive way, as it turns out – of expressing Equation (6.3). Using k for the constant $\pm\sqrt{2mE}/\hbar$ brings us to our final version.

Energy Eigenstates of a Free Particle (Final Version)

The energy eigenstate of a free particle is $\psi(x) = Ce^{ikx}$, with eigenvalue $E = \dfrac{k^2\hbar^2}{2m}$.

- The energy levels are not quantized: k can take any real value, so E can be any non-negative value.
- Watch the signs: e^{3ix} and e^{-3ix} are *different* wavefunctions that correspond to the *same* energy.

Time Evolution of a Free-Particle Eigenstate

Now we're ready to put in time evolution. Any energy eigenstate is therefore multiplied by $e^{-iEt/\hbar}$, so of course we get this:

$$\Psi(x,t) = Ce^{\pm i\sqrt{2mE}x/\hbar}e^{-iEt/\hbar} \qquad \text{(bleah!)}$$

We're never going to write it that way again. Earlier, we wrote the eigenstates more concisely by defining a new constant k; now we will define a second constant called ω, so we can write our wave like this:

$$\Psi(x,t) = Ce^{i(kx-\omega t)} \quad \text{where} \quad \omega = \frac{E}{\hbar} = \frac{\hbar k^2}{2m}. \tag{6.4}$$

You can think of k as a kind of continuous version of the quantum numbers (often called n) that we saw for discrete functions, and ω as a way of expressing the energy or the time evolution with fewer letters. But Equation (6.4) suggests a more important interpretation (and the reason we chose those particular letters):

- The energy eigenstate of a free particle is a traveling wave.
- Its angular velocity is ω, its wave number is k, and it travels with velocity ω/k.
- If k is positive, the wave is traveling to the right. If k is negative, the wave is traveling to the left.

If you don't see all that in Equation (6.4), please review Section 6.1; this is one of the reasons we have emphasized the math of traveling waves.

So in the end, we've concluded that an energy eigenstate of a free particle describes a wave that moves in the same direction at constant velocity forever. It may be reassuring that at least some aspects of classical physics have parallels in the quantum world.[2]

Superpositions of Free-Particle Eigenstates

So now we know the energy eigenstates of the free particle: $\psi(x) = Ce^{ikx}$ and $\Psi(x,t) = Ce^{i(kx-\omega t)}$. If a wavefunction is in one of those states, it has a definite energy. If it's in a superposition of several of those states, then the coefficient in front of each eigenstate tells you the probability of measuring that energy. All familiar, right?

Only there's a problem: as we noted above, the free-particle eigenstates can't be normalized, so it's impossible for a particle to be in one of those states. It's also impossible to normalize the sum of 2 of those states, or of 3 or 17 of them (Problem 12). You can't even normalize an infinite sum of them. How can these impossible states describe physically possible wavefunctions?

Here's another problem. Suppose we could express a wavefunction as a sum of those eigenstates:

$$\psi(x) = \sum_{k=-\infty}^{k=\infty} C_k e^{ikx}. \qquad \text{(doesn't work)} \qquad (6.5)$$

That includes the $k = 0$ and the $k = 3$ and the $k = -3$ states, but what about $k = 2.5$ and $k = \pi$? Remember that k can be any real number. No matter how you sum, you miss most of them!

The solution to both of these problems is the same: instead of a *sum* over free-particle eigenstates, we write a wavefunction as an *integral* over those eigenstates. As it turns out, the mathematical technique for doing precisely that – expressing an arbitrary function of x as an integral over k of e^{ikx} with different coefficients – had been developed in the nineteenth century, and was ready and waiting when quantum mechanics needed it. This technique is called a "Fourier transform."

---

2 Warning: In Section 6.4 you'll see that this parallel is not as simple as it sounds.

Fourier Transforms

For any normalized wavefunction $\psi(x)$ it is possible to find a function $\hat{\psi}(k)$ such that both of the following equations hold:

$$\psi(x) = \frac{1}{\sqrt{2\pi}} \int_{-\infty}^{\infty} \hat{\psi}(k) e^{ikx} \, dk \tag{6.6}$$

$$\hat{\psi}(k) = \frac{1}{\sqrt{2\pi}} \int_{-\infty}^{\infty} \psi(x) e^{-ikx} \, dx \tag{6.7}$$

The function $\hat{\psi}(k)$ is called the "Fourier transform" of $\psi(x)$. The function $\psi(x)$ is called the "inverse Fourier transform" of $\hat{\psi}(k)$.

Before we interpret these functions, make sure you notice a couple of mathematical facts about them:

- Each of these two equations takes as input a function, and gives you back a *completely different function*. Don't be fooled by the fact that these two functions have similar symbols (ψ and $\hat{\psi}$): they have different independent variables (x and k), they have different units, their graphs might look totally different, and they are used for different purposes.
- And speaking of independent variables, those integrals have both x and k in them. But the integral in Equation (6.6) is a definite integral in k, so after you integrate and plug in the limits, you are left with a function of only x. Similarly, the integral in Equation (6.7) gives you a function of only k.

And what does that function of k tell you? You can understand Equation (6.6) by comparing it to Equation (6.5), because every integral is the limit of a sum (a Riemann sum). Equation (6.6) says that you are rewriting your wavefunction $\psi(x)$ as a sum of functions of the form e^{ikx}: that is, free-particle energy eigenstates. Each such eigenfunction has a "coefficient" $\hat{\psi}(k)$ that tells you how much of that energy eigenstate is in this superposition. Seen in this light, Equation (6.7) is the formula for finding those coefficients.

That means $|\hat{\psi}(k)|^2$ gives you energy probabilities. But – as we discussed in Section 4.3 – since k is continuous, $|\hat{\psi}(k)|^2$ is a probability density function.

Energy Probabilities for a Free Particle

If a free particle has wavefunction $\psi(x)$, and $\hat{\psi}(k)$ is its Fourier transform (calculated with Equation (6.7)), then the probability of measuring the energy in a given range is given by

$$P\left(\frac{\hbar^2 k_1^2}{2m} < E < \frac{\hbar^2 k_2^2}{2m}\right) = \int_{-k_2}^{-k_1} |\hat{\psi}(k)|^2 \, dk + \int_{k_1}^{k_2} |\hat{\psi}(k)|^2 \, dk. \tag{6.8}$$

Why two separate integrals? Because $-k_2 < k < -k_1$ represents eigenfunctions in a certain energy range, and $k_1 < k < k_2$ represents *different* eigenfunctions in the *same* energy range.

Aside from that, the fundamental idea should be familiar: we are integrating over a probability density because the variable is continuous.

Wave Packets and Normalization

We've seen that a continuous integral over eigenstates, instead of a discrete sum, allows us to catch all the possible k-values (which are all the real numbers, not just all the integers). Much less obviously, the integral solves our normalization problem too. You cannot normalize any discrete sum of free-particle energy eigenstates, but a Fourier transform can be normalized.

The fact that an integral over unnormalizable traveling waves can be normalized is a remarkable mathematical result that we're not going to prove, although we will look into its consequences a bit more deeply in Section 6.4. At any given moment, the particle's wavefunction is an integral over infinitely many eigenstates. Each eigenstate is an unnormalizable wave, extending forever in both the positive and negative x directions without approaching zero. But the combination of those eigenstates constructively interferes in the neighborhood of one particular x-value, drops quickly to zero as you move away from that x-value, and stays close to zero forever in both directions!

The resulting normalizable function is called a "wave packet" and it tells you exactly what you would expect to find from the wavefunction of an actual particle: the particle's position has some uncertainly in and around this particular x-value, but is all but guaranteed not to be found far away.

The constant in front of Equation (6.7) is chosen so that if your original wavefunction $\psi(x)$ was normalized, then your Fourier transform $\hat{\psi}(x)$ will automatically be normalized: $\int_{-\infty}^{\infty} |\hat{\psi}(k)|^2 dk = 1$.

Example: Superpositions of Free-Particle Eigenstates

A free particle's wavefunction is $\psi(x) = [2/(\pi L^2)]^{1/4} e^{-(x/L)^2}$. (This properly normalized "wave packet" assures us that the particle will be found near $x = 0$.) The following figure shows the position probability distribution $|\psi(x)|^2$.

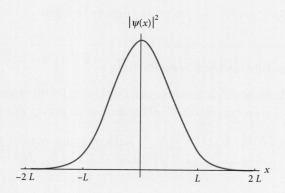

1. Plot $|\hat{\psi}(k)|^2$ for this particle.

2. Just from looking at your plot, estimate the range of energies that you are likely to measure for the particle.

3. What is the probability that the energy is greater than $\hbar^2/(mL^2)$?

Answer:

1. Use Equation (6.7) to find $\hat{\psi}(k)$, using a computer for the integration step:

$$\hat{\psi}(k) = \frac{1}{\sqrt{2\pi}} \int_{-\infty}^{\infty} \left(\frac{2}{\pi L^2}\right)^{1/4} e^{-(x/L)^2} e^{-ikx} dx = \left(\frac{L^2}{2\pi}\right)^{1/4} e^{-L^2 k^2/4}.$$

From this, $|\hat{\psi}(k)|^2 = Le^{-L^2 k^2/2}/\sqrt{2\pi}$. We've plotted it below. The shaded areas represent the probability asked about in Part 3.

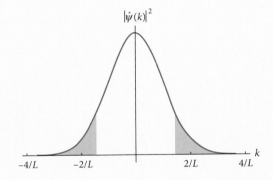

2. We can see that k is very likely to be measured in the range $-2/L$ to $2/L$. That implies that E is most likely to be measured between 0 and $2\hbar^2/mL^2$.

3. For E to be greater than $\hbar^2/(mL^2)$ we need $k^2 > 2/L^2$, which means either $k > \sqrt{2}/L$ or $k < -\sqrt{2}/L$. That probability requires another integral that is best handed to a computer:

$$P\left(E > \frac{\hbar^2}{mL^2}\right) = \int_{-\infty}^{-\sqrt{2}/L} |\hat{\psi}(k)|^2 dk + \int_{\sqrt{2}/L}^{\infty} |\hat{\psi}(k)|^2 dk.$$

Plugging that into a computer, we get 0.16, so there's a 16% chance of measuring $E > \hbar^2/(mL^2)$. (As a rough check, it does at least look plausible that the shaded regions are about 16% of the total area under the curve.)

The preceding Example communicates much of what we want you to learn from this section, so we hope you'll follow it carefully. As you do, it may strike you that the graphs of $|\psi(x)|^2$ and $|\hat{\psi}(k)|^2$ look very similar. That isn't a universal rule, but it isn't entirely a coincidence either.

In general, as we stressed above, $\psi(x)$ and $\hat{\psi}(k)$ are two completely different functions. In some very important cases (Problems 15 and 16, for example), they look entirely different.

But the Gaussian curve e^{-x^2} is a special case in a lot of ways, and one of its special properties is that the Fourier transform of a Gaussian curve is another Gaussian curve. That's why the two curves in the Example have the same overall shape.

Even in that case, however, $\psi(x)$ and $\hat{\psi}(k)$ are different in important ways. We'll explore those differences further in Section 6.3.

6.2.3 Questions and Problems: Free Particles and Fourier Transforms

Conceptual Questions and ConcepTests

1. Briefly define each of the following symbols as it was used in this section, and give its units:

    (a) k

    (b) $\left|\hat{\psi}(k)\right|^2$

    (c) ω

2. A free particle has some wavefunction $\psi(x)$ with Fourier transform $\hat{\psi}(k)$.

    (a) If $|\psi(x_1)|^2 > |\psi(x_2)|^2$, what can you conclude about the probabilities for position measurements of that particle?

    (b) What would have to be true about the values of $\hat{\psi}(k)$ for you to conclude that the particle's energy is more likely to be measured close to E_1 than to E_2?

3. What led us to conclude that energy is quantized for bound states and not quantized for unbound states? (You can focus on a free particle since that's the only unbound state we've discussed so far.)

4. The function $e^{i(kx-\omega t)}$ always represents a traveling wave, but the direction of travel depends on the signs of both k and ω. What led us to conclude, for the energy eigenstate of a free particle, that it is always traveling to the right if $k > 0$ and to the left if $k < 0$?

5. In the Example on p. 274, what is the probability of measuring the energy as exactly $2\hbar^2/(mL^2)$?

6. The solution to Schrödinger's equation for a free particle is the same as it was for a particle inside a square well (infinite or finite). For a free particle we said this solution wasn't normalizable. Why wasn't that a problem for a particle in a square well?

7. You start with a wavefunction $\psi(x)$ and you evaluate its Fourier transform, with the results shown in Figure 6.5. The quantity L is a parameter with units of length. Notice that the plot shows $|\hat{\psi}(k)|^2$, not $\hat{\psi}(k)$.

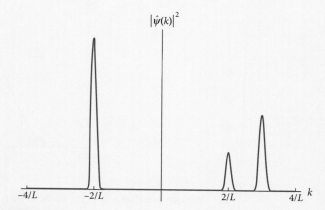

Figure 6.5

If you prepared 100 free particles with precisely that wavefunction, and measured all their energies, estimate what results you would find and in what quantities.

8. In many physical contexts, it is perfectly reasonable for the total energy E of a particle to be negative. Why do we assume throughout this section that our free particle has $E \geq 0$?

9. In general, a wavefunction $\psi(x)$ gives you different probabilities for different energies. The only way in which $\psi(x)$ for a free particle can give you a definite energy is if it has one constant frequency, such as $\psi(x) = Ae^{ikx}$.

    (a) Explain why such a wavefunction is impossible.

(b) What does that impossibility imply about the energy of free particles?

10. If you follow the function $e^{i(kx-\omega t)}$ over time, you will see that ... (Choose one and briefly explain your choice.)

    A. at any given x-value, the function rotates in a circle around the complex plane.

    B. the entire function keeps its shape but moves at constant speed along the x axis.

    C. both A and B are correct; they are actually saying the same thing in this case.

    D. neither A nor B is correct.

For Deeper Discussion

11. When we solved for the energy eigenstates of bound states we discarded any solutions that weren't normalizable, which led us to conclude that energy was quantized. For unbound states, *all* the energy eigenstates are unnormalizable, which led us to conclude that we had to make physical states that were superpositions of them. Why do we treat those states differently? In particular, why didn't we conclude that using unnormalizable energy eigenstates for bound states was fine as long as you only use them to make superpositions of states with those energies?

Problems

12. One of the big differences between a free particle and the bound states we have worked with before is that the free particle's eigenstates cannot be normalized.

    (a) Show that it is impossible to normalize the function $A \sin(kx)$.

    (b) Show that it is impossible to normalize the sum $Be^{ik_1x} + Ce^{ik_2x}$.

    (c) Answer briefly: How do we make use of these impossible wavefunctions to analyze a free particle?

13. Starting from Equation (6.7) (or Equation (6.6), either will work), show that $\int |\hat{\psi}(k)|^2 dk$ has the right units for a probability.

14. Suppose you are given the wavefunction $\psi(x)$ for a free particle.

    (a) Write the integral you would use to find its Fourier transform $\hat{\psi}(k)$.

    (b) Write the integral(s) you would use to find the probability of an energy measurement between E_1 and E_2. (Assume $E_1 < E_2$.)

15. A free particle has the wavefunction $\psi(x) = A$ from $x = -L$ to $x = L$, and 0 everywhere else. (Wavefunctions can't actually be discontinuous but imagine that it rises very rapidly from 0 to A at the two edges of its domain.)

    (a) Find A.

    (b) Find $\hat{\psi}(k)$, the Fourier transform of $\psi(x)$.

    (c) Simplify your answer to Part (b) as much as possible. You may find useful the equations $1/i = -i$ (which follows quickly from $i^2 = -1$), $e^{ix} = \cos x + i \sin x$, $\cos(-x) = \cos(x)$, and $\sin(-x) = -\sin(x)$.

    (d) Draw a graph of $|\psi(x)|^2$, using a wide enough domain to represent its behavior. What can you infer about the results of measurements of this particle by looking at this graph?

    (e) Draw a graph of $\left|\hat{\psi}(k)\right|^2$, using a wide enough domain to represent its behavior. What can you infer about the results of measurements of this particle by looking at that graph?

    (f) Write an integral for the probability of finding the particle's energy between 0 and $2\hbar^2/(mL^2)$.

    (g) Evaluate that integral to find the probability. You can hand this integral to a computer, but make sure to ask for a numerical answer. You may want to first apply the substitution $u = Lk$, which gives you an integral involving only numbers.

16. A free particle has the following wavefunction: $\psi(x) = A \sin(\pi x/L)$ from $x = -L$ to $x = L$ and 0 everywhere else.

    (a) Explain why this is not a physically possible wavefunction. (We can nonetheless treat it as an approximation and work with it.)

    (b) Find the constant A.

(c) Draw a graph of $|\psi(x)|^2$. What can you infer about the results of measurements of this particle by looking at that graph?

(d) Write an integral for the Fourier transform of $\psi(x)$.

The integral you just wrote evaluates to give

$$\hat{\psi}(k) = i\frac{\sqrt{2\pi L}}{k^2 L^2 - \pi^2} \sin(kL),$$ as you can confirm on a computer.

(e) Draw a graph of $|\hat{\psi}(k)|^2$, using a wide enough domain to represent its behavior. What can you infer about the results of measurements on this particle by looking at that graph?

(f) What is the probability of finding that the particle's energy is greater than $2\hbar^2/(mL^2)$?

17. If a free particle begins in the state $\Psi(x,0) = \psi(x)$, write its future state $\Psi(x,t)$. *Hint*: This is going to be messier than it sounds.

18. Particle 1 has wavefunction $\psi_1(x)$ and Particle 2 has wavefunction $\psi_2(x) = \psi_1(x - x_0)$. In other words Particle 2 is simply shifted to the right by x_0 relative to Particle 1.

(a) Prove that $\hat{\psi}_2(k) = e^{-ikx_0}\hat{\psi}_1(k)$.

(b) Assuming both of these wavefunctions represent free particles, what does Part (a) imply about their energy probabilities?

6.3   Momentum Eigenstates

Many texts at this level do not cover eigenstates of any property except energy. But there are a few arguments for going further in an introductory course. First, by discussing two different types of eigenstates – energy and momentum – you get a deeper sense of the overall mathematical framework of quantum mechanics, in which a single wavefunction can be expressed in different bases to provide information about different measurable quantities. Second, momentum eigenstates in particular provide a mathematical justification for the uncertainty principle.

6.3.1   Explanation: Momentum Eigenstates

A quantum mechanical wavefunction is a mathematical function that represents the state of a particle. You know how to use that function to find out (probabilistically) about the particle's position, and also its energy. That same function contains information about the particle's other measurable quantities such as momentum, angular momentum, and kinetic energy.

In this section we're going to discuss momentum in particular. We'll leave most of the other quantities for your advanced quantum mechanics course. Of all these quantities, however, energy has a unique status – because it is the eigenstates of energy, and not of any other quantity, that determine the time evolution of the wavefunction.

What We've Learned So Far about Wavefunctions: A Brief Summary

We're going to start this section with an Example that pulls together much of the content of this chapter so far. As you follow this Example, don't worry about the messy algebra steps that we are skipping. (Full disclosure: we had Mathematica do them for us.) Focus on the way that we use one wavefunction to answer three very different questions about a particle, by expressing that wavefunction in three different (but mathematically equivalent) forms.

Example: Position and Energy from a Wavefunction

A particle has the wavefunction

$$\psi(x) = \frac{1}{5\pi^{1/4}}\left(3 + 4\sqrt{2}\,x\right)e^{-x^2/2}.$$

(For this example we're simplifying our equations by setting constants like m and $\hbar$ equal to 1.)

1. Find the probability of measuring the particle's position between 1 and 3.
2. If the particle is in a simple harmonic oscillator potential $U(x) = (1/2)x^2$, what energies might you measure for it and with what probabilities? (Notice that we've set the harmonic oscillator's ω to 1 also.)
3. If instead this is a free particle with $U(x) = 0$, find the probability density function for its possible energies.

Answer: We're going to skip most of the algebra here and focus on the process.

1. Integrate $|\psi|^2$ from 1 to 3. (Put that in a computer and you get the answer 0.35.)
2. If you look up the energy eigenstates of the simple harmonic oscillator, you can show with a little algebra that

$$\psi(x) = \frac{3}{5}\psi_0 + \frac{4}{5}\psi_1.$$

So there is a 9/25 chance of measuring $E = E_0 = 1/2$, and a 16/25 chance of measuring $E = E_1 = 3/2$.

3. Using the equations for a Fourier transform, you can express this $\psi(x)$ function as

$$\psi(x) = \frac{1}{\sqrt{2\pi}}\int_{-\infty}^{\infty}\left(\frac{3 - 4\sqrt{2}ik}{5\pi^{1/4}}e^{-k^2/2}\right)e^{ikx}dk.$$

The big term in parentheses is $\hat{\psi}(k)$, and its modulus squared is the energy probability density.

Here is a quick summary of the rules illustrated by the preceding Example.

- When you write ψ as a function of x, the modulus squared of that function gives you the probability density function for position measurements.
- For a system with discrete energy levels (bound states), you can express ψ as a sum of energy eigenstates. The coefficients of that sum (modulus squared again) give you the probabilities of all possible energy measurements.
- For a system with continuous energy levels (unbound states), you can express ψ as an *integral* over energy eigenstates. The resulting function (you guessed it, modulus squared again) gives you the probability density function for energy measurements.

Our central point here is that those rules do not involve three different wavefunctions. Rather, they involve three different forms of the same wavefunction, designed to give different information about the same particle.

With that review in mind, we're ready to give you the first new piece of information in this section. For other quantities such as momentum, kinetic energy, and angular momentum, the process is same as the process for energy.

1. Find the eigenstates of that quantity. (This is the step that is different for different observable quantities.)

2. Express the wavefunction as a sum or integral over those eigenstates. (Some beautiful math goes into proving that this is always possible.)

3. Once the wavefunction is re-expressed in that way, the moduli squared of the coefficients give the probability distribution.

In an advanced quantum mechanics course you will learn how to find the eigenstates of any observable quantity. Here we're going to focus on momentum.

Momentum Eigenstates

We're going to tell you the momentum eigenstates and discuss some of their properties. Then we'll talk about where they come from.

Momentum Eigenstates

The momentum eigenstate in quantum mechanics is $\psi(x) = Ce^{ikx}$, with eigenvalue $p = \hbar k$.

If you're thinking "That's the same as the eigenstates of energy!" you're half right. These are the states we discussed in Section 6.2, the energy eigenstates *of a free particle*. But remember that the energy eigenstates depend on the forces acting on a particle: that is, they depend on the potential energy function. (Solving the Schrödinger equation with $U(x) = 0$ led us to the eigenstates of the free particle.) The momentum eigenstates and eigenvalues, by contrast, are always the same.

Because the momentum eigenstates of *any* particle are the same as the energy eigenstates of a *free* particle, many of the conclusions we reached about those energy eigenstates apply to momentum as well:

*   Momentum eigenstates can't be normalized, so a particle can't have a single, definite momentum. (You can reach that same conclusion by thinking about the uncertainty principle.) In fact, no discrete sum of momentum eigenstates is a valid (normalizable) wavefunction.

*   Momentum is a continuous variable, and has a probability density rather than a set of discrete possible values.

*   To find the momentum probability distribution, you rewrite $\psi(x)$ using a Fourier transform. The rule is then:

$$P\left(k_1\hbar < p < k_2\hbar\right) = \int_{k_1}^{k_2} \left|\hat{\psi}(k)\right|^2 dk. \tag{6.9}$$

Note also one of the big *differences* between momentum ($p = \hbar k$) and energy ($E = \hbar^2 k^2/(2m)$), which is the way they work with signs. The functions e^{5ix} and e^{-5ix} are eigenstates of momentum, and they are also eigenstates of energy for a free particle. But those two states represent *different* momentum values (one positive and one negative), while they represent the *same* energy value. That's why Equation (6.9) is a single integral, while its energy equivalent (Equation (6.8) on pp. 273) requires two integrals. (This result should not be too surprising. Classically, a 2 kg particle traveling right at 3 m/s and a 2 kg particle traveling left at 3 m/s have the same kinetic energy, but different momenta.)

Example: Momentum Probabilities

The Example on p. 274 examined a particle with $\psi(x) = [2/(\pi L^2)]^{1/4} e^{-(x/L)^2}$. We found there that $\hat{\psi}(k) = (L^2/(2\pi))^{1/4} e^{-L^2 k^2/4}$. We've reproduced here the plots of $|\psi(x)|^2$ and $\left|\hat{\psi}(k)\right|^2$, with the shading in the second plot changed to reflect the probability question below.

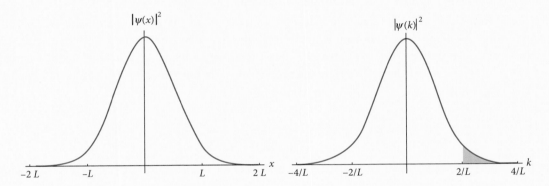

1. Looking at the plot of $\left|\hat{\psi}(k)\right|^2$, estimate the range of momenta that you are likely to measure for this particle.
2. Calculate the probability that the momentum of this particle is greater than $2\hbar/L$.

Answer:

1. You can see that you are very likely to measure the particle in the range $-2\hbar/L < p < 2\hbar/L$, and have almost no chance of measuring $|p| > 4\hbar/L$.
2. This requires another integral that is best handed to a computer:

$$P\left(p > \frac{2\hbar}{L}\right) = \int_{2/L}^{\infty} |\hat{\psi}(k)|^2 dk \approx 0.02.$$

So there's a 2% chance of measuring $p > 2\hbar/L$.

Where Did Those Momentum Eigenstates Come From?

We can offer two different ways of justifying the assertion that Ce^{ikx} is the eigenstate with momentum $p = \hbar k$. The first harkens back to the de Broglie relation.

> ### Active Reading Exercise: Momentum of a Cosine
>
> Suppose that a particle's wavefunction is $\psi(x) = A\cos(kx)$ for some constants A and k.
>
> For the moment, ignore everything we've said in this chapter: eigenstates, normalizability, all of it. Instead, go back to the de Broglie relation $\lambda = h/p$ (which was, as you recall, one of the elements of the old quantum theory that was baked into real quantum mechanics). What does the de Broglie relation tell us about the momentum of this particle?

The wavelength of the particle in that Active Reading Exercise is $\lambda = 2\pi/k$. (If conversions like that aren't second nature then you might want to keep Appendix F handy.) The de Broglie relation tells us that the momentum *magnitude* is $|p| = h/\lambda = hk/(2\pi) = \hbar k$. The de Broglie relation tells us the same about a particle with $\psi(x) = B\sin(kx)$, or about any linear combination of such sines and cosines – including $e^{ikx} = \cos(kx) + i\sin(kx)$.

You can reach a similar conclusion by thinking about the free particle from Section 6.2. Kinetic energy and momentum are *classically* related by the relationship $\mathrm{KE} = p^2/(2m)$. For a free particle, the total energy is just the kinetic energy. So a state of definite energy is also a state of definite kinetic energy, which is in turn a state of definite momentum *magnitude*:

$$\text{For } \psi(x) = Ce^{ikx}, \quad E = \mathrm{KE} = \frac{\hbar^2 k^2}{2m} \quad \rightarrow \quad |p| = \hbar|k|.$$

Both of those arguments establish the magnitude of the momentum, but leave open the question of sign. We can resolve that question by looking at time dependence. For a free particle (Section 6.2 again), each energy eigenstate $\psi(x) = Ce^{ikx}$ evolves in time as $\Psi(x,t) = Ce^{i(kx-\omega t)}$. Positive k means the wave is traveling to the right and negative k means it is traveling to the left,[3] so we can drop the absolute-value bars and write $p = \hbar k$.

Recall that in Section 6.2 we defined the constants k and ω in terms of the energy E. Now that we have them related to the momentum p as well, there is another way we can write those eigenstates:

$$\Psi(x,t) = Ce^{i(px-Et)/\hbar} \qquad \text{Free-particle energy eigenstate written in terms of } p \text{ and } E. \quad (6.10)$$

It shouldn't take long to convince yourself that Equation (6.10) is mathematically equivalent to the $\Psi(x,t) = Ce^{i(kx-\omega t)}$ we wrote previously, since $p = \hbar k$ and $E = \hbar\omega$. But this new form speaks more directly to the energy and momentum associated with a given eigenstate, and it will lead us to some important insights in Section 6.4.

---

3 There's a bit of sleight-of-hand going from "the wave is traveling to the right" to "the particle is traveling to the right." We'll address that in Section 6.4.

Narrow and Wide Distributions

The Example on p. 281 lays out the whole process for finding momentum probabilities. Given a wavefunction, you use Equation (6.7) to find $\hat{\psi}(k)$ and then integrate $|\hat{\psi}(k)|^2$ to find the probability of any given momentum range.

But that Example also illustrates an important property of momentum probabilities.

Active Reading Exercise: Narrow and Wide Bumps

The Example on p. 281 shows a wavefunction that is a bump centered on $x = 0$, and its Fourier transform (which is proportional to its momentum probability distribution). If we had made the wavefunction bump twice as wide, how would that have changed the momentum probabilities?

Hint: Look at how L appears on the horizontal axes of $\psi(x)$ and of $\hat{\psi}(k)$.

As we suggested in our hint, the answer lies in L. We drew ψ so that most of the position probability is in the range $-L < x < L$, without specifying any numerical value for L. We found that most of the momentum probability was in the range $-2\hbar/L < p < 2\hbar/L$.

So what happens if L doubles? The range of positions becomes twice as wide; the range of momenta becomes twice as narrow. This is a specific example of a very general rule:

If you make a function $\psi(x)$ wider, its Fourier transform $\hat{\psi}(k)$ gets narrower.

We've illustrated this rule in Figure 6.6.

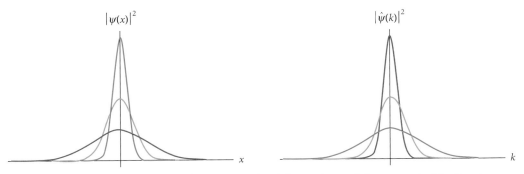

Figure 6.6 The red, green, and blue functions on the right are the modulus squared of the Fourier transforms of the red, green, and blue position probability distributions on the left respectively. As $\psi(x)$ gets wider, $\hat{\psi}(k)$ gets narrower.

At its heart, that is a purely mathematical fact: a narrow function has a broad Fourier transform. More formally, if the standard deviation of a function is low, the standard deviation of its Fourier transform will be high.

But we hope you can see why we're emphasizing this apparent bit of mathematical trivia: that relationship underlies the position–momentum uncertainty principle! A narrowly spiked $\psi(x)$ has a very low position uncertainty. Its Fourier transform is wide, leading to a high uncertainty in momentum.

In conclusion, the Heisenberg uncertainty principle is *not* one of the axioms of quantum mechanics. Rather, it is a mathematically inevitable consequence of the way we use a single wavefunction to calculate the probability distributions of both position and momentum.

6.3.2 Questions and Problems: Momentum Eigenstates

Conceptual Questions and ConcepTests

1. For a certain free particle's wavefunction, $\hat{\psi}(-5) > \hat{\psi}(-2)$.

   (a) What does this inequality tell you about measurements of this particle's momentum?

   (b) Why can't you use this inequality to conclude anything about measurements of the particle's energy?

2. In general, a wavefunction $\psi(x)$ gives you different probabilities for different momenta. The only way $\psi(x)$ can give you a definite momentum is if it has one constant frequency, such as $\psi(x) = Ae^{ikx}$.

   (a) Explain why such a wavefunction is impossible.

   (b) What does that impossibility imply about the momentum of real particles?

3. You start with a wavefunction $\psi(x)$ and you evaluate its Fourier transform, with the results shown in Figure 6.7.

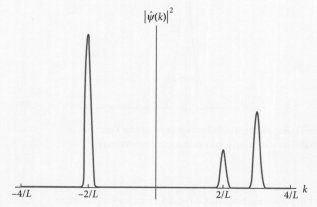

Figure 6.7

If you prepared 100 particles with precisely that wavefunction, and measured all their momenta, estimate what results you would find and in what quantities.

4. Particle 1 has wavefunction $\psi_1(x) = Ae^{-c^2x^2}$ and Particle 2 has wavefunction $\psi_2 = Be^{-2c^2x^2}$. Which of the following describes their momentum probabilities? (Choose one, and briefly explain your answer.)

   A. Particle 1 is more likely than Particle 2 to be found with momentum $0 < p < \hbar c$.

   B. Particle 2 is more likely than Particle 1 to be found with momentum $0 < p < \hbar c$.

   C. They are equally likely to be found with momentum $0 < p < \hbar c$.

   D. We don't have enough information to determine which is more likely.

5. Particle 1 has wavefunction $\psi_1(x) = Ae^{-(x/L)^2}$. The probability of measuring $-\hbar/L < p_1 < \hbar/L$ is 68%. Particle 2 has wavefunction $\psi_2(x) = Be^{-(4x/L)^2}$. (Figure 6.8 shows ψ_1 in blue and ψ_2 in red.) Which of the following is true of the momentum probabilities for Particle 2? (Choose one and explain your choice.)

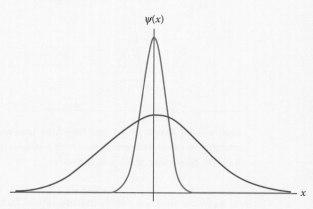

Figure 6.8

   A. There's a 17% chance of finding that $-\hbar/L < p_2 < \hbar/L$.

   B. There's a 68% chance of finding that $-4\hbar/L < p_2 < 4\hbar/L$.

C. Both A and B.

D. Neither A nor B.

6. The Example on p. 274 gave you the wavefunction of a free particle and showed that the probability that you would measure the particle's energy to be greater than $\hbar^2/(mL^2)$ was 16%. What is the probability that you would measure its momentum to be between $-\sqrt{2}\hbar/L$ and $\sqrt{2}\hbar/L$? Explain how you can answer this without evaluating an integral.

A. 16%

B. 8%

C. 32%

D. 84%

E. 42%

For Deeper Discussion

7. Are a particle's position and momentum probabilities independent? In other words, could a particle's wavefunction describe a state such that "If it's found over here then it's likely to have these momenta, but if it's found over there then it's likely to have those other momenta"?

8. We've been treating position differently from all other observable variables, but you can express the position probability rule in the usual way: write the wavefunction as a sum over position eigenstates and the modulus squared of the coefficients will give you the probability distribution. What do the position eigenstates look like?

Problems

9. In Section 6.2 Problem 15 we gave you the wavefunction $\psi(x) = A$ from $x = -L$ to $x = L$, and $\psi(x) = 0$ everywhere else. If you did that problem you should have found that $\hat{\psi}(k) = \sin(kL)/\sqrt{\pi L k^2}$. We've graphed $|\hat{\psi}(k)|^2$ in Figure 6.9.

If you didn't do that problem, that Fourier transform and graph are all you need to keep going with this one.

(a) Qualitatively, what does this graph tell you about momentum probabilities?

(b) The shaded part of this graph has an area $\approx$ 0.9. What does that tell you about the particle's momentum?

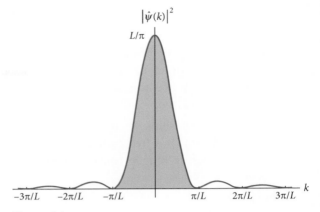

Figure 6.9

(c) Show that the momentum uncertainty in this case is not $\hbar/(2\Delta x)$, but is instead infinite. Why does this not violate the uncertainty principle? *Hint*: You can calculate Δk and multiply by $\hbar$ to get Δp, so you can show this by showing that Δk is infinite.

10. A particle has the wavefunction $\psi(x) = Ax$ from $x = 0$ to $x = L$, and 0 everywhere else. (Actual wavefunctions are continuous and differentiable, but you can think of this as a good approximation to an actual wavefunction.)

(a) Find A.

(b) Find $\hat{\psi}(k)$, the Fourier transform of $\psi(x)$. Simplify as much as possible.

(c) Draw a graph of $|\psi(x)|^2$. What can you infer about the results of measurements of this particle by looking at that graph?

(d) Draw a graph of $|\hat{\psi}(k)|^2$, using a wide enough domain to represent its behavior. What can you infer about the results of measurements of this particle by looking at that graph?

(e) Write an integral for the probability of finding that the particle's momentum lies between $2\hbar/L$ and $3\hbar/L$.

(f) Evaluate that integral to find the probability.

11. **(a)** Write the ground state of a particle in an infinite square well as a sum of momentum eigenstates.

(b) What does your answer tell you about the momentum of the particle?

(c) What does your answer tell you about the momentum uncertainty of the particle?

(d) What does your answer tell you about the kinetic energy of the particle? *Hint*: Think about how momentum and kinetic energy are classically related!

12. Using the relationships we wrote between k, ω, p, and E, show that a free-particle energy eigenstate obeys the de Broglie relations $p = h/\lambda$ and $E = h\nu$.

13. In the Example on p. 281 we found, for $\psi(x) = [2/(\pi L^2)]^{1/4} e^{-(x/L)^2}$, that $P(p > 2\hbar/L) \approx 0.02$.

(a) Write a function that looks very similar to $e^{-(x/L)^2}$, but is *narrower*. In other words, your function (like the original) should peak at $x = 0$, but it should fall to half its peak height closer to zero than the original did.

(b) If you prepared 100 particles with $\psi(x) = [2/(\pi L^2)]^{1/4} e^{-(x/L)^2}$, and measured all their positions, the average would be zero. Your new function should have that property also. Would your new function's positions tend to be closer to zero, or farther, than the original?

(c) Find the normalization constant for your new function.

(d) Was your normalization constant higher or lower than $[2/(\pi L^2)]^{1/4}$? Explain how that could have been predicted without doing the math.

(e) Calculate $\hat{\psi}(k)$ for your function.

(f) On one plot, graph the distributions $|\hat{\psi}(k)|^2$ for the original function and the narrower

$\psi(x)$ that you just defined. How is the new shape different from the original, on both the horizontal and vertical axes? What does this imply about the momentum distributions associated with these two wavefunctions?

(g) Calculate the probability that the momentum is higher than $2\hbar/L$. Did it come out higher, or lower, than 2%?

14. **Exploration: Kinetic Energy Distribution**

In this problem you're going to derive a probability distribution for the kinetic energy of a particle with wavefunction $\psi(x)$.

(a) In order to measure that the particle has kinetic energy between K_1 and K_2, what range of momenta would you have to measure? (You'll need to relate momentum to kinetic energy, and don't forget to include negative momenta.)

(b) What is the probability of measuring the momentum in that range? Because we started by giving you $\psi(x)$, your answer will involve an integral of an integral squared. (Don't use the symbol $\hat{\psi}$ in your answer. Write everything out. Your answer will still have k as an integration variable.)

(c) Use the u-substitution $k = \sqrt{2mK}/\hbar$ to rewrite the probability in terms of K (kinetic energy) rather than k (wave number).

(d) By definition, the probability distribution for kinetic energy is a function $f(K)$ such that the probability of finding the kinetic energy between K_1 and K_2 is $\int_{K_1}^{K_2} f(K) dK$. What is that probability distribution $f(K)$?

6.4 Phase Velocity and Group Velocity

A particle in a lab generally has a "localized" wavefunction. That is, although both its position and momentum are necessarily somewhat uncertain, they aren't very uncertain; the particle is known to be in some small region of space, moving at more or less some particular velocity. Larger objects, of course, approach the classical limit of absolutely specified position and momentum.

It is a remarkable mathematical fact that a reasonably localized wavefunction can be constructed as an integral over infinitely many non-localized functions, such as the energy eigenstates of a free particle. Such a construction is called a "wave packet."

By looking into the mathematical way in which the eigenstates of a particle combine to create its localized wavefunction, we can also gain insight into how the motion of those individual eigenstates leads to the motion of the wave packet – that is, how a physical particle actually moves.

6.4.1   Discovery Exercise: Speed of a Free-Particle Energy Eigenstate

Written in terms of momentum and energy, a free-particle energy eigenstate has the form

$$\Psi(x, t) = Ce^{i(px-Et)/\hbar}. \tag{6.11}$$

As we saw in Section 6.2, Equation (6.11) is a traveling wave.

1. Express this wave's velocity v_{wave} in terms of E and p. (If you need a reminder of the speed of a traveling wave, it's in Appendix F.)

2. Because a free particle has no potential energy, its energy is all kinetic: $E = (1/2)mv^2$. Using that equation and $p = mv$, re-express v_{wave} as a function of the particle's velocity v.

See Check Yourself #7 at www.cambridge.org/felder-modernphysics/checkyourself

You have just found a relationship between the velocity of a traveling wave and the velocity of a particle. But remember that the state of that particle (including its position) is entirely described by that wave! The fact that they are both moving, but with different velocities, is a vital hint to understanding how systems change in quantum mechanics.

6.4.2   Explanation: Phase Velocity and Group Velocity

Two different scientists, Dr. Theory and Dr. Experiment, are watching the same proton make its way along the tunnel of a linear accelerator. You ask them a couple of questions.

Tell me about the proton right now.

Dr. T: "The proton's wavefunction is in a superposition of energy eigenstates, as must always be the case. In this case the proton is essentially a free particle, so each eigenstate is a complex exponential extending infinitely far in both directions, its modulus never changing."

Dr. E: "Oh, the proton is somewhere about here."

What's changing over time?

Dr. T: "Each individual eigenstate is multiplied by $e^{-iEt/\hbar}$ for its particular energy E. That is the only change."

Dr. E: "The proton is flying down the tunnel at 99% the speed of light."

Quantum mechanics works precisely because Dr. T's description correctly gives rise to Dr. E's. You already know the core mathematics that connects the two descriptions, which is a Fourier transform. In this section we'll look at how such an integral can reasonably approximate a classical particle, and how the time-evolution rule gets that particle moving.

Two Waves

Figure 6.10 shows the superposition of two cosine waves with slightly different wavelengths: $\cos(\pi x)$ and $\cos(1.1\pi x)$. As you move along the x axis there are regions where they are in phase and interfere constructively (antinodes), and other regions where they are out of phase and interfere destructively (nodes).

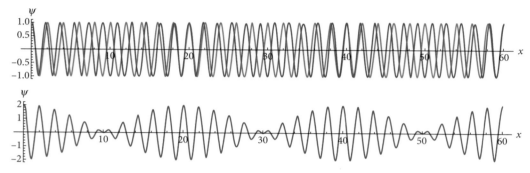

Figure 6.10 The upper panel shows two cosine waves individually. The lower panel shows their sum.

You can describe the lower panel of Figure 6.10 as a high-frequency wave with wavelength ≈ 2, whose amplitude goes up and down with a low-frequency "envelope" with wavelength ≈ 40.

Life gets really interesting when we put in the time evolution. Recall that a free particle's energy eigenstate e^{ikx} evolves as $e^{i(kx-\omega t)}$ where $\omega = k^2\hbar/(2m)$, so it moves at speed $\omega/k = k\hbar/(2m)$. What does that look like in Figure 6.10? The red wave does not change its shape, but moves to the right with speed $v = k\hbar/(2m)$. The blue wave does the same, *but with a different k-value*. The fact that the two waves travel at different velocities shifts the places where they are in and out of phase.

This phenomenon – the change to the overall shape that results from the component waves propagating at different speeds – may not have come up in your earlier math or physics classes, but it is central to the wavefunction behavior we will be describing in this section and the next. So we urge you to spend a few minutes with the following two animations: www.cambridge.org/felder-modernphysics/animations

- "Building a Wave Packet" shows the result of adding multiple waves with slightly different frequencies. Here's the big thing we want you to notice: the more waves you add, the more the result looks like one localized bump. That bump is called a "wave packet."

- "Group Velocity vs. Phase Velocity" shows the sum of only two such waves, but evolves that picture in time. Here's the big thing we want you to notice. The individual waves move with almost, but not quite, the same speed. The individual wave speed is called the "phase velocity." But the wave packet itself moves at a completely different speed, because (as we noted above) the locations of the nodes and antinodes shift in time. The velocity of the overall packet is called the "group velocity."

Now, supposing the sum of all these cosines represents the wavefunction of a particle,[4] consider the physical implications of all this math. It is not the red or the blue wave in Figure 6.10 that determines the probability distribution of the particle's position (or other measurable quantities); it is their sum. Therefore, it is not the phase velocity that determines how fast the particle is moving, but the velocity of the envelope function: the group velocity.

---

4 You can't actually normalize these. We'll take care of that when we start integrating over an infinite number of cosines instead of adding up a finite number.

You know that for a single traveling wave $\cos(kx - \omega t)$, the velocity is ω/k. In Problem 14 you'll show that for a superposition of two waves, the phase velocity (velocity of the high-frequency wave) is ω_{av}/k_{av}, and the group velocity (velocity of the envelope function) is $\Delta\omega/\Delta k$.

Wave Packets

In the "Building a Wave Packet" animation, you saw that adding more cosines gives you a more localized wavefunction: narrower bumps, with more dead space between them. In the limit of an infinite number of cosines, you can approach a single wave packet that represents a localized particle.

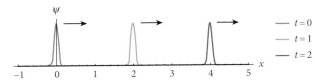

Figure 6.11 One wave packet at three different times.

With that in mind, look at the moving particle represented in Figure 6.11. Make sure you see that this figure, *unlike* Figure 6.10, does not show a periodic function extending out over all space at a single instant. Rather, it shows three different times. The function $\Psi(x, 1)$ peaks at $x = 2$ and is effectively zero for all $x < 1.8$ and all $x > 2.2$. In other words, a measurement at time $t = 1$ is almost guaranteed to find the particle close to $x = 2$.

A second later, a measurement will find this particle around $x = 4$. So this particle is moving to the right at 2 units/s. Of course there is a bit of uncertainty around that number too! But as you may recall from Section 6.2, the Fourier transform of a Gaussian function is another Gaussian function. So both the position and the momentum are sharply peaked curves with relatively little uncertainty.

What you can't see in the drawing is that $\Psi(x, 1)$ is made up of infinitely many e^{ikx} waves, each with its own value of k. It's not easy to picture infinitely many complex exponential functions, which is why we started with the simpler case of two real-valued cosines. So let's see what the simpler example can teach us about the real-life wave packet.

- The overall shape of Figure 6.10 resulted from the two cosines interfering constructively in some regions (such as $x = 20$) and destructively in others (such as $x = 30$). In a similar way, our wave $\Psi(x, 1)$ results from its component waves interfering constructively in some regions (near $x = 2$) and destructively in others (everywhere else).

- The overall shape of Figure 6.10 moved to the right with a "group velocity" based on the fact that its component waves were moving at different speeds, changing the regions of constructive and destructive interference. In a similar way, the motion of our particle at 2 units/s is the group velocity of the wave packet, not the phase velocities of the individual eigenstates.

- Finally, the sum in Figure 6.10 was periodic and could not be normalized. As we have stressed before, no discrete sum of periodic functions is normalizable. Each wave packet in Figure 6.11 is built from a continuous integral, and is normalizable.

Let's look again at those individual eigenstates. Each one has a different wave number k, and each one has a different angular frequency ω. But the two are not independent; they are related by a function $\omega(k)$, based on the mathematical form of the eigenstates. It can be shown that the group velocity of the entire packet is the derivative of that function:

$$v_g = \frac{d\omega}{dk}. \tag{6.12}$$

The proof of this result is beyond the scope of this book, but you can view that formula as a reasonable extrapolation of the result for two cosines, $v_g = \Delta\omega/\Delta k$. The basic logic is the same in both cases: if high-frequency waves move much faster than low-frequency ones, that causes the regions of constructive and destructive interference to shift rapidly.

For a deeper dive into the sums of sine waves, see our online section "Wave Packets" at www.cambridge.org/felder-modernphysics/onlinesections

You don't need that online section to continue from this point. You need to know that, as you add more and more sines, the sum can approach a single localized bump. You need to know that the group velocity of two such waves is $\Delta\omega/\Delta k$, and that the group velocity of many such waves approaches $d\omega/dk$. What you don't need to know, which the online section explores, is why those facts are true.

Of course, you do need to know what all that math tells us about the motions of real particles in the real world! We demonstrate that in the following Example.

Example: Group Velocity for a Free Particle

A free particle is moving with a relatively well defined velocity v (small Δp).

1. Calculate the phase velocities of its eigenstates.
2. Calculate the group velocity of its wave packet.

Answer: A free-particle eigenstate with time dependence can be described by $\Psi = e^{i(px-Et)/\hbar}$. Comparing that to the generic (complex) form of a traveling wave, $e^{i(kx-\omega t)}$, we can see that $k = p/\hbar$ and $\omega = E/\hbar$.

1. Technically the eigenstates all have different phase velocities, but the problem said Δp was small, so they will all have roughly the same momentum, energy, and thus phase velocity.

   The phase velocity for a traveling wave is ω/k:

$$v_p = \frac{\omega}{k} = \frac{E}{p} = \frac{(1/2)mv^2}{mv} = \frac{v}{2}.$$

   This leads to the possibly surprising conclusion that a free particle with velocity v can be made of energy eigenstates that are all moving with velocity close to $v/2$.

   The motion of the particle is represented by the group velocity, however. Recall that when we added our two cosine waves traveling at slightly different speeds their regions of constructive and destructive interference shifted, resulting in a group velocity much faster than the individual phase velocities.

2. To calculate the group velocity we need $d\omega/dk$, so we need to write ω as a function of k:

$$\omega = \frac{E}{\hbar} = \frac{(1/2)mv^2}{\hbar} = \frac{p^2}{2m\hbar} = \frac{\hbar k^2}{2m}.$$

Now we can calculate v_g and then rewrite it in terms of the particle velocity v:

$$v_g = \frac{d\omega}{dk} = \frac{\hbar k}{m} = \frac{p}{m} = v.$$

In other words, a wave packet representing a particle moving at velocity v does indeed move at velocity v. Whew!

You might very reasonably object that "velocity of a particle" is an ill-defined concept in a world where momentum is always uncertain. That's true, but as we mentioned above, particles often have a small enough spread of momenta that we can meaningfully talk about their velocities. In the limit where the probability distribution for p is tightly clustered around a particular value, the particle's wave packet moves at a velocity equal to that value divided by m. More generally, the expectation values of position and momentum are related just as you would expect from classical mechanics (see Question 9):

$$\langle p \rangle = m\frac{d\langle x \rangle}{dt}.$$

Stepping Forward

In Section 6.5 we will look at quantum mechanical particles in motion. The scenarios will be simple: one-dimensional motion, one particle only, simple potential functions. But the math will be unfamiliar, and the resulting behavior will be in some cases strikingly non-classical.

Your ability to follow such scenarios hinges on staying in touch with two results:

1. *A function of the form e^{ikx} (with positive k) represents a wave moving to the right.*

   Why? Because it evolves in time as $e^{i(kx-\omega t)}$, so it moves to the right with speed ω/k.

2. *A free-particle wave packet made up of e^{ikx} functions (with different but positive k-values) represents a particle moving to the right.*

   Why? Because its group velocity is $v_g = d\omega/dk$ (this section), and for a free particle $\omega = \hbar k^2/(2m)$ (Section 6.2). Putting them together, $v_g = \hbar k/m$ tells us that a positive k (waves moving to the right) leads to a positive group velocity (particle moving to the right), and vice versa.

If you become comfortable with those two facts in this section, you can focus on applying them as you learn about scattering and tunneling.

6.4.3 Questions and Problems: Phase Velocity and Group Velocity

Conceptual Questions and ConcepTests

1. Define each of the following:

    (a) Wave packet

    (b) Phase velocity

    (c) Group velocity

2. Figure 6.12 shows an oscillation with a changing amplitude. Estimate the wavelength of the envelope function.

3. Recall that a "node" is a place of destructive interference.

    (a) Looking at Figure 6.10, identify the x-positions of two nodes.

    (b) Over time your two nodes will move to the right. How fast? (Choose one.)

        A. At the phase velocity of the individual waves

        B. At the group velocity of the combined wave

        C. The two nodes will travel at different velocities from each other.

4. Figure 6.13 shows two waves with slightly different wavelengths. If you were to plot their sum, the points where the amplitude was zero would be the nodes and the points with maximum amplitude would be the antinodes.

    (a) Estimate the position of one of the antinodes.

    (b) Assume both waves are traveling to the right. Just to the left of the antinode you identified, is the red wave ahead of or behind the blue wave? What about just to the right of that antinode?

    (c) If the red and blue waves are both free-particle energy eigenstates, which one will move faster? How do you know?

    (d) Putting all your answers together, will the antinode move right or left? How do you know?

5. If you play two musical notes with slightly different frequencies, you can hear "beats," oscillations of the sound from softer to louder and back again. Why? *Hint:* You may find Figure 6.10 helpful here.

6. Which of the following is true about a sum of two cosines? (Choose one.)

    A. Group velocity is always faster than phase velocity.

    B. Group velocity is always equal to phase velocity.

    C. Group velocity is always slower than phase velocity.

    D. Group velocity can be faster than or slower than phase velocity.

7. On p. 287 we presented a discussion with Dr. Theory and Dr. Experiment in which they gave seemingly unrelated answers to two questions about a

Figure 6.12

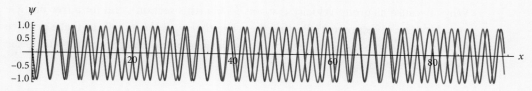

Figure 6.13

proton. Explain how each of Dr. Experiment's answers derives from Dr. Theory's corresponding description.

For Deeper Discussion

8. Elena and Leonicia are arguing about the time evolution of energy eigenstates. Elena says: "When you multiply *any* energy eigenstate by $e^{-iEt/\hbar}$ you're changing its phase but not its modulus. So it just rotates around the complex plane." Leonicia counters: "That may be true for *some* energy eigenstates, but a free-particle eigenstate becomes a traveling wave moving to the right or left." Is Elena right, or is Leonicia, or are they both right, or are they both half-right, or are they both totally wrong? Explain.

9. "Ehrenfest's theorem" says that $md\langle x\rangle/dt = \langle p\rangle$ and $d\langle p\rangle/dt = -\langle U'(x)\rangle$, where angle brackets denote expectation values. Recall that force is related to potential energy by $F = -U'(x)$, so this looks just like the classical definition of momentum and Newton's second law. Ehrenfest's theorem seems to say that the quantum mechanical expectation values $\langle x\rangle$ and $\langle p\rangle$ evolve exactly like the classical values x and p, at least until you make a measurement, but that is not correct. Why not?

Problems

10. Imagine a group of traveling waves, all with different wave numbers k, moving to the right in such a way that $\omega = c/k$ for some positive constant c.

    (a) Calculate the group velocity.

    (b) Is this wave packet moving left, moving right, or staying still? How can you tell?

    (c) Why is it, for traveling waves that look like energy eigenstates of a free particle, that the group velocity v_g always moves in the same direction as the wave?

11. Certain deep-water waves have phase velocities $v_p = \sqrt{g\lambda/2\pi}$, where λ is the wavelength.

    (a) Find the group velocity of a wave packet made of these waves.

    (b) Is the group velocity larger or smaller than the phase velocity? (Is your answer the same or different from what it is for matter waves?)

12. A light beam consists of a wave packet made of light waves with frequencies close to some value ν. Calculate the phase and group velocities of the wave packet. (This will require remembering some properties of light.)

13. Consider the free-particle eigenstates $\psi_1(x) = e^{i\pi x}$ and $\psi_2(x) = e^{1.1i\pi x}$.

    (a) What are $|\psi_1|$ and $|\psi_2|$?

    (b) Let $\psi = \psi_1 + \psi_2$. Calculate $|\psi|^2$, simplifying your answer as much as possible.

    (c) Plot the function $|\psi|^2$ that you found. How is your plot similar to and different from the bottom half of Figure 6.10, which shows this sum with cosines instead of complex exponentials?

14. Consider a superposition of two cosines: $f(x,t) = \cos(k_1 x - \omega_1 t) + \cos(k_2 x - \omega_2 t)$. As in the text, the sum looks like a high-frequency wave whose amplitude is modulated by an envelope function. In this problem you'll calculate the frequencies of the high-frequency wave and the envelope function.

    (a) Use trig identities to prove that $\cos(a+b) \cos(a-b) = (1/2)[\cos(2a) + \cos(2b)]$. (This will require several steps.)

    (b) Apply this result to rewrite $f(x,t)$ as a product of two cosines.

    (c) Which of those two cosines is the high-frequency wave and which one is the envelope function? How do you know?

    (d) Find the velocity of the high-frequency wave and the group velocity (the velocity of the envelope function) for $f(x,t)$.

15. Starting from the free-particle wavefunction $\Psi = e^{i(px-Et)/\hbar}$, we found the phase velocity of an individual eigenstate ($v_p = v/2$) and the group velocity of a wave packet ($v_g = v$).

    (a) Recalculate v_p and v_g using the relativistic definitions of momentum and kinetic energy. *Hint*: The equations for p, E, and how they are related in relativity are in Appendix B.

    (b) You should have found that v_p is *always* greater than c. Why does this not violate causality?

16. Exploration: Evolution of a Wave Packet

 A particle with a spread of momenta centered on p_0 can be modeled with the function $\hat{\psi}(k) = Ae^{-c(k-k_0)^2}$ where $k_0 = p_0/\hbar$.

(a) What is the difference (physically) between a particle of this type with large c and one with small c?

(b) Find A. (It's not equal to 1.)

(c) Find $\psi(x)$.

(d) Now suppose this $\psi(x)$ is the *initial* wavefunction of a free particle. Calculate $\Psi(x, t)$.

Hint: Start by figuring out how each eigenstate $\hat{\psi}(k)$ will evolve in time and then use that time-dependent function to calculate $\Psi(x, t)$ at an arbitrary time t. You can check your answer by making sure it reproduces your initial wavefunction $\psi(x)$ when you plug in $t = 0$.

(e) Take $m = \hbar = c = k_0 = 1$ and plot $|\Psi(x, t)|^2$ at several times. Describe how it's evolving over time. (You should find that it's not just moving; the shape is also changing.)

6.5 Scattering and Tunneling

"Scattering" refers to an interaction in which a particle changes its motion as a result of an interaction. Examples we've talked about include a photon that changes direction after colliding with an electron (Compton scattering), and an alpha particle that changes speed and direction after colliding with a nucleus (Rutherford scattering). In this section we'll explore the scattering in one dimension that occurs when a moving particle encounters a sudden change in potential energy. We'll see that in some cases particles bounce back, when classically they would keep going. We'll also see that particles can pass through barriers that classically would stop them, a phenomenon known as "tunneling."

6.5.1 Discovery Exercise: Scattering and Tunneling

 Each question below gives a potential energy function. In each case a particle is coming in from the left ($x < 0$, $v > 0$) with energy E. Briefly describe how the particle would behave *classically*, assuming all forces are conservative. There is no quantum mechanics in this exercise.

1. $U(x) = 0$ for $x < 0$ and $U(x) = U_0$ for $x \geq 0$. Assume $E < U_0$.

   See Check Yourself #8 at www.cambridge.org/felder-modernphysics/checkyourself

2. The same potential function as Part 1, but this time $E > U_0$.

3. $U(x) = 0$ for $x < 0$ and $x > L$, and $U(x) = U_0$ for $0 \leq x \leq L$. Assume $E < U_0$.

In quantum mechanics the answer to the first question is more or less unchanged, but the other two scenarios come out very different from the classical expectation.

6.5.2 Explanation: Scattering and Tunneling

All the examples in this section involve particles coming in from $-\infty$, which means all the states we consider will be unbound. We'll see how such a particle responds to three different obstacles.

A Potential Step with $E < U_0$

Our first two examples involve a "potential step" (Figure 6.14):

$$U(x) = \begin{cases} 0 & x < 0 \\ U_0 & x \geq 0 \end{cases} \quad \text{where } U_0 > 0$$

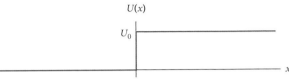

Figure 6.14 A potential step.

If a particle comes in from $-\infty$ with total energy E, what will it do? Classically the solution is trivial. (Classical 1D scattering problems tend to be trivial.) If $E < U_0$ the particle will bounce back and start moving left without changing its speed. If $E > U_0$ the particle will keep moving to the right with a reduced speed. (Remember that all the forces we're considering are conservative, so the particle's total energy remains constant.)

The quantum mechanical picture is more complicated. We'll take those two scenarios one at a time, starting with $E < U_0$.

In the region where $x < 0$, we are solving Schrödinger's equation with $U(x) = 0$. We've solved that one several times, using real-valued sines and cosines (in the infinite square well) or complex exponentials (for a free particle). Here we will opt for the latter form, since it expresses our wavefunction in terms of momentum eigenstates.

But in the region where $x > 0$, Schrödinger's equation looks like this:

$$-\frac{\hbar^2}{2m}\frac{d^2\psi}{dx^2} + U_0\psi(x) = E\psi(x) \quad \rightarrow \quad \frac{d^2\psi}{dx^2} = \frac{2m(U_0 - E)}{\hbar^2}\psi. \tag{6.13}$$

The second derivative of ψ is proportional to ψ, which we've seen many times. But the twist is that the constant of proportionality is now *positive*. (Remember that we're looking at the $E < U_0$ case right now.) That sign change makes a big difference!

Active Reading Exercise: $y''(x) = k^2 y(x)$

1. Find a real non-zero function that satisfies the differential equation

$$\frac{d^2y}{dx^2} = y.$$

Don't try to apply a technique or do any algebra here; just think of a real function that is its *own second derivative*. *Hint*: You won't be able to achieve this with sines or cosines, since we're not allowing i in your solution.

2. Find a *different* real non-zero solution that is not a multiple of your first solution.

3. Write the general solution, with two arbitrary constants, to the following differential equation:

$$\frac{d^2y}{dx^2} = 9y.$$

How far did you get? If you've never seen that equation before, you may have found some but not all of the final answer. If so, take a moment to confirm for yourself that the function $y = Ce^{-3x} + De^{3x}$ fits the bill. Then see if you can apply the same logic to Equation (6.13) to get

$$\psi(x) = Ce^{(-\sqrt{2m(U_0-E)}/\hbar)x} + De^{(\sqrt{2m(U_0-E)}/\hbar)x}. \tag{6.14}$$

So is that what our wavefunction looks like in the $x > 0$ region? It does solve the equation, but it also presents an immediate problem: the second term blows up as $x \to \infty$. (The *first* term blows up as $x \to -\infty$, but we don't care, since this wavefunction only applies on the right-hand side of the world.) The only way we can use this wavefunction is if $D = 0$. (See Question 8 in Section 6.5.3.)

Putting it all together, we end up here:

$$\psi(x) = \begin{cases} Ae^{i(\sqrt{2mE}/\hbar)x} + Be^{-i(\sqrt{2mE}/\hbar)x} & x < 0 \\ Ce^{-(\sqrt{2m(U_0-E)}/\hbar)x} & x \geq 0 \end{cases} \tag{6.15}$$

Equation (6.15) does not represent the wavefunction of a particle. It is one function – one value of ψ for every value of x from $-\infty$ to ∞, at one instant in time – and it represents one energy eigenstate of this system, with eigenvalue E. If you want to know how any actual wavefunction will evolve in time, you express that wavefunction as an integral over infinitely many functions that look like Equation (6.15) with different values of E, and then multiply each individual eigenstate by $e^{-iEt/\hbar}$.

To understand the effect of that transformation, let's look at what $e^{-iEt/\hbar}$ does to each of the three pieces of the eigenstate.

1. The first piece becomes $Ae^{i(x\sqrt{2mE}/\hbar - Et/\hbar)}$, a traveling wave at $x < 0$ moving to the right. This is called the "incident wave." (The word "incident" means "approaching"; this wave is approaching the step.)

2. Similarly, the second piece, with coefficient B, creates a traveling wave at $x < 0$ moving to the left. This is called the "reflected wave."

When discussing these kinds of wavefunctions it's common to describe any function of the form e^{ikx} with positive k as a wave moving to the right, and e^{-ikx} as a wave moving to the left. That shorthand makes sense once you put in the time-dependent piece, as we did above.

3. The third piece, the function in the $x \geq 0$ domain, is not a traveling wave. (Do you see why?) It represents the particle being in the classically forbidden region $x > 0$, with a probability that drops off exponentially as you go farther past the step.

If we took the liberty of arranging these three pieces in chronological order as A, C, and then B, they might tell an almost-classical story: the particle comes in from the left, penetrates a little bit past the step, and then bounces back and heads left toward $-\infty$. But that's not how quantum mechanics works: all three pieces exist at all times, spread out throughout space. How can we interpret that?

The interpretation comes from the wave packet. (As a reminder, the phrase "wave packet" refers to a localized wavefunction constructed as an integral over non-normalizable energy eigenstates.) At any given moment the particle's position is fairly well defined, with a small uncertainty Δx around some particular x-value.

At the particular moment represented in Figure 6.15, the eigenstates that make up the wave packet cancel each other out for almost all x-values, but the incident waves constructively interfere in one small region: the location of the particle. A bit later, everything cancels except for a different small region, closer to the step: the particle has moved. At some still later time everything cancels except the *reflected* waves in one small region, and the packet is moving leftward. The net effect is quite close to the "almost-classical story" we suggested above, but it comes about because of the interference of infinitely many eigenstates.

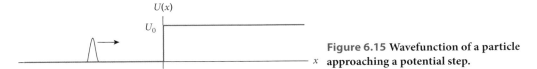

Figure 6.15 Wavefunction of a particle approaching a potential step.

This should all sound familiar. When we interpreted the energy eigenstates of the free particle, we noted that a free-particle energy eigenstate has equal amplitude everywhere. How can that represent a moving particle? In Section 6.4 we concluded that it does so because the motion of all the individual eigenstates that make up a wave packet leads to a group velocity for that wave packet, and that's what we call the motion of the particle. The same logic applies to our current scenario, except that the motion looks different at different times because each energy eigenstate has three different pieces.

Figure 6.16 shows a wave packet reflecting off a potential step. You can find an animated version of this and the other scattering illustrations in this section at www.cambridge.org/felder-modernphysics/animations

We did *not* sit down and draw these wavefunctions by hand, based on imagining what a wavefunction should act like. We took the equations in this book, created initial conditions corresponding to a wave packet moving right, plugged it all into Mathematica, and saw what came out. We're telling you that to make the point that these equations really do lead, mathematically, to those results. Eigenstates add up to localized wave packets, the localized wave packets move and distort and reflect, and all of it comes from Schrödinger's equation.

It is then left to us to see what those results are telling us. In many ways they match our classical expectations (the particle moves to the right, bounces off the step, and moves back to the left with the same velocity). In other ways they don't (during the interaction the particle has a non-zero probability of being detected to the right of the step, even though its energy is less than U_0).

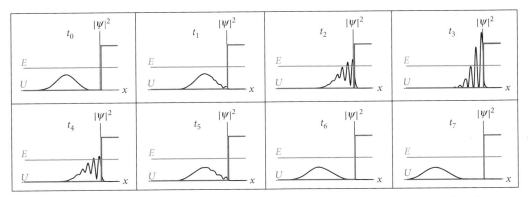

Figure 6.16 Evolution of a wave packet incident on a potential step with $E < U_0$. Notice that, for a brief time while the packet is colliding with the step, it has a small tail at $x > 0$.

A Potential Step with $E > U_0$

Now let's fire a particle with energy greater than U_0 at the same step. Classically it would keep moving to the right; quantum mechanically we have to find its behavior from the energy eigenstates. The energy eigenstates are unchanged for $x < 0$, but they look very different for $x \geq 0$.

Active Reading Exercise: A Potential Step with $E > U_0$

1. Write the energy eigenstate at $x \geq 0$ for a particle with $E > U_0$.

2. Your answer to Part 1 includes two pieces. What does each of the two pieces represent about our particle?

3. Neither of these two pieces blows up as $x \to \infty$, but one of them has to equal zero anyway. Can you guess which one, on physical grounds?

The solution for $x > 0$ now consists of complex exponentials, one piece moving right and the other moving left. We call the right-moving piece the "transmitted wave"; it represents the probability of the particle continuing on past the step. The left-moving piece represents a particle coming in from $+\infty$, which cannot physically happen in this scenario.[5]

So we are left with

$$\psi(x) = \begin{cases} Ae^{i(\sqrt{2mE}/\hbar)x} + Be^{-i(\sqrt{2mE}/\hbar)x} & x < 0 \\ Ce^{i(\sqrt{2m(E-U_0)}/\hbar)x} & x \geq 0 \end{cases} \tag{6.16}$$

The reflected and transmitted waves must be smaller than the incident wave because of normalization. In Figure 6.17 the transmitted wave is larger than the reflected wave: for the parameters we used when we made those pictures, a particle hitting this potential step is more likely to keep going than to bounce back. (As you might imagine, bringing E down toward U_0 increases the probability of reflection.)

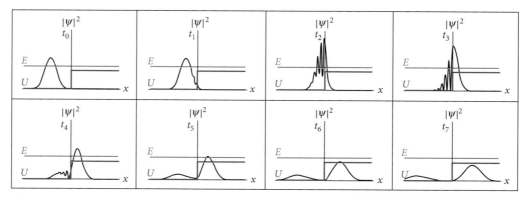

Figure 6.17 Evolution of a wave packet incident on a potential step with $E > U_0$. After the particle hits the step it generates a reflected wave and a transmitted wave.

---

5 That's really a classical argument, and therefore suspect. (Classically there wouldn't be a reflected wave here either.) The rigorous approach is to solve the time-dependent equations, starting with a wave packet approaching the step from the left. At later times you get a wave packet moving left at $x < 0$ (reflected) and one moving right at $x > 0$ (transmitted), but you don't get one at $x > 0$ moving left. The argument above is a way to intuitively understand why that makes sense.

You can calculate the probabilities by comparing the sizes of the incident and reflected waves: $|B/A|^2$ gives you the probability of reflection. That should sound plausible, but it's certainly not obvious: consider that $|C/A|^2$ is *not* the transmission probability! You'll derive the probability formulas for both reflection and transmission in Problem 24.

To calculate the ratio (B/A) you apply the usual boundary conditions that the wavefunction must be continuous and differentiable.

Example: Reflection Probability

A particle collides with a potential step of height U_0. The particle has energy $E > U_0$.

1. Find the reflection probability as a function of U_0/E.
2. Calculate that probability if $E = 2U_0$.

Answer:

1. Equation (6.16) gives an energy eigenstate. The boundary conditions are that it must be continuous and differentiable at $x = 0$:

$$A + B = C$$

$$i\frac{\sqrt{2mE}}{\hbar}A - i\frac{\sqrt{2mE}}{\hbar}B = i\frac{\sqrt{2m(E - U_0)}}{\hbar}C \quad \rightarrow \quad A - B = \sqrt{(E - U_0)/E}\, C.$$

The probability that a particle is reflected is $|B/A|^2$. These equations all came out real, so that's just $(B/A)^2$. With a little algebra you can show that

$$\left(\frac{B}{A}\right)^2 = \left(\frac{1 - \sqrt{1 - U_0/E}}{1 + \sqrt{1 - U_0/E}}\right)^2.$$

So the probability only depends on the ratio U_0/E. As you might guess, the probability approaches zero as $(U_0/E) \rightarrow 0$ (no step, no reflection) and 1 as $(U_0/E) \rightarrow 1$ (guaranteed reflection).

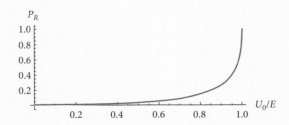

2. Plug in $U_0/E = 0.5$ and you get a probability of 3%. So if a beam of particles of energy E encounters a potential jump of height $E/2$, about 97% will keep going and 3% will bounce back.

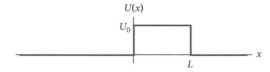

Figure 6.18 A potential barrier.

A Potential Barrier with $E < U_0$

Our final example is a potential barrier that rises up and then goes back to zero: $U(x) = U_0$ for $0 < x < L$ and $U(x) = 0$ elsewhere (Figure 6.18). We'll assume $E < U_0$, so a classical particle coming in from the left would bounce back and never know whether the raised potential went on forever (a step) or came back down (a barrier).

We now have separate solutions in three different regions; for $x > L$ one of those solutions represents a wave coming in from the right, and we will once again throw that one out (our arbitrary constants jump from D to F because for obvious reasons we can't use E):

$$\psi(x) = \begin{cases} Ae^{i(\sqrt{2mE}/\hbar)x} + Be^{-i(\sqrt{2mE}/\hbar)x} & x \leq 0 \\ Ce^{(\sqrt{2m(U_0-E)}/\hbar)x} + De^{-(\sqrt{2m(U_0-E)}/\hbar)x} & 0 < x < L \\ Fe^{i(\sqrt{2mE}/\hbar)x} & x \geq L \end{cases} \quad (6.17)$$

The story from here on is very much like what we said above for the potential step. At any given moment, the particle's wavefunction is an integral over infinitely many of these eigenstates. At early times the eigenstates cancel out except for the incident waves, which create a localized wave packet around some x-value in the left region. That packet moves to the right until it hits the barrier. Things look strange and complicated for a while, with some amplitude in the classically forbidden region (the barrier), and then at later times you have a reflected wave moving left and a transmitted wave moving right (Figure 6.19).

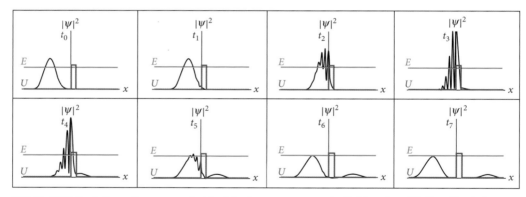

Figure 6.19 Evolution of a wave packet incident on a narrow potential barrier with $E < U_0$. At late times there is a chance of the particle having been transmitted through the barrier.

We want to draw your attention to the transmitted wave at $x \geq L$. A classical particle with $0 < E < U_0$ could *exist* in that region, but it could never *get* there from the left side. But quantum mechanical particles can appear on the far sides of such apparently impassable barriers, and they frequently do. This is called "tunneling" because it seems like the particle somehow burrowed through to the other side.

You can calculate the probability of tunneling from the boundary conditions, as we did for the potential step above, but the algebra is messy (Problem 23). The probability of a particle tunneling through a barrier goes down rapidly as the particle's energy (E/U_0) decreases, and as the barrier width L increases.

Tunneling is not just an arcane result of complicated math; it is a real phenomenon with important consequences. For instance, consider two protons in the Sun. As these protons approach each other, the potential energy of the two-proton system increases because of their mutual Coulomb repulsion. If the two protons get close enough to each other, they fuse into a helium nucleus with a significantly decreased potential energy due to the strong nuclear force. But there is an in-between distance where the Coulomb repulsion creates a high potential energy, and the nuclear force is not yet able to fuse them. This creates a potential hill that is, in general, higher than the energy of the protons. The Sun shines because those protons can tunnel through that potential energy barrier and fuse.

The same principle powers some modern electronic components: they are built with insulating gaps that electrons don't have enough energy to pass through classically, but current flows because the electrons tunnel through.

Stepping Back: Energy Eigenstates and Probabilities

Write down your best answers to the questions below and then compare them with our answers. We're using the potential step with $E > U_0$ as our example, but the ideas we're talking about apply to scattering problems more generally.

Active Reading Exercise: Interpreting the Solutions for a Potential Step

A particle with energy $E > U_0$ is incident on a potential step of height U_0. Here's the formula for an energy eigenstate, and a picture of the wavefunction a while after it hits the barrier. The parts of the wavefunction with coefficients A, B, and C represent the incident wave, reflected wave, and transmitted wave, respectively.

$$\psi(x) = \begin{cases} Ae^{i(\sqrt{2mE}/\hbar)x} + Be^{-i(\sqrt{2mE}/\hbar)x} & x < 0 \\ Ce^{i(\sqrt{2m(E-U_0)}/\hbar)x} & x \geq 0 \end{cases}$$

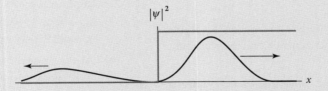

1. How can we keep talking about "the energy of the particle" and writing things like "$E > U_0$" when the particle's wavefunction is an integral over infinitely many eigenstates, each representing a different energy? An unbound particle can never be in a state of definite energy!

2. Is the ratio $|B/A|$ the same for every eigenstate?

3. Does the ratio $|B/A|$ stay the same as the wave evolves through time?

1. When we talk about "a particle with energy E" we mean one with a small ΔE. That is, our wave packet is made almost entirely from eigenstates in a fairly narrow band of E-values. All the other eigenstates, the ones whose energies aren't close to E, have very

small amplitudes (very small values of A, B, and C). Remember, though, that the energy levels aren't discrete; even within a narrow band there are infinitely many different eigenstates!

2. Not exactly. In the Example on p. 299 we found the ratio $|B/A|$. Our answer was a function of E, and every eigenstate has its own eigenvalue E, so every eigenstate has a different $|B/A|$.

   But as we noted above, the wave packet is mostly made up of eigenstates with approximately the same energy. So the ratio $|B/A|$ will be approximately the same for all of them, and we can meaningfully ask what that ratio is for our particle.

3. Yes. Multiplying any function by a given $e^{-iEt/\hbar}$ does not changes its modulus, so evolving an entire eigenstate over time does not change those ratios.

   It's very natural to assume that when the wave is moving to the right, A is high and the other two coefficients are low; after the wave leaves the step, A is low and the other two coefficients are high. But in fact, for each individual eigenstate, A is just as high after the collision as it was before. What changes over time is the interference.

6.5.3 Questions and Problems: Scattering and Tunneling

Conceptual Questions and ConcepTests

1. The last frame of Figure 6.17 on p. 298 shows a wavefunction that consists of two bumps, the "reflected" bump moving left and the "transmitted" bump moving right. (Note that the graph shows $|\psi(x)|^2$, not $\psi(x)$!) Now, suppose you found that the integral under the left-hand bump was 0.2.

   (a) What would that integral tell you about the particle?

   (b) What would be the integral under the right-hand bump, and what would that tell you about the particle?

2. Figure 6.20 shows the wavefunction $\psi(x)$ for a particle that has bounced off a potential step. This wavefunction can be mathematically expressed as an integral over the energy eigenstates of this system. Each such energy eigenstate looks like Equation (6.15) with a different value of E.

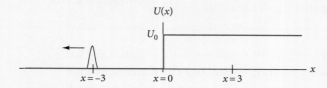

U(x)

U_0

$x=-3$ $x=0$ $x=3$ x

Figure 6.20

At the moment shown, if you look at the "incident" portion of one eigenstate at $x=-3$, it might be almost anything. But if you sum the incident portions of all the eigenstates at $x=-3$, they add up to something very close to zero.

(a) Of the three different pieces of Equation (6.15), which one is relevant at $x=3$? At the moment shown in the drawing, does that piece – summed up over all the eigenstates – necessarily add up to something close to zero at $x=3$, or not? Briefly explain your answer.

(b) At the moment shown in the drawing, do all the reflected waves add up to something near zero at $x=-3$, or do they not? Briefly explain your answer.

(c) At the moment shown in the drawing, do all the reflected waves add up to something near zero at $x=-5$, or do they not? Briefly explain your answer.

(d) How do your answers above change after a few seconds have elapsed?

3. In Equation (6.15), the constant A ... (choose one answer and briefly explain):

   A. must be determined by boundary conditions, but, for any given particle, is a constant

B. could have a different value for every value of x

C. could have a different value for every value of t

D. could have a different value for every value of E.

4. Equation (6.15), an energy eigenstate of a step function with $E < U_0$, has three pieces with coefficients A, B, and C. Because it is an energy eigenstate, the entire function evolves through time by being multiplied by $e^{-iEt/\hbar}$. Explain why this multiplication turns the A and B pieces into traveling waves, but not the C piece. (We're not looking for a discussion of the physics, although that's interesting too. Focus here on the formulas.)

5. In the Example on p. 299 we made a plot of the reflection probability for a particle incident on a potential step. Explain what the left and right limits of that plot tell us physically.

6. In the Example on p. 299 we calculated the probability that a particle of energy E would be reflected from a potential step of height U_0. An unbound particle can't have a definite energy, though, so what does the formula we calculated mean?

7. Define "tunneling" in your own words.

8. **(a)** Section 6.2 discussed the fact that an energy eigenstate of a free particle, Ce^{ikx}, is not normalizable, and therefore can never be the wavefunction of a particle. Briefly explain how these impossible wavefunctions can be relevant in modeling the behavior of real particles.

   (b) But in Equation (6.14), we threw out the constant D because it was the coefficient of an unnormalizable function. Why doesn't the argument you made in Part (a) apply to this case?

9. When a classical particle hits a potential barrier with $E > U_0$, the particle continues to the right more slowly than its original speed. What does that suggest about the behavior of a quantum mechanical wavefunction in the same situation? (Choose one answer and briefly explain.)

   A. The eigenstates on the right side of the barrier, which represent traveling waves, will move

more slowly than the eigenstates on the left side of the barrier.

B. The individual eigenstates will not move more slowly, but their group velocity will be slower.

C. Neither of these is necessarily true.

10. A particle hits a potential barrier of energy U_0. The particle's energy E is just slightly, but measurably, higher than U_0. What behavior would you expect to see in a lab...

   (a) classically?

   (b) quantum mechanically?

11. If a *classical* particle were at rest at $x = 0$ in the potential energy function shown in Figure 6.21, we would call that a "bound state" because it wouldn't have enough energy to leave the well it's in. In quantum mechanics we would say that this system only has unbound states. Why?

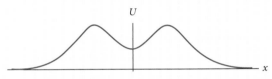

Figure 6.21

12. Figure 6.16 shows the wavefunction of a particle reflecting off a potential step. The figure also shows the particle's total energy. How can the wavefunction at times go much higher than the total energy? *Hint*: This is a trick question! To answer it you have to figure out why it's a trick question.

For Deeper Discussion

13. Figure 6.16 shows the wavefunction of a particle reflecting off a potential step. If you look closely you can see that the reflected wave packet is shorter and wider than the incident one. Why?

14. In the Example on p. 299 we calculated the reflection probability as $|B/A|^2$. The transmission probability is of course one minus the reflection probability. You might reasonably think that this would be equal to $|C/A|^2$, but you can check that it's not. Why not?

15. Critique the statement "Quantum mechanics only matters on microscopic scales."

Problems

16. Equation (6.15) represents an energy eigenstate for a potential step function with $E < U_0$. As with any unbound state, it is impossible for a particle's wavefunction to be a single energy eigenstate. But in this case, unlike other examples we've seen, normalization is not the entire reason.

 (a) Explain why, in order to normalize Equation (6.15), we must choose $A = B = 0$.

 (b) Find the value of C that properly normalizes the resulting function.

 (c) Explain why the resulting eigenfunction is still not a possible wavefunction.

17. Equation (6.15) gives an energy eigenstate for a potential step with $E < U_0$.

 (a) What two boundary conditions must this eigenstate obey at $x = 0$?

 (b) Apply those boundary conditions to prove that $|B| = |A|$.

 (c) What does this imply about the behavior of a particle incident on the potential step?

18. A particle of energy E is coming from $-\infty$ with potential energy $U(x) = 0$ for $x < 0$ and ∞ for $x \geq 0$.

 (a) What is $\psi(x)$ in the region $x \geq 0$? How do you know?

 (b) Write the energy eigenstate of this system. Your answer should involve two arbitrary constants.

 (c) Write the one boundary condition that those functions must obey. Why is there only one?

 (d) Solve that boundary condition to find a relationship between your arbitrary constants. What does that relationship predict about the particle's behavior?

19. A particle of energy E is coming from $-\infty$ with potential energy $U(x) = U_0$ for $x < 0$ and 0 for $x \geq 0$ (where $U_0 > 0$ as usual).

 (a) Sketch $U(x)$.

 (b) How do we know that E must be greater than U_0?

 (c) Write an energy eigenstate of this system. Your answer should involve four arbitrary constants.

 (d) Which one of those four arbitrary constants must equal zero and why?

 (e) Write the boundary conditions that those functions must obey.

 (f) Solve those boundary conditions to find the probability that the particle reflects back from $x = 0$. Your answer should depend only on the ratio U_0/E.

 (g) Evaluate that probability in the limit $E \to \infty$. What does your answer tell you?

 (h) Evaluate that probability in the limit $E \to (U_0)^+$. What physical situation does that correspond to? What would be the behavior of a classical particle in that situation?

20. A particle of energy E is coming from $-\infty$ with potential energy $U(x) = U_0$ (less than E) for $x < -L$, $U(x) = 0$ for $-L \leq x \leq 0$ and $U(x) = \infty$ for $x > 0$.

 (a) Sketch $U(x)$.

 (b) Write the energy eigenstate formulas in each of the regions. For $-L \leq x \leq 0$, use sines and cosines.

 (c) Write (but do not solve) the boundary conditions for this eigenstate. You should, with no algebra, be able to show that one boundary condition leads to one of the arbitrary constants being zero.

 (d) Without having to solve those equations, give the reflection probability. How do you know?

21. Unbound States of a Finite Well

A particle of energy E is coming from $-\infty$ with potential energy $U(x) = U_0$ (less than E) for $x < 0$, $U(x) = 0$ for $0 \leq x \leq L$, and $U(x) = U_0$ for $x > L$.

 (a) Sketch $U(x)$.

 (b) Write the formulas for an energy eigenstate in each of the three regions.

 (c) Which of the six resulting pieces must equal zero and why?

 (d) Write (but do not solve) the four boundary conditions that this eigenstate must satisfy.

(e) What would you have to solve for in those equations to find the probability that the particle is reflected back toward $-\infty$?

22. [*This problem depends on Problem 21.*]

 (a) Find the reflection probability.

(b) For $m = \hbar = U_0 = L = 1$, plot the reflection probability for $1 < E < 20$. Then do it again for $L = 2$.

(c) For $m = \hbar = U_0 = 1$ and $E = 2$, plot the reflection probability for $0 < L < 5$. Explain what the plot is showing you.

23. Tunneling Probabilities

Equation (6.17) gives an energy eigenstate for a particle of energy E coming in from $-\infty$ with potential energy $U(x) = U_0$ (greater than E) for $0 < x < L$ and 0 everywhere else. (You could do this problem by hand but we recommend letting a computer do the algebra for you.)

(a) Write the four boundary conditions that this eigenfunction must satisfy.

(b) Solve to find the tunneling probability.

(c) For $m = \hbar = U_0 = L = 1$, plot the tunneling probability for $0 < E < 1$. Then do it again for $L = 2$.

(d) For $m = \hbar = U_0 = 1$ and $E = 0.5$, plot the tunneling probability for $0 < L < 5$. Explain what the plot is showing you.

24. Exploration: Reflection and Transmission Coefficients[6]

Equation (6.16) gives the energy eigenstates of a wave packet incident on a potential step with $E > U_0$. If you measure the particle's position at a late time, the probability that you will find it was reflected is proportional to the area under the curve of $|\psi|^2$ for the reflected wave packet, which is built from the B piece of the wavefunction. Without bothering with integrals, you can reasonably infer that doubling the constant $|B|^2$ will double the height of the wave packet and therefore double the area. Doubling the width w_B of the wave packet will also double the area. We conclude that

$$\text{Area}_{\text{reflection}} = q|B|^2 w_B,$$

where q is a constant of proportionality that depends on the details of the wave packet, and on the exact way you define the width w_B. The probability of finding that the particle is transmitted is $q|C|^2 w_C$ with the same constant q.

(a) The area under $|\psi|^2$ for the incident wave is $q|A|^2 w_A$. What is the value of that area, and how do you know? (Your answer should be a number.)

(b) In Section 6.4 we showed that a wave packet $\psi = e^{ikx}$ travels with group velocity $v_g = \hbar k/m$. Use that fact to find v_{gC}/v_{gA}, the ratio of the group velocities of the transmitted and incident waves.

(c) The leading edges of the reflected and transmitted waves are created when the leading edge of the incident wave packet hits the step. Their trailing edges are created when the trailing edge of the incident wave reaches the step. Use that fact, and your answer to Part (b), to find w_C/w_A, the ratio of the widths of the transmitted and incident waves.

(d) Find w_B/w_A, the ratio of the widths of the reflected and incident waves.

(e) The probability of reflection is the area under the reflected part of the $|\psi|^2$ curve, divided by the area under the incident part of the $|\psi|^2$ curve. Use your results in this problem to show that the probability of reflection is $|B/A|^2$.

(f) Find the probability of the particle being transmitted. Your answer should not include B.

(g) In the Example on p. 299 we derived the boundary conditions $A + B = C$ and $\sqrt{(E - U_0)/E} = (A - B)/C$. Use those boundary conditions to eliminate C from your answer to Part (f) and check your answers by proving that the probability of reflection plus transmission is 1. (You can simplify your calculations by assuming A, B, and C are all real.)

---

6 Thanks to Travis Norsen, who showed us this derivation.

6.6 The Time-Dependent Schrödinger Equation

The time-dependent Schrödinger equation occupies a place in quantum mechanics comparable to $F = ma$ in classical mechanics. It is the core postulate from which the rest of the system derives.

It is also, like $F = ma$, a second-order differential equation. But Schrödinger's equation is a second-order *partial* differential equation (or PDE), a differential equation built from partial derivatives.

We introduced partial derivatives in Section 6.1, but you may never have seen a PDE before. So this section will serve not only to take us into the heart of quantum mechanics but to introduce a new type of math problem that requires new mathematical techniques.

6.6.1 Discovery Exercise: A Partial Differential Equation

A "partial differential equation" is an equation that involves partial derivatives of a multivariate function; for example,

$$\frac{\partial f}{\partial x} = \frac{\partial f}{\partial t}. \tag{6.18}$$

The function $f(x, t) = 2x + 2t$ is a solution to this equation because $\partial f/\partial x$ and $\partial f/\partial t$ both equal 2. The function $f(x, t) = xt$ is not a solution because $\partial f/\partial x = t$ and $\partial f/\partial t = x$, so they are not equal.

Which of the following are solutions to Equation (6.18)? (Choose all that apply.)

1. $f(x, t) = 5$
2. $f(x, t) = x - t$
3. $f(x, t) = x^2 + t^2$
4. $f(x, t) = x^2 + 2xt + t^2$

6.6.2 Explanation: The Time-Dependent Schrödinger equation

You know the following two facts about energy eigenstates:

- For a particle in a potential energy field $U(x)$, the energy eigenstates are solutions to the "time-independent Schrödinger equation," $- \left[\hbar^2/(2m) \right] \psi'' + U\psi = E\psi$.
- Each energy eigenstate evolves in time according to $\Psi(x, t) = \Psi(x, 0)e^{-iEt/\hbar}$.

As we have presented them so far, those two facts are independent postulates of the system. But in fact they can both be derived from a more fundamental source, the "time-dependent Schrödinger equation." That equation really is a postulate: we will not derive it from anything more basic. Its justification is that it leads to quantum mechanics, and quantum mechanics holds up in the lab.

Before we present this new postulate, we need to distinguish two different types of equations.

- An "ordinary differential equation," or ODE, involves one dependent variable and one independent variable. It therefore involves normal derivatives.

- A "partial differential equation," or PDE, involves one dependent variable with multiple independent variables. It therefore involves partial derivatives. If you have not studied partial derivatives in a math class, make sure you at least go through our introduction in Section 6.1 before going further here.

The time-independent Schrödinger equation in one dimension – the equation we've been solving for the last two chapters – is an ODE for a function $\psi(x)$. In three dimensions it becomes a PDE for a function $\psi(x, y, z)$. But the time-*dependent* Schrödinger equation, the equation we are introducing in this section, is a PDE in any number of dimensions, since it involves both spatial variables and time. This distinction is important because PDEs require different mathematical approaches from ODEs. Much of this section will be devoted to one such technique.

The Time-Dependent Schrödinger Equation in One Dimension

$$-\frac{\hbar^2}{2m}\frac{\partial^2 \Psi}{\partial x^2} + U(x)\Psi = i\hbar\frac{\partial \Psi}{\partial t} \qquad (6.19)$$

Remember that our old friend the time-independent Schrödinger equation says "You give me the potential function, and I'll give you the energy eigenstates." This new equation says "You give me the potential function *and the initial wavefunction* (which could be almost anything), and I'll tell you how that wavefunction will evolve in time." So the wavefunction $\Psi(x, 0)$ must be supplied as an initial condition, and the solution is the multivariate function $\Psi(x, t)$.

Solving the Time-Dependent Schrödinger Equation: Separation of Variables

We're going to solve the time-dependent Schrödinger equation with a technique called "separation of variables." (You may have previously used "separation of variables" to solve *ordinary* first-order differential equations. It's the same name but a completely different technique. Sorry.)

The method begins by assuming a solution of a special form called a "separable function": a function of x only, multiplied by a function of t only. The names of these two functions don't particularly matter, but we will use $\psi(x)$ for our spatial function (because that's what we've been calling spatial wavefunctions), and $T(t)$ for our temporal function (because we have to call it something):

$$\Psi(x, t) = \psi(x)T(t). \qquad (6.20)$$

Equation (6.20) does not represent all, or even most, functions. For example, $(\sin x)(\cos t)$ is separable, but $\sin(x + t)$ is not. Given that $\Psi(x, t)$ is meant to represent all possible wavefunctions, why would we restrict ourselves to such a special case?

It turns out we're not restricting ourselves in the end. Once we find the solutions that look like Equation (6.20), we will add them all with arbitrary coefficients. You will show in Problem 11 that any such sum of solutions is still a solution. What we're not going to prove, although it's also true, is that *all possible solutions of the time-dependent Schrödinger equation can be written as linear combinations of separable solutions*. So the sum we find here will be the general solution.

Active Reading Exercise: Plugging in the Separable Solution

If $\psi(x)$ is a function of x (but not of t), and $T(t)$ is a function of t (but not of x),

1. What is $\dfrac{\partial}{\partial t}[\psi(x)T(t)]$?

2. What is $\dfrac{\partial^2}{\partial x^2}[\psi(x)T(t)]$?

3. Bearing those answers in mind, plug the guess (Equation (6.20)) into the PDE (Equation (6.19)).

The key to that exercise is that when you take a partial derivative with respect to t, you treat x (and therefore any function of x) as a constant, and vice versa.

1. $(\partial/\partial t)[\psi(x)T(t)] = \psi(x)T'(t)$. For example, the derivative of $(\ln x \sin t)$ with respect to t is $(\ln x \cos t)$.

2. Similarly, the second derivative with respect to x of $\psi(x)T(t)$ is $\psi''(x)T(t)$.

3. Knowing that, we can plug our guess into Schrödinger's equation:

$$-\frac{\hbar^2}{2m}\frac{\partial^2}{\partial x^2}[\psi(x)T(t)] + U(x)[\psi(x)T(t)] = i\hbar\frac{\partial}{\partial t}[\psi(x)T(t)]$$

$$-\frac{\hbar^2}{2m}\psi''(x)T(t) + U(x)\psi(x)T(t) = i\hbar\psi(x)T'(t)$$

Now comes the step that gives "separation of variables" its name: get all the x dependence on one side of the equation, and all the t dependence on the other. We achieve that in this case (and in most such cases) by dividing both sides by $\psi(x)T(t)$:

$$-\frac{\hbar^2}{2m}\frac{\psi''(x)}{\psi(x)} + U(x) = i\hbar\frac{T'(t)}{T(t)}. \tag{6.21}$$

The left side of Equation (6.21) has no time dependence, meaning it doesn't change with time. The only way the two sides can stay equal is if the right side doesn't change either. $T(t)$ can be a function of time, but Equation (6.21) implies that $i\hbar T'(t)/T(t)$ is independent of time.

Meanwhile, the right side of Equation (6.21) has no x dependence. For the two sides to be equal at all places, the left side must therefore also be independent of x. Putting it all together, *the only way Equation (6.21) can be satisfied is if both sides equal a constant.*

That point is subtle, but it's the heart of this technique, so we encourage you to sit with it a minute and make sure you're convinced. If we tell you that $f(x) = g(t)$, where f only depends on x and g only depends on t, then they must both equal a constant: $f(x) = g(t) = C$.

Applying that logic to the equation we just derived, we're going to set both sides equal to a constant. We need a name for this constant – you may be able to see where we're going with this – let's call it E:

$$-\frac{\hbar^2}{2m}\frac{\psi''(x)}{\psi(x)} + U(x) = E \tag{6.22}$$

$$i\hbar\frac{T'(t)}{T(t)} = E \tag{6.23}$$

The point of the separation of variables technique is to reduce one partial differential equation to multiple *ordinary* differential equations, which are much easier to solve. That mission is accomplished. And look what we got!

- Equation (6.22) is an old friend. If you don't quite recognize it yet, multiply both sides by $\psi(x)$. As we promised at the beginning, we see that the time-independent Schrödinger equation can be derived from the time-dependent one.

- And what about Equation (6.23)? It may not be as immediately familiar, but it's not hard to solve.

Active Reading Exercise: The Time Equation

Equation (6.23) can be rewritten as

$$\frac{dT}{dt} = -\frac{iE}{\hbar}T(t).$$

(Note that, as usual, we rewrote $1/i$ on the right side as $-i$.)

1. Solve that equation. You can do this by using separation of variables (not the PDE technique we just taught you, but the ODE technique you may have learned in previous math courses). You may also be able to solve it just by thinking about it.

2. Equation (6.22) became a familiar result, the time-independent Schrödinger equation. What familiar conclusion does Equation (6.23) lead to?

The general solution to Equation (6.23) is $T(t) = Ce^{-iEt/\hbar}$. (If you didn't get that, take a moment now to plug it in and convince yourself that it works.) We have derived our second promised result, the time-evolution formula for an energy eigenstate.

So What Did We Find?

Our solution to the time-dependent Schrödinger equation is complete. The separable solutions are all of the form

$$\Psi(x,t) = \psi(x)e^{-iEt/\hbar} \quad \text{where } \psi(x) \text{ is any solution to the time-}independent \text{ equation.}$$

As promised, we were able to use the time-dependent equation to derive both the time-independent equation and the rule for time evolution of its solutions.

Remember that the general solution is a linear combination of all such separable solutions. So the process for any given potential energy function is:

1. Solve the time-independent Schrödinger equation to find the energy eigenstates, each with its own eigenvalue E.

2. Write your initial wavefunction $\Psi(x,0)$ as a linear combination of those eigenstates. (This combination may be a discrete series of bound eigenstates, or a continuous integral over unbound eigenstates.)

3. Multiply each individual eigenstate by $e^{-iEt/\hbar}$ (with its own E) to find the full wavefunction $\Psi(x,t)$.

You already knew that whole process. Now you know that the entire process can be justified based on the solution to a single equation, the time-dependent Schrödinger equation.

It's important to keep track of some assumptions we made along the way.

- The time-dependent Schrödinger equation is not derived from anything else, but is simply a postulate of quantum mechanics.
- We did not prove that the $\psi(x)$ functions we found are energy eigenstates. We called the separation constant E because we know that you can derive that fact from other postulates of quantum mechanics that are beyond the scope of this book.
- We asserted but never proved that you can write any solution $\Psi(x,t)$ as a sum of the separable solutions we found.

We need to also mention a fact that may have been left out of your introduction to differential equations: most of them cannot be solved. At least, most of them can't be solved in the way you're used to, which is "analytically": find a function that works, and then fill in the arbitrary constants based on initial conditions. When analytical approaches fail, we can still make predictions using "numerical" approaches. In Problem 12 you will consider a potential energy function for which you cannot find the energy eigenstates by analytically solving the time-independent Schrödinger equation, and you will predict its time evolution by numerically working with the time-dependent Schrödinger equation.

As a final thought, remember that the time-independent Schrödinger equation describes one type of time evolution for a wavefunction. In the orthodox interpretation there is another form of time evolution – the collapse of the wavefunction when a measurement is performed – that is *not* predicted by this equation, and that introduces an element of randomness to an otherwise predictable process. In the orthodox interpretation, the collapse of the wavefunction must be included alongside Schrödinger's equation as a separate axiom.

A Tale of Two Equations

To summarize, we offer the following side-by-side comparison of the two different Schrödinger equations.

Time-independent Schrödinger equation	Time-dependent Schrödinger equation
$$-\frac{\hbar^2}{2m}\frac{d^2\psi}{dx^2} + U(x)\psi = E\psi$$	$$-\frac{\hbar^2}{2m}\frac{\partial^2\Psi}{\partial x^2} + U(x)\Psi = i\hbar\frac{\partial\Psi}{\partial t}$$
You give it $U(x)$, the potential energy field.	You give it $U(x)$, the potential energy field.
It answers the question: "What are the energy eigenstates of this system?"	Given any initial wavefunction, it answers the question: "How will this wavefunction evolve in time?"
Its answers are functions of space, $\psi(x)$.	Its answers are functions of space and time, $\Psi(x,t)$.
In one dimension it is an "ordinary differential equation" (ODE): its solution is a function of one independent variable only, so it involves an ordinary (second) derivative.	It is a "partial differential equation" (PDE): its solution is a multivariate function, so it involves partial derivatives.

It's easy to get these two equations confused. Their names are similar and the equations themselves look similar. But they are different equations designed to answer different questions, so we urge you to take a moment with the table on the previous page to make sure they are distinct in your mind.

6.6.3 Questions and Problems: The Time-Dependent Schrödinger Equation

Conceptual Questions and ConcepTests

1. Explain in your own words what the solutions to each of the two Schrödinger equations tell you about a particle.

2. In this section we started with the time-dependent Schrödinger equation as a postulate and used it to derive what two results?

3. True or false? All solutions to the time-dependent Schrödinger equation are separable.

4. Dr. McCoy says "I finally solved the time-independent Schrödinger equation for the potential energy function I'm working on. The solution is $Ae^{-x^2}e^{-3it}$." Mr. Spock, who has no idea what the potential energy function is, arches an eyebrow and contemptuously calls the doctor's answer illogical. How does he know Dr. McCoy is wrong?

5. Which of the following represent separable functions? Choose all that apply, and briefly explain your answers (which may require a bit of algebra).

   A. $f_1(x,y) = (x+y)^2 - (x-y)^2$
   B. $f_2(x,y) = e^{2x+3y}$
   C. $f_3(x,y) = \left(e^{2x}\right)^{3y}$
   D. $f_4(x,y) = \ln(xy)$
   E. $f_5(x,y) = \sin(x+y)$

For Deeper Discussion

6. One of the simplest and most important examples of a multivariate function is a string oscillating up and down (in two dimensions only), so its height y is a function of its horizontal position x and the time t. You are watching a particular vibrating string. You don't get told its mathematical function, but you are shown pictures of the string at various times. How could you tell from those pictures whether its height function $y(x,t)$ happens to be a *separable* function $X(x)T(t)$? (Your answer here may not involve any equations. It will involve words, and may also include some pictures.)

Problems

7. Write $\Psi(x,t)$ for a particle in the ground state of an infinite square well. Then plug it in and prove that it is a solution to the time-dependent Schrödinger equation.

8. In this problem you'll find all the separable solutions to the following partial differential equation (PDE):

$$\frac{\partial f}{\partial x} = k\frac{\partial f}{\partial t}.$$

   (a) Before you start, what are the units on the constant k? *Hint*: We didn't tell you the units of f because it doesn't matter.

   (b) Assume a separable solution, $f(x,t) = X(x)T(t)$. Plug it into both sides of the differential equation.

   (c) Separate variables so all the x dependence is on one side and all the t dependence on the other.

   (d) Briefly explain why both sides of the equation you just wrote must equal a constant.

   (e) Set both sides equal to a constant C to get two ordinary differential equations.

   (f) Solve both ordinary differential equations to find $X(x)$ and $T(t)$.

   (g) Multiply those two functions together to find the separable solution $f(x,t)$.

   (h) Show that your final answer is a valid solution to the original differential equation.

9. Find all the separable solutions to the following PDE, where β is a positive constant:

$$\frac{\partial^2 f}{\partial x^2} = \beta\frac{\partial^2 f}{\partial t^2}.$$

When you introduce a separation constant, call it $-k^2$ to indicate that it is negative. *If you're not sure how to go through the process, you can refer to Problem 8 for an example.*

10. Use separation of variables to find a function $f(x, y)$ that solves the following partial differential equation:

$$3\frac{\partial f}{\partial x} + 4y\frac{\partial f}{\partial y} = 5f.$$

Your final solution will include two arbitrary constants: k from separating the variables to create ordinary differential equations, and A from solving those equations. Check your solution by plugging into the differential equation and confirming that it works. *If you're not sure how to go through the process, you can refer to Problem 8 for an example.*

11. **Linearity**

Assume that functions $\Psi_1(x, t)$ and $\Psi_2(x, t)$ have both been shown to solve the time-dependent Schrödinger equation for a given $U(x)$ function. Consider the following function, where A and B are constants: $\Psi_{\text{sum}}(x, t) = A\Psi_1(x, t) + B\Psi_2(x, t)$.

(a) Show that $\Psi_{\text{sum}}(x)$ is a valid solution of Equation (5.2) for the same $U(x)$ function.

(b) What does this result imply about particles and their wavefunctions?

12. **Solving the Time-Dependent Schrödinger Equation Numerically**

In principle you can find how any wavefunction will evolve by writing it as a sum of energy eigenstates and multiplying each one by $e^{-iEt/\hbar}$. In practice, for many potential energy functions it's impossible to find the energy eigenstates analytically. You can still determine the evolution, though, by directly solving the time-dependent Schrödinger equation numerically. To make numerical calculations easier we will set $m = \hbar = 1$ in this problem and use pure numbers without units for everything.

Consider a Gaussian potential energy $U(x) = e^{-x^2}$. You can't find the energy eigenstates analytically. (Try!) Suppose a particle with that potential energy starts in the state

$$\Psi(x, 0) = \frac{1}{(50\pi)^{1/4}}e^{-(x/10)^2}.$$

Numerically solve the time-dependent Schrödinger equation with this initial condition and plot $|\psi|^2(x)$ at various times from $t = 0$ to $t = 6$. Describe how the wavefunction is changing. *Hint:* You may need to set finite limits for x to get a numerical solution. If you evaluate in the range $-60 < x < 60$ you can safely set $\Psi = 0$ on the boundaries without significant loss of accuracy. You should plot your final results on a smaller domain, though.

Chapter Summary

Chapters 5 and 6 together provide an overview of quantum mechanical wavefunctions in one dimension: how to find them, how they evolve in time, and most importantly how to interpret them. These two chapters therefore provide the basic mathematical framework for analyzing the quantum world.

The equations for both chapters are collected in Appendix C. You may want to cross-reference that appendix as you read through this summary.

Section 6.1 Math Interlude **Standing Waves, Traveling Waves, and Partial Derivatives**

If you take a snapshot at one moment in time of either a standing wave or a traveling wave, you will see a sine wave stretched out over the x axis. But the two evolve very differently over time.

- In a "standing wave" the nodes remain fixed while the rest of the wave shrinks and then grows back. In other words the frequency and phase never change, but the amplitude oscillates. See Figure 6.1.

- In a "traveling wave" the entire sine wave moves to the right or left at constant speed. In other words the frequency and amplitude never change, but the phase changes. See Figure 6.2.

- The equations for standing waves and traveling waves are in Appendix F. You will also find there the definitions of, and relationships between, all the constants that define these waves.

- Both standing waves and traveling waves are examples of "multivariate functions": in these cases, y depends on x and t. Multivariate functions require "partial derivatives."

- The notation $\partial y/\partial x$ means "take the derivative of y with respect to x, treating t as a constant"; it gives you the slope of the curve at a particular place and time.

- The notation $\partial y/\partial t$ means "take the derivative of y with respect to t, treating x as a constant"; it gives you the vertical velocity of the curve at a particular place and time.

Section 6.2 Free Particles and Fourier Transforms

A "free particle" means a particle under the influence of no force: in other words, $U(x)$ is a constant (which we generally call zero).

The energy eigenstates of a free particle are of the form $\psi(x) = Ce^{ikx}$, with eigenvalues $E = \hbar^2 k^2/(2m)$.

- With time evolution, $\Psi(x, t) = Ce^{i(kx - \omega t)}$ where $\omega = E/\hbar = \hbar k^2/(2m)$. This is a complex-valued traveling wave with velocity ω/k, traveling to the right if $k > 0$ and to the left if $k < 0$.

- The energy levels are not quantized: k can take any real value, so E can be any non-negative value. (Remember that, for the bound systems in Chapter 5, quantization emerged as a result of boundary conditions.)

- Unbound energy states cannot be normalized, so a system can never be in unbound state with definite energy.

- Rather than expressing an a free particle's wavefunction as a discrete sum of energy eigenstates, we express it as a continuous integral over all possible energy eigenstates: a "Fourier transform." The relevant equations are in Appendix C.

Section 6.3 Momentum Eigenstates

Just as an "energy eigenstate" means a wavefunction for a particle with definite energy, a "momentum eigenstate" means a wavefunction for a particle with definite momentum.

- Unlike energy eigenstates, which depend on the potential energy field (via Schrödinger's equation), the momentum eigenstates are the same for all particles: $\psi(x) = Ce^{ikx}$, with eigenvalue $p = \hbar k$.

- Yes, that's the same as the energy eigenstates of a free particle. So many of the same considerations apply: a particle's wavefunction can never be a pure momentum eigenstate, but it can be expressed as an integral over momentum eigenstates via a Fourier transform.

- A very narrowly spiked function will have a wide Fourier transform (meaning it includes significant contributions from a wide range of k-values). That's a purely mathematical rule, but it provides a justification for the Heisenberg uncertainty principle.

Section 6.4 Wave Packets and Moving Particles

At any given moment, the position of a free particle can be expressed as an integral over infinitely many energy eigenstates e^{ikx}. But the resulting integral can be a single bump around one x-value. (It doesn't have to be, but that's a common situation physically.)

As you move away from that bump, ψ decays rapidly toward zero, *not* because the individual eigenfunctions no longer exist but because they interfere destructively. The resulting "wave packet" represents a physical particle with a fairly (although not exactly) defined position.

- Each individual eigenfunction evolves in time by being multiplied by $e^{-i\omega t}$, which changes its phase but not its amplitude. The result of all these phase changes is to move the wave packet (the region of constructive interference) along the x axis, so the particle moves.

- The velocity of an individual traveling wave is called its "phase velocity" ω/k. The velocity of the wave packet is called its "group velocity" $d\omega/dk$. For a particle with a fairly narrow spread of momenta, the group velocity equals the actual velocity of the particle.

- Hold onto this fact: a free-particle wave packet made up of e^{ikx} functions with positive k-values represents a particle moving to the right. (This makes sense since k is proportional to the momentum.)

Section 6.5 Scattering and Tunneling

This section deals with the behavior of a particle incident on a potential change. In the examples below we assume the particle is initially coming from the left (negative x).

- If a particle encounters a potential step such that $\lim_{x \to \infty} U(x) > E$, the particle's wave packet at late times will be entirely reflected. This is one way in which the quantum world matches the classical one.

- If a particle encounters a potential step or barrier such that $\lim_{x \to \infty} U(x) < E$, the wavefunction at late times will consist of a reflected wave packet and a transmitted one. The relative sizes of these wave packets tell you the reflection and transmission probabilities.

- Even if the potential change involves a barrier with $U > E$, a particle can be transmitted as long as $U < E$ on the far side of the barrier, a purely quantum phenomenon known as "tunneling."

- All of these results can be derived by considering the particle's energy eigenstates and matching boundary conditions across potential changes. In regions with $E > U$, a particle's wavefunction is oscillatory. In regions with $E < U$, a particle's wavefunction is a decaying exponential (not uniformly zero, as you would expect classically).

Section 6.6 The Time-Dependent Schrödinger Equation

The time-dependent Schrödinger equation is one of the starting axioms of quantum mechanics. Two of the results that we presented in Chapter 5 – the time-*independent* Schrödinger equation, and the time evolution of a wavefunction – are derived from this starting point.

Mathematically this section introduces "partial differential equations" (differential equations with partial derivatives) and "separation of variables," the most common technique for solving such equations. (This is not related to the "separation of variables" you used to solve ordinary differential equations in introductory calculus.)

1. Separation of variables involves plugging the "separable solution" $\Psi(x, t) = \psi(x)T(t)$ into the partial differential equation. Then you put all the x dependence on one side and the t dependence on the other and you can set both sides equal to a constant. The result is two ordinary differential equations that you can solve for the functions $\psi(x)$ and $T(t)$.

2. The general solution to a partial differential equation can be built as a linear combination of these separable solutions.

3. For the time-dependent Schrödinger equation, the result is that $\psi(x)$ must satisfy the time-independent Schrödinger equation and $T(t) = e^{-iEt/\hbar}$. In short, we've derived solutions to the time-dependent Schrödinger equation that are energy eigenstates, and they have the time dependence described in Chapter 5. The general solution to the time-dependent Schrödinger equation (for any given potential energy function) is a linear combination of these energy eigenstate solutions.

The table on p. 310 provides a brief summary to help you keep the two different Schrödinger equations straight.

7

The Hydrogen Atom

The story of atoms so far, in three parts:

1. 1911: Rutherford describes an atom as being made of small negatively charged electrons orbiting a large positively charged nucleus, all very analogous to planets orbiting the Sun. Rutherford's model explains his gold-foil scattering, but it cannot explain why these accelerating charges in the atom do not radiate away their energy and fall immediately into the nucleus, and it cannot explain the discrete emission and absorption spectra seen in the lab.

2. 1913: Bohr addresses both of those problems by proposing that the angular momentum of an orbiting electron can only take on certain discrete values, and can jump discontinuously between those values. Like Planck's resolution of the ultraviolet catastrophe and Einstein's explanation of the photoelectric effect, this fits the data but does not provide any fundamental principles.

3. 1926: Schrödinger publishes his wave equation. Eventually, all the ad hoc hypotheses of the old quantum theory are seen to be consequences of Schrödinger's wave mechanics.

This chapter examines how Schrödinger's equation, and the tools we developed in Chapter 5, provide a foundation for many of the conclusions of the Bohr model. Here we focus on the hydrogen atom. Chapter 8 generalizes these conclusions to multielectron atoms, where Schrödinger's equation cannot be solved explicitly but can still be used to make accurate predictions.

7.1   Quantum Numbers of the Hydrogen Atom

An electron orbiting a nucleus is a particle in a bound state. In many ways, this system is just another problem like the bound systems we saw in Chapter 5. But the problems in Chapter 5 were all one-dimensional, so each system was characterized by one quantum number that we generally called n. Because a hydrogen atom is three-dimensional, its eigenstates are characterized by three different quantum numbers.

This section describes those three quantum numbers, and how they can be used to understand the eigenstates of the hydrogen atom. As you go through this section, think about the ways in which these eigenstates are and aren't like the levels described by the Bohr model. Section 7.4

outlines the derivation of those eigenstates from Schrödinger's equation. The two sections in between provide mathematical background for that derivation.

7.1.1 Explanation: Quantum Numbers of the Hydrogen Atom

Suppose you have a particle (mass m) in an infinite square well (length L). You have measured that particle's energy, so the particle is now in an energy eigenstate. A colleague asks you "which state is it in?" You can give a complete answer by saying "$n = 2$." Based on Section 5.3, your colleague now knows that your particle has wavefunction $\psi_2(x) = \sqrt{2/L}\ \sin(2\pi x/L)$ (from $x = 0$ to L) and energy $E_2 = 2\pi^2\hbar^2/(mL^2)$. By identifying the *quantum number* $n = 2$ you have communicated the exact energy of the particle, the probability distribution of its position, and the values or probability distributions of other measurable quantities.

However, suppose your colleague has never read Section 5.3, and wants to know how particles behave in infinite square wells. Your best quantum mechanical answer would be a description of the energy eigenstates: here is how the system behaves when $n = 1$ (the ground state), here's what changes for $n = 2$, and so on. Each state is tagged by its quantum number n, and is described by the wavefunction and energy associated with that quantum number.

Of course, both you and your colleague should keep in mind that the actual wavefunction at any given moment might be pretty much anything! But we analyze an arbitrary wavefunction by decomposing it into a sum of energy eigenstates, so those states are still the key to understanding the system (not to mention predicting its time evolution).

With all that in mind, we turn our attention to the hydrogen atom. The particle is an electron, and the potential energy comes from the electric force exerted by the nucleus:[1] $U = -k/r$, where k is a positive constant and r is the distance from the nucleus.

We will see in Section 7.2 that the energy eigenstates of a three-dimensional system are characterized by three different quantum numbers. For a hydrogen atom these numbers are generally called[2] n, l, and m_l. This section discusses what each of these quantum numbers means for the eigenstate ψ_{nlm_l}. Section 7.4 will discuss those states in more mathematical detail and consider how they are derived from the Schrödinger equation, and Chapter 8 will extend the analysis to atoms with more than one electron.

Much of the information in this section is summarized in Appendix G.

The "Principal Quantum Number" n

The quantum number n can be any positive integer.

The term "principal quantum number" refers to the fact that n determines energy: each eigenstate ψ_{nlm_l} has energy $E_n = -(1/n^2)$ Ry, where 1 Ry $\approx$ 13.6 eV is a unit of energy called a "rydberg." So the ground state of hydrogen has energy -1 Ry ≈ -13.6 eV, the first excited state has energy $-(1/4)$ Ry ≈ -3.4 eV, and so on.

As we discussed in Section 4.1, these energies are negative because the negative potential energy has a higher magnitude than the positive kinetic energy, meaning the electrons are bound to the atom. The energy differences tell you how much energy is required to kick an electron

---

1 We are using the word "nucleus" rather than "proton" because many of the ideas presented in this section scale up to larger atoms.

2 Some sources call the third quantum number m without the subscript, but we'll follow the sources that use m_l to distinguish it from the electron mass m_e and from another quantum number m_s that we'll introduce in Section 7.5.

from a low-energy state to a high-energy one, or how much is released if an electron goes the other way. It takes about 10 eV of energy to knock a hydrogen atom from the ground state to the first excited state, but from there it only takes about 2 eV more to kick it up to the second excited state. If you start with a hydrogen atom in the ground state, it takes about 13.6 eV to knock out the electron completely and ionize the atom.

Note that one energy eigenstate ψ_{nlm_l} of a hydrogen atom is characterized by several different quantum numbers, but the actual energy E_n associated with that state is just a function of n. Different states that have the same energy – in this case, that means the same n but different values of the other quantum numbers – are said to be "degenerate" states.

The "Angular Momentum Quantum Number" l

The quantum number l can be any integer from 0 to $n-1$, inclusive. In the ground state ($n = 1$) the only option is $l = 0$; each step up in n allows for one more possible l.

As you might guess from its name, l controls the angular momentum of the state. Specifically, the magnitude of the total angular momentum of the electron about the nucleus is $L = \sqrt{l(l+1)}\,\hbar$.

This is the first time we've discussed the angular momentum of a quantum mechanical state – it won't be the last – and it's worth pausing to consider how this differs from a classical description.

A classical particle has angular momentum $\vec{L} = \vec{r} \times \vec{p}$. So you can find its angular momentum around a given axis based on its position $\vec{r}$ from that axis and its momentum $\vec{p}$, which are of course both well defined. An object can have zero angular momentum (even if it's moving), but an *orbiting* object never has zero angular momentum around the orbital center.

Now consider a quantum mechanical electron in energy eigenstate ψ_{nlm_l}. Some properties of this electron can be described only as probability distributions: the electron does not have a specific position, momentum, kinetic energy, or potential energy, for instance. But other properties can be specifically quantified: this electron *does* have a well-defined total energy (dictated by its quantum number n), and it also has a well-defined angular momentum magnitude (dictated by l). To put it another way, if you measured 100 hydrogen atoms in the state $\psi_{3,1,0}$, you would find they all have exactly the same energy and exactly the same angular momentum magnitude.

Remember also that a given atom may exist in a superposition of eigenstates. The coefficients of those states tell you the probabilities of finding different energies and different angular momentum magnitudes.

Finally, it's worth noting (from $L = \sqrt{l(l+1)}\,\hbar$) that the $l = 0$ state has no angular momentum. This makes no sense if you imagine the electron as a classical orbiting particle. But if you imagine the electron as an unchanging cloud of probabilities, it may come as less of a surprise that some clouds have angular momentum and others don't.

The "Magnetic Quantum Number" m_l

The quantum number m_l can be any integer from $-l$ to l, inclusive.

While l determines the magnitude of the angular momentum vector, m_l is related to its direction. Specifically, the z component of the angular momentum is $L_z = m_l \hbar$. The name comes from the fact that the simplest way to measure a particular component of the angular momentum of a charged particle is to measure the magnetic field it creates.

This means that, if an atom is in an energy eigenstate ψ_{nlm_l}, its angular momentum has a well-determined magnitude and a well-determined z component. Because (for any three-vector) $|\vec{L}|^2 = L_x^2 + L_y^2 + L_z^2$, you can calculate a precise value for $L_x^2 + L_y^2$. But there is an uncertainty principle that dictates that L_x and L_y are not fully determined in such a state. (The one exception is the state $l = m_l = 0$, in which all components of $\vec{L}$ equal zero with no uncertainty.)

For example, suppose $l = m_l = 1$, which means $L = \sqrt{2}\,\hbar$ and $L_z = \hbar$. You now know that $\vec{L}$ lies somewhere on the circle shown in perspective in Figure 7.1. You know L_z exactly but there is still uncertainty in L_x and L_y.

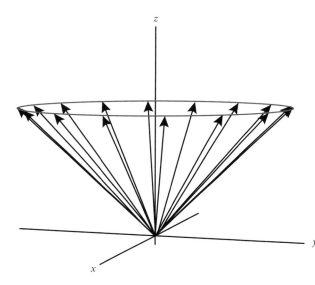

Figure 7.1 Angular momentum $\vec{L}$ for a hydrogen eigenstate with $l = m_l = 1$.

You may also be wondering what defines the "z axis". The answer is, as always, that you can pick any direction you want to be the z axis. Your choice doesn't affect the atom in any way, but it affects the equations you use to describe the atom.

Think about that classically. You measure the angular momentum to have a magnitude of 5 and to point in a certain direction. If you choose to call that the $+z$ direction, then your angular momentum vector is $5\hat{k}$. Someone else with different axes might describe the same vector as $4\hat{j} + 3\hat{k}$. Different descriptions, same atom.

So, quantum mechanically, suppose you measure the energy, the magnitude of the angular momentum, and one particular component of the angular momentum, all at the same time. If you choose to call that component the z component, then you will describe the atom as being in one of its eigenstates ψ_{nlm_l}. Someone else with different axes would describe the same atom as being in a superposition of eigenstates, corresponding to a definite E and a definite $|\vec{L}|$ but an uncertain L_z. Different descriptions, same atom.

So What Are the Hydrogen Atom Eigenstates?

In all the abstract discussion of the meanings of eigenstates and the interplay between them, it's easy to lose sight of the fact that they are just mathematical functions. For instance, the energy eigenstates of a particle in an infinite square well are sines. So, you may wonder, why do we keep talking about the hydrogen atom's eigenstates without actually writing them down?

Well, since you asked . . .

$$\psi_{nlm_l}(r,\theta,\phi) = \sqrt{\left(\frac{m_e e^2}{2\pi\epsilon_0\hbar^2 n}\right)^3 \frac{(n-l-1)!}{2n(n+l)!}}\, e^{-m_e e^2 r/(4\pi\epsilon_0\hbar^2 n)}$$

$$\times \left(\frac{m_e e^2 r}{2\pi\epsilon_0\hbar^2 n}\right)^l L_{n-l-1}^{2l+1}\left(\frac{m_e e^2 r}{2\pi\epsilon_0\hbar^2 n}\right) Y_l^{m_l}(\theta,\phi).$$

Don't panic! In Section 7.3 we will introduce those possibly unfamiliar independent variables, and in Section 7.4 we will start to break that formula down. We will see that some factors have important implications for the measurable behaviors of the hydrogen atom, and other factors can be ignored for our purposes. We will also outline the process of deriving that formula by solving Schrödinger's equation.

Corrections and Generalizations

The quantum mechanical description of the hydrogen atom has been wildly successful. Recall that Bohr's model was able to predict the spectrum of hydrogen by listing the energy levels, but Bohr's quantization had no theoretical basis. Schrödinger's equation enables us to derive those energies E_n and thus correctly predict the spectrum. Quantum mechanics also allows us to derive many properties of hydrogen that the Bohr model can't explain, such as the splitting of spectral lines (Section 7.7) and the lifetimes of excited states (found in a more advanced quantum course). Nonetheless, we should note a couple of limitations to the model as we've discussed it so far.

The first is simple and easy to correct. You may recall from Section 4.1 that Bohr brought his original model closer in line with experiment by replacing the electron mass with the "reduced mass,"

$$\mu = \frac{m_e M}{m_e + M}. \tag{7.1}$$

This idea comes from classical mechanics, and takes into account that the electron (mass m_e) and the nucleus (mass M) are both rotating around their common center of mass. Because the nucleus is thousands of times more massive than the electron, ignoring nuclear motion is a reasonable approximation. (Note that Equation (7.1) predicts almost exactly $\mu = m_e$ because $M \gg m_e$.) However, using μ makes the equations just a bit more accurate (see Problem 12).

The other caveat is a much bigger issue. Everything we've said in this section applies to a hydrogen atom: one proton and one electron. What about other atoms?

Adding neutrons to the nucleus has almost no effect on the energy eigenstates. Adding more protons doesn't change the picture too much. An ionized helium atom, or a doubly ionized lithium atom, are examples of "hydrogen-like" atoms: Z protons with one electron. In the wavefunction given above, you replace e^2 with Ze^2. The energy is then multiplied by Z^2:

$$E = -\frac{Z^2}{n^2}\,\text{Ry}.$$

Adding more electrons, on the other hand, complicates the picture considerably because the electrons interact with each other as well as with the nucleus. In Chapter 8 we'll start to explore how those interactions lead to all the richness of the periodic table.

7.1.2 Questions and Problems: Quantum Numbers of the Hydrogen Atom

Conceptual Questions and ConcepTests

1. For each quantity give a brief description of what it represents in this section, and units that could be used to measure it.

    (a) l

    (b) L_z

    (c) ψ_{nlm_l}

2. The restriction $-l \leq m_l \leq l$, along with the formulas for $|\vec{L}|$ and L_z, describe the relationship between the magnitude of the angular momentum vector and its z component.

    (a) If the z component of a vector were bigger than the magnitude of the vector, that would be bad. (We can all agree on that, right?) Show that this contradictory situation can never happen, based on the equations given in the Explanation.

    (b) The rules do allow for L_z to precisely equal $|\vec{L}|$. What l and m_l values lead to this situation, and what does it say about L_x and L_y?

3. Suppose a hydrogen atom is in the state $\psi_{3,2,1}$. You measure L_z, then measure L_x, and then measure L_z again. Are you guaranteed to get the same answer for L_z both times? Why or why not?

4. The function $(1/2)\psi_{4,3,1} + (\sqrt{3}/2)\psi_{4,1,0}$ is an eigenstate of energy, but not of angular momentum magnitude. Is it possible to have the reverse: an eigenstate of angular momentum magnitude that is not an eigenstate of energy? If so, write one. If not, why not?

For Deeper Discussion

5. Suppose a hydrogen atom is in the state $\psi_{3,2,1}$ and then you measure its angular momentum in the x direction. After that measurement, will it be in an energy eigenstate?

6. Make a list of the main ways in which the Bohr model matches the quantum mechanical understanding of the hydrogen atom, and the main ways in which it is different. Try to get at least two items on each list. Your list of differences should include at least one conceptual difference and at least one quantitative difference.

Problems

7. Consider a hydrogen atom in the state $\psi_{4,3,-3}$.

    (a) If you measure the energy, what will you find?

    (b) If you measure the magnitude of the angular momentum, what will you find?

    (c) If you measure the z component of the angular momentum, what will you find?

    (d) If the electron drops into a different state with $n = 2$, how much energy will be released?

8. Consider a hydrogen atom in the following state, where C is a constant:

$$\frac{1}{\sqrt{5}} \psi_{10,5,0} + \sqrt{\frac{2}{5}} \psi_{4,3,2} + C\, \psi_{4,2,1}.$$

    (a) Find the constant C.

    (b) What are all the energies you might measure, and with what probabilities?

9. Consider a hydrogen atom known to be in the state $\psi(x, y, z) = \psi_{7,5,3}$.

    (a) What are the magnitude and z component of the angular momentum vector?

    (b) You do not know L_x or L_y, but you can find $L_x^2 + L_y^2$. What is it?

10. Let θ_L represent the angle of the angular momentum vector from the z axis. Calculate θ_L for an arbitrary hydrogen atom eigenstate ψ_{nlm_l}. Your answer will of course be a function of some or all of the quantum numbers!

11. Recall that the "degeneracy" of an energy level refers to how many independent quantum states correspond to the same energy.

    (a) How many different states ψ_{nlm_l} correspond to the ground state ($n = 1$) of a hydrogen atom, and therefore have the same energy?

    (b) Repeat Part (a) for the $n = 2$, $n = 3$, and $n = 4$ energy levels.

(a) Write a general function for the degeneracy of the nth energy level of the hydrogen atom. (We're not looking for a mathematical proof. Just spot the pattern.)

12. The Earth doesn't actually orbit about the Sun; both bodies orbit about their mutual center of mass. The Sun is so heavy that the center of mass is pretty close to the center of the Sun, so to a good approximation you can pretend that the Earth revolves around a non-moving Sun. But instead of approximating, you can solve for the motion by treating the two-body pair as a single orbiting particle with "reduced mass" $\mu = m_1 m_2/(m_1 + m_2)$. The same trick works in quantum mechanics, and you can treat an electron–proton pair as a one-particle system with a reduced mass.

(a) Calculate the reduced mass of a hydrogen atom. (You'll need to look up some constants for this, and you'll need to hold onto quite a few decimal places in order to see any difference between μ and m_e.)

(b) The energies of the eigenstates ψ_{nlm_l} are proportional to the mass m of the particle for which you are solving. Calculate the percentage error introduced by using the electron mass instead of the more correct reduced mass.

7.2 The Schrödinger Equation in Three Dimensions

One-dimensional problems make a useful playground for developing the techniques of quantum mechanics, and they do have some important applications, but the world we[3] live in is three-dimensional. In this section we're going to introduce three-dimensional quantum mechanics through the example of a particle in a box. Both this section and Section 7.3 provide mathematical background for Section 7.4, in which we outline the solution of Schrödinger's equation for the hydrogen atom.

7.2.1 Discovery Exercise: The Two-Dimensional Infinite Square Well

The two-dimensional version of the infinite square well is a square box of side length L. A particle is free to move anywhere inside the box, but can never leave it (Figure 7.2).

The time-independent Schrödinger equation in 2D is just like the 1D version except it has spatial derivatives with respect to both x and y. Inside the box (where $U = 0$) it looks like

$$-\frac{\hbar^2}{2m}\left(\frac{\partial^2 \psi}{\partial x^2} + \frac{\partial^2 \psi}{\partial y^2}\right) = E\psi. \qquad (7.2)$$

The Explanation (Section 7.2.2) will show you how to solve such equations. Here we are going to skip to the solution:

$$\psi(x, y) = A \sin\left(\frac{a\pi}{L}x\right) \sin\left(\frac{b\pi}{L}y\right). \qquad (7.3)$$

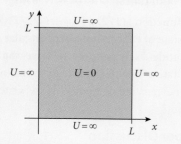

Figure 7.2 A particle in a 2D box has potential energy $U = 0$ in the region $0 \le x \le L$, $0 \le y \le L$, and potential energy $U = \infty$ outside that region.

---

3 except string theorists

Recall that at an infinite potential jump ψ may not be differentiable, but it must still be continuous. Because $\psi(x, y)$ must be zero outside the box, continuity gives us four boundary conditions, including "$\psi(x, y) = 0$ when $x = 0$" and "$\psi(x, y) = 0$ when $y = 0$." (We've already built those two into Equation (7.3); make sure you see that.)

1. Write the other two boundary conditions, similarly to how we just wrote the first two.

2. What must be true of the numbers a and b in order for Equation (7.3) to match the boundary conditions you wrote in Part 1?

3. Plug the solution (Equation (7.3)) into the differential equation (Equation (7.2)). Solve the resulting equation to find the energy (E) in terms of the quantum numbers a and b. *There should be no x or y in your answer.*

   See Check Yourself #9 at www.cambridge.org/felder-modernphysics/checkyourself

4. Based on your solutions to Parts 2 and 3, the lowest possible energy of this particle is $\pi^2 \hbar^2/(mL^2)$. What are the next two energy levels?

7.2.2 Explanation: The Schrödinger Equation in Three Dimensions

In one-dimensional problems, a time-dependent wavefunction is a function of two variables, $\Psi(x, t)$. Separation of variables on the time-dependent Schrödinger equation gives us a general formula for the time dependence and gives us the time-independent Schrödinger equation, an "ordinary differential equation" (only one independent variable) for the spatial wavefunction $\psi(x)$.

In three dimensions, where the wavefunction can be written as $\Psi(x, y, z, t)$, the process of separating variables for the time-dependent Schrödinger equation is very similar to the one-dimensional version (Chapter 6). The resulting $T(t)$ function is the same as before, indicating that once again each energy eigenstate is multiplied by $e^{-iEt/\hbar}$. But the spatial result now involves three independent spatial variables. So unlike the 1D case, the time-independent Schrödinger equation is now another *partial* differential equation. For some potential energy functions you can then keep using separation of variables to get ordinary differential equations in all three spatial variables.

The Probability Interpretation of the Wavefunction in 3D

In 1D, as you know, $|\psi|^2$ gives the probability per unit length of finding a particle in a particular region. If $|\psi|^2$ is constant over some finite interval of length L, then the probability of finding the particle in that interval is $|\psi|^2 L$. For non-constant wavefunctions, that multiplication becomes an integral:

$$P(x_1 \le x \le x_2) = \int_{x_1}^{x_2} |\psi|^2 dx.$$

For a properly normalized wavefunction, $\int_{-\infty}^{\infty} |\psi|^2 dx = 1$.

In 3D, $|\psi|^2$ is the probability per unit *volume*. (Notice that this means that ψ in 3D has different units from ψ in 1D.) So if $|\psi|^2$ is constant in a finite region of volume V, then the probability of finding the particle in that region is $|\psi|^2 V$. For non-constant wavefunctions, that multiplication becomes a triple integral, a mathematical operation that will be familiar if you have studied multivariate calculus:

$$P(x_1 \leq x \leq x_2, y_1 \leq y \leq y_2, z_1 \leq z \leq z_2) = \int_{z_1}^{z_2} \int_{y_1}^{y_2} \int_{x_1}^{x_2} |\psi|^2 \, dx \, dy \, dz.$$

You normalize the wavefunction by setting that integral over all of space equal to 1.

Some of the problems at the end of this section require multiple integrals. Those problems are clearly labeled. You can do the rest of the section without that skill.

The Time-Dependent and Time-Independent Schrödinger Equations in 3D

The three-dimensional Schrödinger equation looks much the same as the 1D version, but it has derivatives with respect to y and z in addition to x:

$$-\frac{\hbar^2}{2m} \left(\frac{\partial^2 \Psi}{\partial x^2} + \frac{\partial^2 \Psi}{\partial y^2} + \frac{\partial^2 \Psi}{\partial z^2} \right) + U\Psi = i\hbar \frac{\partial \Psi}{\partial t}$$

The time-dependent Schrödinger equation in 3D. (7.4)

Much as we did in 1D, we start by guessing $\Psi(x, y, z, t) = \psi(x, y, z)T(t)$. In Problem 21 you'll show that this leads to $T(t) = e^{-iEt/\hbar}$ (exactly as in 1D), and to a 3D version of the time-independent Schrödinger equation.

The Time-Independent Schrödinger Equation in Three Dimensions (in Cartesian Coordinates)

A particle subject to the potential energy function $U(x, y, z)$ has energy eigenstates

$$\Psi(x, y, z, t) = \psi(x, y, z)e^{-iEt/\hbar},$$

where the spatial function $\psi(x, y, z)$ obeys the time-independent Schrödinger equation,

$$-\frac{\hbar^2}{2m} \left(\frac{\partial^2 \psi}{\partial x^2} + \frac{\partial^2 \psi}{\partial y^2} + \frac{\partial^2 \psi}{\partial z^2} \right) + U\psi = E\psi.$$ (7.5)

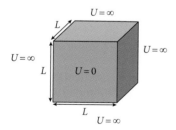

Figure 7.3 A particle in a 3D box has potential energy $U = 0$ in the region $0 \leq x \leq L$, $0 \leq y \leq L$, $0 \leq z \leq L$, and potential energy $U = \infty$ outside that region.

Section 7.4 will outline the process of solving the time-independent Schrödinger equation to find the energy eigenstates of the hydrogen atom. This section illustrates the process on a simpler (and still important) 3D problem.

A Particle in a 3D Box

Just as we did in one dimension, we will start with an infinite potential energy well. A particle is free to move within a cubical box of width L. The particle cannot leave that box, but, aside from when it collides (elastically) with the walls, it experiences no forces inside the box. We can say the same thing by saying that U is zero inside the box and infinite outside it (Figure 7.3).

Where $U = \infty$, the math demands that $\psi = 0$ (Problem 22). This tells us that the particle can never go outside the box. (We hope you knew that anyway.)

Inside the box Equation (7.5) becomes

$$-\frac{\hbar^2}{2m}\left(\frac{\partial^2\psi}{\partial x^2}+\frac{\partial^2\psi}{\partial y^2}+\frac{\partial^2\psi}{\partial z^2}\right)=E\psi. \tag{7.6}$$

We have gone from the time-dependent Schrödinger equation to the time-independent one by separating $\Psi(x,y,z,t)$ into $\psi(x,y,z)$ and $T(t)$. Now we separate variables a second time: plug $\psi(x,y,z)=X(x)Y(y)Z(z)$ into Equation (7.6), divide both sides by ψ, and separate x from the other two variables. You'll carry out these steps in Problem 23 to get the following equation:

$$-\frac{X''(x)}{X(x)}=\frac{Y''(y)}{Y(y)}+\frac{Z''(z)}{Z(z)}+\frac{2mE}{\hbar^2}. \tag{7.7}$$

Remember that the energy E is treated as a constant in this equation. (It was introduced as a separation constant when we derived the time-independent Schrödinger equation.)

The next step follows the same idea we used on the way to Equation (7.5): the left side of Equation (7.7) depends only on x, and the right side only on y and z, so both sides must equal the same constant. In Problem 24 you'll show that this constant must be positive, and, knowing that, we'll call it k^2.

Setting the left side of Equation (7.7) equal to our new constant allows us to solve for $X(x)$:

$$-\left(\frac{X''(x)}{X(x)}\right)=k^2 \quad \rightarrow \quad X''(x)=-k^2X(x). \tag{7.8}$$

This is the same equation we solved for $\psi(x)$ in the 1D infinite square well, and it is subject to the same boundary conditions. (The only difference is the name of the constant.) In Problem 25 you'll find that $X(x)$ is given by the following equation (note that we have replaced the constant k with a new constant a; they are related by $k=a\pi/L$):

$$X(x)=A\sin\left(\frac{a\pi}{L}x\right)\quad\text{where }a\text{ is an integer.} \tag{7.9}$$

What makes this different from the 1D problem is that this isn't the whole energy eigenstate; we have to find $Y(y)$ and $Z(z)$ as well.

To do that we return to Equation (7.7). If the left side of that equation equals k^2, then the right side must do so as well:

$$\frac{Y''(y)}{Y(y)}+\frac{Z''(z)}{Z(z)}+\frac{2mE}{\hbar^2}=k^2. \tag{7.10}$$

Active Reading Exercise: The Equations for Y and Z

Separate variables to turn Equation (7.10) into two ordinary differential equations, one for $Y(y)$ and one for $Z(z)$. Both equations will have a new constant p^2 that you'll introduce in the same way we introduced k^2 above. (One of the equations will also still have k^2 in it.)

The Y equation is $Y''=-p^2Y$ and its solution is $Y(y)=C\sin(py)$, where $p=b\pi/L$ and b must be an integer.

The Z equation ends up as

$$Z''(z)=-\left(\frac{2mE}{\hbar^2}-k^2-p^2\right)Z(z).$$

This looks more complicated than the X and Y equations, but it's actually the same. Like them, it says that Z'' equals minus a constant times Z. We can make our algebra simpler by defining a new name for that constant: $q^2 = 2mE/\hbar^2 - k^2 - p^2$. Now the Z equation looks identical to the X and Y ones and the solution is $Z(z) = F\sin(qz)$, where $q = c\pi/L$ and c must be an integer.

We can now construct the energy eigenstates for this system as $\psi(x, y, z) = X(x)Y(y)Z(z)$. The arbitrary constants are determined by normalization; see Problem 26. We can also solve for E in the equation $q^2 = 2mE/\hbar^2 - k^2 - p^2$ to find the energy of each of those eigenstates. In both cases we write our final answers in terms of the quantum numbers a, b, and c rather than k, p, and q.

Eigenstates of a Particle in a 3D Cubical Box

A particle that is trapped in a cube that goes from 0 to L in all three directions and experiences no forces inside the cube has the following energy eigenstates:

$$\psi_{abc}(x, y, z) = \sqrt{\frac{8}{L^3}} \sin\left(\frac{a\pi}{L}x\right) \sin\left(\frac{b\pi}{L}y\right) \sin\left(\frac{c\pi}{L}z\right) \tag{7.11}$$

$$E_{abc} = \frac{\pi^2\hbar^2}{2mL^2}\left(a^2 + b^2 + c^2\right) \tag{7.12}$$

We put subscripts on ψ and E to indicate that each combination of integers a, b, and c corresponds to a different energy eigenstate with its own energy.

Example: Eigenstates of the Particle in a 3D Box

A particle in the box described above is in the following state:

$$\psi(x, y, z) = \frac{1}{\sqrt{3}}\sqrt{\frac{8}{L^3}} \sin\left(\frac{\pi}{L}x\right) \sin\left(\frac{2\pi}{L}y\right) \sin\left(\frac{\pi}{L}z\right)$$
$$+ \sqrt{\frac{2}{3}}\sqrt{\frac{8}{L^3}} \sin\left(\frac{2\pi}{L}x\right) \sin\left(\frac{\pi}{L}y\right) \sin\left(\frac{3\pi}{L}z\right).$$

1. If you measure the energy of this particle, what will you find?

   Answer: This wavefunction is $1/\sqrt{3}$ times $\psi_{1,2,1}$ plus $\sqrt{2/3}$ times $\psi_{2,1,3}$. So there is a $1/3$ probability that you will find $E = (1^2+2^2+1^2)\pi^2\hbar^2/(2mL^2) = 3\pi^2\hbar^2/(mL^2)$, and a $2/3$ probability that you will find $E = (2^2 + 1^2 + 3^2)\pi^2\hbar^2/(2mL^2) = 7\pi^2\hbar^2/(mL^2)$.

2. If you measure the energy of this particle a second time, what will you find?

   Answer: Your first measurement collapsed the wavefunction into an eigenstate, so your second measurement is guaranteed to find the same answer you found the first time.

Particle in a Box, Stepping Back 1: What Did We Do?

There was enough algebra in that solution that you may have lost the forest for the trees, so let's review what we did.

1. As always, we plugged the problem's potential energy function into the time-independent Schrödinger equation. In this case $U = 0$ gave us Equation (7.6).

So far this is just like 1D problems. But Equation (7.6) is a *partial* differential equation in three variables that we had to separate out one at a time.

2. We plugged a separable solution $\psi(x, y, z) = X(x)Y(y)Z(z)$ into Equation (7.6).

3. We pulled all the x dependence to one side and the y and z to the other, and set both sides equal to a constant. This gave us two equations: one for $X(x)$, and the other for $Y(y)$ and $Z(z)$.

4. We solved the ordinary differential equation for $X(x)$ and applied its boundary conditions. This led us to introduce an integer quantum number a.

5. Then we turned to the partial differential equation for Y and Z. Separating variables a final time led us to ordinary differential equations for Y and Z.

6. Finally we solved those Y and Z equations. Each solution led to the introduction of another quantum number. The three quantum numbers were not entirely independent; they were related to the constant E in the original differential equation.

The final result was Equations (7.11) and (7.12).

When we solve for the energy eigenstates of the hydrogen atom in Section 7.4, the math will be somewhat more complicated and you won't be able to solve all the equations by hand, but you should nonetheless recognize the same steps as those we described above.

Particle in a Box, Stepping Back 2: What Did We Get?

A common mistake is to think that we have found "the wavefunction that our particle will have" (wrong), or "all the possible wavefunctions that our particle might have" (more sophisticated but still wrong). A particle in our 3D box could have any wavefunction inside the box that is continuous everywhere, differentiable everywhere except possibly at the infinite potential jump at the boundaries, and normalized.

What we have found is the energy eigenstates of a particle in this particular box. Let's review what that means.

- If a particle's wavefunction looks like Equation (7.11) for some positive integers a, b, and c, then the particle is in a state of definite energy. That energy is given by Equation (7.12).

- The energy eigenstates are also "stationary states." If a particle is in one such state, then the probability distributions for measurable quantities such as position and momentum will not change over time.

- If a particle is not in an energy eigenstate, you can write its wavefunction as a sum of such eigenstates:

$$\psi(x, y, z) = c_{1,1,1}\psi_{1,1,1}(x, y, z) + c_{1,1,2}\psi_{1,1,2}(x, y, z) + c_{1,2,1}\psi_{1,2,1}(x, y, z) + \cdots .$$
$$(7.13)$$

For each energy eigenstate, $|c_{abc}|^2$ is the probability of finding the particle with that energy. The sums of the probabilities $|c_{abc}|^2$ must equal 1. (When you measure the energy, the probabilities of getting *something* have to add up to 1.)

- Finally, the energy eigenstates are the key to predicting time evolution. You rewrite your wavefunction as a sum of energy eigenstates (as we just discussed), and then multiply each individual eigenstate, with eigenvalue E, by $e^{-iEt/\hbar}$.

The lowest possible energy of the particle is $3\pi^2\hbar^2/(2mL^2)$, which occurs if the particle is in the state $\psi_{1,1,1}$. This is called the "ground state" of the particle.

The next level up, referred to as the "first excited state," is $\psi_{2,1,1}$ with energy $3\pi^2\hbar^2/(mL^2)$. But the states $\psi_{1,2,1}$ and $\psi_{1,1,2}$ also have that same energy. We say that the first excited state of this particle has a "degeneracy" of 3. These degenerate states still represent different possible states of the particle. Each one, for example, has a different probability distribution for where you would find the particle (Problem 16). But a particle that is in one of these states, or any combination of these states, has a well-defined energy. You'll explore the degeneracies of the different energy levels for this box in Problem 18.

7.2.3 Questions and Problems: The Schrödinger Equation in Three Dimensions

Conceptual Questions and ConcepTests

1. For each quantity below, give its units and a brief description of what it represents in this section.

   (a) $\psi(x,y,z)$

   (b) $X(x)$ (You don't need to give units for this one.)

   (c) ψ_{abc}

   (d) E_{abc}

   (e) c_{abc}

2. Given the equation $f(x,y) = g(x,t)$, which of the following can you conclude? (Choose one.)

   A. This equation is impossible for any functions.

   B. Both sides can be functions of x, but not of y or t.

   C. Both sides can be functions of y and t, but not of x.

   D. Both sides must equal a constant.

   E. None of the above.

3. Box A contains a particle in the state $\psi_{1,2,1}$ and Box B contains a particle in the state $\psi_{1,3,1}$. In which box are you more likely to find the particle very close to $y = L/2$?

4. Why do the energy eigenstates of the particle in a box have to equal zero at the boundaries of the box?

5. You have two cubical boxes, each containing a particle in its ground state. Box A is one meter across

and Box B is two meters across. In which box is $|\psi|^2$ larger in the middle of the box? (Choose one.)

   A. Box A

   B. Box B

   C. The values of $|\psi|^2$ are equal.

6. Imagine working in four spatial dimensions with some potential energy function $U(w,x,y,z)$.

   (a) How many times would you have to separate variables to solve the time-dependent Schrödinger equation?

   (b) How many times would you have to separate variables to solve the time-independent Schrödinger equation?

   (c) Of the five functions $W(w)$, $X(x)$, $Y(y)$, $Z(z)$, and $T(t)$, which one do you know right now? What is it?

7. Which of the following are possible wavefunctions for a particle in a 3D cubical box of side length L? (Choose all that apply. Assume the constants A and B are chosen to normalize the functions correctly.)

   A. $\psi(x,y,z) = A\sin(\pi x/L)\sin(\pi y/L)\sin(\pi z/L)$

   B. $\psi(x,y,z) = A\sin(\pi x/L)\sin(\pi y/L)\sin(\pi z/L)$ $+ B\sin(\pi x/L)\sin(2\pi y/L)\sin(\pi z/L)$

   C. $\psi(x,y,z) = A\sin(\pi x/L)\sin(\pi y/L)\cos(\pi z/L)$

   D. $\psi(x,y,z) = A\sin(\pi x/L)\sin(\pi y/L)z(z-L)$

For Deeper Discussion

8. The function $\psi_{1,1,1}$ represents the unique ground state of a particle in a box, meaning this is the only possible wavefunction the particle can have that has a definite energy equal to $3\pi^2\hbar^2/(2mL^2)$. The next energy level has degeneracy 3, meaning that $\psi_{2,1,1}$, $\psi_{1,2,1}$, and $\psi_{1,1,2}$ are three different states with the same definite energy $3\pi^2\hbar^2/(mL^2)$. Write another state for that particle with that same energy. How many other states could you write with energy $3\pi^2\hbar^2/(mL^2)$? *Hint*: No other combination of a, b, and c will give you this same energy.

9. How would Equations (7.11) and (7.12) change if the box had a constant potential energy U_0 inside, instead of $U = 0$?

10. What would the energy eigenstates be for a particle confined to an infinitely long square tube, $0 \le x \le L$, $0 \le y \le L$, $-\infty \le z \le \infty$, with $U = 0$ inside that region?

Problems

11. Suppose you have calculated a wavefunction $\psi(x, y)$ for a particle confined to a two-dimensional region. Like most functions we tend to work with, this wavefunction looks constant if you look at a sufficiently small region.

   (a) Over one square of side length 0.01 m, you find that $\psi \approx 4\,\text{m}^{-1}$. Estimate the probability of finding the particle in that square.

   (b) Over a different square of side length 0.01 m, you find that $\psi \approx -4\,\text{m}^{-1}$. Estimate the probability of finding the particle in that square.

   (c) Over a circle of radius 0.01 m, you find that $\psi \approx (3 + 4i)\,\text{m}^{-1}$. Estimate the probability of finding the particle in that circle.

12. Suppose you have calculated a wavefunction $\psi(x, y, z)$ for a particle in some three-dimensional region. Like most functions we tend to work with, this wavefunction looks constant if you look at a sufficiently small region.

   (a) Over one cube of side length 0.01 m, you find that $\psi \approx 2\,\text{m}^{-3/2}$. Estimate the probability of finding the particle in that cube.

   (b) Over a different cube of side length 0.01 m, you find that $\psi \approx -2\,\text{m}^{-3/2}$. Estimate the probability of finding the particle in that cube.

   (c) Over a sphere of radius 0.01 m, you find that $\psi \approx (5 - 12i)\,\text{m}^{-3/2}$. Estimate the probability of finding the particle in that sphere. (You may need to Google a formula here.)

13. Figure 7.4 shows a wavefunction for a particle in 2D.

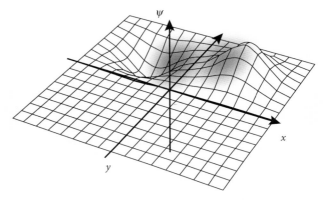

Figure 7.4

   (a) Are you more likely to find the particle in the region $x < 0$ or in the region $x > 0$, or are you equally likely to find either?

   (b) Are you more likely to find the particle in the region $y < 0$ or in the region $y > 0$, or are you equally likely to find either?

   (c) You ask your friend Cecelia to measure the particle's position and just report the sign of y. She comes back and tells you that she found the particle at positive y. Based on that, do you think she's more likely to have found it at $x > 0$, or at $x < 0$, or is she equally likely to have found it in either region?

14. Figure 7.5 shows a wavefunction for a particle in 2D. The figure shows a deep, narrow trench going from the lower left to the upper right of the plot.

   (a) Are you more likely to find the particle in the region $x < 0$ or in the region $x > 0$, or are you equally likely to find either?

   (b) Are you more likely to find the particle in the region $y < 0$ or in the region $y > 0$, or are you equally likely to find either?

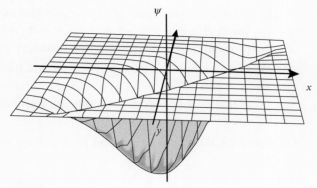

Figure 7.5

(c) You ask your friend James to measure the particle's position and just report the sign of y. He comes back and tells you that he found the particle at positive y. Based on that, do you think he's more likely to have found it at $x > 0$, or at $x < 0$, or is he equally likely to have found either?

15. A particle is in a cubical box of width L in the ground state. In this problem we'll take $L = \hbar = 1$ to keep all the numbers simple.

(a) Plot $\psi(x, y, z)$ as a function of x and y, at $z = 0$, $z = 1/4$, $z = 1/2$, $z = 3/4$, and $z = 1$. Make sure you show five different plots.

(b) Around which of those z-values are you most likely to find the particle?

(c) Based on your five graphs, around what spot in the box are you most likely to find the particle?

(d) Would your answer change if the particle were in the state $\psi_{2,1,1}$? You can answer this by drawing new graphs, but you can also answer it just by looking at the function. Either way, you must justify your answer. If you say it would change, where would you be most likely to find the particle in this new state?

16. Consider the states $\psi_{1,2,1}$ and $\psi_{1,1,2}$ in the three-dimensional box discussed in the Explanation.

(a) Write down the wavefunctions for each of these states.

(b) Calculate the energy of each of these states.

(c) Box A contains a particle in the state $\psi_{1,2,1}$ and Box B contains a particle in the state $\psi_{1,1,2}$. In which box are you more likely to find the particle very close to $y = L/2$?

17. A particle in our 3D cubical box is known to be in the following state:

$$\psi(x, y, z) = \sqrt{\frac{2}{L^3}} \sin(3\pi x/L) \sin(\pi y/L) \sin(4\pi z/L)$$
$$+ \sqrt{\frac{4}{L^3}} \sin(2\pi x/L) \sin(\pi y/L) \sin(2\pi z/L)$$
$$+ \sqrt{\frac{2}{L^3}} \sin(\pi x/L) \sin(4\pi y/L) \sin(3\pi z/L).$$

(a) Rewrite this as a set of coefficients times eigenfunctions, similar to Equation (7.13). Remember that each eigenfunction includes a factor of $\sqrt{8/L^3}$, so, for example, the first coefficient will *not* be $\sqrt{2/L^3}$.

(b) If you measure the energy of this particle, what are all the possible energies you might find, and with what probabilities? *Hint*: If your probabilities don't come out unitless, that means you did something wrong.

(c) If you then make a second measurement of the same particle, what energy will you find?

18. Equation (7.12) gives the energy associated with the eigenstate ψ_{abc} in our infinite square well.

(a) The state $\psi_{2,1,1}$ has energy $3\pi^2\hbar^2/(mL^2)$. Name two other eigenstates with that same energy level.

(b) Because there are three independent eigenstates with the same energy as $\psi_{2,1,1}$, we say that those states have a "degeneracy" of 3. What is the degeneracy of $\psi_{2,2,1}$?

(c) Find an energy level with a higher degeneracy than $\psi_{2,2,1}$ has. Demonstrate that your answer works by listing all the states with that energy.

19. Consider the following wavefunction:

$$\psi(x, y, z) = \sqrt{\frac{8}{L^3}} \sin\left(\frac{2\pi}{L}x\right) \sin\left(\frac{3\pi}{L}y\right) \cos\left(\frac{4\pi}{L}z\right)$$

(a) Demonstrate that this function is a valid solution to Equation (7.6), Schrödinger's equation inside our three-dimensional box.

(b) Explain why this is *not* a possible wavefunction for a particle in our three-dimensional box. (Don't say "because the solution is Equation (7.11) and this isn't." The question is why, on our way to Equation (7.11), we threw away all the cosines.)

20. Equation (7.11) gives the energy eigenstates ψ_{abc} for a particle in our three-dimensional box.

(a) Write the state $\psi_{1,1,1}$.

(b) For the state you wrote in Part (a), at what point is ψ maximized?

(c) If you measured the position of this particle many times, would you tend to find it near the middle, or near the edges?

(d) Write the state $\psi_{2,2,2}$. Is a particle in this state more or less likely to be in the middle of the box than one in the state $\psi_{1,1,1}$?

(e) How much energy would be required to kick the particle from $\psi_{1,1,1}$ to $\psi_{2,2,2}$?

In Problems 21–26 you will fill in missing steps in the Explanation (Section 7.2.2).

21. In this problem you're going to use separation of variables to derive the time and spatial dependence of energy eigenstates in 3D.

(a) Plug the separable solution $\Psi(x,y,z,t) = \psi(x,y,z)T(t)$ into Equation (7.4).

(b) The next step is to get all the spatial dependence on one side and all the time dependence on the other. You accomplish this by dividing both sides of the equation by ψT. Write the resulting equation.

(c) Explain how you know that both sides of that equation must equal a constant.

(d) Set both sides equal to a constant called E. One of the resulting equations should be Equation (7.5). If it isn't, try re-doing some of the algebra until you get that.

(e) The other side should be an ordinary differential equation for $T(t)$. Solve that equation to find $T(t)$.

22. Explain why, in a region where $U \to \infty$, Equation (7.5) can only be satisfied if $\psi \to 0$. You should answer this from the equation, not from physical arguments.

23. Fill in the steps required to get from Equation (7.6) to Equation (7.7).

24. In Equation (7.8) we set up an ordinary differential equation for $X(x)$ with a constant that arose from separation of variables. We called that constant

k^2 to indicate that it must be positive. Consider the opposite, a negative constant that we will call $-k^2$. (We will assume throughout this problem that everything – the wavefunction, the variables, and the constants like k – are real.)

(a) Write the differential equation for $X(x)$. (This will look almost identical to the equation we wrote.)

(b) Write the general solution. (This will look nothing like our solution – in fact, it will not involve sines or cosines.)

(c) Explain why this solution cannot possibly meet the boundary conditions of going to 0 at $x = 0$ and $x = W$. That's why we insisted on a positive constant!

25. Write the general solution to Equation (7.8). (Because it is a second-order equation, this will have two arbitrary constants.) Then write the boundary conditions on $X(x)$ and show how they lead to Equation (7.9). Make sure to note the relationship between k and the quantum number a, and show mathematically that a must be an integer. *Hint*: At any value of x where ψ is zero at all times, $X(x)$ must equal zero.

26. [*This problem requires multiple integrals, a topic from multivariate calculus.*] The energy eigenstates of the particle in a box are $\psi_{abc} = A_{abc} \times \sin(a\pi x/L)\sin(b\pi y/L)\sin(c\pi z/L)$, where a, b, and c can be any positive integers and A_{abc} is an arbitrary constant. When we say "arbitrary" we mean that any value of A_{abc} would satisfy Schrödinger's equation and the boundary conditions. Only one value of A_{abc} is consistent with normalization, however. In this problem you'll derive that value.

(a) The integral of $|\psi|^2$ over any region tells you the probability of finding the particle in that region. Suppose the particle is in the energy eigenstate $\psi_{1,1,1}$. Set up and evaluate a triple integral for the probability of finding the particle anywhere inside the box. Your answer will include the constant $A_{1,1,1}$.

(b) The total probability of finding the particle inside the cube must equal 1. Set your answer to Part (a) equal to 1 and solve for $A_{1,1,1}$.

(c) Repeat Parts (a) and (b) for the eigenstate $\psi_{1,1,2}$ to find the constant $A_{1,1,2}$ in front of it.

27. Find all the separable solutions for the differential equation

$$\frac{\partial^2 f}{\partial x^2} + \frac{\partial f}{\partial y} + z\frac{\partial f}{\partial z} = 0.$$

Because there are three independent variables, you will need to separate variables twice.

28. **Particle in a 2D Box**

The Explanation (Section 7.2.2) finds the energy eigenstates for a particle trapped in a cube. Repeat the process for a particle trapped in a square box. That is, $U=0$ when $0 \le x \le L$ and $0 \le y \le L$, and $U=\infty$ elsewhere. Find the energy eigenstates $\psi_{ab}(x,y)$ and their associated energies E_{ab}.

29. **Particle in a Rectangular Solid Box**

The Explanation (Section 7.2.2) finds the energy eigenstates for a particle trapped in a cube. Repeat the process for a particle trapped in a rectangular solid. That is, $U=0$ when $0 \le x \le L$ and $0 \le y \le W$ and $0 \le z \le H$, and $U=\infty$ elsewhere. Find the energy eigenstates ψ_{abc} and their associated energies E_{abc}.

30. A particle is trapped in a cubical box just like the one in the Explanation (Section 7.2.2), except this time it experiences a potential energy function $U=mgz$, where mg is a constant. (Outside the cube, U is still ∞.) Write the time-independent Schrödinger equation and separate variables to get ordinary differential equations for $X(x)$, $Y(y)$, and $Z(z)$. You do *not* need to solve these equations.

7.3   Math Interlude: Spherical Coordinates

You have learned that some two-dimensional problems lend themselves to polar coordinates rather than to the more familiar x and y. A three-dimensional problem with spherical symmetry is best approached with "spherical coordinates." So before we turn our attention to the hydrogen atom, we pause to discuss this possibly unfamiliar coordinate system.

7.3.1   Discovery Exercise: Polar Coordinates

If you want to specify the location of a point on a plane, you can use the "Cartesian coordinates" x and y. Alternatively, you can use "polar coordinates": ρ is the distance to the origin and ϕ is the angle measured from the positive x axis (Figure 7.6).[4]

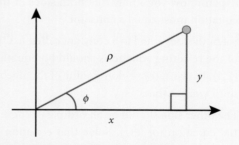

Figure 7.6 Coordinates of a point in Cartesian and polar coordinates.

1. Write functions for finding x and y if you are given ρ and ϕ. (You should be able to see these functions quickly from the diagram.)

2. Write functions for finding ρ and ϕ if you are given x and y. (Same comment.)

   See Check Yourself #10 at www.cambridge.org/felder-modernphysics/checkyourself

---

4 You may have learned polar coordinates with r and θ instead of ρ and ϕ. The letters aren't important; the meaning is the same.

3. Draw the set of all points for which $2 \leq \rho \leq 3$.

4. Draw the set of all points for which $0 \leq \phi \leq \pi/2$.

5. The point $(5, \pi/2)$ is the same as the point $(5, 9\pi/2)$ in polar coordinates. Give one other (ρ, ϕ) combination that identifies this same point.

7.3.2 Explanation: Spherical Coordinates

The Cartesian coordinates x, y, and z are the most common way to identify a point in three-space, but other systems are also used. The system that will prove most useful for the hydrogen atom is called "spherical coordinates" (or sometimes "spherical polar coordinates"):

- r: the distance from a point to the origin

- θ: the angle down from the positive z axis

- ϕ: the angle around the xy plane, starting at the positive x axis and moving toward the positive y axis (counterclockwise when viewed from above)

Figure 7.7 shows a point (r, θ, ϕ) with its Cartesian and spherical coordinates labeled. We have also labeled ρ, the distance you would walk on the xy plane to find yourself directly under our point. The variable ρ is part of a different coordinate system, but it is useful to us as a stepping-stone in finding the conversions between Cartesian and spherical coordinates, which you will partly do in Problem 9. The results are shown in Table 7.1.

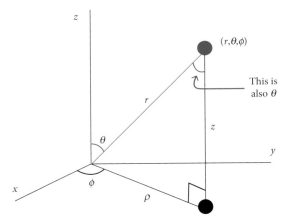

Figure 7.7 Coordinates of a point in Cartesian and spherical coordinates.

Table 7.1 Conversions between Cartesian and spherical coordinates

Cartesian-to-spherical	Spherical-to-Cartesian
$x^2 + y^2 + z^2 = r^2$	$x = r \sin\theta \cos\phi$
$\cos\theta = z/r$	$y = r \sin\theta \sin\phi$
$\tan\phi = y/x$	$z = r \cos\theta$

You may find it helpful to think of ϕ and θ as longitude and latitude, respectively. (If you're not used to longitude and latitude, just ignore this paragraph.) The only difference is that latitude is 0 at the Equator and $\pm \pi/2$ (or $\pm 90°$) at the poles. The spherical θ is defined to be 0 at the North Pole (the positive z axis) and π at the South Pole (the negative z axis).

Active Reading Exercise: Covering All of Space with Spherical Coordinates

In Cartesian coordinates you cover all of space by letting x, y, and z range from $-\infty$ to ∞. What range of values do r, θ, and ϕ need in order to cover all of space? Make sure that you try to answer this yourself before reading on, but we will start you off with the hint that none of the three variables needs to go from $-\infty$ to ∞.

There is more than one possible answer, but the simplest way to do this is to let r go from 0 to ∞, θ go from 0 to π, and ϕ go from 0 to 2π. It makes sense to not let r be negative because it's defined as a distance. For θ, 0 is the positive z axis and as you swing down from there you reach your highest value at the negative z axis, where $\theta = \pi$. For ϕ, going from 0 to 2π rotates you all the way around the z axis.

Example: Spherical Coordinates

What are the spherical coordinates of the point $x = -1, y = -1, z = 0$?

Answer: The right approach to a problem like this is to start with a picture. We have opted to start with two: one in three dimensions and one down on the xy plane:

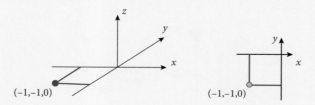

Because the point is on the xy plane it's at an angle of $\pi/2$ down from the z axis, which tells us θ. The angle ϕ starts at the positive x axis and goes around counterclockwise, reaching π at the negative x axis and continuing 45° further to $\phi = 5\pi/4$. Finally, r is the distance from the origin, $\sqrt{1^2 + 1^2 + 0^2} = \sqrt{2}$:

$$(r, \theta, \phi) = \left(\sqrt{2}, \frac{\pi}{2}, \frac{5\pi}{4} \right).$$

You can check these answers using the conversions above. For example, $\tan(5\pi/4) = 1$, which does equal y/x. However, if you had just used these conversions blindly and plugged in $\tan^{-1}(y/x)$ for ϕ, you would have found the wrong answer because there's more than one angle that has that same tangent. (Try it, and see Problem 10 for more on this issue.) You should understand what the coordinates mean and always look to make sure your answers make sense.

Our goal in this section is not a thorough treatment of this mathematical topic, but a summary of the information you need before you tackle the hydrogen atom. After the basic information above, there are three other questions we need to answer:

1. When is it helpful to use spherical coordinates instead of Cartesian?

2. Given that our goal is to analyze the hydrogen atom quantum mechanically, how do you express Schrödinger's equation in spherical coordinates?

3. I've seen non-Cartesian coordinate systems before, and those don't look like the right letters. What's up with that?

The answers, in order, are:

1. Spherical coordinates are generally useful for physical situations with spherical symmetry – that is, things that radiate outward from a point. Consider the density of the Sun. More to the point, consider the electric potential of the nucleus of a hydrogen atom. In Cartesian coordinates these are functions of x, y, and z. In spherical coordinates they are functions of r only.

2. Recall that the time-independent Schrödinger equation in 3D Cartesian coordinates is

$$-\frac{\hbar^2}{2m}\left(\frac{\partial^2\psi}{\partial x^2}+\frac{\partial^2\psi}{\partial y^2}+\frac{\partial^2\psi}{\partial z^2}\right)+U\psi=E\psi. \tag{7.14}$$

The Cartesian coordinates are inside the parentheses, so that's where the conversion happens. But you can't just replace $\partial^2\psi/\partial x^2$ with $\partial^2\psi/\partial r^2$ and so on. Instead you have to use the conversions we gave above, along with the multivariate chain rule, to convert derivatives with respect to x, y, and z to derivatives with respect to r, θ, and ϕ. The good news is that you don't have to do that over and over; it's been done once and for all. When you plug in the results, the Schrödinger equation becomes:

$$-\frac{\hbar^2}{2m}\left(\frac{\partial^2\psi}{\partial r^2}+\frac{2}{r}\frac{\partial\psi}{\partial r}+\frac{1}{r^2}\frac{\partial^2\psi}{\partial\theta^2}+\frac{\cos\theta}{r^2\sin\theta}\frac{\partial\psi}{\partial\theta}+\frac{1}{r^2\sin^2\theta}\frac{\partial^2\psi}{\partial\phi^2}\right)+U\psi=E\psi. \tag{7.15}$$

For any given function ψ the terms in parentheses in Equations (7.14) and (7.15) will give the same results, as you'll show for one example in Problem 11. We can therefore introduce a new piece of notation and write Schrödinger's equation in a coordinate-independent way:

$$-\frac{\hbar^2}{2m}\nabla^2\psi+U\psi=E\psi.$$

That possibly unfamiliar $\nabla^2\psi$ is called the "Laplacian" of ψ. It stands for the term in parentheses in either equation.

3. If you learned spherical coordinates in a prior class, you may have expressed the same variables with different letters (switching ρ and r, and/or switching θ and ϕ). We are presenting the letters as they are generally used in physics, and we will consistently use them as they appear in Figure 7.7. We apologize for any (completely understandable) confusion. (We should also note that there are two common forms of the letter phi: ϕ and φ. We will consistently use ϕ.)

7.3.3 Questions and Problems: Spherical Coordinates

Conceptual Questions and ConcepTests

1. The Explanation lists two systems that are most easily represented in spherical coordinates: the density of the Sun, and the electric potential around a hydrogen atom. Give two more.

2. Draw and briefly describe the shape described by the spherical equation $r = 10$.

3. Draw and briefly describe the shape described by the spherical equation $\theta = \pi/6$.

4. Draw and briefly describe the shape described by the spherical equation $\phi = \pi/6$.

5. Write the equations in spherical coordinates to describe the bottom half (below the xy plane) of the sphere of radius 2 centered on the origin. (Not a filled-in sphere, just the surface.)

6. "The gravitational force of the Sun varies inversely with the square of your distance to the Sun." Express that sentence using spherical coordinates with the Sun at the center of your coordinate system.

Problems

7. The points below are given as Cartesian coordinates (x, y, z). Express them in spherical coordinates (r, θ, ϕ). You should be able to answer by picturing where each point is, rather than by plugging into formulas.

    (a) $(3, 3, 0)$

    (b) $(1, 0, 1)$

8. The points below are given as spherical coordinates (r, θ, ϕ). Express them in Cartesian coordinates (x, y, z). You should be able to answer by picturing where each point is, rather than by plugging into formulas.

    (a) $(\sqrt{2}, \pi/2, 3\pi/4)$

    (b) $(5, 3\pi/4, \pi)$

9. The conversions between Cartesian and spherical coordinates come from two right triangles.

    (a) Draw a point on the xy plane, and draw a right triangle that shows its x and y coordinates.

    Label the hypotenuse of your triangle ρ, and label the angle as ϕ.

    (b) Whenever you draw a right triangle and label an angle, you can immediately write down four relationships: the sine, the cosine, the tangent, and the Pythagorean theorem. Based on the triangle you drew in Part (a), write down all four relationships between x, y, ρ, and ϕ.

    (c) The second right triangle is shown in Figure 7.7 on p. 333. Use this triangle to write down all four relationships between ρ, z, r, and θ.

    (d) Now use the relationships you have written to write x, y, and z in terms of r, θ, and ϕ. Remember that you don't want ρ to appear in your final answers.

10. When you convert from spherical to Cartesian coordinates you can safely plug into the formulas given in this section. When you go the other way the formulas say $r = \sqrt{x^2 + y^2 + z^2}$, $\theta = \cos^{-1}(z/r)$, and $\phi = \tan^{-1}(y/x)$. Unfortunately that last formula can sometimes give you wrong results. You always have to think about where the points are graphically to make sure you are getting the right answer. We'll show you what we mean.

    (a) Consider the point $(x, y, z) = (1, 1, 0)$. Plug into these formulas to find r, θ, and ϕ.

    (b) Now do the same thing for the point $(x, y, z) = (-1, -1, 0)$.

    (c) Which of your answers was wrong? What are the correct spherical coordinates for that point?

11. Equations (7.14) and (7.15) are both forms of Schrödinger's equation. They are equivalent because their spatial derivatives express the "Laplacian" operator $\nabla^2 \psi$ in their respective coordinate systems. We're not going to prove that they are equivalent in general, but in this problem you'll show that they give same answer for one simple example, the function $f(r) = r^4$.

(a) Calculate $\nabla^2 f$ using the spherical coordinates formula for the Laplacian. This is the part in parentheses in Equation (7.15), but acting on a function called f instead of ψ.

(b) Rewrite the function $f(r) = r^4$ as a function of x, y, and z.

(c) Calculate $\nabla^2 f$ in Cartesian coordinates. This time the Laplacian is the part in parentheses in Equation (7.14).

(d) Demonstrate that your spherical and Cartesian answers are the same. (If they aren't, go back and find your mistake!)

7.4 Schrödinger's Equation and the Hydrogen Atom

The wavefunction of the electron in a hydrogen atom can be described as some combination of eigenstates ψ_{nlm_l}. Each such eigenstate is a state of definite energy, angular momentum magnitude, and angular momentum z component, which is specified by the quantum numbers n, l, and m_l, respectively. Section 7.1 discussed these quantum numbers and their meanings; this section will look at the actual wavefunctions for these eigenstates, and outline their derivation.

7.4.1 Discovery Exercise: Schrödinger's Equation and the Hydrogen Atom

An electron orbiting a nucleus feels a potential energy $U = -k/r$.

1. Write the time-independent Schrödinger equation for the electron in spherical coordinates. (See Equation (7.15) on p. 335.)

2. Make a guess $\psi(r,\theta,\phi) = R(r)\Theta(\theta)\Phi(\phi)$, and separate variables to derive an ODE for $R(r)$.

7.4.2 Explanation: Schrödinger's Equation and the Hydrogen Atom

The energy eigenstates of any system are the solutions to the time-independent Schrödinger equation with the appropriate potential energy. For the hydrogen atom, the potential energy is $U = -k/r$, where r is the distance from the nucleus and $k = e^2/(4\pi\epsilon_0)$ is a constant from Coulomb's law. Choosing the nucleus as the origin of our coordinate system means that r is also the radial variable in spherical coordinates. As you saw in Section 7.3, the time-independent Schrödinger equation in those coordinates is

$$-\frac{\hbar^2}{2m_e}\left(\frac{\partial^2\psi}{\partial r^2} + \frac{2}{r}\frac{\partial\psi}{\partial r} + \frac{1}{r^2}\frac{\partial^2\psi}{\partial\theta^2} + \frac{\cos\theta}{r^2\sin\theta}\frac{\partial\psi}{\partial\theta} + \frac{1}{r^2\sin^2\theta}\frac{\partial^2\psi}{\partial\phi^2}\right) - \frac{k}{r}\psi = E\psi. \qquad (7.16)$$

As usual, our approach to a partial differential equation is separation of variables: plug in $\psi(r,\theta,\phi) = R(r)\Theta(\theta)\Phi(\phi)$, rearrange to get the r dependence on one side and the θ and ϕ dependences on the other, and then set both sides equal to a constant. You'll go through the details in Problem 16. The resulting equations are

$$r^2\frac{R''}{R} + 2r\frac{R'}{R} + \frac{2m_e}{\hbar^2}(Er^2 + kr) = l(l+1) \qquad (7.17)$$

$$\frac{\Theta''}{\Theta} + \frac{\cos\theta}{\sin\theta}\frac{\Theta'}{\Theta} + \frac{1}{\sin^2\theta}\frac{\Phi''}{\Phi} = -l(l+1) \qquad (7.18)$$

The first equation gives the radial dependence of the energy eigenstates and the second one gives their angular dependence. Solving these equations is beyond the scope of this book, but we will describe some of the mathematical features of their solutions below.

The Radial Function $R(r)$

The full solution of Equation (7.17), given in Appendix G, is an intimidating concoction of polynomials, fractions, a square root, a factorial or two, and a decaying exponential. Here is a somewhat boiled-down version:

$$R_{nl}(r) = A_{nl} r^l e^{-r/(a_0 n)} \times \left(\text{a polynomial in } \frac{r}{a_0 n} \text{ of degree } n - l - 1 \right). \qquad (7.19)$$

The constant a_0 is called the "Bohr radius," a distance roughly equal to half an angstrom. The A_{nl} in front is a normalization factor that is given explicitly in Appendix G.

The decaying exponential $e^{-r/(a_0 n)}$ guarantees that $R \to 0$ at large radii. If it didn't, then ψ wouldn't be normalizable. The polynomial at the end causes the wavefunction to oscillate as r increases. In other words, as you move out from the origin you alternate between radii where the electron is relatively likely to be found and radii where it is unlikely to be found. How many times it oscillates depends on l. The lower l is (relative to n), the more times the probability oscillates. For $l = n - 1$, the polynomial is "zeroth-order," meaning a constant, and (except for $n = 1$) the radial wavefunction simply rises to a peak and then falls monotonically.

If you compare Equations (7.17) and (7.19), you might notice a number of constants that appear in the problem but not in the solution. The constants m_e, k, and $\hbar$ got absorbed into a_0, which is defined as $\hbar^2/(m_e k)$. But the more conspicuous omission is that the differential equation has an E that the solution doesn't have. Pause on that. Look at the equations, and think back to what you've previously learned about the hydrogen atom. Can you see how to identify the energy associated with a particular $R_{nl}(r)$ state?

Congratulations if you noticed that Equation (7.19) has an n that wasn't in the differential equation. In the course of applying the boundary conditions you can show that the energy can only take on values $E = -(m_e k^2)/(2\hbar^2 n^2)$, where n is a positive integer. Equation (7.19) is expressed in terms of n rather than E for convenience, but it still assigns a particular energy to each state. (Recall that Section 7.1 said $E = -1/n^2$ Ry; you can now see that $1 \text{ Ry} = m_e k^2/(2\hbar^2) \approx 13.6 \text{ eV}$.)

We're skipping a lot here. In later courses you may learn how to derive and work with the "associated Laguerre polynomials" and other specialized functions and fill in those gaps. But you've been through simpler problems such as the infinite square well (in both 1D and 3D) in detail. In each case the energy E appeared as a constant in the time-independent Schrödinger equation. Then the boundary conditions restricted E to an infinite-but-discrete set of values, and those values were expressed by introducing a quantum number (often called n). In outline form, the solution to the hydrogen atom – and every other bound system – follows the same pattern as those earlier problems, even though the specific differential equations get more obscure as the potential energy functions get more complicated.

The Angular Functions $\Theta(\theta)$ and $\Phi(\phi)$

As you may recall from Section 7.2, the next step is to turn to Equation (7.18) and separate variables a second time. You'll do that in Problem 16 to get ordinary differential equations for $\Theta(\theta)$ and $\Phi(\phi)$. The latter ends up as follows:

$$\Phi''(\phi) = -m_l^2 \Phi(\phi). \qquad (7.20)$$

(In this context the letter m_l here is just a new constant that got introduced in the separation of variables.) You may recognize this as the simple harmonic oscillator equation. You can write the solution with sines and cosines, but it's more convenient to use complex exponentials:[5]

$$\Phi(\phi) = e^{im_l\phi}. \qquad (7.21)$$

As we said above, bound systems lead to quantized solutions. You will show in Problem 18 that the boundary conditions on Φ require that m_l in Equation (7.21) be an integer.

The $\Theta(\theta)$ equation is harder to solve, but the answer ends up being polynomials in $\sin\theta$ and $\cos\theta$. You'll show that a few of them work in Problems 16 and 17.

In general you will not see $\Phi(\phi)$ and $\Theta(\theta)$ written separately. Those particular functions, multiplied by each other, occur often enough to merit a special name and symbol[6] *together*:

$$\Theta(\theta)\Phi(\phi) = \Theta(\theta)e^{im_l\phi} = Y_l^{m_l}(\theta,\phi) \quad \text{Spherical harmonics.}$$

For example, $Y_3^{-1} = \left[\sqrt{21}/(8\sqrt{\pi})\right]\sin\theta\left(5\cos^2\theta - 1\right)e^{-i\phi}$. The general formula and the first several spherical harmonics are listed in Appendix G. Each one is a polynomial of order l in $\sin\theta$ and $\cos\theta$, multiplied by $e^{im_l\phi}$.

You can begin to see why spherical harmonics have such general importance when you consider this: when we separated variables, the potential energy function went to the r side of the equation. If you replace $U = -k/r$ with a different function of r, you end up with a different $R(r)$ differential equation, but with the exact same equations – and therefore the same solutions – for $\Theta(\theta)$ and $\Phi(\phi)$. That tells us that spherical harmonics don't just apply to the hydrogen atom. They are the angular part of the energy eigenstates for any $U(r)$, which is to say, for any physical situation with a spherically symmetric potential energy function.

Electron Probability Distributions

We've said that a quantum mechanical electron isn't a planet-like object orbiting around a nucleus, but rather a cloud of probability described by a wavefunction. We have now talked about all the pieces of that wavefunction,

$$\psi_{nlm_l}(r,\theta,\phi) = R_{nl}(r)\Theta_{lm_l}(\theta)\Phi_{m_l}(\phi) = R_{nl}(r)Y_l^{m_l}(\theta,\phi).$$

This form allows us to analyze the probability distributions one variable at a time.

- $R(r)$ goes to zero in the limit $r \to \infty$, but exhibits polynomial behavior on its way there. So if you start at the nucleus and move outward, ψ may oscillate a finite number of times before it decays.

- $\Theta(\theta)$ has sines and cosines. This may cause ψ to oscillate as you rotate from the positive z axis down to the negative z axis.

- Finally, $\Phi(\phi) = e^{im_l\phi}$. Let's consider what that implies.

---

5 You might object that we should also include $e^{-im_l\phi}$. Since we haven't specified whether m_l is positive or negative, though, that is included in the form we wrote. Put another way, all we've said so far is that the solutions are all of the form "e raised to an imaginary number times ϕ" (or combinations of such functions).

6 We could write Y_{lm_l}, similarly to how we wrote R_{nl} above, but with spherical harmonics it's conventional to write the m_l as a superscript.

Active Reading Exercise: The ϕ Probability Distribution

For any values of the variables m_l and ϕ, it is always the case that $|\Phi(\phi)| = \left|e^{im_l\phi}\right| = 1$.

1. What does that fact tell us about the ϕ dependence of position probabilities for an electron that is in an eigenstate?

2. What does that same fact tell us about the ϕ dependence of position probabilities for an electron that is in a superposition of different eigenstates?

You can see our answers at www.cambridge.org/felder-modernphysics/activereadingsolutions

You'll explore this question further in Problem 19. Here we will turn our attention from ϕ to r. In many cases it's useful to know the probability per unit radius of finding the electron at a given distance from the nucleus.

How Far is the Electron?

The formula $r^2 R^2(r)$ gives the probability per unit r of finding the electron at distance r from the nucleus. Therefore the probability of finding the electron between r_1 and r_2 is given by an integral:

$$P(r_1 \le r \le r_2) = \int_{r_1}^{r_2} r^2 R^2(r)\,dr. \tag{7.22}$$

Equation (7.22) says that the probability does not just scale like $R^2(r)$, as you might expect. That's because $|\psi|^2$ gives you a probability *per unit volume,* and there is a lot more volume between (say) 5 and $5 + dr$ than there is between 1 and $1 + dr$. You'll formalize that argument a bit in Problem 20.

Notice that a hydrogen atom doesn't have a well-defined "size." The probability of finding the electron asymptotically approaches zero at large radii, but never reaches it. You can roughly define a size based on the expectation value of r, and this size generally increases as n increases. See Problems 8–11.

Example: Electron Probabilities

A hydrogen atom is in the state $\psi_{3,2,2}$. What is the probability of finding the electron in the region $r > 15a_0$?

Answer: We looked up the relevant wavefunction information in Appendix G. Then we integrated $r^2 R^2(r)$ on a computer:

$$R_{3,2}(r) = \frac{4}{81\sqrt{30}} a_0^{-3/2} \left(\frac{r}{a_0}\right)^2 e^{-r/(3a_0)}$$

$$P(r > 15a_0) = \int_{15a_0}^{\infty} r^2 R^2\,dr = 0.13 = 13\%$$

You can build some intuition for the different eigenstates by visualizing the probabilities they represent. First, let's consider the ground state $\psi_{1,0,0}$. For any state with $l = m_l = 0$ the spherical harmonic Y_0^0 is just a constant, and for $n = 1$ the radial wavefunction is just a constant times e^{-r/a_0}. So the probability distribution $|\psi|^2$ is spherically symmetric, largest at the origin and falling steadily toward zero as you move away from it. The radial probability distribution $r^2 R^2$ is zero at $r = 0$, and rises to a peak before going back to zero, as shown in Figure 7.8 (see Question 5).

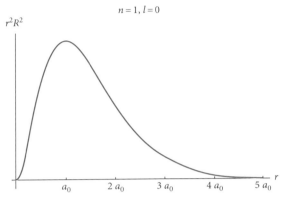

$n = 1, l = 0$

r^2R^2

$a_0 \qquad 2\,a_0 \qquad 3\,a_0 \qquad 4\,a_0 \qquad 5\,a_0$

r

Figure 7.8 Probability per unit r of finding the electron at a given radius in the ground state $\psi_{1,0,0}$ of the hydrogen atom.

Higher energy states tend to be more complicated, and states with $l \neq 0$ are generally not spherically symmetric. However, every eigenstate ψ_{nlm_l} is symmetric around the z axis, so you can get a sense of what the probabilities look like by looking at a cross-section. The left panel of Figure 7.9 shows a contour plot of the probability distribution for the state $\psi_{4,2,0}$, viewed on the xz plane.

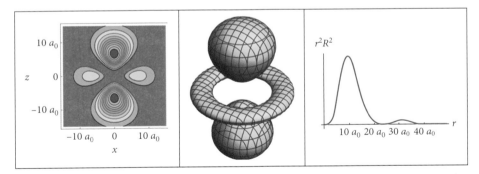

$10\,a_0$

$z \quad 0$

$-10\,a_0$

$-10\,a_0 \quad 0 \quad 10\,a_0$

x

r^2R^2

$10\,a_0 \ 20\,a_0 \ 30\,a_0 \ 40\,a_0$

r

Figure 7.9 Probability distributions for the state $\psi_{4,2,0}$. The left panel shows contours of probability on the xz plane. The middle panel shows one such contour in 3D. The right panel shows the radial probability distribution. (We've exaggerated the second bump to make it more visible.)

Each of the curves on the plot is a contour of equal probability, and the colors range from blue for the highest to red for the lowest probability. Because this is a 2D slice, you should imagine all of these contours and regions being rotated about the vertical axis to get the full 3D distribution. Unlike the ground state, where the most likely spot was the origin, the electrons in this state are most likely to be found somewhat above or below the origin on the z axis or somewhat away from the origin on the xy plane. The middle panel shows one such contour. Finally, the right panel shows the radial probability distribution. It rises and falls a few times before approaching zero at large r (see Problem 13).

7.4.3 Questions and Problems: Schrödinger's Equation and the Hydrogen Atom

Conceptual Questions and ConcepTests

1. Consider a particle – not an electron in a hydrogen atom, just some particle – with the wavefunction $\psi(r,\theta,\phi) = R(r)\Theta(\theta)\Phi(\phi)$. Describe briefly what each of the following hypothetical parts would tell you about this hypothetical particle. (For example, if in Cartesian coordinates we told you that $\psi(x,y,z) = X(x)Y(y)Z(z)$ and $X(x) = e^{-x^2}$, you would say that the probability of finding this particle dropped off rapidly as you moved away from the yz plane.)

   (a) $R(r) = e^{-r}$

   (b) $R(r) = (\sin r)/r$

   (c) $R(r) = k$ (in other words the function has no r dependence)

   (d) $\Phi(\phi) = \begin{cases} 0 & 0 \le \phi \le \pi \\ 1 & \pi < \phi < 2\pi \end{cases}$

   (e) $\Theta(\theta) = \cos^2\theta$

2. An electron is a cloud of probability hovering around a nucleus, and that probability does not drop to zero no matter how far from the nucleus you go. Yet you will often see the statement that a hydrogen atom in its ground state has a radius of about 1 angstrom (10^{-10} m). What does that mean?

3. We said that Equation (7.18) and its solution $Y_l^{m_l}(\theta,\phi)$ describe the angular dependence of the energy eigenstates for any spherically symmetric potential energy function $U(r)$. Since the potential energy doesn't show up at all in Equation (7.18), you might think that this would be valid for *any* potential energy function. Why is this not correct? Think about which step in our derivation of this equation would be invalid if U depended on θ and/or ϕ.

4. We said that the probability distribution is independent of ϕ because the modulus of $e^{im_l\phi}$ is always 1. Does this mean the $e^{im_l\phi}$ term is unimportant? Why can't we just leave it out of the equations entirely?

5. In the ground state of the hydrogen atom, the largest value of the probability distribution $|\psi|^2$ occurs at $r = 0$, meaning the single most likely place to find the electron is at the nucleus. But the radial probability r^2R^2 is zero there and peaks at about $r = a_0$, meaning that is the most likely radius at which to find the electron. How can you resolve this apparent contradiction?

6. Label each of the following as "possible" or "impossible" for a hydrogen atom in one of the eigenstates ψ_{nlm_l}. If there is any eigenstate in which the statement would be true you should answer "possible."

   (a) The probability distribution is independent of time.

   (b) The probability distribution is independent of ϕ.

   (c) The probability distribution depends on ϕ.

   (d) The electron is more likely to be at positive x than at negative x.

   (e) The probability per unit increase in radius of finding the electron in a given range of radii is highest at $r = 0$.

7. [*This question depends on Question 6.*] Which, if any, of your answers in Question 6 would change if we considered superpositions of the states ψ_{nlm_l} in addition to pure states? Of course any statements that you labeled as possible will still be possible. But for any statement that you labeled impossible, explain why that would or wouldn't still be true when you include superpositions.

Problems

8. Consider a hydrogen atom in the state $\psi_{2,1,1}$. Remember that the probability per unit radius of finding the electron at a particular radius is given by $P(r) = r^2R(r)^2$.

   (a) Find the r-value that maximizes $P(r)$, that is, the distance from the nucleus where you are most likely to find the electron.

   (b) If you measure the electron's distance to the nucleus many times, the average of all your results will be the expectation value $\langle r \rangle = \int_0^\infty rP(r)dr$. Calculate this.

   (c) Did your two answers give the same r-value? What does that tell you?

9. The "size" of a hydrogen atom grows more or less proportionally to $a_0 n^2$. Let $l = 0$ and plot the radial probability distribution $r^2 R^2$ for $1 \leq n \leq 5$. For each n estimate the radius at which the probability becomes vanishingly small. (We're not asking for calculations here. Just estimate by eye when the probability distribution is almost hitting the horizontal axis.) Express your answer as a multiple of $a_0 n^2$. For example, for $n = 1$ we estimate that the probability becomes negligible at about $r \approx 5 a_0 n^2$. *Hints*: We recommend that you use a computer to calculate the radial wavefunctions as well as plotting them; doing it by hand gets tedious. You can easily make these plots numerically by setting $a_0 = 1$.

10. One way to define the size of a hydrogen atom is by the expectation value of the radius: $\langle r \rangle = \int_0^\infty r P(r) dr$. Calculate this expectation value for $l = 0$ for all values of n from 1 to 10. How does this size grow as you increase n?

11. Roughly speaking, the size of a hydrogen atom is about $a_0 n^2$. How much energy would you have to give to a hydrogen atom to make it about a meter wide? Why do we never encounter atoms that large? *Hint*: Think about how much energy it takes to ionize a hydrogen atom.

12. We said that in the ground state of the hydrogen atom the probability distribution $|\psi|^2$ is largest at the origin, which is of course inside the nucleus! The nucleus, however, is very small, with a radius of roughly 10^{-15} m. If a hydrogen atom is in the ground state, what is the probability of finding the electron inside the nucleus?

13. Figure 7.9 plots the probability distribution for the state $\psi_{4,2,0}$ in several ways, showing how the radial probability distribution $r^2 R^2$ rises and falls as you move to higher radius.

    (a) Write the radial probability distribution for this state. You will need to calculate $R_{4,2}(r)$, either by hand or by computer, starting with the general formula given in the Appendix G.

    (b) At how many positive r-values will this distribution equal 0?

14. For each n from 1 to 4, make a plot showing the radial probability distribution $r^2 R^2$ for each pos-

sible value of l. The first plot will have just one curve on it, the second one will have two, and so on. In each plot be sure to clearly label which curve corresponds to which value of l. Using your plots, describe how n and l each affect the radial probability distribution.

15. **The Radial Solution** $R(r)$

    Look up the definition of $R(r)$ in Appendix G and use it to write out $R(r)$ for $n = 1, l = 0$. You can then use the same appendix to check your answers.

In Problems 16–20 you will fill in missing steps in the Explanation.

16. **Separating the Angular Equations**

    (a) Plug the separable solution $\psi(r, \theta, \phi) = R(r)\Theta(\theta)\Phi(\phi)$ into Equation (7.16) and separate variables to obtain Equations (7.17) and (7.18). At the point where you introduce a separation constant, you will need to call it $l(l+1)$ to match the way in which we wrote the equations. (You could just call it p or something like that, but this choice makes the solution of the angular equation simpler.)

    (b) Separate variables in Equation (7.18) to get all the θ dependence on one side and the ϕ dependence on the other. Set both sides equal to a constant and write the resulting differential equation for $\Theta(\theta)$. (You don't need to solve it.)

17. [*This problem depends on Problem 16.*] In Problem 16 you separated the angular equations and found the differential equation for $\Theta(\theta)$. Look up the spherical harmonics in Appendix G. Each one is a solution to the above differential equation, multiplied by a factor of $e^{im_l \phi}$. Use that list to write down $\Theta(\theta)$ for $l = 2$, $m_l = 0$ and show that it is a valid solution to the differential equation you found.

18. **The Boundary Conditions for** $\Phi(\phi)$

    We showed in the text that the solutions to Equation (7.20) for $\Phi(\phi)$ are of the form $e^{im_l \phi}$. Next we need to apply the boundary conditions to find the allowed values for m_l. But ϕ doesn't have boundaries; it just keeps going around the z axis. The boundary condition is that the function $\Phi(\phi)$ must be periodic. In other words, since ϕ and $\phi + 2\pi$ are two ways of representing the same place, it must be true that $\Phi(\phi) = \Phi(\phi + 2\pi)$.

(a) Find a value of m_l that would cause $\Phi(\phi)$ to have period 2π. *Hint:* Remember that $e^{ix} = \cos x + i \sin x$, so e^{ix} has the same period as $\cos x$ and $\sin x$.

(b) Find $\Phi(\pi/3)$ and $\Phi(\pi/3+2\pi)$ using the value of m_l you found in Part (a) and verify that they come out the same. If they don't, go back and fix the mistake you made in Part (a).

(c) Find a value of m_l that would cause $\Phi(\phi)$ to have period π.

(d) Find $\Phi(\pi/3)$ and $\Phi(\pi/3+2\pi)$ using the value of m_l you found in Part (c) and verify that they come out the same.

(e) Find a value of m_l that would cause $\Phi(\phi)$ to have period 4π and find $\Phi(\pi/3)$ and $\Phi(\pi/3+2\pi)$ using that value.

(f) Which of the three m_l values you just found would satisfy the boundary conditions for Φ? (It might be more than one of them.)

(g) Would any negative values of m_l work? Why or why not?

(h) What are all of the allowed values of m_l for which $\Phi(\phi)$ obeys the boundary condition $\Phi(\phi) = \Phi(\phi + 2\pi)$?

19. When ϕ Matters

An eigenstate of energy for the hydrogen atom can be expressed as $R(r)\Theta(\theta)\Phi(\phi)$. The first two functions are complicated, but the third is just $\Phi = e^{im_l\phi}$ for some integer m_l. In this problem you'll consider what that means.

(a) Find the modulus of $e^{im_l\phi}$. (Remember that, for any complex number z, the modulus $|z|$ is given by $\sqrt{zz^*}$ where z^* is the complex conjugate.)

(b) Your answer to Part (a) tells us that if a hydrogen atom is in an eigenstate of energy, its probability density is independent of ϕ. Describe visually what kind of symmetry that implies.

(c) Find the modulus of $e^{i\phi} + e^{2i\phi}$. (*Note:* You can't normalize that function without also knowing

$R(r)$ and $\Theta(\theta)$, but that isn't important for our purpose here.)

(d) Your answer to Part (c) is *not* ϕ-independent. What does that fact tell you about hydrogen atoms?

20. The Radial Probability Distribution

In this problem you will calculate the probability of finding a particle between the radii r and $r + dr$ (Equation (7.22)).

(a) How much volume is contained between r and $r + dr$?

(b) You should be able to write your answer to Part (a) in the form

$$< something > dr+ < something \ else >$$
$$dr^2+ < a \ third \ thing > dr^3.$$

In the limit $dr \to 0$ the last two terms become arbitrarily smaller than the first, so you can drop them. Write the resulting volume as $< something > dr$.

(c) The probability that the electron is in that region is the volume times $|\psi|^2$, or equivalently the volume times $R^2(r)Y^2(\theta,\phi)$. Use your result from Part (b) to explain why the probability of finding an electron between r and dr is proportional, not to $|\psi|^2$, but to $r^2|\psi|^2$.

(d) Once you're convinced that the probability is proportional to r^2R^2, what's the constant of proportionality? The answer is: the constants out in front of our $R(r)$ functions are carefully chosen to make that constant 1, leading to Equation (7.22). As an example, show that the ground state function $R_{1,0}(r)$ is properly normalized. (That is, that the probability of finding the electron at *some* distance from the nucleus is 1.)

Problems 21–25 are all about probabilities for hydrogen atoms in different energy eigenstates. You can find all the information you need about the wavefunctions in Appendix G. Some of these problems will require integration; you can do the integrals by hand but some get quite tedious. Feel free to do them on a computer but clearly show what you are integrating (and why), and then what result the computer gave you for that integral.

21. Consider a hydrogen atom in its ground state.

   (a) What is the probability of finding the electron somewhere in a sphere of radius a_0 centered on the nucleus?

   (b) What is the probability of finding the electron somewhere in the region $z > 0$? *Hint*: No integration is required.

22. [*This problem requires multiple integrals, a topic from multivariate calculus.*] Consider a hydrogen atom in the state $\psi_{2,1,1}$.

   (a) What is the probability of finding the electron in the region $0 < r < a_0$, $0 < \theta < \pi/4$, $0 < \phi < \pi/2$?

   (b) What is the probability of finding the electron in the region $0 < \theta < \pi/2$, $0 < \phi < \pi/2$, at any radius? *If you think about it, this part is easier than the previous one.*

23. Consider a hydrogen atom in the state $\psi_{2,1,1}$. You should be able to answer all of the following questions *without* evaluating any integrals.

   (a) What is the probability of finding the electron in the region $z > 0$?

   (b) What is the probability of finding the electron in the region $x > 0$?

   (c) If you pick a small region around $r = a_0$ on the positive z axis and an identical region at $r = a_0$ on the positive x axis, in which of those regions are you more likely to find the electron?

24. Suppose a hydrogen atom is in a state with $n = 2$, $l = 1$. For what value(s) of m_l is the electron most likely to be found near the positive z axis? Don't just give a number; explain how you know.

25. [*This problem requires multiple integrals, a topic from multivariate calculus.*] Like other systems, a hydrogen atom doesn't have to be in an energy eigenstate. It can be in any properly normalized combination of them. For each wavefunction below, calculate the probability of finding the electron in the region $0 < \phi < \pi/4$.

   (a) $\psi_{2,1,1}$ (*Hint*: You do not need to evaluate an integral to answer this.)

   (b) $\psi_{2,1,0}$ (Nor this one either.)

   (c) $(1/2)\psi_{2,1,1} + (\sqrt{3}/2)\psi_{2,1,0}$

   (d) $(i/2)\psi_{2,1,1} + (\sqrt{3}/2)\psi_{2,1,0}$

26. Exploration: The Infinite Spherical Well

In Section 7.2 we solved for the energy eigenstates of the infinite square well in 3D, also called a "particle in a box." In this problem you are going to do the same thing for a spherical well, meaning a particle that is confined to the region $r < L$ for some constant radius L. Inside that sphere the potential energy is zero; outside, it is infinite.

   (a) Write the three-dimensional time-independent Schrödinger equation in spherical coordinates for the region $r < L$.

   (b) Suppose you plug in a separable guess $\psi(r, \theta, \phi) = R(r)\Theta(\theta)\Phi(\phi)$. What would the angular solution $\Theta(\theta)\Phi(\phi)$ be? *Hint*: You should be able to write the answer with no calculation.

   (c) What is the differential equation for the radial part $R(r)$? (It's once again possible to read this off from the text, but if you don't see how, go ahead and plug in the guess and separate variables to find it. You will still want to call the separation variable $l(l+1)$ as we did when we solved for the energy eigenstates of the hydrogen atom.)

   (d) Use a computer to find the general solution to the differential equation you just wrote. Because it's a second-order equation, the solution should have two arbitrary constants, each multiplying a different independent solution. Don't be worried if the solutions involve functions you've never heard of.

   (e) Plug in $r = 0$ to both of the independent solutions. You should find that one of them is infinite. We reject that solution (which is to say the arbitrary constant in front of that one must equal zero). Write the resulting solution for $R(r)$, which should now just have one arbitrary constant. In addition to that constant, your function should have the energy E and the quantum number l in it, as well as the constants m (mass) and $\hbar$.

   (f) Explain why $R(L)$ must equal zero.

To finish the problem, you are going to use that boundary condition $R(L) = 0$ to find the ground state energy of the system. To show you how this works, let's take the familiar example of the 1D infinite square well. At this point in that calculation you would have found that the wavefunction was $\psi(x) = A \sin(x\sqrt{2mE}/\hbar)$ and you would know that the boundary condition was $\psi(L) = 0$. Of course we know that the first positive value of α for which $\sin(\alpha) = 0$ is π, but, if you didn't know that, you could just plot $\sin(\alpha)$ on a computer and estimate that the first zero-crossing was at $\alpha \approx 3.1$. That

would tell you that the lowest energy that would satisfy the boundary condition would be the one that satisfies $L\sqrt{2mE}/\hbar \approx 3.1$, and you could then estimate the ground state energy of the 1D infinite square well as $(3.1)^2\hbar^2/(2mL^2)$.

(g) Take $l = 0$ in the solution you found for $R(r)$ and then apply the same logic we just described to estimate the ground state energy of the infinite spherical well. (You can find the first zero-crossing of the function graphically as we described, or by looking up the properties of the function and finding it exactly.)

7.5 Spin

The description we've given above, based on Schrödinger's equation, explains many observed features of hydrogen and other atoms. But in the 1920s both experimental evidence and theoretical arguments began to suggest another degree of freedom – another quantum number – beyond those we've described so far.

7.5.1 Explanation: Spin

The original Bohr model predicted that the magnitude of angular momentum was quantized. A later extension of the model by Ernst Sommerfeld predicted that the z component of angular momentum was quantized as well. In 1922 Otto Stern and Walter Gerlach did an experiment to look for that quantization. (Stern was a skeptic and was trying to disprove the quantization idea.)

The Stern–Gerlach Experiment

The Stern–Gerlach experiment experiment involved passing a beam of atoms through a magnetic field. You can treat each electron as a current loop.[7] The following results arise from classical electromagnetic theory. (You'll demonstrate the first two in Problem 7.)

- A loop of current in a uniform magnetic field experiences a torque, but no net force.
- A loop of current in a non-uniform magnetic field experiences a net force.
- That net force is proportional to the angular momentum of the charged particles in the current loop.

Stern and Gerlach passed a beam of silver atoms through a strong, non-uniform magnetic field. Assuming $(\partial B/\partial z) > 0$, an atom whose electrons have a net angular momentum upward would

---

[7] A single, classically orbiting particle is different from a current loop because at each moment there's only "current" in one spot. If it's orbiting fast enough, however, then averaged over one full orbit it acts like a loop of current. Quantum mechanically the electron is spread out around the nucleus so this model is even more accurate.

be deflected down, and vice versa. (The direction is reversed because electrons have negative charge.) See Figure 7.10.

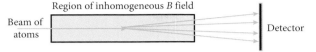

Figure 7.10 A Stern–Gerlach apparatus. Atoms are sent through a region with an inhomogeneous magnetic field that deflects them up or down proportionally to the z component of their electrons' angular momentum.

Active Reading Exercise: The Stern–Gerlach Experiment

Imagine a Stern–Gerlach experiment with hydrogen atoms. Each atom is bent up or down by an amount proportional to the z angular momentum of its electron. Then each atom strikes a detector on the other side of the apparatus and makes a mark where it hits. Draw what you would expect the pattern on the detector to look like in each of the following cases:

1. A classical model, in which the electron in each atom is orbiting with a random orientation.

2. The Bohr–Sommerfeld model, in which the electron in each atom can have $L_z = \pm\hbar$.

3. A quantum mechanical hydrogen atom in the ground state.

In a classical view, each atom has a random value of L_z ranging from $-|\vec{L}|$ to $|\vec{L}|$, so there should be a continuous spread of heights on the detector. (That's what Stern was expecting.) In the Bohr–Sommerfeld model, there should be spots on the detector at two heights, with blank space in between. In the quantum mechanical model presented in Sections 7.1 and 7.4, the hydrogen atom has zero angular momentum in its ground state, so all the atoms should hit the detector in one tiny spot in the middle.

The results of the original Stern–Gerlach experiment are shown in Figure 7.11, a postcard sent by Gerlach to Bohr.

The Stern–Gerlach experiment clearly showed that L_z is quantized. Silver atoms are complicated, but in 1927 T. E. Phipps and J. B. Taylor reproduced the experiment with hydrogen atoms and got the same result. While this result ruled out a classical model and fitted the Bohr–Sommerfeld model, it doesn't fit our quantum picture of the hydrogen atom as presented so far. Somehow an electron in a state with zero orbital angular momentum was being deflected by the magnetic field.

We'll come back to Stern and Gerlach in a moment.

A New Quantum Number

A couple of years after Stern and Gerlach did their work, Wolfgang Pauli proposed a solution to an apparently unrelated theoretical problem.

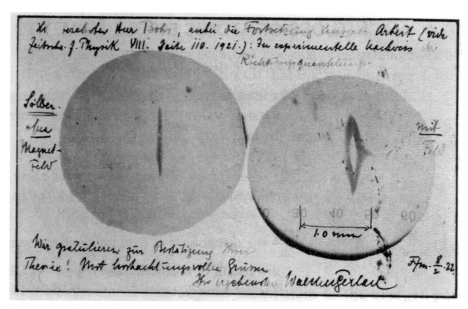

Figure 7.11 Results of the original Stern–Gerlach experiment. The left image is from a weaker field in which no splitting is visible. As the field was increased the atoms split into two separate beams, each with a different quantized value of L_z. (What we are calling the z axis corresponds to right and left in this image.) *Source: Alamy © 2005.*

Each energy eigenstate in hydrogen is characterized by the three quantum numbers n, l, and m_l, with $0 \leq l < n$ and $-l \leq m_l \leq l$. The energy is determined only by n, and from the ranges of l and m_l you can calculate that each energy level has degeneracy n^2. However, that degeneracy can be measured using the Zeeman effect (which we'll discuss in Section 7.7) and from the structure of the periodic table (Chapter 8), and both of those methods show that in fact each energy level n has degeneracy $2n^2$.

In 1924 Pauli suggested that each electron in an atom has a fourth quantum number that can only take on two possible values. Each combination of n, l, and m_l corresponds to two states, one for each value of this new quantum number, thus explaining the doubled degeneracy.

Pauli gave no physical explanation for this new electron property, which he called a "two-valuedness not describable classically." In 1925 two graduate students, George Uhlenbeck and Samuel Goudsmit, proposed that Pauli's new quantum number could be explained if the electron were spinning about its own axis.[8] Pauli's new quantum number would then be related to the spin angular momentum of the electron, which we now just call its "spin."

So every electron has two sources of angular momentum – orbit and spin – and the two are mathematically analogous in many ways:

---

8 Shortly after writing the paper, Uhlenbeck spoke to Lorentz, who criticized the idea, and Uhlenbeck told his teacher Paul Ehrenfest that they shouldn't publish. Ehrenfest replied that he had already submitted the paper on their behalf and it was due to be published in two weeks. Thus is history made.

The orbital angular momentum has magnitude $\sqrt{l(l+1)}\hbar$.	The spin angular momentum has magnitude $\sqrt{s(s+1)}\hbar$.
The quantum number m_l ranges in integer steps from $-l$ to l.	The quantum number m_s ranges in integer steps from $-s$ to s.
The orbital angular momentum has z component $m_l\hbar$.	The spin angular momentum has z component $m_s\hbar$.

All very parallel, right? But here's the key difference: l for an electron in an atom can be any integer between 0 and $n-1$. For an electron in any circumstances, s is *always* 1/2. Take a moment to plug $s = 1/2$ into the three facts above. You should conclude that . . .

- For every electron, in every circumstance, the magnitude of spin is $\sqrt{s(s+1)}\ \hbar = \sqrt{3/4}\ \hbar$.
- The quantum number m_s can have two values, $-1/2$ or $+1/2$.
- The z component of the spin is therefore $\pm(1/2)\hbar$.

A complete description of an electron in an atom therefore requires four numbers: the three quantum numbers that arise when solving Schrödinger's equation, and a completely separate quantum number m_s. The two states represented by m_s are often called "spin-up" ($m_s = 1/2$) and "spin-down" ($m_s = -1/2$).

Putting It Together: Spin and the Stern–Gerlach Experiment

Do you see how Pauli's 1924 proposal explains Stern and Gerlach's 1922 result? The ground state of hydrogen (or of silver) has no orbital angular momentum; what they were actually measuring was spin.

A hydrogen atom has two degenerate ground states. Both have $n = 1, l = 0$, and $m_l = 0$; they are distinguished by whether m_s equals $+1/2$ or $-1/2$. When you pass these atoms through a Stern–Gerlach apparatus you physically separate the spin-up atoms from the spin-down atoms.

Stern and Gerlach originally used silver, which has 47 electrons. A more complete analysis of their results requires some information about multielectron atoms that we will discuss in Chapter 8, but here's the short version: the first 46 of those electrons have orbital angular momenta that cancel perfectly, and spins that cancel perfectly, for a total $L_z = 0$. The outermost electron is in an $l = 0$ state, meaning it has no orbital angular momentum. So Stern and Gerlach measured the only angular momentum that *doesn't* cancel, the spin of the 47th electron, which has equal probabilities of pointing up or down. (Their choice of silver was a stroke of luck. The Bohr model predicted that silver in its ground state should have orbital angular momentum, which we now know it doesn't. We also now know that it does have a net spin, which was unknown at the time.)

So What is Spin?

It is a useful picture for many purposes to think of an electron as a little ball spinning on its own axis. Like the Bohr model, this classical image explains some of the key features of spin: in this case it associates with every electron both an angular momentum and a magnetic field that are not predictable from the electron's orbit. Also like the Bohr model, this picture is not actually accurate. (If it were literally true, the surface of the electron would be moving faster than the speed of light: see Problem 8.) An electron is "really" a quantum mechanical collection of probabilities for properties that include both orbital and spin angular momentum.[9]

Spin is unlike any other particle property because the spin angular momentum always has the same magnitude for a given particle, and in macroscopic objects the spins almost all cancel. When we talk about energy, momentum, or orbital angular momentum in quantum mechanics, they are all familiar from classical physics. Spin is an intrinsically quantum phenomenon that doesn't appear on a macroscopic scale. (Of course a spinning ball has angular momentum, but that's actually the *orbital* angular momentum of all the small parts of the ball about the axis of rotation.)

For our purposes here, spin has to be introduced as a new postulate to quantum mechanics. It cannot be deduced from Schrödinger's equation or derived from the wavefunction, so a complete description of an electron's state has to include its wavefunction *and* the orientation of its spin. In 1928 Paul Dirac developed a relativistic version of quantum mechanics, and showed that spin is a necessary consequence of that theory rather than a new postulate. Spin also arises naturally in quantum field theory, our most complete and accurate theory of how particles behave.

A Technical Note: The Gyromagnetic Ratio

A magnet in a non-uniform magnetic field experiences a force with $F_z = \mu_z(\partial B_z/\partial z)$, where $\vec{\mu}$ is the "magnetic moment" of the magnet. A charged particle orbiting with angular momentum $\vec{L}$ has magnetic moment $\vec{\mu} = q\vec{L}/2m$. Plugging in $S_z = (1/2)\hbar$ and $q = e$ you would expect μ_z for an electron to be $-e\hbar/(4m)$, but Stern and Gerlach found μ_z to be $-e\hbar/(2m)$, a quantity now known as the "Bohr magneton." That suggests that $\vec{\mu}$ for spin is $q\vec{L}/m$, twice the value that it has for orbital angular momentum. (This is another way in which an electron with spin is not like a classically spinning ball, for which the ratio would be the same as it is for orbital angular momentum.)

All of this is summarized in the equation $\vec{\mu} = gq\vec{L}/(2m)$, where g is called the "gyromagnetic ratio."[10] So experimentally $g = 1$ for orbital angular momentum and $g \approx 2$ for spin. How well does the theory line up with the measured g for spin?

- Non-relativistic quantum mechanics makes no prediction. Since spin is introduced as a new postulate, the theory can simply define g as its measured value.

- In Dirac's relativistic quantum mechanics, spin is a consequence of the fundamental axioms instead of being itself a new axiom. Dirac's equations predicted $g = 2$ for spin, suggesting that he was on the right track. But precise modern measurements show $g \approx 2.0023$, a small difference that Dirac's theory cannot account for.

- Quantum field theory correctly predicts the measured value to more than 10 decimal places.

---

9   At least it is in the orthodox interpretation. There is no interpretation of quantum mechanics in which the electron is a classical spinning ball.

10 Some sources define the "gyromagnetic ratio" as μ/L and call this the "g-factor."

7.5.2 Questions and Problems: Spin

Conceptual Questions and ConcepTests

1. What would Stern and Gerlach have found if they had measured the component of angular momentum in a direction perpendicular to the one they used (e.g. the x or y direction)?

2. List the values of all four quantum numbers for each of the two electron states in the ground state of helium.

3. What would have happened if Stern and Gerlach had used helium atoms instead of silver?

4. Protons have charge equal and opposite to electrons, and have the same spin ($s = 1/2$) as electrons. So why don't the protons in the nucleus contribute significantly to the Stern–Gerlach result? *Hint*: The Technical Note on p. 350 is relevant here.

For Deeper Discussion

5. To a very good approximation, the states $(n, l, m_l, m_s) = (2, 1, 1, 1/2)$ and $(2, 1, 1, -1/2)$ have the same energy, but in fact one does have a slightly higher energy than the other. Which one, and why?

Problems

6. An atom has all of the $n = 1$ and $n = 2$ electron states filled.

   (a) How many electrons does the atom have?

   (b) What would you see if you did a Stern–Gerlach experiment with this type of atom? Explain.

7. [*This is purely a classical electromagnetism problem. Its purpose is to demonstrate what the Stern–Gerlach results imply about atoms.*]

   Consider a current going around a loop of wire. When a magnetic field is applied, each electron in the current feels a magnetic force $q\vec{v} \times \vec{B}$. (Recall throughout this problem that $q < 0$ for electrons.) Figure 7.12 shows three scenarios. In each one you are viewing the current loop edge-on, so that the electrons are moving into the page on the right side and moving out of it on the left side.

   In the parts of the problem where you are asked to copy one of these drawings, you can just copy the line for the loop and arrows for the $\vec{B}$ field; you don't have to copy the labels.

   (a) Copy drawing 1, which shows a uniform magnetic field perpendicular to the loop. Draw force vectors showing the direction of the magnetic force on the moving electrons at the left and right edges of your drawing. Explain why the net force on the loop is zero.

   (b) Copy drawing 2, in which the loop has been tilted. Draw force vectors showing the direction of the magnetic force on the moving electrons at the right and left edges of your drawing.

      i. Explain why the net force on the current loop is still zero.

      ii. The net torque on this current loop is non-zero. From your point of view looking at the page, would the loop tend to rotate clockwise or counterclockwise?

   (c) Explain, using the idea of field lines, why a field with $(\partial B/\partial z) \neq 0$ cannot point in the z direction at every point in space.

   (d) Copy drawing 3, in which the loop is once again horizontal but the magnetic field is now non-uniform. Draw force vectors showing the

Current loops in magnetic fields

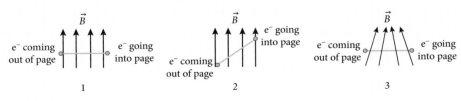

Figure 7.12

direction of the magnetic force on the moving electrons at the right and left edges of your drawing. In which direction is the net force on the loop?

8. In quantum field theory the electron is viewed as a point particle, with zero extent in space. High-energy collision experiments have put an upper limit on the electron size; if it isn't infinitely small, its radius is smaller than 10^{-16} m.

   (a) Assume that a uniform sphere of that size, and with the mass of an electron, is rotating with angular momentum $\sqrt{3/4}\,\hbar$. Calculate the fastest speed at which any point on the sphere would be moving. Find the ratio of that speed to the speed of light.

   (b) Remember that 10^{-16} m is an *upper* limit on the electron radius. What happens to your calculated speed if the radius is smaller than that?

   Remember our point here. An electron acts in many ways like a spinning ball – its spin gives it both angular momentum and a magnetic moment, as you would expect – but it is not actually a spinning ball.

7.6   Spin and the Problem of Measurement

According to the orthodox interpretation of quantum mechanics, measuring a particle instantly changes its state in a way that is not described by Schrödinger's equation:

$$\psi(x) = c_1\psi_1 + c_2\psi_2 \quad \rightarrow \textit{ you measure the energy and find } E = E_2 \rightarrow \quad \psi(x) = \psi_2.$$

Our treatment of quantities such as position, momentum, and energy is not complete without specifically discussing measurement. But that's a difficult discussion to have without getting bogged down in the algebra. Spin, on the other hand, can illustrate the effects of measurement without all the baggage of complex-valued multivariate functions.

So in this section we will revisit the Stern–Gerlach experiment and learn more about spin. But the lessons of this section apply equally well to more complicated properties such as position, momentum, and energy; our real topic is measurement in the quantum world.

Throughout this section we will describe measurements using the orthodox interpretation of quantum mechanics, in which a measurement collapses a particle's state from a mix of probabilities into one particular value. Other interpretations describe this process differently, but predict the same experimental results.

7.6.1   Explanation: Spin and the Problem of Measurement

We are considering only particles with $s = 1/2$, such as electrons. Such particles are called "spin-1/2 particles." When you measure the value of a spin-1/2 particle's spin in any particular direction you always find $\pm(1/2)\hbar$. We call a particle with $S_z = (1/2)\hbar$ "spin-up" and write its state as ↑, while a "spin-down" state has $S_z = -(1/2)\hbar$ and is written ↓. We similarly use → and ← for states of positive and negative S_x.

A composite system such as a silver atom in its ground state can act like a spin-1/2 particle because all of the spins cancel except the spin of one electron. That brings us back to the Stern–Gerlach experiment. Their apparatus divided an incoming stream of silver atoms into two outgoing streams, based (we now know) on the spins of the outermost electrons *in a particular direction.* If you choose to call that direction the *z* axis, then we can describe the apparatus like this:

- Input: a stream of atoms with randomly oriented spins
- Output: one stream of ↑ atoms and a different stream of ↓ atoms

Repeated Spin Measurements

We're going to start with simple scenarios and build toward more interesting, and perhaps surprising, predictions of quantum mechanics.

Active Reading Exercise: Two Identical Spin Measurements

A stream of silver atoms heading in the *y* direction passes through a Stern–Gerlach apparatus that separates it into two separate streams, one ↑ and one ↓. The initial stream was random, so the atoms split roughly 50/50 between the two outgoing streams.

Now the ↑ stream passes through a second apparatus identical to the first. It splits *that* stream based on positive and negative S_z-values.

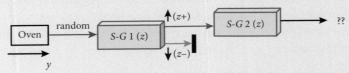

What comes out of the second device? (This is *not* a trick question.)

The answer is that the second device will output only one stream. All the atoms coming into that device have positive S_z, so they will come out together as they came in. A thousand subsequent measurements of that same property will all yield the same results.

Now we rotate the second apparatus 90°.

Active Reading Exercise: Two Orthogonal Spin Measurements

We begin once again with an apparatus that separates atoms into two streams based on S_z. But this time the ↑ stream goes into an apparatus that redirects atoms based on S_x, the *x* components of their spins.

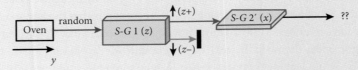

What comes out of the second device?

The answer to that one is that the outgoing stream is split 50/50. We have no knowledge of the *x* components of the spins, so any given atom is as likely to be positive as negative.

Things get more interesting when we now bring back our second *z*-splitter.

Active Reading Exercise: Three Alternating Spin Measurements

We begin once again by segregating atoms into two streams based on S_z. Then the ↑ stream goes into the S_x-splitter as before, and comes out 50/50.

But now we focus on one of the two streams coming from that second device. Every atom in this stream has now been sorted twice: first based on having a positive S_z, and then by having a positive S_x. We feed this stream into a third Stern–Gerlach apparatus; this one – like the first – sorts the stream on the basis of the z component of spin.

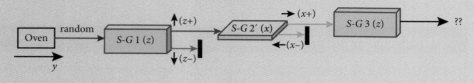

What comes out of the third device?

It's tempting to conclude that the outcoming streams from Apparatus 2′ have both definite S_z *and* definite S_x. But that's not how measurement works in quantum mechanics. A particle can be in state ↑ or state →, but not both. When we measured S_x we *changed* the state of the ↑ atoms into either → or ←. When you measure either of those output streams in the z direction you are equally likely to find ↑ or ↓. Apparatus 3 therefore splits the beam 50/50.

The Quantum Interpretation of Repeated Measurements

The result of our last experiment demonstrates a spin-related version of the uncertainty principle. You know that a precise measurement of position creates a state of high momentum uncertainty, and vice versa. Similarly, when you put an atom into a state of definite S_x it has maximum uncertainty in S_y and S_z.

Suppose, however, that your friend Al doesn't believe in all this quantum weirdness and uncertainty. "Each particle has a particular spin," Al claims. "Your measurements are just finding out what that spin is. There's nothing about the process of measurement that inherently has to change the state."

What might Al say when confronted with the z-x-z measurements described above? Didn't the measurement of S_x in the middle clearly erase the S_z state of the atom? Take a moment to think about Al's response.

One possible answer is that your Apparatus 2′ (the S_x-measuring device) jostled the atoms, shifting their S_z-values in unpredictable ways. Al contends that the act of measuring S_x doesn't inherently have to erase S_z information; it was just this particular S_x-measuring device that did something to mess them up. If we're clever enough we can design a better Apparatus 2′ that measures S_x while leaving S_z undisturbed. Might someone come up with such a device next month?

If quantum mechanics is correct, then the answer is no. If you observe the particle's spin in the x direction, it ceases to have a definite spin in the z direction, no matter how you do the measurement.

To summarize, here is how the orthodox interpretation of quantum mechanics describes an electron's spin:

- When you measure an electron's spin angular momentum in any particular direction, you will always get $\pm\hbar/2$.

- At any moment, each electron has one direction along which a spin measurement is guaranteed to give $\hbar/2$. (That fact doesn't follow from what we've told you so far, but it's true.) If you measure it in the opposite direction, you'll get $-\hbar/2$. Measurement along any other axis will have non-zero probabilities of getting positive or negative results.

- If you measure the spin along an axis perpendicular to its orientation you are equally likely to get a positive or negative result. (Problem 6 discusses measurements along non-perpendicular directions.)

- Once you've done that measurement, the electron's spin is now oriented along the direction you measured it. Any subsequent measurements in that direction will give the same result, and any subsequent measurement along its original orientation will be split 50/50.

Recombined Streams and Quantum Erasers

At this point you may find yourself sympathetic with Al's position. The orthodox interpretation of quantum mechanics says that there's something intrinsic about the act of measuring one quantity that can change a particle's state so that it no longer has a definite value of another quantity. Al, meanwhile, says that there's just something poorly designed about your measuring device. That seems like a more sensible, intuitive conclusion.

But it's not a tenable one. We want to discuss three experiments that conclusively rule out the classical view that particles have definite states of all quantities and that we simply discover them by measurement.

First Experiment: The One-Particle-at-a-Time Double Slit: Do you remember the first quantum experiment we described in the book? When you send particles through a double slit one at a time, they show an interference pattern on the back wall, which can only be explained by saying they went through both slits at once.

If you put a measuring device at one slit (call it Slit A), the interference pattern disappears. This means that some particles go through Slit B and then hit regions of the back wall that would have been empty when you didn't do the measurement. Somehow having a measuring device at Slit A allowed a particle at Slit B to do something it otherwise couldn't have done, *even though the measuring device showed that the particle never touched it*. That seems to suggest that the act of measuring, rather than any particular features of the measuring device, changed the particle's state from a superposition of two positions into a single position.

Second Experiment: Combining the Streams: The following Active Reading Exercise considers an experiment discussed by Richard Feynman in the *Feynman Lectures on Physics*.

Active Reading Exercise: Recombining the Streams

We use a Stern–Gerlach apparatus to produce a stream of atoms known to have positive S_z. That is then sorted into two streams, one with positive S_x and one negative. We know that if we measured either of those two streams for its S_z-values, the results would come out 50/50.

But instead we *recombine* the two streams before performing any further measurements. There is no way to look at an atom in the resulting stream and determine whether it came from the → stream or the ← one.

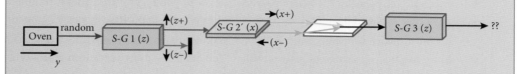

What will we find if we now measure S_z for this final stream of atoms?

Quantum mechanics predicts that we will find that all the atoms in this final stream have positive S_z-values! *It was not the details of Apparatus 2′ that disrupted S_z. Rather, it was the certain determination of S_x that forced S_z to be uncertain.* If we recombine the streams well enough that no possible measurement could determine how each atom came out of the second Stern–Gerlach apparatus, then we have erased the x measurement and the atoms still have definite z orientation.

To the best of our knowledge, as of this writing nobody has performed this experiment. (It turns out to be technically difficult to recombine the streams perfectly enough without disturbing the atoms.) However, many experiments measuring other quantities have been performed that rule out the explanation that our measuring devices somehow "jiggled" the particles and messed up our results.

Third Experiment: Quantum Erasers: Possibly the most dramatic demonstration of this idea is called a "quantum eraser" experiment. We're not going to explain such experiments here, but we want to say something about what they show.

The basic idea is that you create two particles in a specially prepared state and then send them in different directions. You send one of the particles (call it Particle A) through a double slit. You can show that a measurement done on Particle B, *arbitrarily far away*, can create or destroy the double-slit interference pattern created by Particle A.

A description of these experiments (including why they don't allow faster-than-light communication) is in the online section "The Quantum Eraser" at www.cambridge.org/felder-modernphysics/onlinesections

Unlike the Feynman thought experiment, quantum eraser experiments have been done, and seem to completely rule out any explanations in which the measuring device somehow disturbs the particles in a classical way.

Stepping Back: Schrödinger's Cat

Long before you learned how to solve differential equations or use the phrase "angular momentum" in a sentence, you may have heard stories about Schrödinger's cat. Schrödinger first described this thought experiment in 1935; it was based on an earlier idea from Einstein. Although the infamous cat makes a great conversation topic among philosophers of varying scientific backgrounds, you need to learn some quantum mechanics before you can fully appreciate Einstein's and Schrödinger's point.

And now you have, so now you can.

The experiment posits a cat and a radioactive substance in a steel chamber. Radioactive decay is a classic example of quantum mechanical randomness: after one half-life, any given atom has a 50/50 chance of having decayed. Schrödinger's experiment is jury-rigged so that after exactly one hour there is a 50% chance that an atom will have decayed, releasing a flask of poison that kills the cat.

Following the logic of quantum mechanics, the atom is now in the state

$$\psi_{\text{atom}} = \frac{1}{\sqrt{2}}(\text{not decayed}) + \frac{1}{\sqrt{2}}(\text{decayed}).$$

This in turn means the cat is in the state

$$\psi_{\text{cat}} = \frac{1}{\sqrt{2}}(\text{alive}) + \frac{1}{\sqrt{2}}(\text{dead}). \tag{7.23}$$

At that point you open the lid and look. There is a 0% chance that you will see the state described by Equation (7.23). Instead will see a living cat, blissfully unaware of the fate that it narrowly escaped. Or perhaps you will see the opposite. Either way, the cat is now in an eigenstate of alive-ness, which means the atom is now in an eigenstate of decayed-ness. The fact that the Schrödinger equation predicts a cat in a superposition of alive and dead, but we always see one or the other, is an example of what's called the "measurement problem."

Much has been written about the measurement problem over the past 100 years.[11] Schrödinger's cat is the most recognized icon of quantum mechanics because it boils the problem down to a simple image and a specific question: what state was the cat in before you looked? More generally, does measuring something reveal to you what it already was, or cause it to be in the state you measure? Here are three possible answers.

- *Measurement just reveals states.* This is the classical interpretation (which we associated with Al, above). The atom either decayed or it didn't, and the cat was either alive or it wasn't. Your act of opening the lid didn't change anything in the box; it just informed you of the state.

  The problem with that interpretation goes all the way back to the one-photon-at-a-time double-slit experiment. If we say that each photon goes through the left slit (but you don't know it), or else it goes through the right slit (but you don't know it), then we cannot explain the interference pattern on the back wall. We can only explain the dark patches by postulating that the wavefunction goes through both slits at once and interferes with itself. The state of the photon when it goes through the first wall is "maybe this slit, maybe that one."

- *Measurement changes states.* Before you open the box the atom is in a state of "maybe decayed, maybe not," which means the cat is in a state of "maybe alive, maybe dead." When you open the box, you *cause* the wavefunction to collapse into one state or the other. Now the cat is either alive, or dead:

$$\frac{1}{\sqrt{2}}(\text{alive}) + \frac{1}{\sqrt{2}}(\text{dead}) \quad \rightarrow \text{measurement} \rightarrow \quad 1(\text{alive}) + 0(\text{dead}).$$

---

11 For an excellent summary we recommend Norsen, Travis, *Foundations of Quantum Mechanics*, Springer, 2017. Norsen is an advocate of the "pilot-wave" interpretation, but he also provides an overview of the issues involved in reconciling the mathematics of quantum mechanics with the physics of reality, and describes the pros and cons of various interpretations.

This interpretation is what quantum mechanics seems to push us toward. But neither Einstein nor Schrödinger believed it; rather, they argued that the manifest absurdity of this conclusion proves that quantum mechanics is not a complete description of reality. As Einstein wrote to Schrödinger, "Nobody really doubts that the presence or absence of the cat is something independent of the act of observation."

If you do adopt the position that the cat is in an ambiguous superposition of states, you are led to the question of when the ambiguity ends. If your act of opening the box collapsed the cat's wavefunction, why couldn't the cat's observation of itself do the same thing earlier? That kind of question leads many people to a third position:

- *It depends on the size.* You will often hear that the most counterintuitive aspects of quantum mechanics – inherent randomness, uncertainty, and so on – apply to the microscopic world, while the macroscopic world is fundamentally classical. Sure, atoms may exist in these weird superpositions, but cats are always either alive or dead. Schrödinger's cat is designed to undermine this small/large distinction by explicitly linking the two worlds. If the atom is in a superposition, and the cat's state is tied to the atom's, then the cat must be both alive and dead at the same time.

  But many modern physicists say that the collapse of the wavefunction is not caused by observation but by macroscopic interaction. Schrödinger's atom goes into a superposition of decayed and not-decayed, but as soon as it causes a macroscopic flask to break or not break, the state has been measured and therefore collapses. This is how people often understand the orthodox interpretation. The word "macroscopic" is not precisely defined, leaving open the question of exactly when this collapse occurs, but many proponents of the orthodox interpretation of quantum mechanics would still say that a cat can't be simultaneously alive and dead.

Many working physicists dismiss such questions as philosophy rather than science. Different interpretations of the state of Schrödinger's cat have enormous bearing on how we understand the nature of reality, but they have no known effect on predictions of experiments.

Nonetheless, many physicists today – like Einstein and Schrödinger before them – do find these questions important, and there are many debates about the right way to interpret quantum mechanics. We offer a brief introduction to these debates in the online section "Interpretations of Quantum Mechanics," at www.cambridge.org/felder-modernphysics/onlinesections

However, everyone agrees that tiny Newtonian billiard-ball-like particles do not match the data. When we talk about the "quantum world" vs. the "classical world," what we really mean is the actual world vs. the one we used to believe in.

7.6.2 Questions and Problems: Spin and the Problem of Measurement

Conceptual Questions and ConcepTests

1. A million spin-up ($\uparrow$) silver atoms pass through an x-oriented Stern–Gerlach apparatus, and then all the ones that came out positive are passed through a z-oriented apparatus. Roughly how many come out as $\downarrow$?

   A. A million
   B. Half a million
   C. A quarter million
   D. None

2. A stream of silver atoms enters an x-oriented Stern–Gerlach apparatus and comes out as two streams,

one of which has "right" spins and the other of which has "left" spins. The "right" spins are then put through a z-oriented apparatus. Half come out up and half down. A classical analysis would say: "Even before the atoms entered the second apparatus, half of them were spin-up and the other half were spin-down; the device simply helped us figure out which was which."

(a) Replace the quoted sentence above with an analysis based on the orthodox interpretation of quantum mechanics.

(b) Now make an argument, based on the experiments, that the quantum analysis is correct and the classical analysis is not.

3. Consider a classically minded physicist looking at our Active Reading Exercise entitled "Three Alternating Spin Measurements" (p. 354).

(a) Explain how such a physicist could explain the final result of this experiment without recourse to the uncertainty principle.

(b) If someone were able to do the experiment "Recombining the Streams" (p. 355) and the results matched the quantum mechanical predictions we described, how would that refute such a classical explanation?

4. We compared the set of Stern–Gerlach experiments described in this section to the one-particle-at-a-time Young double-slit experiment, as two ways of illustrating core principles of quantum mechanics.

(a) What fundamental ideas about quantum mechanics can be derived from both experiments?

(b) What's different about the principles shown by the two experiments, and how they demonstrate them?

For Deeper Discussion

5. In the 1935 paper in which he first described his cat experiment, Schrödinger wrote: "There is a difference between a shaky or out-of-focus photograph and a snapshot of clouds and fog banks." Science, like a camera, views reality and creates a model of that reality. What point is Schrödinger making about quantum mechanics as a model of reality?

Problems

6. If an electron's spin is oriented in a particular direction and you measure it along an axis at an angle θ away from that direction, the probability of getting a positive result is $\cos^2(\theta/2)$.

(a) If an electron is spin-up (positive z orientation) and you measure it along the z axis, what does this formula predict about the probabilities?

(b) If an electron is spin-up (positive z orientation) and you measure it along the x axis, what does this formula predict about the probabilities?

(c) If an electron is spin-up (positive z orientation) and you measure it first along an axis at $45°$ between the x and z axes, and then along the x axis, what is the probability that your final measurement shows it to be positive along the x axis? *Hint*: You'll have to consider the two possible results of the first measurement separately.

7. [*This problem depends on Problem 6.*] You start with a ↑ particle and take N measurements of its spin: first at an angle $\pi/(2N)$ down from the z axis, then at twice that, and so on until the Nth measurement is on the x axis.

(a) What is the probability that all N of your measurements come out positive?

(b) Take the limit as $N \to \infty$ of your answer to Part (a).

(c) The limit you evaluated in Part (b) corresponded to an experiment. What experiment does it describe, and what result does your answer predict?

8. **Exploration: Spin Basis Vectors**

It's common to write the spin state of an electron (or any spin-1/2 particle) as a "column vector." In this context you can think of that as a fancy term for a set of numbers lined up vertically: $\binom{a}{b}$. The following column vectors represent electrons pointing in the up, down, right, and left directions. (We're not proving any of this, but it's true.)

$$\uparrow (z+) = \begin{pmatrix} 1 \\ 0 \end{pmatrix} \qquad \downarrow (z-) = \begin{pmatrix} 0 \\ 1 \end{pmatrix}$$

$$\rightarrow (x+) = \begin{pmatrix} 1/\sqrt{2} \\ 1/\sqrt{2} \end{pmatrix} \qquad \leftarrow (x-) = \begin{pmatrix} 1/\sqrt{2} \\ -1/\sqrt{2} \end{pmatrix}$$

If you measure an electron's spin along the x axis, you can predict the results by writing its column vector in the form

$$\begin{pmatrix} a \\ b \end{pmatrix} = c_1 \begin{pmatrix} 1/\sqrt{2} \\ 1/\sqrt{2} \end{pmatrix} + c_2 \begin{pmatrix} 1/\sqrt{2} \\ -1/\sqrt{2} \end{pmatrix}.$$

Then $|c_1|^2$ gives the probability of a positive result and $|c_2|^2$ gives the probability of a negative result. You can do the same thing to get probabilities for a z measurement.

(a) Suppose an electron is in the state $\uparrow$. Write its column vector and rewrite it in the form above to predict the results of a measurement in the x direction. What are c_1 and c_2?

(b) Using the c_1 and c_2 you found, what are the probabilities of measuring $\rightarrow$ and $\leftarrow$ for an $\uparrow$ electron? (The results should not be surprising.)

(c) Now suppose an electron is in the state $\begin{pmatrix} 1/2 \\ \sqrt{3}/2 \end{pmatrix}$. What are the probabilities of getting $\uparrow$ and $\downarrow$ if you measure it along the z axis? What are the probabilities of getting $\rightarrow$ and $\leftarrow$ if you instead measure it along the x axis? (This is instead of the z measurement, not after it.)

7.7 Splitting of the Spectral Lines

Theory and experiment agree that each energy level of the hydrogen atom has degeneracy $2n^2$. Careful measurements show, however, that this degeneracy is not exact: different states at the same value of n do not have precisely the same energy. That effect is directly measurable as a splitting of each spectral line into multiple, closely spaced lines.

7.7.1 Discovery Exercise: Splitting of the Spectral Lines

Here is the time-independent Schrödinger equation for a single electron orbiting a single proton:

$$-\frac{\hbar^2}{2m}\nabla^2\psi - \frac{e^2}{4\pi\epsilon_0 r}\psi = E\psi. \tag{7.24}$$

Unlike equations that represent more complicated systems (such as *two* electrons orbiting a nucleus), Equation (7.24) can be solved analytically. The math leads to the eigenstates and eigenvalues in Appendix G, and those formulas lead to predictions that hold up very well in the lab.

Very well... but not perfectly. Photons emitted when electrons drop down to other levels, and other experimental evidence, point to very small but consistent deviations from the energy levels $E_n = -(1/n^2)$ Ry.

Why? Can you think of an approximation we have made in this chapter, or a property of protons and electrons that we have not taken into account? Can you think of two or three?

7.7.2 Explanation: Splitting of the Spectral Lines

Recall from Chapter 4 that the spatial frequencies of the emission lines of hydrogen follow the Rydberg formula, $f = 1/\lambda = R_H \left(1/n_1^2 - 1/n_2^2\right)$. Each line represents a photon emitted when a hydrogen atom transitions from energy level n_2 to energy level n_1. Since the energy of each level mostly depends on n, the quantum numbers l and m_l don't appear in this formula.

In practice, however, there are a number of effects that cause states to have slightly different energies based on their values of l and m_l. You can think of this as splitting each major energy level (corresponding to a value of n) into multiple, closely spaced but not identical energies. As a result, each spectral line is split into multiple, closely spaced lines. In this section we briefly describe the most important of these mechanisms and their effects on the hydrogen energy levels.

The Zeeman Effect

The Zeeman effect refers to the changes in atomic spectra caused by an external magnetic field. To help understand that, here are a few facts that you may remember from classical electricity and magnetism:

- A rotating charge produces a magnetic field.
- Magnets tend to align with magnetic fields. Or, to put the same fact a different way, a magnet pointing along an external magnetic field has less energy than one pointing opposite to the external field. (The direction in which a magnet "points" is the direction of its "magnetic moment." For a current loop that's given by the right-hand rule.)
- Therefore a rotating charge in an external magnetic field experiences a torque, twisting it to align with the external field.

Active Reading Exercise: The Zeeman Effect

Suppose you put a gas of atoms in a magnetic field pointing in the positive z direction ($\vec{B} = B\hat{k}$).

1. Would the external magnetic field cause the energy of the eigenstates to depend on l, on m_l, or on both?

2. For whichever one(s) you chose, would the energy be higher when that quantum number was high or low?

Ignore spin for now.

You need to stack up a few insights to fully answer that Active Reading Exercise. Give yourself a pat on the back if you thought of all or most of the following:

- Both l and m_l are related to the angular momentum of the electron, which in turn correlates with its magnetic moment. That's why they matter in the presence of a magnetic field.
- Because the external magnetic field in this exercise points in the $+z$ direction, the z component of the electron's magnetic moment matters here, and its x and y components do not.
- Recall that l controls the angular momentum's magnitude, m_l its z component. If two electrons have different l-values but the same m_l, they have the same L_z but different L_x

and/or L_y. They therefore react the same way to the field $\vec{B} = B\hat{k}$. This means m_l matters and l does not.

- Because an electron is negatively charged, its magnetic moment points opposite to its angular momentum. So a positive m_l means its magnetic moment is pointing down, opposite to this particular external field. A negative m_l means the electron's magnetic field is pointing with the external field, which is the preferred (low-energy) orientation.

The bottom line is, in an external magnetic field pointing in the positive z direction, higher m_l leads to higher energy. For weak magnetic fields (neglecting spin), the Zeeman contribution to the energy of a state is

$$\Delta E = m_l \mu_B B \quad \text{where} \quad \mu_B = \frac{e\hbar}{2m_e} \qquad \text{The Zeeman effect for weak } B.$$

That constant μ_B is called the "Bohr magneton," as you may remember (but probably don't) from Section 7.5.

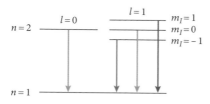

Figure 7.13 The Zeeman effect splits one spectral line into three.

Consider the transition $n = 2$ to $n = 1$. The ground state, $n = 1$, can only have $m_l = 0$, but the $n = 2$ state can have $m_l = -1, 0$, or 1. So there are now three transitions, $\Delta m_l = -1, 0$, and 1, with slightly different photon energies (Figure 7.13). This is observed as a splitting of the Lyman alpha spectral line into three closely spaced lines.

In a stronger magnetic field, the effects get more complicated and ΔE depends non-linearly on B.

Our discussion so far has ignored spin. If an atom transitions between states in which all its electron spins cancel, then you can ignore spin, as we have been doing. In that case the splitting caused by a magnetic field is called the "normal Zeeman effect." When the initial or final states have a net spin the calculations are more complicated, and the effect is called the "anomalous Zeeman effect."

The Zeeman effect can be used to measure magnetic fields. For example, by measuring the splitting of spectral lines in the Sun's atmosphere scientists can plot variations in magnetic field across the Sun's face (see Figure 7.14).

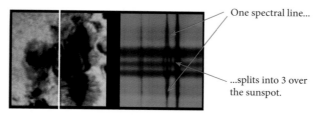

Figure 7.14 The image on the left is a picture of a sunspot (dark region). The entire green plot on the right is a spectrum taken along the white line through the middle of that picture. On the right, the vertical axis is position (the same as on the left), and the horizontal axis is frequency. In the region over the sunspot, one spectral line is split into three by Zeeman splitting. *Source: NSO/AURA/NSF © 2020.*

Fine Structure

In the absence of a magnetic field, m_l has no effect on an electron's energy. But for an electron with $l \neq 0$ there is a magnetic field in the atom even if you don't apply one externally. Consider things from the (non-inertial) frame of the electron. In that frame the electron is stationary and the proton orbits around it, creating a magnetic field. That field interacts with the electron's spin. The result is a change in the electron's energy depending on the orientation of its spin (Problem 12). This effect is called "spin–orbit coupling." The effect is similar to the Zeeman effect, and it also breaks the degeneracy of the energy levels, but it occurs without any external field.

Another correction that breaks this degeneracy is due to relativity, which says the kinetic energy of an electron is slightly less than $p^2/(2m)$. Taking this into account lowers all the energy levels, but the effect is strongest for low-l states, which spend more time on average near the nucleus (where they have higher speeds). This relativistic splitting of energy levels is, coincidentally, of the same order of magnitude as the spin–orbit coupling (Problem 13). Taken together, these two corrections are called the "fine structure" of an atom. The energy changes due to these effects are smaller than the overall energy of the states by roughly a factor of α^2, where α is a unitless number called the "fine structure constant":

$$\alpha = \frac{e^2}{4\pi \epsilon_0 \hbar c} = 0.00730 \qquad \text{The fine structure constant.}$$

Selection Rules

We showed above that, in the presence of an external magnetic field, the normal Zeeman effect splits the $n = 2$ to $n = 1$ transition into three separate transitions based on Δm_l. You might expect that the transition $n = 3$ to $n = 1$ would be split into five levels since $n = 3$ can have $-2 \leq m_l \leq 2$, but in fact the normal Zeeman effect still only shows three lines. That's the result of a "selection rule" that limits which electron transitions generally occur. The two most important selection rules for atomic transitions are

$$\Delta l = \pm 1$$
$$\Delta m_l = -1, 0, \text{ or } 1$$

Deriving these rules is complicated, but you can at least partially understand them with a simple argument. A photon is a spin-1 particle so it can have angular momentum states $m_s = -1, 0,$ or 1. By conservation of angular momentum the electron's z angular momentum should change by the opposite amount. Hence the rule for Δm_l. The argument for the Δl rule is similar (Question 10).

The term "selection rule" is somewhat misleading, and the term "forbidden transition" that is usually used for transitions that violate those rules is even worse. Such transitions do occur, either through collisions or through simultaneous emission of more than one photon. For example, the selection rules imply that the $n = 2, l = 0$ state can never decay to $n = 1$. In fact that state does decay, after a mean lifetime of about 0.1 s. That may sound fast, but the "allowed" atomic transitions typically occur after about a nanosecond, so the $n = 2, l = 0$ state lasts a lifetime by comparison. States that last extremely long (on atomic scales) before decaying are called "metastable."

Levels of Splitting

We've discussed some of the most important effects that alter Rydberg's simple formula for spectral lines. There are other, smaller effects as well. Here is a list designed to convey some sense of scale:

- As we noted above, fine structure (spin–orbit coupling and relativistic corrections) creates energy corrections about 20,000 times smaller than the bare energy levels ($\Delta E \sim \alpha^2 E$).

- The "Lamb shift" comes from treating the electric field of the proton quantum mechanically. This produces corrections about 140 times smaller than fine structure ($\Delta E \sim \alpha^3 E$).

- "Hyperfine" splitting comes from coupling between the spin magnetic moments of the electrons and of the protons. This produces corrections about 15 times smaller than the Lamb shift ($\Delta E \sim \alpha^2 (m_e/m_p)E$).

Spectroscopic measurements are more than accurate enough to detect all of these effects.

A Parting Note on the Fine Structure Constant and Other Unitless Constants

Quantum mechanics and our understanding of the structure of the hydrogen atom allow us to derive the formula we gave earlier for the fine structure constant:

$$\alpha = \frac{e^2}{4\pi \epsilon_0 \hbar c} \approx \frac{1}{137}.$$

It's a mix of constants, many of which have units, but the result is a unitless quantity. In other words "$\alpha \approx 1/137$" seems to be a fundamental fact of nature, a fact that affects atomic behavior. You could imagine a different universe that obeys all the same equations but with a slightly different value of α. Such a universe would not sustain life, or even matter, as we know it.

Fundamental unitless quantities such as the fine structure constant are called "free parameters" of a physics theory. Other examples include the number of types of quarks (there are 6 of them) and the ratio of the proton mass to the electron mass (roughly 2000).

One of the goals of many high-energy theoretical physicists is a theory that predicts the many unitless constants we measure in nature. Other theorists have suggested that those "constants" actually vary over space and time, in which case the values we see for things like α are just features of our part of the universe. At present these are open questions.

7.7.3 Questions and Problems: Splitting of the Spectral Lines

Conceptual Questions and ConcepTests

1. Ignoring spin, into how many spectral lines does the $n = 6$ to $n = 3$ transition split as a result of an external magnetic field? (Choose one and explain.)

   A. 2

   B. 3

   C. 4

   D. 5

   E. >5

2. In the presence of an external magnetic field in the positive z direction, the Lyman alpha line ($n = 2$ to $n = 1$) is split into three separate lines. For the highest-frequency of those three lines, specify all three quantum numbers for the $n = 2$ state in which the electron starts. How do you know?

3. An electron is in a state with $l = 1$, $m_l = 1$, and $m_s = 1/2$. Which of the following effects does spin–orbit coupling have on the state? Explain how you know. (Choose one.)

   A. It increases the state's energy.

   B. It decreases the state's energy.

   C. It doesn't change the state's energy.

   D. The answer depends on n.

4. True or False? If you want to predict a hydrogen atom's energy levels to about 1% accuracy in the absence of an external magnetic field, you only need to consider n.

5. Which spectral line is wider? (Choose one answer.) *Hint*: This section discussed why each spectral line can be higher or lower than you would naïvely expect, but didn't talk about their widths. What would cause a particular spectral line to have non-zero width?

    A. $n = 2, l = 0$ goes to $n = 1$.

    B. $n = 2, l = 1, m_l = 0$ goes to $n = 1$.

    C. These two lines are equally wide.

6. We listed the relative strengths of splitting caused by fine structure, Lamb shift, and hyperfine splitting. Why couldn't we include Zeeman splitting on this list?

7. For each of the following, specify on which of the quantum numbers n, l, m_l, and m_s the energy depends. Your answer to each one will be a list of quantum numbers, plus a short explanation. (We're leaving out spin–orbit coupling because it's more complicated.)

    (a) The Bohr model

    (b) The Zeeman effect, for a weak magnetic field (assume you define the direction of the field as the z direction)

    (c) Relativistic corrections

For Deeper Discussion

8. Why is the Zeeman effect formula $\Delta E = m_l \mu_B B$ only valid for weak magnetic fields?

9. From the inertial reference frame in which the nucleus is at rest and the electron is orbiting, why is there a spin–orbit coupling? *Hint*: Just as a rotating charge produces a magnetic dipole, a rotating magnet produces an electric dipole.

10. In the text we made an argument, based on conservation of angular momentum, that an electron transition that emits a photon should obey the selection rule $\Delta m_l = -1, 0$, or 1. Make a similar argument

for why $\Delta l = -1, 0$, or 1. (The fact that Δl can't equal 0 is harder to demonstrate.)

Problems

11. The magnitude of the Earth's magnetic field is roughly 45 microteslas.

    (a) Calculate the contribution of this magnetic field, based on the Zeeman effect, for a transition with $\Delta m_l = 1$.

    (b) How does that contribution compare to the fine structure correction?

12. **Exploration: Spin–Orbit Coupling**

    You're going to use a series of approximations to get an order-of-magnitude estimate for the effect of spin–orbit coupling. We'll consider the reference frame of the electron, in which the proton is a spinning ball of charge.[12] The energy shift associated with the electron's spin magnetic moment in the magnetic field of the proton is $-\vec{\mu}_s \cdot \vec{B}$, where μ_s is the magnetic moment associated with the electron's spin.

    (a) Find the z component of μ_s, the electron's magnetic moment. Express your answer in terms of the quantum number m_s. *Hint*: You might want to look at the discussion of gyromagnetic ratio in Section 7.5.

    (b) Treating the proton as a circular current loop in the xy plane, its magnetic field is $\vec{B} = \mu_0 I/(2r)\hat{k}$, where the current I equals e (the proton charge) over the period T. Write an expression for B_z in terms of the proton's radius and velocity (without the period T).

    (c) The proton's speed in this non-inertial frame equals the electron's speed in the usual (proton-centered) frame. Write that speed in terms of L_z, the electron's z angular momentum. Use that to write an expression for B_z that depends on L_z and r, but not on v.

    (d) Plug in the radius from the Bohr model for r. (This will introduce the quantum number n into your formula for B_z.)

---

12 Using a non-inertial reference frame as if it were inertial is one of our approximations. A careful treatment shows that this introduces a factor-of-2 error. We'll have other errors of roughly that same size.

(e) Expressing L_z in terms of the quantum number m_l, put everything you've done together to come up with an expression for the spin–orbit energy correction ΔE as a function of n, m_l, and m_s.

(f) Use your expression to numerically estimate the splitting between the two spin states of the hydrogen eigenstate $\psi_{2,1,1}$. Express your answer in eV, and also as a fraction of that state's uncorrected energy.

(g) At the beginning of the problem we mentioned that treating the electron frame as if it were inertial is an approximation. Identify at least two other approximations we made in this calculation.

13. **Exploration: The Relativistic Correction**

You're going to estimate the relativistic correction to the ground state energy of hydrogen.

(a) Using the relativistic expression for kinetic energy, $KE = \sqrt{p^2c^2 + m^2c^4} - mc^2$, find the first two non-zero terms of a Maclaurin series expansion for KE. You should find that the first term is the classical $KE = p^2/(2m)$. The second term is the first-order relativistic correction. If you're familiar with the binomial expansion, that can save you some algebra.

To estimate that correction to the electron's energy, we'll need to estimate p, and to do that we'll need to estimate the value of KE before the relativistic correction. In classical mechanics the "virial theorem" says that for an orbiting particle the time-average of kinetic energy is $-(1/2)$ times the time average of potential energy. In quantum mechanics the same formula holds for expectation values of an electron in orbit.

(b) Using the virial theorem, what's the expectation value of kinetic energy for the ground state of hydrogen, without a relativistic correction?

(c) Using that expectation value, estimate the magnitude of the electron's momentum. (You can use the non-relativistic expression to get a good order-of-magnitude estimate.) Then, use that result to estimate the relativistic correction to the electron's kinetic energy. You should find an answer roughly four orders of magnitude smaller than the ground state energy.

Chapter Summary

This chapter discusses the quantum mechanical picture of the hydrogen atom: the physical meaning of its quantum numbers, the derivation of its eigenstates from the Schrödinger equation, and a quantum mechanical property called "spin" that is not represented in the electron's wavefunction.

Much of this information is concisely laid out in Appendix G. You may want to cross-reference that appendix as you read through this summary.

Section 7.1 Quantum Numbers of the Hydrogen Atom

The energy eigenfunctions of a single electron in orbit around a nucleus are designated as ψ_{nlm_l}. For instance, $\psi_{3,2,0}$ is the eigenfunction with $n = 3$, $l = 2$, and $m_l = 0$. A quantum mechanical investigation of the hydrogen atom centers on these three "quantum numbers."

- The principal quantum number n, which can be any positive integer, is the primary determinant of the electron's energy.

- The angular momentum quantum number l, which can be any integer from 0 to $n - 1$, determines the *magnitude* of the electron's angular momentum.

- The magnetic quantum number m_l, which can be any integer from $-l$ to l, determines the z component of the electron's angular momentum.

Section 7.2 The Schrödinger Equation in Three Dimensions

This section and the next are not specifically about the hydrogen atom, but build physical and mathematical tools we need to analyze it. This section considers wavefunctions and Schrödinger's equation in three dimensions.

- For a continuous one-dimensional position, $|\psi|^2$ gives a position probability per unit length. For a continuous three-dimensional position, $|\psi|^2$ gives a position probability per unit volume.
- In one dimension the time-independent Schrödinger equation is an ordinary differential equation (ODE), while the time-dependent Schrödinger equation is a partial differential equation (PDE). In more than one dimension they are both PDEs.
- Therefore we solve them both using the technique of separation of variables, as presented in the previous chapter.
- As an example of this process, this section solves for the energy eigenstates and eigenvalues of a particle trapped in a 3D box. You can find the results in Appendix C.

Section 7.3 Math Interlude: Spherical Coordinates

Spherical coordinates are a three-dimensional alternative to the Cartesian coordinate system (x, y, z) (just as polar coordinates are an alternative to Cartesian in 2D). Spherical coordinates are easier to work with for systems that have spherical symmetry, such as the electric potential field generated by an atomic nucleus.

- r is the distance from a point to the origin; θ is the angle down from the z axis; ϕ is the angle around the z axis.
- Figure 7.7 illustrates spherical coordinates. Table 7.1 gives the conversion equations between spherical and Cartesian coordinates.

Section 7.4 Schrödinger's Equation and the Hydrogen Atom

The energy eigenstates described in Section 7.1 come from solving Schrödinger's equation for an electron orbiting a nucleus. Without going through all the details, this section outlines the process.

- Write Schrödinger's equation for the potential energy of an electron near an atomic nucleus, using spherical coordinates.
- Separate variables to turn that one PDE into two equations: an ODE for r, and a PDE for θ and ϕ.
- The result is a decaying oscillatory function $R(r)$, multiplied by the "spherical harmonic" function $Y_l^{m_l}(\theta, \phi)$.
- The probability of finding the electron at a given radius is proportional to $r^2 R^2(r)$. This probability approaches zero as $r \to 0$ and $r \to \infty$, reaching a maximum in the middle.

Section 7.5 Spin

An electron has an intrinsic angular momentum called "spin" with two eigenstates: one pointing up (along whichever direction you choose to measure it), and the other pointing down.

An electron's spin causes it to act something like a ball of charge spinning around its own axis, with angular momentum and a magnetic moment, but an electron is not literally a physically spinning charge.

- In a Stern–Gerlach experiment you measure the spin of an electron along any axis, and you always find its component along that axis to be $\pm(1/2)\hbar$.

- The spin state of an electron is described by a quantum number m_s, which can only be $1/2$ if it's pointing in the positive direction or $-1/2$ for the negative direction.

- In an atom, m_s becomes a fourth quantum number specifying the state of an electron, in addition to n, l, and m_l.

Section 7.6 Spin and the Problem of Measurement

This section describes variations of the Stern–Gerlach experiment that highlight the subtle and important issue of measurement in quantum mechanics.

- When you measure a property of a particle, you get a specific answer. In the orthodox interpretation of quantum mechanics, measurement causes the wavefunction to "collapse" discontinuously in a way not predicted by Schrödinger's equation. It goes from a superposition of eigenstates to a single eigenstate of the quantity you measured.

- There is an uncertainty relationship between different components of spin: an eigenfunction of S_x has maximum uncertainty for S_y and vice versa.

- So measuring S_x puts a particle into a state in which S_x is certain, and therefore S_y and S_z are completely uncertain.

- The section describes a thought experiment proposed by Richard Feynman, involving recombining streams. This experiment highlights the view that it is *the fact of measuring* S_x, not the details of the measurement apparatus, that erases all certainty in S_z.

- Schrödinger's famous "cat" thought experiment vividly illustrates the problem of measurement in quantum mechanics. If the microscopic world exists in a superposition of states until someone observes it, that implies that macroscopic objects do as well.

Section 7.7 Splitting of the Spectral Lines

The formula presented in Appendix G, that the energy of an electron in a hydrogen atom is $-(1/n^2)$ Ry, is very close but not exact. Deviations from that value can be measured as a widening or splitting of the lines in the hydrogen emission spectrum. This section examines some of the causes of this small variance.

- The "Zeeman effect" applies to an atom in an external magnetic field. If the electron's magnetic moment lines up with the external field, that lowers the energy; if it points the opposite way, that raises the energy.

- "Spin–orbit coupling" slightly reduces the energy if the magnetic moments of the electron's orbit and its spin are aligned with each other, and increases the energy if they are opposite to each other.

- Relativistic corrections to the electron energy slightly lower the total energy of all the hydrogen atom states, most strongly for states with lower l.

- Spin–orbit coupling and relativistic corrections are together referred to as the "fine structure" of an atom; their contributions are very small and roughly comparable to each other.

- "Selection rules" indicate which electron transitions are likely to happen, and which will happen much less frequently. In general, electron transitions will obey $\Delta l = \pm 1$ and $\Delta m_l = -1, 0,$ or 1.

- The section also briefly discusses two other corrections, the "Lamb shift" (due to quantization of the electric field) and "hyperfine splitting" (due to coupling between the magnetic moments of the electron and proton). Both effects are orders of magnitude smaller than fine structure effects.

8

Atoms

In 1869 Dmitri Mendeleev presented to the Russian Chemical Society a periodic table and a set of laws that laid the foundation for modern chemistry. He showed that the elements could be placed in an order, corresponding loosely but not perfectly to their atomic weights, and this order could be used to classify and predict their properties. He was even able to predict the existence and properties of elements (such as gallium and germanium) that had not yet been discovered.

It would be another half-century before the mechanism behind this regularity was understood. By the end of this chapter you will understand *why* the periodic table is laid out in the way it is and how that layout explains the basic properties of the elements, all based on fundamental principles of quantum mechanics.

But you won't get there in the way you might expect, by extending the solution of Schrödinger's equation beyond hydrogen. Just as classical equations offer no analytical solution for a multi-planet solar system, quantum mechanics offers no analytical solution for a multi-electron atom. What it does offer is numerical simulations, qualitative arguments, and mathematical approximations that make very successful predictions – and explain the periodic table.

8.1 The Pauli Exclusion Principle

In 1925, along with his introduction of the spin quantum number, Pauli proposed his "exclusion principle." This principle provides a vital link between quantum mechanics and chemistry.

8.1.1 Explanation: The Pauli Exclusion Principle

Before we dive into the structure of multielectron atoms we need to introduce a new postulate to our description of quantum mechanics. The "Pauli exclusion principle" says that no two electrons can be in the same quantum state as each other. For example, if you put 17 electrons in an infinite square well they can't all occupy the ground state together. They have to spread out into different states.

This principle is important because systems tend to settle into their lowest-energy states. If you look at 80 different hydrogen atoms at room temperature you will find virtually all of them in their ground states (meaning $n = 1$). Without the Pauli exclusion principle, if you looked at a single mercury atom you would find 80 electrons all in the $n = 1$ state. That is quantum mechanically forbidden, however, so in many-electron atoms the electrons fill in lots of different energy levels (starting at the bottom). The result is a wide range of properties for different types of atoms.

Before we talk about how this affects atoms, we have to clarify three things about this principle.

1. *What exactly do we mean by a quantum state?*

    In this case, a "quantum state" means a set of quantum numbers. Remember, however, that in addition to the three quantum numbers we discussed in Section 7.1 (n, l, and m_l), there is also the fourth quantum number m_s for spin. So a multielectron atom can have two electrons in the ground state ($n = 1, l = 0, m_l = 0$), one with spin up and one with spin down. The third electron, however, needs a different value of n, l, and/or m_l.

2. *Is this just about electrons, or do all particles obey this?*

    It's about more than just electrons, but not all particles obey it.

    Every particle can be classified as either a "fermion," meaning it obeys the Pauli exclusion principle, or a "boson," meaning it does not. Fermions include electrons, protons, and neutrons, while bosons include photons and some more obscure particles like "gluons" and "Higgs."

    Whether a particle is a fermion or a boson depends on its spin. Any particle with half-integer spin is a fermion. (Electrons, protons, and neutrons are all spin-1/2 particles.) Any particle with integer spin is a boson. (Photons have spin 1.)

    Composite particles can also be classified as fermions or bosons depending on their total spin. For example, a ^{4}He atom (2 electrons, 2 protons, 2 neutrons) is a boson.[1] So within a ^{4}He atom the electrons must be in different states, but you can put arbitrarily many of these atoms in the same state as each other.

3. *Where does this principle come from?*

    We described the Pauli exclusion principle as an additional postulate of quantum mechanics, but that's not entirely true. When you solve for the wavefunctions of multiparticle systems (a topic beyond the scope of this book) you can show that the wavefunctions of certain particles have properties that lead to the exclusion principle, but you can't tell which particles. (For more details, see Problem 6.) The result that being a fermion or boson depends on a particle's spin can be derived from quantum field theory (a topic way beyond the scope of this book). For our purposes we'll simply treat the exclusion principle as a new postulate.

---

[1] The "4" in ^{4}He refers to the total number of protons plus neutrons.

Example: Multiple Particles in an Infinite Square Well

Suppose you have three particles of mass m in an infinite square well of width L, and assume they exert no forces on each other. Find their total ground state energy (the lowest possible energy for the entire three-particle system) . . .

1. if they are spin-1 bosons.
2. if they are spin-1/2 fermions.

Answer: The energy of a particle in an infinite square well is $n^2\pi^2\hbar^2/(2mL^2)$ (Section 5.3).

1. The three bosons will all go into the $n=1$ state, so $E_T = 3\pi^2\hbar^2/(2mL^2)$.

2. Two of the fermions can go into the $n=1$ state, one with spin up and the other with spin down. The third one will go into the $n=2$ state, so the total energy is $E_T = (1+1+4)\pi^2\hbar^2/(2mL^2) = 3\pi^2\hbar^2/(mL^2)$.

(The boson spins are irrelevant since they can all go into the same state.)

We can extend the preceding Example by asking: what if there were three protons and three electrons in the same square well? The answer is, different particles don't exclude each other. So two of the protons, and two of the electrons, would settle into the ground state, leaving one of each in their respective $n=2$ states. You might also throw 100 bosons into the same well, and the exclusion principle would not have anything to say about them. What is forbidden is two or more of the *same type of fermion* in the *same state*.

8.1.2 Questions and Problems: The Pauli Exclusion Principle

Conceptual Questions and ConcepTests

1. How many electrons can occupy the $n=2$ state of an atom?

2. Your friend Simplicio tells you that the Pauli exclusion principle makes perfect sense because electrons are negatively charged and repel each other, so it would take infinite energy to jam two of them into precisely the same spot. How would you critique your friend's explanation?

3. Critique the following argument. Three electrons are evenly spaced in an infinite square well 10 light-years wide. If one of them on each side falls into the ground state the one in the middle can't be affected by it for at least five years, so there's nothing to prevent it from also falling in and violating the Pauli exclusion principle.

For Deeper Discussion

4. If an electron is in the state "$n=2, l=1, m_l=0$, spin-up" then the Pauli exclusion principle says that no other electron around the same atom can occupy that particular state. But suppose an electron is in the state "n might be 2, but then again it might be 3."

What does Mr. Pauli have to say about the other electrons in that atom?

5. *You should answer this after doing Problem 6.* If a million ^{4}He atoms are all in the same quantum state as each other, aren't their individual electrons all in the same quantum state as each other? Why does this not violate the Pauli exclusion principle?

Problems

6. **Exploration: Antisymmetric Wavefunctions and the Pauli Exclusion Principle**

The Pauli exclusion principle actually says something more general than "two electrons can't be in the same quantum state as each other." In this problem you'll explore the more general formulation and how it gives rise to that simpler conclusion. To keep things simple we'll ignore spin, so we'll pretend the wavefunction is the full description of the electron.

Consider an electron and a neutron in a 1D infinite square well. This system does not have one position, but two positions x_e and x_n. The state of the system is therefore a single wavefunction of two independent variables: $\psi(x_e, x_n)$. Any such function you write is a possible state of the system, provided it meets the boundary conditions and is normalized.

We'll call the one-particle energy eigenstates of the square well ψ_1, ψ_2, and so on. If the electron is in the ground state $\psi_1(x_e)$ and the neutron is in the first excited state $\psi_2(x_n)$, then you can write the wavefunction for the whole system as a product of the two: $\psi(x_e, x_n) = \psi_1(x_e)\psi_2(x_n)$.

Put two *electrons* in the same well, however, and your options are much more restricted. The Pauli exclusion principle says that the wavefunction $\psi(x_1, x_2)$ describing two electrons must be "antisymmetric," which means that if you switch x_1 and x_2 you get back the same function times a minus sign.

(a) Which of the following functions are antisymmetric? (Choose all that apply.)

A. $f(x_1, x_2) = x_1 x_2$

B. $f(x_1, x_2) = -x_1 x_2$

C. $f(x_1, x_2) = x_1 - x_2$

D. $f(x_1, x_2) = \sin(x_1)\, e^{x_2} - e^{x_1} \sin(x_2)$

If one electron is in the ground state and the other is in the first excited state, the Pauli exclusion principle prohibits $\psi(x_1, x_2) = \psi_1(x_1)\psi_2(x_2)$ because it isn't antisymmetric. But there is a clever workaround:

$$\psi(x_1, x_2) = A\left[\psi_1(x_1)\psi_2(x_2) - \psi_1(x_2)\psi_2(x_1)\right]. \tag{8.1}$$

The first part of this wavefunction says that one electron is in state ψ_1 and the other is in state ψ_2. The second part says the same thing. It switches which electron is in which state, but since every electron is just like every other one it still describes the same physical state. Mathematically, though, it satisfies the antisymmetry requirement of the Pauli exclusion principle.

(b) Why did we have to include an extra constant A in Equation (8.1)? *Hint:* The answer has nothing to do with antisymmetry or the Pauli exclusion principle.

(c) Try to write an antisymmetric wavefunction that describes the state "Both electrons are in the ground state." Explain why this isn't a physically meaningful wavefunction. (The answer is not the Pauli exclusion principle.)

What you should have shown in Part (c) is how the requirement of an antisymmetric wavefunction leads to the requirement that you can't put two electrons in the same state. (You can take spin into account by simply including the spin as part of your definitions of the states ψ_1 and ψ_2. Then you would conclude that there are two states with the same energy and you can put one electron in each, but you still can't put them both into any one state.)

In quantum mechanics you can show that any pair of identical particles must have either an antisymmetric or a symmetric wavefunction. In quantum field theory you can show that particles with half-integer spin have antisymmetric wavefunctions and particles with integer spin have symmetric wavefunctions.

(d) Photons have integer spin and are therefore bosons. Write a symmetric wavefunction expressing the statement "Two photons are in the ground state of my infinite square well."

7. **Exploration: Neutron Stars and Degeneracy Pressure**

[*This problem requires multiple integrals, a topic from multivariate calculus.*] When a star larger than our Sun ends its life it can collapse inward and form a "neutron star," a sphere of neutrons with a mass larger than our Sun and a radius of about six miles. The gravitational pull inward is enormous. Neutrons do not repel each other, so what keeps the neutron star from collapsing inward? The short answer is the Pauli exclusion principle. Now you'll work out the longer version of that answer.

Each neutron is trapped inside a spherical region of radius R. We will approximate that with an easier-to-work-with cube of side length $2R$. We solved *that* problem in Section 7.2 and (with $L = 2R$) we found energy levels $E = \left(n_x^2 + n_y^2 + n_z^2\right)\pi^2\hbar^2/(8mR^2)$.

(a) Write the energy in the star as a triple sum. Each of the sums will go up to an as-yet-unknown value n_{max}. Don't forget that because of spin each triplet (n_x, n_y, n_z) can be occupied by two neutrons.

(b) That triple sum isn't easy to evaluate, but you can approximate it as a triple integral: $\sum_{n=1}^{n_{max}} f(n) \approx \int_0^{n_{max}} f(n) dn$. (Normally when you approximate a sum with an integral like this you have to multiply out front by $1/\Delta x$ to compensate for the extra dx inside the integral. Do you see why you didn't have to do that this time?) Evaluate the resulting triple integral to find an expression for the energy. Your expression should still include n_{max}.

(c) To find n_{max}, write an equation relating the total number of neutrons N in the star to n_{max}. Then, since the mass of the neutron star is just N times the neutron mass m_n, turn it into an equation relating M to n_{max}.

(d) Plug the expression you found for n_{max} in Part (c) into the expression you found for energy in Part (b).

(e) The expression you just found for the energy is only the kinetic energy. (The particle-in-a-box

problem assumes no potential energy.) Your result should show that, the larger the star is, the smaller its kinetic energy. Explain why that makes sense physically.

(f) In addition to this kinetic energy, though, the neutron star also has gravitational potential energy $-3GM^2/(5R)$. Add the two energies to write the total energy.

(g) At equilibrium the neutron star will have whatever radius minimizes its total energy. Find the value of R that minimizes the total energy you wrote in Part (f).

(h) Plugging in a total mass of 1.5 times the mass of our Sun, find the equilibrium radius of a typical neutron star.

(i) What would that equilibrium radius be if neutrons were bosons? *Hint*: This does require going back through your calculation, but this time it should be much easier and won't require any integrals.

(j) In words, since the neutrons exert no forces on each other aside from gravity, why doesn't the neutron star collapse in on itself?

8.2 Energy Levels and Atomic States

The ground state of hydrogen is very simple: the one electron is in the state $n = 1, l = 0$. Atoms with more electrons have to fill in higher energy levels, which tends to make those new electrons less tightly bound to the nucleus. But those atoms also have more protons, which bind the electrons more tightly. As a result of these two competing effects, the energy required to pull an electron out of an atom oscillates up and down as the atomic number increases. That alternating pattern has profound consequences for how atoms interact with each other.

8.2.1 Discovery Exercise: Energy Levels and Atomic States

A lithium atom has three protons and three electrons. The first two electrons are in the state $n = 1, l = 0$, while the third one has $n = 2, l = 0$.

1. In a hydrogen-like atom (only one electron), the energy of an eigenstate is $-(13.6 \text{ eV})Z^2/n^2$, where Z is the number of protons in the nucleus. If the $n = 2$ electron were the only electron in the lithium atom, how much energy would it have?

2. In the actual lithium atom, the $n = 2$ electron feels forces from the other electrons as well as from the nucleus. Those $n = 1$ electrons act like a spherical cloud of charge at a smaller radius than the $n = 2$ electron (Figure 8.1). Taking into account the force from those inner electrons, would you expect the actual energy of the $n = 2$ electron to be higher (less negative), or lower (more negative), than your answer to Part 1? Why?

See Check Yourself #11 at www.cambridge.org/felder-modernphysics/checkyourself

Figure 8.1 Lithium energy levels.

8.2.2 Explanation: Energy Levels and Atomic States

The atoms of each element are distinguished by their "atomic number" Z, the number of protons in the nucleus. An atom with one proton in the nucleus ($Z = 1$) is called hydrogen, one with $Z = 2$ is called helium, and so on.

If you change the number of neutrons in the nucleus you have a different "isotope": still the same element, but a different atomic mass. If you change the number of electrons orbiting the nucleus you have an "ion": still the same element, but now electrically charged. Here we will primarily consider neutral atoms (so Z will be the number of electrons as well as the number of protons), and we will not pay much attention to neutrons at all. Our interest will be how the electrons fill the available states as you go to higher Z-values.

Spectroscopic Notation

For many purposes the energy and total angular momentum of a state are more important than its orientation, so an atomic state is often designated by its values of n and l. For historical reasons, however, the value of l is specified by a letter rather than a number. The first four values are s, p, d, and f.[2] From there they proceed alphabetically, except they skip j because it looks too much like i (Table 8.1).

Table 8.1 Labels for values of l

l-value	0	1	2	3	4	5	6	7	8	and so on . . .
letter	s	p	d	f	g	h	i	k	l	and so on . . .

For example, $2p$ refers to $n = 2$, $l = 1$. There are six states in $2p$, corresponding to the three allowed values of m_l and the two spin orientations.

Each value of n is called a "shell," while each combination of n and l is called a "subshell." For example, the $n = 2$ shell consists of the subshells $2s$ and $2p$. The letter is often given a superscript to indicate how many electrons occupy that subshell. For example, the ground state of nitrogen ($Z = 7$) is written $1s^2 2s^2 2p^3$ to indicate that it has two electrons in the state $n = 1, l = 0$, two electrons in the state $n = 2, l = 0$, and three electrons in the state $n = 2, l = 1$.

---

2 These letters come from properties of spectral line series and stand for "sharp," "principal," "diffuse," and "fundamental." Don't ask.

The shells themselves also have associated letters, which start with K for $n = 1$ and go up alphabetically. We will not use these often, although we will refer to them when we discuss X-ray spectroscopy. We will make frequent use of notation like $1s$ and $3p$, however, so it's worth getting used to. You can keep Appendix H handy for this purpose.

Electron Screening

The simplest approximation for a multielectron atom might be to ignore forces between electrons and treat each electron as occupying a state in a hydrogen-like atom with nuclear charge $+Ze$. That would be a terrible approximation, though, because for each electron the collective force from the other $Z - 1$ electrons is comparable to the force from the nucleus.

But we can get a lot of insight into multielectron atoms with a slightly more complicated model. Consider lithium ($Z = 3$), which in its ground state has two electrons in the $1s$ subshell and one in the $2s$ subshell. Rather than trying to solve Schrödinger's equation for all the electrons (which we can't), let's just treat the two inner electrons as uniform spherical shells of negative charge at the Bohr radius a_0.

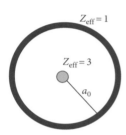

As you may recall from classical electricity and magnetism, a uniform shell of charge acts like a point at its center, as far as any charges outside it are concerned. So outside the shell ($r > a_0$) the lithium atom is just like a hydrogen atom; the three protons and two inner electrons combine to create an "effective atomic number" $Z_{eff} = 1$. What about $r < a_0$? A uniform spherical shell exerts no force on charges inside it, so for $r < a_0$ the effective potential is that of the lithium nucleus with its three positive charges (Figure 8.2).

So what does this mean for the third electron in a lithium atom? It is on average at a radius higher than a_0, so to first approximation it feels a net charge of $+e$ coming from the nucleus. We say that the inner electrons "screen" the nuclear charge.

Figure 8.2 The effective nuclear charge felt by the outer electron in lithium ($Z = 3$). The thick shell represents the two inner electrons. Inside that shell the field is just that of the nucleus. Outside the shell the first two electrons screen the nuclear charge so the electric field is that of a single positive charge.

According to this model, the outer electron in lithium should be just like a $2s$ electron in hydrogen. We noted in Chapter 4 that for a hydrogen-like atom (one electron) the energy of a level-n electron is $-(13.6 \text{ eV})Z^2/n^2$. So the model predicts that the "ionization energy" of lithium – that is, the amount of energy required to remove its outermost electron – should be $(13.6 \text{ eV})/2^2 = 3.4 \text{ eV}$.

Active Reading Exercise: Electron Screening

Jot down your answers to the following questions before reading on. (If you're stuck on one of them you should still try to answer the other; they are independent of each other.)

1. We said above that the electron screening model predicts an ionization energy of 3.4 eV for lithium, but that model isn't exact. Would you expect the actual value to be higher or lower than that? Why?

2. What does the electron screening model predict for the effective charge felt by the outer electron of beryllium ($Z = 4$)?

The answer to the first question above is "higher than 3.4 eV," and here's why. We said above that the third electron is outside the shell of the inner electrons. But, really, the third electron is a wavefunction of position probabilities. It has a high probability of being outside the shell ($r > a_0$ so $Z_{eff} = 1$), and a low – but not zero! – probability of being inside the shell (where $Z_{eff} = 3$). That's why that electron is more tightly bound than our simple screening model predicts. Experimentally the ionization energy is 5.4 eV. Using $E = -(13.6 \text{ eV})Z^2/n^2$, this represents an effective Z of 1.3.

For the beryllium question you need two insights. The simple one is that the nucleus has one more proton than lithium. In addition, however, the screening is more complicated. The inner electrons shield the two outer ones about as well as they do in lithium, but how much do the outer ones screen each other from the nucleus? As a rough estimate you can say that each electron feels the other one screening it about half the time. Then you would estimate that each outer electron feels 4 protons screened by 2.5 electrons, for a Z_{eff} of 1.5. Experimentally the ionization energy is 9.3 eV, which corresponds to $Z_{eff} = 1.7$.

Example: Using a Simple Model to Predict Ionization Energy

The ground state of nitrogen ($Z = 7$) is $1s^2 2s^2 2p^3$. Use the simple model discussed above to predict the Z_{eff} felt by one of the outermost electrons, and, based on that, the ionization energy of a nitrogen atom.

Answer: The nucleus has 7 protons. The 4 inner electrons provide full shielding, and the other two electrons in the $2p$ state provide half-shielding:

$$Z_{eff} = 7 - 4 - 2(1/2) = 2$$

$$E = \frac{13.6 \times Z_{eff}^2}{n^2} = \frac{13.6 \times 2^2}{2^2} = 13.6 \text{ eV}.$$

The experimental value is 14.5 eV.

Predicting the exact energies (or equivalently Z_{eff}) for electrons in different atoms requires large numerical calculations, taking into account the radial distributions of both the outer and inner electrons. (For instance, we've been treating the inner electrons as perfect spherical shells, but of course their wavefunctions spread over a range of radii.) Experimental ionization energies can differ quite significantly from the calculations we're doing here,[3] especially at high Z. But these simple calculations correctly predict which atoms are easy to ionize and which are very difficult, and that distinction in turn predicts a great deal about atomic bonding in chemistry.

Pause to Remember Which Way is Up

As we describe energy levels and effective Z, it's easy to get confused about what it means for each of those to be high or low. Here's a quick reminder.

- Every electron in an atom has negative total energy because $E = 0$ is defined as the energy required for the electron to escape the atom. For instance, if an electron has $E = -4$ eV, that means it would take 4 eV of energy to pull the electron out of the atom.

---

3 Perhaps the greatest inaccuracy in our model is that as Z gets large there are more inner electrons, so the small part of their wavefunctions outside the outer electrons becomes more important. Numerical calculations and experimental results agree that the effective Z of the outer electrons grows roughly proportionally to the n-value of the outer shell.

- The lower the energy (i.e. the more negative it is), the more tightly is the electron bound.

- The ionization energy is the absolute value of the energy of the outermost electron. So if the outermost electron has energy $-4\,\text{eV}$, then the ionization energy of the atom is $+4\,\text{eV}$. This is where signs can get particularly confusing: if that electron is kicked into a *higher* shell, that will *raise* its energy (less negative), thus *lowering* the ionization energy of the atom (less positive).

- The higher Z is, the more tightly bound the electrons are, and the lower (more negative) their energies. Because inner electrons partially screen the nucleus, the "effective Z" felt by most electrons is lower than the actual Z of the atom.

Which Energy Levels Fill Up First?

For an atom in its ground (or unexcited) state, the electrons fill the lowest energy levels available. As you may recall, in a hydrogen atom the energy is solely determined by the quantum number n. If all atoms were like hydrogen the electrons would occupy the lowest shells (lowest n-values), and within each shell would fill in the states in any order. (That's not exactly true, but it's a pretty good approximation.) In multielectron atoms, however, the value of l has a significant effect.

To see why that is, let's go back to lithium. It has two electrons in the $1s$ subshell. Why does the third electron go into $2s$ and not $2p$? The answer has to do with their radial probability distributions. We've plotted the radial probability distributions for those subshells in hydrogen, based on the formulas in Appendix G. The distributions for multielectron atoms can be calculated numerically and are qualitatively similar to the hydrogen distributions shown in Figure 8.3.

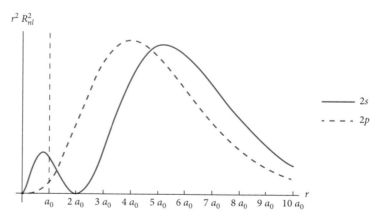

Figure 8.3 Radial probability distributions for the $2s$ and $2p$ subshells of hydrogen. The vertical line at a_0 shows the radius at which the $1s$ subshell peaks.

Active Reading Exercise: Subshell Energies

A lithium atom has its $n = 1$ shell filled by the first two electrons. The third electron will be either in subshell $2s$ or in $2p$.

1. Based on Figure 8.3, which of these two subshells has a higher probability of the electron being found at $r < a_0$?

2. Explain how your answer to Part 1 tells us which subshell has lower energy (and will therefore be occupied first). *Hint*: Your answer should include some form of the verb "to screen."

What Figure 8.3 shows is that the $2p$ wavefunctions lie almost entirely beyond a_0, so the $1s$ electrons effectively screen the nuclear charge felt by the $2p$ electrons. If the third electron is in $2p$ then it feels an effective Z of roughly 1. The $2s$ wavefunction, however, has a much higher (though still small) probability of being inside the radius of the $1s$ electrons. The $2s$ electron therefore experiences less screening and so feels a higher Z_{eff}. Remember that high Z_{eff} means tighter binding and therefore lower energy, so the state $1s^2 2s^1$ is the ground state and $1s^2 2p^1$ is an excited state.

This conclusion is not specific to $2s$ and $2p$. In general, states with lower l are more likely to be found at small radii than states with higher l. So they experience less screening and have lower energy than the higher-l states.

The result is that, within a shell, the subshells fill up in order of increasing l. As you keep adding more electrons (and thus more screening) the effect of l becomes more pronounced, so eventually you don't go in order by n. The first time this happens is between argon (Ar, $Z = 18$) and potassium (K, $Z = 19$). The ground state of argon has the $n = 1$ and 2 shells filled, plus the subshells $3s$ and $3p$. You might expect that potassium would add an electron in $3d$, but instead it adds one in $4s$. The additional screening felt by the $3d$ electrons is enough to raise their energy above that of the $4s$ electrons. The next element, calcium, continues this by filling up the $4s$ subshell, and only then does $3d$ begin to fill.

So, when n and l both have effects on the energy level of a state, how can you tell the order in which the subshells will fill? Fortunately there is a handy rule that works most of the time.[4]

The $n + l$ Rule for Filling Subshells (which works for most but not all elements)

Given two states ψ_{nl}, the subshell with the lower sum $n + l$ will fill up first.

In the case of a tie, the lower-n state fills first.

The states $3p$ and $4s$ both have $n + l = 4$, so the $3p$ state (lower n) fills up first. But $3d$ has $n + l = 5$, so that fills up after $4s$.

Hund's Rule

We have seen how the quantum numbers n and l affect the energy of an electron, which explains why certain shells and subshells fill before others. *Within* a particular subshell, the electrons still end up in the states that minimize the energy level. This order is based on a subtler effect called "Hund's rule."

---

4 In addition to the "$n + l$ rule" this is sometimes called the "Madelung rule," the "Janet rule," the "Klechkowsky rule," the "aufbau approximation," the "diagonal rule," and – we are not making this up – the "Uncle Wiggly path." We'll stick with the "$n + l$ rule."

Since we don't cover multi-particle quantum mechanics in this book, we cannot fully justify the following rule. It can be shown, however, that two nearby electrons whose spins are identical will be, on average, farther apart from each other than if they have opposite spins. The result is that electrons with identical spins shield each other from the nucleus less than electrons with different spins, so same-spin is a lower energy state than opposite-spin.

So, imagine a subshell that can contain up to six electrons, as illustrated in Figure 8.4. If it contains only one electron, Hund's rule is irrelevant. (That electron will either be spin-up or spin-down; it doesn't matter.) If it contains all six, Hund's rule is again irrelevant. (There will be three with each spin). But if there are two or three electrons, Hund's rule says they will all have the same spin. Only after the subshell is half-filled will the opposite spin states be occupied.

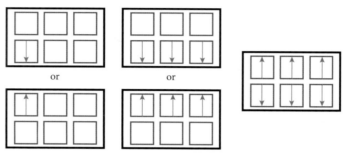

Figure 8.4 Three examples of how the lowest energy levels of a p subshell fill according to Hund's rule. With one electron it can have spin up or down. With three, they can be up or down but they will all be in the same direction. With six, all of the states are filled.

Stepping Back

When we studied hydrogen, we solved Schrödinger's equation to find the energy eigenstates and their energies, and everything else followed from those solutions. We can't do that for multielectron atoms. Instead we can give qualitative arguments that justify some general rules for the energies of different states. Those rules have been confirmed by numerical solutions to Schrödinger's equation, and more importantly by experiment. In the rest of this chapter we'll derive some of their consequences for the properties of elements.

Here are the general rules.

- The Pauli exclusion principle says no two electrons can occupy the same quantum state. That means that, in the ground state of a multielectron atom, the Z electrons fill up the Z lowest-energy states.
- Because each electron can have spin up or spin down, two electrons can be in each combination of n, l, and m_l. That in turn means each subshell consists of $2(2l+1)$ states.
- Those subshells fill up in order of $n + l$, with lower n going first for ties.
- Within a subshell the electrons tend to fill in with as many spins as possible pointing in the same direction.

There are some exceptions to the $n + l$ rule scattered throughout the periodic table. (You can see them in Appendix H. The first is chromium, which "should" have a full $4s$ subshell and four electrons in the $3d$ subshell, but instead has one in $4s$ and five in $3d$.) Ignoring those exceptions and using this rule, however, we will be able to explain most of the basic properties of the periodic table.

8.2.3 Questions and Problems: Energy Levels and Atomic States

Conceptual Questions and ConcepTests

1. Is there a $2d$ subshell? If so, how many electrons can be in that subshell simultaneously? If not, how do you know it doesn't exist?

2. Explain in your own words why the $4s$ subshell fills before the $3d$ one. (Don't say "because the sum $n+l$ is lower." Say something about physics.)

3. Figure 8.3 shows that the $2s$ state is more likely to be at small radii than the $2p$ state, but it shows that $2s$ is also more likely to be at large radii. It's simply more spread out. Why does being spread out with high probabilities at both low and high r cause the $2s$ state to feel a higher effective Z than the more concentrated $2p$ state?

4. Nitrogen has filled $1s$ and $2s$ subshells, and three electrons in the $2p$ subshell. What does Hund's rule tell us about the values of m_l for those three electrons when a nitrogen atom is in its ground state?

5. If you give a cadmium atom just enough energy to push one electron into a higher energy state, what subshell will it go into?

6. In a hydrogen atom, the energy (to a very good approximation) only depends on n, and is independent of l and m_l. In multielectron atoms, the energy of an electron also depends on l. Why does it still (mostly) not depend on m_l?

7. An "unpaired" electron spin is one that isn't canceled out by another electron. For example, if four electrons had spins up–down–up–up we would call that two unpaired electrons. How many unpaired electrons does carbon ($Z = 6$) have in its ground state? (Choose one.)

   A. 0

   B. 1

   C. 2

   D. 4

   E. 6

8. In helium, in its ground state, the total spin angular momentum is zero. If you excite it to a higher energy state, will that still be true? Briefly explain.

   A. It will definitely still be true.

   B. It will definitely no longer be true.

   C. It depends on the excited state.

For Deeper Discussion

9. We've based our discussion of electron screening on the hydrogen wavefunctions and their associated radial probability distributions. How do you think the radial probability distribution of the $2s$ electron in lithium might differ from a $2s$ electron in hydrogen, and why?

10. We've said that an inner electron screens the outer ones, pushing outward enough to cancel the pull of one nuclear proton. But the outer electrons do *not* exert a net inward push on the inner ones. How can that be true without violating Newton's third law? *Hint:* You can answer this with entirely classical arguments. Imagine two concentric spherical shells, one inside the other, and think about the forces they exert on each other.

Problems

11. We've presented a model in which the outermost electron in an atom feels fully screened from the nucleus by all the inner electrons and roughly half-screened by each of the other electrons in the same shell. While this model becomes numerically inaccurate at high Z, it nonetheless correctly reproduces many of the patterns in ionization energies seen experimentally.

    (a) As an example, let's consider magnesium ($Z = 12$), with outer shell $n = 3$.

       i. How many electrons are in that shell?

       ii. Considering one of those electrons to be screened from one proton by each of the inner electrons and from half a proton by each of the other $n = 3$ electrons, what effective Z does that electron feel?

iii. Use that Z_{eff} to estimate the ionization energy for magnesium.

(b) Using that same model, estimate the ionization energy of the outermost electron in elements 1 through 11.

(c) Of all of those, which is the easiest to ionize and which is the hardest?

(d) Why does the ionization energy you calculated go up and down rather than simply increasing or decreasing as Z increases?

(e) What column on the periodic table should contain elements that are easiest to ionize? Briefly explain why, based on your work in this problem.

12. Write the ground state of silicon in spectroscopic notation.

13. List all of the subshells in the order they fill up, through $7p$ (the highest subshell in the ground state of any known element as of this writing).

14. Which is easier to ionize, helium or lithium? Explain.

15. Which is easier to ionize, sodium or magnesium? Explain.

16. Rubidium has 37 protons.

(a) What is the outermost occupied subshell?

(b) Estimate the ionization energy of the outermost electron in rubidium, assuming that it is screened by all of the electrons in other shells. (There are no other electrons in this same subshell, so you don't have to worry about that.)

(c) The actual ionization energy is 4.2 eV. Was your estimate high or low? Explain why it makes sense that it should be.

(d) In the Explanation (Section 8.2.2) we explained why this rough model didn't quite give the right answer for lithium, but it was pretty close. You should have found that it gives a much less accurate answer for

rubidium. Why should the estimate be less accurate for rubidium than for lithium?

17. **Screening in Lithium**

The effective screening felt by an outer electron depends on the radial probability distributions of both the outer and the inner electrons. In this problem you will numerically simulate the screening of the outer electron in lithium in both the $2s$ and $2p$ states. For simplicity we're going to take $a_0 = 1$ throughout this problem.

(a) Write the radial probability distributions $P(r) = r^2 R_{nl}^2$ for the $2s$ state of hydrogen. You can find the energy eigenstates in Appendix G.

(b) Generate a random electron from that probability distribution. To do that you can generate a random number r from 0 to 15, and another random number y from 0 to 1. (Make sure to generate random real numbers, not just integers.) If $y < P(r)$ then r is your randomly generated radius. If it isn't then generate a new pair of numbers r and y until you get one that works. (Think about why this gives you random numbers that obey this probability distribution. The upper limit 15 is chosen because the probability distribution is negligible beyond $r = 15a_0$.)

(c) Use that procedure to generate 10,000 random radii drawn from the $2s$ probability distribution. For each one, let $Z = 1$ if $r > 1$ and $Z = 3$ if $r < 1$. Add all of the resulting Z-values and divide by 10,000 to get their average, Z_{eff}. (If you don't get a final average Z_{eff} between 1 and 3 then you did something wrong.)

(d) Repeat the entire procedure using the $2p$ distribution instead of $2s$.

(e) For which one did you find the larger value of Z_{eff}? Why does this make sense?

(f) The Z_{eff} you found shouldn't equal the actual one we said has been measured for lithium. Give at least two reasons why not.

8.3 The Periodic Table

In the previous section we described the energy levels occupied by electrons in multielectron atoms. In this section we talk about how those energy levels determine the behavior of different elements, and how that behavior leads to the structure of the periodic table.

8.3.1 Discovery Exercise: The Periodic Table

Section 8.2 presented a simple model that is not numerically accurate but good for understanding qualitative trends. In this model the outermost electron in an atom feels the electric pull of the nucleus "screened" by other electrons. The electrons in its own subshell screen half a proton each, and the electrons in lower subshells screen a full proton each. The resulting "effective" charge goes into the formula for the energy required to liberate the outermost electron: $E = -(1\ \text{Ry})Z_{\text{eff}}^2/n^2$, where n is the principal quantum number of that electron.

In this exercise you will use this model to compare fluorine (ground state $1s^2 2s^2 2p^5$), neon (ground state $1s^2 2s^2 2p^6$), and sodium (ground state $1s^2 2s^2 2p^6 3s^1$).

1. Would you expect fluorine, neon, or sodium to be most likely to give up an electron?

You can also use this model to estimate how likely an atom is to accept an electron from another atom. If the extra electron would have a very low energy (very negative), the atom is likely to absorb that electron.

2. Would you expect fluorine, neon, or sodium to be most likely to accept an extra electron? Explain how you can answer this from our simple model.

8.3.2 Explanation: The Periodic Table

Appendix H shows the periodic table of the elements. All the properties of these elements come, in one way or another, from the quantum mechanical wavefunctions of the atoms (and their spins).

In this section we will focus on how atoms of different elements interact with each other. To a good approximation, the reactions that occur between atoms are those that bring the whole system to its lowest possible energy. If one atom has a very loosely bound electron (high energy), and giving that electron to another atom would drop it into a tightly bound state (low energy), then the transfer will tend to happen.

That suggests that an atom's tendency to interact with other atoms depends on two things: having loosely bound outer electrons, and having tightly bound available states (sometimes called "holes"). We'll see below how each of those properties depends on the electron configuration (how many electrons are in each subshell). Then we'll see how the periodic table groups together atoms with similar interaction tendencies.

Ionization Energies

Section 8.2 discussed the "ionization energy": the amount of energy required to remove the outermost electron from an atom. We wrote the ionization energy as $(13.6\ \text{eV})Z_{\text{eff}}^2/n^2$. The effective Z is a result of strong screening by the inner electrons and weaker screening by the other outer electrons.

Active Reading Exercise: Ionization Energy

Argon ($Z = 18$) in its ground state has all of the states up through $n = 3$ filled. Potassium ($Z = 19$) has all of those states plus one $4s$ state filled, and calcium ($Z = 20$) has the same plus a second $4s$ electron. Rank these three atoms in order from lowest ionization energy to highest. You do not need to do any calculations for this problem. Just explain in words which ones you would expect to be higher and lower, and why.

The outermost electron in argon shares the $n = 3$ shell with seven other electrons, all of which weakly shield the nucleus. So that electron feels a large effective Z and thus has a large ionization energy. In potassium, on the other hand, the outermost electron is very well screened so it feels a low effective Z. Plus, it is at $n = 4$, which has a much lower ionization energy than $n = 3$. For both of those reasons (low Z_{eff} and high n), potassium is much easier to ionize than argon. When you go from potassium to calcium you add one more proton and do not effectively shield it, so the binding is slightly stronger for calcium than for potassium.

The correct ranking is potassium (4.3 eV), calcium (6.1 eV), argon (15.8 eV). As you would expect from our arguments above, there's a huge drop from argon to potassium when you simultaneously lower Z_{eff} and raise n, and then a small increase from potassium to calcium when you leave n the same and slightly increase Z_{eff}.

That pattern repeats throughout the periodic table. Figure 8.5 shows the actual (experimental) ionization energies. Apart from some bumps and wiggles (some of which can be explained by considering the effects of subshells), the overall trend is for the ionization energy to rise as you add electrons within a shell, and then drop precipitously when you first add an electron to a new shell.

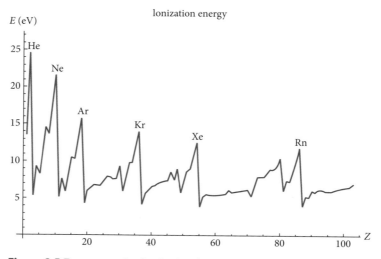

Figure 8.5 Energy required to ionize the outermost electron, by element. (A high value on this plot means it is hard to ionize the atom.) The high peaks represent the noble gases, with filled shells. The big drop in ionization energy after each one represents the alkali metals, atoms with a single electron at a new value of n.

Electron Affinity

We just discussed how much energy it takes to remove an electron from an atom. For example, if you hit a chlorine atom with a photon with 13 eV (or more) you can knock out one of its electrons.

The flip side is the energy you get out of a system when you *add* an electron to an atom. As an example, if a chlorine atom captures a free electron, it can emit a photon with 3.61 eV (or more, depending on how much kinetic energy the electron had before it was captured). It emits the photon because the bound electron has less energy than it did when it was unbound. (That energy can be released in other forms, such as kinetic energy of the atom. The point is that if a chlorine atom captures an electron, that makes 3.61 eV available in some other form.)

The amount of energy an atom could release by capturing an electron is the atom's "electron affinity." So what determines the electron affinities of neutral elements?

Active Reading Exercise: Electron Affinity

Fluorine $(Z = 9)$ in its ground state has five of the six $2p$ states filled. Neon $(Z = 10)$ has all six of those states filled.

1. If you were to add an electron to each of these two atoms, which one would release more energy and why?

2. Therefore, which of these two elements is more likely to bond with another element by absorbing an electron?

For the first question above, the reasoning closely mirrors what we said above about ionization energies. If you give a fluorine atom an extra electron, it will fill that last $2p$ state. Because the $2p$ electrons screen each other weakly, that electron will be tightly bound to the nucleus. So it will have a large negative energy and release a lot of energy when it is captured. If you add an electron to neon, however, that electron will go into the $3s$ state. It will be weakly bound, have a small negative energy, and release very little energy during its capture.

Generalizing, an atom with a small number of unfilled states in its outer shell will have a high electron affinity, while an atom with a completely full outer shell will have a very low electron affinity.

What does that tell us about how the atom behaves? That brings us to the second question above. Because systems tend to evolve toward lower energy, atoms with high electron affinities (such as fluorine) are much more likely to absorb electrons than atoms with low affinities (such as neon).

For some atoms (including all the noble gases) the electron affinity is actually negative. If you add an electron to neon, that electron's small negative energy is more than compensated by the weak extra screening it adds to the other electrons, and the neon's energy goes up. So that electron will promptly fall back out.

A Tour of the Periodic Table

Much of what we have presented in Section 8.2 and this section can be summarized in two rules:

- If an atom has only one or two electrons in its outermost shell, those outer electrons feel a low Z_{eff}, so the atom has a low ionization energy. Such an atom is likely to give up an electron.

- If an atom has an almost-but-not-quite-full outermost shell, a newly added electron would feel a high Z_{eff}. Such an atom has a high electron affinity, so it is likely to absorb an electron.

Those two rules explain how different regions in the periodic table react with each other. Keep Appendix H handy as you go through this section. You can find a printable version at www.cambridge.org/felder-modernphysics/figures

The Noble Gases (Column 18)   The far right column of the periodic table contains the elements with completely filled p subshells. Those elements are called the "noble gases" (or sometimes "inert gases"). Our arguments above show why these atoms have high ionization energies (so they don't tend to give electrons up) and low electron affinities (so they don't tend to absorb electrons). The result is that they almost never react with other atoms. They exist at room temperature as "monatomic gases," gases consisting of single unbound atoms.

A gas condenses into a liquid when the temperature is low enough to get the molecules to bond to each other. Because noble gases bond so weakly, they only become liquids at temperatures close to absolute zero.

The Alkalis and Alkaline Earths (Columns 1 and 2)   Columns 1 and 2 form the "s block" of the periodic table. Each "alkali" element (Column 1) has a single electron in a new shell (value of n). Those electrons have very low ionization energies and low electron affinities, so the alkalis tend to react by giving up electrons. The "alkaline earth" elements (Column 2) have somewhat higher ionization energies because each has two s electrons in its outer shell. They tend to form bonds in which they give up both of their outer electrons.

The Halogens (Column 17)   Back on the other side of the periodic table, the "halogens" in Column 17 have high ionization energies and high electron affinities. In other words, their electrons are tightly bound and they each have room for one more tightly bound electron, so they tend to react by absorbing one extra electron.

The Transition Metals (Columns 3–12), the Lanthanides, and the Actinides   Columns 3–12 form the "d block," known collectively as the "transition metals." The $n + l$ rule (p. 379) tells us that the $4s$ subshell will be filled before $3d$, and $5s$ before $4d$, and so on. So each atom in this block is working its way through a d subshell, but already has electrons in the s subshell of a *higher n-value*. (You can tell that the energies of these s and d subshells are very close because a number of the transition metals "steal" an electron from the s subshell to fill an extra spot in d. Those exceptions to the $n + l$ rule are labeled in Appendix H.)

The s electrons are on average at much higher radius than the d electrons. This leads to the seemingly paradoxical result that even though the s electrons are lower energy and get added before the d electrons as you increase Z, they are almost always the ones that get pulled off when a transition metal interacts with another atom. Because it is the s electrons that dominate chemical reactions, the chemical properties of the transition metals tend to all be pretty similar. Like the alkalis and alkaline earths, they mostly react by giving up electrons.

The "lanthanides" and "actinides" form the f block, which is put below the rest of the table without column numbers. (This is simply so the table can fit on a page. Logically they should occupy 14 columns between the *s* and *d* blocks, down at Rows 6 and 7.) Like the transition metals, they are similar to each other chemically because they tend to give up their *s* electrons rather than their *f* electrons.

Columns 13–16, from Metals to Non-metals   So far we've discussed all of the periodic table except Columns 13–16 (boron through oxygen on the top row). To the left of these columns everything is a metal[5] (except hydrogen – more on that in a moment). That implies many things about the properties of those elements, but it primarily means that they tend to react by giving up electrons rather than absorbing them. To the right of these columns everything is a non-metal:[6] one column tends to absorb electrons, and one tends not to have any chemical reactions.

Within these four columns there is a somewhat irregular diagonal line going from the upper left to the lower right that distinguishes metals on the lower left from non-metals on the upper right, with a number of so-called "metalloids" occupying ambiguous positions between the two. We're not going to get into a detailed discussion of the properties of these elements, or of the reasons for this strange dividing line. We will simply note that this is a region of the periodic table with intermediate values of both ionization energy and electron affinity, so it defies the simple descriptions that we can give for other regions.

It's not a coincidence that organic compounds – the building blocks of life – are largely formed from the elements in these columns. Simply considering all the common forms of carbon (coal, graphite, diamond, . . .) gives you a sense of how complex the behaviors of these atoms can be.

Hydrogen   Ironically, the simplest atom to study quantum mechanically is perhaps the most complicated chemically. Although hydrogen is most commonly shown on the far left with the alkalis, it is unique. Some periodic tables show it in Column 14 above carbon, some with the halogens in Column 17, and some in multiple places. It has a single electron in its outer shell (alkali) and is one electron short of filling a shell (halogen). But hydrogen has a much higher ionization energy than any of the alkalis, and a much lower electron affinity than any of the halogens.

We'll discuss how hydrogen reacts when we talk about covalent bonds in Chapter 9.

Stepping Back: From Quantum Mechanics to Chemistry

When the authors took chemistry we were given rules that govern atomic bonding, the most important being that atoms "want to have full shells." The reason for this strange desire was "because of quantum mechanics." You have now learned enough quantum mechanics to understand much of why atoms behave the way they do.

An atom with a nearly full shell has one or more available states for tightly bound electrons, i.e. a high electron affinity, so it tends to absorb electrons. If there are one or two electrons beyond the last full shell, those electrons are effectively screened, weakly bound, and easy to remove. So those atoms tend to release electrons. All of this comes from what we described about electron screening in Section 8.2, and all of this leads to a tendency toward a full outer shell.

---

5 Astronomers use a different definition of "metal," meaning any element other than hydrogen or helium. Their use of the term has nothing to do with chemical properties.

6 Elements 117 (Ts) and 118 (Og) don't occur naturally and are short-lived in laboratories, so it's hard to characterize their chemical properties.

The periodic table reflects this tendency by lining up in columns all the atoms that are filling one type of subshell. For example, all of the atoms in Column 15 (starting with nitrogen) have an outer, half-filled p subshell. Thus atoms in each column tend to have similar chemical properties.

There are two terms chemists often use to describe these properties:

- "Electronegativity" measures whether an atom tends to absorb electrons (high electronegativity) or release them (low electronegativity). There are many different definitions that lead to different numbers, but in general the alkalis (Column 1) have the lowest electronegativities and the halogens (Column 17) have the highest.

- "Valence" can roughly be thought of as how many electrons are available to be released. The alkali atoms have a valence of 1 and the alkaline earths have valence 2. You could assign fluorine a valence of 7 for the 7 electrons in its $n = 2$ shell, but it is often more useful to think of its valence as -1 since it is most likely to react by absorbing one electron.

The simple rules and trends we've discussed obviously do not explain all of chemistry. Appendix H lists exceptions to the $n + l$ rule. The division into metals, non-metals, and metalloids is imprecise, with some borderline elements listed differently in different periodic tables. Periodic tables also differ in regard to which elements are included in the lanthanides and actinides.[7] For elements with $Z \gtrsim 100$, relativistic effects become important and some of our simple descriptions break down. The elements in the $7p$ row are so unstable that when you synthesize them they decay in a tiny fraction of a second, so it's not all that meaningful to talk about their chemical properties. And all of the ionization energies, electron affinities, and even electron configurations we've quoted are for free atoms, and can be significantly different when the atoms are bound into solids. We'll discuss atomic bonding in more detail in Chapter 9 (molecules) and Chapter 11 (solids).

Despite all these caveats, if you come out of this chapter understanding how quantum mechanics leads some columns of the periodic table to release electrons easily and others to absorb them (and the noble gases to do neither), you'll have a great start toward understanding the physical basis of chemistry.

---

7 La and Ac are sometimes left out and Lu and Lr are sometimes included. The convention we've followed, which is what it would look like if the $n + l$ rule were followed perfectly, is the one used in most physics books.

8.3.3 Questions and Problems: The Periodic Table

Conceptual Questions and ConcepTests

1. An element's "electronegativity" is high if it is likely to react by absorbing electrons, low if it is likely to release them. Which of the following makes a good definition of electronegativity? We're using IE and EA here to mean ionization energy and electron affinity, respectively. (Choose one. The correct choice is one of the definitions sometimes used for it.)

   A. $(IE + EA)/2$

   B. $(IE - EA)/2$

   C. $(-IE + EA)/2$

   D. $(-IE - EA)/2$

2. Although we said hydrogen reactions are complicated, you can, to a first approximation, just imagine that hydrogen tends to give up one electron like the other alkalis. How many hydrogen atoms would you expect a nitrogen atom to bind with? Why?

3. Why are Columns 1 and 2 called the "s block"? What would you call the block consisting of Columns 13–18?

4. List each element that has a p subshell with one electron in it.

5. (a) Name an element that is likely to give up an electron, but not likely to absorb one.

   (b) Name an element that is likely to absorb an electron, but not likely to give one up.

   (c) Name an element that is unlikely to give up *or* absorb an electron.

6. Many problems in classical mechanics can be approached two ways: either by considering forces, or by considering energy. Explain why fluorine (ground state $1s^2 2s^2 2p^5$) has a strong electron affinity by discussing forces rather than discussing energy. *Hint*: It's still about screening.

7. The lanthanides are considered to be in Row 6 of the periodic table because their s electrons are in the 6s subshell. If more elements are discovered, what will be the first row with a g subshell? (Assume the $n + l$ rule continues to hold.)

8. Transition metals often react chemically as if they have a valence of 2. Explain.

9. There are numerous chemical techniques that separate elements from each other, but it is very hard to chemically separate rare earth elements (lanthanides) from each other. Why?

10. **Hund's Rule and Half-Filled Shells**

    We've explained how electron screening leads to the general rule that an atom "wants" to have a full shell. In other words, alkali atoms with one electron in a new shell will tend to give up that electron, while halogens with a nearly full shell will tend to absorb an extra electron. Chemists also talk about a general rule that atoms "like" configurations with half-filled shells. For example, using the $n + l$ rule you would expect chromium to have $4s^2 3d^4$ in its ground state, but instead it has $4s^1 3d^5$. (We're not bothering to write out the full subshells below these.) Explain why. *Hint*: Imagine that 4s and 3d are exactly equal in energy and explain why Hund's rule predicts that $4s^1 3d^5$ is a lower energy state than $4s^2 3d^4$.)

Problems

11. Make a rough sketch of electron affinity as a function of Z, similar to the one we made for ionization energy (Figure 8.5). Label the elements that represent key turning points in the graph.

12. In Section 8.2 we presented a simple model for the screening felt by the outermost electron in an atom. Each electron in a lower shell screens one proton, and each electron that shares the outer shell screens half a proton. While this model is not particularly numerically accurate, it does correctly reproduce the trends we see in the periodic table.

    (a) Using this simple model, estimate the electron affinity of oxygen (ground state $1s^2 2s^2 2p^4$).

    (b) Now estimate the electron affinity of fluorine (ground state $1s^2 2s^2 2p^5$).

    (c) Now estimate the electron affinity of neon (ground state $1s^2 2s^2 2p^6$).

    (d) Generalize your results to explain why certain columns in the periodic table have strong electron affinity and others very weak.

13. In Section 8.2 we presented a simple model for the screening felt by the outermost electron in an atom. Each electron in a lower shell screens one proton, and each electron that shares the outer shell screens half a proton. Use that model to produce a graph of ionization energy as a function of Z, similar to Figure 8.5. You should find that this simple model doesn't match the experimental values well, but does reproduce some important features of the graph. What features does the model accurately represent, and in what ways is it not accurate?

8.4 X-Ray Spectroscopy and Moseley's Law

Multielectron atoms have their own absorption and emission spectra, different from hydrogen in complicated ways. Detailed numerical solutions of Schrödinger's equation can accurately predict the energy levels and characteristic spectra of many of these atoms. For the innermost electrons, however, simple analytical models are fairly accurate. Henry Moseley's measurements of those spectra provided a crucial step toward our modern understanding of the atom.

8.4.1 Explanation: X-Ray Spectroscopy and Moseley's Law

Atomic transitions in which outer electrons are knocked into higher states or fall back down into their ground state configurations tend to involve energies on the order of a few eV, so the corresponding spectral lines are usually in the visible or ultraviolet range. Such processes are often called "optical transitions."

Electrons in low-energy states with little screening, however, often have transitions in the keV range and beyond, which correspond to X-ray photons. When an inner electron is ejected from an atom, a higher level electron will fall into its spot, and another one will fall into that spot, until the atom reaches the ground state (not of the neutral atom, but of the ion).

If you bombard a gas of atoms with either high-energy particles or photons, the atoms radiate photons with frequencies that characterize a particular element. Each emission line – each transition from a higher energy level to a lower level – is designated by a two-letter naming scheme:

- You begin with the spectroscopic notation for the *lower* of the two n levels, the state the electron dropped into: K for $n = 1$, L for $n = 2$, and so on alphabetically. (Just as with hydrogen emission spectra, this is the level that defines a particular series of emission lines.)
- Then you use a subscripted Greek letter for how many levels the electron dropped: α for a drop of one level, β for two, and so on (Table 8.2).

Table 8.2 Notation for emission lines

Dropped into \ Dropped from	$n = 2$	$n = 3$	$n = 4$	$n = 5$
$n = 1$	K_α	K_β	K_γ	K_δ
$n = 2$		L_α	L_β	L_γ
$n = 3$			M_α	M_β

Each of those spectral lines can actually correspond to several closely spaced lines because the energy depends on l as well as n, but deep in the lower orbitals that effect isn't as strong as it is for the outer electrons.

In 1913 Henry Moseley measured the X-ray emission spectra of many different elements.

Theoretical Prediction

We're going to cheat a bit and predict Moseley's results based on more knowledge about atoms than he had.

Consider the K_α transition, from $n = 2$ to $n = 1$. The emitted photon has an energy equal to the energy difference between those two states:

$$E = (13.6 \text{ eV}) \left(\frac{Z_{\text{eff}, 1}^2}{1^2} - \frac{Z_{\text{eff}, 2}^2}{2^2} \right).$$

The terms $Z_{\text{eff}, 1}$ and $Z_{\text{eff}, 2}$ refer to the effective Z-values felt by an electron in the $n = 1$ and $n = 2$ shells, respectively, accounting for screening of the nucleus by other electrons. For low-Z atoms like lithium and beryllium we estimated earlier that the $n = 1$ electrons would be roughly half-screened by each other and not screened by the outer electrons, so $Z_{\text{eff}, 1}$ should be $Z - 0.5$. That's not very accurate even for low-Z atoms, and it becomes wildly inaccurate for atoms with many electrons.

Here's why. The wavefunction of an "inner" electron has a small but non-zero magnitude at very high radii; the wavefunction of an "outer" electron has a small but non-zero magnitude very close to the nucleus. Therefore an outer electron actually does provide some screening, lowering the Z_{eff} felt by an inner electron. This small effect starts to add up when there are a lot of outer electrons. The end result, based on both numerical calculations and experimental measurements, is that for most atoms $Z_{\text{eff}, 1} \approx Z - 2$ and $Z_{\text{eff}, 2} \approx Z - 10$.

Now we make a significant approximation: because of the $1/n^2$ dependence of electron energy, and also because of the different Z_{eff}-values at the different levels, $E_{n=2}$ is very close to zero compared to $E_{n=1}$. We will therefore ignore $E_{n=2}$ in our calculation, and predict that Moseley should find

$$E_{K_\alpha} \approx (13.6 \text{ eV}) (Z - 2)^2 . \tag{8.2}$$

By this logic we expect K_α, K_β, K_γ, and so on to produce very similar photons, but the L series to produce much lower energies.

Moseley's Plot

Moseley did not have the theoretical background we have. (Actually we owe some of that background to his experiment!) But he did have the Bohr model of the atom, which had come out earlier that year. Based on that model, he made a plot of the square root of photon frequency vs. atomic number Z.

Active Reading Exercise: The K_β Plot

Suppose you could do an experiment using many different elements (aka many different Z-values), but all of them undergo K_α transitions only. Based on Equation (8.2), what would be the shape of your $\sqrt{\nu}$ vs. Z plot? You can start your answer with just one word describing the shape, but then you can make several other predictions about the details of the graph.

One way to approach the exercise above is to replace E with $h\nu$ and then rearrange Equation (8.2) as

$$\sqrt{\nu} \approx \sqrt{\frac{13.6 \text{ eV}}{h}} (Z - 2) = \sqrt{3.29 \times 10^{15} \text{ Hz}} (Z - 2).$$

So we expect a line, and we can predict its slope. We can also predict that if you extrapolated that line toward the horizontal axis ($v = 0$) it should hit at $Z = 2$. (Of course the frequency of a K_α photon for helium is not 0; our approximations are not valid for very low Z.)

That's an approximation for the K transitions. If you start over and calculate for the L transitions (Problem 7) you get a different equation, which predicts a different $\sqrt{v}$ vs. Z line with a different slope and x-intercept.

Moseley of course could not isolate individual transitions in this way. His experiment generated photons of many different frequencies from a variety of elements, but he was able to organize the resulting data points into straight lines. His plot is shown in Figure 8.6. (He plotted Z on the *vertical* axis and $\sqrt{v}$ on the horizontal one, so his K_α slope is the inverse of what we calculated.)

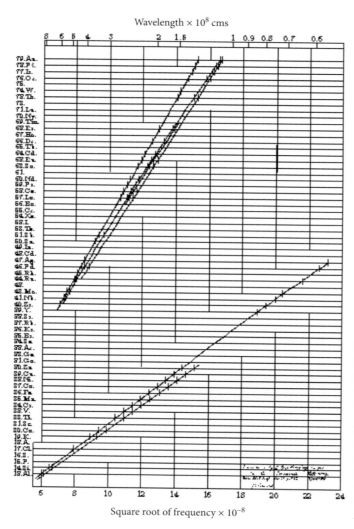

Figure 8.6 Moseley's original plot, showing the square root of photon frequency on the horizontal axis and atomic number on the vertical axis. The plot includes the K series of X-ray emission (a group of almost-but-not-quite identical lines), and the L series (a different group of almost-but-not-quite identical lines). *Source: H.G.J. Moseley 1914.*

"Moseley's Law" for X-ray emission spectra is $\sqrt{v} = k_1(Z - k_2)$, where k_1 and k_2 are constants that depend on which spectral line you are plotting (K_α, L_γ, etc.). Plots such as Figure 8.6 are called "Moseley plots."

When Moseley did his work, the structure of the atom wasn't well understood. The periodic table was (mostly) arranged by atomic mass.[7] "Atomic number" referred to an element's position in the table, rather than to any known physical property. Moseley's work established the idea of a measurable atomic number. During this work he noticed some gaps in his plot and correctly predicted the existence of three as-yet-unknown elements with atomic numbers 43, 61, and 75. The third, rhenium, was discovered in 1925. The first two, technetium and promethium, do not occur in nature and were synthesized in labs in 1937 and 1945, respectively.

---

7 Atomic mass mostly follows the same order as atomic number, but not perfectly. Even as far back as Mendeleev, people knew that to group elements with similar chemical properties together you sometimes had to arrange them out of order of their measured masses. This was often assumed to reflect errors in the mass measurements.

8.4.2 Questions and Problems: X-Ray Spectroscopy and Moseley's Law

Conceptual Questions and ConcepTests

1. Every point on Moseley's plot represents a transition *from* a certain n-level *to* a lower n-level, *in* a certain element (that's three distinct independent variables), thus producing a photon of a certain frequency (that's the dependent variable, which he put on the horizontal axis). Figure 8.7 shows an unlabeled but colored Moseley plot.

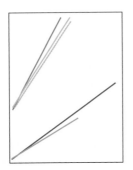

Figure 8.7

(a) All the red points lie on one line, in the top left of the plot. Of the three independent variables listed above, which ones do these points have in common, and which one makes them different from each other?

(b) The orange points lie on a different line, close to but separate from the red line. Which of the three independent variables distinguishes the red line from the orange line?

(c) The blue and green lines lie far away in the lower part of the plot. Which of the three independent variables makes the blue and green lines so different from the red, orange, and yellow lines?

2. Figure 8.6 shows a group of K transitions and a group of L transitions. One is a pair of nearby lines near the bottom of the plot and the other is a set of three nearby lines near the top. Which is which? How do you know?

3. Based on the simple model we presented in Section 8.2, each atom at the $n = 1$ level of a many-electron atom experiences $Z_{eff} = Z - 0.5$.

(a) Based on that same simple model, what Z_{eff} is experienced by each atom at the $n = 2$ level of a many-electron atom? (Remember that there are two electrons in the $n = 1$ shell and eight in the $n = 2$ shell.)

(b) Experiment shows that the actual $Z_{eff,2}$ in a many-electron atom is much lower than your answer to Part (a). Why?

4. Make a rough sketch of λ vs. Z for the K_α line. You don't have to include numbers, but you should show the shape of the plot.

5. In this section we've been talking about high-energy (X-ray) transitions. If you were to plot $\sqrt{\nu}$ vs. Z for photons emitted in the lowest-energy transition in a multielectron atom, would the results be linear? Why or why not?

Problems

6. What is the frequency of the K_α emission line for nickel?

7. The Explanation (Section 8.4.1) analyzes the K_α transition. Here you will similarly analyze the L_α transition.

   (a) L_α represents a transition from which (higher) energy level, to which (lower) energy level?

   (b) What is the energy of the lower state into which your electron transitions?

   (c) What is the energy of the higher state from which your electron transitions? Assume $Z_{\text{eff},3} \approx Z - 18$.

   (d) Subtract to calculate the energy of the photon emitted by this transition.

   (e) For the K transitions we approximate the photon energy by approximating the energy of the upper level as zero. Using $Z = 45$, calculate the percentage error in using this approximation for the L_α transition.

   (f) Finally, rearrange your approximation to predict the slope and x-intercept of a plot of $\sqrt{\nu}$ vs. Z for an L transition.

8. Make a Moseley plot of Z vs. $\sqrt{\nu}$ for the K_α line. Be sure to include numbers (with appropriate units) on both axes.

9. Suppose you were to plot $\sqrt{\nu}$ vs. Z for the L_α line. (Note that your axes will be reversed from Moseley's plot.)

   (a) Would the slope be higher or lower than it is for K_α? Explain.

   (b) Would the x-intercept be higher or lower than it is for K_α? Explain.

10. In the Explanation (Section 8.4.1) we estimated the energy of a K_α photon by neglecting the energy of the $n = 2$ state.

   (a) Using the values we gave for Z_{eff}, estimate the percentage error that the above approximation introduces for calcium ($Z = 20$) and for gold ($Z = 79$).

   (b) Would the formula we derived be a better estimate for K_α or for K_β? Explain.

11. Figure 8.8 shows a simulated spectrum obtained by bombarding cesium atoms with high-energy particles. The continuous spectrum is from bremsstrahlung radiation and the four spikes are emission lines.

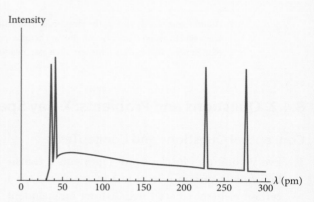

Figure 8.8

   (a) The four spectral lines are K_α, K_β, L_α, and L_β. Make a rough copy of the figure and label which line is which. Briefly explain your reasoning.

   (b) Use Equation (8.2) to estimate the wavelength of the K_α line. Make sure it roughly agrees with what's shown on the plot. (Note that the wavelengths are plotted in picometers.)

12. [*This problem depends on Problem 11.*]

   (a) Use Figure 8.8 to estimate the energy of the particles used to bombard the cesium target.

   (b) If the goal of an experiment is to measure the X-ray emission lines of a target, the bremsstrahlung radiation is noise. Explain why such experiments are often done using protons or alpha particles to bombard the target instead of electrons.

Chapter Summary

Chapter 7 dealt with atoms with only one electron. When more than one electron orbits the same nucleus, the electrons interact with each other, complicating the picture significantly. Although we can't analytically solve Schrödinger's equation for such atoms, we can apply quantum mechanical principles to them, to explain much of the structure of the periodic table.

Section 8.1 The Pauli Exclusion Principle

The Pauli exclusion principle says that no two electrons can occupy the same quantum state.

- The Pauli exclusion principle applies to all particles that are "fermions," meaning they have half-integer spins. It does not apply to "bosons," particles with integer spins. Protons, neutrons, and electrons are all fermions, while photons are bosons.

- Composite particles are classified in the same way, on the basis of their total spin. For example, ^{3}He nuclei are fermions (they can't occupy the same quantum state together), while ^{4}He nuclei have integer spin and are thus bosons (they can).

- Because fermions can't crowd together into one state, an atom in its ground state has electrons occupying the N_e lowest-energy eigenstates, where N_e is the number of electrons. (For a neutral atom, N_e equals the atomic number Z.) Each combination of the quantum numbers n, l, and m_l can be occupied by two electrons, one with spin up and the other with spin down.

Section 8.2 Energy Levels and Atomic States

The energy that binds an electron to its nucleus is affected by the number of protons in the nucleus, and also by the other electrons orbiting the same nucleus. As a result of these two competing effects, the energy required to free an electron – its "ionization energy" – oscillates as the atomic number increases.

- "Spectroscopic notation" uses a number to represent an n-value (the shell), and a letter for the l-value (subshell). For example, $1s$ is the $n = 1$, $l = 0$ state and $2p$ is the $n = 2$, $l = 1$ state. Each subshell contains more than one state with different values of m_l and m_s.

- Electrons in low orbits "screen" electrons in high orbits, effectively reducing the nuclear attraction they feel.

- In hydrogen, the energy depends only on n (to a very good approximation). In multielectron atoms the energy also depends strongly on l because electrons with different radial distributions feel different amounts of screening.

- As a general but imperfect rule, subshells fill in order of ascending $n + l$.

- "Hund's rule" says that electrons within a subshell fill same-spin states before they begin to fill opposite-spin states.

Section 8.3 The Periodic Table

The goal of this section was to show how the properties of different elements illustrated in the periodic table arise from the quantum mechanical properties of atoms we have discussed in previous sections.

- Because of the screening effects discussed in Section 8.2, an atom with a full outer shell will tend to have a high ionization energy (because its outermost electrons are weakly screened). An atom with only one or two electrons in its outer shell will tend to have a low ionization energy (because its outermost electrons are strongly screened).

- An atom's "electron affinity" reflects its tendency to gain an electron. An atom with an almost-full outer shell will tend to have a high electron affinity – again, because a newly added electron would be weakly screened.

- These two effects explain the columns of the periodic table, with some elements inclined to give up electrons, some inclined to absorb electrons, and some inclined to do neither. The short version is that elements near the left of the periodic table tend to give up electrons, elements near the right (but not all the way there) tend to absorb electrons, and elements in the far right column don't tend to react at all.

- To review how these effects lead to the structure of the periodic table in more detail, go through the "tour" of these columns that starts on p. 385, cross-referencing each category with the periodic table (Appendix H).

- The term "electronegativity" is used to measure whether an atom tends to absorb electrons (high electronegativity) or release them (low electronegativity). High ionization energy and high electron affinity both contribute positively to electronegativity.

- The term "valence" refers to the number of electrons an atom is inclined to release; a negative valence indicates a tendency to absorb electrons.

Section 8.4 X-Ray Spectroscopy and Moseley's Law

The "optical spectra" of atoms are produced by transitions of the outer electrons, and they vary in complicated ways among different elements. The "X-ray spectra" involve transitions of lower electrons and follow a more predictable pattern.

In 1913 Henry Moseley used X-ray spectra to demonstrate regular patterns among the elements, leading to the notion of "atomic number" (Z), which we now associate with the number of protons in the nucleus.

- Emission lines are described with a two-letter naming scheme: the spectroscopic notation for the lower of the two n-levels, followed by a subscripted Greek letter that indicates how many levels the electron dropped. For example, L_β refers to a transition to $n = 2$ (because that's what L indicates) from $n = 4$ (because β means n went down by 2).

- Moseley found that a plot of Z for the element vs. $\sqrt{\nu}$ for the emitted photon produced a series of lines. Each such line represents the same atomic transition for many different elements.

- Moseley's work led to the prediction and subsequent discovery of several elements with atomic numbers between those of the then-known elements.

9

Molecules

Chapter 8 talked about atoms in isolation. But most of the atoms around you are joined together, a fact that can dramatically change their physical and chemical properties. In this chapter we explore atoms that are joined together in molecules. Chapter 11 will describe solids, large collections of atoms or molecules bound into a macroscopic size.

9.1 Ionic and Covalent Bonds

You may remember from a bygone chemistry class that "ionic" bonding involves one atom giving up an electron to another atom, and "covalent" bonding involves two atoms sharing an electron. In physics we ask why all this electron-swapping happens, and why it bonds atoms together. The answer is pretty much always the same: "because it lowers the total energy of the system." With Chapter 8 as background we can consider the question of why – and under what circumstances – ionic and covalent bonds are energetically favorable.

9.1.1 Discovery Exercise: Ionic Bonds

When one atom gives an electron to another, they acquire opposite charges and attract each other. That attraction creates an "ionic bond."

As you answer the following questions about ionic bonds, keep your eye on Appendix H, and keep in mind what you learned in Chapter 8 about why some elements tend to give up electrons and others are inclined to absorb them.

1. Which of the following pairs of elements are likely to form an ionic bond? (Choose all that apply and briefly explain your answers.)

   A. Li and Na

   B. Li and F

   C. F and Cl

2. Sometimes an atom gives one electron each to two other atoms and ionically bonds to both of them. Which of the following molecules could form this way? (Choose one.)

A. LiS_2

B. Li_2S

C. $BeCl_2$

D. Be_2Cl

9.1.2   Discovery Exercise: Covalent Bonds

Consider a simple model of a hydrogen atom as a proton surrounded by a thin spherical shell of negative charge at one Bohr radius. Nearby is a single bare proton. See Figure 9.1.

The lone proton   Hydrogen atom

Figure 9.1 A bare proton and a hydrogen atom.

1. Does the bare proton feel a net force, and if so in which direction?
2. Does the hydrogen nucleus feel a net force, and if so in which direction?
3. Does the electron cloud feel a net force, and if so in which direction?
4. One moment later, sketch how the picture would have changed because of the forces you predicted.
5. In your new picture, does the bare proton feel a net force, and, if so, in which direction?

9.1.3   Explanation: Ionic and Covalent Bonds

To introduce bonding, we will initially discuss "ionic bonds" (one atom gives up an electron to another) and "covalent bonds" (multiple atoms share an electron) as if they were wholly separate categories. We'll see later, however, that most actual atomic bonds fall on a spectrum between the two extremes.

Ionic Bonds

Chapter 8 explained why some atoms have low ionization energies and are thus likely to give up electrons, and why some atoms have high electron affinities and therefore tend to absorb electrons. The lowest ionization energies occur in Column 1 of the periodic table, (the alkali elements) and the highest electron affinities in Column 17 (the halogens). As you might expect, this means that alkali atoms and halogens tend to react with each other.

Consider the familiar case of sodium (Na) and chlorine (Cl), which combine to make salt. Both are in Row 3 of the periodic table, sodium in Column 1 and chlorine in Column 17. Sodium has an ionization energy of 5.14 eV, one of the lowest in the periodic table. Chlorine has an electron affinity of 3.61 eV, which is the highest of any element.

You're probably expecting us to tell you that the system lowers its energy if sodium gives an electron to chlorine, but if you just look at those numbers, that does not seem to be true. The lowest ionization energy on the periodic table (cesium, 3.89 eV) is still higher than the highest electron affinity, so if you have two atoms floating in empty space it is *always* energetically favorable for them to keep their own electrons, unless they can do something else to lower their energy.

Which, in fact, they can. When a sodium atom gives an electron to a chlorine atom, the sodium acquires a net positive charge and the chlorine a net negative charge. So now they attract each other, and can reduce their potential energy by coming closer together. If they get close enough, the overall energy will be less than it was when they were far apart and neutral. They are now a molecule, held together by an "ionic bond."

What keeps the nuclei from crashing into each other? The nuclei themselves are positively charged. Once the ions are so close that their electron wavefunctions overlap, the nuclei are less shielded and repel each other. Meanwhile the electrons experience a subtler effect called "Pauli repulsion": as the electron wavefunctions overlap, their energies go up. The details of this are beyond the scope of this book, but you can see that this must be true by considering the limit in which the two nuclei merge completely. In that case you would just have one atom and the combined electrons would fill the lowest 28 energy levels, rather than just the 10 lowest of the sodium ion and 18 lowest of the chlorine ion.

So the nucleus of each atom attracts the electrons of the other, while their nuclei repel each other and their electron clouds repel each other. The balance of these forces leads to the energy being minimized at a particular separation between the nuclei.

For Na and Cl that distance is 0.236 nm, giving the system a total energy that is 4.26 eV below what the two atoms would have separately. That 4.26 eV is called the "binding energy" of the molecule. It is also called the "dissociation energy" because it's the energy you have to supply to pull the molecule apart. The energy of Na and Cl as a function of their distance is shown in Figure 9.2.

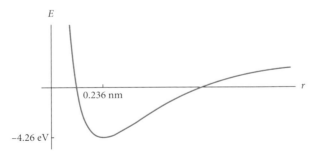

Figure 9.2 Energy of an NaCl molecule as a function of the separation between the nuclei. (At large r the energy doesn't approach 0: see Question 2).

A molecule like NaCl is called "diatomic" (two atoms). In some cases, more than two atoms join together ionically. Each Column 2 atom (the alkaline earths) has two electrons with low ionization energies; one such atom can bind together with two halogens to form a molecule such as $MgCl_2$. You can call this "triatomic," but the term "polyatomic" is often used for any molecule with more than two atoms.

Covalent Bonds

Imagine a fairly empty universe. Somewhere in this fairly empty universe is a lone proton. And somewhere else, very far away, is a hydrogen atom: a single electron in its ground state orbit around a single proton. See Figure 9.3. The energy of that ground-state electron, as you know, is -13.6 eV. Nothing else has much energy to speak of.

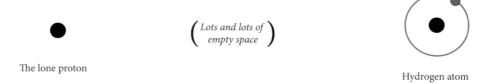

The lone proton

$\begin{pmatrix} \text{\textit{Lots and lots of}} \\ \text{\textit{empty space}} \end{pmatrix}$

Hydrogen atom

Figure 9.3 A bare proton and a hydrogen atom far away from each other.

Now we begin to bring this hydrogen atom closer to the lone proton. For a while, this has no discernible impact on the energy of the system. An electron getting closer to a proton loses energy, and a proton getting closer to a proton gains the same amount of energy. To put it another way, we're bringing a neutral object closer to a positively charged object, and who cares?

Active Reading Exercise: The Proton and the Hydrogen Atom

When the proton and hydrogen atom get closer together (but not yet so close that the lone proton is inside the electron cloud), there *is* a net force between them. Which way (attractive or repulsive) and why? *Hint*: Try sketching how our picture of the hydrogen atom will change as it gets close to the proton.

The lone proton

Hydrogen atom

Figure 9.4 A proton and a hydrogen atom approach each other.

Figure 9.5 A proton and a hydrogen atom merge to a form a molecule.

When our two objects get close enough, attraction to the lone proton deforms the electron's orbit so that it is no longer spherically symmetrical. At this point the electron is, on average, closer to the lone proton than its proton is. (In a word, the hydrogen atom is "polarized." See Figure 9.4.) The hydrogen atom is thus attracted to the proton. Or, to put it another way, it loses energy by coming closer.

So the hydrogen atom comes closer and closer. But eventually it gets so close that the electron orbit envelopes both protons (Figure 9.5). Those protons no longer want to get closer together.

At some point the system reaches a stable equilibrium: bringing the two protons either closer together or farther apart would increase the total energy. Numerical calculations and experiments show that this minimum energy is -16.3 eV, obtained by a proton separation of 0.106 nm. The resulting covalently bonded molecule is called H_2^+, a singly ionized hydrogen molecule. The binding energy is 2.7 eV, the difference between the bound state energy and the energy that the particles would have if you pulled the protons apart and left the electron bound to one of them.

Some of the details of this calculation are illustrated in Figure 9.6. In the limit where the two protons are right on top of each other, you have a helium nucleus, and we know the ground state electron energy is -54.4 eV. In the limit where the two protons are infinitely far apart, the ground state electron energy is just -13.6 eV, the energy of an electron bound to a single proton. In between those two limits, the energy depends in a complicated way on the exact shape of the electron cloud, so we can't write an analytical formula for it, but we know it goes smoothly from one of those limits to the other. The exact form can be determined through numerical calculations. The proton–proton potential energy is much simpler: $e^2/(4\pi\epsilon_0 r)$. The total energy is the sum of the two and it has a minimum at 0.106 nm.

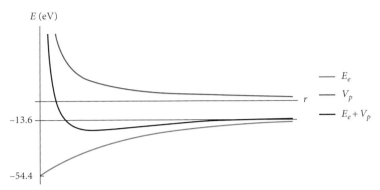

Figure 9.6 Ground state energy of an electron bound to two protons as a function of proton–proton separation. The energy (black) is the sum of the proton–proton potential energy (blue) and the electron total energy (red).

The H$_2$ Molecule

The most common molecule in the universe is H$_2$, two protons and two electrons. When you add a second electron to the H$_2^+$ described above, it falls into the same ground state as the first electron, but with opposite spin. The details are more complicated, but the basic argument illustrated in Figure 9.6 still holds: if the protons get closer together the energy of the electron states goes down and the energy of the proton–proton repulsion goes up, with the net result that there is an equilibrium radius that minimizes the total energy. Because the electron density between the protons is higher than it is for the H$_2^+$ molecule, the protons are more effectively screened from each other, and the equilibrium radius is smaller: 0.074 nm.

Other Covalently Bonded Molecules

The alkali elements can form diatomic molecules in the same way we just described for hydrogen. Each one has a single s electron in its outer shell and when two of them get together they can share those electrons to form Li$_2$, LiH, KNa, and so on.

For atoms farther to the right in the periodic table, the rules for forming covalent bonds depend on the spins of their outer electrons. Consider oxygen, whose outer shell looks like $2s^2 2p^4$ in its ground state. Hund's rule says the outermost subshell will fill all its states with electrons of the same spin before allowing any of the opposite spin. Because a p subshell has three wavefunction states available with each spin, it will fill its first three electrons with the same spin and the fourth with the opposite.

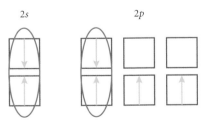

Figure 9.7 The outer shell of oxygen, with six of the eight states filled. The red ovals indicate electrons that are paired together in the same wavefunction. (We arbitrarily put three up and one down in the 2p subshell; it could be three down and one up.)

The outer shell of oxygen is illustrated in Figure 9.7. When two electrons have the same wavefunction (which must mean they have opposite spins), we say they are "paired." So the outer shell of oxygen has four paired electrons and two unpaired electrons.

With that terminology in mind, we can give the rule that determines how many covalent bonds an atom can form:

Only the unpaired electrons take part in covalent bonds.

That rule is not at all obvious, and we will explain where it comes from in Section 9.2. But by following that rule you can predict the kinds of molecules that are likely to form covalently.

For instance, consider a halogen such as fluorine. Its outer shell has $2s^2 2p^5$, meaning it has six paired electrons and one unpaired. It can therefore form a single covalent bond.

Figure 9.8
Structure of H_2O and O_2. Each O atom has two lines coming from it, indicating that it is covalently sharing two of its electrons. This makes a total of four electrons buzzing around the O_2 molecule. The other electrons remain attached to one or the other of the oxygen nuclei.

Oxygen, as we have seen, has two unpaired electrons. It can therefore form two covalent bonds. That enables it to form the most famous covalent molecule, H_2O, in which each unpaired oxygen electron is shared with one hydrogen atom. An oxygen atom can also share both electrons with a single other oxygen atom, leading to an O_2 molecule (see Figure 9.8).

The molecule OH can also form (and does), but the oxygen atom is only sharing one of its two available electrons. The resulting molecule is therefore highly reactive: it will lower its energy further by sharing that second electron, so it generally doesn't last long.

The same basic logic continues as we move left to nitrogen, which has three unpaired p electrons and can form three covalent bonds, forming molecules such as NH_3 and N_2 (Question 5).

You might think that carbon with its two p electrons would form two covalent bonds. But if you've studied chemistry you might already know that carbon forms four covalent bonds. Those bonds lower a carbon atom's energy enough to make it worth stealing an s electron into the p subshell. That makes the outer shell $2s^1 2p^3$, with four unpaired electrons with which to form CH_4 or C_2. For similar reasons boron can form three bonds (Question 6).

These are guidelines, and there are exceptions, but these rules will allow you to understand many common molecules, including complicated arrangements such as CCl_4, $SiHF_3$, and C_6H_6. As we mentioned in Chapter 8, the rich complexity of molecules formed by the atoms in the p block is one of the main reasons these elements are so essential to life.

A Spectrum from Ionic to Covalent

Thus far we've presented molecular bonds as either ionic or covalent. That's a useful way to start understanding them, but in reality most bonds lie somewhere on a spectrum between the two. In NaCl the shared electron is closer to the chlorine nucleus than the sodium nucleus, but its wavefunction still extends to the sodium nucleus and beyond. In H_2O the shared electrons mostly live between the hydrogen and oxygen molecules, but they are closer to the oxygen than to the hydrogens. There are no completely ionic bonds, and the only completely covalent bonds are symmetric ones, as in H_2 or O_2. Nonetheless, most bonds lie close enough to one end of the spectrum to be usefully described as ionic or covalent.

One way to understand where a molecule falls on this spectrum is through electronegativity, sometimes defined as the average of the ionization energy and electron affinity. Atoms with high electronegativity are eager to accept electrons and loath to give them up, while low electronegativity implies the opposite. A bond between atoms with very different electronegativities will be mostly ionic, whereas one between atoms with similar electronegativities will be mostly covalent.

9.1.4 Questions and Problems: Ionic and Covalent Bonds

Conceptual Questions and ConcepTests

1. An NaCl molecule is at its equilibrium separation. Name at least one force that prevents the two atoms from coming closer together. Name at least one force that prevents them from moving farther apart.

2. Figure 9.2 shows the energy of an NaCl molecule as a function of the separation between the two nuclei. The energy is defined as 0 for an Na and Cl atom far from each other. So why doesn't the energy in the plot approach 0 in the limit $r \to \infty$?

3. Why are the protons in an H_2 molecule in its ground state closer together than those in an H_2^+ molecule?

4. How many covalent bonds can a chlorine atom ($Z = 17$) form?

5. Figure 9.8 shows the molecules H_2O and O_2. Note that each H atom is connected by one line (one covalently shared electron), each O atom by two.

   (a) Explain why a nitrogen atom can form three covalent bonds.

   (b) Draw the structures of NH_3 and N_2 molecules.

   (c) Explain why we don't see a lot of NH floating around.

6. (a) Write the spectroscopic notation for a boron atom in its ground state. Assume it is filling up its subshells in order of ascending energy.

   (b) How many electrons would such a boron atom have available for covalent bonding?

   (c) But in fact boron can form three covalent bonds, because it does not fill in its subshells in perfect order. Write the modified spectroscopic notation to show how it fills its subshells to make three electrons available.

7. Draw a diagram showing all the bonds between the atoms in benzene (C_6H_6), making sure each hydrogen has one and each carbon has four. *Hint*: Start by putting the carbons in a ring.

8. A water molecule is electrically neutral. Explain why two nearby water molecules will attract each other. (When the molecules stick together in this way it's called a "hydrogen bond.")

For Deeper Discussion

9. In equilibrium an NaCl molecule is small enough that the electron wavefunctions of the two ions overlap significantly. In that situation how is it meaningful to say that the Na lost an electron and the Cl gained one?

10. Figure 9.6 shows the ground state energy of an H_2^+ molecule. The electron energy includes potential and kinetic energy, but the proton–proton energy only shows potential energy. Why?

11. Figure 9.6 shows how the positive potential energy of the proton–proton repulsion and the negative energy of the proton–electron attraction combine to lead to an equilibrium radius for a hydrogen molecule. (The figure is actually for H_2^+, but an H_2 figure would be qualitatively similar.) Helium occurs when the two protons get much closer together than that equilibrium radius.[1] How is that possible? Put another way, what feature of the forces between the particles is not included in Figure 9.6?

Problems

12. An ionic bond has four sources of energy, some positive and some negative. In this problem you will consider the contributions of these four sources to an NaCl molecule. Remember that a change that

---

1 Helium also needs at least one neutron in addition to the protons in order to be stable.

tends to happen on its own is associated with a negative energy; a change that you have to force is associated with a positive energy.

(a) The ionization energy of sodium is 5.14 eV. Is that a positive or a negative contribution to the formation of an NaCl molecule?

(b) The electron affinity of chlorine is 3.61 eV. Is that a positive or a negative contribution?

(c) The average distance between the two ionically bonded molecules is 0.236 nm. Calculate the electrostatic energy associated with bringing the two atoms from infinity to that separation. Is this energy positive or negative?

(d) The final contributing factor is repulsion: the electric repulsion of the two nuclei, and the Pauli repulsion of the electron clouds. Does this final factor contribute positive or negative

energy to the molecule? Briefly explain your answer (qualitatively).

(e) The binding energy of an NaCl molecule is 4.26 eV. Based on this number and your three previous answers, calculate the energy associated with the repulsion discussed in Part (d).

13. Lithium has an ionization energy of 5.39 eV and chlorine has an electron affinity of 3.61 eV.

(a) Neglecting repulsive forces, how close would a lithium ion and a chlorine ion have to get in order to have a net energy lower than they would have as separated, neutral atoms?

(b) The equilibrium separation of an LiCl molecule is 0.2 nm and the dissociation energy is 4.86 eV. What is the total energy contribution of the nuclear and Pauli repulsions at that distance?

9.2 Bonding and Antibonding States

If you followed Section 9.1, you can generally predict what types of molecules are likely (or unlikely) to form covalently. But the conclusions of that section follow from a rule – "only unpaired electrons form covalent bonds" – that we have yet to justify. Here we will examine the quantum mechanical wavefunctions that underlie covalent bonding, and in doing so we will see where that rule comes from.

9.2.1 Explanation: Bonding and Antibonding States

We began our discussion of covalent bonding in the last section by bringing a hydrogen atom gradually closer to a proton to create an H_2^+ molecule. Let's walk through that process again, starting when the two protons are very far from each other. The electron can be in a $1s$ state bound to the first proton, or in a $1s$ state bound to the second proton. Let's call those wavefunctions ψ_1 and ψ_2. (Figure 9.9 depicts both situations in one dimension only; you can think of this as plotting the 3D wavefunctions along the line connecting the two nuclei.)

Figure 9.9 The state ψ_1 represents an electron bound to one proton, ψ_2 to the other proton.

Both ψ_1 and ψ_2 are energy eigenstates of our two-distant-protons-and-one-electron system (Question 1). That is, they are both solutions to the time-independent Schrödinger equation $-\hbar^2/(2m)\nabla^2\psi + U(\vec{r})\psi = E\psi$ for the same potential energy function U and the same energy E

(−13.6 eV). Because Schrödinger's equation is linear and homogeneous, any linear combination of ψ_1 and ψ_2 is *also* a solution to that same Schrödinger equation with that $U(\vec{r})$ and E. The two combinations we will focus on here are $\psi_S = A(\psi_1 + \psi_2)$ and $\psi_D = B(\psi_1 - \psi_2)$, where the constants A and B are chosen for normalization. (The subscripts S and D stand for "sum" and "difference.") These combinations are shown in Figure 9.10.

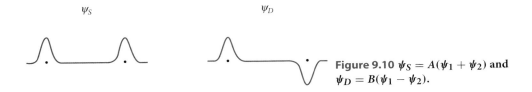

ψ_S ψ_D

Figure 9.10 $\psi_S = A(\psi_1 + \psi_2)$ and $\psi_D = B(\psi_1 - \psi_2)$.

Without going through any math, we can answer two important questions about these states:

- *Where is the electron?* Wavefunction ψ_S represents a state in which the electron has equal probabilities of being close to the left proton or close to the right proton. ψ_D represents exactly the same probabilities. These are two different states, but their position probability distributions $|\psi_S|^2$ and $|\psi_D|^2$ are identical.
- *What is the energy?* We said above that any linear combination of ψ_1 and ψ_2 solves the same time-independent Schrödinger equation that each does, so ψ_S and ψ_D are both energy eigenstates of this system with energy $E = -13.6$ eV.

All four states ψ_1, ψ_2, ψ_S, and ψ_D are degenerate. They represent the ground state – the lowest energy level – of this system.

Incidentally, you can create infinitely many other ground states for this system, such as one that says "the electron has a 3/4 probability of being bound to the left proton and a 1/4 probability of being bound to the right." All such states have the same energy, and all of them can be written in the form $F\psi_1(x) + G\psi_2(x)$, with the constants F and G chosen for normalization. You'll show in Problem 8 that all such states can also be written in the form $H\psi_S(x) + J\psi_D(x)$.

Now Let's Bring the Atoms Closer

Now, just as we did in the previous section, let's see what happens when our two protons approach each other.

Active Reading Exercise: Sum and Difference Wavefunctions

Consider two protons close enough to each other that ψ_1 and ψ_2 overlap. Sketch ψ_S and ψ_D, similarly to how we did above for the case of two widely separated protons. In which state is the electron more likely to be found close to the midpoint between the two protons?

In the region in between the two nuclei, the two parts of ψ_S interfere constructively while the two parts of ψ_D interfere destructively. As a result, an electron in the state ψ_S is most likely to be found in between the nuclei, while an electron in the state ψ_D is more likely to be found outside the two nuclei (Figure 9.11).

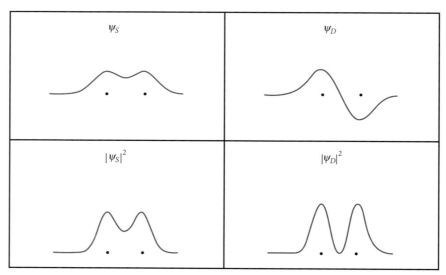

Figure 9.11 In between the two nuclei, ψ_S shows constructive interference and ψ_D shows destructive interference.

That fact has significant consequences for the energy associated with these two states. An electron in between the nuclei screens their charges from each other, reducing their repulsion and lowering their total energy. An electron outside this middle region doesn't provide that screening. Moreover, an electron in between the two nuclei feels a very low potential energy, while an electron on the far side of either nucleus feels a much higher one.

So we can see that ψ_S lowers the total energy of the system and therefore helps bind two atoms together, while ψ_D has the opposite effect (Figure 9.12). For this reason ψ_S is called the "bonding state" and ψ_D the "antibonding state."[2]

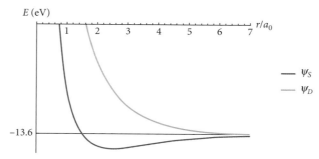

Figure 9.12 The total energy (electron energy plus proton repulsion energy) for an H_2^+ molecule as a function of nuclear separation. If the electron is in the bonding state, there is an equilibrium radius that minimizes the energy. If the electron is in the antibonding state, the nuclei will move apart.

---

2 Technically the bonding and antibonding states aren't exactly equal to ψ_S and ψ_D. The bonding and antibonding states are defined as energy eigenstates of the system, and to find those energy eigenstates you would have to solve the Schrödinger equation for the potential energy function created by two nuclei. The solutions would look close to, but not exactly equal to, ψ_S and ψ_D. They wouldn't look at all like ψ_1 and ψ_2. Since we are making qualitative arguments here, we will use "ψ_S and ψ_D" interchangeably with "bonding and antibonding states."

Recall that, when the two protons were far apart, ψ_1 and ψ_2 represented two different energy eigenstates with the same energy, and you could build any other eigenstate with that energy as a combination of those two. ψ_S and ψ_D *also* represented two different energy eigenstates with that same energy, and you could build all those same eigenstates by combining those two.

With the nuclei close to each other, the picture is very different. The bonding and antibonding states are two independent energy eigenstates with different energies:

- The bonding state ψ_S is the new ground state, with an energy lower than the ground state energy when the protons were widely separated.
- The antibonding state ψ_D is an energy eigenstate, but not the ground state. It has an energy higher than the ground state energy when the protons were widely separated.
- ψ_1 and ψ_2, the states that represent the electron centered on one nucleus or the other, are no longer energy eigenstates. To find the energies associated with these states, and the probability of each energy, you can build the states as combinations of energy eigenstates.

As you might imagine, therefore, ψ_1 and ψ_2 will no longer be of much interest to us. The correct way to analyze this covalently bonded system is by thinking about ψ_S and ψ_D.

So Now – Finally – Why Do Only Unpaired Electrons Contribute?

Consider two helium atoms coming together, their $1s$ subshells filled.

When the protons were far apart there were two $1s$ states (one around each proton), each of which could be occupied by two electrons (one spin-up and one spin-down). Now that the protons are closer, there are still two states around the twin protons whose energies are close to the $1s$ energy of a helium atom. One is the bonding state and the other is the antibonding state, each still capable of holding two electrons.

Those four electrons would love to all crowd into the low-energy bonding state, but Pauli forbids it. Two of them (with opposite spins) occupy that state, and that leaves the other two (also with opposite spins) in the antibonding state. The bonding state lowers the energy of the system, the antibonding state *increases* the energy of the system, and in the final analysis, no energy is saved. This is why you're not likely to run into an He_2 molecule any time soon. (Remember that the electron energy has to go *down* far enough to more than compensate for the increased repulsion energy of the nuclei, as you pull them closer together.)

In the language of Section 9.1, the helium outer subshell has no unpaired electrons available for covalent bonds.

On the other hand, consider fluorine, whose outermost shell is $2p^5$. Two distantly separated fluorine atoms have a total of six $2p$ states (three around each nucleus). Bring them close together and they still have six $2p$ states, three bonding and three antibonding, capable of holding two electrons each.

Active Reading Exercise: F_2

Two individual fluorine atoms have, between them, ten $2p$ electrons. How will those ten electrons fill the available states in the two-proton system?

Figure 9.13 Ten electrons in six 2p states.

The answer is that the electrons will fill the lowest-energy states, the bonding states, first. Only when those are full will the remaining electrons land in the antibonding states. The result is depicted in Figure 9.13.

As we have noted before, a filled bonding state and a filled antibonding state more or less cancel each other out in terms of their effect on the total energy. That leaves the two electrons in the third bonding state to lower the energy of the system. In the language of Section 9.1, each fluorine outer subshell has four paired electrons that do not contribute to the covalent bond, and one unpaired electron that does.

9.2.2 Questions and Problems: Bonding and Antibonding States

Conceptual Questions and ConcepTests

1. Figure 9.9 shows $\psi_1(x)$ and $\psi_2(x)$, two possible ground states of a system comprising two protons (far away from each other) and one electron.

   (a) What is the energy of these states? (Here you can just write down a number.)

   (b) What is the energy of the state $\psi_S(x) = A(\psi_1 + \psi_2)$, with the constant A chosen to properly normalize the function?

2. In this question you're going to walk carefully through what a bonding state is, and how it gets its name.

   (a) Draw two protons relatively close to each other. Draw $\psi_1(x)$, a wavefunction representing an electron attached to the left-hand proton. The wavefunction should be centered on the left-hand proton, but should *not* go to zero at the right-hand proton.

   (b) Draw the two protons again and draw $\psi_2(x)$, the wavefunction representing an electron attached to the right-hand proton.

   (c) Draw the function $\psi_S(x) = \psi_1(x) + \psi_2(x)$.

   (d) How can you tell, based on your drawing, that an electron in the $\psi_S(x)$ state is fairly likely to find itself between the two protons?

   (e) Why does having an electron between the two protons tend to lower the energy of the entire system?

   (f) Finally, why is $\psi_S(x)$ sometimes called the "bonding state"?

3. Figure 9.9 shows $\psi_1(x)$, a wavefunction that says "This electron is attached to the left-hand proton." The drawing of course takes place entirely in one dimension. Describe (you do not need to draw) what $\psi_1(x, y, z)$ would look like in three dimensions.

4. For the case of two protons far away from each other and a single electron, we showed that ψ_1, ψ_2, ψ_S, and ψ_D all represent possible ground states of the system. That is, they all have the same energy as each other, and that is the lowest energy the system can have.

   (a) Describe a possible state of this system that is *not* the ground state: one that has a higher energy. You don't need to do any math here, just tell us about the electron.

   (b) Can your state be built as a linear superposition of ψ_1 and ψ_2? How do you know?

5. Two argon atoms are not likely to form a covalent bond. In Section 9.1 we would have predicted that result by saying that argon's 3p subshell has no unpaired electrons. Explain the same result by showing how two nearby argon atoms would fill up the bonding and antibonding states in their 3p subshells.

6. Draw diagrams showing how the 2p states are filled in N_2 and O_2 that are similar to Figure 9.13 for F_2.

For Deeper Discussion

7. In Figure 9.9 we showed pictures of ψ_1 and ψ_2, and then slightly below that we showed pictures of their superpositions ψ_S and ψ_D. We said ψ_S and ψ_D had to have equal energy because of the general rule that any superposition of two eigenstates with equal energy also has that same energy. In Figure 9.11 we showed versions of those pictures in which the atoms are much closer together, and we argued that, because of screening of the proton–proton repulsion, ψ_S has lower energy than ψ_D. Why does that not violate the general rule that led us to previously conclude that they had to have equal energies?

Problems

8. Remember the definitions $\psi_S = A(\psi_1 + \psi_2)$ and $\psi_D = B(\psi_1 - \psi_2)$, where A and B are non-zero constants. We are concerned here with the set of all combinations $F\psi_1 + G\psi_2$, where F and G are any constants. Show that any such combination can also be written in the form $H\psi_S + J\psi_D$. (If you want the fancy terminology, you are proving here that ψ_S and ψ_D form a "basis" for the same set of functions as ψ_1 and ψ_2.) *Hint*: The way you prove this can always be done by finding G and H as functions of E and F.

9. **Exploration: Bonding and Antibonding State Energies**

   [*This problem requires multiple integrals, a topic from multivariate calculus.*] Consider two protons separated by a distance L. Let ψ_1 be the ground state of one hydrogen atom and ψ_2 the ground state of the other. (It will be the same function, but centered on a different point.)

   (a) Write the potential energy function $U(x, y, z)$ for the electron in the presence of the two protons.

   (b) Write the wavefunctions ψ_1 and ψ_2 in Cartesian coordinates. They are just hydrogen atom ground states, but they are centered on different places. *Hint*: You can start by looking at Appendix G.

   (c) Use a numerical integral to normalize the sum and difference wavefunctions $\psi_S = A(\psi_1 + \psi_2)$ and $\psi_D = B(\psi_1 - \psi_2)$. For the rest of this problem use $L = 0.106$ nm, the equilibrium separation for H_2^+. *Hint*: To get a numerical answer you may need to integrate over a finite region of space. Just make sure it's big enough that the wavefunctions are negligible outside it.

   (d) The expectation value of energy for an electron with a wavefunction ψ is the integral over all space of

   $$\psi\left[-\frac{\hbar^2}{2m_e}\left(\frac{\partial^2\psi}{\partial x^2} + \frac{\partial^2\psi}{\partial y^2} + \frac{\partial^2\psi}{\partial z^2}\right) + U(x, y, z)\psi\right].$$

   (Normally the first ψ in this formula would be ψ^*, the complex conjugate. We can ignore that here because our wavefunctions are real.)

   Have a computer numerically integrate to find the expectation value of energy of the states ψ_S and ψ_D.

   (e) Calculate the (positive) potential energy of the two protons at this separation.

   (f) Using your answers to Parts (d) and (e), calculate the dissociation energy of an H_2^+ molecule, assuming the electron is in the bonding state. What would you get if you assumed the antibonding state instead, and what does that imply?

   (g) You should have found a dissociation energy close to but not exactly equal to the empirical value, 2.7 eV. Aside from taking numerical integrals, what approximation(s) did we make here that would cause our answer to not come out exactly correct?

9.3   Vibrations, Rotations, and Molecular Spectra

Molecules, like individual atoms, have unique patterns of emission and absorption lines. Studying the spectrum of an object can allow us to determine its molecular composition, what isotopes are present, what fraction of the molecules are ionized, and more. Molecular spectroscopy is used in applications ranging from studying interstellar dust clouds to monitoring pollution concentrations in the atmosphere.

9.3.1   Explanation: Vibrations, Rotations, and Molecular Spectra

Like an atom, a molecule can gain or lose energy when its electrons move between different quantum states. But a molecule has two forms of energy that an individual atom does not: vibration and rotation. Each of these involves motion of the nuclei as well as the electrons, and each leads to a characteristic set of spectral lines in addition to the electronic ones.

Vibrations

Consider a diatomic molecule. (We will use the example here of a purely ionic bond, but the argument applies equally well to a covalent bond; as we have mentioned, most molecules have characteristics of both types of bonding.) As we discussed in Section 9.1, the two separate ions are attracted to each other by their opposite electrical charges. But if they get too close, they are repelled by their positively charged nuclei and by Pauli repulsion. So they settle into an equilibrium radius that minimizes the total energy of the system.

> **Active Reading Exercise: The Not-Quite-Equilibrium Radius**
>
> Two atoms in perfect bonded equilibrium are jostled. They are now a bit too far apart (and are therefore pulled toward each other), or else a bit too close together (and are therefore pushed apart). What do they do now? *Hint*: "They come to rest at their equilibrium radius" is a nice, simple answer that doesn't conserve energy.

The answer to the exercise above is that the two molecules do return to their equilibrium radius, but they don't stop there. Their built-up kinetic energy carries them beyond that radius until they stop at the other side, at which point they are pulled in the other direction. In short, they oscillate – or vibrate – about the equilibrium radius.

So the whole system is analogous to a mass-on-a-spring scenario from mechanics. Small oscillations in almost any system can be well approximated by a spring force $F = -kx$, resulting in simple harmonic oscillation with angular frequency $\omega = \sqrt{k/m}$. The "x-position" in this analogy is the distance r between the two nuclei, the "spring constant k" depends on the forces between the atoms (which are complicated), and the "mass m" is the reduced mass of the system: $m_1 m_2/(m_1 + m_2)$. These oscillations result from the potential energy shown in Figure 9.14.

A classical harmonic oscillator can vibrate with any amplitude. If there is a damping force it loses energy gradually until the amplitude decays down to zero. (Picture a squeaky spring.) But as we saw in Chapter 5, a *quantum* simple harmonic oscillator can only have energies $E = (n + 1/2)\hbar\omega$, where $n = 0, 1, 2, \ldots$ is the "vibrational quantum number." Such a system makes discrete jumps ("vibrational transitions") between quantum numbers, giving off or

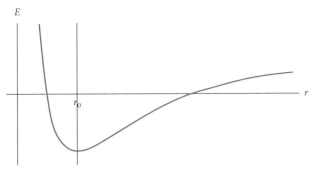

Figure 9.14 The separation between two nuclei in a diatomic molecule can oscillate about its equilibrium value r_0. For small oscillations the potential energy is mathematically equivalent to that of masses on a spring.

absorbing photons in the process. There is a selection rule (beyond our scope to prove) that says such transitions only change n by ±1, so vibrational transitions nearly always[3] involve photons of energy $\hbar\omega$.

The formula $E = (n + 1/2)\hbar\omega$ also implies that molecules have a ground state vibrational energy of $(1/2)\hbar\omega$. A molecule can't just sit at its equilibrium state (which would violate the uncertainty principle), but must oscillate about it with at least that much energy. This ground state oscillation is measurable because it reduces the dissociation energy required to pull a molecule apart.

Rotations

In addition to vibrating about equilibrium, a molecule can also rotate. We said in Chapter 7 that the angular momentum of an electron orbiting a proton is quantized as $L^2 = l(l+1)\hbar^2$, where $l = 0, 1, 2, \ldots$ is a rotational quantum number. The same rule applies to rotation of molecules (or anything else). The energy associated with the rotation is $L^2/(2I)$, where I is the moment of inertia of the rotation, so the possible energies are $\hbar^2/(2I)$ times 0, 2, 6, 12, 20, ...

Once again there is a selection rule that says $\Delta l = \pm1$. Since the rotational states don't have evenly spaced energies, this selection rule does not imply that all the emission (or absorption) photons have the same energy. If we define $E_0 = \hbar^2/(2I)$ then the possible transition energies are $2E_0$, $4E_0$, $6E_0$, and so on.

Molecular Spectra

Because the energies of the allowed rotational transitions are evenly spaced, their emission and absorption spectra consist of evenly spaced lines (in frequency, not wavelength). Pure vibrational transitions all have the same energy. Some molecules can undergo pure rotational transitions or pure vibrational transitions, but in many cases transitions are only allowed that change both quantum numbers by 1. (They can go in the same or opposite directions.)

Lines involving vibrational and rotational transitions produce photons with energy $\hbar\omega + q\hbar^2/I$, where q can be any non-zero integer (positive or negative). See Figure 9.15.

---

3 Remember: selection rules are strong tendencies, not unbreakable laws.

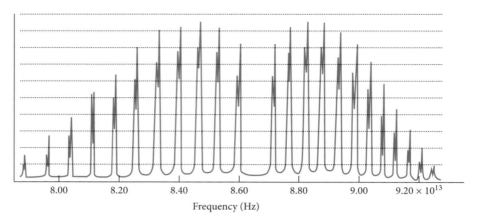

Figure 9.15 Infrared absorption spectrum of HCl. The spectrum involves a transition from the vibrational ground state to the first excited state. Meanwhile each peak corresponds to a different rotational transition. All lines represent transitions where n and l each change by 1.

We've summarized below the rules governing vibrational and rotational transitions.

Vibration	Rotation
$E = \left(n + \dfrac{1}{2}\right)\hbar\omega$	$E = \dfrac{L^2}{2I}$
where $\omega = \sqrt{\dfrac{k}{m}}$	where $L^2 = l(l+1)\hbar^2$
Quantum number: $n = 0, 1, 2, \ldots$	Quantum number: $l = 0, 1, 2, \ldots$
Selection rule: $\Delta n = \pm 1$	Selection rule: $\Delta l = \pm 1$
therefore $|\Delta E| = \hbar\omega$	therefore $|\Delta E|$ for a transition from $l-1$ to l is $l\dfrac{\hbar^2}{I}$

As a rough guide, electronic transitions are typically[4] on the order of a few eV (visible to ultraviolet), vibrational transitions are tenths of an eV (near-infrared), and rotational transitions are hundredths of an eV (far-infrared). At room temperature, collisions between molecules don't have enough energy to excite electronic or vibrational transitions, so molecules at room temperature are virtually all in their electronic and vibrational ground states. Molecular collisions at room temperature can excite rotational transitions, so the air molecules in your room are rotating with a range of energies.

Our discussion of spectral lines applies to all molecules except "homopolar" ones, such as N_2 or O_2, which consist of two identical atoms. Because these are perfectly symmetric they have zero electric dipole moment. That means they have essentially no interaction with electromagnetic fields, so they cannot absorb or emit photons. They rotate and vibrate just like all molecules, and they can still make transitions in their rotational and vibrational energies via collisions or other forms of energy exchange, but they don't produce vibrational or rotational spectra.

---

4 assuming they only involve outer electrons.

9.3.2 Questions and Problems: Vibrations, Rotations, and Molecular Spectra

Conceptual Questions and ConcepTests

1. At very low temperatures a sample of diatomic molecules will all be in their rotational and vibrational ground states. Assuming they can only make transitions involving both rotation and vibration, how many lines would these molecules have in their absorption spectra?

2. If you measure the rotational spectrum of a molecule and then mix in a small amount of a heavier isotope of that molecule (effectively the same but with a higher mass), how does that change the measured spectrum? How can molecular spectroscopy be used to determine the ratios of different isotopes in a sample?

3. Explain why there is a gap in the otherwise evenly spaced peaks in Figure 9.15. What transition(s) would a peak at that frequency represent?

4. These questions are based on Figure 9.15.

    (a) Give the initial and final rotational quantum numbers for the first three peaks to the right of the gap in the middle.

    (b) What rotational state was most highly occupied in this HCl sample?

5. Deuterium (D) is an isotope of hydrogen with a single electron orbiting a proton and a neutron.

Because the neutron has no charge, deuterium nuclei exert nearly the same force on each other as hydrogen nuclei. So a vibrating D_2 molecule has the same spring constant k as a vibrating H_2 molecule. Which molecule is easier to dissociate (break apart) and why? *Hint*: Think about the ground state energy.

For Deeper Discussion

6. The effective potential energy felt by a diatomic molecule is not actually a simple harmonic oscillator potential. Instead it is asymmetric, such that vibrations have an average R greater than the equilibrium value. Based on that fact, how might you expect the rotational spectrum of a molecule to change when it is in an excited vibrational state?

Problems

7. Derive the formula $|\Delta E| = l\hbar^2/I$ for a rotational transition that obeys the selection rule $\Delta l = \pm 1$.

8. The moment of inertia of a diatomic molecule is mR^2, where $m = m_1 m_2/(m_1 + m_2)$ is the reduced mass and R is the separation between the two nuclei. Looking at Figure 9.15, estimate the equilibrium separation of the two atoms in an HCl molecule.

Chapter Summary

Chapter 8 showed how quantum mechanics explains the tendency of some atoms to give up electrons, and other atoms to absorb them. This chapter explores how those tendencies determine the ways in which atoms combine into molecules.

Section 9.1 Ionic and Covalent Bonds

Molecular bonds are traditionally categorized as "ionic" (one atom gives an electron to another) or "covalent" (atoms share electrons). Most bonds fall on a spectrum, with electrons neither fully donated nor fully shared. But the binary model still provides a useful framework for exploring the mechanisms of atomic bonding.

- An ionic bond occurs when one atom gives up an electron (raising the energy of the total system), another atom gains that electron (lowering the energy of the total system), and then the two resulting ions approach each other (lowering the energy of the system to less than its original level).

- The two ions do not get too close because of the mutual repulsion of their nuclei, and also because of "Pauli repulsion": as the two electron clouds merge, electrons are forced by the Pauli exclusion principle into higher-level states. So the two atoms settle into an equilibrium radius.

- The "binding energy" or "dissociation energy" of the resulting molecule is the difference between its total energy and the energy of the original two atoms.

- In a "covalent bond" the orbit of an electron surrounds multiple nuclei.

- Only an "unpaired" electron – a spin-up electron in the outer shell that has no corresponding spin-down electron, or vice versa – takes part in a covalent bond. As an example, the outermost shell of an oxygen atom has six electrons but only two unpaired electrons, so it can form two covalent bonds.

Section 9.2 Bonding and Antibonding States

This section takes a deeper dive into the quantum mechanical wavefunctions that underlie covalent bonding.

- When two hydrogen atoms are far away from each other, there are two ground states of the whole system that are available to an electron: "1s orbit around this nucleus" and "1s orbit around that nucleus." (Remember that, because of spin, each of these states is capable of holding two electrons.)

- As the atoms approach each other, these states add (linear superposition). One resulting state gives the electron a high probability of appearing between the two nuclei; this is called a "bonding state" because it lowers the energy of the system. The opposite, the "antibonding state," raises the energy of the system.

- The end result lowers the energy of the entire system – and therefore bonds the two atoms together – if the electrons can occupy more bonding states than antibonding states. This explains the rule presented in the previous section that only unpaired electrons form covalent bonds.

Section 9.3 Vibrations, Rotations, and Molecular Spectra

Vibrations and rotations represent energetic modes of a molecule that are based on the interactions between its nuclei rather than the states of its electrons. They therefore represent new physical principles – and new spectral lines – that emerge only when atoms combine into a molecule.

- Vibrations involve two nuclei oscillating around their equilibrium radius. The resulting motion is a quantum harmonic oscillator, so its energy levels are evenly spaced.

- Molecular rotational energies are proportional to $l(l+1)$ for integer l, so they are not evenly spaced.

- Both vibrations and rotations obey a selection rule that says that their quantum numbers generally only change by ± 1 in a transition. Since the vibrational states are evenly spaced, they always contribute the same amount of energy to an emission photon, while the rotational transition energy depends on the quantum numbers involved.

- The table on p. 412 concisely lays out the equations governing molecular vibration and rotation.

10

Statistical Mechanics

How would you model the air in your room?

You might try to measure the states of the 10^{27} or so molecules floating around, and then write equations to predict their behavior. But whether you use Newton's laws or Schrödinger's equation, such an approach would be impossible both physically and computationally. And even if it weren't, what would you do with the predictions you got? You don't care if this molecule bumps into that one; you care about aggregate properties that you can measure, such as air pressure. (Incidentally, have you ever considered that an individual molecule does not have a "pressure," or a "temperature," or a "color"? Those exist only as emergent properties of large collections of molecules.)

So, for macroscopic objects we use statistical methods. We talk about the average properties of the molecules, and the distributions of those properties. In this chapter you'll learn techniques for analyzing large collections of particles statistically, and you'll apply those techniques to problems ranging from the rate of fusion in the Sun to the energy required to heat up a room by a degree.

10.1   Microstates and Macrostates

Probability and statistics are powerful predictive tools, but only when applied to large numbers of events.

- If you roll two dice, you should not be surprised by any result.
- If you roll 100 dice, you should expect the average number on a die to fall between 3 and 4.
- And if you roll a million fair dice, you can bet the rent money that the average will be almost exactly 3.5.

In physics it is not uncommon to work with systems of 10^{23} molecules, sometimes far more. With those orders of magnitude, predictions based on probability can very reasonably be expressed as guarantees. This is one of the reasons that deterministic Newtonian mechanics makes such a good predictor for systems that are, at their lowest levels, full of quantum randomness.

10.1.1   Discovery Exercise: Microstates and Macrostates

🔍 You roll two dice, one at a time, and write the results as an ordered pair. For example $(1, 5)$ means you rolled a 1 and then a 5.

1. The results $(1, 6)$ and $(6, 1)$ each have a sum of 7. How many combinations (including those two and others) have a sum of 7?

2. How many combinations have a sum of 2?

   See Check Yourself #12 at www.cambridge.org/felder-modernphysics/checkyourself

3. After a while you have filled a sheet of paper with ordered pairs. Now you count how many pairs sum to 2, and how many pairs sum to 7. Which of these two sums do you expect to see more often? How much more often?

10.1.2   Explanation: Microstates and Macrostates

Consider a canister filled with a gas of N molecules. (A typical number for N would be 10^{23}.) Classically, a complete description of the gas would specify the position and velocity of all N molecules. A quantum mechanical description would be the wavefunction for all of those particles. Either description would be what we call the "microstate" of the system: a full account of every part of the system at the microscopic level.

You can see that describing a room at the microstate level is impossible. It would be impossible to do the measurements, impossible to do the calculations, and impossible to use them even if you had them. So why even talk about it? Because thinking about microstates on a *statistical* level can lead to important conclusions. (That's a quick summary of this entire chapter.)

The "macrostate" of a system comprises all the things that you could actually measure on a macroscopic level: the mass, the density, the temperature, and so on. With good measuring instruments you can include less obvious characteristics such as chemical composition and heat capacity. The quantities that make up the macrostate of our canister come from sums or averages of the properties of the molecules that make up the gas.

A "microstate" is a well-defined notion, but the definition of a "macrostate" depends on the sensitivity of your measuring devices. You could separately measure the mass of each half of our canister and see whether the gas is evenly distributed between them or not. Maybe you could divide the canister into 100 small volumes and measure the mass and energy inside each one, and then the macrostate would include all of those. If your measuring devices were sufficiently sensitive you could measure every particle, and then there would be no distinction between microstate and macrostate, but in practice we cannot come near that for any macroscopic system.

A Simple Example: A Paramagnet

In a paramagnet, each atom has a small magnetic moment that can point up or down. (The difference between a paramagnet and the more familiar "ferromagnet" is that in a ferromagnet the atoms tend to align with their neighbors. In a paramagnet they are far enough apart that each atom points up or down independently of the others.)

The system we will consider here is a paramagnet with N atoms. Our system has no external magnetic field biasing the results between up and down.

- The *microstate* of our system is a complete list of the orientations of all the atoms.[1]
- The *macrostate* – what we actually measure – is the total magnetic field. That tells us how many atoms are pointing up, but not which ones. So we can identify our macrostate as an integer, ranging from 0 (all down) to N (all up).

At random intervals, each atom might suddenly flip its state. So, regardless of the initial condition, after enough time the state of the system is entirely random; each atom has a 50% chance of pointing up, independently of all the other atoms.

Here's an easy question (most people get it right unless they overthink it), and a hard question (most people don't).

Easy question: Which is a more likely value for our macrostate, $N/2$ or N?

Answer: The state $N/2$ is more likely than any other state.

Hard question: Why?

Most people answer "because it's the average" which only begs the question: why is the average more likely than other outcomes? How can we say that the state of the system is completely random, and in the same breath say that some states are more likely than others?

Let's tackle that question by looking at a small value of N that we can get our heads around.

Active Reading Exercise: Microstates of a Paramagnet

Suppose you have a paramagnet with four atoms. One possible microstate is UUDD, meaning the first two are up and the last two are down.

1. How many possible microstates are there? (There are several ways to figure this out, but if you're stuck you can just list them all.)
2. How many microstates have all four atoms pointing up?
3. How many microstates have exactly two atoms pointing up?
4. If you were to measure the total magnetic field of this paramagnet at some instant, what is the probability that you would find all four atoms pointing up? What is the probability that you would find exactly two atoms pointing up (i.e. zero magnetic field)?

We hope you answered that Active Reading Exercise carefully, because we're going to walk through the whole thing now, and you'll learn this a lot better by checking your answers than by just reading ours.

1. There are two possible states for the first atom. For each of those there are two possible states for the second atom, so there are four possible states for the first two:

---

1 The full microstate would also include how the atoms are vibrating and, well, everything else about them, but here we are only interested in magnetic orientation.

All two-atom states:　UU, UD, DU, DD.

For each of those four states there are two possible states for the third atom, so there are eight possible states for the first three. (You can make them by sticking first a U and then a D after each of the four we just wrote.) Continuing in this way we conclude that there are 2^4, or 16, possible microstates for our four-atom system.

2. Only one of those states, UUUU, has all four atoms pointing up.

3. The simplest way to find how many microstates have two "up" atoms is to list them all. The key to making a list like this is to have a system such that you know you haven't missed any. We'll start with all the ones that have a U in the first position, and we'll list all the places the second U could be. Then we'll move to ones where the first U that appears is in the second position, and so on:

   All two-up-and-two-down states:　UUDD, UDUD, UDDU, DUUD, DUDU, DDUU.

So there are six of these microstates.

4. The probability at any given moment of having all four atoms pointing up is 1/16. The probability of exactly two atoms pointing up is 6/16, or 3/8.

Now we have the answer to the hard question we posed before the exercise. The macrostate $N/2$ is more likely than the macrostate N because $N/2$ arises from *more possible microstates* than N. If you want to make predictions for this system, place your bets on the macrostates that arise from the most possible microstates.

You won't always win those bets. It's true that our four-atom system has a pretty good chance of ending up 50/50, and that result is more likely than another other. But it's also true that different results – including "all four up" – are not all *that* unlikely, and they will come up from time to time. Statistical reasoning can't predict with confidence the behavior of a small number of particles.

On the other hand, for large numbers, statistics make very accurate predictions. In Problem 15 you'll show that, as the number of atoms goes up, the probability that you'll find an answer very close to 50% rises quickly. For a million atoms the percentage pointing up is effectively guaranteed to be between 49% and 51%. For a macroscopic number of atoms the fraction pointing up will be so close to 50% that any nett magnetic field of the paramagnet will be undetectable.

Combinations and Two-State Systems

For our four-atom paramagnet you can list all the microstates, but for larger systems you need other ways to calculate probabilities. To illustrate one helpful tool for such calculations, consider a bag with five tiles labeled A, B, C, D, and E. You reach into the bag and pull out two tiles. How many possible draws can you get?

We can answer that question by listing all the possible draws, being systematic as always. We get a total of 10:

AB, AC, AD, AE, BC, BD, BE, CD, CE, DE.

This mathematical topic – "how many different ways can you choose m items from a group of n items?" – is called "combinations." (It is distinguished from "permutations," which ask the same question except that order matters. In our combinations question above, we did not count A B and B A as two different draws.)

The math symbol for "how many ways to choose two things from a set of five" is written $\binom{5}{2}$, and is pronounced "5 choose 2." The general formula is

$$\binom{n}{m} = \frac{n!}{m!\,(n-m)!}. \tag{10.1}$$

The exclamation points are factorials. We can check this with the example we just worked out:

$$\binom{5}{2} = \frac{5!}{2! \times 3!} = \frac{120}{2 \times 6} = 10.$$

You'll justify Equation (10.1) in Problem 18. You will find that formula very useful in the problems in this section (and the rest of the chapter) – especially when working with "two-state systems" such as the atoms of our paramagnet, each of which had only two available states.

Above we asked how many microstates in our four-atom paramagnet had exactly two atoms pointing up. That's a combinations question: how many pairs of two can you find in a group of four? You can use Equation (10.1) to confirm that $\binom{4}{2}$ is 6, the same result we got by listing them all.

The Fundamental Assumption of Statistical Mechanics

In that paramagnet example, we calculated probabilities based on how many *micro*states would lead to any given *macro*state. In the rest of this chapter we're going to use that method of calculating probabilities to derive a number of non-obvious results about macroscopic systems. First, however, we have to step back and identify the underlying assumption we've been using in these calculations.

Active Reading Exercise: Identifying Our Assumption

We calculated above that a 4-atom paramagnet has 16 possible microstates, and that 6 of those involve equal numbers of up and down atoms. From those numbers we concluded that the probability of the system being in a state with equal numbers of up and down atoms was 6/16.

What assumption did we have to make to go from those two numbers to that conclusion about probability? (We're not asking what assumptions went into finding the numbers 16 and 6; that was just counting. We are asking what assumption went into the next step, where we said that their ratio equals the probability of finding that particular macrostate.)

The assumption we had to make is that all of the microstates are equally likely. If you flip a fair coin four times, you have a 3/8 chance of getting equal numbers of heads and tails. If you flip a weighted coin that is more likely to give heads than tails, then you will get a different result.

All of statistical mechanics flows from this one assumption.

The Fundamental Assumption of Statistical Mechanics

Over a long enough span of time, an isolated system will spend equal times in each of its accessible microstates.

This principle is sometimes called the "ergodic hypothesis."

For the four-atom paramagnet, this means each microstate such as U U D D or U U U U will occur equally often. For a macroscopic paramagnet with 10^{23} atoms, virtually all of the microstates have between 49.9999% and 50.000001% up atoms,[2] so you could wait the lifetime of the universe and never measure a deviation from the 50/50 up and down macrostate.

Note the two key words "isolated" and "accessible." The first means that this assumption only applies to systems that are not able to exchange energy, particles, or anything else with their environment. (At least not anything significant to the measurements you are making. So it's OK if you bounce low-energy photons off the system to observe it, for example, as long as they don't noticeably perturb the system.)

The word "accessible" means that you only count microstates that it's possible for the system to reach. For instance, an isolated system has a fixed amount of energy, so it can only be in a microstate with that amount of energy. There is zero chance of an isolated gas canister evolving to a state where all its molecules have twice as much energy as they started with. Accessibility imposes other restrictions as well; you can imagine a microstate in which some of the molecules of gas are outside the canister, but if the walls are impermeable then that microstate isn't accessible. (The time it would take a milligram of gas to quantum-tunnel through a macroscopic canister wall is much longer than the lifetime of the universe.)

With those two restrictions, the fundamental assumption is a powerful statement about how systems behave. It's not obvious that it should be true, but its consequences have been successfully tested in so many different contexts that many physicists regard it as perhaps the most broadly applicable postulate in all of physics.

---

2   We just stopped because we were tired of typing 9s and 0s, but we could have kept going.

Example: Four Oscillators

A particular kind of particle can have energy 0, 1, 2, etc. (It might be a quantum simple harmonic oscillator where the energy unit is $\hbar\omega$ and we've subtracted off the ground state energy for simplicity.) Suppose you have four of these particles lined up in a row.

- The microstate of this system is a list of the energies of each particle. For example, here is one possible microstate:

$$2102.$$

  The first and fourth particles have two energy units each, the second has one, and the third has none.

- The macrostate in which we are interested is the total energy in each half of the system (the first two particles, and the last two). In the example in the previous bullet, the macrostate would be "three energy units in the first half and two in the second."

Question: Suppose the total system has two energy units. What is the probability at any given moment that both of those energy units are in the first half?

Answer: First we have to count the accessible microstates, meaning all the ones that have two energy units. The key is to be systematic so you don't miss anything. We chose to organize our list into three rows based on the energy of the first particle. The "winning" states are indicated in red.

20 00					
11 00	10 10	10 01			
02 00	01 10	01 01	00 20	00 11	00 02

There are 10 accessible microstates and 3 of them have both energy units in the first half, so the probability is 3/10.

10.1.3 Questions and Problems: Microstates and Macrostates

Conceptual Questions and ConcepTests

1. For each term below, give a brief description of what it means in your own words.

   (a) Microstate

   (b) Macrostate

   (c) The fundamental assumption of statistical mechanics

2. In which of the following situations will a paramagnet obey the fundamental assumption of statistical mechanics? (Choose all that apply and explain in each case.)

   A. It is in a hermetically sealed container interacting with nothing.

   B. It is in a constant, external magnetic field, but otherwise interacting with nothing.

   C. It is in a constant, external magnetic field while sitting out on the counter in a hot room.

3. At any given moment each molecule of air in a room is roughly equally likely to be found anywhere inside that room. Using the fundamental assumption, explain why you never find all the air in a room bunched up in one corner.

4. An isolated paramagnet (with no external field) is more likely to have half of its atoms pointing up than it is to have all of them pointing up. Why is that not a violation of the fundamental assumption?

5. You roll two normal six-sided dice, one red and one black. Figure 10.1 gives all the possible sums (the macrostates in which we are interested), and the number of microstates to which each corresponds.

Sum	2	3	4	5	6	7	8	9	10	11	12
Ω (number of microstates)	1	2	3	4	5	6	5	4	3	2	1

Figure 10.1

For instance, the table tells us that there are 3 different ways to add up to 10. (They are 4–6, 5–5, and 6–4.) This number is sometimes referred to as the "multiplicity" of a macrostate. What mathematical operation would you perform to turn this table of multiplicities into a table of probabilities?

6. Suppose you flip two coins, but they are weighted coins; each one is twice as likely to show up heads as tails. So there are four possible microstates of this system (HH, HT, TH, and TT), but they are not all equally probable. Is this a violation of the fundamental assumption of statistical mechanics? Why or why not?

For Deeper Discussion

7. When a system isn't isolated, it often tends to evolve into its lowest-energy state. For example, a ball rolling inside a bowl will quickly settle down to rest at the bottom of the bowl. Macroscopically we say this occurs because of friction, but at a microscopic level that just means that in the many collisions between atoms of the ball and atoms of the bowl, the ball's atoms give up their energy to the bowl much more often than the reverse happens. Explain why that is true based on the fundamental assumption. *Hint*: You can't apply the fundamental assumption to the ball by itself because it's not isolated.

Problems

8. You roll two normal six-sided dice, one red and one black. One resulting microstate is "red die gets a 5, black die gets a 1."

    (a) How many microstates are there?

    (b) How many of these microstates result in the macrostate "the sum of the two dice is six"? (The microstate we gave above is one of them.)

    (c) What is the *probability* that the sum of the two dice is six?

9. You are holding two cups, each containing two coins. You shake the cups, randomizing the coins, and then you look into each of them. What is the probability that the two cups have the same number of heads as each other?

10. In the presence of an external magnetic field, the up and down states of the atoms in a paramagnet have different energies. For simplicity we'll call those energies 0 (for down) and 1 (for up). Suppose you have a paramagnet with six atoms, and you're able to measure the total energy in each half of the

paramagnet: the first three and the last three. For example, if the microstate is UUD DUD then we would say the macrostate is "two up in the first half and one up in the second," or equivalently "two energy units in the first half and one in the second." Now suppose the total system has two energy units. What is the probability at any given moment that both of those energy units are in the first half?

11. The "Four Oscillators" Example on p. 420 describes a system of simple harmonic oscillators, each of which can have any integer number of energy units from 0 up to the total energy of the system.

    (a) Suppose you have a system with three oscillators and two total energy units. What is the probability at any given moment that both of the energy units are in the first oscillator?

    (b) Suppose you have a system with four oscillators and two total energy units. What is the probability at any given moment that the first oscillator has exactly one energy unit?

    (c) Suppose you have a system with four oscillators and two total energy units. What is the probability at any given moment that the first two oscillators between them have exactly one energy unit?

To answer Problems 12–16 you can use Equation (10.1).

12. On p. 417 we calculated the number of states of a four-atom paramagnet that have exactly two atoms pointing up. We did it there by just listing them. Show how we could have done it using Equation (10.1) instead.

13. Suppose you flip a coin 15 times.

    (a) How many possible outcomes are there? (This does not require Equation (10.1).)

    (b) How many of those outcomes have exactly three heads? (This does.)

    (c) What's the probability that you will get exactly three heads when you do this experiment?

14. Suppose you have a paramagnet with 10 atoms.

    (a) What is the probability at any given moment that none of the atoms is pointing up?

(b) What is the probability at any given moment that exactly half of the atoms are pointing up?

15. In this problem you are going to calculate the probability of a paramagnet with N atoms having somewhere between 40% and 60% of those atoms pointing up.

   (a) Find this probability for $N = 10$. This will require summing three probabilities, for $N_{up} = 4$, 5, and 6.

   (b) Find this probability for $N = 100$. This will require summing all 21 probabilities from $N_{up} = 40$ through 60.

   (c) Find this probability for $N = 1000$. (You may need to ask a computer to find the difference between this number and 1 to see any digits past the decimal.)

   (d) What do your answers imply about the state of an actual macroscopic paramagnet that you can hold in your hand? Explain.

16. A system consists of 20 particles, each of which can have energy 0 or 1. (This could be, for example, a paramagnet in an external magnetic field.) The system has a total energy of 5. The amount of energy in the first 10 particles could be anywhere from 0 to 5. Calculate the probability of each of those six possibilities.

17. The Example on p. 420 describes a system of simple harmonic oscillators, each of which can have any integer number of energy units from 0 up to the total energy of the system. If a system consists of 10 such oscillators and has total energy 5, what is the probability that all of the energy is in the first 5 oscillators?

18. **Exploration: The Formula for Combinations**

   In this problem you will justify Equation (10.1), which answers the question: "A group of n distinct items contains how many subgroups of m items?" We will focus here on a specific example: *In a room with* 10 *people, how many* 3-*person committees can you form?* (Each person can belong to more than one committee.)

   (a) You begin by selecting the first member of your committee. How many options do you have?

   (b) After you pull that first person out, you select the second member. How many options do you have?

   (c) With both of those people pulled out, how many options do you have for the third?

   (d) How many different ways, in total, could you have made those three selections?

   (e) Show that your answer to Part (d) can be written as $(10!)/(7!)$. (*Do not* do this by multiplying everything out.)

   We're not done yet, and here's why. One of the committees you created was Rack–Shack–Benny. Another was Rack–Benny–Shack. But those are the same committee!

   (f) List all the possible orders in which you could create the committee that includes Rack, Shack, and Benny. How many are there?

   (g) Explain why your answer to Part (f) can be expressed as $(3!)$. (*Hint*: How many choices do you have for who goes first? And from there ...)

   (h) Explain how all this leads to Equation (10.1).

10.2 Entropy and the Second Law of Thermodynamics

You may have heard that entropy is a measure of the "disorder" or "chaos" of a system. Such terms can be helpful up to a point, but they are ambiguous and can even be quite misleading. In this section we will see the precise meaning of the term entropy. (You may want to think about how it does, and how it does not, map to more common terms such as "disorder.") We'll also see how the second law of thermodynamics, which says "entropy tends to increase over time," follows from the fundamental assumption of statistical mechanics.

10.2.1 Discovery Exercise: Entropy and the Second Law of Thermodynamics

There are N air molecules in a room. Imagine that the molecules don't affect each other at all (a reasonable approximation), and that in each millisecond each molecule has a 10% chance of moving from the side of the room where it is at that moment to the other side. Initially all the air is on the left side of the room.

1. A millisecond later, about how many molecules are on the right side of the room?

2. A second later, about how many molecules are on the right side of the room?

   See Check Yourself #13 at www.cambridge.org/felder-modernphysics/checkyourself

3. How will you see the number of molecules on the right side change if you keep watching for several hours?

10.2.2 Explanation: Entropy and the Second Law of Thermodynamics

In Section 10.1 we said all the possible microstates of an isolated system are equally likely. That means macrostates corresponding to many microstates are more likely than ones corresponding to very few microstates. For example, suppose you flip a coin 10 times. You are just as likely to get HHHHHHHHHH as HTTHHTHTTH; those are each microstates of the system. For the macrostate "total number of heads," however, HHHHHHHHHH is the only possible microstate corresponding to the macrostate "10 heads," while there are hundreds of microstates corresponding to "5 heads," so the macrostate "5 heads" is much more likely than the macrostate "10 heads."

The Second Law of Thermodynamics

> #### Multiplicity and Entropy
>
> For each possible macrostate of a system, we define the "multiplicity" Ω as the number of corresponding microstates. We define the "entropy" S of a macrostate as
>
> $$S = k_B \ln \Omega, \tag{10.2}$$
>
> where "Boltzmann's constant" $k_B = 1.38 \times 10^{-23}$ J/K.
>
> Multiplicity is a unitless quantity; entropy has the same units as Boltzmann's constant (energy over temperature).

A few notes about those definitions:

- We haven't normalized any probabilities here. If we tell you that a particular macrostate has multiplicity $\Omega_1 = 10^{20}$, and therefore entropy $S_1 = 6 \times 10^{-22}$ J/K, you still have no idea how probable or improbable that particular macrostate is. But you do know that that macrostate is 10 times as likely as a different state for which $\Omega_2 = 10^{19}$.

- The point of the logarithm is to create an "extensive" variable: if one system has entropy S_1 and another has entropy S_2, then the combined system has entropy $S_1 + S_2$. You'll show in Problem 19 that entropy has this property and multiplicity doesn't.

- If we were creating statistical mechanics from scratch today, we might define entropy as a unitless quantity just equal to $\ln \Omega$. But Equation (10.2) matches historical usage, which was developed before Ludwig Boltzmann figured out that this quantity is related to microstates and multiplicity.

- This definition of multiplicity (and therefore of entropy) is predicated on discrete counts of macrostates and microstates. The examples we need in this book will generally fall into that category. But in case you're curious, the general approach for continuous variables is to break their possible values into discrete bins and use those to define the multiplicity. In the limit of sufficiently small bins, the exact bin size drops out of the measurable results.[3]

Example: A Five-Atom Paramagnet

A paramagnet consists of five atoms, each of which can point up or down. What is the entropy of the macrostate "two atoms pointing up"?

Answer: We can get the multiplicity by listing the corresponding microstates and counting them:

$$UUDDD, \quad UDUDD, \quad UDDUD, \quad UDDDU, \quad DUUDD,$$
$$DUDUD, \quad DUDDU, \quad DDUUD, \quad DDUDU, \quad DDDUU.$$

Alternatively, we can note that the number of ways you can "choose" two atoms out of five to point up is

$$\binom{5}{2} = \frac{5!}{2!\ 3!}.$$

Either way you get $\Omega = 10$, which leads to

$$S = k_B \ln \Omega = \left(1.38 \times 10^{-23} \text{ J/K}\right) (\ln 10) = 3.2 \times 10^{-23} \text{ J/K}.$$

Probability problems tend to be framed in terms of individual experiments: "If you randomized each of the five atoms in your five-atom paramagnet, what is the probability that you would get exactly two of them pointing up?" But here we are less interested in sudden bursts of randomness, and more interested in gradual evolution.

Active Reading Exercise: An Evolving Paramagnet

Suppose the five-atom paramagnet from the previous example begins with all five atoms pointing down. But every millisecond, each atom has a 1% chance of randomly flipping its state. What do you expect to see after 5 ms? What do you expect to see after a year? (We are asking for some thought and a general answer; we are not asking for calculations.)

---

3 That kind of binning is what Planck was doing when he quantized the energy in cavity radiation. In that case he found that he could match the experimental data by tuning the bin size to a specific value rather than taking the limit where it becomes infinitesimal. Thus was quantum mechanics born.

That Active Reading Exercise is a simple model of the kind of behavior we see with large-scale systems: over time, they change randomly. After 5 ms you will still see your initial condition (all down), or something very much like it. After a year, when every atom has randomized itself many times, you are as likely to see any *microstate* as any other. Which means you are most likely to see the *macrostates* with the highest entropy: two or three atoms up.

After that, the system will occasionally, by random chance, slip back into a lower entropy state. One out of every 32 times you measure it, this paramagnet will be in the state "all down." But that's only because this system is so small. We ran a computer simulation of 10,000 atoms with the condition above: they all start down, and each ms each one has a 1% chance of flipping. Figure 10.2 shows the results. We see the number of "up" atoms rise rapidly to 50% of the total. From then on there are only small fluctuations away from a 50/50 split of up and down.

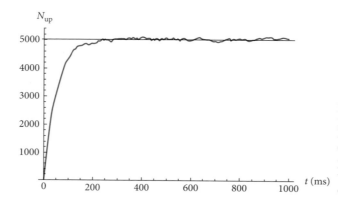

Figure 10.2 A 10,000-atom paramagnet starts out with all spins pointing down. Each ms each spin has a 1% chance of flipping. After 0.2 s, about half the spins are pointing up, and from then on the number of "up" atoms never changes much.

If this were a macroscopic system with 10^{23} atoms, the "all-down" state would quickly evolve into the 50/50 state, but the 50/50 state would never spontaneously evolve back into a state noticeably different from 50/50. Again, this is not because the atoms are conspiring to stay in balance; it is because the overwhelming majority of microstates fall within 0.00001% of that 50/50 ratio.

All of this is summed up in one of the most important laws of physics.

The Second Law of Thermodynamics

An isolated macroscopic system will never decrease its entropy.

Just as in our discussion of the fundamental assumption in Section 10.1, the word "isolated" is crucial here. You can certainly put a system into a low-entropy state. For example, a canister with all the air on one side of it (and a vacuum on the other) has a lower entropy than the same canister with the air spread out evenly. Therefore, left to itself, the air will never spontaneously migrate to an all-on-one-side state. But you can use a piston to force the air to one side, leaving a vacuum on the other. If you do that slowly at a constant temperature, the canister's entropy will decrease. In that case, though, the canister of air isn't isolated. If you pull back your view to define a system comprising the canister of air, the piston, you, and the room around you, you'll find that the total entropy of that system went up in the operation.

Irreversible Processes

Imagine a computer simulation of billiard balls bouncing around a billiard table. In this simulation the balls roll perfectly along the table, with no sliding and no air resistance, so they don't slow down as they move. Furthermore, all the collisions are perfectly elastic, so the balls don't lose energy when they collide with each other or the sides of the table.

Question: If you watch a video of this simulation, how can you determine whether the video is running forward or backward?

Answer: You can't! The laws that govern each collision – conservation of energy and conservation of momentum – are perfectly reversible. If you make the "billiard table" a three-dimensional room and add gravity to the simulation, you *still* can't tell forward time from backward time; a ball speeds up as it heads downward, bounces elastically off the floor, and then slows down as it heads upward, and in the backward video it does the exact same thing.

Do the conditions of our scenario – perfectly elastic collisions, no friction, and no air resistance – sound too contrived to be meaningful? They shouldn't, because those are precisely the conditions of the microscopic world. Energy is always conserved. If you could watch a video of air molecules bouncing around a room, that movie would describe the same laws of physics if you watched it forward or backward.

Why, then, in a universe built from time-symmetric laws, is time itself so asymmetric? Why don't ruins un-crumble into castles? Why don't oak trees un-grow into acorns? Why does your phone right now contain all the pictures you took yesterday, but none of the pictures you will take tomorrow? The answer is that all irreversible processes involve a net increase in entropy.[4]

Consider a block sliding along a table. Friction gradually brings the block to rest. At a macroscopic level we say that the block lost energy to the non-conservative force of friction, and that's why the same motion would never run in the reverse direction. Where did that energy go? It went into "thermal energy," meaning the energy associated with how hot an object is. A block moving in a cold room becomes a block at rest in a (slightly) warmer room, and energy is conserved.

But a microscopic description of the same event looks quite different. At this level the kinetic energy of the block became kinetic energy of the air molecules. The difference is that the molecules of the block were initially all moving in the same direction, but the air molecules that now have that kinetic energy are all moving in random directions. That's why at a macroscopic level we see thermal energy rather than kinetic energy.

The macrostate "all the molecules of the block are moving to the left at 3 m/s" has one corresponding microstate. The macrostate "there's a lot of energy in the random motions of the air molecules" has a lot of corresponding microstates. For that process to reverse would require that all of the randomly moving air molecules happen to hit the block over and over in the same direction, imparting a net macroscopic velocity to it. You could wait countless lifetimes of the universe and never see that happen.[5]

---

4 OK, technically there are some obscure particle interactions in quantum field theory that aren't reversible for reasons having nothing to do with entropy, but everything irreversible you see in daily life is so because of the second law.

5 Temperature is not the same thing as energy, as we'll discuss in Section 10.3. A hotter object doesn't always have more energy than a colder one, but for any particular object, adding more thermal energy (random molecular motion) heats the object up.

We're making two separate claims here. One claim is that, left to itself, a sufficiently large-scale system will move toward the state of maximum entropy and then never go back the other way. That claim follows directly from the fundamental assumption of statistical mechanics. The microstates will of course keep changing, but the vast majority of microstates correspond to high-entropy macrostates, so you won't see a return to the low-entropy states.

Our second claim – which does not obviously follow from the first – is that this tendency for entropy to increase underlies *all* the irreversible time evolution we see in nature. Castles crumble because "ruins" is a higher-entropy state than "castle." The Earth rotating around the Sun neither increases nor decreases its entropy, so that video would play backward as well as forward. Any time you see something happening that could never physically happen in reverse, you're seeing the second law of thermodynamics at work.

Entropy is Not a Force (But It Acts Like One)

To explain that heading about entropy and force, consider an analogy. In his 1976 treatise on evolution, *The Selfish Gene,* Richard Dawkins writes:

> It is good to get into the habit, whenever we are trying to explain the evolution of some characteristic, such as altruistic behaviour, of asking ourselves simply: "what effect will this characteristic have on frequencies of genes in the gene pool?" At times, gene language gets a bit tedious, and for brevity and vividness we shall lapse into metaphor. But we shall always keep a sceptical eye on our metaphors, to make sure they can be translated back into gene language if necessary.

Dawkins' point is that we often discuss evolution in terms of purpose: "Mothers give food to their babies so that their genes will propagate." Such language is only a metaphor; no one believes that a mother is literally thinking about her genes when she feeds her children! But the metaphor provides a useful shorthand for the longer, more accurate description: "Over millions of years, genes that tend to promote maternal child care have been passed down more frequently than genes that discourage such care, which is why the first sort of genes are more common today."

A similar shorthand is often applied to discussions of entropy. You may hear the phrases "entropic force" and "entropic pressure." The metaphor is that the second law of thermodynamics is *pushing* a system toward a higher-entropy macrostate. Just as with purposeful-evolution language, this metaphor can be useful as long as you keep in the back of your mind that it is a shorthand for a blind statistical tendency.

For example, human cells often have more potassium ions (K^+) inside them than outside, compensated for by an equal number of negatively charged ions. Since potassium can flow through cell membranes, this can result in a net flow of K^+ ions outward. That flow of positively charged ions sets up a potential difference that pushes the potassium back toward the cell (Figure 10.3). In sloppy but useful language, we might say that "the flow stops when the electric force pushing the ions into the cell is balanced by the entropic force pushing them outward." Try to restate that equilibrium condition in more careful language that describes what's actually happening microscopically.

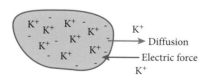

Figure 10.3 In equilibrium the cell still has an excess of K^+ ions, which means they randomly diffuse outward more than inward, but the electric force is pulling them inward.

Here's our answer. The K^+ ions do feel a net electric force toward the cell (because it is now negatively charged). But they also feel lots of forces in random directions because of collisions with other molecules. Because there are more K^+ ions inside the cell than outside, it is more common for those random collisions to push a molecule out of the cell than it is for them to push one in. In equilibrium the net flow of K^+ ions out of the cell due to that random motion is equal to the net flow into the cell due to the combination of the electric force and that random motion.

As long as you keep that more accurate description in the back of your mind, it is often convenient to summarize all of that by saying that in equilibrium the "entropic force" is equal and opposite to the electric force.

Energy and Entropy

At each moment, any given system has both a certain energy and a certain entropy. They are not the same thing! But the Example below points to an important connection between these two quantities.

Example: Entropy of a Collection of Oscillators

Consider a collection of 300 simple harmonic oscillators, each of which can have energy 0, ϵ, $2\epsilon, \ldots$ This could be, for example, atoms vibrating in a crystal lattice.

1. Calculate the entropy of the system with no energy, and then with total energy ϵ, 2ϵ, and 3ϵ.

2. What is $\lim_{E \to \infty} S$?

3. Use your results to make a sketch of entropy vs. energy for this system.

Answer:

1. • With no energy, every oscillator must be in the ground state, so $\Omega = 1$ and $S = 0$.

   • With one unit of energy, one oscillator is at energy ϵ. That could be any of the 300 oscillators, so $\Omega = 300$ and $S/k_B = \ln 300 = 5.7$.

   • With two units of energy you have two possibilities: one oscillator with 2ϵ or two oscillators with ϵ each. The multiplicity is

$$300 + \binom{300}{2} = 45{,}150.$$

   So $S/k_B = \ln 45{,}150 = 10.7$.

   • With three units of energy you have three possibilities. There are 300 ways to put all the energy in one oscillator. If you have 2 units of energy in one and 1 in another, there are 300 options for which one has 2, and for each of those there are 299 options for which one oscillator has 1 energy unit. Finally, there are $\binom{300}{3}$ options for putting 1 unit each into 3 oscillators. Put all that together and you get $\Omega = 4{,}545{,}100$ and $S/k_B = 15.3$.

2. With an arbitrarily large amount of energy to distribute, the system will have arbitrarily many ways to distribute it, so the entropy approaches ∞.

3. From the results above, we can see that the entropy grows with increasing energy, but the rate of growth slows, so the curve is concave down. But it still grows without bound. That's enough information for a qualitative sketch.

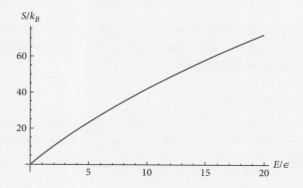

This Example illustrates an important general point that we will need in the next section:

Increasing the energy of a system generally increases its entropy.

The more energy a system has, the more different ways in which it can distribute that energy. There are exceptions to this rule, but they are almost entirely restricted to the kind of small artificial examples in textbooks like this one ("Consider a paramagnet with four atoms . . . "). For real systems, S virtually always increases with increasing energy. See Problem 18.

10.2.3 Questions and Problems: Entropy and the Second Law of Thermodynamics

Conceptual Questions and ConcepTests

1. Suppose you compare two macrostates of a system, and you find State A has a higher multiplicity than State B. Is it possible for State B to have a higher entropy than State A? Explain.

2. For each of the following events, does the total entropy increase, decrease, or remain the same? For each, give a one-sentence explanation of your answer.

   (a) Two cars have a head-on collision.

   (b) An astronaut drops a ball on the Moon. (Include only the time the ball is falling, not its collision with the ground.)

   (c) A pot of water on the stove boils.

   (d) The Earth revolves around the Sun.

3. Consider a system which – unlike macroscopic systems – has only 100 possible microstates. Of these microstates, 94 correspond to Macrostate 1, and the other 6 correspond to Macrostate 2. The system changes its microstate randomly 10 times every year. Which of the following is fairly certain to be true? (Choose one, and explain your choice.)

   A. The system will start in Macrostate 1.

   B. After 1000 years, the system will be in Macrostate 1.

   C. Over a 1000-year period, the system will spend about 94% of its time in Macrostate 1.

   D. Once the system is in Macrostate 1, it will never revert to Macrostate 2.

4. A canister of water initially has red dye on the left side and blue dye on the right, separated by a barrier. When you remove the barrier the dyes start to mix and soon the whole mixture is purple. Your friend says that this shows that the "entropy force" was pushing the red dye right and the blue dye left. Was there a net force on each type of dye? If so, what was the source of that force? If not, how could they have started moving?

5. When you leave a cup of hot coffee on a counter it cools off, thus lowering its internal energy. With less energy it has fewer ways to arrange that energy, so its entropy goes down. Why does this not violate the second law?

6. Choose one irreversible process (other than the ones we've already discussed) and explain how it involves an increase in net entropy.

7. Give at least one example of a reversible process.

8. Explain why, as a general rule, giving more energy to a system increases its entropy.

9. In this section we discussed the second law of thermodynamics. There are three others: the first, the third, and (for historical reasons) the "zeroth law." Look them up and briefly describe what they say.

For Deeper Discussion

10. Maxwell's Demon

In 1867 James Clerk Maxwell proposed a thought experiment to show how net entropy could be decreased. Start with a gas uniformly spread out throughout a box. Put a wall in the middle of the box that divides it into two parts, and place a molecule-sized door in the wall. Imagine a super-fast, tiny demon standing by the door. Whenever a fast molecule approaches from the left side, the demon opens the door and lets it through. Whenever a slow-moving molecule approaches from the right, the demon lets it through. After a while a lot of the fast-moving molecules will start to accumulate on the right and the slow-moving ones will build up on the left, thus separating the box into a high-temperature half and a low-temperature half. That state has lower entropy than a well-mixed box of gas, so the net entropy has gone down. Does

this thought experiment disprove the second law of thermodynamics? Explain.

11. A macroscopic system will spontaneously tend to evolve toward states of higher entropy. What about non-spontaneous events, though? If you put a human and some other physical systems in a box, could that human decide to interact with those systems in a way that lowered the net entropy? Discuss.

Problems

12. You throw a penny, a nickel, a dime, and a quarter into the air.

   (a) What is the multiplicity of the result "all four heads"?

   (b) What is the entropy of the result "all four heads"?

   (c) What is the multiplicity of the result "three heads and one tail"?

   (d) What is the entropy of the result "three heads and one tail"?

13. You throw 20 pennies into the air.

   (a) What is the multiplicity of the result "exactly three heads"? *Hint*: You can write down a formula (using "choose" notation which becomes factorials) and then punch it quickly into a calculator.

   (b) What is the entropy of the result "exactly three heads"?

   (c) What is the multiplicity of the result "exactly 10 heads"?

   (d) What is the entropy of the result "exactly 10 heads"?

14. A system contains four objects, each of which can have energy 0, 1, 2, 3, or 4. The total energy of the system is 4.

   (a) Calculate the entropy of the state "One object has all 4 units of energy." (We're not specifying which object, just that all the energy is in one of them.)

   (b) Calculate the entropy of the state "Each object has 1 unit of energy."

(c) Suppose the system changes every second to a completely random microstate (still always with a total energy of 4). How long would you expect to wait before seeing the state "One object has all 4 units of energy"?

15. A hundred ordinary six-sided dice are placed on a rubber sheet, each one with the number 1 facing up. The rubber sheet starts to ripple in such a way that every minute each individual die has a 1% chance of flipping to a random state. The microstate of this system corresponds to the number facing up on each die. The macrostate is the average value of all those numbers.

(a) What is the initial macrostate?

(b) What will be the macrostate (approximately) one hour later?

(c) What would you guess will be the macrostate (approximately) one year later?

(d) Assume that at some point in the future, the macrostate of the system is 3.5. Suddenly two dice flip from 1 to 6. What is the new macrostate?

(e) How many dice would have to spontaneously flip from 1 to 6 to change the macrostate from 3.5 to 4?

(f) Now imagine a million dice instead of a hundred. How many dice would have to spontaneously flip from 1 to 6 to change the macrostate from 3.5 to 3.6?

(g) Explain what your conclusions imply about the behavior we observe for macroscopic systems.

16. Create a computer simulation of the scenario in Problem 15 (with a hundred dice, not a million). Draw a graph of the macrostate over time for the first 1000 minutes. (You do not need to have solved Problem 15 to solve this one.)

17. Consider a system with 100 particles, each of which can have energy $E = 0$ or $E = 1$.

(a) How many microstates are possible if the total energy of this system is 0?

(b) How many microstates are possible if the total energy of this system is 1?

(c) How many microstates are possible if the total energy of this system is 5?

(d) How many microstates are possible if the total energy of this system is 20?

18. We said in the text that entropy almost always increases with energy, but we also noted that there are exceptions.

(a) Consider a paramagnet with four atoms, each of which can have energy 0 or ϵ. Calculate S/k_B for each possible system energy from 0 to 4ϵ and plot your results.

(b) Under what circumstances (if any) does adding energy to the four-paramagnet system *decrease* its entropy?

(c) Sketch a plot of S/k_B vs. energy for a 300-paramagnet system. *Hint:* If you calculate it for $E = 150\epsilon$ and put that point in your plot, you should be able to figure out the shape of the rest of the plot without any other calculations.

(d) Compare your plot to the one we made on p. 430 of a 300-oscillator system. What is it about the two systems that causes their plots to look so different?

In general it is only possible for S to decrease with E in a system with a maximum possible energy that corresponds to a very high-energy state. You'll see in Section 10.4 why this never spontaneously occurs, no matter how much you heat up the system. For a macroscopic system you are safe assuming that S vs. E is an increasing function.

19. **Why Use a Log?**

In this problem you'll show that entropy is an extensive variable, meaning the entropy of two systems together is just the sum of their individual entropies. We'll consider a simple system (System A) made of three coins that is in the macrostate "two coins are showing heads."

(a) What is the multiplicity of this macrostate?

(b) Suppose System A is in a box, and next to it is a box labeled B that also contains three coins in the macrostate "two conis are showing heads." List all of the possible microstates of the combined system and give their combined multiplicity. We'll start you off by listing one of them: A(HTH), B(HHT).

(c) In general, if one system has multiplicity Ω_A and another has multiplicity Ω_B, what is the multiplicity of the combined system? Make sure your answer reproduces your result from Part (b).

(d) Repeat Parts (a) and (b) with entropy instead of multiplicity. (You can use the multiplicity results you found so the calculations should be quick.) Write your answers in decimal form.

(e) Using your answer to Part (c) and the definition of entropy, prove that entropy is extensive.

20. **The Gibbs Paradox**

Imagine a box with N non-interacting particles. We define a microstate of the system as a complete list of which quarter of the box each individual particle is in, and the macrostate as the total number of particles in each quarter. Assuming N is large, we can assume that in equilibrium each of those four quarters has almost exactly $N/4$ particles in it.

(a) Find the entropy of the macrostate "each quarter of the box has exactly $N/4$ particles."

(b) Now slide a partition into the middle of the box, dividing it into a left half and a right half. Each half is made of two parts, each containing $N/4$ particles. Calculate the entropy of each half of the box.

(c) After you put in the partition, what's the total entropy of the system?

(d) Assuming $N = 100$, what's the ratio of the original entropy to the entropy after you put in the partition?

You should find that simply sliding a partition into the box reduced its entropy, thus violating the second law. That result is known as the Gibbs paradox. You'll resolve it in Section 10.6.

10.3 Temperature

If entropy feels like a strange new acquaintance, temperature is more like a familiar old friend. We all know the pleasures of a hot meal, a warm blanket, a cool breeze, and an ice cold soda.

But if you try to *define* temperature you might equate it to an amount of energy, or an amount of kinetic energy, or perhaps an energy density – and those are all incorrect. Temperature is closely related to energy, but they are not the same thing, and it is certainly possible for a cold object to have more energy than a hot object.

It turns out that a proper definition of temperature takes us back to a discussion of entropy.

10.3.1 Discovery Exercise: Temperature

System S_1 has 5 objects, and System S_2 has 50 objects. Each object can have energy 0 or 1.

The initial macrostate of this system is "S_1 has a total energy of 1, and S_2 has a total energy of 3."

Then one unit of energy flows from S_2 to S_1.

1. Does the entropy of S_1 increase or decrease? By how much?
2. Does the entropy of S_2 increase or decrease? By how much?

3. Overall do you expect to see energy flowing from S_2 to S_1 as we described, which is from the higher-energy system to the lower? Or do you expect to see it flow the other way? *Hint*: Your answer to this part should be based on your answers to the other parts!

See Check Yourself #14 at www.cambridge.org/felder-modernphysics/checkyourself

10.3.2 Explanation: Temperature

Figure 10.4 A coffee cup.
Source: Swigart © 2010.

If you put a cold spoon into a hot drink (Figure 10.4), energy will flow from the drink to the spoon.

That fact presumably comes as no surprise, but think about what it implies. Does it mean that the drink has more energy than the spoon? Not at all! (Energy is also flowing from the coffee to the room, which presumably has much more total energy than the coffee.) This is not about total amount of energy, but there is a property of those two objects – the drink is hot, the spoon is cold – that makes the energy flow in that direction.

That property is what we call "temperature."

An Operational Definition of Temperature

We define Object A as hotter than Object B ($T_A > T_B$) if energy flows spontaneously from A to B when you put the two objects in contact.

This isn't really a definition: it just tells us how to rank-order objects' temperatures, not how to assign numbers to them. Nonetheless, this is the key property we are referring to when we talk about temperature. So any quantitative definition must lead to the conclusion that energy flows from hot things to cold things.

In the rest of this section we will see how this idea relates to entropy and the second law of thermodynamics, and we'll be led to a more fundamental definition of temperature.

Moving from "Operational Definition" to Real Definition

"Energy flow from a hot object to a cold object" is a classic example of irreversibility. Leave our spoon and drink together for a while and they will end up at the same temperature. After that they will never spontaneously revert to a state of significantly different temperatures.

In Section 10.2 we claimed that the only reason why any process is irreversible is that the process increases entropy. So if you want to understand why energy flows in the direction it does, follow the entropy.

Recall that increasing the energy of a system increases its entropy. (That's because the more energy a system has, the more different ways it can arrange that energy.) So this particular energy flow is increasing the entropy of the spoon, and simultaneously decreasing the entropy of the coffee. But those two effects cannot cancel each other perfectly; if they did, and thus led to $\Delta S = 0$ for the overall system, then the process would not be irreversible.

So the spoon and the coffee must react to the *same* change in energy with *different* changes in entropy. That difference is the key to understanding why energy flows at all, and to defining temperature. The following Active Reading Exercise is designed to help you put those pieces together.

Active Reading Exercise: Entropy vs. Energy

Two systems called "System B" and "System R" are in contact, so they can exchange energy with each other. The blue curve below shows the entropy of System B as a function of its total energy, and the red curve shows the same for System R. Both systems start at the energy E_0 shown on the plot.

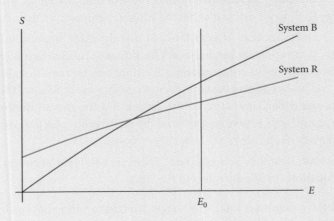

1. Imagine that energy flows from System B to System R. From these plots, which would be bigger in magnitude, B's decrease in entropy or R's increase?

2. Would that energy flow spontaneously happen?

Here are the steps to thinking through that exercise:

- A slight decrease in B's energy would decrease its entropy by a lot (large slope).
- The same increase in R's energy would only increase its entropy a little (small slope).
- The result would be a net decrease in total entropy, which will never spontaneously happen.

Conversely, energy flow from R to B would cause R's entropy to decrease a little and B's to increase a lot, so that is what will happen. Notice that the direction of energy flow has nothing to do with which system has more energy (corresponding to the horizontal position on the plot); it is related to the *slopes* of the curves.

Do you see how this points toward a mathematical definition of temperature? Energy will tend to flow spontaneously from R to B, which is our definition of the statement "R has a higher temperature than B." We know this is true because the E vs. S graph has a *lower slope* for R than for B. That leads us to the following mathematical definition of temperature. (We'll explain the note about "heat flow" next to the equation in a moment.)

The Quantitative Definition of Temperature

$$T = \frac{1}{dS/dE} \qquad dE \text{ refers to energy change due to heat flow.} \qquad (10.3)$$

This definition is only valid for temperatures measured in kelvins (or some other scale that starts at absolute zero).

Make sure you see how that definition reproduces our operational definition on p. 434. System R in the Active Reading Exercise has a smaller dS/dE than System B, which makes it likely to give up energy to B, so we *define* it as having a higher temperature than B.

As always, remember that "energy flows in that direction because it increases entropy" is shorthand for "there are more random ways for things to happen that way than the other way." Going back to our spoon-in-coffee example, there are constant collisions between the molecules of the spoon and of the coffee at the boundary where they meet. Some of those collisions transfer net energy to the coffee, and some to the spoon. But the overall motions are such that the collisions are more likely to give energy from the coffee molecules to the spoon molecules than vice versa, so macroscopically that's how energy flows.

The note about "heat flow" is important. There are two types of energy transfer, "work" and "heat," and Equation (10.3) only refers to the latter.

- If a piston compresses a gas, or a weight stretches a spring beyond its natural length, or a cue ball slams into a motionless billiard ball, one object imparts energy to another object. Those are examples of "work," and they transfer energy regardless of temperature.

- If your hand touches a hot stove, or the air inside the refrigerator is suddenly mixed with the air at room temperature, or sunlight reaches the surface of Mercury, one object imparts energy to another object. Those are examples of "heat," and they happen spontaneously because of a temperature difference. (If Mercury were as hot as the Sun, it would be radiating back as much energy as it absorbs – by definition.)

A change in energy ΔE can be caused by heat, work, or a combination of the two, but Equation (10.3) applies to energy changes caused by heat flow only.

The letter Q is often used to designate the energy flowing into a system specifically as heat. Heat flowing out is a negative contribution to Q.

Heat Capacity

Let's return to our coffee. We said many times above that the hot coffee gives up energy to the cold spoon. But we also slipped in a very different claim that you may not have noticed: over time, as the coffee and spoon change their energy levels, they also change their temperatures. The coffee cools down, and the spoon warms up.

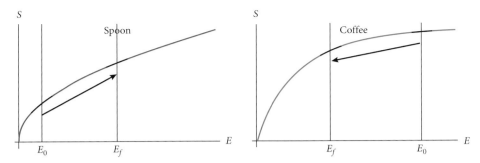

Figure 10.5 Initially the coffee is hotter (lower slope) than the spoon (higher slope). Heat flows from the coffee to the spoon until they have the same slope, meaning they have the same temperature.

That fact, although certainly familiar, does not follow obviously from anything we've said. The process is illustrated in Figure 10.5. We have said that as the spoon's energy E increases, that causes its entropy S to increase; now we're saying that this change also causes its derivative dS/dE to *decrease*. Meanwhile the coffee's energy loss is increasing the slope dS/dE for it. When the two systems reach the same dS/dE, the energy transfer will be equal in both directions: thermal equilibrium. That does not mean they have the same energy, or the same entropy. It means the slope is the same for both.

That new fact begs a new question: when the coffee's energy changes, *how much* does the temperature change in response? We answer that question with a variable called "heat capacity." You might think we would define that variable as dT/dE (change in temperature per unit change in energy), but it turns out to be more useful to define it the other way around.

Heat Capacity

$$C = \frac{dE}{dT}$$

One way to think about this definition is that C is roughly the amount of energy you would need to provide to increase a system's temperature by one degree.

Section 10.5 will give some general guidelines for estimating heat capacity, and Section 11.7 will look at how different models of atomic motion lead to increasingly accurate predictions of the heat capacities of various solids. A more complete treatment would have to factor in not only the properties of the object being heated but also the way you heat it. If you heat a gas at constant pressure its volume will increase; if you heat it at constant volume its pressure will increase. The former turns out to require more energy, so "heat capacity at constant pressure" is greater than "heat capacity at constant volume." That's an important distinction in many areas of chemistry and physics, but we don't need to worry about it here.

Heat capacity also depends on how much stuff you're heating up. All other things being equal, raising the temperature of two cups of coffee takes twice as much energy as raising the temperature of one cup. So it's often useful to talk about the heat capacity of an object per unit mass, or per number of molecules. Those are both called "specific heat capacity" and written c.

When you see a specific heat capacity you have to figure out from the context whether it's per unit mass or per number of molecules, etc.

Entropy Changes

Above we defined Q as the energy that flows into a system as heat, and we defined temperature (Equation (10.3)) as the entropy change that results from such heat flow. Putting them together, we can say the entropy change of a system *in response to heat flow* will be $\Delta S = Q/T$. If work is also done on that system, that won't generally change its entropy. The system's entropy may change spontaneously – for instance, a drop of dye in water will spread out over time – but such changes will always increase the entropy, never decrease it. So we can write the expression for entropy change very generally as

$$\Delta S \geq \frac{Q}{T}.$$

Historically, entropy was defined by this relationship long before Boltzmann figured out that he could define it in terms of microstates. *Warning*: Don't try to apply this formula using temperatures in degrees Celsius or Fahrenheit. It only works for scales like the Kelvin scale that start at absolute zero.

This equation is another way of seeing why heat flows in the direction it does. If heat flows from an object at a high temperature T_H to an object at low temperature T_C then the hot object will lose entropy Q/T_H and the cold object will gain entropy Q/T_C. Since $T_H > T_C$ the net entropy of the entire physical universe will increase. If heat flowed the other way then that would violate the second law of thermodynamics.

Everything you do causes an increase in entropy. When you do work on a system, entropy increases by at least a little because of friction, drag, or other irreversible effects, but you can make that increase arbitrarily small. When you put two objects in contact so heat can flow between them, heat flow from the hotter to the colder object causes a net increase in entropy.

Refrigerators

A refrigerator extracts heat from its cold interior and dumps heat into the comparatively warm kitchen.

Question: How do we know that such an energy transfer will never happen spontaneously?

First answer: Because it violates the definition of temperature, which says that heat flows from hot objects to cold objects.

That's perfectly correct, but if you're trying to invent a refrigerator, that answer doesn't suggest a solution. If we go deeper into the fundamental physics, we can find the loophole.

Second answer: When the refrigerator interior loses energy, its entropy goes down. When the kitchen gains energy, its entropy goes up. But because the refrigerator is cold, it has a high dS/dE: the exchange of energy lowers its entropy *more* than it raises the entropy of the kitchen. So the entropy of the overall refrigerator-plus-kitchen system will decrease, which will never happen spontaneously.

The second law of thermodynamics *underlies* the rule that heat flows from hot objects to cold ones, so our second answer is just a more careful restatement of the first. But the second answer also suggests what we need to do if we want to force this heat transfer: we need to add *more* energy to the kitchen air than we remove from the refrigerator. Then we can arrange things so that the entropy of the total system will rise.

And where will that extra energy come from? From the outlet, which supplies electrical work to the refrigerator. So the energy balance for this system looks like

Energy supplied by outlet + Energy removed from the interior of the refrigerator
= Energy gained in kitchen air.

Any refrigerator must satisfy this energy conservation equation *and* lead to a net increase in entropy. In Problem 25 you'll use those two facts to derive a limit on how efficient a refrigerator can be.

Similarly, the second law puts limits on the efficiencies of engines. The things we call engines are typically "heat engines," meaning they make something very hot (gasoline, coal, nuclear fuel, . . .) and then use the heat flow from that fuel to produce work (moving a car, turning an electrical turbine, . . .). But when heat flows out of the fuel, that reduces its entropy, and work doesn't generally affect entropy. So engines must use up some of their energy by transferring heat to something else, usually a surrounding reservoir of air or water. In Problem 27 you'll use that fact to derive a limit on how efficient an engine can be.

10.3.3 Questions and Problems: Temperature

Conceptual Questions and ConcepTests

1. For each quantity give a brief description of what it represents, and units that could be used to measure it.

    (a) Temperature

    (b) Heat

    (c) Heat capacity

2. Explain in your own words how the quantitative definition of temperature (p. 436) gives rise to the operational definition (p. 434).

3. Explain why the definition $T = 1/(dS/dE)$ can only be valid for a scale like the Kelvin scale, which starts at absolute zero. *Hint*: Think about what it would imply to say that $T = 0$ in this definition.

4. Harold Hill, the traveling salesman, tries to sell you a new air conditioner he's invented. "It doesn't use any electricity!" he tells you. "It just takes energy from your room, so that makes your room cool down. The energy is dumped as excess heat outside." What's the flaw in Mr. Hill's proposal?

5. Give at least one example of two systems where the one with more energy has a lower temperature.

6. For almost all systems, $dS/dE > 0$. That tells us physically that the more energy the system has the more entropy it has, presumably because it has more ways it can arrange all that energy.

    (a) It is also true that for almost all systems $d^2S/dE^2 < 0$. Similarly to how we described the positive slope, what does this negative concavity tell us about these systems? Your answer should be in terms of entropy and energy, and *not* use the word "temperature."

    (b) Now reframe your answer to Part (a) in terms of temperature. What does the fact that $d^2S/dE^2 < 0$ for most systems tell you about the relationship between energy and temperature?

7. Why can't you cool down your kitchen by leaving your refrigerator door open?

8. Many older cars were sold without air conditioning, but all cars had heaters as a free option. Why does it cost virtually no extra energy for a car to heat its interior?

9. Figure 10.6 shows entropy as a function of energy for two systems, both of which are at energy E_0.

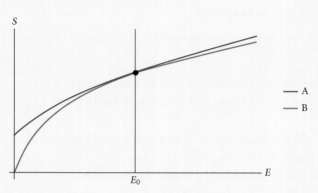

Figure 10.6

(a) Which system has a higher temperature? (Choose one and explain.)

   A. A

   B. B

   C. They have the same temperature.

   D. There's not enough information in the graph to tell.

(b) Which system has a higher heat capacity? (Choose one and explain.)

   A. A

   B. B

   C. They have the same heat capacity.

   D. There's not enough information in the graph to tell.

For Deeper Discussion

10. One of the many strange properties of black holes is that they have negative heat capacity. Discuss what this implies about how black holes interact with their environments. *Hint*: Because of quantum mechanical effects, black holes can emit radiation and give energy to their surroundings.

11. Figure 10.7 shows entropy vs. energy for a paramagnet consisting of N atoms, each of which can have energy 0 or ϵ. If $E = 0$ or $E = N\epsilon$ there is only one possible microstate, so the entropy is zero in both cases. In between, the entropy rises to a maximum.

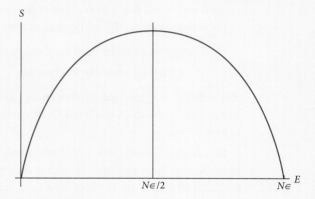

Figure 10.7

(a) Draw a graph of temperature as a function of energy for this system.

You should have seen that the temperature, as defined in Equation (10.3), is *negative* for $N\epsilon/2 < E < N\epsilon$. You can't get a paramagnet to the negative temperature state by just heating it up or cooling it down. First you cool it down so most of the paramagnets are in the low-energy state. Then you suddenly reverse the magnetic field so that the orientation of most of the atoms is the high-energy one, thus putting you on the right side of the plot.

(b) Consider two objects, one with a negative temperature and one with a positive temperature, that are free to exchange energy with each other. Will energy tend to flow from the positive to the negative, or the other way around? Explain your answer in terms of the effect on the overall entropy of the system.

(c) Consider two objects, one with a negative temperature and one with a *more* negative temperature, that are free to exchange energy with each other. Which way will energy flow in this system? Explain your answer in terms of the effect on the overall entropy of the system.

(d) Explain why, outside very artificial situations such as a suddenly flipped external magnetic field, negative temperatures never happen.

Problems

12. The "Sackur–Tetrode equation" gives the entropy of a monatomic ideal gas (such as helium) as a function of the number of molecules N, the volume V, the energy E, and the mass m of an atom:

$$S = Nk_B \left[\ln \left(\frac{V}{N} \left(\frac{4\pi mE}{3Nh^2} \right)^{3/2} \right) + \frac{5}{2} \right].$$

Find the temperature. Simplify your answer as much as possible. (You may recognize the result.)

13. Figure 10.8 shows entropy as a function of energy for a system of 100 oscillators, each of which can have energy 0, ϵ, 2ϵ, etc.

E/ϵ	100	200	300	400	500	600	700	800	900	1000
S/k_B	135	187	220	245	265	282	296	309	320	330

Figure 10.8

(a) Find (approximately) the temperature at $E = 300\epsilon$ and $E = 600\epsilon$. *Hint:* When estimating the derivative at a certain point from a set of values, it's usually more accurate to use values on both sides of that point rather than just on one side.

(b) Find (approximately) the heat capacity at $E = 300\epsilon$ and $E = 600\epsilon$.

(c) Based on your answers, would it be easier to heat the system up by one degree if it started with 300ϵ of energy or with 600ϵ?

14. A paramagnet has N atoms, each of which can have energy 0 or ϵ. In this problem you'll consider the high-temperature limit of this system analytically. In Problem 15 you'll reach the same final result graphically. Neither problem depends on doing the other.

(a) Write the entropy of the paramagnet as a function of its total energy E.

(b) Use the Stirling approximation, $\ln(x!) \approx x \ln x - x$, to help you estimate the paramagnet's temperature as a function of E.

(c) Invert your formula to write E as a function of T.

(d) What is $\lim_{T\to\infty} E$? In that limit, how many of the atoms (on average) are in their excited state?

15. A paramagnet has N atoms, each of which can have energy 0 or ϵ. In this problem you'll consider the high-temperature limit of this system graphically. In Problem 14 you reached the same final result analytically. Neither problem depends on doing the other.

(a) What is the entropy of the system when it has zero energy?

(b) What is the entropy when it has $E = N\epsilon$, the most energy it can possibly have?

(c) Make a qualitative sketch of entropy vs. energy for this system. You can draw this graph without doing any calculations beyond Parts (a) and (b).

(d) What should the slope of that sketch be in the limit $\lim_{T\to\infty}$? Mark the point on your graph where this occurs.

(e) Based on your answer to Part (d), argue that in the limit where you keep heating a paramagnet toward $T \to \infty$ you do not approach a point where all the atoms are pointing up.

16. When a solid melts it absorbs energy but does not change its temperature. For example, you have to add 333 J/g to ice at $0\,°C$ to change it into water at $0\,°C$. Suppose a 10 g ice cube melts while sitting in a room at $27\,°C$.

(a) How much does the ice cube's entropy change? (Make sure to get the sign right.)

(b) How much does the room's entropy change? (Make sure to get the sign right.)

(c) How much does the universe's entropy change? Does it increase or decrease?

17. A house uses 10^7 J to stay at $70\,°F$ for one day while the outside temperature is $50\,°F$. How much does the entropy of the universe increase as a result?

18. Estimate the energy you would need to add to a room to heat it by one degree. Explain any assumptions or approximations you use, and any numbers you need to look up.

19. Derive a formula for heat capacity that only depends on E and S. (It will include one or more derivatives.) You can assume that the only changes to the system you are considering are heat flows, so there is no work and $dS = dE/T$.

20. Consider a system of N distinguishable particles, each of which can be in one of three energy states: 0, ϵ, and 2ϵ. The system has q energy units, meaning the total energy of the system is $E = q\epsilon$. In short, this system is just like a paramagnet except that each particle has three possible states instead of two. For $N \gg q \gg 1$ the multiplicity can be approximated as $\Omega \approx N^q/q!$. Find expressions for the entropy, temperature, and heat capacity of this system. The heat capacity should be expressed as a function of temperature. In this problem you may use "Stirling's approximation": $\ln(x!) \approx x \ln x - x$.

21. [*This problem depends on Problem 20.*] Take the limits of the heat capacity you found as $T \to 0^+$ and $T \to \infty$. Explain why your answer makes sense in the $T \to \infty$ limit.

22. System A has 4 objects, and System B has 20 objects. Each object can have energy 0 or ϵ. The macrostate of this system – the quantity you measure – is "the amount of energy in A, and the amount of energy in B." The initial state is one unit of energy in A and two in B (Figure 10.9).

Figure 10.9

Now imagine that one unit of energy (ϵ) flows from B to A.

(a) Calculate the change in the entropy of B. (You can keep your numbers simpler by expressing your answer as a number times k_B rather than plugging in the value for k_B.)

(b) Calculate the change in the entropy of A.

(c) Take $\epsilon = 1$ eV. Estimate the temperature of B. Note that the energy of this system is discrete, so you will estimate dS/dE by finding $\Delta S/\Delta E$ over a small discrete change.

(d) Estimate the temperature of A.

(e) Explain which direction energy will generally flow in this system, and why.

(f) A friend of yours argues that energy always flows from higher-energy to lower-energy systems. Briefly explain why this is sometimes but not always true. Do not use the word "temperature" in your explanation.

23. Consider two systems with two distinguishable particles each. The particles in System A can each be in one of two possible states with energies 0 or ϵ. (This could be a very small paramagnet or any other two-level system.) The particles in System B can each be in one of five possible states with energies 0, ϵ, 2ϵ, 3ϵ, or 4ϵ (Figure 10.10).

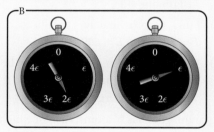

Figure 10.10

(a) Initially the two systems are not in contact and they each have energy 2ϵ. What are the entropies S_A and S_B of the two systems?

(b) Now the two systems are put in contact with each other (i.e. they are able to exchange energy). Once they have had time to exchange energy, what is the most likely way in which they will divide the 4ϵ of energy between them?

(c) In Part (b) you found the division of energy that is most likely. If you keep observing for a long time, what fraction of the time will the system spend with that division?

(d) Take $\epsilon = 0.01\,\mathrm{eV}$ and estimate the temperature of System B in the equilibrium state. (A note of caution: If you likewise estimate the temperature of System A in equilibrium you will not find them equal. You may find it useful to think about why that is, but you don't need to address that for this problem.)

24. A 5 kg cup of water at $10\,^\circ\mathrm{C}$ warms up to the room temperature $27\,^\circ\mathrm{C}$.

(a) The specific heat capacity of water is $4186\,\mathrm{J/(kg\,K)}$. How much heat flows into the water? (Ignore any heat flowing into or out of the cup itself.)

(b) How much does the entropy of the room change? *Hint*: You don't want to use degrees Celsius. Why not?

(c) How much does the entropy of the water change? *Hint*: You will have to set up and evaluate an integral.

(d) Does the total entropy of the universe increase, decrease, or stay the same?

25. A refrigerator removes heat Q_C from its interior at some cold temperature T_C, dumps heat Q_H into a kitchen at temperature T_H, and extracts work W from an outlet. We will define all of those quantities to be positive.

(a) What relationship between Q_C, Q_H, and W is implied by conservation of energy?

(b) If you assume the total entropy change of the universe is zero, what relationship is implied between Q_C and Q_H?

(c) The "coefficient of performance" (COP) of a refrigerator is defined as Q_C/W (the cooling that you want, divided by the work it costs to get it). Use your answers to Parts (a) and (b) to derive the COP of a refrigerator in the limit where the entropy of the universe doesn't change. Your answer should only depend on T_H and T_C.

(d) If your refrigerator is at $4.5\,^\circ\mathrm{C}$ and your kitchen is at $27\,^\circ\mathrm{C}$, what is the smallest possible amount of electrical energy you need for each joule of heat that it extracts from the interior?

26. [*This problem depends on Problem 25.*] A heater can be 100% efficient, turning every joule of work it uses into heat. A "heat pump" can achieve efficiencies *greater than* 100% by extracting heat from the cold exterior.

(a) If you are heating a room at temperature T_H while the outside temperature is T_C, what is the maximum possible efficiency (Q_H/W) that a heat pump can achieve?

(b) A heat pump does the same thing as a refrigerator; it extracts heat from a cold place and dumps even more heat into a hot place, using some electrical work to make up the difference. So why is the COP of an ideal refrigerator different from the efficiency of an ideal heat pump? Which is higher?

27. A heat engine extracts heat Q_H from a fuel at temperature T_H, produces work W, and dumps waste heat Q_C into a "cold reservoir" at temperature T_C. We will define all of those quantities to be positive.

(a) What relationship between Q_C, Q_H, and W is implied by conservation of energy?

(b) If you assume the total entropy change of the universe is zero, what relationship is implied between Q_C and Q_H?

(c) The efficiency e of a heat engine is defined as W/Q_H (the work that you want, divided by the heat input it costs to get it). Use your answers to Parts (a) and (b) to derive the efficiency of a heat engine in the limit where the entropy of the universe doesn't change. Your answer should only depend on T_H and T_C.

(d) If your fuel burns at $2700\,^\circ\mathrm{C}$ and your car dumps excess heat into the surrounding air at $27\,^\circ\mathrm{C}$, how many joule of waste heat must you expel for each joule of work your car does in moving forward? (In practice car engines are much less efficient than the theoretical limit.)

28. A Rubber Cord

Figure 10.11 shows a mass m hanging from a rubber cord of length L.

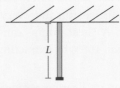

L

Figure 10.11

Figure 10.12 peeks inside the rubber cord, which we imagine as a linked chain of molecules. The left side shows that chain with each link (length d) going straight down from the top. On the right side, one of those links has flipped upward, shortening the cord.

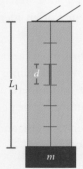

L_1, d, m

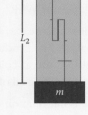

L_2, m

With one link pointing up, the cord is shortened to $L = 4d$

At the cord's maximum extent (all links going straight down), $L = 6d$

Figure 10.12

The effect of one link in the chain pointing up is to lift the mass at the bottom a distance $2d$, adding $2mgd$ to the potential energy of the mass. If a total of N_{up} links are pointing up, the total energy E of the system relative to its lowest-energy state is $2mgdN_{up}$.

To specify the macrostate of this system, we must say how long the cord is, or how many links are pointing up, or how much total energy the cord has. All three are equivalent. To specify the microstate, we must say which particular links are pointing up.

(a) For a cord with N links, find the multiplicity as a function of N_{up}.

(b) Using the result of Part (a) and the relationship between E and N_{up}, write the entropy S (not the multiplicity!) as a function of E.

(c) Use the result of Part (b) to find the temperature of the cord as a function of its energy. You can use the approximation

$$\frac{d}{dx} \ln \binom{N}{x} \approx \ln \left(\frac{N - x}{x} \right).$$

(Be careful! The left side of that equation involves a "choose" symbol and the right side involves a fraction; they look similar until you notice the horizontal line in the middle.)

(d) If you heat the cord, should it become shorter or longer? Justify your answer.

10.4 The Boltzmann Distribution

Our discussion so far has focused on the simple-but-important case of an "isolated system," one that is unaffected by other systems. Now we turn to a different case, also simple and also important: a small system in contact with a much bigger one.

10.4.1 Discovery Exercise: The Boltzmann Distribution

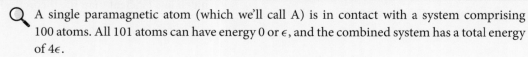

A single paramagnetic atom (which we'll call A) is in contact with a system comprising 100 atoms. All 101 atoms can have energy 0 or ϵ, and the combined system has a total energy of 4ϵ.

1. How many accessible microstates does the entire combined system have?

2. In how many of those microstates does Atom A have energy 0, and in how many does it have energy ϵ?

    See Check Yourself #15 at www.cambridge.org/felder-modernphysics/checkyourself

3. Based on your answer, you can conclude that A is far more likely to have energy 0 than ϵ. Why does this not violate the fundamental assumption of statistical mechanics?

10.4.2   Explanation: The Boltzmann Distribution

A system so large that it can gain or lose energy with no significant change to its temperature is called a "reservoir." For example, when a glass of cold water is left on the table it warms up to room temperature. In the process it gives energy to the room, but the room is so much larger than the glass that this has no noticeable effect on the room, so the room counts as a reservoir for the water glass. Meanwhile, if the window is open, the outside atmosphere acts as a reservoir for the room.

The microstate distribution for our water glass is quite different from the "fundamental assumption" in Section 10.1.

- *For an isolated system*: The total energy is constant, and any individual microstate with that energy is just as probable as any other individual microstate.

- *For a small system in contact with a reservoir*: The energy can fluctuate, and an individual microstate with low energy is more probable than a microstate with higher energy.

We have emphasized throughout this book that systems tend to move into low-energy states. For example, an ionic bond forms if and only if the state "two bound atoms" has lower energy than "two unbound atoms."

But with statistical mechanics we can take the next step and ask *why* systems tend toward lower-energy states. If you did the Discovery Exercise above, you found that answer for yourself: given two different microstates of our system, the microstate with lower energy leaves more energy – and therefore more entropy – for the reservoir. In other words, a given low-energy state of our system corresponds to *more possible microstates* of the reservoir than does a given high-energy state. (This logic cannot apply to an isolated system, for which all available microstates have the same energy.)

In Problem 22 you'll formalize that argument into a derivation of the "Boltzmann distribution."

The Boltzmann Distribution

Let System S be in equilibrium with a reservoir at temperature T. The probability of finding System S in a particular microstate with energy E is given by

$$P = \frac{1}{Z} e^{-E/(k_B T)}. \tag{10.4}$$

The quantity $e^{-E/(k_BT)}$ is called the "Boltzmann factor" for that particular microstate. The unitless proportionality constant $1/Z$ is the same for all possible microstates of the system. Its denominator Z is called the "partition function" for the system.

This formula only works if you measure temperature in kelvins (or some other scale that starts at absolute zero).

(It might look a bit odd to call the proportionality constant a fraction, $1/Z$, instead of just giving it a letter. It turns out that a number of equations end up simpler because we define it this way.)

The Boltzmann distribution is the most important equation in this chapter. The rest of this chapter will be spent drawing out the consequences of that formula for the air in a room, a strand of DNA, protons in the core of the Sun, beams of light inside a cavity, and a wide variety of other systems.

You can often predict the behavior of such systems *without* going through calculations if you keep the following points in mind. (Keep your eye on Equation (10.4) and Figure 10.13 and all these points should be clear.) Remember that these points apply to a system in thermal equilibrium with a reservoir.

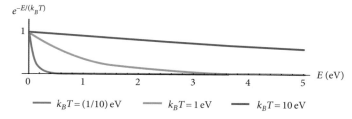

Figure 10.13 The Boltzmann factor as a function of energy at three different temperatures.

- The Boltzmann factor $e^{-E/(k_BT)}$ always decreases as E increases. This means that any particular low-energy microstate is more probable than any particular high-energy microstate.

- However, that does not guarantee that any system is more likely to have a low energy than a high energy. There may be many microstates with the same energy. All of these microstates have the same probability as each other. The probability of finding the system with this energy depends on the probability of each of its microstates, but also on how many of them there are. This subtle but important distinction is illustrated by the Example "Degeneracy" on the following page.

- For small values of T, the exponential decrease is very pronounced. ($y = e^{-10x}$ decays much faster than $y = e^{-x}$.) So in a system at low temperature, the high- and even medium-energy states have very low probabilities; the particles tend to fall into the lowest-energy states available.

- For high values of T, the exponential decrease is very gradual. This does *not* mean that a high temperature kicks all particles into the highest-energy states possible. Rather, it flattens the probability curve. So in the (physically unreachable) limit $T \to \infty$, all available microstates are equally likely. At an (actual) finite high temperature, you find a fairly even distribution up to some high energy level before the probabilities tail off.

Those last two points can be summarized by saying that, at any temperature, microstates with energies less than $\sim k_B T$ are roughly equally probable, and microstates with energies much higher than $k_B T$ are very improbable. We will return to that point in the next section.

In the meantime, the following Example illustrates the fact that the Boltzmann distribution gives the probability of finding *any particular microstate* with energy E; you can use that to find the (often very different) probability that your system has energy E.

Example: Degeneracy

System S has 11 possible microstates: 1 state at energy 0, and 10 states at energy $\epsilon = 0.05\,\text{eV}$. Its environment is at $27\,^\circ\text{C}$ (a typical room temperature). Is it more likely to be at energy 0 or ϵ?

Answer: First convert the temperature to kelvins. To convert from degrees Celsius to kelvins you add 273, so $27\,^\circ\text{C}$ is 300 K. Using the given values for ϵ and T and looking up the value for k_B, the ground state is more likely than any of the excited microstates by a factor of

$$\frac{e^{0/(k_B T)}}{e^{-\epsilon/(k_B T)}} = \frac{1}{0.1445} = 6.92.$$

Since the ground state is therefore about 7 times more likely than each excited state, and there are 10 excited states, the system is most likely to be found with energy ϵ.

Here's a time-saving tip. Suppose, in the preceding Example, the first state had energy $\epsilon_1 = 0.1\,\text{eV}$ and the other states had energy $\epsilon_2 = 0.15\,\text{eV}$. That would make the calculations uglier. But you could subtract $0.1\,\text{eV}$ from both numbers, and that would turn this new problem into exactly the problem we just solved. The math would be simpler, but the final result would be the same.

We're going to use that trick almost every time we apply the Boltzmann distribution. The general rule is: *if you subtract the same number from every energy level in your system, the Boltzmann distribution is unchanged.* We will consistently subtract off the ground state energy from every energy level. Then the new ground state energy will be 0, and its Boltzmann factor will be 1.

Why is that legitimate? You can think of it in purely physical terms as choosing a different point to be the zero of potential energy (which is always an arbitrary decision). Mathematically, you will show in Problem 21 that such a shift multiplies the Boltzmann factors and the partition function Z by the same factor, which makes no difference in the resulting probabilities.

Normalizing the Boltzmann Distribution

The preceding Example shows how you can use the Boltzmann distribution to calculate relative probabilities, such as "the ground state is seven times more likely than each of the higher-energy microstates." If you want to know the actual probability of each state, you need to find the partition function Z. As you might guess, it's a matter of normalization: add up all of the probabilities for all the microstates, with the unknown constant Z in them, and set the total equal to 1.

Active Reading Exercise: Normalizing the Boltzmann Distribution

In the preceding Example ("Degeneracy"), find the value of the partition function and the probability of finding the system in its ground state.

You can find our answers at www.cambridge.org/felder-modernphysics/activereadingsolutions

The general rule that you can see from solving that problem is that the partition function is always the sum of $e^{-E/(k_BT)}$ over all of the possible microstates. (This is why it is convenient to define the Boltzmann distribution in terms of a constant called $1/Z$ rather than Z.)

Example: A Three-Level System

A system has one state with energy 0, two states with energy 2 eV, and four states with energy 5 eV. If its surroundings are at temperature 1200 K, what is the probability of finding it with energy 2 eV?

Answer: First, you can calculate $k_BT = 0.103$ eV. The ground state has Boltzmann factor $e^0 = 1$. (If you call the ground state energy of a system 0 then its Boltzmann factor always equals 1.) The 2 eV state has a Boltzmann factor equal to $e^{-(2\text{ eV})/(0.103\text{ eV})} = 0.380$. The 5 eV state has a Boltzmann factor equal to $e^{-(5\text{ eV})/(0.103\text{ eV})} = 0.00794$.

The partition function is the sum of seven Boltzmann factors:

$$Z = 1 + 2(0.380) + 4(0.00794) = 1.792.$$

The probability of each 2 eV state is its Boltzmann factor divided by the partition function. The probability that it is in *either* of the two 2 eV states is twice that:

$$P = \frac{2e^{-E/(k_BT)}}{Z} = \frac{2(0.380)}{1.792} = 0.42.$$

So there's a 42% chance of finding it at that energy. (You can also check that there's a 56% chance of finding it in the ground state and a 2% chance of finding it at 5 eV, so the probabilities do all add up to 1.)

Where Does the Boltzmann Distribution Come From?

We said above that low-energy states of our water glass are more likely than high-energy states because the glass can give up energy to the reservoir around it. In the following Active Reading Exercise you will flesh out that argument.

Active Reading Exercise: Small System, Big Reservoir

Consider a small system S in contact with a reservoir. The reservoir consists of 1000 atoms, each of which can have energy 0, or ϵ, or 2ϵ, etc. The entire system (S and the reservoir together) has a combined energy of 2ϵ.

1. Suppose System S is in a microstate (which we'll call M_1) with energy ϵ. How much energy is left for the reservoir? How many possible microstates are available in the reservoir?

2. Now suppose System S is in a microstate (M_2) with energy 2ϵ. How much energy is left for the reservoir? How many possible microstates are available in the reservoir?

When answering the following two questions, do *not* use the Boltzmann distribution. Instead use the fact that the combined system (S and the reservoir together) obeys the fundamental assumption of statistical mechanics.

3. Based on your answers above, can you reasonably conclude that System S is more likely to be in Microstate M_1 than Microstate M_2?

4. Based on your answers above, can you reasonably conclude that System S is more likely to have energy ϵ than energy 2ϵ?

You can find our answers to the first two questions at www.cambridge.org/felder-modernphysics/activereadingsolutions

Below we explain why the answers to the last two questions are "yes" and "no," respectively.

Part 3: When you compare two specific microstates of System S, the one with lower energy is more probable. Why? Because that leaves more energy for the reservoir. More energy levels always implies more possible microstates, and more microstates implies higher probability.

Part 4: You cannot therefore conclude that lower energy levels are always more likely, because System S may have *more* microstates at a higher energy level than at a lower one.

Part 4 makes the same point as the "Degeneracy" Example on p. 447, in which a lower-energy microstate of a system was 7 times more likely than any of the high-energy microstates, but there were 10 times as many of the high-energy states. The Boltzmann factors tell you the relative probabilities of individual microstates of a system. The Boltzmann factor for each energy times the degeneracy of that energy tells you the relative probability of the system having that much energy.

Stepping Back: When Does the Boltzmann Distribution Apply?

The Boltzmann distribution can give microstate probabilities for a glass of water, a block of copper, or a building, each of which is a system in contact with a much bigger reservoir.

If you have a gas in a sealed, insulated container, you cannot apply the Boltzmann distribution to the microstates of the gas. You can, however, treat each molecule of the gas as a system that is in contact with a bigger reservoir, namely the rest of the gas. One of the most common uses of the Boltzmann distribution is for individual particles, atoms, or molecules inside a macroscopic system.

The other caveat is the word "equilibrium" that we included in the definition of the Boltzmann distribution on p. 445. If I put a rock into a hot fire it will reach a state in which the average energy of its atoms is much higher than it was before it went into the fire. If I then pull it back out into a cold room, it is now in contact with a reservoir at a cold temperature, but the rock itself is still in a fairly high-energy microstate. I cannot apply the Boltzmann distribution to the rock until it comes to thermal equilibrium with the room. This should sound very much like our discussion of entropy; you can start a system in a low-entropy state, but given enough time it will drift toward its highest-entropy available macrostate.

When you apply the Boltzmann distribution to a single atom or molecule, the time scales for it to sample its available states are usually measured in tiny fractions of a second. A rock, on the other hand, might take hours or days to reach equilibrium.

10.4.3 Questions and Problems: The Boltzmann Distribution

Conceptual Questions and ConcepTests

1. For each quantity give a brief description of what it represents, and units that could be used to measure it.

   (a) $e^{-E/(k_B T)}$ (the "Boltzmann factor")

   (b) Z (the "partition function")

2. In the Boltzmann factor $e^{-E/(k_B T)}$,

   (a) of what is E the energy?

   (b) of what is T the temperature?

3. Under what circumstances does the Boltzmann distribution apply to a system?

4. A hot glass of water and a cold glass of water are sitting in the same room, which is at temperature T. Which of the following describes the probabilities of their microstates? (Choose one.)

   A. Because they are both in contact with the same reservoir, we know from the Boltzmann distribution that the microstates of the hot glass all have the same probabilities as the corresponding microstates of the cold glass.

   B. The probability distribution for the hot glass is more highly weighted toward high-energy states than the probability distribution for the cold glass, thus indicating that one or both of them don't obey the Boltzmann distribution with temperature T.

   C. They both have the same probability distribution, but the fact that we see that one glass is hotter than the other tells us that the hot glass randomly landed in a higher-energy microstate than the cold glass.

   D. It is meaningless to talk about probabilities of microstates until the glasses have reached equilibrium with the surroundings.

5. A system in equilibrium with a reservoir at temperature T has 10 particles, each of which has 30 available microstates, all with different energies.

   (a) Describe the state that this system will approach as $T \to 0^+$.

   (b) Describe the state that this system will approach as $T \to \infty$.

   (c) Is there any temperature at which the first excited state of the entire system would be more probable than the ground state?

   (d) How would your answers change if these particles were subject to the Pauli exclusion principle?

6. A paramagnet consists of N atoms, each of which can have energy 0 or ϵ. What is $\lim_{T \to \infty}$ of the number of atoms in the excited state? (Choose one and explain your answer using the Boltzmann distribution.)

   A. 0

   B. Less than $N/2$ but more than 0

   C. $N/2$

   D. More than $N/2$ but less than N

   E. N

7. The air in a room can be considered a small system, in contact with the much larger reservoir of air outside. Assume the room isn't being heated or cooled except by contact with the outside, and assume the outside air maintains a constant temperature. Why

doesn't the air in the room all sink down to the bottom of the room, which is its lowest possible energy state?

8. When a small system can exchange energy with a large reservoir, the low-energy microstates of the small system are more likely than its high-energy ones. Does that mean that the high-energy microstates of the *reservoir* are more likely than the low-energy ones?

For Deeper Discussion

9. In a hydrogen atom, there are two ground states at $E = -13.6$ eV and there are *infinitely* many excited states with energy asymptotically close to 0. The Boltzmann distribution implies that, accounting for degeneracy, all hydrogen atoms should be in those excited states at any non-zero temperature. In reality virtually all of the hydrogen atoms at room temperature are in the ground state. Why?

Problems

10. A system has one state with energy 0, two states with energy 0.1 eV, and four states with energy 0.5 eV. Its surroundings are at temperature 1200 K.

    (a) Calculate the Boltzmann factor $e^{-E/(k_B T)}$ for each of the seven possible microstates of this system.

    (b) The probability of any given microstate is its Boltzmann factor divided by the partition function Z. Calculate the partition function to normalize your probability distribution.

    (c) What is the probability of the system being in a particular *one* of the microstates with $E = 0.1$ eV?

    (d) What is the probability of the system having energy $E = 0.1$ eV?

11. A paramagnet is in an external magnetic field. Each atom can have energy 0 or 0.02 eV. If the paramagnet is at 300 K, what fraction of the atoms are at the higher energy? *Hint*: The fraction that are in a particular state equals the probability of any individual atom being in that state.

12. For a system in equilibrium with a reservoir, sufficiently high-energy microstates are so unlikely that we can consider them unreachable. The higher the temperature, the higher the unreachable energy levels are. How high does the energy have to be for the Boltzmann factor to drop below 0.001 . . .

    (a) at $T = 100$ K (very cold)?

    (b) at $T = 300$ K (room temperature)?

    (c) at $T = 1000$ K (very hot)?

    (d) What can you physically say about a microstate whose Boltzmann factor is 0.001? *Hint*: The answer is not "the probability of the system being in this microstate is 0.001."

13. The following model of DNA was proposed by Charles Kittel.[6] A strand of DNA is like a zipper that can be pulled apart at the end. The ground state of DNA is completely closed (zipped). When one link is open the energy goes up by about 0.3 eV. This opening allows many possible orientations of the unzipped link, so the degeneracy of this level is about 10,000. In sum, DNA has a non-degenerate ground state and a 10,000-fold degenerate excited state, whose energy is 0.3 eV above the ground state. Neglecting energy states above this first excited energy level, what percentage of DNA molecules are in the ground state (completely zipped) at 37° Celsius (body temperature)?

14. A hydrogen atom has two ground states (one for each electron spin) and eight first excited states with an energy roughly 10 eV higher than the ground state. At what temperature would a hydrogen atom be equally likely to be in the first excited state or in the ground state?

15. A new molecule called Felderonium has been discovered. The ground state is not degenerate. The first excited state is three-fold degenerate, meaning there are three states of the molecule at that energy. The energy of the first excited state is 0.10 eV above the energy of the ground state. The next excited state is six-fold degenerate and has an energy 0.15 eV above the ground state. There are no higher energy states.

---

6 Kittel, C., *Am. J. Phys.* **37**, 917 (1969).

(a) What fraction of Felderonium molecules at room temperature (298 K) are at their first excited energy level?

(b) An astronomer observes an interstellar dust cloud and notes from its emission lines that there's one singly excited Felderonium molecule (i.e. a molecule at the first excited energy level) for every five ground state Felderonium molecules. What is the temperature of this dust cloud?

16. A hydrogen atom has two ground states (one for each electron spin). The next energy level consists of eight states at 10.2 eV above the ground state, and the next after that consists of 18 states at 12.1 eV above the ground state.

(a) A hydrogen atom is surrounded by gas at 1000 K. Neglecting all states above the first excited state, find the probability that the atom is in the first excited state.

(b) Would your answer to Part (a) have changed significantly if you had included the next energy level? How do you know?

(c) Repeat Part (a) assuming the gas is at 300 K (room temperature) and again at 5000 K. How do the properties of hydrogen atoms change with temperature?

17. System A consists of two particles, each of which can be in one of three possible states with energies 0, ϵ, or 2ϵ. System B consists of three particles, each of which can be in one of two possible states with energies 0 or 2ϵ. See Figure 10.14.

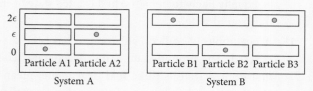

Figure 10.14

You may write the energy of each system as $E = q\epsilon$. The variable q is thus a unitless measure of how many energy units the system has. For example, in the state pictured above, $q_A = 1$ ($E_A = \epsilon$) and $q_B = 4$ ($E_B = 4\epsilon$). (Do not actually assume $q_A = 1$ and $q_B = 4$ for this problem; that's just an example.)

(a) The two systems are put in contact with each other (i.e. they are able to exchange energy) and the total energy of the combined system is 4ϵ. If you wait a long time and then look at the two systems, what is the probability that the energy will be distributed evenly between them (i.e. two energy units each)?

(b) Now suppose that you remove system A and put system B in contact with a thermal reservoir at a temperature of $T = 10,000$ K. Once it has had time to reach equilibrium, what is the probability that system B will have two energy units? Assume $\epsilon = 1$ eV.

18. A water molecule can vibrate by having the hydrogen atoms "flex" toward and away from each other. Quantum mechanically this vibration is like a simple harmonic oscillator – evenly spaced energies with no degeneracy – with an energy spacing of roughly 0.2 eV. In this problem we will assume for simplicity that a water molecule can only be in one of its first three states: the ground state, the first excited state, or the second excited state.

(a) Roughly what percentage of water molecules in a typical room are in the first excited state?

(b) Was it a good approximation to neglect states above the second excited state? Justify your answer.

(c) Suppose we had asked you about water molecules in the core of the Earth, where the temperature is estimated to be roughly 5000 K. Would it still be a good approximation to neglect states above the second excited state? Your justification for this answer must include a calculation.

19. Consider a paramagnet made of three atoms, each of which (because of an external field) can be at energy 0 (down) or 0.05 eV (up). Let q be the number of atoms at the higher energy. For example, if $q = 2$ then the whole system has energy 0.1 eV.

(a) If $q = 0$, the only possible state of the system is DDD. In a similar way, list all the possible microstates for each possible value of q.

(b) If the surrounding environment is at 20 °C, what is the probability that the paramagnet has $q = 1$?

20. A simple harmonic oscillator can have energy 0, $\hbar\omega$, $2\hbar\omega$, and so on. (Usually you add the ground state energy $(1/2)\hbar\omega$ to each of these, but we're defining the ground state as zero for simplicity here.) Use the fact that the infinite series $1 + x + x^2 + x^3 + \cdots$ converges to $1/(1-x)$ for all $|x| < 1$ to write an exact formula for the partition function for this system. Then use it to calculate the probability of finding a simple harmonic oscillator in the second excited state $(E = 2\hbar\omega)$ if $\hbar\omega = 0.01$ eV and the surrounding temperature is 400 K.

21. Shifting the Ground State

A quantum mechanical system has only two states, a ground state E_0 and a first excited state E_1. The system is in contact with a reservoir at temperature T.

(a) Write a formula for the probability that the system is in the excited state.

(b) Now you decide to define the ground state energy as 0, which makes the first excited state energy $E_1 - E_0$. Recalculate the probability of the system being in the first excited state using these new energies and prove that the answer is equivalent to what you derived in Part (a).

22. Exploration: The Derivation of the Boltzmann Distribution

Consider a small system S in contact with a large reservoir at temperature T, and consider two microstates of System S with energies E_A and E_B. Your goal is to show that the probabilities of finding the small system in either of those microstates is proportional to $e^{-E/(k_B T)}$, which is equivalent to saying that the ratio of the probabilities is $P_A/P_B = e^{-E_A/(k_B T)}/e^{-E_B/(k_B T)}$.

(a) Let Ω_A be the multiplicity of the *reservoir* when the system is in State A, and let Ω_B be the multiplicity of the reservoir when the system is in State B. Express the ratio P_A/P_B in terms of Ω_A and Ω_B, and then rewrite it in terms of entropies of the reservoir, rather than multiplicities. Using the properties of exponentials, express your answer in terms of $\Delta S_R = S_{R,B} - S_{R,A}$, the entropy difference between the two reservoir states.

(b) Temperature is defined as $T = 1/(dS/dE)$, which we can approximate as $T = \Delta E/\Delta S$. Explain why this approximation becomes exact in the limit of an infinitely large reservoir. *Hint*: Remember that this limit corresponds to the reservoir having a constant temperature.

(c) Using $T = \Delta E_R/\Delta S_R$, express the ratio P_A/P_B in terms of ΔE_R and T.

Pause for a moment to remember what all these variables mean: P_A and P_B are the probabilities of finding the small system in microstates A and B. Most of the other variables we're using refer to the reservoir. For example, ΔE_R is the change in the energy of the *reservoir* when the *small system* transitions from state A to state B.

(a) Let $\Delta E = E_B - E_A$ be the energy difference between the two states of the small system. How is ΔE related to ΔE_R?

(b) Express the ratio P_A/P_B in terms of ΔE. (You'll still have the reservoir temperature T in your answer.)

(c) Finally, re-express P_A/P_B in terms of the individual energies E_A and E_B rather than the difference ΔE. You should get the Boltzmann distribution.

10.5 Some Applications of the Boltzmann Distribution

The Boltzmann distribution predicts the probabilities of individual microstates based on the energy levels they represent. But, as we emphasized at the beginning of this chapter, you could never measure the microstate of all the air in a room. The Boltzmann distribution is useful (and credible) because it leads to larger-scale predictions. In this section we will see how the

Boltzmann distribution leads to the answers to three questions: answers that can be verified experimentally.

For a given gas of particles at a given temperature, . . .

1. What is the average thermal energy of the particles?
2. What is the heat capacity of the gas?
3. What is the distribution function of molecular speeds?

Average thermal energy will be an important idea in later sections and later chapters of this book. Heat capacity will come up again in Chapter 11, but if you're not covering that chapter it won't be needed for anything else in this chapter or subsequent ones. Nothing else in this book depends on the distribution of molecular speeds.

10.5.1 Discovery Exercise: A Perfectly Even State Density

Consider a system that has three possible energy levels: $E = 0$, $E = 3k_BT$, and $E = 6k_BT$. Each of those levels represents exactly one microstate, and their probabilities follow the Boltzmann distribution $P = (1/Z)e^{-E/(k_BT)}$.

1. Recall that the formula for expectation value is:

$$\langle E \rangle = \sum_E EP(E).$$

Calculate the expectation value of energy for this system. Your answer should be in the form of a decimal times k_BT.

See Check Yourself #16 at www.cambridge.org/felder-modernphysics/checkyourself

2. Explain briefly what your result means, in a sentence that starts "If you measured a million of these particles, . . ."

10.5.2 Explanation: Some Applications of the Boltzmann Distribution

The applications in this section are all based on applying the Boltzmann distribution to one particle, atom, or molecule in a large system.

Average Energy

Consider a collection of particles such as a gas full of molecules. Each individual molecule is a "system" that obeys the Boltzmann distribution, with the gas itself acting as the reservoir. What is the *average thermal energy* of those particles? (Remember that "thermal energy" refers to the internal energy, which macroscopically looks like warmth, rather than the potential or kinetic energy.)

We're going to offer two different answers to that question. The first answer gives a rough order-of-magnitude approximation for any system that meets some fairly broad conditions that we will spell out. The second gives a more accurate answer under more narrowly defined conditions.

You can arrive at the first, rougher answer by looking at a plot of the Boltzmann factor. Figure 10.15 shows how the probabilities of individual microstates decrease as their energies increase. Note that we have defined the ground state to be $E = 0$; as we discussed in Section 10.4, you can subtract off the ground state energy from every energy level, and this bit of simplification does not change the calculated probabilities.

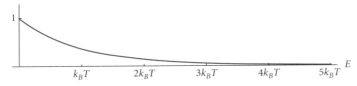

Figure 10.15 The Boltzmann factor $e^{-E/(k_BT)}$.

Figure 10.15 suggests that microstates with energies close to or below k_BT occur relatively often, while microstates with energies much higher than k_BT are rare. That leads to the following guideline:

The average thermal energy of a particle is typically of order k_BT above its ground state.

We write $E \sim k_BT$, using the symbol $\sim$ to designate an order-of-magnitude estimate.

This general guideline doesn't always apply.

Active Reading Exercise: The $E \sim k_BT$ Guideline

Suppose a system of particles obeys the Boltzmann distribution, meaning that, for each particle, each of its states has a probability proportional to $e^{-E/(k_BT)}$. What additional assumptions must we make to conclude that each particle's average energy will be roughly k_BT?

Below we're going to give two important answers to that question. We hope you spot at least one of them before we say it, because it's a point that we emphasized in Section 10.4. (Yes, that's a hint.)

The first assumption is that every energy level corresponds to the same number of microstates (at least approximately). Remember that the Boltzmann distribution gives the probability for each *microstate,* not for each energy level. If there are many microstates with low energy, and only a few with high energy, that will tend to push the average down, and vice versa. This is the point we emphasized in Section 10.4.

Figure 10.16 illustrates this point by drawing a horizontal blue line to represent every microstate in two different systems. System A has many microstates with low energy levels, and fewer with high levels; System B has the opposite. If both systems follow the Boltzmann distribution, System A will have a lower energy on average than System B. Neither one may have an average energy at or near k_BT.

The second assumption is illustrated by the two systems in Figure 10.17. Can you see which one will have an energy level close to k_BT, and which will not?

System C will have an energy roughly equal to k_BT. If you're looking at the drawing of System C's energy levels and thinking "the average looks much higher than k_BT," you're forgetting the Boltzmann distribution; all those higher energy levels have very low probabilities!

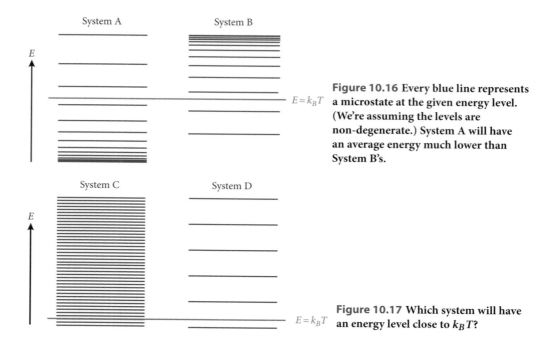

Figure 10.16 Every blue line represents a microstate at the given energy level. (We're assuming the levels are non-degenerate.) System A will have an average energy much lower than System B's.

Figure 10.17 Which system will have an energy level close to $k_B T$?

And System D? Its ground state energy has a Boltzmann factor of 1, as with all our systems. But its first excited state has a much higher energy, and therefore a near-zero probability. So System D is almost guaranteed to have energy 0, making its average energy very low. (You should have seen this come out if you did the Discovery Exercise for this section.)

Is this starting to sound a bit familiar? This is how Planck solved the ultraviolet catastrophe, all the way back in Chapter 3! He did not modify the Boltzmann distribution; instead he proposed discrete energy levels. For low-frequency radiation his levels were closely spaced like System C above, leading to the same average energy as classical models. But for high frequency his levels were widely spaced like System D, leading his model to predict a very low average energy – which matched the data.

This issue also arises in many more mundane circumstances. The thermal energy at room temperature typically isn't enough to excite either molecular vibrations or electron orbitals above their ground states. Those systems and many others therefore spend virtually all their time in their ground states.

We can summarize both assumptions above by introducing the term "density of states," the number of microstates per unit energy, as a function of energy:

A system that obeys the Boltzmann distribution will obey the rule $E \sim k_B T$ if its density of states is reasonably high and reasonably uniform.

Systems A and B above have non-uniform densities of states, and System D has a density of states that is uniform but low. Of the four, only System C can be assumed to have an average energy around $k_B T$.

Near the start of Section 10.5.2 we promised two different rules for estimating average energy. The rule $E \sim k_B T$ is very rough but applies to a broad range of systems. The more precise rule that applies under more narrow circumstances is called the "equipartition theorem."

In order to state that theorem, we need to introduce another new term: "degrees of freedom." Although a rigorous definition turns out to be complicated, you can think of degrees of freedom as the variables you need to specify in order to give the state of a system at any given moment. For example, a particle moving through space is classically described by six variables: x, y, z, v_x, v_y, and v_z. In the absence of external forces its energy is $(1/2)mv_x^2 + (1/2)mv_y^2 + (1/2)mv_z^2$. So the particle has six degrees of freedom, but its energy depends on only three of them.

In that example we call v_x, v_y, and v_z "quadratic degrees of freedom" because the energy is a quadratic function of each of them. And with that terminology in place, we can present the equipartition function.

The Equipartition Theorem

For a system in equilibrium with a reservoir at temperature T, the average value of any energy term that depends quadratically on one of the system's degrees of freedom is $(1/2)k_B T$.

We can illustrate how this works through some examples. (The results below are non-relativistic except the one where we specify otherwise.)

- A free particle (as we discussed above) has three quadratic degrees of freedom, so it has average thermal energy $(3/2)k_B T$.

- A gas made of N free non-relativistic particles has $3N$ quadratic degrees of freedom, so it has average thermal energy $(3/2)Nk_B T$.

- A one-dimensional simple harmonic oscillator has an energy that depends quadratically on two degrees of freedom: $E = (1/2)mv^2 + (1/2)kx^2$. So its average thermal energy is $k_B T$.

- A hydrogen molecule has an energy that depends on five degrees of freedom, three translational and two rotational: $E = (1/2)mv_x^2 + (1/2)mv_y^2 + (1/2)mv_z^2 + (1/2)I\omega_x^2 + (1/2)I\omega_y^2$. (We'll talk below about why there are only two possible rotations.) So its average thermal energy is $(5/2)k_B T$.

- A free relativistic particle has three degrees of freedom, but they are not quadratic. (KE $= (1/2)mv^2$ is a Newtonian formula.) The guideline $E \sim k_B T$ still applies to each degree of freedom, but the exact coefficient $(1/2)$ doesn't. (You can show that the thermal energy of a free relativistic particle in 3D is $3k_B T$ rather than $(3/2)k_B T$.)

Example: Air Molecules

Estimate the average speed of nitrogen molecules in a typical room.

First (rough) answer: Assume that the molecules have kinetic energy on the order of $k_B T$:

$$\frac{1}{2}mv^2 \sim k_B T \quad \rightarrow \quad v \sim \sqrt{\frac{2k_B T}{m}}.$$

Typical room temperature is about 27 °C, which is 300 K. The mass of a nitrogen molecule is about 5×10^{-26} kg, so this gives $v \sim 400$ m/s.

Second (more precise) answer: Each molecule is essentially a free particle moving in three-dimensional space, so a more precise estimate of its average thermal energy is $(3/2)k_B T$. Plugging that in and using the more precise $m = 4.65 \times 10^{-26}$, the same equations now give $v = 517$ m/s. (This isn't the average value of v but rather the square root of the average value of v^2, often called the "root-mean-square" velocity.)

The derivation of the equipartition theorem is *not* quantum mechanical: it assumes that all the relevant variables are continuous. When quantization becomes important, the equipartition theorem breaks down.

You can understand the equipartition theorem qualitatively as another result of the fundamental assumption of statistical mechanics. Imagine a free particle and a hydrogen molecule exchanging energy. The energy can go into eight buckets: the three degrees of freedom of the particle, and the five degrees of freedom of the molecule. (Some energy might go into "the particle moving in the x direction" and some might go into "the molecule rotating in the y direction" and so on.) Over time each bucket will average the same amount of energy as every other bucket, which is why the hydrogen molecule will average 5/3 times as much energy as the particle. (This argument also depends on the density of states in each bucket, which is what makes the energy distribution different for types of energy that don't depend quadratically on the degrees of freedom.)

If the hydrogen molecule were to break apart into two separate hydrogen atoms, they would between them have more degrees of freedom and thus take more energy than the molecule by itself.

Heat Capacity

In Section 10.3 we defined heat capacity as the energy per unit temperature required to heat an object: $C = dE/dT$. Based on what we've said about average energy, you're now in a good position to estimate heat capacities.

Active Reading Exercise: Heat Capacity

Estimate the order of magnitude of the heat capacity per molecule of most objects at room temperature. *Hint: Think about what we've said about the average energy per molecule.*

Roughly speaking, most objects have a thermal energy on the order of $E \sim k_B T$ per molecule (or atom). More accurately, we said that applies to each type of energy (translational kinetic energy, rotational kinetic energy, etc.), so we expect E to equal a few $k_B T$ per molecule. The heat capacity per molecule should therefore be a few k_B.

At standard temperature and pressure, the heat capacity per molecule of air is roughly $2.5k_B$, of mercury $3.4k_B$, and of lead $3.1k_B$. Those are three very different substances (a gas, a liquid, and a solid), but they all have macroscopic properties that can be explained by the Boltzmann distribution.[7]

We'll talk more about heat capacities of solids in Chapter 11, but here we want to say a bit more about gases. We can use the equipartition theorem to make our prediction for the heat capacity of gases more precise. Consider a gas of helium atoms. According to the equipartition theorem, the average kinetic energy of a molecule is $(3/2)k_BT$, so the heat capacity per molecule of helium gas should be $1.5k_B$. The measured value is $1.505k_B$.

Now consider hydrogen gas. Helium is a noble gas, so it doesn't form molecules, but hydrogen at room temperature is in the form H_2. Each of those molecules still has (on average) a translational kinetic energy of $(3/2)k_BT$, but each molecule can also rotate. You can think of a hydrogen molecule as a thin rod with the hydrogen nuclei at the two ends. If we call the axis of that rod the z axis then it can rotate in the xz plane or the yz plane (Figure 10.18).

The hydrogen molecule's total kinetic energy is the sum of its three translational and two rotational pieces: $(1/2)mv_x^2 + (1/2)mv_y^2 + (1/2)mv_z^2 + (1/2)I\omega_x^2 + (1/2)I\omega_y^2$. The equipartition theorem says that each piece should have on average $(1/2)k_BT$ of thermal energy, so $E = (5/2)k_BT$ and $c = (5/2)k_B$. The measured value is $2.47k_B$.

At $T = 40$ K, however, the energy required to excite one quantum of rotational energy is higher than k_BT. We say the rotational degrees of freedom are "frozen out": that is, the rotational energies look like System D in Figure 10.17, with even the first excited state too high

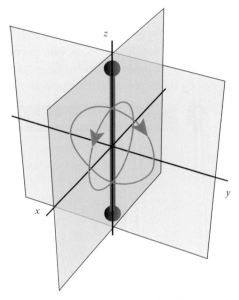

Figure 10.18 A hydrogen molecule can rotate in two planes. Any other possible rotation can be written as a sum of rotations in these two planes.

to reach. That leaves three degrees of freedom, so our predicted heat capacity per molecule drops to $(3/2)k_B$. The measured value is $1.51k_B$.

Conversely, vibrations of the molecule are frozen out at room temperature but become significant at much higher temperatures.

We should mention that the values you will typically see in tables of heat capacity will look different from the values we have presented here, for two reasons. First, tables often list values per mole while we are calculating per molecule; second, table values are often heat capacity at constant pressure (which is easier to measure), while we are calculating heat capacity at constant volume (which is easier to calculate).

---

7 We don't mean to leave you with the impression that every substance has virtually the same specific heat capacity. Water, for example, has $9k_B$, one of the highest for any common substance. That's because water molecules are held together by hydrogen bonds whose bonding energy is in the right range to be broken by thermal motions, so the number of bonds and thus the amount of thermal energy is strongly dependent on temperature. Most common substances fall between $1.5k_B$ and $4k_B$, however.

The Maxwell Speed Distribution

We've seen that in many circumstances the average thermal energy of a molecule is of order $k_B T$. In some cases, however, knowing the average is not enough.

Consider the core of the Sun. The Sun's core is mostly ionized hydrogen, meaning individual protons and electrons. When two protons collide strongly enough they can fuse, and those fusion reactions produce the energy that makes the Sun shine. The core of the Sun is about 1.5×10^7 K, so an average proton has energy on the order of $k_B T = 10^{-16}$ J. That turns out to be less than 1% of the energy required for a collision to result in fusion.

So why does the Sun shine? Because it doesn't need an *average* proton to have enough energy to fuse. It just needs *some* protons to have enough energy. To analyze fusion in the Sun we need to explore the probabilities of a proton having various speeds. Those probabilities are given by the Boltzmann distribution.

Active Reading Exercise: Boltzmann Factor for Speed

A molecule of mass m is in a gas at temperature T. Each possible velocity the molecule could have is a microstate of that molecule. (We're ignoring position.)

1. What is the Boltzmann factor for a microstate with speed v?

2. At room temperature the Boltzmann factor for a nitrogen molecule to be moving at 300 m/s is about five times as large as the Boltzmann factor for it to be moving at 600 m/s, but it is *not* true that a nitrogen molecule is five times as likely to be moving at 300 m/s as it is at 600 m/s. Why not?

The Boltzmann factor depends on the energy, which is $(1/2)mv^2$, where $v = |\vec{v}|$ is the molecule's speed. So the Boltzmann factor is $e^{-mv^2/(2k_B T)}$. You can plug in the mass of a nitrogen molecule and $T = 300$ K to confirm that this is about five times bigger at $v = 300$ m/s than it is at 600 m/s. So why is that not the ratio of probabilities?

Remember that the probability of a particle having a certain energy depends both on the Boltzmann factor and on the degeneracy at that energy. Only now we hit a complication. Our earlier examples involved discrete energy states, but speed is a continuous variable. So instead of degeneracy we have to talk about the density of states $g(E)$.

We defined density of states above as the "number of microstates per unit energy." That's not a rigorous definition when you have a continuum of microstates, but the basic idea still applies; certain energies have more available arrangements than others, and that gets counted in the probabilities.[8]

For a continuous variable we write a continuous distribution of probability densities, and the probability is an integral. If a small system is in contact with a reservoir at temperature T and the energy of that system is continuously distributed, the probability of the system having energy

---

8 In many cases the underlying energy eigenstates are truly discrete. For example, the air molecules in the atmosphere are bound – they don't have enough energy to go flying into space – so their energy eigenstates are technically discrete. But their possible states are so closely spaced that you can think of them as continuous, with a higher density at some energies than at others.

between E_1 and E_2 is

$$P(E_1 < E < E_2) = \int_{E_1}^{E_2} g(E)e^{-E/(k_B T)}\,dE.$$

You normalize the distribution by setting its integral from 0 to ∞ to 1.

Coming back to molecular speeds, you can show in Problem 26 that the density of states is proportional to v^2, so the probability of a molecule having speed v is proportional to $v^2 e^{-mv^2/(2k_B T)}$. Normalizing then leads to the "Maxwell speed distribution."

The Maxwell Speed Distribution

The probability that a molecule of mass m in an ideal gas at temperature T has speed between v_1 and v_2 is given by

$$P = 4\pi \left(\frac{m}{2\pi k_B T}\right)^{3/2} \int_{v_1}^{v_2} v^2 e^{-mv^2/(2k_B T)}\,dv. \tag{10.5}$$

That phrase "ideal gas" means that the molecules experience no forces except during very brief collisions.

Figure 10.19 shows the probability distribution for molecular speeds. It goes to zero at $T = 0$ and $T \to \infty$ and has a peak in the middle at $v = \sqrt{2k_B T/m}$. That implies (not surprisingly) that typical speeds go up with increasing temperature.

We'll finish off this section by using the Maxwell speed distribution to calculate the probability of an electron escaping from a block of copper. In Problem 18 you can apply the same technique to the scenario we began with: protons fusing in the Sun.

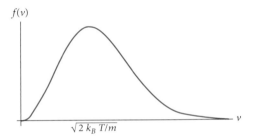

Figure 10.19 The Maxwell speed distribution.

Example: Escaping Copper

Inside a block of copper you can consider the outer electrons to be approximately an ideal gas. When an electron approaches the edge of the block, it needs about 5 eV of kinetic energy to leave the copper. (You may recall from our discussion of the photoelectric effect that this energy is called the "work function" of copper.) If you heat a block of copper to 1000 K (a large bonfire), what is the probability of an electron having enough energy to escape?

Answer: The question is asking for the probability that an electron has an energy equal to or greater than the work function. Using $E = (1/2)mv^2$, the required speed is $v_{min} = 1.325 \times 10^6$ m/s, and

$$P = 4\pi \left(\frac{m}{2\pi k_B T}\right)^{3/2} \int_{v_{min}}^{\infty} v^2 e^{-mv^2/(2k_B T)}\,dv.$$

Plugging in $T = 1000$ K and evaluating this integral numerically gives $P = 6 \times 10^{-25}$: very close to zero. (By the way, copper melts at about 1350 K, long before it gets hot enough for electrons to escape by random thermal motion.)

10.5.3 Questions and Problems: Some Applications of the Boltzmann Distribution

Conceptual Questions and ConcepTests

1. The formula $E \sim k_B T$ can be used to approximate ... (Choose all that apply, and briefly explain.)

   A. the average energy of all the particles in an isolated, macroscopic system.

   B. the maximum energy of all the particles in an isolated, macroscopic system.

   C. the average energy of all the particles in a small macroscopic system in contact with a reservoir.

   D. the maximum energy of all the particles in a small macroscopic system in contact with a reservoir.

2. The text talked about using the Boltzmann distribution to analyze solar fusion. What general type of system does the Boltzmann distribution apply to, and why does that system meet that criterion?

3. We said that microstates with energies much higher than $k_B T$ occur rarely, but you are always free to redefine the zero of energy. If you define the ground state energy to be much higher than $k_B T$, that doesn't make it less likely to occur. Explain how that fact can be consistent with the Boltzmann distribution.

4. Which would you expect to be moving faster in the air around you? (Choose one and explain.)

   A. A hydrogen molecule

   B. A nitrogen molecule

   C. They would on average move at the same speed.

5. (a) Why is the heat capacity of hydrogen lower at 40 K than it is at room temperature?

   (b) At temperatures above about 1000 K, the heat capacity of hydrogen rises above its room temperature value. Why?

   (c) Keeping in mind the explanation you just gave, what do you expect the heat capacity per molecule of hydrogen to be at high temperatures? (Choose one and briefly explain your answer.) *Hint*: You can neglect electronic excitations and relativistic effects.

   A. $3k_B T$

   B. $(7/2)k_B T$

   C. $4k_B T$

   D. $> 4k_B T$

6. Rank-order the following by heat capacity per molecule, from highest to lowest. (For a noble gas, read "atom" for "molecule.") For some the answer might be that they are approximately equal. Explain your reasoning.

   A. Helium at room temperature

   B. Hydrogen at room temperature

   C. Helium at 10 K

   D. Hydrogen at 10 K

7. A water molecule is made up of three atoms that are *not* all in a line with each other. Based on only that information, which of the following is the best prediction for the heat capacity, per molecule, of water vapor at room temperature? (Choose one, and briefly explain your answer.)

   A. $(1/2)k_B$

   B. k_B

   C. $(3/2)k_B$

   D. $(5/2)k_B$

   E. $3k_B$

8. Bart tells Lisa that he calculated that the Maxwell speed distribution at 200 m/s for a particular gas equals 1/3, so he knows that the molecules of that gas have a 1/3 chance of moving at 200 m/s. "Oh, Bart," Lisa sighs, "That's not what it means at all." What was wrong with Bart's assertion, and how should he have interpreted his result?

9. Explain why the Maxwell speed distribution goes to zero at $v = 0$ and $v \to \infty$. Don't just note that the formula says so; explain why it occurs physically. One or two short sentences for each limit should be plenty.

10. Why does the Maxwell speed distribution only apply to ideal gases (ones in which the molecules

feel no forces)? In other words, what step in the arguments we made for this formula would be invalid if there were forces on the molecules, either from each other or from an external source?

11. In an ideal gas of molecules, rank-order the following speeds, and explain how you know. Note that $\langle x \rangle$ means the average value of x. You can answer this question just by looking at Figure 10.19, with no calculations. *Hint*: No two are equal.

   A. The peak speed in the Maxwell speed distribution

   B. $\langle v \rangle$

   C. $\sqrt{\langle v^2 \rangle}$

12. On p. 458 we wrote: "If the hydrogen molecule were to break apart into two separate hydrogen atoms, they would between them have more degrees of freedom and thus take more energy than the molecule by itself." Why doesn't that violate conservation of energy?

For Deeper Discussion

13. A gas of 10^{23} molecules will have a typical thermal energy of $10^{23} k_B T$. But that entire canister of gas still obeys the Boltzmann distribution for its microstates. So why doesn't the same argument that said a molecule should have roughly $k_B T$ of energy say that the gas as a whole should have energy $k_B T$?

14. A gas of hydrogen atoms will ionize above a certain temperature T_I. Which of the following is true about the temperature T_I? (Choose one and explain your choice.)

   A. $k_B T_I \ll 13.6\,\text{eV}$

   B. $k_B T_I \approx 13.6\,\text{eV}$

   C. $k_B T_I \gg 13.6\,\text{eV}$

Problems

15. Inside a metal there are electrons that are not bound to any one atom but instead can move freely throughout the metal. The following questions refer to a block of iron in a kiln at $700\,°\text{C}$. You can estimate your answers to an order of magnitude, but you should show calculations to support them.

   (a) What's the typical energy of an electron in that block of iron?

   (b) What's the typical speed of an electron in that block of iron?

16. In the Example "Air Molecules" on p. 457 we used the equation $E \sim k_B T$ to estimate the average speed of nitrogen molecules in a typical room.

   (a) We predicted a typical speed of 517 m/s. What is the probability of a given molecule's speed being within 50 m/s of that value?

   (b) What is the probability of a given molecule's speed being more than twice the average speed?

17. You have a large collection of electrons, each confined in its own infinite square well of width 10^{-10} m. (You can find relevant equations about infinite square wells in Appendix C.) At what temperature would you expect a noticeable percentage of those electrons to be above the ground state? You can estimate your answer to an order of magnitude, but you should show calculations to support it.

18. A proton in the core of the Sun strikes another proton. For simplicity you can take the second proton to be at rest. The core of the Sun is at 1.5×10^7 K. In order to get close enough to fuse, the moving proton will have to have at least 1.24×10^{-14} J of kinetic energy. You can estimate your answers in this problem to an order of magnitude, but you should show calculations to support them.

   (a) Estimate the probability that the two protons will fuse.

   (b) A typical proton in the Sun experiences about 10^8 collisions per second. Assuming your answer to Part (a) gives the probability of fusion occurring in any given collision, estimate how long it will take before a typical proton fuses.

   (c) Re-do your calculations for a nuclear reactor with a temperature of 10^8 K. (It's true that some fusion reactors achieve this temperature for short periods, but the reactions are complicated and you shouldn't put too much stock in the numbers you get out of this simple analysis. What you will get is a sense for the temperature dependence of these formulas.)

19. In the Discovery Exercise on p. 454, you calculated the average energy for a system with three evenly spaced energy levels with $\Delta E = 3k_B T$. You should have found roughly $0.16 \times k_B T$.

    (a) Repeat the exercise, still using three energy levels, but this time with $\Delta E = 6k_B T$. Briefly explain why the answer comes out the way it does.

    (b) Repeat the exercise, this time using 60 energy levels, with $\Delta E = (1/10)k_B T$. Briefly explain why the answer comes out the way it does.

    (c) In the limit as $\Delta E \to 0$, the sum becomes an integral. Evaluate that integral, letting E range from 0 to ∞.

20. (a) Write an equation for the total energy of a classical simple harmonic oscillator.

    (b) How much thermal energy does the equipartition theorem predict such an oscillator will have on average?

    (c) If you hang a mass on a spring and then heat up the room, you *do not* see it spontaneously vibrating. How can you reconcile that with your answer to Part (b)?

    (d) Under what circumstances would you expect your prediction in Part (b) to be invalid?

21. In this problem you're going to determine whether the density of states of a hydrogen atom increases or decreases with increasing n. The answer depends on two factors. (You may want to keep Appendix G handy for this problem.)

    (a) Write down the energy gap between the ground state and the first excited state, and then the energy gap between the first and second excited states, and then second to third. Based on that, is the density of states increasing or decreasing?

    (b) Now write down the number of microstates represented by the ground state, first excited state, and second excited state. Based on that, is the density of states increasing or decreasing?

    (c) Now answer the question, and justify your answer.

    (d) Based just on that, would you expect the average energy to be above, or below, $k_B T$?

    (e) In fact, at room temperature the answer to whether $\langle E \rangle$ is larger or smaller than $k_B T$ is the opposite of what you should have concluded in Part (e). Why?

22. Equation (10.5) involves the function $4\pi \left(\frac{m}{2\pi k_B T} \right)^{3/2} v^2 e^{-mv^2/(2k_B T)}$. In this problem we will use $f(v)$ to refer to that function (*not* to its integral).

    (a) Calculate the speed at which $f(v)$ peaks.

    (b) An incorrect interpretation of your answer to Part (a) would be "that's the most likely speed for a particle to have." Explain why this is incorrect, and write a better interpretation.

    (c) Write an equation that represents the normalization condition for this function. (The constants were chosen to satisfy that condition.)

    (d) Is this function perfectly symmetrical about the v-value you found in Part (a)? What does your answer imply about the speeds of particles close to that value?

    (e) Write an integral that represents the expectation value $\langle v \rangle$. Briefly explain in words what the result would represent. (If you don't remember how to calculate expectation values, you can review Section 4.3.) Your final answer to this problem will be an integral; you will evaluate that integral in Problem 23.

23. (a) Using the Maxwell speed distribution defined on p. 461, write and evaluate an integral to find the expectation value $\langle v \rangle$. *Hint*: If you don't remember how to calculate expectation values you can review Section 4.3.

    (b) Set up and evaluate an integral to find the "root-mean-square" speed, defined as $\sqrt{\langle v^2 \rangle}$.

    (c) Which of your answers matches the prediction of the equipartition theorem? How could you have predicted which one would match equipartition without doing the calculations?

24. Equation (10.5) gives the Maxwell speed distribution. It's a complicated mess of constants and variables, but it can be greatly simplified with a u-substitution.

 (a) Define a new unitless variable x that lets you write it in the form $P = \int_{x_1}^{x_2} f(x)\,dx$ with no other letters.

 (b) In the Example "Escaping Copper" on p. 461 we calculated the probability of an electron escaping from a block of copper at 1000 K. Show how that integral looks using your substitution from Part (a). Then evaluate your integral and make sure you get the same probability that we found!

25. Spectral lines appear broadened because some of the atoms that emit them are moving toward the detector and others are moving away, and the resulting Doppler shifts result in a range of measured frequencies for the lines. Estimate the width in frequencies of a Lyman-alpha emission line (the transition from $n = 2$ to $n = 1$ in a hydrogen atom) caused by Doppler broadening at room temperature.

26. Density of States for the Maxwell Speed Distribution

The state of a free classical particle can be specified with six numbers: x, y, z, v_x, v_y, and v_z. Of those, only the velocity components affect the energy, so those are the degrees of freedom we'll consider. Our goal is to calculate the density of states, or number of states per unit energy, as a function of speed. The problem is that speed is a continuous variable, so we're going to have to break all the possible values of v_x, v_y and v_z into discrete bins of width Δv.

 (a) We're going to start with the simple case of a particle in 1D. Sketch the v_x axis broken up into small bins of width Δv.

 (b) Assume a particle has some specified energy E. Sketch the possible states for that particle on your v_x axis. How many bins might the particle be in?

 (c) Now consider a particle moving in 2D. To specify its state we need two axes: v_x and v_y. Sketch that two-dimensional space broken into small square bins of width and height Δv.

 (d) On your 2D plot, sketch a curve representing all the possible states with some particular energy E. What is the shape of that curve?

 (e) In the limit where the bins are extremely small, the number of bins intersected by that curve is proportional to the length of the curve. Give that length as a function of the speed v.

 (f) Now imagine a particle moving in 3D, with a state specified in a (v_x, v_y, v_z) space. We won't ask you to draw it, but what 2D shape would represent the set of all states with a specified energy E?

 (g) Give the area of that 2D shape as a function of speed v.

 (h) If a particle is moving in 3D, what is the density of states $g(v)$? Give your answer in the form of an unknown constant times v raised to a power, and explain how you know this, based on your previous answers.

10.6 Quantum Statistics

Every electron in the universe is *exactly* the same as every other electron. That fact leads electrons to behave in ways that are measurably different from how they would behave if they could in any way be distinguished from each other.

In this section we'll discuss the statistical behavior of collections of identical particles, a topic known as "quantum statistics."

10.6.1   Discovery Exercise: Quantum Statistics

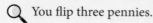 You flip three pennies.

1. List all of the possible outcomes. For example, HHT is one and HTH is another. Assuming all of these outcomes are equally likely, what is the probability of getting three heads?

2. List all the possible outcomes again, but this time only list how many heads and tails you got. For example, don't list HHT and HTH as separate states. Now, assuming all of *these* outcomes are equally likely, what is the probability of getting three heads?

   See Check Yourself #17 at www.cambridge.org/felder-modernphysics/checkyourself

3. Which of your two answers is the correct probability for getting three heads when you flip three pennies?

10.6.2   Explanation: Quantum Statistics

Figure 10.20 Two particles, each with two available states.

We're going to start with one of the simplest systems you can imagine. This system has only two particles. Each particle has only two available states, A and B (Figure 10.20). And these two states have equal energy, so the Boltzmann distribution predicts that all possible microstates of this system are equally likely.

Question: What is the probability that the two particles are in the same state as each other?

Not-exactly-an-answer: There are three possible answers, depending on what types of particles they are.

We said in Chapter 8 that there are two types of particles. Fermions (such as electrons, protons, and neutrons) obey the Pauli exclusion principle, which says no two identical fermions can be in the same state. Bosons do not obey this principle. With that in mind, here are the three cases.

- *Two different particles.* First let's assume the two particles are different, such as a proton and a neutron. In this case there are four possible states: both in A, both in B, proton in A and neutron in B, or neutron in A and proton in B. Two of these four microstates of the system have both particles in the same state, so the probability is 1/2.

- *Two identical fermions.* If the particles are two protons or two neutrons, then the probability that the two are in the same state is 0, because that's impossible for identical fermions.

- *Two identical bosons.* Finally, let's assume the particles are two photons, which is an example of the more general case of two identical *bosons*. This system has three possible states: both in A, both in B, or one in each. So the probability of matching states is now 2/3.

The Pauli exclusion principle is weird, but that's a bit of quantum weirdness that you're used to by now. The new weirdness here is that our "two photons" system acts mathematically differently from our "one proton and one neutron" system. That's not how probability works for any classical system! If you flip two quarters then "first quarter heads and second quarter tails" is a state and "first quarter tails and second quarter heads" is a *different* state. It doesn't matter if you flipped them at the same time, or if they look identical: they are two different quarters whether you can tell them apart or not, and the probability that they both come out the same is $1/2$.

But bosons obey different rules. "This photon is in A and that photon is in B" is one possible state, and "this photon is in B and that photon is in A" is the *same* state, because – other than the states they are in – any two photons are completely interchangeable. The two-photon system has only three states, so the fundamental assumption of statistical mechanics predicts that each of these states will have a $1/3$ probability. That prediction turns out to be correct.

Generalizing, identical bosons are always statistically more likely to be in the same state as each other than they would be if they weren't identical. That fact can lead to dramatic physical consequences such as superconductivity and superfluidity, which we'll discuss in Section 10.8.

Example: Two Particles and Three States

Two non-interacting particles can each be in one of three states, A, B, or C, all with the same energy. What is the probability of finding one particle in State A and one in State B if they are . . .

1. distinguishable particles?
2. identical bosons?
3. identical fermions?

Answer: We will list the possible states of the system, coloring the "winning" states **red**.

1. We'll use 1A to mean the first particle is in State A, and so on. The possible states are:

$$(1A\ 2A), (1A\ 2B), (1A\ 2C),$$
$$(1B\ 2A), (1B\ 2B), (1B\ 2C),$$
$$(1C\ 2A), (1C\ 2B), (1C\ 2C).$$

2. This time we can't label which particle is in which state, just which states are occupied. That means we don't list A B and B A as two separate states, so the possibilities are

$$(AA), (AB), (AC), (BB), (BC), (CC).$$

3. This is like the previous case, except we leave out states like A A that violate the Pauli exclusion principle:

$$(AB), (AC), (BC).$$

So the probabilities are $2/9$, $1/6$, and $1/3$, respectively.

It may seem contradictory to call particles "non-interacting" when they do change each other's statistics. It's a common term, however, meaning that they don't exert forces on each other, so they don't interact in any way *except* through the statistics of identical particles.

The Fermi–Dirac and Bose–Einstein Distributions

The Example on p. 461 looked at the electrons in a block of copper. Electrons, like photons, are indistinguishable from each other. So when we list the microstates of such a system, we never say "this particular electron is in that state"; such an assertion would be meaningless. Instead we list the total *number* of electrons in each available state. That number is called the "occupation number," or just "occupancy," of the state.

By applying the Boltzmann distribution to the whole ensemble, you can in principle calculate the average occupation number of any given single-particle state for a system of fermions or bosons. In practice that's very hard for large systems, but the problem has been solved very generally using another method. You can learn that method and derive the following results in Problems 23 and 24.

The Quantum Distribution Functions

In a system of identical particles at temperature T, the average number of particles in a single state with energy E is given by one of the following:

$$\bar{n} = \frac{1}{e^{(E-\mu)/(k_B T)} + 1} \qquad \text{Fermi–Dirac distribution, applies to fermions} \qquad (10.6)$$

$$\bar{n} = \frac{1}{e^{(E-\mu)/(k_B T)} - 1} \qquad \text{Bose–Einstein distribution, applies to bosons} \qquad (10.7)$$

As always, begin by making sure you understand what these two formulas do.

- The formulas answer the question "how many particles in this system are in this particular state?" The answer is a function of the energy associated with that state. So if eight different states all represent the same energy, these formulas will predict the same occupancy for all eight of those states. Multiply that occupancy by eight and you will predict the number of particles at that energy level.

- The notation $\bar{n}$ designates the *average* you would get if you measured the occupancy of that state many times. Each individual measurement would be an integer, but $\bar{n}$ generally is not. For example, in a system of fermions you will always find 0 or 1 particles in any given state, so $\bar{n} = 1/3$ would mean that for 1/3 of the time there's a particle in this particular state.

- The constant μ (with units of energy) is called the "chemical potential." It serves as a normalization constant, and is the same for all of the states of the system.[9] For a system of bosons μ is always less than the ground state energy; for a system of fermions it can be higher or lower than the ground state energy. In both cases μ depends on temperature.

---

9   If you study more thermodynamics you will learn that the chemical potential of a system measures how strongly its entropy depends on its number of particles, similarly to how the temperature measures how the entropy depends on energy. It is true, but far from obvious, that the chemical potential appearing in the quantum distributions is the same quantity. For our purposes, it's just a normalization constant.

To normalize these distributions, set the sum of the occupation numbers for all states equal to the total number of particles in the system: $N = \sum_{\text{states}} \bar{n}$. For large systems, that kind of sum is difficult to work with, and we usually replace it with an integral. Assuming we've called the ground state energy 0,

$$N = \int_0^\infty \bar{n} g(E) \, dE.$$

Recall that the density of states $g(E)$ is defined so that $\int g(E) dE$ is the number of states in a given energy range, so this is equivalent to the sum over states. (Replacing a sum with an integral is an approximation, but if $\bar{n} g(E)$ doesn't change much from one state to the next, it's a good approximation.)

In a similar vein, you can find the total energy of the system by multiplying the number of particles in each state by the energy of that state:

$$E_{\text{system}} = \int_0^\infty \bar{n} E g(E) \, dE.$$

One example that we'll use several times in the rest of this chapter is a gas of identical particles in a box. You can show in Problem 31 that the density of states for a gas of particles of mass m in a box of volume V is

$$g(E) = \frac{m^{3/2} V}{2^{1/2} \pi^2 \hbar^3} \sqrt{E} \qquad \text{Density of states for particles in a box of volume } V. \qquad (10.8)$$

We'll talk about some applications of the Bose–Einstein distribution in Sections 10.7 and 10.8. Here we want to say a bit more about the Fermi–Dirac distribution, particularly in the limit of low temperatures.

Degenerate Fermi Gases

If you plug $E = \mu$ into the Fermi–Dirac distribution (Equation (10.6)) you get $\bar{n} = 1/2$. So, for fermions, states with energies lower than μ are likely to be occupied and states with energies above μ are likely to be unoccupied. The exercise below probes what that implies at very low temperatures.

Active Reading Exercise: The Fermi–Dirac Distribution at Low Temperatures

Answer the first two questions below based on Equation (10.6), the Fermi–Dirac distribution.

1. What is $\lim_{T \to 0^+} \bar{n}$ for $E < \mu$?
2. What is $\lim_{T \to 0^+} \bar{n}$ for $E > \mu$?
3. Based on both of your answers, describe the occupancy of a low-temperature system of fermions. Explain briefly why this answer comes as no surprise.

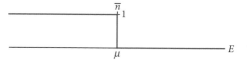

Figure 10.21 Fermi–Dirac distribution at $T = 0$. For a system of N fermions, the N lowest states have energy less than μ and are all occupied. No states above that are occupied. (At higher temperatures, μ is lower and the distribution decreases more gradually.)

In the limit of low temperature, $\bar{n}$ approaches 1 for all $E < \mu$ and 0 for all $E > \mu$ (Figure 10.21). You can (and hopefully just did) derive that from Equation (10.6), but it also makes physical sense. In the limit of zero temperature, the fermions will fill up *all* of the N lowest-energy states and none above that. Put another way, the system as a whole will be in its ground state in the limit $T \to 0^+$, which we already knew had to be true from the Boltzmann distribution.

For a familiar example, think of a multielectron atom. A neon atom has 10 electrons, and at very low temperatures they fill up the 10 lowest available states, one to a state as required by Pauli. So $\bar{n} = 1$ for the $n = 1$ and $n = 2$ states (10 states in all), and $\bar{n} = 0$ for the $n = 3$ states and above. The energy halfway between the 10th and 11th energy levels is called the "Fermi energy" ϵ_F of a neon atom. (Typical thermal energies at room temperature are a few percent of an eV, which is too small to excite atomic transitions, so room temperature counts as "very low temperature" for this example.)

The box below defines the Fermi energy, and also explains how it is related to the chemical potential μ.

Fermi Energy

For a system of N fermions, the Fermi energy ϵ_F is halfway between the energies of the Nth and the $(N + 1)$th particle states.

The Fermi energy is independent of temperature, but in the limit $T \to 0^+$, the chemical potential μ approaches ϵ_F. In that limit, all the states with energy up to ϵ_F are filled and all states with higher energies are unfilled.

This is a good time to remind you that all of these probability distributions – the fundamental assumption, the Boltzmann distribution, and the quantum distributions – refer to systems in equilibrium. You can excite an electron to a higher state by hitting it with a photon, but at normal temperatures it will quickly fall back down, and it will almost never get knocked up to that state by random interactions with the atoms around it.

A system of fermions at equilibrium that are all in their ground states is called a "degenerate Fermi gas." In Problem 32 you'll analyze a degenerate Fermi gas for an unlikely sounding application: calculating the radius of a burned-out star!

State Distribution and the Correspondence Principle

As you may recall, the "correspondence principle" states that quantum rules must approach classical rules for high quantum numbers. If a light beam is sufficiently intense, you can ignore the fact that it is made of individual photons; if a macroscopic mass is vibrating on a spring, its energy is so much larger than the spacing between its energy levels that you can treat those levels as continuous. How does the correspondence principle apply to the material in this section?

We have seen that for distinguishable particles you can use the Boltzmann distribution to ask "What is the probability that this particular particle is in that particular microstate?" For identical particles, you can never properly use the phrase "this particular particle." Instead you apply the quantum distributions (Fermi–Dirac or Bose–Einstein) to find the occupation number for each state available to the individual particles. But if you naïvely make predictions for a system of fermions or bosons based on the equations for distinguishable particles, it turns out you get the right answers anyway, if you are working at high enough temperatures. That's the correspondence principle in this case.

To see what that looks like mathematically, we have to use the Boltzmann distribution to predict the occupation number $\bar{n}$ of a particle state. If we apply the Boltzmann distribution to one individual particle (which is valid for distinguishable particles), it tells us that the particle has probability $e^{-E/(k_B T)}/Z$ of being in a state with energy E. For a system of N particles, therefore, the occupation number of that state should be $Ne^{-E/(k_B T)}/Z$.

Now consider what happens to Equations (10.6) and (10.7) in the limit $(E - \mu) \gg (k_B T)$. In that limit the exponential term in the denominator is very large compared to 1 and both equations become

$$\bar{n} \approx \frac{1}{e^{(E-\mu)/(k_B T)}} = e^{\mu/(k_B T)} e^{-E/(k_B T)} \qquad (\text{for } E - \mu \gg k_B T).$$

That looks just like the Boltzmann distribution: a normalization constant times the Boltzmann factor. Normalization ensures that the proportionality constants will come out equal in both cases, so the quantum distributions approach the classical prediction in the limit of high energy.

Figure 10.22 makes the same point graphically: at a constant temperature, the three formulas give completely different answers for low energy levels but approach each other in the high-energy limit.

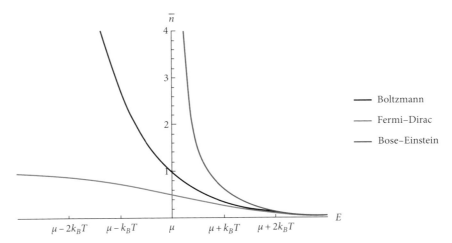

Figure 10.22 Three distribution functions. Moving to the right, the three curves approach each other.

We said this would be a valid approximation for high temperatures, but the argument above showed that it's valid for high energies. The connection has to do with μ, which decreases as temperature increases. At a high enough temperature essentially all the occupied states obey the condition $(E - \mu) \gg (k_B T)$, and the Boltzmann distribution gives accurate results.

We can summarize much of that with the following Q&A.

Question: When is the Boltzmann distribution valid for predicting the state of one individual particle?

Answer (in three parts):

- For a system of identical particles at low temperature, it's only accurate for high-energy states.
- For a system of identical particles at high temperature, it becomes accurate for all states.
- For a system of distinguishable particles, it's always valid.

This result makes sense if you think of quantum statistics as fundamentally changing the probability of finding more than one particle in the same state. That change makes a big difference for low energy levels at low temperatures: there is a certain probability of finding a shared state if the particles are distinguishable, a higher probability if they are bosons, and zero probability if they are fermions. But you're unlikely to find multiple particles sharing a very high-energy state no matter what kinds of particles they are. And at high temperatures there are so many accessible states that the probability becomes low for all states. So, in those situations, the three different distributions don't look all that different from each other.

As an example of a system of identical particles, consider the air molecules in your room. The molecules in a macroscopic gas are always identical. At extremely low temperatures their quantum behavior becomes significant, but at any normal temperature you can predict their behavior quite well using non-quantum statistics.

As an example of a system of distinguishable particles, consider a paramagnet. Each atom can be identified by its unique and unchanging position. There's a difference between "the one over here points up and the one over there points down," and the reverse state. So the statistics of a paramagnet are those of distinguishable particles. At any energy level or temperature, you can legitimately use the Boltzmann distribution on each atom to find its probabilities.

10.6.3 Questions and Problems: Quantum Statistics

Conceptual Questions and ConcepTests

1. For each quantity, give a brief description of what it represents in this section, and units that could be used to measure it.
   - (a) $\bar{n}$
   - (b) μ (for a collection of identical fermions)

2. You have to be careful when comparing the Boltzmann distribution to the Fermi–Dirac and Bose–Einstein distributions, because they measure different things. Suppose you have a system of particles, and there is a state with energy E_0 that a particle in this system could be in.
   - (a) If the particles are distinguishable and you plug $E = E_0$ into the Boltzmann distribution and

   find that $e^{-E_0/(k_B T)}/Z = 0.1$, briefly explain what that result tells you about this system.
   - (b) Now suppose the particles are bosons and you plug $E = E_0$ into the Bose–Einstein distribution and find that $\bar{n} = 0.1$. Briefly explain what that result tells you about this system. *Hint*: It's not the same as Part (a)!
   - (c) Now go back to the system in Part (a). If that system has 20 particles, what is $\bar{n}$ for a microstate with energy E_0?

3. We said that for a system of identical fermions $\bar{n} = 1/3$ means that the probability of finding a particle in that state is 1/3. Can you similarly use $\bar{n} = 1/3$ to determine the probability of finding a particle in

that state for a system of bosons? If so, what is the probability? If not, why not?

4. The text made the claim that identical bosons are always more likely to be in the same state as each other than they would be if they weren't identical. Make a very general argument about why that's true.

5. What is the Fermi energy of a hydrogen atom?

6. Sketch the Fermi–Dirac distribution twice.

   (a) First, in the limit $T \to 0$

   (b) Second, at a non-zero T

   (c) What's different about the two plots?

7. A system of identical particles is in contact with a reservoir. The Boltzmann distribution applies to . . . (Choose one.)

   A. individual particles in the system

   B. the system as a whole

   C. both

   D. neither

8. Problem 20 in Section 10.2 presented the "Gibbs paradox," in which splitting a box of gas into two parts seemingly reduced its entropy. Read over that problem. (You don't need to solve it to answer this question, but it might help.) Explain how to resolve the paradox.

For Deeper Discussion

9. We said that the quantum distributions approach the Boltzmann distribution if $E - \mu \gg k_B T$. That seems to imply that the classical approximation works better at low temperatures. But in fact, low temperatures are precisely where the quantum difference is most important! Discuss the behavior of both the Bose–Einstein and Fermi-the Dirac distributions for very low temperatures, and explain why the resulting behavior is *not* classical.

Problems

10. A gas at room temperature (say $T = 300$ K) has a state 0.01 eV above μ. Calculate the average occupation number of that state if the gas is made of . . .

   (a) fermions

   (b) bosons

(c) Would it be possible for a measurement to find no particles at all in that state? Answer for both fermions and bosons.

(d) Would it be possible for a measurement to find five particles in that state? Answer for both fermions and bosons.

11. A gas of fermions is at room temperature (say $T = 300$ K). Calculate the occupation numbers for states with the following energies:

   (a) $\mu - 0.01$ eV

   (b) μ

   (c) $\mu + 0.01$ eV

   (d) The three examples above were chosen to make a very general point about μ in the Fermi–Dirac distribution. Summarize that point in a brief sentence.

12. In a gas of bosons at 100 K, at what energy will the occupation number drop to 0.1? Your answer will be in the form μ plus a number (with units).

13. For the scenario in the Example "Two Particles and Three States" on p. 467, calculate the probability of finding both particles in the same state as each other if the particles are . . .

   (a) distinguishable

   (b) identical bosons

   (c) identical fermions

   (d) The text made the general claim that identical bosons are always more likely to be in the same state as each other than they would be if they weren't identical. Does this work?

14. Consider five identical fermions in a simple harmonic oscillator whose states have energies $(n + 1/2)\hbar\omega$ for $n = 0, 1, \ldots$

   (a) What is the value of the Fermi energy for this system?

   (b) At room temperature, would you expect μ for this system to be the same, higher, or lower, than your answer to Part (a)? Briefly explain why.

15. Suppose you have three particles, each of which can be in one of three states, A, B, and C, all with equal energies. What are the probabilities of having one in each state if they are . . .

(a) distinguishable?

(b) bosons?

(c) fermions?

16. Consider a system with three particles, each of which can have energies 0, ϵ, or 2ϵ. The system as a whole has energy 3ϵ. Find the entropy of the system if it is made of ...

(a) distinguishable particles.

(b) fermions.

(c) bosons.

17. In addition to bosons and fermions we have just invented a new kind of particle: trions. There can be zero, one, or two identical trions in a state, but no more.[10] If you have five identical trions in a system with three possible states (A, B, and C) of equal energy, what is the probability of having two particles in State A?

18. Suppose you have two particles in a system with k possible single-particle states, all of equal energy. What is the probability of finding the two particles in the same state if they are ...

(a) distinguishable?

(b) bosons?

(c) fermions?

19. If you have a gas of fermions in equilibrium with a reservoir at some temperature T, which of the following is more likely? Show the calculations that lead to your answer.

A. A state with energy $\mu - \Delta E$ is unoccupied.

B. A state with energy $\mu + \Delta E$ is occupied.

C. The two previous answers are equally likely.

D. There isn't enough information to answer the question.

20. Consider a gas of photons in a 1D container of length $L = 0.3\,\text{m}$. (This could be a very thin wave guide, for example, such as a fiber optic cable.) The possible energies for particles in this potential are

$$E_n = \frac{\pi \hbar c}{L} n,$$

where n can be any positive integer. (This is different from the usual infinite square well energies because the photons are relativistic.) A true fact that we are not going to derive here is that, for massless particles such as photons, $\mu = 0$. If the container is in equilibrium with its surroundings at temperature $T = 300\,\text{K}$, find the energy level (i.e. the value of n) for which the expected number of photons is $1/2$.

21. [*This problem depends on Problem 20.*] Calculate the number of photons in Problem 20.

22. Consider two identical bosons with two available states, one at $E = 0$ and the other at $E = \epsilon$, in contact with a reservoir at temperature T. You're going to calculate the ground state occupation number using the Boltzmann distribution with the two particles as your system (not either one separately).

(a) List all of the possible microstates of the system. For each one, give its energy and the number of particles in the ground state.

(b) Calculate the probability of each of those microstates using the Boltzmann distribution.

(c) Take a weighted average of those probabilities to find the average number of particles in the ground state.

23. **Deriving the Fermi–Dirac Distribution**

The Boltzmann distribution is based on a system exchanging energy with an effectively infinite reservoir. Such a system is called a "canonical ensemble." Different microstates of the system have different energy levels, and the probability of finding the system in any given microstate is a function of the total energy of the system when it is in that microstate.

Now consider a "grand canonical ensemble": a system that can exchange both energy *and particles* with a reservoir. Each microstate is now characterized by the energy E of the system and the number of particles N in the system. The probability of any given microstate is given by

$$P = \frac{1}{\mathcal{Z}} e^{-(E - N\mu)/(k_B T)}. \tag{10.9}$$

($\mathcal{Z}$ here is not identical to the Z in the Boltzmann distribution, but it plays a similar role. In the case

---

10 You can make quantum mechanical arguments for why such a particle would be impossible, but run with it.

where N is the same for all microstates, you can absorb $e^{N\mu/(k_BT)}$ into the proportionality constant $1/\mathcal{Z}$, reducing this to the Boltzmann distribution.)

One example of a grand canonical ensemble is a single-particle state: for instance, one particular state inside a hydrogen atom is free to exchange its one electron with the rest of the atom. We can derive the Fermi–Dirac distribution by applying Equation (10.9) to such a single-particle state. Its possible microstates are "occupied" ($N = 1$, and the energy is some value E) and "unoccupied" ($N = 0$ and $E = 0$).

(a) Write the probabilities for the state being occupied and for it being unoccupied.

(b) Set the sum of the two probabilities equal to 1 and solve for $\mathcal{Z}$.

(c) The average occupation number of the state is simply equal to the probability of it being occupied. Explain why.

(d) Write a formula for the average occupation number of the state. You may have to do some algebra to show that it equals the Fermi–Dirac distribution.

24. **Exploration: Deriving the Bose–Einstein Distribution**

[*This problem depends on Problem 23.*] You're going to repeat the process you used in Problem 23, only this time for bosons. Once again we'll consider a single-particle state to be our system, a grand canonical ensemble governed by Equation (10.9). The difference is that, instead of just 0 or 1, the occupation number N can now be any non-negative integer. The energy of any given microstate is NE, where E is the energy of one particle in our state.

(a) Normalize the probability distribution to express $\mathcal{Z}$ as an infinite series.

(b) The average occupation number of the state is $\bar{n} = \sum_{N=0}^{\infty} NP(N)$, where $P(N)$ is the probability of having occupation number N. Use this to write the average occupation number. Your answer will be an infinite sum that includes (among other things) the constant $\mathcal{Z}$.

(c) Show that if you define $x \equiv (E - \mu)/(k_BT)$, your answers together imply that $\bar{n} = -(1/\mathcal{Z})(d\mathcal{Z}/dx)$.

(d) Use the fact that $\sum_{i=0}^{\infty} r^i = \frac{1}{1-r}$ (for any $|r| < 1$) to express $\mathcal{Z}$ (your answer from Part (a)) in closed form.

(e) Put it all together to derive the Bose–Einstein distribution.

25. A helium atom is in a gas at 10,000 K. Ignore all states except the ground state (with degeneracy 2) and the first excited state (with degeneracy 8). We'll call the energy of the ground state 0, which makes the energy of the first excited state about 20 eV. (This is the energy per electron. If both electrons are in the first excited state the total energy will be 40 eV above the ground state.) Remember throughout this problem that electrons are identical fermions.

(a) How many states does the helium atom have in which both electrons are in the ground state?

(b) How many states does the helium atom have in which one electron is in the ground state and one in the first excited state?

(c) How many states does the helium atom have in which both electrons are in the first excited state?

(d) Using your answers to Parts (a)–(c) and the Boltzmann distribution, what is the probability of finding both electrons in the first excited state?

Problems 26–30 all concern the shapes of the Fermi–Dirac and Bose–Einstein distributions. You might find it helpful to keep Equations (10.6) and (10.7) in front of you.

26. All of the questions in this problem refer to the Fermi–Dirac distribution (Equation (10.6)).

(a) As E increases, does $\bar{n}$ increase monotonically, does it decrease monotonically, or does it have any local maxima or minima?

(b) Find $\lim\limits_{E \to -\infty} \bar{n}$.

(c) Find $\lim\limits_{E \to \infty} \bar{n}$.

(d) Find $\bar{n}$ when $E = \mu$.

(e) Based on all your answers, describe in words the kind of distribution that this formula indicates at any fixed temperature. (A good example to keep in mind is an atom with many electrons.)

27. All of the questions in this problem refer to the Fermi–Dirac distribution (Equation (10.6)).

    (a) Find $\lim_{T\to\infty} \bar{n}$ assuming μ is a constant.

    (b) Explain why your answer to Part (a) can't be correct physically.

    (c) What can you conclude about the behavior of μ in the limit $T \to \infty$ for a system of identical fermions?

    (d) Given your answer to Part (c), what does the Fermi–Dirac distribution predict about a system of fermions at high temperatures?

28. All of the questions in this problem refer to the Bose–Einstein distribution (Equation (10.7)).

    (a) Explain why Equation (10.7) implies $E > \mu$. *Hint*: A math professor looking at that equation would not automatically conclude that $E < \mu$ is out of domain.

    (b) As E increases, does $\bar{n}$ increase monotonically, does it decrease monotonically, or does it have any local maxima or minima?

    (c) Find $\lim_{E\to\mu^+} \bar{n}$.

    (d) Find $\lim_{E\to\infty} \bar{n}$.

29. All of the questions in this problem refer to the Bose–Einstein distribution (Equation (10.7)).

    (a) Find $\lim_{T\to 0^+} \bar{n}$ if $E > \mu$. (Assume μ is constant.)

    (b) Keeping in mind that $E > \mu$ for all states of a bosonic system, explain why your answer to Part (a) can't be correct physically.

    (c) What can you conclude about the behavior of μ in the limit $T \to 0^+$ for a system of identical bosons?

    (d) Now describe the behavior you expect to find if you look at a large system of bosons at low temperature.

30. Make plots of the Fermi–Dirac distribution at several different temperatures. For simplicity you can take $\mu = 0$ (even though in reality it depends on

temperature). Include a large enough range of negative and positive E-values to see the behavior of the function.

31. **Density of States for an Ideal Gas in a 3D Cavity**

    [*This problem requires multiple integrals, a topic from multivariate calculus.*] Many important formulas rely on knowing the density of states $g(E)$ for a system. So how do you find that? One approach is to use the fact that the total number of particles in a system is given by $N = \int \bar{n}g(E)dE$. If you can find an expression for N in the form $N = \int \bar{n}\langle\text{some function of energy}\rangle dE$, you can conclude that the function of energy in the integral is $g(E)$. In this problem you'll apply that approach to particles in a cube of width L. The energy eigenstates are defined by three quantum numbers n_x, n_y, and n_z, and the energy eigenvalues are

    $$E = \frac{\pi^2\hbar^2}{2mL^2}\left(n_x^2 + n_y^2 + n_z^2\right).$$

    The total number of particles N is the sum of $\bar{n}$ over all those states:

    $$N = \sum_{n_x=1}^{\infty}\sum_{n_y=1}^{\infty}\sum_{n_z=1}^{\infty} \bar{n}.$$

    Our first step is to approximate that sum with an integral:

    $$N = \int_0^{\infty}\int_0^{\infty}\int_0^{\infty} \bar{n}\,dn_x dn_y dn_z.$$

    (a) Answer the following two questions about this approximation. First: if you approximate $\sum f(x)$ with $\int f(x)dx$ you normally have to multiply by $1/\Delta x$ outside the integral, so why do you not have to do that in this case? Second: why is it okay to start your integrals at 0 even though your sums started at 1?

    (b) You can think of n_x, n_y, and n_z as Cartesian coordinates in a 3D space. Convert your integral to spherical coordinates, which you can call n (instead of the usual r), θ, and ϕ. *Don't forget the Jacobian.*

    (c) Evaluate the angular integrals to reduce your formula for the energy of the box to a single integral over n.

(d) How is n related to the energy E of an energy eigenstate? Use that relationship to do a u-substitution and rewrite your formula as an integral over E. You should have L^3 in your formula; replace it with V, the volume of the box.

(e) Comparing the equation you've derived to the formula $N = \int \bar{n} g(E) dE$, read off the density of states.

(f) Does the formula you just derived for $g(E)$ apply only to fermions, or to any type of particle? Briefly explain your reasoning.

(g) What other approximation(s) did you make? Under what circumstances would the formula you derived be accurate?

32. **Exploration: Equilibrium Radius of a White Dwarf Star**

The Sun exists in a state of equilibrium between the inward pull of gravity and the outer push from the nuclear reactions in its core. When it uses up its nuclear fuel it will collapse under its own weight into an incredibly dense remnant known as a "white dwarf." To a good approximation we can model the white dwarf as a degenerate Fermi gas, meaning the particles fill all the available states up to the Fermi energy.

(a) Using the density of states for particles in a box (Equation (10.8)), write an integral for the total number of electrons in the white dwarf. (You'll see soon why we're ignoring the protons and neutrons.) Your integral will have the Fermi–Dirac distribution in it, but for now you can just leave that as $\bar{n}$ rather than writing it out. You will have to multiply the density of states by 2 to account for the spin degeneracy of the electron states.

(b) Plug in the $T \rightarrow 0^+$ limit of the Fermi–Dirac distribution. Explain why this changes the upper limit of your integral to the Fermi energy ϵ_F.

(c) Evaluate the integral and solve the result to find ϵ_F as a function of N_e, the number of electrons in the white dwarf.

(d) The total thermal energy of the white dwarf is given by $E_{thermal} = \int_0^\infty \bar{n} E g(E) dE$. Explain how you can tell that this energy is all kinetic. (We'll introduce gravitational potential energy later.)

(e) Write and evaluate an integral for $E_{thermal}$ at $T = 0$, using the density of states and Fermi energy you already found.

(f) Looking at your formula for $E_{thermal}$, explain why we can neglect the thermal energy of the protons and neutrons.

(g) Your formula for $E_{thermal}$ should have the volume V and the number of electrons N_e in it. Rewrite it in terms of the radius ($V = (4/3)\pi R^3$) and the total mass of the white dwarf. For the latter, assume there are roughly equal numbers of protons, neutrons, and electrons. You can neglect the mass of the electrons.

(h) The gravitational potential energy of a uniform sphere is $U = -(3/5)GM^2/R$. Add this to the thermal energy to find the total energy. Find the value of R that minimizes the total energy. This is the equilibrium radius of the white dwarf.

(i) When our Sun collapses it will first explode, shedding about half its mass, and then the remaining half will collapse into a white dwarf. How big will it be?

In the 1910s, astronomers began to observe that there are objects in the universe with mass comparable to a star and size comparable to the Earth. An understanding of these objects had to wait for the development of quantum mechanics. Some other questions you might want to think about include: What's the source of the force that opposes gravity and keeps a white dwarf from collapsing? How would this look different if the particles making up the Sun were bosons? For large enough stars the immense pressure causes the electrons and protons to all fuse into neutrons. What does that do to their radii?

10.7   Blackbody Radiation

Section 3.4 introduced blackbody radiation, and explained how Planck's solution to the "ultra-violet catastrophe" was the first step on the road to quantum mechanics. But that section had a lot of take-our-word-for-it moments. Now we are in a position to explain more fully what blackbody radiation is, how its properties arise from statistical mechanics, and how it is used in applications ranging from improving the energy efficiency of houses to seeing in the dark.

If you haven't read Chapter 3 (or you read it a long time ago), you should still be able to understand this section. We will mention things we discussed in that chapter here, but we will redefine any terms we need, so the discussion here is self-contained.

10.7.1   Explanation: Blackbody Radiation

All objects emit radiation. That's pretty obvious for the Sun or a light bulb, but rocks and tables and people continually emit radiation too. That fact comes as a surprise to many people, and two questions often come up immediately.

1. *Question*: Why do all objects emit radiation?

    Answer: The radiation we're talking about here is called "thermal radiation" and is due to the never-ending thermal motion of the molecules or particles that make up an object.

2. *Question*: Well, why can't I see it, then?

    Answer: The radiation emitted by most of the objects around us is primarily in the infrared region of the spectrum, and our eyes can't see infrared light. With an infrared camera you can go into a completely sealed room and see the people around you glowing.

As an object heats up, the radiation it emits gets stronger, and the average frequency gets higher. For example, as you heat up coals in a fire you see them glow a dull red, and as they get hotter they start to look brighter and more orange.

As we study thermal radiation in this section, we will be treading some of the same ground that we covered in Chapter 3. Once again our goal is to explain and derive the Planck spectrum. But this time we approach the task with the right tools in hand: the second law of thermodynamics, the definition of temperature, the Boltzmann distribution, and the Bose–Einstein distribution.

We begin with a description of what a "blackbody" is, and why it is the simplest case for studying thermal radiation.

Blackbodies

When light strikes an object, it can be absorbed, reflected, or transmitted. Since most objects are opaque, we're going to ignore transmission and talk about reflection and absorption. The kinds of light that get absorbed determine an object's color. Leaves are green, for example, because they reflect most green light but absorb a significant amount of light of other colors. White objects reflect nearly all of the visible light that hits them. At the other extreme, an object that absorbs all electromagnetic radiation is called a "blackbody."

The name makes sense, but it can be misleading. The Sun, for example, is an almost perfect blackbody: if you shine a beam of light at the Sun it will not reflect back. The Sun looks yellow, not because of reflected light, but because of the light that it emits.

And that brings us back to where we started this section: "all objects emit radiation." In Section 3.4 we claimed that the radiation emitted by a blackbody depends, not on its chemical or physical properties, but only on its temperature. Now we're ready to talk about why.

Active Reading Exercise: Two Blackbodies

Imagine two blackbodies of the same size, shape, and temperature sitting near each other. For simplicity we'll assume that nothing else around them is emitting significant amounts of radiation. As each one absorbs radiation emitted by the other one, energy is being transferred back and forth. That transfer counts as heat, not work.

1. Based on the definition of temperature, what can you conclude about the amount of energy transfer between the two systems?

Now suppose we were to tell you that Blackbody A emits more radiation than Blackbody B.

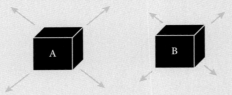

2. Some of the radiation emitted by Blackbody A strikes Blackbody B. What happens to that radiation?
3. Based on your answer to Part 2, does Blackbody A absorb more radiation than Blackbody B, or the other way around, or the same?
4. Your answers should have uncovered a logical contradiction. What can you therefore conclude?

In the scenario we described above, Blackbody B receives more photons than Blackbody A. (Remember that the two objects are the same size and shape.) Both blackbodies absorb all the light that they receive (that's the definition of a blackbody), so there is a net transfer of energy from A to B. By the definition of temperature, though, two objects of the same temperature should exchange energy equally, so there's the contradiction.

The conclusion is that our premise must be wrong. If A and B are blackbodies with the same size, shape, and temperature, they *must* emit the same amount of radiation. If they didn't, then you would have a net transfer of heat between two objects at the same temperature.

We based that argument on the definition of temperature, but all such arguments are ultimately grounded in the second law of thermodynamics. If Blackbody A radiates more energy than it absorbs, its energy and its temperature will go down. That means dS/dE will now be higher for Blackbody A than for Blackbody B. Any net energy transfer from Blackbody A to Blackbody B would then decrease the entropy of the entire system, so it won't happen.

The Planck Spectrum for Cavity Radiation

Section 3.4 analyzed the spectrum of energy inside a fully enclosed cavity. That's the same question we're addressing here, because the spectrum of radiation inside a cavity is also the spectrum emitted by a blackbody.

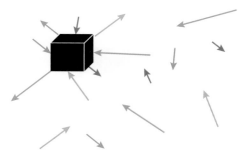

To see why these are the same, consider a blackbody inside a cavity (Figure 10.23). Once the whole system reaches thermal equilibrium, the blackbody must be emitting the same spectrum that it's absorbing. The spectrum emitted by a blackbody must therefore be the same as the spectrum of radiation inside a cavity (up to a proportionality constant that we'll discuss a little later).

Figure 10.23 In equilibrium a blackbody inside an enclosed cavity must emit the same spectrum that it absorbs.

So we want to mathematically characterize the radiation inside a cavity at equilibrium. Just as Planck did, we will start with the answer: the data that our theory needs to fit.

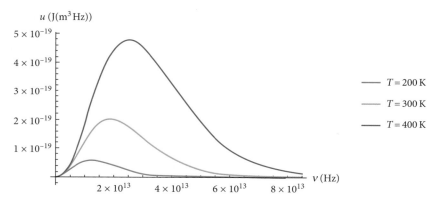

Figure 10.24 The Planck spectrum of cavity radiation.

Figure 10.24 shows the radiation spectrum at three different temperatures. At a quick glance you can see that the higher-temperature radiation has more total energy, and that it peaks at a higher frequency (although all three of these curves peak in the infrared range).

But it's important to notice that our dependent variable $u(v)$ is not simply a measure of energy: its units are (J/(m^3 Hz)), meaning energy *per unit frequency, per unit volume*. Integrating over a frequency range gives you energy per unit volume:

$$\int_{v_1}^{v_2} u(v)\, dv \quad \text{The energy per unit volume in the frequency range } v_1\text{--}v_2.$$

Integrating from $v = 0$ to ∞ gives the total energy per unit volume in the cavity, which we call ρ. Multiplying ρ by the volume of the cavity gives the total energy.

Deriving the Planck Spectrum

Remember that when Planck began his work, light was conceived of in entirely wave-like terms. Classical calculations, based on Maxwell's equations and the Boltzmann distribution, predicted that higher frequencies should always lead to higher energy. Planck modified the theory by proposing quantized energy levels $E = h\nu$, choosing the constant h to make his predicted spectrum match the curves in Figure 10.24.

With the benefit of the theoretical advances that followed from Planck's work, we can approach the same question from a different starting point. Planck treated the light inside a cavity as an electromagnetic wave – one that happened to allow only certain discrete energy levels – and did calculations involving wavelengths and amplitudes. We will consider light as a gas of photons, and count them. After that, we will sketch out the connection between the two models.

There are two ways in which you can calculate the number of photons with a given frequency ν. One is to apply the Boltzmann distribution directly: the system is "all the light at a particular frequency ν" and the reservoir is "the walls of the cavity." (You can explain why this is a valid approach in Question 6, and do the math in Problem 19.)

The other approach, which is quicker and easier, is to use the Bose–Einstein distribution. The key piece of information you need for the second approach is that, for a gas of photons, μ always equals 0. (That isn't at all obvious, but Problem 19 effectively proves it.) Plugging that into the Bose–Einstein distribution gives

$$\bar{n}(\nu) = \frac{1}{e^{h\nu/(k_B T)} - 1}. \tag{10.10}$$

Here are the steps leading from the Bose–Einstein distribution to the Planck spectrum:

- Equation (10.10) gives an *occupancy per state*. You specify a particular state (by identifying its energy E, or equivalently its frequency ν), and the resulting $\bar{n}$ is the number of photons you should expect to find in that particular state.

- Now imagine that you want to know *how many photons* there are in some small range of frequencies between a particular ν and $\nu + d\nu$. You would multiply $\bar{n}$ (from Equation (10.10)) times the number of states in that range. That number of states is $g(\nu)d\nu$, where $g(\nu)$ is the density of states expressed in terms of frequency rather than energy.

- But what you actually want to know is *how much energy* lies in that small range of frequencies between ν and $\nu + d\nu$. So you multiply the number of photons by the energy per photon, which is of course $h\nu$.

- Integrate to find the total energy in any given frequency range:

$$E = \int_{\nu_1}^{\nu_2} \bar{n}(\nu)\, g(\nu)\, h\nu\, d\nu. \tag{10.11}$$

Note that the density of states, $g(\nu)$, grows in proportion to the volume of the cavity. (The occupancy of each state, $\bar{n}(\nu)$, does not.) So you have to divide out that volume before you can see $u(\nu)$, the spectrum function, hiding inside Equation (10.11).

In Problem 20 you can apply the process we just described to a cube of width L. We already found the density of states in such a cube: that was Equation (10.8). But that derivation was non-relativistic, and now we're working with photons. So we will back up to the energy eigenstates for a particle in a 3D box, and fill in the steps from the occupancy per state (Equation (10.10) above) to the spectrum (Equation (10.12) below).

> ## The Planck Spectrum for Cavity Radiation
>
> In an enclosed cavity in equilibrium at temperature T, the electromagnetic field energy per unit volume in the frequency range v_1 to v_2 is
>
> $$\int_{v_1}^{v_2} u(v)\,dv \quad \text{where} \quad u(v) = \frac{8\pi h}{c^3}\frac{v^3}{e^{hv/(k_B T)} - 1}. \qquad (10.12)$$

Connecting the Two Models

In Section 3.4 we derived the Planck spectrum based on a wave model of light (albeit discretized), and here we have derived the same result from a particle model. Remember that these are not two different kinds of light; they are two different descriptions of the same radiation. This wave/particle duality lies at the heart of quantum mechanics.

- Section 3.4 opens with a variable called $E_w(v)$, the energy of one wave. The spectrum u is based in part on E_w, but also on the number of waves at any given frequency.

- The derivation in this section starts with $\bar{n}(v)$, the occupancy of one photon state. The spectrum u is based in part on $\bar{n}$, but also on the density of states.

The connection between the two models is this: *each possible standing wave in the classical picture is a possible state that a photon could be in.* The number of possible standing waves becomes the density of quantum mechanical states (more and more at higher energy levels). The energy E_w of a given wave is proportional to the number of photons in that state (fewer and fewer at higher energy levels.) As you consider states of increasing energy, the combination of that rising density of states and falling occupation number causes the spectrum to rise to a peak and then fall.

Blackbody Radiation

We argued above that the spectrum emitted by a blackbody should be proportional to the spectrum of radiation in a cavity in equilibrium. Why proportional, and not equal? The simplest answer is that the units are different. In a cavity the spectrum is the energy per unit frequency, per unit volume. For a blackbody the spectrum is energy emitted per unit frequency, per unit surface area, per unit time. When we imagined a blackbody sitting in a cavity, our argument really showed that the spectrum emitted by the blackbody per unit time has to equal the spectrum of cavity radiation *that hits the surface* of that blackbody per unit time.

To convert from one to the other, you need to do a geometry problem in which you have a certain density of photons, all moving at speed c in randomly distributed directions, and figure out how many of them per unit time will hit a given surface area.

You can simplify that geometry a bit by reducing your surface area to a tiny region dA. One way to think about that is to poke a tiny hole in your cavity. The spectrum that leaks out of that hole will be proportional to the spectrum of radiation in the cavity. In fact the hole itself is a blackbody! Any radiation that hits the hole from outside will be absorbed (fall into the cavity), so the spectrum emitted by the hole has to be a blackbody spectrum. In Problem 21 you can do the geometry to convert the formula for cavity radiation into a formula for blackbody emission.

You can mostly predict the result with almost no calculation. If you look at the units we gave above, you can see that to get from cavity radiation to blackbody emission you have to multiply by a distance over a time, i.e. a speed. The only speed in the problem is c, so you know the blackbody spectrum is the cavity radiation spectrum times c times a number. When you do the geometry you find that number equals $1/4$.

Blackbody Spectrum

The intensity (energy per time per unit surface area) emitted by a blackbody is

$$I = \frac{2\pi h}{c^2} \int_0^\infty \frac{v^3}{e^{hv/(k_B T)} - 1}\, dv. \tag{10.13}$$

Two important results of Equation (10.13) are:

- *Wien's law*: The blackbody spectrum peaks at $hv = 2.82\, k_B T$.
- *Stefan's law*: The total integrated intensity of a blackbody is σT^4, where $\sigma = 5.67 \times 10^{-8}$ W/(m^{2}K^4) is the "Stefan–Boltzmann constant."

Equation (10.13) says that the radiation emitted by a blackbody approaches zero at both low and high frequencies, and reaches a peak in the middle. Recall that this is not what classical physics predicted, and Planck originally postulated quantization to make the theory match this experimentally verified fact.

Wien's law says that the typical energy of a thermally emitted photon is a few $k_B T$, but you already knew that, right? Typical thermal energies are almost always of order $k_B T$. Stefan's law says that the total power emitted by a blackbody is proportional to T^4, so a small change in temperature has a large effect on the emission. You can derive these laws in Problems 22 and 23.

Equation (10.13) applies to blackbodies, but the same basic behavior also governs non-blackbodies: that is, objects that reflect some light. The difference is that a blackbody emits *more* radiation than any other object at the same temperature. Specifically, an object that absorbs half the radiation that hits it at a particular color will emit half as much radiation of that color as a blackbody. In practice the only perfect blackbodies are black holes,[11] but most objects emit a sizeable fraction of the energy that a blackbody would emit, so studying blackbody radiation gives us a good idea of the radiation emitted by nearly all the objects around us.

The temperature dependence in Equation (10.13) allows the blackbody spectrum to be used as a thermometer. We measure the temperatures of stars, for example, by measuring their spectra. At normal Earth temperatures, most objects emit primarily in the infrared. By measuring the infrared emission from a house on a cold day, people can detect where the house is leaking heat and needs more insulation. Night-vision goggles are actually infrared cameras; they allow people to see other people and animals in the dark because people and animals are generally warmer than their surroundings, and therefore emit more radiation.

Remember, though, that the blackbody spectrum only describes "thermal radiation" from molecular motion. When a source like a laser or LED light emits radiation for other reasons, that radiation is added on top of the object's thermal emission.

---

11 Black holes seem to disprove our argument since they absorb everything that hits them and emit nothing. In 1974 Stephen Hawking resolved this dilemma by proving that, when you take quantum field theory into account, black holes do emit a perfect blackbody spectrum of radiation.

10.7.2 Questions and Problems: Blackbody Radiation

Conceptual Questions and ConcepTests

1. Blackbody A and Blackbody B are the same shape and size, but A is hotter than B. Which of the following are true? (Choose all that apply.)

   A. A emits more total power than B.

   B. The radiation emitted by A is, on average, higher frequency than B's.

   C. At low frequencies B emits more than A.

   D. At high frequencies A emits more than B.

   E. A absorbs a higher percentage of the radiation hitting it than B does.

2. Imagine you had an object that was pitch black but which emitted no radiation. Explain why this would violate the second law of thermodynamics.

3. (a) Can an item look black and not be a very good blackbody? Explain.

   (b) Can an item be a (nearly perfect) blackbody and not look black? Explain.

4. In the text we argued that two blackbodies of the same size, shape, and temperature would have to emit equal amounts of radiation to stay in thermal equilibrium. How would that argument change if Blackbody A were bigger than Blackbody B? Would they still emit the same amount of radiation as each other? How can you get your answer from the argument we made?

5. The function $u(\nu)$ represents the energy per unit frequency, per unit volume, inside a closed cavity at equilibrium. To find the total energy, you have to integrate across all frequencies, but you don't have to integrate through the volume: you can just multiply by the volume of the cavity. Why don't you need to use calculus in that step?

6. We have applied the Boltzmann distribution to a glass of cold water, the air in a kitchen, and many individual particles. In Problem 19 you can approach the problem of cavity radiation by applying the Boltzmann distribution to "the radiation of a given frequency ν inside the cavity." What is the reservoir in this system? What does it mean to exchange energy between the system and the reservoir? Why does this system meet the conditions for using the Boltzmann distribution?

7. In the text we argued that two blackbodies of the same size, shape, and temperature would have to emit equal amounts of radiation to stay in thermal equilibrium. Explain how you can extend that argument to say that they must emit the same spectrum of radiation, i.e. the same amount of each color. *Hint*: Imagine putting colored filters between them.

For Deeper Discussion

8. We argued that no object can emit more thermal radiation than a blackbody at the same temperature without violating the second law of thermodynamics. But we also said that objects such as LEDs can emit non-thermal radiation that is more intense than a blackbody would emit at the same temperature. Why did our argument not apply to these other types of radiation?

Problems

9. Figure 10.25 shows a measured spectrum of a star.[12] Estimate the star's temperature.

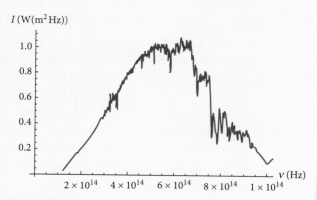

Figure 10.25

---

12 Data courtesy of Suzan Edwards.

10. The early universe was a hot plasma of ionized gas that emitted radiation as a blackbody. That radiation now fills the entire universe with blackbody radiation at 2.7 K. This pervasive sea of radiation is called the "cosmic microwave background," or CMB.

   (a) What is the energy density of the CMB?

   (b) What's the peak frequency of the CMB?

   (c) The average density of matter in the universe is equivalent to about 1.5 protons per cubic meter. How does the energy of that matter compare to the energy in radiation?

11. Calculate the total energy of radiation inside an oven while it is cooking. Clearly state any assumptions and approximations you make.

12. To a very good approximation, the human body can be treated as a blackbody.

   (a) Estimate the power radiated by your body. Clearly explain any assumptions and approximations you make.

   (b) Night-vision goggles work by detecting the thermal radiation emitted by warm objects. To detect human beings, to roughly what frequency should night vision goggles be sensitive?

13. Sketch on one plot the blackbody spectra for an object at 1000 K and for one at 2000 K.

14. Figure 10.26 shows a (simulated) spectrum of an incandescent light bulb.

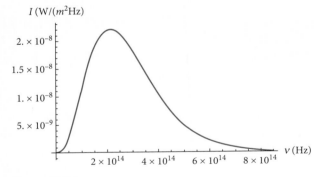

Figure 10.26

   (a) What is the temperature of the bulb filament?

   (b) What percentage of the light bulb's emitted energy is in the visible range? (You may want to use Appendix D for reference.)

   (c) Water molecules have non-degenerate vibrational states with energies 0, ϵ, 2ϵ, and so on (up to ∞), where $\epsilon = 0.2$ eV. In the vicinity of the light bulb filament, what percentage of water molecules are in their first excited state (the one with energy ϵ)? (If you make an approximation, note explicitly what approximation you are making and why it is justified.)

15. Suppose a cubic meter of water at 350 K is floating in empty space. (Don't ask us why.) The specific heat capacity of water is 4.2×106 J/(m^3 K).

   (a) How long would it take for it to cool down by 1 K? (You do not need calculus for this part, because it's reasonable to assume that the radiation intensity is roughly constant.)

   (b) How long would it take for the water to cool down from its original 350 K down to its freezing temperature of 273 K? (Now you need to take into account the varying intensity.)

16. The Earth's upper atmosphere receives 1370 W/m^2 of sunlight, and about half of that passes through the atmosphere and reaches the Earth's surface.

   (a) Use that fact to estimate the temperature of asphalt on a hot summer day. *Hint*: Once it has reached equilibrium the asphalt will emit the same amount of energy that it receives.

   (b) Why is black asphalt hotter than white pavement on a summer day?

17. The Sun has a radius of 7×10^8 m and its surface is at about 6000 K. Mercury orbits at a distance of about 6×10^{10} m. Assuming the Sun-facing surface of Mercury is in equilibrium, estimate its temperature.

18. Equation (10.13) gives the formula for the blackbody spectrum as an integral over frequency.

   (a) Use a u-substitution to rewrite it as an integral over wavelength.

   (b) Sketch the resulting shape.

   (c) At what wavelength does the spectrum have a peak? You should have found that the peak wavelength is *not* at $\lambda_{\text{peak}} = c/\nu_{\text{peak}}$. Just to take a concrete example, if you plot the Sun's spectrum in terms of frequency it peaks in the near-infrared, but if you plot it in terms of wavelength it peaks in visible light. How is that not a contradiction?

In Problems 19–23 you will fill in missing steps in the Explanation (Section 10.7.1).

19. **From Boltzmann Distribution to Occupation Number**

    In the text we found the number of photons in a cavity at a given frequency by using the Bose–Einstein distribution, but we asserted without proof that $\mu = 0$. You can derive this occupation number without that assumption using the Boltzmann distribution. (The fact that these two methods give the same answer is one way to prove that $\mu = 0$ in this case.)

    Our "system" is an electromagnetic wave of a given frequency, inside an enclosed cavity. This wave obeys the Boltzmann distribution because it is free to exchange energy with the walls of the cavity. Classically we would describe our wave as thus increasing or decreasing its amplitude; quantum mechanically we describe it as increasing or decreasing its number of photons, with each photon having energy $h\nu$.

    (a) Using the Boltzmann distribution, what is the probability of the wave being made up of N photons? Your answer will include an unspecified constant Z.

    (b) Express the constant Z as an infinite series (using a summation sign Σ).

    (c) You can find the average energy of the wave by summing its energy for each possible number of photons times the probability that it has that number of photons. Write an expression for that average energy as an infinite series.

    (d) You can evaluate those sums analytically using the method from Section 10.6 Problem 24, or you can ask a computer to do it. Either way, write a closed-form expression for the average energy of the wave. Your final answer will not be expressed as a sum, and it will not involve an unknown Z.

    (e) How many photons are represented by that energy?

20. **From Occupation Number to Energy Density**

    [*This problem requires multiple integrals, a topic from multivariate calculus. The problem depends on Problem 31 in Section 10.6.*] In Section 10.6 Problem 31 you calculated the density of states of a non-relativistic ideal gas. In this problem you'll modify that calculation for an ideal gas of photons and use the result to derive the energy density of cavity radiation.

    (a) Find the density of states for photons in a cubical box of width L. Because photons are relativistic, their energy eigenvalues are

    $$\epsilon = \frac{hc}{2L} \sqrt{n_x^2 + n_y^2 + n_z^2}.$$

    The process will closely mirror the non-relativistic one, except for the different energy eigenstates. Each photon can have two polarization states, so just like an electron it can be in two possible states for each combination of the quantum numbers n_x, n_y, and n_z.

    (b) Write the total energy in the box as an integral over particle energies involving the density of states and the occupation number. Plug in the density of states you found and the occupation number for bosons with $\mu = 0$. Divide your answer by the volume of the box to find the energy density.

    (c) Convert your integrand from energy to frequency using the de Broglie relation $E = h\nu$. You should get the Planck spectrum for cavity radiation.

21. **From Energy Density in a Cavity to Power Emitted by a Blackbody**

    [*This problem requires multiple integrals, a topic from multivariate calculus.*] A cavity is filled with a density (energy per volume) ρ of photons. If we put a small hole of size dA in the side of the box, how much energy will leak out of that hole in a time dt?

    Figure 10.27 represents the cavity (above), the small hole, and the outside world (below).

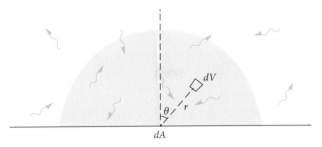

Figure 10.27

(a) The shading in the drawing represents a hemispherical region in the cavity. All the photons that emerge from the hole during some finite time period dt must have come from that shaded region. What is the radius of that hemisphere?

(b) If we consider a small volume dV in that region, how much energy is contained in that region? (If this takes you more than 20 seconds, you're overthinking it.)

For a photon in that small volume to reach the hole, it must be pointing in the right direction. The hole is at some distance r from the photon. Once the photon has traveled that distance, it will be somewhere on a sphere of surface area $4\pi r^2$ surrounding its original position. The probability of it hitting the hole is the effective area of the hole, $\cos\theta\, dA$, divided by $4\pi r^2$.

(a) Explain why the effective area of the hole depends on θ, and in particular why it is bigger at small θ than at large θ. (If you're stuck, think about a photon that starts at $\theta = \pi/2$.)

(b) Putting everything together, set up and evaluate a 3D integral for the amount of energy that escapes through the hole in a time dt.

(c) Using your answer to Part (b) and Equation (10.12), write a formula for the intensity (energy per time per area) emitted by a blackbody.

22. Wien's Law

Starting from the blackbody spectrum in Appendix I, derive Wien's law (also in that appendix). At the end of the calculation you'll need to numerically solve an equation that you can't solve analytically. *Hint*: At that point, and not before then, you might find it helpful to define a combination of T and some constants as a new variable x to simplify your equation.

23. Stefan's Law

Starting from the blackbody spectrum in Appendix I, derive Stefan's law (also in that appendix). *Hint*: Simplify the integral with a u-substitution that takes all the constants out. You can then evaluate the integral numerically or ask a computer to take it analytically.

10.8 Bose–Einstein Condensation

The statistical difference between bosons, fermions, and distinguishable particles has to do with the probabilities of particles being in the same state together; if the particles in a system have enough energy that the particles can occupy far more states than there are particles, that's not likely to happen anyway. So at high temperatures the behaviors of these different types of particles are similar. For example, we can use the Maxwell speed distribution for gases at room temperature without worrying about whether the air molecules are bosons or fermions. (They're mostly bosons, in case you were wondering.)

At low temperatures, however, the distinctions become important as more of the particles try to crowd together into the lowest-energy states. At very low temperatures, a system made of bosons can enter a phase called "Bose–Einstein condensation," in which a large fraction of the particles fall into the ground state together, and in many ways they start acting like a single particle. This can cause effects such as fluids that flow with zero viscosity and currents that flow with zero resistance.

10.8.1   Discovery Exercise: Bose–Einstein Condensation

Note: You can answer the questions below with no calculations.

System S_D comprises 10 distinguishable particles, each of which can have energy 0, ϵ, 2ϵ, etc.

1. How many microstates of this system have a total energy of 0?
2. How many microstates of this system have a total energy of ϵ?

System S_B comprises 10 bosons, each of which can have energy 0, ϵ, 2ϵ, etc.

3. How many microstates of this system have a total energy of 0?
4. How many microstates of this system have a total energy of ϵ?

See Check Yourself #18 at www.cambridge.org/felder-modernphysics/checkyourself

5. Assuming both systems obey the Boltzmann distribution, which system would be more likely to have a total energy of zero? Briefly explain your answer.

10.8.2   Explanation: Bose–Einstein Condensation

At very low temperatures, quantum statistics can lead to remarkable physical phenomena *at the macroscopic level* that are impossible to explain classically. Two dramatic examples are superconductivity and superfluidity.

In 1911 Kamerlingh Onnes found that when he cooled mercury below about 4 K it lost all electrical resistance and became what he later called a "superconductor." Scientists have observed current loops flowing in superconductors for years with no measurable decrease. Superconductors also repel magnetic fields, and can be used to levitate magnets (Figure 10.28, left).

In 1938 *Nature* published two back-to-back articles about the low-temperature behavior of helium, one by Pyotr Kapitsa and the other by John Allen and Donald Misener. The two groups had independently discovered that, below about 2 K, helium lost all viscosity (resistance to flow) and became what Kapitsa termed a "superfluid." Superfluid helium can flow through pores of only a few atomic widths, can sustain rotating vortices indefinitely without dissipation, and can climb up the walls of any container in which it is placed (Figure 10.28, right).

Superconductivity and superfluidity are both examples of "Bose–Einstein condensation." The word "condense" in this context refers to a collection of low-temperature bosons dropping into a state where almost all the particles are in their ground states. (This is in some ways analogous to the more common use of "condense" to mean vapor collecting into droplets of liquid: in both cases, atoms collect together into a new state of matter.)

Question: Why does the "condensed" state – meaning all the particles in their ground states – lead to superconductivity and superfluidity?

Figure 10.28 Left: A magnet levitates over a (solid) superconductor. *Source: Bobroff and Bouquet © 2010.* **Right: A small container has been dipped into superfluid liquid helium. The helium in the container flows up and around the top of the bowl and forms a drop below it (visible in the figure) that falls back into the pool of superfluid below it. This process continues until the bowl is empty.** *Source: Leitner 1963.*

Answer: Resistance to current flow and fluid flow arise from collisions. In a Bose–Einstein condensate, all of the atoms are in the ground state of the system together. They are described by one stationary wavefunction, so there are no collisions and no scattering.

Question: Why do collections of bosons display this behavior at low temperatures?

Answer: That's what this section is about. We will first present an argument for why bosons behave this way and distinguishable particles do not. Then we will outline a derivation of the critical temperature at which a bosonic gas condenses.

A Toy Model of Bose–Einstein Condensation

To understand the basic mechanism of Bose–Einstein condensation, we're going to start with a relatively simple scenario. (You will extend this scenario to more realistic cases in Problems 9 and 10, and discover that the oversimplified version actually works quite well.)

Active Reading Exercise: Condensation in a Two-State System, Part 1 (compare p. 490)

Consider a system composed of N non-interacting particles at temperature T. Each individual particle has only two possible states, one with energy $E = 0$ and the other with energy $E = \epsilon$.

1. In the limit $T \to 0^+$, what will the particles do? (This part requires no calculations, and the answer is the same for both distinguishable particles and bosons.)

2. Assuming the particles are distinguishable, about how many will be in the ground state if $T = \epsilon/k_B$? *Hint*: Apply the Boltzmann distribution to a single particle.

3. Now assuming the particles are indistinguishable bosons, about how many will be in the ground state if $T = \epsilon/k_B$? *Hint*: Apply the Boltzmann distribution to the system as a whole. (So, also assume the existence of an external reservoir.) The system has $N + 1$ possible states, but you should find that they become irrelevant after the first two or three.

This might take you five or ten minutes – longer than most of our Active Reading Exercises – but we encourage you to take the time. If you work through all three parts, or try your best and get stuck before you read our answers, you will understand why Bose–Einstein condensation happens better than you could by just reading a derivation.

1. We know from the Boltzmann distribution that in the limit $T \to 0^+$ higher-energy states of the entire system become infinitely less probable than the ground state, so the whole system will go into its ground state, meaning all the particles will have energy 0.

2. At temperature $T = \epsilon/k_B$ the Boltzmann distribution says that each distinguishable particle is more likely to be in the ground state than the first excited state, by a factor of e. Roughly 3/4 of the particles will be in the ground state.

3. For bosons it's important to remember that the system has only one state at any given energy level. (For instance, the state with $E = 5\epsilon$ is "Five particles have $E = \epsilon$ and the others have $E = 0$." You cannot meaningfully ask "Which five?") The probability of the system being in the ground state is e times higher than the probability of its being in the first excited state, and states above that have negligible probabilities. So there's about a 3/4 chance of the system being in the ground state. The math looks very similar to the preceding calculation for distinguishable particles, but this result applies to the state of the entire system.

To make sure you see what we have just concluded, consider a system with 10^{23} particles. Our results above suggest that, if these are distinguishable particles, you should expect about 2×10^{22} of them to be excited at this temperature. But if these particles are bosons, *there is about a 3/4 chance that all the particles are in their ground states*. There is a 1/4 chance that one particle (out of the 10^{23}) is excited, and a much smaller chance of finding more than that.

The argument above demonstrates why bosons at low temperatures "condense": that is, all drop into their ground states at low temperatures. It also explains why distinguishable particles don't display the same behavior, but you may feel that we cheated by applying the Boltzmann distribution on two different levels for the two types of systems. You can use the same approach for both systems, and – if you're careful – you will reach the same conclusion.

Active Reading Exercise: Condensation in a Two-State System, Part 2

For the system of distinguishable particles above, we applied the Boltzmann distribution to a single particle; for the bosons, we applied it to the whole system. Now, still supposing the existence of an external reservoir, suppose we applied the Boltzmann distribution to the entire system of distinguishable particles. How would those calculations differ from the calculation for bosons, and what effect would that difference have on the results?

You can find our answer at www.cambridge.org/felder-modernphysics/activereadingsolutions

The moral of this story is that, because you count the states of a bosonic system differently for distinguishable particles, there is a strong statistical tendency for bosons to cluster together in one state. At high temperatures this tendency doesn't matter because they have so many states available that it's unlikely for two or more particles to occupy the same state as each other. (Our toy example had only two states, but most real systems have infinitely many.) At low temperatures the tendency for bosons to cluster together makes an enormous difference.

You might think that with 10^{23} particles a macroscopic system would surely have particles sharing states, but the number of available states grows with the size of the system. For a macroscopic system at room temperature, the number of available states is often much larger than 10^{23}, so classical statistics work fine. We discussed this in Section 10.6 as an example of the correspondence principle: at high energy levels, quantum behavior starts to look classical.

Bose–Einstein Condensation in an Ideal Gas

Our argument above was based on a toy model of particles with only two allowed energy levels. We will now consider the more general (and more realistic) scenario of an ideal gas of bosons, and derive the critical temperature at which condensation occurs.

Remember that an ideal gas means a system of non-interacting, freely moving particles. For simplicity we'll put our particles in a cubical box of width L, and as usual we'll call the ground state energy 0.

The number of particles in a given state is given by the Bose–Einstein distribution:

$$\bar{n} = \frac{1}{e^{(E-\mu)/(k_B T)} - 1}.$$

We know the energy eigenstates of a particle in a box. In principle we could sum $\bar{n}$ over all those states, set the result equal to N to find μ, and plug that back into $\bar{n}$ with $E = 0$ to find how many particles are in the ground state at any given temperature. That approach is pretty much hopeless analytically. (You can do it numerically for a small system in Problem 14.)

So instead, we will approximate that sum with an integral:

$$N = \sum_{\text{all states}} \bar{n} \approx \int_0^\infty \bar{n} g(E) dE \qquad \text{(not a great approximation).}$$

Remember that we define the density of states $g(E)$ as the number of states per unit energy interval, so $\int_{E_1}^{E_2} g(E) dE$ equals the number of states with energy between E_1 and E_2. That means the integral above is roughly equivalent to the sum over all states of $\bar{n}$, the number of particles per state.

There's a catch, though. It's a reasonable approximation to replace a sum with an integral, provided the function you are summing over changes very little from one point to the next in your sum (see Figure 10.29). But in our condensate we expect a large number of particles to be in the ground state. So $\bar{n}$ is going to end up being much bigger for the ground state than the other states, and it will be a terrible approximation to include it in our integral.

We can correct for that by adding n_0, the number of particles in the ground state, outside the integral. Above the ground state the integral is a good approximation:[13]

$$N = n_0 + \int_0^\infty \bar{n} g(E) dE.$$

---

13 You might think we're over-counting the ground state by having the integral still go down to zero. The energies from the ground state to the first excited state make a negligibly small contribution to the integral, particularly because $g(0) = 0$.

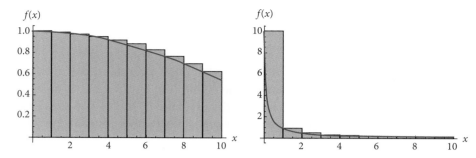

Figure 10.29 The sum represented by the area of the gray boxes on the left is well approximated by the integral of the curve in that figure. On the right the approximation is not good because the function changes rapidly from the first value to the next. After that the integral is a reasonable approximation.

Next we need $g(E)$. We derived the density of states for a particle in a box in Section 10.6. Plugging that formula into our integral gives

$$N = n_0 + \frac{m^{3/2}}{2^{1/2}\pi^2\hbar^3}V\int_0^\infty \frac{\sqrt{E}}{e^{(E-\mu)/(k_BT)} - 1}dE. \tag{10.14}$$

You can read Equation (10.14) as "The total number of particles is some unknown number in the ground state, plus a distribution in the other states given by the density of states $g(E)$ times the Bose–Einstein occupation number $\bar{n}$." But we still have an unknown constant μ. You can show in Problem 12 that at low temperatures $\mu \approx -k_BT/N$, so for a macroscopic system μ is almost exactly zero. (Remember that we're taking zero to be the ground state energy.) For the ground state itself, the tiny difference between μ and 0 is important. (Otherwise $\bar{n}$ would be infinite.) But for all the other states, $E - \mu$ is indistinguishable from E. Since the integral is only over excited states, we can substitute $\mu = 0$:

$$N = n_0 + \frac{m^{3/2}}{2^{1/2}\pi^2\hbar^3}V\int_0^\infty \frac{\sqrt{E}}{e^{E/(k_BT)} - 1}dE. \tag{10.15}$$

Now there are no unknowns left. You can do a u-substitution to get all the constants out of the integral and evaluate it numerically (Problem 13). The final result is

$$\frac{n_0}{N} = 1 - \left(\frac{T}{T_C}\right)^{3/2}, \tag{10.16}$$

where the critical temperature is

$$T_C = 3.3\frac{\hbar^2}{mk_B}\left(\frac{N}{V}\right)^{2/3}. \tag{10.17}$$

This formula suggests that, below the critical temperature, the fraction of the particles in the ground state rapidly approaches 1. A system begins to condense at $T \approx T_C$, and for $T \ll T_C$ essentially all of the particles are in the same quantum state.

The first observations of Bose–Einstein condensation in a gas were made in 1995 by two independent groups, Eric Cornell and Carl Wieman using Rb (Figure 10.30) and Wolfgang Ketterle using Na. The three of them shared the Nobel prize for this achievement in 2001.

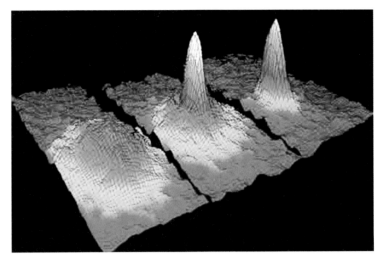

Figure 10.30 The velocity profile of Rb atoms at three successively lower temperatures in Cornell and Wieman's experiments. In each of the three plots the horizontal axes are components of velocity v_x and v_y, and the height (also marked with color) indicates how many of the atoms had that particular velocity. In the left plot the atoms are distributed among many states. In the middle plot a substantial fraction are in the central state, but there's still a noticeable bump around the central peak showing atoms in other states. On the right, at the lowest temperature, essentially all of the atoms are in the ground state. *Source: Royal Swedish Academy of Sciences.*

10.8.3 Questions and Problems: Bose–Einstein Condensation

Conceptual Questions and ConcepTests

1. Define "Bose–Einstein condensate" in your own words.

2. Why does a system of bosons behave differently from a system of distinguishable particles at low temperatures, but not at high temperatures?

3. The text describes a system of N particles, each of which has two energy levels: 0 and ϵ. Let n_0 be the occupation number of the ground state. Parts (a)–(d) are all multiple-choice questions with the following choices: 0, 1, $N/2$, N, and "None of the above." Briefly explain your thinking in each case.

   (a) What is $\lim_{T \to 0^+} n_0$ for a system of distinguishable particles?

   (b) What is $\lim_{T \to 0^+} n_0$ for a system of identical bosons?

   (c) What is $\lim_{T \to \infty} n_0$ for a system of distinguishable particles?

   (d) What is $\lim_{T \to \infty} n_0$ for a system of identical bosons?

   (e) For temperatures between these two limits, is n_0 always bigger for the system of bosons than for the distinguishable particles, always bigger for the distinguishable particles, or neither? Explain.

4. Equations (10.16) and (10.17) together represent an important fact about a bosonic ideal gas at a low temperature. In the context of those formulas, briefly explain what each of the following variables represents, and what units might be used to represent it.

   (a) V

   (b) T_C

   (c) n_0/N

5. Equation (10.17) gives the critical temperature at which a bosonic ideal gas condenses. Consider a container filled with identical bosons.

(a) If you double the number of bosons in the container but keep the volume of the container constant, what happens to the critical temperature?

(b) If you double the number of bosons in the container but keep the density constant, what happens to the critical temperature?

(c) We defined a quantity as "intensive" if two copies of the system have the same value as one copy. For example, density is intensive because two identical boxes with density $3 \, kg/m^3$, when viewed together, still have density $3 \, kg/m^3$. An "extensive" quantity doubles when you have two identical copies. For example, mass is extensive because two 10 kg boxes are, together, 20 kg. Is T_C intensive, extensive, or neither?

6. Equation (10.16) gives the fraction of particles in the ground state for a bosonic ideal gas. At temperatures above T_C this formula says n_0 is negative! Explain.

For Deeper Discussion

7. One of the consequences of the second law of thermodynamics is that it is impossible to build a "perpetual motion" machine. Since every interaction in the world increases the net entropy, all macroscopic repetitive motions must receive energy input from outside or wind down eventually. We said in the text, however, that rotating vortices in a superfluid or currents in a superconductor dissipate no energy and can continue indefinitely. How do these systems not violate the second law?

8. An atom of ^{4}He is made of two protons, two neutrons, and two electrons. The less common isotope ^{3}He is made of two protons, *one* neutron, and two electrons. ^{4}He becomes superfluid at about 2 K, but ^{3}He doesn't become superfluid until you cool it down to a few mK.

(a) Why is it easier to make a superfluid of ^{4}He than of ^{3}He?

(b) Given your answer to Part (a), how is it possible to have a ^{3}He superfluid at all? *Hint*: Think

about what ^{4}He is made of, and how it can form a superfluid.

Problems

9. On p. 489 we calculated the occupation number of the ground state for a system of N particles with two energy levels, 0 and ϵ. In this problem you will re-do those calculations for a system of particles with three energy levels: 0, ϵ, and 2ϵ. This system, like the one in the text, is at temperature $T = \epsilon/k_B$.

(a) Calculate $\bar{n}_0$, the expected occupancy of the ground state, assuming the N particles are all distinguishable.

(b) Now assume the N particles are indistinguishable bosons. List all the microstates of this system with $E \leq 3\epsilon$. (*Hint*: There aren't many, but there are more than four.) For each microstate, give the occupancy of the ground state n_0 and the total energy of the system E.

(c) Our goal is to calculate the expected occupancy of the ground state. Explain briefly why it is reasonable to neglect all the microstates you did *not* list in Part (b).

(d) Calculate $\bar{n}_0$, the expected occupancy of the ground state, for the bosonic system.

10. [*This problem depends on Problem 9.*]

(a) Find $\bar{n}_0$ for both distinguishable particles and bosons, this time for a system in which every particle can have infinitely many energy levels 0, ϵ, 2ϵ, 3ϵ, ... Just as in Problem 9, you will want to cut off your calculations when the Boltzmann factors become negligible.

(b) What do your results suggest about the behavior of these systems?

11. On p. 489 we calculated the occupation number of the ground state for a system of N particles, each of which could have energy 0 or ϵ, at a temperature $T = \epsilon/k_B$. For this problem assume $N = 5$. *You don't need a computer for this problem, but it does involve a number of repetitive calculations so you might find it easiest to set up the calculation in a spreadsheet, or with some other system for doing calculations.*

(a) Calculate $\bar{n}_0$, the occupation number of the ground state, for a system of distinguishable particles at a range of temperatures from $T = \epsilon/(2k_B)$ to $T = 10\epsilon/k_B$. (You're going to plot $\bar{n}_0$ as a function of T, so choose enough points to see the shape of the plot clearly.) Now assume the five particles are identical bosons. List all the possible microstates of the system. For each microstate, specify n_0 (the number of particles in the ground state) and E (the total energy of the system.)

(b) Find $\bar{n}_0$, the expected ground state occupancy of the bosonic system, at all the same temperatures as those you used in Part (a). Once again it is not surprising to see that, as the temperature rises, the expected occupancy of the ground state decreases.

(c) Plot $\bar{n}_0$ vs. temperature for both systems on the same set of axes. How are they similar and how are they different?

12. **Approximating μ for a Low-Temperature Bosonic Gas**

Consider a system of N non-interacting bosonic particles, where $N \gg 1$. (For a typical macroscopic gas, $N > 10^{23}$, so $N \gg 1$ is a pretty good assumption.) We'll call the ground state energy 0 for convenience.

(a) In the limit $T \to 0^+$, what is the occupation number $\bar{n}$ of the ground state?

(b) Explain how your answer to Part (a) and the formula for $\bar{n}$ for the ground state from the Bose–Einstein distribution imply that μ must be almost exactly equal to 0 in the limit $T \to 0^+$.

(c) Now consider μ at some small, non-zero temperature. Using the fact that $e^x \approx 1 + x$ for small x, solve for μ in terms of N and T.

(d) For a ^{4}He atom in a $1\,\text{cm}^3$ box, the difference between the ground state energy and the first excited state energy is about $10^{-18}\,\text{eV}$. Assuming a gas of 10^{23} atoms, how does that energy difference compare to the value of μ

at 1 K, using the approximation you derived in Part (c)?

13. Starting with Equation (10.15), use a u-substitution and a numerical integral to end up with Equation (10.16).

14. **Exploration: Numerical Calculation of Bose–Einstein Condensation**

We noted in the text that the rigorous way to calculate μ is to set the sum of $\bar{n}$ over all energy eigenstates equal to N. Once you solve the resulting equation for μ, you can plug it back in to $\bar{n}$ to find the occupation number of any particular eigenstate, and in particular of the ground state. In this problem you're going to carry that out numerically for a system of 10 identical bosons in a box. The energy levels are

$$E = \epsilon \left(n_x^2 + n_y^2 + n_z^2 - 3 \right).$$

The quantum numbers n_x, n_y, and n_z can each be any positive integer. We've inserted the -3 to set the ground state energy to 0. (You don't have to do that, but it makes it easier to interpret the values of μ that you get.) Initially you're going to consider $T = 0.5\epsilon/k_B$.

(a) Write a formula for $\bar{n}$ as a function of μ and the quantum numbers.

(b) Set $\mu = -0.04\epsilon$ and calculate the sum of $\bar{n}$ for all values of the quantum numbers from 1 to 10. (Technically the sum goes up to ∞, but taking it up to 10 will be enough to ensure a high level of accuracy, here and in all the later parts of this problem.)

(c) Repeat Part (b) for different values of μ until you find the correct value of μ for this temperature. (How will you know which is the "correct" one? One way to do this is to have the computer calculate a long list of results for different values of μ and plot them to see where the correct value is. You can have the computer interpolate between those values to find an accurate solution for μ).

(d) Use your value of μ to find the occupation number of the ground state at $T = 0.5\epsilon/k_B$.

(a) Repeat the process to find the ground state occupation number for $T = \epsilon/k_B$.

(b) Keep doubling the temperature and re-doing the calculations until the ground state occupation number drops below 1. Make a plot of n_0 vs. $k_B T/\epsilon$.

(c) Estimate the critical temperature at which this system becomes a Bose–Einstein condensate. (People wouldn't normally apply the term to such a small system, but you can still see the same basic process at work.)

Chapter Summary

Statistical mechanics predicts the behavior of macroscopic objects (such as a roomful of air) by applying statistical predictions to large numbers of microscopic objects (such as 10^{23} molecules). Although an individual atom is subject to quantum mechanical randomness, predictions based on such large numbers can be stated with statistical certainty.

The equations for this chapter are collected in Appendix I. You may want to cross-reference that appendix as you read through this summary.

Section 10.1 Microstates and Macrostates

The "microstate" of a system is a complete description of the state of every particle in the system. The "macrostate" of a system only refers to properties that you can measure macroscopically, such as temperature and pressure.

- The "fundamental assumption of statistical mechanics" asserts that, over a long enough span of time, an isolated system will spend equal time in each of its accessible microstates.

- Therefore such a system will spend most of its time in macrostates that correspond to many different microstates. For instance, there are many microstates that correspond to the macrostate "the air is distributed pretty evenly around your room," and comparatively few microstates that correspond to the macrostate "all the air in your room is clustered toward the floor." This is why you are unlikely to ever measure the latter.

- Many of the systems we study in this chapter are "two-state systems." For instance, an atom in a magnet can point up or down. When analyzing such systems it is often useful to calculate "combinations": the notation $\binom{n}{m}$ represents the number of groups of m items you can make from a total list of n distinct items.

Section 10.2 Entropy and the Second Law of Thermodynamics

The fact that entropy tends to increase is a natural result of the fundamental assumption of statistical mechanics, once you understand what entropy actually means.

- The "multiplicity" Ω of a macrostate is the number of microstates to which it corresponds.

- The "entropy" of a macrostate is $S = k_B \ln \Omega$.

- The second law of thermodynamics asserts that an isolated macroscopic system will never decrease its entropy. This law follows directly from the fundamental assumption of statistical mechanics presented in Section 10.1.

- This tendency toward increasing entropy underlies all the time-irreversible processes we see around us.
- Increasing the energy of a system generally increases its entropy because, with more energy, there are more ways to distribute the energy.

Section 10.3 Temperature

Temperature is closely related to energy, but it is *not* a measure of the total energy of a system, nor of its energy density. A proper understanding of temperature helps explain why many physical processes occur in the directions that they do.

- Temperature can be operationally defined by the sentence "Object A is hotter than Object B ($T_A > T_B$) if energy flows spontaneously from A to B."
- The quantitative definition of temperature is $T = 1/(dS/dE)$.
- You can relate the two definitions by considering what happens when energy flows from A to B, assuming $T_A > T_B$. The entropy of A decreases, but the entropy of B increases *more,* so the entropy of the entire system increases. A flow in the other direction would decrease the net entropy, so this will not spontaneously occur.
- The dE in the definition of temperature is energy transfer through "heat" (often designated Q), meaning a spontaneous transfer of energy that occurs when two objects are put in contact. Other forms of energy transfer (e.g. via macroscopic forces) are considered "work."
- Conservation of energy, and the relation $\Delta S \geq Q/T$, allow us to calculate limits on the efficiency of devices such as refrigerators and heat pumps.
- The "heat capacity" of a system is defined as $C = dE/dT$. It is roughly the amount of energy required to heat the system by one degree. The heat capacity of a system per molecule or per kilogram is called "specific heat capacity."

Section 10.4 The Boltzmann Distribution

The fundamental assumption of statistical mechanics applies to an isolated system: any microstate is as likely as any other microstate. However, the Boltzmann distribution applies to a small system in contact with a much larger system (the "reservoir"): a low-energy microstate of the small system is more likely than a high-energy microstate.

- The reason why a low-energy state is more likely is because it corresponds to a higher energy level – and therefore a higher entropy – for the reservoir.
- The equation for the Boltzmann distribution is $P = (1/Z)e^{-E/(k_B T)}$, where $1/Z$ is a normalization constant. Z is called the "partition function" and $e^{-E/(k_B T)}$ is called the "Boltzmann factor."
- Remember that this formula is not the probability that your system will be in energy E; it is the probability that your system will be in any given microstate with energy E. Multiply that by the total number of microstates with energy E (the "degeneracy") to find the total probability of that energy level. For this reason, it is possible for a high energy level to be more probable than a low level.

- The Boltzmann distribution predicts that, for very low temperatures, all the particles crowd into the lowest-energy microstates possible. For very high temperatures, the probabilities even out somewhat; all states with energy much less than $k_B T$ are roughly equally likely.

Section 10.5 Some Applications of the Boltzmann Distribution

The Boltzmann distribution can be used to predict (among other things) the average thermal energy of a system of particles, the heat capacity of a gas, and the distribution function of molecular speeds.

- As a general guideline, the average thermal energy of a particle is typically of order $k_B T$ above its ground state.

- That guideline applies when the "density of states" – the number of available microstates per unit energy – is reasonably uniform, and reasonably high. A very non-uniform density of states can push the average energy much higher or much lower than $k_B T$. (This should make sense if you think about it.) A very low density of states can push the average energy much lower, because the first excited state may be beyond reach.

- A more accurate estimate of thermal energy is provided for some systems by the "equipartition theorem": the average value of any energy term that depends quadratically on one of the system's degrees of freedom is $(1/2)k_B T$. Appendix I lists several important examples of the equipartition theorem.

- Once you have the average thermal energy (either the simple answer $E \sim k_B T$ or the equipartition theorem), you can use the definition $C = dE/dT$ to predict heat capacity.

- The "Maxwell speed distribution" (Appendix I) gives the probability distribution for the speeds of molecules in an ideal gas.

Section 10.6 Quantum Statistics

Any two electrons are identical. Therefore "one of these two electrons is in the ground state and the other is in the first excited state" is one microstate of a system, not two. Because statistical mechanics is all based on counting microstates, this difference in counting leads to important differences in statistical outcomes.

- Fermions obey the "Fermi–Dirac distribution"; bosons obey the "Bose–Einstein distribution." Both formulas (see Appendix I) give the "occupation number" $\bar{n}$ of a state – the average number of particles you will find in that state, over many measurements – as a function of the energy of that state.

- Both formulas include a normalization factor, the "chemical potential" μ, which in many circumstances depends only weakly on temperature.

- For a system of N fermions, the "Fermi energy" is defined as $\epsilon_F = \lim_{T \to 0} \mu$, which falls halfway between the energies of the Nth and $(N+1)$th particle states. At very low temperature all the states with $E < \epsilon_F$ will be occupied, and all the states with $E > \epsilon_F$ will be unoccupied. Such a system is called a "degenerate Fermi gas."

- For most temperatures, the quantum distributions lie very close to each other and to the classical distribution. Only at very low temperatures do the three types of systems act very differently.

Section 10.7 Blackbody Radiation

When Planck did his work on blackbody radiation in 1900, he brought to the task an understanding of temperature, entropy, and the Boltzmann distribution. We hadn't yet covered any of those topics when we introduced Planck's quantum hypothesis in Chapter 3. So in this section we shine a more coherent light on the subjects we dimly illuminated in that chapter.

- When light strikes an object, it can be absorbed, reflected, or transmitted. An object that absorbs all incident radiation is called a "blackbody."

- All objects emit thermal radiation, and blackbodies emit more than any other objects. The thermal radiation emitted by a blackbody depends only on its size and temperature.

- The spectrum of blackbody emission is, with a properly chosen proportionality constant, the same as the spectrum of radiation in a fully enclosed cavity in equilibrium. This spectrum experimentally goes to zero for very high and very low frequencies.

- The section outlines a step-by-step derivation of the Planck spectrum. In brief: the Bose–Einstein distribution gives the number of photons at any given frequency, you multiply that by the density of states to get the number of photons in any frequency range, and then multiply that by $E = h\nu$ to get the amount of energy in any given frequency range.

- From the spectrum for cavity radiation you can derive the spectrum emitted by a blackbody. The peak frequency is proportional to temperature ("Wien's law"), and the total radiated intensity is proportional to T^4 ("Stefan's law"). All these formulas are in Appendix I.

Section 10.8 Bose–Einstein Condensation

At low temperatures, particles tend to cluster in the lowest available energy states. But this tendency applies far more to bosons than to distinguishable particles, leading to some of the more visible consequences of quantum mechanics. A system in which a large percentage of bosons are in the ground state together is called a "Bose–Einstein condensate."

Bose–Einstein condensates can display remarkable behaviors including dropping to zero electrical resistance and repelling magnetic fields ("superconductivity"), or losing all viscosity and climbing up the walls of containers ("superfluidity"). The section does not primarily focus on these phenomena, but on why bosons (and not distinguishable particles) condense at sufficiently low temperature.

11
Solids

You probably learned in school that matter comes in three phases: solid, liquid, and gas. (A fourth phase called "plasma" only tends to occur in extreme environments like the center of the Sun or physics laboratories, so your teachers can be forgiven if they left it out.) Gases can flow and conform their shapes to their containers, and can also compress or expand; liquids can also flow and conform shape, but they cannot compress or expand; solids can't really flow, conform, compress, or expand. These differences arise from the interactions of the atoms and/or molecules that make up the substances.

- In a gas, the separations between molecules are much larger than their sizes. The forces between the molecules are much weaker than the forces holding the molecules together, and they are too weak to bind molecules to each other.
- In a liquid, the molecular separations are comparable to the sizes of the molecules. The forces between molecules are strong enough to keep them close together, but weak enough that they continually break and re-form, which is why a liquid has no fixed structure.
- In a solid, the molecular separations are also small, and the intermolecular forces hold the atoms or molecules together in a fixed configuration.

This chapter is about solids. We had some chapters about liquids and gases too, but they flowed away. Sorry about that.

11.1 Crystals

We can divide solids into "amorphous solids" and "crystals."

An amorphous solid such as glass or plastic looks like a liquid on the molecular level, with all the molecules randomly thrown together. (It's a solid because, whatever random configuration they are in, they stay there.) In a crystal the atoms or molecules are arranged in a regular, repeated pattern.

This chapter discusses only crystals. We could have named it "Crystals" but that could be misleading since many students might assume it was all about rubies and diamonds. In fact the category "crystal" includes granite, copper, and many other solids that you might not expect.

11.1.1 Explanation: Crystals

A crystal is composed of a regular lattice of atoms. You specify the lattice's structure by describing a "unit cell," the smallest group of atoms from which you can make the whole lattice by repeating it. (That idea is not as simple as it sounds; we'll come back to it at the end of this section.)

The simplest 3D structure is a "simple cubic lattice," for which the unit cell is a cube with atoms at each of the six corners. But this simple shape is not typically the most efficient way to pack atoms, so it is not very common. The only element that forms a simple cubic lattice at standard temperature and pressure is polonium.

Two more common types of lattice are shown in Figure 11.1. Both of them start with the simple cubic lattice described above, but the "face-centered cubic" (fcc) lattice adds an atom at the center of each face, and the "body-centered cubic" (bcc) lattice adds an atom at the center of the cube. Elements with the fcc structure include aluminum, copper, gold, silver, and lead; the bcc structure is seen in lithium, sodium, potassium, chromium, and tungsten.

Remember that this pattern repeats in a grid. So the top face of one cube (five atoms in the fcc case, four in the bcc) is also the bottom face of an adjacent cube.

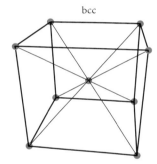

Figure 11.1 Unit cells of a "face-centered cubic" lattice (fcc) and a "body-centered cubic" lattice (bcc).

A crystal such as diamond or quartz has a regular atomic structure throughout the entire object. But such large-scale regularity is the exception; much more common are "polycrystals" such as ice, ceramics, and most metals and rocks. A polycrystal is composed of "crystallites" that are typically less than a millimeter wide. The crystallites themselves are packed together with random orientations, but within each crystallite we see the regular atomic lattice.

Consider salt. Chapter 9 discussed an NaCl molecule as the prototype of an ionic bond: the sodium atom gives an electron to the chlorine atom (which raises the energy of the entire system), and the two oppositely charged atoms then draw near each other (which lowers the energy of the system and binds the molecule).

The salt on your table is different. It *is* made of positively charged sodium atoms and negatively charged chlorine atoms, lowering their energy through their mutual attraction. However, these ions are *not* grouped into pairs that you could identify as molecules. Instead they form a much larger structure called an "ionic crystal," a lattice of alternating negative and positive ions. (The term "ionic solid" is also common.) Figure 11.2 shows the repeating structure of a salt crystal, with face-centered cubic cells of sodium (fcc) interlaced with identical cells of chlorine.

Figure 11.2 The lattice structure of NaCl. The figures in the center and on the right highlight the unit cells of the Na and Cl lattices, each of which is a face-centered cubic.

Cohesive Energy

The "cohesive energy" of a solid is the energy required to break it apart into its constituent atoms. You can think of this as a larger-scale version of the "dissociation energy" required to pull apart a molecule.

The most obvious contribution to the cohesive energy applies to crystals like NaCl that consist of ions; it's the electric potential energy of the ions. A positive ion in a crystal has a negative potential energy associated with its attraction to all the negative ions, and a positive potential energy associated with its repulsion from all the other positive ions. The net result is a potential energy per ion that looks like

$$U = -\alpha \frac{e^2}{4\pi\epsilon_0} \frac{1}{R}. \tag{11.1}$$

That constant α is called the "Madelung constant" for the crystal, and it depends only on the geometry of the crystal.

As an example, consider a sodium ion buried somewhere in Figure 11.2. Its six neighboring chlorine ions contribute $-6e^2/(4\pi\epsilon_0 R)$ to the potential energy, so they add 6 to the Madelung constant. (Equation (11.1) has a negative sign in it, which is why this negative energy constitutes a positive contribution to α.)

The next-closest neighbors are the 12 sodium ions at a distance $\sqrt{2}R$; their positive contribution to the potential energy adds $-12/\sqrt{2}$ to the Madelung constant. After that come 8 chlorine ions at a distance $\sqrt{3}R$, and so on. All this becomes an infinite series that you will often see written as follows:

$$\alpha = 6 - \frac{12}{\sqrt{2}} + \frac{8}{\sqrt{3}} - \cdots \qquad \text{The } \textit{wrong} \text{ way to calculate } \alpha \text{ for an fcc crystal.}$$

The problem is that this series diverges! To get a series that converges, and matches experimental data, you have to expand out in successive cubes, not successive spheres (see Problem 14). The result is that for an fcc lattice $\alpha \approx 1.75$.

Active Reading Exercise: Cohesive Energy

Imagine a large-but-finite collection of sodium atoms and chlorine atoms – not ions, regular atoms – all very far away from each other. Over time they get together, exchange a few electrons, and end up in the lattice in Figure 11.2.

There are many energetic differences between the original configuration (individual atoms) and the final configuration (salt). Equation (11.1) represents two of those differences: the attraction between oppositely charged ions is a negative contribution to the final potential energy, and the repulsion between same-charged ions is a positive contribution. List as many *other* differences as you can, classifying each one as negative or positive.

How many did you come up with? Our list includes the ionization energy of sodium (positive), the electron affinity of chlorine (negative), and the effects that alter the forces between ions when they get very close, namely the Pauli repulsion caused by overlapping electron clouds (positive), and repulsion between the unscreened nuclei (positive). All of these contribute to the cohesive energy of the crystal.

The last one we listed, inter-proton repulsion, is the force that keeps the ions from collapsing into each other. Recall from our discussion of molecules that, when atoms are close enough that their electron clouds overlap, the nuclei are less screened from each other, leading to a net repulsion in addition to the overall ion–ion repulsion (or attraction) that we've already included. That repulsive energy can be modeled as AR^{-n}, where R is the distance between the nuclei and A and n are free parameters (see Problem 12).

Above we defined "cohesive energy" as the energy required to break a solid apart. Unfortunately that term is used slightly differently in different sources. First of all, it is sometimes calculated per atom, sometimes per mole, and sometimes per a small group of atoms (e.g. per pair of Na and Cl ions, for salt). You can usually tell this from the units (eV/atom or kJ/mol, say). It is also sometimes used to mean the energy used to break a solid into neutral atoms, and sometimes to mean the energy required to break it into individual ions. We will always use it to mean the energy required to break the solid into neutral atoms (sometimes called the "atomic cohesive energy").

Types of Bonding

You can divide crystals into four types based on how they are held together.

- *Ionic crystals*

   An ionic crystal is a large-scale structure such as the NaCl crystal we discussed above, built from positively charged ions (atoms that have given up an electron) and negatively charged ions (atoms that have gained an electron). Our discussion of Madelung constants above only applies to ionic crystals.

   Because the ions in an ionic bond have full outer shells, they pack together like spheres. (Figure 11.2 distorts sizes to show the crystal structure clearly; for a more accurate picture, imagine the blue and red dots as electron clouds packed together so closely that they overlap.) Not all ionic crystals follow the interlaced fcc lattice of NaCl. Different crystals have different structures because of the relative sizes of the ions, which determine what packing structure is most efficient.

   Ionic crystals are generally soluble in water because the polar water molecules (positive on one side and negative on the other) can pull apart ionic bonds.

- *Covalent solids*

  As you might guess, covalent crystals are held together by covalent bonds. For example, diamond is a covalent crystal in which each carbon atom is bonded with four others. The structure of the lattice is determined by the geometry of those bonds; in the case of diamond, the carbon bonds form at equal angles and make a tetrahedral structure. Each pair of adjacent carbon atoms shares two electrons between them.

- *Metals*

  In metallic crystals like copper and gold, some of the outer electrons from each atom are effectively shared by the entire lattice, somewhat like one enormous covalent molecule. Those collective electrons are referred to as an "electron gas." The solid is held together by the attraction between the positively charged lattice and the negatively charged electron gas.

  Because the bonds in a metal aren't directly between atoms, it's usually possible to insert different kinds of atoms into the lattice. So it's easier to make alloys – different materials mixed together with various proportions – from metals than from other types of solids. For example, the bronze age in human history was marked by the discovery that melting copper with a small admixture of tin led to a stronger metal than either one by itself.[1]

- *Molecular crystals*

  The crystals we've been discussing so far consist of lattices of individual atoms, but some crystals such as ice are made of molecules. The molecules themselves are held together by strong bonds, and are held to each other by much weaker bonds (but still stronger than those in a gas or liquid).

  The bonds between molecules arise from the electric dipole moments of the molecules; such electric attractions between dipole molecules are called "van der Waals forces." The resulting bonds are relatively weak, so these materials have fairly low melting points.

  In ice, each molecule has a permanent dipole moment because the shared electrons are on average closer to the hydrogen than to the oxygen. This allows the slightly negative oxygen of one molecule to bind to the slightly positive hydrogen of another. When a bond between two molecules with permanent dipole moments involves a positively charged hydrogen, it's called a "hydrogen bond."

  The weakest van der Waals bonds are the bonds between molecules with no permanent dipole moments. If you put two hydrogen molecules near each other, for example, they induce opposite dipole moments in each other and thus attract. The resulting force is very weak and falls off with distance as $1/R^6$, so these types of solids are extremely easy to pull apart. Hydrogen, for example, has a melting point of 14 degrees above absolute zero.[2]

---

1 Interesting historical sidenote: the first bronze was made by mixing copper and arsenic. People then figured out that tin is a lot safer.

2 Some authors only use "van der Waals" to refer to bonds between induced dipoles. The International Union of Pure and Applied Chemistry (IUPAC), however, defines "van der Waals forces" to include all forces between dipole molecules, permanent or induced. The van der Waals forces between induced dipoles are also called "London forces."

Properties of the Different Types of Crystals

You can make generalizations about the properties of crystals based on how they are bonded. For example:

- *Melting point.* Melting a crystal requires pulling apart the bonds that hold it in its fixed lattice structure (without completely separating the atoms or molecules from each other). The stronger the bonds holding a solid together, the harder it is to pull them apart. Generally speaking, ionic and covalent crystals have very high melting points, metallic crystals somewhat lower ones, and molecular crystals lower still. Among molecular crystals those with no permanent dipole moments have the lowest melting points of all. The same strong bonds that make ionic and covalent crystals have high melting points also tend to make them hard (although in fact ionic crystals are often brittle).

- *Conductivity.* We'll talk about electrical conduction later in the chapter, but here's a spoiler: because metals have a ready supply of electrons that are free to move, they tend to conduct both electricity and heat better than other types of solids do.

- *Opacity.* When a photon hits a solid it can pass through with little or no interaction (transmission). Alternatively it may excite electrons inside the solid (absorption), which may then fall back and emit new photons (reflection). So we see a solid as transparent if its electrons are unlikely to be excited by photons in the visible light spectrum.

  Those visible photons are generally too weak to cause the high-energy electron leaps from one shell to another (such as 2s to 3s), but they can cause the much lower-energy leaps within a shell (such as 2s to 2p). Ionic crystals are likely to have full outer shells, with no low-level excitation available, and therefore to be transparent. Covalent and metallic crystals tend to have excited states available for electrons, and are therefore usually opaque.

You shouldn't take this list of properties too seriously. It's a useful guide, and thinking about it can help you understand the nature of the molecular bonds in solids, but there are many exceptions. NaCl, the most common and familiar ionic crystal, has a melting point of 1074 K, considerably lower than most metals (1358 K for copper, for example). Most covalent solids are opaque, but diamond is transparent.

Moreover, not all crystalline solids fit neatly into these categories. Just as molecular bonds fall on a spectrum from ionic to covalent, solids can be bound by various combinations of these mechanisms. Also, many common solids such as wood are mixtures of different solids (cellulose, lignin, . . .) that are bound in a variety of ways to themselves and to each other.

Nonetheless, understanding these four basic mechanisms of bonding takes you a long way toward understanding how solids are put together.

Final Thoughts on Repeating Patterns

We defined a "unit cell" earlier as "the smallest group of atoms from which you can make the whole lattice by repeating it." With that in mind, take another look at Figure 11.2, the lattice structure of NaCl. How many atoms compose its unit cell?

You might reasonably (but incorrectly, as it turns out) answer "two": one sodium ion and one chlorine ion separated by a distance R. You can build the entire lattice by copying *that*

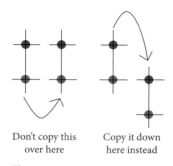

Don't copy this   Copy it down
over here   here instead

Figure 11.3 The right and wrong ways to build an NaCl crystal.

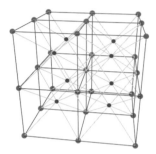

Figure 11.4 The lattice structure of cesium chloride (CsCl).

unit over and over, but it requires a little care. Each time you make a copy, you have to slide over by a distance R from the previous copy (Figure 11.3).

That little two-ion piece is not technically the unit cell, but it is the smallest unit of our crystal, and therefore an important guide to predicting the behavior of the crystal.

If you were to draw each of those pairs as a single dot, an NaCl crystal would look like a face-centered crystal of those dots. So we consider the unit cell of NaCl to be a face-centered crystal with 28 atoms: 14 pairs of an Na ion and a Cl ion. The entire crystal can be built by repeating that unit cell in a simple stack in all three directions.

As another example, consider the CsCl crystal shown in Figure 11.4.

You might think of this as a bcc lattice because each cube of cesium ions (red) has a chlorine ion (blue) in the middle. It's not considered bcc, though, because the atom in the middle is different from the ones on the cube. Like NaCl, you can consider CsCl to be made of two interlaced lattices of two different types of ions. For CsCl, though, each of those is a simple cubic lattice, so that's what we consider CsCl to be. Equivalently, you can imagine drawing each CsCl pair as a single dot. (For example, you could pair each blue dot with the red one just up, back, and to the right of it.) The lattice you would make out of those dots would be a simple cubic lattice.

11.1.2 Questions and Problems: Crystals

Conceptual Questions and ConcepTests

1. What distinguishes "crystals" from other solids?

2. Equation (11.1) shows the potential energy of an ion in a crystal lattice, where the Madelung constant α is calculated by summing over the contributions from all other ions in the lattice. If you multiplied Equation (11.1) by the number of ions to get the potential energy of the whole lattice, the result would be too big. By what factor, and why? *First note*: That potential energy is negative, so "too big" means "too negative" in this case. *Second note*: We're not talking about the fact that there are other contributions to the energy; we're saying this would overestimate the contribution of the ion–ion potential energy term.

3. Figure 9.6 on p. 401 shows the ground state energy of a covalent bond as a function of the separation between the nuclei. (Focus on the black curve.)

In equilibrium, the nuclei will rest at the stable minimum (or as close as the uncertainty principle allows). As you raise the temperature, however, the vibrations of the nuclei will cause their average energy to rise above the minimum.

(a) Looking at the figure, will the average separation be... (Choose one and explain.)

   A. smaller than the equilibrium separation?

   B. equal to the equilibrium separation?

   C. larger than the equilbirum separation?

(b) If this represents the covalent bond between atoms in a crystal, what will happen to the overall crystal when you heat it (but not enough to melt it)? Explain how your answer follows from your answer to Part (a).

4. Briefly describe what happens at an atomic or molecular level when a crystal melts. What general

property makes some solids easier to melt (i.e. have lower melting points) than others?

5. Why can't you have an ionic crystal made entirely of one type of element?

6. Why can't hydrogen form a covalent crystal?

7. How many atoms are in a unit cell of CsCl (Figure 11.4)? (Choose one and explain.)

   A. 2

   B. 9

   C. 16

   D. 18

   E. 28

8. The "ionic cohesive energy" is the energy required to break a crystal into the charged ions of which it is made (still charged, but far away from each other). The "atomic cohesive energy" is the energy required to break a crystal into neutral atoms far away from each other. Which one is higher? What sources of energy does one of these take into account that the other ignores?

Problems

9. Figure 11.5 shows a 1D ionic crystal whose ions all have charge q (blue) or $-q$ (red). Neighboring ions are separated by a distance R.

Figure 11.5

In this problem you will calculate the Madelung constant for this crystal.

(a) Choose an arbitrary blue point (positive charge) on the grid. How many nearest neighbors does it have, and how far away are they? Write down the total potential energy associated with the attraction or repulsion between your blue point and those neighbors.

(b) How many *next*-nearest neighbors are there, and how far away are they? Write down the total potential energy associated with the attraction or repulsion between your blue point and those neighbors.

(c) Repeat for the third-nearest and fourth-nearest sets of neighbors.

(d) Write down an infinite series for the total potential energy associated with the attraction and repulsion of all neighbors to your one ion. Give your answer in the form $-\alpha q^2/(4\pi\epsilon_0 R)$, where α is an infinite series.

(e) You should be able to look up (or possibly recognize) the infinite series you just wrote. Do so and find the Madelung constant α for this crystal.

10. Consider a two-dimensional crystal built by placing alternatively positive and negative charges at every integer-pair point on the coordinate plane. Each point has a charge of q or $-q$, and the horizontal and vertical separations are R (Figure 11.6).

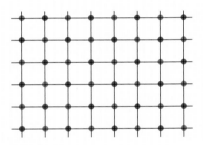

Figure 11.6

You can find the Madelung constant for this crystal by expressing the potential energy of a single ion as a sum over the ions in one square around it, the ions in the next square around that, and so on.

(a) Write the Madelung constant including all the ions in the first two concentric squares. Your final answer should only include numbers (no letters). If you're stuck you might find it helpful to work through Problem 9 first.

(b) For an infinite 2D crystal like this, the Madelung constant is 1.6. What percentage error did you have with just two concentric squares?

11. Equation (11.1) shows the first few terms of the Madelung constant for an fcc crystal.

(a) Estimate this Madelung constant using just those three terms. What is the percentage error?

(b) What would the fourth term be? What is your percentage error using that term? (You should find that adding the fourth term makes the sum much farther from the correct answer, the first hint that this series actually diverges.)

12. There are three contributions to the cohesive energy of an ionic crystal such as NaCl: the ion–ion negative potential energy, the repulsive potential energy between the nuclei, and the energy difference between free ions and free neutral atoms ($I_{Na} - \text{Aff}_{Cl}$). There's also Pauli repulsion between the electron clouds, but we can include that with the repulsive potential energy and model that whole term as $U_{rep} = AR^{-n}$, where R is the separation length between the ions. Putting all that together, the potential energy per ion pair (Na and Cl) is

$$U = I_{Na} - \text{Aff}_{Cl} - \alpha \frac{q^2}{4\pi\epsilon_0 R} + AR^{-n}.$$

(a) Set $dU/dR = 0$ and solve for R to find an expression for the equilibrium separation length. Your answer should have A, n, and the Madelung constant α in it.

(b) For NaCl the equilibrium separation length is 2.81×10^{-10} m and the Madelung constant is 1.75. The ionization energy of sodium is 5.14 eV, and the electron affinity of chlorine is 6.45 eV. The cohesive energy per ion pair is 7.98 eV. Estimate the parameter n.

13. [This problem depends on Problem 12.] Plot the potential energy of an NaCl crystal as a function of the ionic separation R, using the values you calculated for A and n.

14. **Exploration: The Madelung constant for NaCl, Two Ways**

Consider an NaCl crystal with an ion at each point (iR, jR, kR), where i, j, and k are all integers. To find the potential energy of the ion at the origin, you can sum over all the contributions of the other ions, $\pm q^2/(4\pi\epsilon_0\sqrt{i^2+j^2+k^2}\, R)$, where the sign $\pm$ depends on whether the ion is positive or negative.

(a) Calculate the potential energy of the ion at the origin from all the ions for which i, j, and k go from -1 to 1. Find the contribution of these 26 ions to the Madelung constant. Then repeat, letting the indices go from -2 to 2, and then from -3 to 3, and so on. Make a plot of α vs. r where each calculation of α includes all the points from $-r$ to r in each direction, up to $r = 20$. Describe what α is doing as you increase r. Hint: As a check, you should find that for $r = 1$ you get $\alpha = 2.134$. As you go to higher r, make sure to add the old and new contributions, so that for $r = 20$ your value of α includes *all* the ions from -20 to 20 in each direction.

(b) Now re-do your calculation, but expanding out in spheres instead of cubes. Your first value of α will use the ions closest to the origin, the next one will use the next-closest, and so on. You should find that the first few terms reproduce the beginning of the series we wrote on p. 502. Make a plot of α vs. r, where r is the radius of the sphere. Describe what α is doing as you increase r. Hint: One way to find which ions to include in each new sphere is to start by calculating the distance from the origin to every ion in a $20 \times 20 \times 20$ cube and then sort the resulting list by radius.

(c) The calculation in Part (a) converges to the correct value for α. Many texts present the calculation in Part (b). Based on your results, explain briefly how you can tell this is wrong!

11.2 Band Structure and Conduction

You know by now that one of the most important properties of any physical system is the structure of its energy levels. This section begins with a qualitative discussion of the energy values availabel to an electron in a crystal: the crystal's "band structure." The rest of this section explains how the band structure can be used to predict the electrical and heat conduction properties of a solid. Sections 11.3 and 11.4 focus on the properties and uses of semiconductors, once again based on their band structures, and Section 11.5 discusses *why* the lattice structure of a crystal leads to allowable energy levels in "bands."

11.2.1 Explanation: Band Structure and Conduction

The energy levels of a free particle are continuous: it can always absorb just a little more energy, or give away a little. The energy levels of a bound particle are discrete: it can gain or lose only specific amounts of energy.

An electron in a crystal is a bound particle, but its energy levels are organized into "bands." For instance, Figure 11.7 depicts a band of allowable energy levels between 2 and 3 eV. The energy levels in this band are technically discrete, but their density of states is so high that they are effectively continuous. Between 4 and 5 eV is another band; an electron could be measured with essentially any energy in this region. But you will *never* measure an electron in this crystal with energy between 3 and 4 eV. Bands of effectively continuous energy levels alternate with bands of forbidden levels.

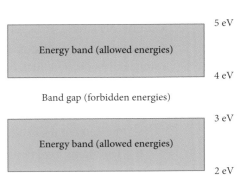

Figure 11.7 A simple example of bands. The electron can have any energy between 2 and 3 eV, or between 4 and 5 eV, but cannot have energy between 3 and 4 eV.

Imagine an electron in such a structure with exactly 4 eV of energy. That electron can gain 1/2 eV, or 1/10 eV, or 1/1000 eV. But it cannot give up energy unless it gives up a full 1 eV or more.

This section and Sections 11.3 and 11.4 explore how this band structure explains conduction properties in general, and semiconductor properties in particular. Section 11.5 discusses *why* the lattice structure of a crystal leads to allowable energy bands.

The Flow of Current

Figure 11.8 shows a simple circuit with a voltage source and a resistor. When the switch is closed, a constant current $I = V/R$ will begin to flow continuously around the circuit.

That current consists of a flow of charges (electrons) through a metal (the wire). The battery supplies an electric field that pushes the electrons around the loop. But think about that description more carefully. An electric field exerts a force on an electron, so the electrons should keep accelerating all the way through the circuit. That's not what happens. A steady current requires that the electrons have, on average, a constant *velocity*. What's going on?

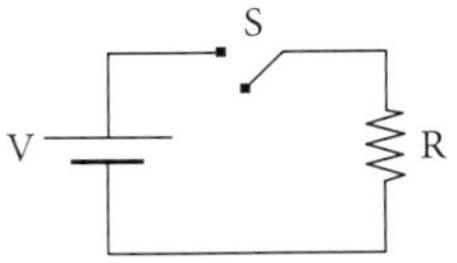

Figure 11.8 A simple circuit.

Classically, the resolution is simple. The classical electron is like a tiny bullet traveling through a lattice populated with ionized atoms. Each time the electron collides with one of those ions, its forward momentum slows. It then begins accelerating again, and this alternation of speeding up and colliding leads to an unchanging average velocity as it moves through the wire. (The situation is analogous to pushing an object through a medium with drag: the balance of the pushing force and the drag from constant collisions with molecules of the medium can lead to a constant velocity.)

To treat the same scenario quantum mechanically, we have to calculate the electron's wavefunction. There is a potential field applied by an external source (which you can think of as linearly increasing as you move along the wire), and also a potential field created by the lattice of ions in the wire (alternating dips and bumps). You might imagine that scattering off those dips and bumps would slow down the electron's wave packet, just as classical collisions do. But it doesn't work that way.

A full treatment of this quantum mechanical problem is beyond the scope of this book, so you'll have to take our word for the following result: for an infinite, perfectly periodic lattice, the probability of scattering becomes zero. So once again it seems like the electrons should accelerate to arbitrarily large speeds, rather than flowing in a steady current.

In Section 11.5 we'll look a bit more closely at some properties of Schrödinger's equation in a simple periodic potential. For now, we want to note that real crystals have resistance because their lattices aren't perfectly periodic. That occurs for two main reasons:

1. Every lattice has imperfections, such as atoms of different elements interspersed throughout the crystal.

2. The atoms have random thermal vibrations that break the symmetry of the lattice.

At very low temperatures, when thermal vibrations are minimal, the lattice imperfections dominate; at room temperature, thermal vibrations typically contribute more to electron scattering. We're not going to analyze this scattering process in detail, although you'll show in Problem 8 that a simple model of thermal vibration effects can correctly predict the temperature dependence of resistance for metals. In a more careful treatment, you can think of these quantized lattice vibrations as particles called "phonons" and model the scattering as a set of collisions between electrons and phonons propagating separately through the lattice. We'll discuss phonons in Section 11.7.

The result of all this acceleration and scattering is that the electrons acquire an average velocity. That "drift velocity" is typically much smaller than the random thermal motion of the electrons. So you shouldn't picture an army of electrons moving in lockstep in the direction of the current. A more accurate image would be a cloud of electrons moving in all directions, with a slight statistical preference for one direction over the other.

This whole picture of electrons propagating and scattering assumes that the solid *has* electrons that are free to move about, as opposed to being rigidly attached to one or two nuclei. Whether a solid has such electrons, and is thus capable of conducting electricity, depends on its band structure.

Insulators, Conductors, and Semiconductors

In a solid at low temperatures the N electrons fill the N lowest energy states. We define the "Fermi energy" ϵ_F as halfway between the Nth and $(N + 1)$th lowest-energy state, so at very low temperatures all the states with $E < \epsilon_F$ are filled and all the ones with higher energy are unfilled. At room temperature some of the electrons near the Fermi energy may be thermally excited into higher energy states, but the N lowest states are nearly all filled and the ones above that are almost all empty. For most materials, if you raised the temperature to a point where the previous statement stopped being true, you would vaporize them, so we can safely say that virtually all solids remain close to the zero-temperature Fermi–Dirac distribution.

Remember, however, that in solids those available energy states come in bands.

Active Reading Exercise: Available Electrons in a Crystal

Imagine a crystal whose band structure includes the one "forbidden" and two "allowed" bands portrayed in Figure 11.7 on p. 509.

1. Suppose $\epsilon_F = 2.5$ eV, so the highest-energy filled level is in the middle of the bottom (allowed) band. If an external source supplies $1/10$ eV, are there electrons that can absorb it?

2. Now suppose $\epsilon_F = 3.5$ eV, meaning that the bottom band is entirely filled and the top band entirely empty. If an external source supplies $1/10$ eV, are there electrons that can absorb it?

If you thought through that Active Reading Exercise properly, you can see the difference between a conductor and an insulator.

- A conductor, like silver or copper, has a partially filled band of electron states. Electrons near the Fermi energy are thermally excited into higher states, so the region from $\epsilon_F - k_B T$ to $\epsilon_F + k_B T$ contains electrons moving up and down in energy through random collisions.

  Now suppose you apply an electric field. Some of those electrons near the Fermi energy get pushed into higher energy states with momentum in the direction of the force being exerted on the electrons. The result is that the electron gas acquires a net drift velocity.

  From there you can use the description we gave above: the applied field continues to give energy to the electrons, pushing them to higher energy states, but they keep losing energy to scattering off of lattice imperfections. The result is a net "drift velocity." Current flows.

- An insulator, such as diamond, has a full band; the Fermi energy is in the middle of a forbidden region so the band below is full and the one above is empty. Virtually no electrons have enough energy to jump the gap to the next allowable band (see Problem 5). The states in that fully occupied band have an equal mix of momenta in all directions. A modest electric field cannot change the electron states, so no current flows. A sufficiently strong field can push electrons across the gap, but for most insulators the required field is enormous.[3]

- There is a third category of crystals: semiconductors, such as silicon and germanium. These are like insulators in that their top bands are fully occupied. The difference is that they have a relatively small gap up to the next band, so a small but measurable number of electrons are able to jump the gap and conduct at room temperature.

---

3 For diamond, for example, it requires a field of roughly 10 MV/cm. So if you want to put current through a diamond ring, attach it to 11 million 9-volt batteries in series. We are not responsible for any marital strife that may arise from such an experiment.

The band structures of conductors, insulators, and semiconductors are shown in Figure 11.9.

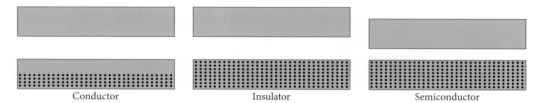

Conductor   Insulator   Semiconductor

Figure 11.9 A solid with a partly filled electron band is a conductor. A solid whose highest occupied band is full is either an insulator or a semiconductor, depending on how wide the gap is before the next band.

There is no universal definition of the difference between an insulator and a semiconductor, but generally a gap of less than about 2 eV qualifies a solid as a semiconductor. Silicon has a gap of 1.14 eV, germanium 0.67 eV.

Semiconductors are terrible at conducting electricity; a silicon wire that's 1 cm long and 1 mm wide would have a resistance of 20 million ohms! But semiconductors can be manipulated in ways that dramatically increase or decrease their resistance, which means they can be used to construct controllable electronic devices. We'll discuss semiconductor devices in Sections 11.3 and 11.4.

It's easy to think that "conductor" is synonymous with "metal." While it's true that metals are generally conductors, it's not true that all conductors are metals. For example, graphene is a form of carbon that has a half-filled band and therefore can conduct electricity, but it is not a metal because it is principally held together by covalent bonds rather than by attraction between the lattice and the electron gas.

To finish our discussion of electrical conduction, let's return to that circuit we discussed at the beginning. Conductors such as copper allow current to flow with very little resistance, so that accounts for the wires in the circuit. Insulators and semiconductors allow very little current to flow, so you can use them to block unwanted flow (e.g. by coating your wires in case they touch each other). So how do you make a resistor, which allows moderate amounts of current to flow?

There are many different answers to that question, but they mostly involve conductors and insulators in various combinations. For example, many resistors are made by grinding together a carbon-based conductor and an insulating fill. By altering the ratio of conductor to insulator you can make a device with whatever resistance you want.

Heat Conduction

To think about electrical conduction, again picture a wire. If you apply a potential difference across a copper wire, current flows, because copper is a good conductor. A wire made of rubber would not produce the same effect.

For a comparable image of heat conduction, imagine a thermos flask. A thermos flask made entirely of metal would be a terrible idea: heat would flow through the walls, and it wouldn't be long before your hot soup or your cold drink reached room temperature. But a good insulating material, such as Styrofoam, is much more resistant to heat transfer, and your lunch is saved. (In a typical thermos flask, the Styrofoam is surrounded with another material, quite possibly metal – not for insulation, but for rigidity.)

In short, heat conductivity is a measure of how easily heat flows through a material, just as electrical conductivity measures how easily current flows through a material. We're not going to analyze the mechanisms of heat conduction, but they follow many of the same ideas as electrical conduction. When you apply heat to a metal, you excite its electrons into higher energy states. Unlike electrical conduction, this doesn't give the electrons a net velocity, but it does still require that the electrons have available states to jump into. So, generally speaking, materials that are good conductors of electricity are also good conductors of heat and vice versa.

This tendency is formalized in the "Wiedemann–Franz law," which says that the ratio $K/(\sigma T)$ is roughly the same for all metals, where K is the thermal conductivity and σ is the electrical conductivity. The thermal conductivity grows faster with temperature because it depends on the heat capacity of the electron gas, which is proportional to T.

11.2.2 Questions and Problems: Band Structure and Conduction

Conceptual Questions and ConcepTests

1. What's the difference between a conductor, an insulator, and a semiconductor? Answer by saying how they behave differently, and also what causes those different behaviors. A few short sentences are plenty.

2. Imagine that Figure 11.7 on p. 509 shows part of the band structure for the fictitious solid Vibranium.

   (a) Suppose that in its ground state, Vibranium has enough electrons to fill energy levels up to 2.8 eV. (In other words, $\epsilon_F = 2.8$ eV.) Is Vibranium an insulator, conductor, or semiconductor?

   (b) Now suppose that in Vibranium's ground state, the 2–3 eV band is entirely full and the 4–5 eV band is entirely empty. (In other words, $\epsilon_F = 3.5$ eV.) Is this version of Vibranium an insulator, conductor, or semiconductor?

   (c) To make Vibranium an insulator you need to change the energy labels on Figure 11.7 *and* choose the right value of ϵ_F. Copy the figure and do so.

3. An insulator has no electrons that can be excited (unless you *really* excite them). So if you heat up an insulator by a few degrees, where is that energy stored?

4. If you think of the electrons in the electron gas as (mostly) free, then their eigenfunctions look like $\psi(x) = e^{ikx}$ with energy eigenvalues $E = \hbar^2 k^2/(2m)$. If you change the sign of k you change the eigenfunction, and you change the direction of the wave's time evolution, but you do not change the energy.

So what if, instead of representing all the electrons by their energy levels (as we have been doing in this section), we represented them by their k-values?

Figure 11.10 shows one solid – a conductor – at three different times. At one of these three times the conductor is in its ground state; at another, it is being excited by a heat source; and at a third time,

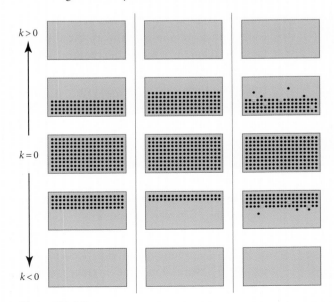

Figure 11.10

it is being excited by an externally applied voltage. Identify which is which, and explain how you know.

Problems

5. The highest occupied electron band in diamond is full, and the gap to the next band is about 5.5 eV.

   (a) Use the Fermi–Dirac distribution (see Appendix I) to estimate the fraction of states at the bottom of that upper band that are filled at room temperature. *Hint*: The Fermi energy for diamond, and therefore its chemical potential μ, lies in the middle of the band gap.

   (b) A 20 carat diamond has about 2×10^{23} atoms. Assuming each atom contributes one state near the bottom of the upper band, estimate how many of those states are filled on average at any given moment (and thus are able to conduct electricity). Your answer should give you some idea why diamond is such a good insulator.

6. **The Classical Model of Conduction**

   Consider a block of silver, which we can model as a lattice of singly ionized silver atoms surrounded by a gas of electrons, one per silver atom. Let ρ be the density of silver (mass per volume) and M_S be the mass of a silver atom. (Express your answers using letters until we tell you to plug in numbers.)

   (a) Find the volume per atom and take the cube root of that volume to estimate the distance between silver ions. Assuming the atoms completely fill the space (which is more or less true since their electron clouds overlap), you can take that distance as your estimate of the mean free path traveled by an electron between collisions.

   (b) Use the equipartition theorem (Appendix I) to estimate the thermal speed of an electron, and use that and the mean free path to find the average time between collisions.

   (c) A potential V is applied across a silver wire of length L, producing an electric field V/L within the wire. Find the average velocity imparted to an electron by that field in the time between collisions.

   (d) Find the current in the wire – charge per unit time – assuming the wire has a cross-sectional area A. *Hint*: The number density of electrons is the same as the number density of silver atoms, which you can express in terms of the letters we've defined.

   (e) Resistance is voltage over current. Resistivity is defined as resistance times cross-sectional area, divided by length. Find an expression for the resistivity of silver.

   (f) Plug in numbers to get a value for the resistivity of silver at 20 °C. You'll need to look up ρ and M_S, as well as the electron mass and electron charge.

   (g) Experimentally the resistivity of silver is 1.59×10^{-8} Ω m. You should find that your prediction using this classical model was more than an order of magnitude too large. Give at least one reason why this model gives too large an answer.

7. [*This problem depends on Problem 6.*]

   (a) Using the resistivity of silver (the real number, not the wrong prediction) and the number density of its electrons, find the drift velocity of silver electrons for an applied field of 100 V/m.

   (b) The Fermi energy of silver is 5.49 eV. What is the typical thermal speed of a conduction electron in silver? (Again, give the real answer and not the incorrect classical prediction.)

   (c) What must the mean free path of those electrons be? How many lattice spacings does that mean free path correspond to?

8. **The Temperature Dependence of Conduction**

   The conductivity of a metal is proportional to the mean free path and inversely proportional to the average speed of the electrons. Other factors that affect it are mostly temperature-independent.

   (a) Explain why the conductivity depends on mean free path and average electron speed in those ways.

(b) In a classical model the mean free path is the distance between ions and is approximately constant. If you treat the electrons as a classical gas (ignoring the fact that they are fermions), how does their average speed scale with temperature? How does that imply that the conductivity scales with temperature?

(c) In a quantum model the average speed of the conduction electrons does not depend on temperature. Why not?

(d) In a quantum model the mean free path at room temperature is typically determined by thermal lattice vibrations, which cause the mean free path to be inversely proportional to temperature. How does conductivity scale with temperature at room temperature?

(e) At much lower temperatures, defects dominate the mean free path, which becomes temperature-independent. How does conductivity scale with temperature in this regime?

11.3 Semiconductors and Diodes

Left to itself, a semiconductor acts similarly to an insulator. But because the gap between the highest filled band and the lowest empty band is smaller than the gap in a true insulator, we can engineer a semiconductor to be more like a conductor. In this section we'll discuss the physics underlying such manipulation, and see how it can be used to create an important circuit element called a diode. In Section 11.4 we'll see how the same technology can be used to create an even more important circuit element, the transistor.

11.3.1 Discovery Exercise: Semiconductors

Consider a wire made from a semiconductor. As we explained in Section 11.2, that word implies three things: there is an energy band that is completely full, the band immediately above it is completely empty, and the gap between these two bands is relatively small (Figure 11.11).

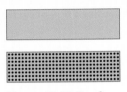

Figure 11.11 Bands in a semiconductor.

   1. If a voltage difference is applied across this wire, very little current will flow. Briefly explain why.

Now imagine that the electricity fairy sprinkles a few extra electrons into the wire. (The fairy also increases the positive charge of the lattice so the wire remains electrically neutral.)

   2. Where in Figure 11.11 will those electrons go? You can copy the figure and draw them in or explain in words where they will end up.

   3. How does this change the resistance of the wire? Explain briefly why.

11.3.2 Explanation: Semiconductors and Diodes

A semiconductor, left on its own, acts much like an insulator. That is, it has a resistivity so high that very little current flows through it even in response to a strong electric field.

What makes a semiconductor so useful is that we can customize its resistivity. We will begin this section by looking at the properties of semiconductors at the atomic level, and how we can exploit those properties to lower their resistivity. Then we will discuss how that technology allows us to create two important electronic components: diodes (later in this section) and transistors (in the next section).

Current Flowing through a Semiconductor

In Section 11.2 we defined a semiconductor as a solid with a relatively small (less than ~2 eV) gap between a full electron band and an empty one. The full band is called the "valence band" and the empty one above it is called the "conduction band." That's because electrons that jump up to the conduction band have lots of available states close to their energy, so they can conduct current.

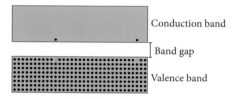

Conduction band

Band gap

Valence band

Figure 11.12 A small number of electrons in a semiconductor have enough thermal energy to jump the gap and fill states in the conduction band, leaving behind holes in the valence band. If an electric field is applied to the right, the conduction electrons will move to the left, and the holes in the valence band will move to the right, both causing a flow of current to the right.

Each electron that jumps the gap also leaves a "hole" behind, an unfilled state in the valence band (Figure 11.12). A nearby electron might jump into that hole, leaving behind a different hole – which might in turn be filled by a different electron. Since the hole "moves" in a direction opposite how the electrons move, you can think of the hole as a positive charge that moves along the valence band in response to an applied electric field.

The band gap of a semiconductor is typically much larger than $k_B T$. For pure silicon at room temperature, fewer than one in a billion atoms contribute electrons to the conduction band. The resistivity of silicon (resistance per unit length, times cross-sectional area) is therefore thousands of ohm meters, many orders of magnitude higher than that of conductors.

What makes semiconductors so interesting, and useful, is that we can *change* their resistance by a process known as "doping."

Doping

Phosphorus and silicon have similar band structures, but phosphorus sits just to the right of silicon on the periodic table: that is, it has one more electron per atom. That electron sits at the bottom of the conduction band, making phosphorus a conductor (although a weak one compared to some other elements).

Now imagine that you take a block of silicon and replace one in every 10,000 atoms with a phosphorus atom. Each of those atoms contributes an electron to the conduction band. Recalling that pure silicon contributes less than one electron per billion atoms to the conduction band at room temperature, this tiny sprinkling of phosphorus increases the conductivity of the silicon by more than five orders of magnitude!

Suppose instead that you replace those silicon atoms with aluminum. Aluminum lies just to the *left* of silicon on the periodic table – it has one fewer electron per atom – so each aluminum atom adds a hole, an unfilled state, to the valence band. Once again you have dramatically increased the silicon's conductivity.

The process of adding impurity atoms to a semiconductor to increase its conductivity is called "doping." (You may recall from Section 11.1 that it is relatively easy to mix metals in this way, because a metal is not held together by bonds between individual atoms.) A pure semiconductor is called "intrinsic," while one that has been doped is called "extrinsic."

When a semiconductor is mixed with a small number of atoms with extra electrons like phosphorus, we say it is an "n-type" extrinsic semiconductor. The "n" stands for negative, because you have added negative charge carriers that can conduct electricity. A semiconductor with extra holes (e.g. silicon doped with aluminum) is "p-type" because the added holes act like positive charge carriers. See Figure 11.13.

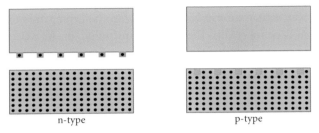

n-type   p-type

Figure 11.13 An n-type semiconductor has atoms that contribute extra electrons to the conduction band. A p-type semiconductor has atoms that contribute extra holes to the valence band. The number of charge carriers in a doped semiconductor is typically many orders of magnitude larger than the number that arises from thermal fluctuations.

Diodes

Figure 11.14 shows two circuits, identical except for the orientation of the battery. If these circuits had nothing in them but the battery and the resistor, that change in orientation would make no difference in the strength of the resulting current. But the circuit element called a "diode" (represented in circuit diagrams as a triangle) allows current to flow freely in one direction, while blocking current in the opposite direction.

The triangle representing a diode is drawn pointing in the direction in which it allows current to flow. Hence the circuit on the left will have a constant current $I = V/R$; the circuit on the right will have no current. (As physics textbooks so often do, we are assuming ideal electronic components. We'll explain in a little bit why both of these statements are approximately but not perfectly correct for real, physical diodes.)

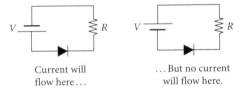

Current will flow here...   ...But no current will flow here.

Figure 11.14 The diode on the left acts as a closed circuit, the one on the right like an open circuit, because the diode only allows current to flow in one direction.

How do you construct such a unidirectional circuit element? The answer, for reasons that we're about to explain, is that you push a p-type semiconductor up against an n-type semiconductor, creating what's called a "p-n junction."

Active Reading Exercise: A p-n Junction

The figure below shows two p-n junctions connected to batteries. In which of these two arrangements will current flow through the junction?

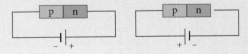

Even if you don't know the answer, write down a few thoughts about how the two configurations might differ. Remember that the difference is *not* that the n-side of the junction is negatively charged or the p-side positive; they are both electrically neutral! The difference, as we explained above, has to do with available charge carriers.

The correct answer is that current can only flow in the arrangement on the right, where the negative terminal is connected to the n-type semiconductor. In a circuit diagram, this diode would be an arrow pointing to the right.

It's tempting to pass over that result too quickly: electrons flow from negative to positive, and not the other way around, right? But explaining the diode's behavior is not that simple, because an n-type semiconductor does *not* have a net negative charge. What it *does* have is electrons in its conduction band, available to carry current, so we need to see how that arrangement of electrons leads to the unidirectional behavior we see. (It's equally true and important that the p-type semiconductor has holes in its valence band. We will focus our discussion on the electrons, but make sure you see how all the same arguments apply to the holes, and lead to the same current direction.)

In the circuit on the right side, the voltage pushes the conduction electrons from the n-type region into the p-type region. As in any circuit, the electrons that flow away are replaced by new ones from the battery, and you get a steady current.

In the circuit on the left the voltage tends to push electrons from the p-side to the n-side, but the p-side doesn't have any free electrons to push. The electrons that are available on the n-side get pulled *toward* the positive terminal of the battery rather than flowing across the junction. Soon the region near the p-n boundary is devoid of any free charge carriers. In short, there's nothing to carry the current in the direction that the battery is trying to push it.

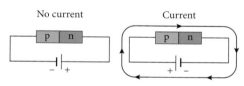

Figure 11.15 The diode on the right is forward-biased and allows current to flow. The diode on the left is reverse-biased and acts like an insulator.

The result of all this is that current can only flow from p to n, as shown in Figure 11.15. So a diode is drawn in a circuit as an arrow pointing from p to n.[4]

When you connect a voltage source to a diode in the direction that allows current to flow through it, you "forward-bias" the diode. When you connect a voltage across it the other way you "reverse-bias" it and no current flows.

The argument we just gave is a reasonable hand-waving explanation, but the details are somewhat more complicated, and more interesting, than this simple picture would suggest.

Diodes and p-n Junctions: A Deeper Look

Consider a p-n junction that is *not* attached to a battery. As we emphasized above, both sides of the junction are electrically neutral. (For each extra electron in an n-type semiconductor, for example, there's an extra proton balancing the charge.) So you might suppose that all the electrons will stay exactly where they are, and for the most part, you would be right.

---

4 The arrow symbol for a diode points in the p-to-n direction because we imagine current as a flow of positive charge, even though it is really electrons moving. This is Benjamin Franklin's fault, but there was honestly no way he could have known any better.

But there is still random thermal motion, and every now and then an electron on the n-side wanders over to the p-side for no particular reason. Once there, it quickly falls into a hole in the valence band. Such random migration is much less likely to happen in the other direction, since a p-side electron would have to jump *up* from the valence band to the conduction band before it could find a home on the n-side.

This asymmetrical diffusion of electrons leaves the n-side with a net positive charge and the p-side negative. That charge imbalance causes an electric field that tends to push electrons *back* to the n-side. The effect of that field is to raise the potential energy of electrons on the p-side relative to the n-side, so in equilibrium the energy diagram looks like Figure 11.16.

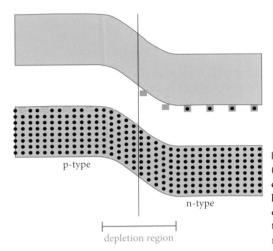

Figure 11.16 A p-n junction in equilibrium (with no external voltage applied). The charge carriers near the junction have diffused across, leaving a "depletion region" with no free charge carriers. This creates an electric field that raises the potential energy of electrons on the p-side relative to the n-side.

The green squares on the right represent conduction-band levels that were originally filled on the n-side; the yellow squares on the left represent valence-band levels that were originally empty in the p-side. Far away from the junction, those green squares are still filled with electrons and the yellow squares are still vacant. But in a small region around the junction, known as the "depletion region," both valence bands are full and both conduction bands are empty.

Figure 11.17 shows the two currents that are continually flowing across the junction in opposite directions.

1. The "recombination current": random motion leads to a net diffusion of electrons in the conduction band from the n-side to the p-side. On the p-side they quickly fall across the gap into holes in the valence band.

2. The "thermal current": when electrons on the p-side thermally jump the band gap into the conduction band, the electric field pushes them to the n-side.

(As a reminder, we're focusing on the motion of electrons, but there is also a recombination current of holes from p to n and a thermal current of holes going the other way.)

The recombination current tends to widen the depletion region and the thermal current tends to narrow it. In equilibrium those two effects cancel and the depletion region stays at a constant width.

Now we can consider what happens when an external voltage is applied. The thermal current is essentially unaffected. The reason is that, with or without that external voltage, an electron

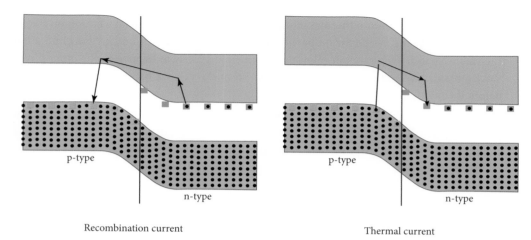

Recombination current   Thermal current

Figure 11.17 In equilibrium the "recombination current" is a flow of electrons from the n-side to the p-side via diffusion. The "thermal current" is a flow of electrons in the other direction driven by the electric field.

has to jump up to the conduction band *before* it can cross to the n-side. The probability of such a jump is simply determined by the temperature, and virtually every electron that jumps the gap flows across to the n-side.

The recombination current, however, is affected by an external voltage. First consider forward-biasing, in which the current is trying to push electrons from the n-side to the p-side. The effect of the applied voltage is to reduce the energy hill that electrons have to climb to move in that direction. The stronger you make the voltage, the more electrons will be able to climb up the potential energy hill and flow. So for small applied voltages the circuit roughly follows Ohm's law, with current proportional to applied voltage. At a certain point the applied voltage overcomes the internal electric field of the p-n junction and current can flow with very little resistance.[5] (See Figure 11.18.) In a typical p-n junction the voltage at which this occurs might be less than a volt.

Next consider reverse-biasing the circuit. This time the applied voltage increases the size of the hill and reduces the recombination current, so it once again creates a net flow of current. You can't reduce the recombination current below zero, though, so once you make the hill high enough that no electrons can flow up it, the current saturates at the level of the thermal current. That thermal current is quite small, so you get a very small current even for large applied voltages.

All this explains why a p-n junction acts like a diode, but also why it's not an ideal diode. For very small voltages in either direction it approximately obeys Ohm's law. For reverse-biasing, the current saturates at a very small value. For forward-biasing, it rapidly rises to a very large value.[6] See Figure 11.19.

---

5 Of course the resistance isn't zero at that point, but is limited by electron collisions with thermally excited ions, as in a regular conductor. See Section 11.2.

6 Generally reverse-biasing a p-n junction produces a very small current, but there is a breakdown voltage beyond which a reverse-biased p-n junction will become conducting. For typical devices the breakdown voltage tends to be about 50–100 V or more, so this is not a concern in most circuits. There are some electronic components designed to have smaller breakdown voltages, so that a small change in applied voltage leads to a large change in the current.

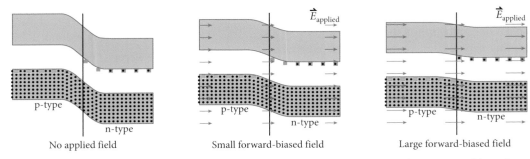

No applied field — Small forward-biased field — Large forward-biased field

Figure 11.18 Forward-biasing a p-n junction lowers the potential hill so more electrons are able to flow from the n-side to the p-side. For small applied fields, the stronger the field, the larger the current. Once the applied field is large enough, there is essentially no potential barrier to overcome and the p-n junction acts like a conductor. (Any applied field stronger than the one shown on the right would reverse the slope of the hill across the junction so the p-side was at lower potential than the n-side.) Notice that increasing the field also decreases the size of the depletion region. When the slope of the hill reverses, there is no depletion region left.

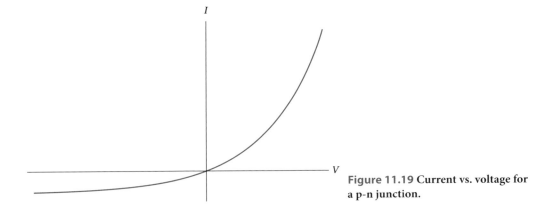

Figure 11.19 Current vs. voltage for a p-n junction.

We can explain the same phenomenon in a different way by focusing on the depletion region. Forward-biasing shrinks the depletion region, bringing charge carriers closer to the junction. With a relatively small amount of forward-biasing there is no more depletion region and the diode is a conductor. Reverse-biasing widens the depletion zone, quickly leading to a situation in which very few charge carriers of either type can reach the junction to cross it. We will come back to this perspective in Section 11.4, because it provides the clearest explanation of transistors.

11.3.3 Questions and Problems: Semiconductors and Diodes

Conceptual Questions and ConcepTests

1. For each term below, give a brief description of what it represents.

   (a) Intrinsic semiconductor

   (b) Conduction band

   (c) p-type semiconductor

   (d) p-n junction

   (e) Recombination current

   (f) Depletion region

   (g) Diode

2. When you dope silicon with phosphorus, each phosphorus atom contributes one excess electron

to the conduction band. But each phosphorus atom also has one more proton than a silicon atom, so the resulting n-type semiconductor is still electrically neutral. Write an equivalent description for silicon doped with aluminum.

3. Figure 11.16 shows a p-n junction in equilibrium. Which side has a higher electric potential, the p-side or the n-side? How do you know? *Hint*: Remember that electrons are negatively charged.

4. We speak of two different types of current flowing through a semiconductor: electrons drift in one direction, and holes drift in the opposite direction. (These two movements do not cancel each other out; both of them represent current flowing in the same direction!) But really, a "hole" isn't a thing that moves; when we say a hole moved to the right, we actually mean an electron that used to be on its right jumped into the hole. So what is the difference between these two forms of current?

5. At what energy is the Fermi energy in an n-type semiconductor? (Choose one and explain.)

   A. Near the top of the valence band

   B. Near the middle of the band gap

   C. Near the bottom of the conduction band

   D. None of the above

6. Figure 11.17 shows the recombination and thermal currents of electrons across a p-n junction. Draw similar pictures showing the recombination and thermal currents of holes flowing across.

Problems

7. The band gap of silicon is about 1.1 eV.

   (a) Use the Fermi–Dirac distribution to estimate the fraction of states at the bottom of the conduction band that are filled at room temperature. *Hint*: The Fermi energy for silicon lies in the middle of the band gap.

   (b) What fraction of the states at the top of the valence band of silicon would you expect to be unoccupied (aka holes)? Briefly explain how you know.

   (c) Now imagine doping the silicon, replacing one out of every 10,000 atoms with aluminum. This increases the number of available valence holes by what factor?

8. A "photodiode" is a diode in which light can cause electrons to jump the gap into the conduction band and flow across the junction. Photodiodes have many applications, from light sensors to solar cells.

   (a) Would this photo-current contribute to the thermal current or to the recombination current? Explain.

   (b) The band gap in germanium is about 0.67 eV. Estimate the wavelength of light to which a germanium photodiode would be sensitive. What range of the electromagnetic spectrum does this fall into?

   (c) Is the wavelength you calculated in Part (b) an upper limit or a lower limit on the wavelengths that would excite current in a germanium diode? Explain.

11.4 Transistors

In 1947 three researchers at Bell Labs, William Shockley, John Bardeen, and Walter Brattain, built the first working transistor. Their invention, for which they won the 1956 Nobel Prize in Physics, led to the computer revolution. Today many of us walk around with phones that have billions of transistors in them, enabling the computations that have reshaped modern society. All of these advances are possible because of the unique properties of semiconductors.

11.4.1   Explanation: Transistors

Figure 11.20 shows a device with three leads labeled A, B, and C. Current can flow along the wire between A and B, but *not* to or from C. That doesn't mean C is irrelevant, though. Small, controllable changes in the voltage at C can change the resistance between A and B. That's a transistor: a device that controls the current flowing between two points via a voltage applied at a third point.

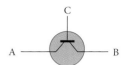

Figure 11.20 A transistor has three connections, here labeled A, B, and C. When a voltage is applied between leads A and B, current flows. Increasing or decreasing the voltage applied at C changes the resistance and thus the current between A and B.

You can see that a transistor is effectively a switch: you can open or close a circuit without physically disconnecting or connecting wires. And indeed, transistors are often used as switches because they require very little time or power to turn on and off. But they also have many more uses, as we'll see below.

Because they are cheap to make and use, because they switch states rapidly, and can be made microscopically small, transistors are at the core of all modern electronics. As mentioned above, a single smartphone has billions of transistors. It has been estimated that from their invention in 1947 until the early twenty-first century, over 10^{22} transistors have been made! (That's very difficult to put in context, but if it helps, the total number of grains of sand on Earth is probably a smaller number than that.)

First we're going to describe how to make a transistor, and then we'll talk about some of the ways transistors are used in circuits.

Most transistors are field effect transistors, or FETs. (The most common type is the metal oxide semiconductor field effect transistor, or MOSFET.) The basic architecture of a FET is shown in Figure 11.21. Note that the "gate" here corresponds to the label C in Figure 11.20; the "source" and "drain" are leads A and B.

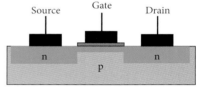

Figure 11.21 A field effect transistor. The current between the source and drain flows through the narrow channel below the gate. A small voltage applied at the gate can greatly increase or decrease this current. The source and drain are electrically connected to the n-type semiconductor (green), while the gate is separated from it by an insulating layer (gray).

The current flows between the source and the drain, each of which is connected to one side of an n-type semiconductor region. The middle of that region is a narrow channel resting on a p-type substrate. That substrate creates a depletion region: in Figure 11.21 that region is at the bottom of the green zone. The depletion region has no charge carriers, further narrowing the channel through which current can flow between the source and drain.

Active Reading Exercise: A Transistor

Suppose a positive voltage is applied to the gate in Figure 11.21, creating an electric field that points downward through the channel below it. Does that increase or decrease the current between the source and drain? Why?

The answer to that Active Reading Exercise is that the electric field *reduces* the flow, quite possibly almost cutting it off entirely. This is the core insight into transistor behavior, so it's worth some time to make sure you understand it.

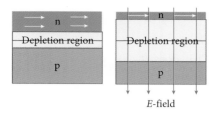

Figure 11.22 This is a zoom into the n-type channel in Figure 11.21. Current (white arrows) flows between the source off to the left and the drain off to the right, but it can't flow in the depletion region because there are no charge carriers there. When a positive voltage is applied at the gate (above the figure) the resulting E-field reverse-biases the p-n junction and widens the depletion region, shrinking the region where current can flow.

Figure 11.22 shows a blow-up of the narrow channel through which current (white arrows) flows from source to drain. Because the bottom of the channel is a p-n junction, it has a depletion region with no charge carriers. The lines in the middle of the depletion region indicate the boundary between the n-doped and p-doped materials.

The physical situation is similar to the diodes we discussed in Section 11.3, but note the differences between the two scenarios:

- In a diode, the current flows perpendicularly to the p-n junction. The conduction-band electrons in the n-type semiconductor are available to flow through the depletion region into the p-type.

- In an FET, the current flows parallel to the p-n junction. That means conduction-band electrons flow through the n-type region. The depletion region represents a dead zone.

So what happens when you reverse-bias the junction? The depletion region is widened. We discussed that fact in our discussion of diodes in Section 11.3, but let's briefly revisit the argument. Reverse-biasing pulls electrons in Figure 11.22 upward, but the p-region and the depletion region have no electrons that are available to flow (that is, electrons in the conduction band). So the electrons at the bottom of the n-region are the only ones that go anywhere: they go upward, widening the depletion band. The n-region, the region available for current flow, is therefore constricted. (You can go through this argument again with a different picture in Question 1.) The net result is that the resistance between the source and drain increases.

As promised, the FET is a device in a which a very small change to the gate voltage can lead to a large change in the current between the source and drain. Many modern transistors are designed so that at one easily reachable gate voltage the channel is effectively wide enough to act as an open wire, and at another easily reachable gate voltage the channel is effectively entirely closed, allowing no current to flow. By setting the gate voltage to that high or low voltage you can turn the current on or off, providing a binary switch.

We've discussed an n-type channel on a p-type substrate, and concluded that a high gate voltage reduces current between the source and drain. You can instead put a p-type channel on an n-type substrate and reverse the effect of the gate voltage. And there are other, very different ways to make transistors. The defining characteristic of a transistor is that a small change at one terminal dramatically changes the resistance between the other two.

Logic Gates

The core of a computer is built out of logic gates, circuit elements that produce "on" or "off" results depending on the values of certain inputs.[7] For example, an AND gate has two input

---

7 The word "gate" in "logic gate" does not refer to the gate region of a field effect transistor, a fact made all the more confusing by the fact that logic gates are usually made from transistors. You have to pay attention to which use of the word "gate" is meant in any given context.

wires and one output wire. The output wire is at high voltage ("on") if and only if both the inputs are at high voltage, while an OR gate has an output that is on if either or both of the inputs are on. See Figure 11.23.

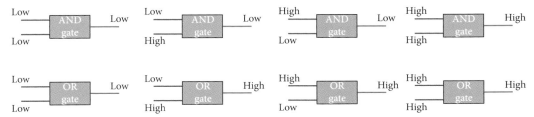

Figure 11.23 The top row shows the four possible states of an AND gate. The inputs on the left (in blue) can each be at high or low voltage. The output on the right (red) is at high voltage if the first input AND the second input are both high. The bottom row shows an OR gate, whose output is high if the first input OR the second input is high.

Other common logic gates include XOR ("exclusive OR": output is high if exactly one of the inputs is high but not if they both are), NAND ("not AND": output is high *unless* both inputs are high), and NOT (only one input, and the output is the opposite of the input).

All of these logic gates can be constructed using transistors. In this discussion we will adopt the convention that a high input voltage to a transistor allows current to flow while a low input voltage blocks current.

Active Reading Exercise: A Logic Gate

The figure below shows a circuit with two inputs and one output. Each input is connected to the gate of one transistor. Using what we've said about transistors, is this circuit one of the logic gates listed above (AND, OR, XOR, or NAND), or something different? To answer, work out what will happen in the circuit for each possible combination of high and low voltages for the inputs. Remember that a resistor with current flowing through it has a voltage difference across it, while a resistor with no current has essentially the same voltage on both sides.

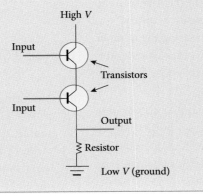

You can find our answer to that exercise at www.cambridge.org/felder-modernphysics/activereadingsolutions

11.4.2 Questions and Problems: Transistors

Conceptual Questions and ConcepTests

1. Figure 11.18 on p. 521 shows the effect of forward-biasing on a p-n junction.

   (a) Draw similar illustrations to show the effect of reverse-biasing, paying particular attention to the change in the potential drop between the p-type and n-type sides. (Your answer here will be two drawings: "no applied field" and "strong reverse-biased field.")

   (b) Explain, based on your drawing, why reverse-biasing widens the depletion region.

   (c) Suppose an FET has a potential difference between the source and drain, driving current (See Figure 11.21). Explain why applying a voltage at the gate that reverse-biases the p-n junction decreases the source–drain current.

2. We said that in a field effect transistor a current between the source and drain has to flow through the narrow channel connecting them. Why can't current flow between them through the p-type substrate below?

3. In the Active Reading Exercise on p. 523 and the discussion that followed, we said that reverse-biasing the p-n junction in a field effect transistor reduces the number of available charge carriers in the channel and thus reduces the current flow. Your friend Simplicio proposes this counterargument: "When the p-n junction is in equilibrium with no applied voltage at the gate, electrons will have diffused across the gap from the n-side to the p-side. When you apply that positive voltage you should pull them back, thus increasing the number of charge carriers in the narrow n-type channel and *increasing* the current flow." Explain to Simplicio the flaw in his logic.

4. Figure 11.24 shows a circuit similar to the one in the Active Reading Exercise on p. 525. Figure out how this circuit will behave for each possible combination of inputs, and determine which type of logic gate this circuit represents.

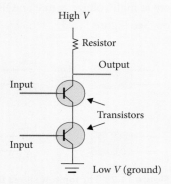

Figure 11.24

5. Figure 11.25 shows a circuit similar to the one in the Active Reading Exercise on p. 525. Explain why this circuit cannot be used as any possible logic gate.

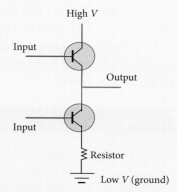

Figure 11.25

Problems

6. Draw a circuit diagram for each of the following logic gates. You may want to use the Active Reading Exercise on p. 525 as a guide.

   (a) NOT

   (b) AND

   (c) OR

7. In formal logic we distinguish between the "logical OR" and the "exclusive OR." The former, designated as OR, is satisfied if at least one of the propositions is true. (You will get wet if you jump into either a bath or a shower.) The latter, XOR, is satisfied if exactly one of the propositions is true, but not both. (You

did your civic duty if you voted for either the incumbent or the challenger.) Draw a circuit diagram showing how to construct an XOR gate using other gates (NOT, AND, OR, and/or NAND).

8. A NAND gate has a high voltage output for any combination of inputs *except* both high. All of the other logic gates can be built from NAND gates. As an example, draw a circuit for an OR gate built out of NAND gates. Your diagram shouldn't have a high voltage source or ground or any transistors or resistors (although all of those things are found inside the NAND gates). You should only show NAND gates and wires (including two input wires and one output wire).

9. By combining transistors you can make gates for the logical operators AND, OR, XOR, and NOT (among others). We showed one example in an Active Reading Exercise in the text, and you may have gone through others in some of the Conceptual Questions above. Now you will take the next step: design a circuit to add two two-digit binary numbers. You will have four inputs, two for each input number. For example if the first two are high (1) and low (0) that would represent the number 2 in binary, and if the next two are low (0) and high (1) that would represent 1. So the sum would be 3. Since the highest possible sum you could produce is 6, you will need three output wires to indicate the output as a binary number. Your circuit would ultimately be built from transistors, but you don't need to show them; instead you can show the logic gates that are built from those transistors.

11.5 Why Do Crystals Have a Band Structure?

The ideal way to explain band structure would be to mathematically analyze a network of electrons in the potential field created by an effectively infinite, mostly periodic lattice of positively charged ions, solve Schrödinger's equation in this field, and show that the band structure emerges from the solutions.

You probably won't be too surprised to hear that we're not going to do that, given that we can't actually solve Schrödinger's equation for an atom with two electrons. But we can lay out an argument – actually two very different arguments – for why such a solution *would* lead to energy bands. More rigorous (although still not complete) calculations, numerical solutions, and experimental data all confirm that energy levels in solids do occur in bands.

11.5.1 Discovery Exercise: Why Do Crystals Have a Band Structure?

The first few questions in this Discovery Exercise review bonding and antibonding states. If you have trouble with these questions, review that material in Section 9.2.

Figure 11.26 shows the four lowest energies for an electron in the vicinity of a single proton (aka a hydrogen atom). The electron has ground state energy $-13.6\,\text{eV}$ (the bottom line in the drawing). Because of spin, there are two states available at that energy.

Now consider an electron in the vicinity of two protons.

1. In the limit where those two protons are very far from each other, what is the ground state energy for the electron and how many states are available at that energy?

2. If the two protons are roughly 10^{-10} m apart (a typical molecular separation), how many states are available to the electron at roughly -13.6 eV? Is the energy of those states . . .

   A. higher than -13.6 eV?

   B. lower than -13.6 eV?

   C. higher for some of the states and lower for others?

3. Sketch an energy level diagram similar to Figure 11.26 for an electron orbiting around two protons 10^{-10} m apart.

   See Check Yourself #19 at www.cambridge.org/felder-modernphysics/checkyourself

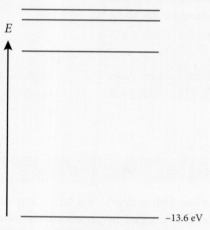

Figure 11.26 The first few energy levels of hydrogen.

These last questions extend beyond Chapter 9.

4. Now imagine an electron in the vicinity of three protons, all far from each other. How many ground states are available to that electron?

5. If you bring three hydrogen atoms very close together (don't worry for the moment about how you do this), what will the energy level diagram for an orbiting electron look like? It's OK if you get this one wrong, but make your best guess and write a sentence or two explaining your reasoning.

11.5.2   Explanation: Why Do Crystals Have a Band Structure?

We have seen that electrical conductivity depends on the "band structure" of electron levels in a crystal. Why do electron levels come in bands, and why do those bands have the particular energies and widths that they do?

In this section we will answer that question in two different ways. First we'll consider the energy eigenvalues of individual atoms, and see how those eigenvalues change as you bring the atoms close together. Then we'll directly consider eigenstates in a periodic potential.

Bringing Atoms Together

In Section 9.2 we saw what happens to the available electron energy states when you bring two nuclei close together. When the nuclei are far apart there are two ground states, one around each nucleus. (Throughout this discussion we're going to ignore spin; it has no impact on the arguments we make here, except that each state has room for two electrons rather than one.) As our two nuclei approach each other, those two states with the same energy change into two states with slightly different energies: a bonding state with slightly lower energy, and an antibonding state with slightly higher energy. We say the energy level "splits" as the nuclei get closer.

Now imagine *three* nuclei. When they are far apart, these nuclei offer three different ground states – different eigenfunctions with the same eigenvalue – that an electron might occupy. As the nuclei approach each other the wavefunctions become more complicated, but those three states must still exist in some form. Without going through all the math, it is reasonable to suppose that their energy levels will be close to their original energy, but not identical to each other: the ground state splits into three closely spaced energy levels. In a similar way, each excited state splits into three states with energies close to, but not identical to, what they were in the individual atoms.

Now – since this is a chapter about solids – imagine bringing together 10^{23} nuclei, give or take a few. Each energy level splits into so many closely spaced energies that it's useful to think of them as a continuous band of energy (see Figure 11.27).

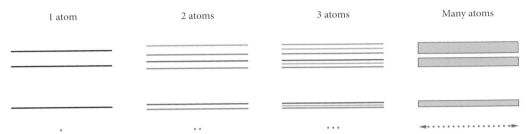

Figure 11.27 Available electron states around a single nucleus and around multiple nuclei. In the limit of many nearby nuclei, the states widen into almost continuous bands.

The number of states in any particular band is proportional to the number of atoms. But the width of the band – that is, the energy difference between its top and its bottom – is just a function of the lattice spacing, and doesn't change as you add more atoms.

In Figure 11.27 each energy eigenstate of the individual atoms turns into a single band in the crystal, but the reality can be more complicated. As the nuclei get closer together the bands widen, so for tightly packed lattices different atomic subshells can overlap and merge into a single band. Conversely, in a single atom the spherically symmetric potential means the energy doesn't depend on m_l, but that symmetry can be broken in a crystal and a single subshell can split into two or more separate bands. So the original energy levels of the atom do *become* the bands, but there isn't a one-to-one correspondence between those levels and the resulting bands.

Because of such issues, predicting the band structure of a solid based on the properties of its atoms is a complicated matter. Experimentally, however, the energies and widths of the bands can be observed through spectral lines.

Schrödinger's Equation for a Periodic Lattice

We're going to back up now, starting over with the same situation (an electron in a crystal) and arriving at the same result (bands of allowable energy levels). But instead of approaching the problem indirectly, by bringing distant atoms together, we will apply Schrödinger's equation to our lattice of nuclei.

We begin of course with potential energy. The left side of Figure 11.28 shows the potential energy function felt by an electron in a lattice of atoms. The right side shows a simpler function that approximates this potential, the "Kronig–Penney" model. It's possible to solve exactly for the eigenstates of the Kronig–Penney potential, and you can draft the beginning of that process in Problem 10. Here we're going to talk qualitatively about the form of those eigenstates to see why their eigenvalues come in bands.

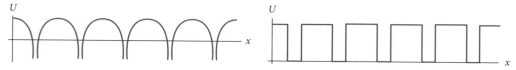

Figure 11.28 The plot on the left shows the potential energy function felt by an electron in a lattice. The plot on the right shows the Kronig–Penney model, which can be fully solved to find the allowed energies.

Because a macroscopic solid is so much bigger than the lattice spacing between its atoms, we can think of that periodic lattice as going on infinitely and ignore the boundary conditions at the edges of the crystal. (We made the same approximation when we calculated Madelung constants in Section 11.1. Did you notice?) In between the potential barriers, the electron is a free particle. As you may recall from Chapter 6:

- The energy eigenstates of a free electron are of the form e^{ikx}, each one representing a traveling wave moving to the right if $k > 0$ and to the left if $k < 0$.
- The energy of that wave is $E = \hbar^2 k^2/(2m)$.

We also saw that a potential step or dip splits the wave into a transmitted piece and a reflected piece. It's tempting, and sometimes useful, to think of this in the language of time dependence: "The wave moves, then hits a barrier, then splits." Recall, though, that each eigenstate includes all three parts (incident, transmitted, and reflected waves) at all places and all times, each part moving at speed ω/k without changing its shape or amplitude. The story that unfolds over time – a particle hits a barrier as an incident wave and then heads out as a transmitted wave and a reflected wave – is the story of a wave packet built from the interference patterns of all those eigenstates.

Chapter 6 discussed all that in scenarios with only one or two potential barriers. Now imagine infinitely many potential barriers, spaced periodically. That means there is no distinction between the incident wave and the transmitted wave; there is just a wave moving in one direction. Meanwhile, moving in the opposite direction, the reflected wave is the superposition of all the waves reflected from all the potential barriers. In general those reflected waves have different phases and interfere with each other destructively, so the electron behaves essentially as a free particle with $\psi = e^{ikx}$.

But for eigenstates with certain specific frequencies, the reflected waves interfere constructively with one another. That occurs when $k = n\pi/R$ for integer n (Problem 8).[8] The nett reflected wave and the incident wave combine to form a standing wave.

That reflected wave alters the energy of those specific eigenstates. To see why, remember that the total energy associated with any eigenstate is a combination of kinetic and potential energy.

---

8 The same formula is used for a very similar reason in "Bragg scattering," when a beam is fired perpendicularly to the surface of a crystal; see Problem 16 on p. 126.

- The kinetic energy associated with a free-particle eigenstate is $E = \hbar^2 k^2/(2m)$. (This is the eigenvalue of Schrödinger's equation with $U = 0$: all kinetic energy, no potential energy.)
- The potential energy in the Kronig–Penney potential field is 0 in between the barriers, but has a higher value at the barriers. Therefore, the potential energy of any given wavefunction is proportional to the probability of finding the particle inside a barrier. (If you have a 4/5 probability of being where $U = 0$, and a 1/5 probability of being where $U = 100$, then your expected potential energy is 20. Make sure that makes sense!)

For the k-values that have few or no reflected waves, the wavefunction $\psi = e^{ikx}$ gives an equal probability of finding the particle at any position, in or out of the barriers. So the barriers do contribute to the total energy of such wavefunctions, but they contribute the same amount for all k-values.

But the k-values that lead to standing waves correspond to perfect resonance: an integer number of waves fit between the potential barriers. If those potential barriers are at the nodes of a wave, then the probability of finding the particle at the high-energy barrier is relatively low, so the total potential energy is low. Conversely, if the potential barriers are at antinodes, then the total potential energy is high. These two situations are represented in Figure 11.29.

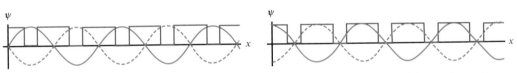

Figure 11.29 Two possible standing waves on a periodic lattice. The one on the left has nodes at the points of high potential energy, and thus a lower average energy than the one on the right, which has antinodes there.

It turns out that for k-values close to but slightly less than $n\pi/R$, you get a standing wave whose nodes are at the potential barriers. That pushes their total energy down. For k-values slightly higher than $n\pi/R$, the potential barriers are at the antinodes, which pushes the energy up. (See Figure 11.30.) We are not in any way justifying this result, which comes from calculations of the eigenstates in a periodic potential, but we are going to explain how this result gives rise to band structure.

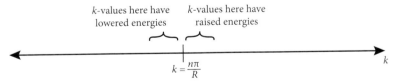

Figure 11.30 k-values above and below resonance are affected differently by the periodic potential.

The resulting plot of E vs. k is shown in Figure 11.31.

- When k is far from the resonance values, the plot lies on the same parabola as it did for free particles: $E = \hbar^2 k^2/(2m)$, plus a constant contribution from the potential energy.
- When k is just under $n\pi/R$, the standing wave lowers the potential energy.
- When k is just above $n\pi/R$, the standing wave raises the potential energy.

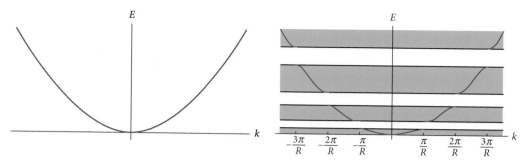

Figure 11.31 Energy as a function of wave number for a free particle (left) and for a particle in a periodic lattice (right). For a free particle, E depends quadratically on k. The plot for a periodic lattice looks like the free-particle plot far from the frequencies $n\pi/R$, but the energy jumps discontinuously at those values. Particles can still have any value of k, but energy can only be in the allowed bands (shaded).

The result is a discontinuous jump in energy at $k = n\pi/R$. Study the right side of Figure 11.31 to see that we have arrived through a new route at a familiar endpoint: bands of allowed energy, alternating with bands of forbidden energy.

11.5.3 Questions and Problems: Why Do Crystals Have a Band Structure?

Conceptual Questions and ConcepTests

1. Give two reasons why the allowed energy bands in a crystal don't correspond one-to-one with the energy levels of the individual atoms.

2. When we considered a free particle hitting an energy barrier in Chapter 6, we broke the energy eigenstates into three parts: an incident wave, a transmitted wave, and a reflected wave. When we considered the energy eigenstates of a particle in a periodic lattice we only talked about two parts. Why the difference?

3. If two crystals are made of different types of atoms, but both have the same lattice spacing, will they have the same band structure? Why or why not?

4. Copy the Kronig–Penney potential energy function from Figure 11.28 and mark on it the locations of the lattice ions. How do you know?

5. How can two different wavefunctions, both defined over all of space and over the same potential energy function, have different expected values for potential energy?

6. In the text we made arguments for why we would expect allowed and forbidden bands of energy for an electron in a periodic potential. In which situations would those arguments hold? (Choose one and explain.)

   A. The total energy is lower than the top of the potential barrier (so classically the particle would be trapped in a single well).

   B. The total energy is higher than the top of the potential barrier (so classically the particle would be able to propagate freely).

   C. It applies in both of these scenarios.

For Deeper Discussion

7. The electrons we are calling "free" are actually bound because they don't have enough energy to escape the crystal. That means their eigenstates should be identified with a discrete quantum number. So why do we say that their wave number k can take on any real value?

Problems

8. A traveling wave $\psi = e^{i(kx-\omega t)}$ is moving to the right. At $x = 0$, part of the wave reflects back to the left. We'll call that part of the wave W_1. Another part of the wave keeps going to $x = R$, where it is

reflected back. We'll call that wave W_2. (Presumably part of that wave is also transmitted at $x = R$, but it won't matter for what we're doing here.) The waves W_1 and W_2 have the same wavelength.

(a) At a point $x < 0$, how much farther has W_2 traveled than W_1?

(b) What is the largest wavelength the waves could have that would cause them to interfere constructively at a point $x < 0$?

(c) Why didn't the answer to Part (b) depend on which negative x-value you look at?

(d) Write a formula for *all* the possible wavelengths that would cause the two waves to interfere constructively at $x < 0$. Your answer should have an arbitrary integer n in it.

(e) Rewrite your formula for wavelength as a formula for k. *Hint*: The relationship between wavelength and wave number is in Appendix F. You should reproduce the formula in the text for wave numbers that produce large-amplitude reflected waves.

9. Consider an electron in a periodic lattice with lattice spacing R. The electron's wavefunction has wave number k and momentum $\hbar k$. In this problem δ represents a small change in k, with $\delta \ll \pi/R$.

(a) Is the *kinetic* energy associated with $k = \pi/R - \delta$ less than, the same as, or greater than, the kinetic energy associated with $k = \pi/R + \delta$? Briefly explain how you know.

(b) Is the *potential* energy associated with $k = \pi/R - \delta$ less than, the same as, or greater than, the potential energy associated with $k = \pi/R + \delta$? Briefly explain how you know. (This will not involve any math, just an argument based on the text.)

(c) Does your result point toward a band of *allowable* energy values clustered around $\hbar^2\pi^2/(2mR^2)$, or a band of *forbidden* energy levels clustered around that value?

10. **The Kronig–Penney Potential**

In this problem we're going to start the solution for the energy eigenstates of a particle in the Kronig–Penney potential, $U(x) = 0$ for $0 < x < R - a$, $U(x) = U_0$ for $R - a \le x \le R$, repeating periodically outside that range. Assume $E > U_0$. (You will be able to simplify some of your formulas if you define $\alpha = \sqrt{2mE}/\hbar$ and $\beta = \sqrt{2m(E - U_0)}/\hbar$.)

(a) Write the form for an energy eigenstate in the region $0 < x < R - a$.

(b) Write the form for an energy eigenstate in the region $R - a \le x \le R$.

(c) Felix Bloch proved that the eigenstates for a particle in a periodic potential with period R must be of the form $\psi(x) = u(x)e^{ikx}$, where $u(x)$ also repeats every R. Write $u(x)$ in each of the two regions by multiplying $\psi(x)$ by e^{-ikx}, and simplify as much as possible. (Treat k as an unknown parameter.)

(d) Your wavefunctions should have four arbitrary constants between them. Using the fact that $u(x)$ has period R and the requirement that wavefunctions be continuous and differentiable, write four equations relating your four arbitrary constants. You do not need to solve the four equations.

In principle you could solve the equations that you just wrote to find three of the four arbitrary constants, and a condition relating k and E. That condition would only have real solutions for certain values of E, which determine the allowed energies.

11.6 Magnetic Materials

In an introductory class you may have learned that an electric current creates a magnetic field (e.g. $B = \mu_0 I/(2\pi r)$ around a long straight wire), and that a moving charge outside that current feels a force from that magnetic field ($\vec{F} = q\vec{v} \times \vec{B}$). At least, we hope you learned those things, because we're not going to explain or review them here.

What we are discussing here is matter. Matter is made up of atoms, and those atoms can produce magnetic fields. Our concern is how and when those magnetic fields align to create large-scale magnetic phenomena.

By the way, although our focus in this chapter is on solids, the material in this section also applies to the magnetic properties of liquids and gases.

11.6.1   Explanation: Magnetic Materials

Why aren't all objects magnets?

That question may sound silly, but consider: an electron is a charged particle with angular momentum (spin and/or orbital), so every electron is a magnet. The reason why most objects don't generate macroscopic magnetic fields is that the magnetic fields of the electrons point randomly in all directions, canceling each other out.

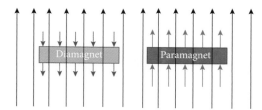

In the presence of an external magnetic field, however, the electrons in an object can align and cause the object to emit a measurable field of its own. If the induced field opposes the external field we call that object a "diamagnet." If the induced field reinforces the external field we call the object a "paramagnet." See Figure 11.32.

Figure 11.32 In an external magnetic field (black), a diamagnet produces a small field pointing the opposite way (blue), while a paramagnet produces a small field pointing in the direction of the external field (red). The field produced by a paramagnet is typically larger than that produced by a diamagnet.

This section will discuss these two effects, and then the rarer but more familiar case of "ferromagnetism," in which a solid produces a magnetic field in the absence of an applied external field. First we'll give a qualitative overview of why these effects happen, and then a slightly more quantitative discussion of the strengths of these effects and how they depend on temperature. These discussions are intended to give a general idea of the mechanisms involved in these phenomena, but full quantum mechanical treatments are beyond the scope of this book.

Atomic Magnetism: A Brief Reminder

An electron in an atom produces a magnetic field for two different reasons: orbit and spin. You can picture the classical images "an electron circling a nucleus" and "an electron rotating around its own axis," but the quantum behavior is significantly different in some ways from what classical mechanics would predict. The magnetic fields are tied to the quantum numbers m_l and m_s respectively, they are quantized along whatever axis you choose to measure them on, and a precise measurement of one component precludes certainty of the others. (Also, the magnetic field associated with spin is twice as large as you would predict classically for a spinning sphere of charge; see Chapter 7.)

For our purposes here, we don't need to distinguish between the orbital and spin contributions. What we need you to know is this: each electron in a solid produces a magnetic field, but in most solids measurements in any direction would show equal amounts of up and down fields, so the net field would be zero.

If you wanted a solid to produce a net magnetic field upward, you could bias the result in either of the two following ways:

- Change the eigenfunctions so that the states with "up" fields point up a little more and the "down" states point down a little less.
- Change the probabilities so that more of the atoms point up than down.

Those two changes correspond to diamagnetism and paramagnetism, respectively.

Diamagnetism

If you introduce a magnetic field to a wire loop, a current will be induced in the loop. Lenz's law says that this current will flow in the direction that generates a magnetic field *opposite* to the newly applied field. If the loop already has a current flowing, that current will increase or decrease to oppose the newly applied field.

In an analogous way, applying a magnetic field to a solid shifts the dipole moments of its electrons so as to oppose the field. The result is a (typically very slight) reduction of the field in and around the solid.

Some sources discuss diamagnetism as if it were a straightforward application of Lenz's law. While the underlying mechanism is related, there are important differences. The applied field isn't inducing classical currents, but instead is slightly altering the energy eigenstates of the electrons. The atomic magnetic fields that support the external field are slightly weakened, the opposite fields slightly strengthened.

Diamagnetism differs from classical induction in another crucial way: the macroscopic effects of Lenz's law occur only as the field is *changing*. When you stop increasing the applied field and leave it at a steady value, the induced current almost instantly dissipates. (The induced field described by Lenz's law is proportional to the time derivative of magnetic flux, not to the flux itself.) By contrast, the diamagnetic field persists as long as the external field is applied. When the external field is removed, the electrons revert to their usual state and the induced field vanishes.

Paramagnetism

Imagine an atom such as lithium or aluminum that has a completely filled subshell plus one extra electron in a subshell all by itself. The electrons in the full subshells have equal numbers of up and down spins, and equal amounts of up and down orbital angular momentum. In short, their magnetic fields completely cancel each other out. But the spin of that one extra electron gives each atom a permanent dipole moment. (This may remind you of what Stern and Gerlach unwittingly discovered about silver atoms!)

Now imagine a block of material made of such atoms. Each atom has a dipole moment, but under normal circumstances they all point in random directions, so the block produces no net magnetic field. When you apply an external field, however, that field exerts a torque on each atom. (Recall that a uniform magnetic field exerts a torque on a magnetic dipole.) The atoms tend to line up with the external field, so their internally generated fields no longer add up to zero. That's paramagnetism.

In terms of the internal mechanisms, as we have said before, diamagnetism and paramagnetism change the atoms in very different ways. Diamagnetism changes the magnetic field associated with the "up" and "down" wavefunctions. Paramagnetism changes their probabilities: an internal magnetic field aligned with the external field has lower energy, and is therefore more probable. (Recall from Chapter 10 that a lower energy state of a system is more likely than a higher energy state, as is described by the Boltzmann distribution.)

In terms of its effect, paramagnetism differs from diamagnetism in three notable ways:

1. The diamagnetic field opposes the external field; the paramagnetic field reinforces it.
2. Diamagnetism affects all atoms; paramagnetism only affects atoms that have intrinsic magnetic moments of their own.
3. Finally, the paramagnetic field is typically a much *larger* effect than the diamagnetic.

The net result of all this is that if you apply an external magnetic field to an atom such as helium, you're going to measure a small opposing (diamagnetic) field. If you apply an external magnetic field to an atom such as lithium, you will measure a reinforcing (paramagnetic) field, only slightly weakened by the diamagnetic response that is *also* going on. We characterize lithium as a paramagnet.

You will often see the guideline that atoms with all their electrons paired are diamagnetic, while atoms with any unpaired ones are paramagnetic, because even a single electron contributing to paramagnetism generally overwhelms the effects of diamagnetism. That's all correct. For example, you can use that to explain why all the noble gases are diamagnetic.

But if you use that guideline to predict the properties of elements based on the periodic table, you'll be wrong as often as you're right. The problem is that, aside from the noble gases, atoms don't generally float about individually. A single nitrogen atom has three unpaired electrons, but in an N_2 molecule, which is how nitrogen actually occurs around us, the electrons are all paired. So nitrogen is diamagnetic. Oxygen molecules do have unpaired electrons, so oxygen is paramagnetic. And the electron structure in solids gets even more complicated and hard to predict: lithium is paramagnetic, but boron with its single electron in the 2p subshell is nonetheless diamagnetic.

The bottom line is that the magnetic properties of different materials are determined experimentally, and predicting them theoretically is a messy business.

Ferromagnetism

When most of us think about magnetism we don't picture objects that show slight magnetic properties in certain laboratory experiments. We picture *magnets*, picking up paperclips and sticking to refrigerators and pointing North in the wilderness. Physicists call those "ferromagnets," objects that produce their own magnetic fields in the absence of external applied fields.

A ferromagnet is like a paramagnet in that each atom has a permanent dipole moment. But in a ferromagnet these atoms are close enough to each other that their dipole moments interact, and tend to line up with each other.

We need to head off one common student misconception about ferromagnetism: the dipoles in a ferromagnet do *not* interact by a classical Lorentz force, in which the magnetic field of one atom tends to rotate neighboring atoms in that same direction. That force is orders of magnitude too weak to have any noticeable effect (Problem 9). The tendency of atoms in a ferromagnet to align with their neighbors comes from a purely quantum mechanical effect called an "exchange interaction." We'll outline that mechanism at the very end of this section, but focus here on the result: for some closely spaced atoms, it's a lower energy state to have electrons' dipole moments aligned than to have them anti-aligned.

Taking that fact as a given, let's pursue its consequences. If you zoom in on a tiny region inside a block of iron, you will generally find each atom lined up with its neighbors. Nearby you'll find another group of atoms, also aligned with each other, but in a different direction from the first group. These "domains" are typically a fraction of a millimeter across, and each one is oriented randomly, independent of the others (Figure 11.33). The result is that most blocks of iron, like most other materials, produce no net magnetic field.

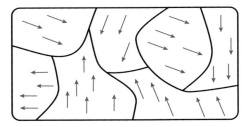

Figure 11.33 Magnetic domains in a block of iron.

If you put your block of iron in an external magnetic field, the domains start lining up with that field. This happens in two ways: some entire domains rotate to line up with the field, and some individual atoms at the boundary of two domains rotate so that one domain grows at the expense of its neighbor. The result is that the iron now produces its own magnetic field, aligned with the external field.

The effect of ferromagnetism as we've described it so far is similar to paramagnetism, although the induced field is typically much stronger. The qualitative difference comes when you remove the external field. In a paramagnet the atoms quickly reorient in essentially random directions. But the ferromagnet's atoms are held in their orientations by their interactions with their neighbors, so the magnetization of the ferromagnet persists. You have just created a magnet.

This is an easy experiment to do. Take a strong magnet and use it to pick up a paper clip. Then take the magnet away and you'll find that the first paper clip can now pick up a second paper clip. You have magnetized the metal by putting it in the field of the first magnet.

Numbers and Formulas

We have said in general terms that ferromagnetism is stronger than paramagnetism, which is stronger than diamagnetism. How do you quantify the strength of these effects?

A small loop of area A with current I has a "magnetic dipole moment"[9] with magnitude $|\mu| = IA$, pointing in a direction given by the right-hand rule. This definition of μ means that the energy of a current loop in an external magnetic field is $E = -\vec{\mu} \cdot \vec{B}$. The dipole moment of a solid is the sum of its atomic dipole moments, and we define the "magnetization" M of an object as its magnetic dipole moment per unit volume.

Roughly speaking, that magnetization M is a measure of the internal magnetic field produced by the material: $B_{\text{induced}} \sim \mu_0 M$. See Figure 11.34. (The exact magnetic field depends on the geometry of the object. For example,

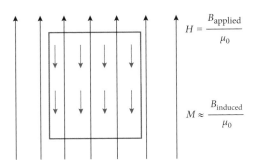

$$H = \frac{B_{\text{applied}}}{\mu_0}$$

$$M \approx \frac{B_{\text{induced}}}{\mu_0}$$

Figure 11.34 When an external magnetic field is applied around an object, we define H as the applied field divided by μ_0. The field magnetizes the object, which causes it to produce its own internal field, and the magnetization M is roughly equal to this induced field divided by μ_0. In this image the induced field opposes the applied field, indicating that the object is a diamagnet.

---

9 Don't confuse the magnetic moment μ with the constant μ_0.

the internal magnetic field is $\mu_0 M$ inside a uniform cylinder and $(2/3)\mu_0 M$ inside a uniform sphere.)

For many diamagnetic and paramagnetic materials, the magnetization is proportional to the external applied field. We define H as the applied field divided by μ_0, and define the "magnetic susceptibility" χ of a material by the relation $M = \chi H$.

For most paramagnetic and diamagnetic materials, $|\chi| \ll 1$. That means the applied field $(H\mu_0)$ is roughly the same as the total field (B), so $M \approx \chi B/\mu_0$.

> ## Active Reading Exercise: Magnetic Susceptibility
>
> 1. What is the sign of χ for a diamagnet?
> 2. What is the sign of χ for a paramagnet?
> 3. Why can't we meaningfully ask the same question about a ferromagnet?

1. For a diamagnet the induced field opposes the applied field, so $\chi < 0$. Typical diamagnetic susceptibilities are of the order -10^{-5}.

2. For a paramagnet the induced field is aligned with the applied field, so $\chi > 0$. Typical paramagnetic susceptibilities are of the order 10^{-4}.

3. For a ferromagnet the magnetization isn't a function of the applied field; it depends on the entire history of fields applied to the object. So χ is not defined for a ferromagnet.

In a paramagnet, the lowest-energy state is for all the atoms to be aligned with the field. But thermal fluctuations prevent them all from aligning perfectly. As a result, χ decreases with increasing temperature. "Curie's law" says that $\chi \propto 1/T$ (Problem 10). This works well until you get to very low temperatures where the paramagnet starts to saturate, meaning nearly all the dipoles align with the external field.

At very high temperatures the dipoles in a ferromagnet stop aligning with their neighbors. This occurs at the "Curie temperature" for a substance, which is roughly when $k_B T$ becomes comparable to the interaction energy between neighboring dipoles. Above the Curie temperature ferromagnets become paramagnets, but often with very high susceptibilities. (The susceptibility can be as high as 10^5 for some materials.)

Stepping Back: Different Types of Magnetism

We've described the three principal types of magnetic materials. In order from the typically weakest to strongest effects they are:

- Diamagnets. In a diamagnet, an external applied field alters the magnetic dipole moments of the electrons, causing them to produce fields that oppose the applied field. All materials experience diamagnetism.

- Paramagnets. When a material has unpaired electrons, the atoms or molecules have permanent dipole moments. An external applied field will orient those dipoles to align with the field, reinforcing the applied field and (typically) overwhelming the effects of diamagnetism.

- Ferromagnets. In a ferromagnet each atom is a permanent dipole, and the atoms are close enough together that they tend to align with their neighbors. A ferromagnet will tend to produce a field that reinforces an external field, and that magnetization will persist when the external field is removed.

This is not an exhaustive list of all possible types of magnetism. In an "antiferromagnet" the atoms are permanent dipoles that are close enough to interact, but the details of that interaction tend to make each atom anti-align with its neighbors. A "ferrimagnet" is like an antiferromagnet – neighboring atoms tend to oppose each other – but the material is made of different types of atoms with different dipole moments. The strong and weak dipoles pointing in alternating directions produce a net field. And many solids are made of combinations of materials with different magnetic properties.

A Final, Brief Word about the Exchange Interaction

The atoms of a diamagnet or paramagnet are influenced by an external magnetic field, but not by each other; each one reacts to the external field independently. In a ferromagnet, by contrast, the atoms exert a force on each other that tends to align them. So if an external field biases them toward one direction, their mutual interaction sustains that bias after the external field is removed.

A classical understanding of magnets tells us that nearby magnets can influence each other to line up. That sounds like a perfect explanation of ferromagnetism until you do the math. Then you discover that the magnetic force of each atom on its neighbors, classically analyzed, is far too weak to cause the effect we see. The effect is actually caused by an "exchange interaction."

All you need to know, in order to follow this section, is the *effect* of the exchange interaction; you don't need to understand its *cause*. But below we offer a brief gloss on the cause, in case you're curious. If nothing else, it reminds us that almost every property of a solid – electrical conductivity, transparency, magnetism, and so on – finds its roots in quantum mechanical behavior that cannot be explained classically.

Section 8.1 Problem 6 discussed the Pauli exclusion principle. That principle *leads to* the conclusion that two electrons in the same atom cannot occupy the same quantum mechanical state, but it begins with a more fundamental rule: the wavefunction $\psi(x_1, x_2)$ describing two electrons must be "antisymmetric." That means that if you swap the individual states of the two electrons, the overall wavefunction is multiplied by -1.

Remember that the overall state of an electron is described by a spatial wavefunction and a spin. What does it mean for the overall two-electron state to be antisymmetric? If the two electrons have the same spin state, it means the spatial wavefunction is antisymmetric; if the spin states are opposite, the spatial wavefunction must be symmetric.

So what? Well, two electrons with an antisymmetric spatial function are on average farther apart from each other than if they have a symmetric wavefunction. Being farther apart reduces their Coulomb energy, so the state with an antisymmetric spatial wavefunction has a lower energy than the opposite state. Hence a tendency toward parallel spins, aka ferromagnetism.

11.6.2 Questions and Problems: Magnetic Materials

Conceptual Questions and ConcepTests

1. For each term below, give a brief description of what it represents and the units you would use for it.

    (a) Magnetization

    (b) Magnetic susceptibility

    (c) Curie temperature

2. Define "diamagnet," "paramagnet," and "ferromagnet" in your own words.

3. If you put some solids near a permanent magnet, which will be attracted to the magnet? (Choose one and explain.)

    A. A diamagnet

    B. A paramagnet

    C. Both

    D. Neither

4. Are the noble gases diamagnetic or paramagnetic? Explain how you know. ("I looked it up" is not an acceptable explanation.)

5. Inside a block of iron, where is the energy density typically higher: within a domain, or on the boundary between domains? Explain.

6. Put a strong magnet over a piece of iron (a ferromagnet) and the iron will acquire a magnetic field and thus be lifted up toward the magnet. Put that same magnet over a piece of aluminum (a paramagnet) and it will also acquire a magnetic field in the same direction. But you can't pick up aluminum with a magnet. (Try!) Why not?

For Deeper Discussion

7. Explain antiferromagnetism in terms of bonding and antibonding states. In other words, explain why it would in some cases be energetically favorable for neighboring electrons to have opposite spins.

Problems

8. **A Semiclassical Model of Diamagnetism**

    An electron (mass m_e, charge $-e$) executes a perfectly circular counterclockwise orbit (radius R, speed v) around a nucleus (Figure 11.35).

Figure 11.35

All the calculations in this problem are classical, except that they assume the electron magically emits no radiation. (You may recognize this set of conditions from the Bohr model.)

(a) Calculate the current represented by this single electron moving in a circle, and use that to calculate the magnetic field it produces at the position of the nucleus. Your final answer will be the internally generated magnetic field B_{int} as a function of R, v, and constants.

(b) Does B_{int} point into, or out of, the page? *Hint*: On this and several other parts of this problem, keep reminding yourself that the electron charge is negative!

(c) The only force acting on the electron, at this point in the problem, is the electric attraction of the nucleus. That force imparts the centripetal acceleration that keeps the electron orbiting. Write the relationship between that force F and the orbital v and R.

Now suppose a magnetic field B_{ext} is applied, pointing into the page. For simplicity we'll assume the electron continues to orbit at the same radius.

(d) Will this external field exert an outward force (thus decreasing the net force on the electron) or an inward force (thus increasing the net force)? Explain.

(e) Stepping through the equations you've written: will that change in net force increase or decrease v? Will that, in turn, increase or decrease B_{int}?

(f) Explain how your final result matches the predicted direction of a diamagnetic field.

9. Iron has an atomic spacing of approximately 3×10^{-10} m and a permanent magnetic dipole

moment per atom of about 2×10^{-23} A m^2 (or equivalently 2×10^{-23} J/T). The formulas you need for this problem can be found in any introductory electricity and magnetism textbook.

(a) Calculate the magnetic field produced by one iron atom at the location of its neighbor atom. For simplicity we'll assume that the atoms are lined up along the axis along which the first atom's magnetic moment points (Figure 11.36).

First atom   Neighbor

Figure 11.36

(b) What is the energy difference between that neighbor atom being aligned with the field vs. being aligned against it?

(c) How does that energy difference compare to $k_B T$ at room temperature? Comment on the implications of your answer for ferromagnetism.

10. Curie's Law

A paramagnet with N dipoles is in an external field, whose direction we'll call "down." As a consequence, each dipole can be in a state with energy 0 (pointing down, with the field) or energy ϵ (pointing up, against the field). The magnetization of the paramagnet is $m(N_{down} - N_{up})/V$, where m is the magnetic moment of each dipole and V is the volume of the dipole.

(a) Using the Boltzmann distribution, derive an expression for the magnetization of the paramagnet at a temperature T.

(b) Show that for $k_B T \gg \epsilon$ your formula gives Curie's law.

(c) Find $\lim\limits_{T \to 0} M$. Explain why your answer makes sense physically.

11. Exploration: Hysteresis

We have seen that, for a diamagnetic or paramagnetic material, the induced field is a function of the externally applied field. But for a ferromagnetic material the relationship is more complicated, because the tendency for domains to align with each other gives the material a kind of "memory" of previous states. A ferromagnet is therefore characterized, not by a proportionality constant χ, but by a "hysteresis loop." In this problem you will sketch out part of that curve, plotting M (the magnetization, or induced magnetic field) on the vertical axis, and H_{ext} (the externally applied field) on the horizontal. Begin your diagram at the origin, representing the lack of any external or induced magnetic field.

(a) As you apply an external field, the ferromagnet starts to magnetize. As the external field continues to increase, the increase of the induced field slows, approaching a maximum value: the "saturation point." Briefly explain why, and sketch the relevant part of the curve.

(b) If the external field is now gradually reduced back down to zero, the induced field also reduces – but *not* all the way back down to zero. Briefly explain why, and sketch the relevant part of the curve.

(c) You can find a picture of a hysteresis loop at www.cambridge.org/felder-modernphysics/figures

You have now stepped through an explanation of the first two parts of this curve. Walk through the remaining parts, explaining the behavior at each step.

11.7 Heat Capacity

Section 10.5 offered a very general rule that the heat capacity of most objects should be on the order of a few times Boltzmann's constant k_B, and used the equipartition theorem to make more specific predictions for several different gases. Here we will investigate the heat capacities of solids. We will begin with a simple model and build toward more complexity and greater accuracy.

11.7.1 Discovery Exercise: Heat Capacity

 A crystal contains N nuclei arranged in a lattice structure. Each nucleus has a fixed position in the lattice, but can make small vibrations around that position. Because those vibrations can occur in all three directions, these N nuclei can be considered as $3N$ simple harmonic oscillators.

1. According to the equipartition theorem (Appendix I), what is the total *thermal* energy of such a collection of oscillators?

   See Check Yourself #20 at www.cambridge.org/felder-modernphysics/checkyourself

2. Remember that "heat capacity" can be approximately defined as the amount of energy required to raise the temperature of an object by one degree. Based on your answer to Part 1, what is the heat capacity of this crystal?

11.7.2 Explanation: Heat Capacity

In Section 10.3 we introduced heat capacity as follows:

Heat Capacity

$$C = \frac{dE}{dT}$$

One way to think about this definition is that C is roughly the amount of energy you would need to provide to increase a system's temperature by one degree.

Heat capacity is an extensive property, meaning that the heat capacity of two identical bricks is twice the heat capacity of either brick alone. Another term you will commonly see is "specific heat capacity" (often signified by a lowercase c): the heat capacity per unit, and therefore a property of a particular substance. You have to gauge from the context whether c means the heat capacity per atom, per mole, or per kilogram.

The Law of Dulong and Petit

In 1819 Pierre Dulong and Alexis Petit noticed that the molar heat capacity (specific heat capacity defined as the heat capacity per mole) is essentially the same for all solids. We can express their result as

$$c = 3R \quad \text{where } R \text{ is the "ideal gas constant": } R = 8.3 \text{ J/(mol K).} \tag{11.2}$$

If you did the Discovery Exercise 11.7, then you already know why this law holds in general. (If not, we'll go through the argument below.) As Table 11.1 shows, the "law of Dulong and Petit" works pretty well at room temperature – but not always. (We'll come back to diamond in Question 4.)

We can understand Equation (11.2) if we think of each atom in a lattice oscillating in three dimensions. The equipartition theorem says that each degree of freedom in a crystal has an

Table 11.1 Room temperature molar heat capacities, along with the prediction of the law of Dulong and Petit

	c at 298 K $\left(\dfrac{\text{J}}{\text{mol K}}\right)$
Predicted	**24.9**
Al	24.2
Cu	24.5
C (diamond)	6.1
Au	25.4
Fe	25.1
Ag	24.9
Pb	27.1
Zn	25.2

average thermal energy of $(1/2)k_B T$. No matter what different crystals are made of or how their atoms are arranged, they all have the same number of degrees of freedom per atom. So they all have the same thermal energy per atom (at any given temperature), and thus the same heat capacity. You can fill in the details to show why that heat capacity usually comes out to $3R$ in Problem 11.

The law of Dulong and Petit predicts that heat capacity should be independent of temperature. As you can see from Figure 11.37, that works pretty well at normal temperatures. (Remember that 150 K is close to −200 degrees Fahrenheit!) But when the temperature drops low enough, the heat capacity approaches zero. To understand that behavior we need to factor in quantum mechanics.

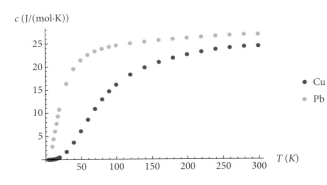

c (J/(mol·K))

• Cu
• Pb

Figure 11.37 Experimentally measured values of specific heat capacity for copper and lead at various temperatures.

The Einstein Model of Heat Capacity

Einstein realized that at low temperatures it's important to treat atomic vibrations quantum mechanically.[10] A quantum mechanical simple harmonic oscillator has evenly spaced energy levels $(1/2)\hbar\omega$, $(3/2)\hbar\omega$, ... At high temperatures ($k_B T \gg \hbar\omega$) you can treat the oscillator classically, and it obeys the equipartition theorem. But when $k_B T$ is comparable to or smaller than $\hbar\omega$, the equipartition theorem no longer holds.

---

10 Sounds almost obvious, right? Except that Einstein figured this out in 1907.

Active Reading Exercise: The Einstein Model of Heat Capacity

1. For $k_B T < \hbar\omega$, will the average energy per oscillator be higher or lower than the equipartition theorem predicts? Why?

2. Explain how that will affect the heat capacity.

The first question in that Active Reading Exercise revisits an issue we discussed in Chapter 3 when we talked about blackbody radiation, and again in Chapter 10 when we talked about the Boltzmann distribution. The answer is "lower" because there is not enough energy to kick any oscillator into its first excited state, so almost all the atoms stay in their ground states. We say that the higher-level states are "frozen out": they are inaccessible at this low temperature.

And that brings us to the second question. Because E is pretty much flat at zero for all temperatures below this threshold, the heat capacity $C = dE/dT$ must be small. This explains why Figure 11.37 shows a relatively constant heat capacity at high temperatures (Dulong–Petit, equipartition theorem), but a dramatic plunge toward zero at low temperatures (frozen-out excited states, energy nearly independent of temperature).

In Problem 12 you can calculate[11] that the time-averaged energy of a quantum oscillator is

$$E = \frac{1}{2}\hbar\omega + \frac{\hbar\omega}{e^{\hbar\omega/(k_B T)} - 1} \qquad \text{Average energy of a quantum oscillator.} \qquad (11.3)$$

The energy of a crystal is $3N$ times this (because each atom is essentially three quantum oscillators). Taking the derivative gives you the Einstein formula for heat capacity (Problem 13). Expressing the result as a heat capacity per mole,

$$c = 3R \left(\frac{T_E}{T}\right)^2 \frac{e^{T_E/T}}{\left(e^{T_E/T} - 1\right)^2} \quad \text{where} \quad T_E = \frac{\hbar\omega}{k_B} \qquad \text{Einstein formula for specific heat.}$$

$$(11.4)$$

We've written this in terms of the "Einstein temperature" of a crystal: $T_E = \hbar\omega/k_B$. Of course $\hbar$ and k_B are constants, but ω – the angular frequency of vibration – is a property of a particular crystal. For any particular type of crystal, the Einstein temperature specifies how quickly the heat capacity rises with temperature. For $T \gg T_E$, Einstein's formula approaches the classical prediction $c = 3R$ (you can show this in Problem 14), while as $T \to 0^+$ it approaches zero (Problem 15). Both of these limits match the predictions we made qualitatively above, and (more importantly) the data.

But while the Einstein model correctly predicts that $\lim\limits_{T \to 0^+} c = 0$, it does not correctly predict the function $c(T)$ at very low temperatures. The Einstein model predicts that the heat capacity approaches zero exponentially as $T \to 0^+$, but the data show that $c \propto T^3$ at low temperatures. To further refine the model, we need to include interactions between atoms.

The Debye Model of Heat Capacity

In 1912 Peter Debye significantly improved on Einstein's model. (And really, how often does anyone get to say that?)

---

11 You did the same calculation in a different way if you did Problem 38 of Section 3.4.

We will demonstrate Debye's idea using a simple one-dimensional model. Figure 11.38 depicts a horizontal row of identical atoms, each free to vibrate vertically. You can treat each atom as a classical simple harmonic oscillator (Dulong and Petit) or a quantum harmonic oscillator (Einstein). But both of these approaches treat each atom as an independent system. Because the atoms are identical they all oscillate with the same frequency, but their amplitudes and phases are entirely uncorrelated.

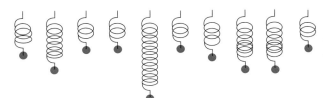

Figure 11.38 Each atom in this array can be thought of as a simple harmonic oscillator, identical to – but independent of – each other atom. But that turns out to be the wrong model for a crystal.

Debye recognized that the atoms are *not* independent of each other. When one atom vibrates, that causes its neighbors to vibrate. The differential equations are "coupled," meaning that the forces on each atom depend on the positions of its neighbors. The solution is therefore not N different functions to represent N different atoms; it is one complete description of the system.

You can analytically solve coupled differential equations using linear algebra.[12] Alternatively, you can approximate a solution by letting the lattice spacing approach zero, so the row of atoms acts like a continuous oscillating rope, and the coupled differential equations become one partial differential equation. (You'll look at that equation in Problem 18.) Either way, you find a set of solutions corresponding to atoms linked in a wave.

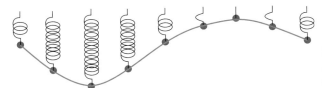

Figure 11.39 A standing-wave solution for atomic vibrations in the crystal.

Figure 11.39 represents one moment in the life of our one-dimensional crystal. In the next moment, the peaks move down and the troughs move up, until the system has flipped itself upside down, and then the process starts over again. In other words, the system describes a standing wave. (You may want to briefly review standing waves in Section 6.1. The relevant equations are in Appendix F.) This standing wave is formed by the coordinated positions of all the atoms in the lattice, much like a group of sports fans doing "the wave" in a stadium.

We're not saying that if you peeked inside a crystal, you would see atoms lined up in a beautiful sine wave. The pattern at any given moment, and the time evolution of that pattern, would be so complicated that they would appear to be almost random. But this seemingly chaotic motion can always be written as a *superposition* of these standing waves, each with a characteristic wavelength and frequency. Each such standing wave – each "vibrational mode" of the crystal – has the following properties:

- The edges of the crystal must be nodes of the wave. This restricts the vibrations to certain discrete allowable frequencies, defined by the length of the entire crystal. The longest

---

12 You can see this process in upper-level textbooks on condensed-matter physics or statistical mechanics, or in our text *Mathematical Methods in Engineering and Physics.*

possible wavelength (and therefore the wave with the lowest allowed frequency) is twice the crystal length.

- The shorter the wavelength λ, the higher the (temporal) frequency ω. In other words, vibrational modes with very long wavelengths take a long time to complete a full oscillation.

- Finally, if the individual atoms are spaced a distance R apart, the shortest possible wavelength (and therefore the wave with the highest allowed frequency) is $2R$. So the minimum frequency is set by the crystal length, and the maximum frequency by the lattice spacing.

So far we've been talking about this system classically. If this were a classical system, then each of these modes could vibrate with any amplitude, and thus any energy. Quantum mechanically, each standing wave has a discrete set of allowed vibrational energies. In fact, each standing wave acts as a quantum oscillator, with allowed energies $n\hbar\omega$ for integer n.[13] Because ω is largest for short-wavelength modes, those modes have the largest spacing between allowed energies.

In one sense the Debye model is quite like the Einstein model because it describes the vibrations of the crystal as a set of simple harmonic oscillators. The difference is that Einstein's oscillators are individual atoms, all with the same frequency, while Debye's oscillators are standing waves with a wide range of frequencies.

The Debye model may sound familiar if you remember our discussion of cavity radiation in Section 3.4 or Section 10.7. There's an allowed set of standing waves with discrete frequencies set by the size of the cavity. Each wave has a frequency inversely proportional to its wavelength, and the energy of each wave is quantized as $n\hbar\omega$. In cavity radiation we say that n is the *number of photons* of that frequency in the cavity. By analogy, we call the excitations of Debye's standing waves "phonons." (Remember that sound is a vibration in matter, and the prefixes "photo" and "phono" refer to light and sound, respectively.) For each frequency ω, the integer n is the number of phonons at that frequency.

One difference between phonons and photons is that phonons in a solid have a maximum possible frequency set by the lattice spacing, while photons of arbitrarily large frequency can exist inside a cavity. But in normal circumstances this is not as big a difference as it may sound: probability drops exponentially with frequency, so you won't find too many photons with $\hbar\omega \gg k_B T$.

Just as we calculated the total radiation energy in a cavity, you can sum over the expected thermal energies of all the standing waves in a solid. We're not going to go through the details of that calculation, but the result is that the total energy in these lattice vibrations is

$$E = \frac{9Nk_BT^4}{T_D^3} \int_0^{T_D/T} \frac{x^3}{e^x - 1} dx. \tag{11.5}$$

The Debye temperature T_D is a parameter that depends on the properties of a particular crystal. You can differentiate Equation (11.5) with respect to T to get heat capacity (Problem 17), but the answer is complicated and not particularly enlightening. Instead we'll present a few properties

---

13 Technically the allowed energies are $(n + 1/2)\hbar\omega$ but the ground state energies are irrelevant to the heat capacity, which is what we're interested in here.

of this formula. (All of these properties come from Equation (11.5), but none obviously. You can verify the high- and low-temperature limits in Problem 16, and see the T_D/T dependence in your final answer to Problem 17.)

- At high temperatures Equation (11.5) reproduces the classical prediction $c = 3R$.
- At low temperatures this reproduces the experimental formula $c \propto T^3$.
- The Debye heat capacity is only a function of T_D/T. Because all the important differences between one solid and another (for this purpose) are represented in their different T_D constants, this means that the $c(T)$ curves for all solids look the same, but are stretched by different factors along the horizontal axis.

We can plot the experimental values for different solids as a function of T/T_D. The elements shown have very different heat capacities as a function of temperature, but plotted as a function of the ratio T/T_D they all fall very close to the Debye prediction (Figure 11.40).

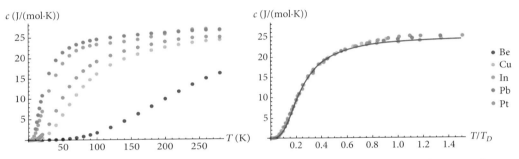

Figure 11.40 Heat capacity per unit mole as a function of temperature (left) and as a function of T/T_D (right). The black curve on the right is the Debye formula prediction.

The Electron Gas

We're going to add one final correction to our model. So far we've only talked about the energy of the nuclei, vibrating in their lattice structure. What about the energy of the electrons? If the number of electrons is of the same order as the number of atoms, and if each particle's thermal energy is of order $k_B T$, then it seems reasonable to guess that the electronic energy would be comparable to the lattice energy. But it's not.

To see why, let's talk about what those electrons are doing. Remember that any such collection of fermions obeys the Pauli exclusion principle and the Fermi–Dirac distribution (Section 10.6). Do you remember what that looks like at normal temperatures? The ground state has an occupancy close to $\bar{n} = 1$ ("I'm always occupied"). Higher-level states have smaller and smaller occupancy numbers, dropping to $\bar{n} = 1/2$ at energy μ, and approaching zero ("we're never occupied") for high-energy states.

But at low temperatures the Fermi–Dirac distribution starts to look like a step function: something very much like $\bar{n} = 1$ for all $E < \mu$, and $\bar{n} = 0$ for all $E > \mu$. This state – all the fermions in their lowest possible energy states pretty much all the time – is called a "degenerate Fermi gas," and it turns out that's the normal state of solids until they get far above room temperature.

(In fact most solids would vaporize before they got hot enough to depart significantly from this distribution.)

So consider an electron occupying the very lowest-energy state in the solid. That electron does have energy, but it doesn't have *thermal* energy: it can't give up its energy by dropping down to a lower level, and it can't easily accept energy since the levels above it are full, so it doesn't participate in energy exchanges. For an insulating solid, that's pretty much the whole story: the electrons are all tightly bound and cannot easily be excited. When you give energy to the solid (heat it up), that energy goes into the nuclei, not into the electrons, so the electrons don't contribute to the heat capacity.

But for a conductor the story looks quite different. Recall that a conductor has a half-filled band, so the electrons in that band have available states that they can easily be excited into. When you heat up the conductor, some of that energy goes into exciting those electrons, so they *do* contribute to the heat capacity. How much? Roughly speaking, the number of electrons in the top-level band that get thermally excited is proportional to T, and the energy they acquire is of order $k_B T$, so the thermal energy of the electron gas is proportional to T^2, and the heat capacity grows linearly with T.

Even though the heat capacity of the electron gas is small, it shrinks at low temperatures more slowly than the lattice contribution. At a low enough temperature it starts to dominate, so, near absolute zero, metals transition from $c \propto T^3$ to $c \propto T$ (Figure 11.41).

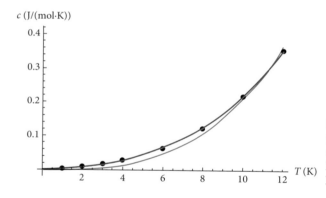

Figure 11.41 Heat capacity for platinum (points) along with a best-fit cubic function (red) and a best-fit linear plus cubic function (blue). The latter fits the data perfectly.

Stepping Back: Models of Heat Capacity

In this section we've taken you through four successively more accurate models for the heat capacity of solids.

- The law of Dulong and Petit uses the equipartition theorem to predict that all solids have the same heat capacity per mole, independent of temperature. This model neglects quantum effects. It works pretty well for most solids at room temperature, but not for all of them, and it fails for all solids at very low temperatures.

- The Einstein model treats each atom as a 3D quantum oscillator, which obeys equipartition at high temperatures but has a heat capacity that falls to zero at low temperatures. This model neglects interactions between atoms. It accurately models the drop in heat capacity as you lower the temperature, but has the wrong functional shape at very low temperatures.

- The Debye model considers modes of coordinated vibrations across the lattice, aka phonons. This model works well at all temperatures for many crystalline solids, but for conductors it predicts the wrong temperature dependence at very low temperatures.

- Our most accurate predictor is the Debye model, modified to include the contribution from the electron gas.

The Einstein and Debye models each give molar heat capacities as a function of T divided by a reference temperature: T_E or T_D, respectively. Roughly speaking, T_E and T_D are the temperatures above which you can use the law of Dulong and Petit. Specifically, we have $c(T = T_E) = 2.76R$ in the Einstein model and $c(T = T_D) = 2.86R$ in the Debye model. The reference temperatures T_E and T_D depend on the material. For most materials they are below room temperature, so the law of Dulong and Petit works pretty well in everyday situations.

Given that the Debye model plus the electron gas is the most accurate of these, why bother learning the others? There are two main reasons.

The first is that the simpler models are much easier to use. If you want to know the heat capacity of a solid at room temperature to within about 10%, the answer for most solids is "about 25 J/(mol K)" and you're done. If you want to know the temperature dependence over a wide range of temperatures that doesn't get too close to absolute zero, the Einstein model gives a simple formula that works pretty well. A great deal of physics is about picking the approximations that will make your calculations as easy as possible while still maintaining enough accuracy for whatever you want to do.

The second reason is that each of these models gives you a different physical insight. If you understand why almost all crystals approach the same specific heat capacity at high temperatures, it means you have a good grasp of the equipartition theorem. If you understand why it fails at low temperature, it means you understand the difference between classical and quantum thermal energy distributions. And if you understand why the Debye model is more accurate than the Einstein model, it means you understand quantized vibration modes. For a student learning quantum mechanics, those fundamental ideas are as, or more, important than understanding heat capacity.

11.7.3 Questions and Problems: Heat Capacity

Conceptual Questions and ConcepTests

1. The heat capacity of a solid approaches zero as the temperature approaches zero. Heat capacity can be approximately defined as the amount of energy required to increase a system's temperature by one degree. Using that definition, what does it mean to have a heat capacity approaching zero?

2. The equipartition theorem says that the thermal energy per degree of freedom of a system depends on temperature. So why did we say that a classical calculation using that law implies that heat capacity should be independent of temperature?

3. Why does the heat capacity of a crystal lattice approach zero at low temperatures? We're not looking for formulas or calculations. Physically, why does increasing the temperature have almost no effect on the lattice's energy when the temperature is close to absolute zero?

4. At room temperature, diamond has a much lower specific heat capacity than the law of Dulong and Petit predicts. That implies that ... (Choose one and explain your choice.)

   A. diamond has a low Debye temperature.

   B. diamond has a high Debye temperature.

   C. diamond is not well described by the Debye model.

5. One key feature of the Debye model is that the sum over all the oscillations cuts off at a maximum frequency. Is that cutoff more important at high temperatures or low temperatures? Explain.

6. We said that for metals the electron gas contributes to the heat capacity. Why don't the more tightly bound electrons also contribute?

7. The spectroscopic notation for an aluminum atom is $1s^2 2s^2 2p^6 3s^2 3p^1$. Which of the following dominates the heat capacity of aluminum at temperatures very close to absolute zero? (Choose one and explain.)

   A. The protons and neutrons in the nucleus

   B. The two electrons in the $1s$ shell

   C. The electron in the $3p$ shell

   D. All of the electrons contribute significantly.

   E. All of the electrons and the nuclei contribute significantly.

8. Suppose you predict the specific heat capacity of a metal at very low temperature using each of the models below. Rank-order the predictions you would make from lowest c to highest c.

   A. The law of Dulong and Petit

   B. The Einstein model

   C. The Debye model without the electron gas contribution

   D. The Debye model with the electron gas contribution

Problems

9. The measured specific heat capacity of copper at 100 K is 16.1 J/(mol K). Use that fact to predict its specific heat capacity at 200 K using ...

(a) the law of Dulong and Petit.

(b) the Einstein model of heat capacity.

(c) the Debye model of heat capacity.

For comparison, the measured value is 22.6 J/(mol K).

10. At 1 K platinum has a specific heat capacity of 0.004749 J/(mol K). At 10 K it's 1.4384 J/(mol K).

(a) At low temperatures the Debye model (with the electron-cloud correction) predicts that a metal's heat capacity can be modeled as $aT + bT^3$. Use these two data points to find a and b.

(b) If you cool a block of platinum, at what temperature will its heat capacity start to be dominated by its electrons?

11. **Deriving the Law of Dulong and Petit**

(a) As we saw in Chapter 10 on statistical mechanics, the average thermal energy of a classical simple harmonic oscillator in 1D is $k_B T$. So if a crystal lattice comprises N atoms, each acting as a simple harmonic oscillator in three different directions independently, what is the total thermal energy of the crystal?

(b) Take the derivative of that energy with respect to temperature to get the heat capacity of the crystal.

(c) A mole consists of Avogadro's number (N_A) of atoms. What is the heat capacity per mole of the crystal? Express your answer in terms of the ideal gas constant R, using the relation $R = N_A k_B$.

12. **Average Energy of an Oscillator**

The energy eigenvalues for a simple harmonic oscillator are $(n + 1/2)\hbar\omega$. Ignoring the ground state energy for now, we can define $\epsilon = \hbar\omega$ and write these energies as $n\epsilon$, where n can be any non-negative integer. The Boltzmann distribution says that at a temperature T the probability of an oscillator having energy $n\epsilon$ is $e^{-n\epsilon/(k_B T)}/Z$. The average value of the energy over time is $\sum_n n\epsilon e^{-n\epsilon/(k_B T)}/Z$. We can make one more simplification by defining $\alpha = \epsilon/(k_B T)$. Then

$$\langle E \rangle = \frac{\alpha k_B T}{Z} \sum_{n=0}^{\infty} n e^{-\alpha n}.$$

We're going to show you a clever trick for evaluating that sum. (This trick can be used for a variety of tricky sums.)

(a) Let $S_1 = \sum_n n e^{-\alpha n}$. Write S_1 as the derivative with respect to α of a different sum, S_2.

(b) S_2 is a "geometric series," meaning each term is equal to a constant r times the previous term. The sum of an infinite geometric series is $a_1/(1 - r)$, where a_1 is the first term, as long as $|r| < 1$. Use this formula to evaluate S_2.

(c) Differentiate your answer for S_2 to find S_1.

(d) Normalization requires that $Z = \sum_n e^{-\alpha n}$. Evaluate that sum to find Z. *Hint:* You've already done the hard work for this.

(e) Putting it all together, calculate $\langle E \rangle$. Plug in the formulas for α and ϵ so your answer is in terms of ω and T. Don't forget to add back in the ground state energy $(1/2)\hbar\omega$.

13. Fill in the steps to get from Equation (11.3) to Equation (11.4).

14. Take the limit of the Einstein formula for specific heat (Equation (11.4)) as $T \to \infty$. (There are several ways you can do this.) Explain why your answer makes sense physically.

15. Take the limit of the Einstein formula for specific heat (Equation (11.4)) as $T \to 0^+$. *Hint:* In this limit you can replace $(e^{T_E/T} - 1)$ with $e^{T_E/T}$. Do you see why?

16. In this problem you're going to calculate the high- and low-temperature limits of the Debye energy formula, and then differentiate it to find the Debye heat capacity in those limits. The starting point will be Equation (11.5).

(a) In the limit $T \gg T_D$, the upper bound of the integral is very small, meaning x only takes on very small values. Using the approximation $e^x \approx 1 + x$ for small x, simplify the integral, and then evaluate it to find a high-temperature approximation for E.

(b) Using your answer to Part (a), take the derivative dE/dT to find the heat capacity C. Then divide by N to find the heat capacity per atom. Then multiply by Avogadro's number N_A to find the heat capacity per mole. (*Hint:* $N_A k_B = R$.). How could you have predicted your answer without doing the calculations?

(c) Now consider the limit $T \ll T_D$. Explain why E is proportional to T^4 in that limit, which in turn implies that c is proportional to T^3.

17. Starting from Equation (11.5), derive the heat capacity per mole of a solid using the Debye model. *Hint:* You may find it helps to define a new variable $u = T_D/T$.

18. Exploration: The Wave Equation

Debye considered a crystal lattice as an array of coupled harmonic oscillators. In the limit where the lattice spacing approaches zero, such a system classically obeys the "wave equation":

$$\frac{\partial^2 y}{\partial x^2} = \frac{1}{v^2} \frac{\partial^2 y}{\partial t^2} \qquad \begin{array}{l} \text{The one-dimensional} \\ \text{wave equation.} \end{array} \qquad (11.6)$$

To form a simpler mental picture, imagine a guitar string vibrating up and down. The letter v is a constant for any particular string, but it is a different constant for a different string.

(a) Show that the function $y(x, t) = x$ is a valid solution to Equation (11.6).

(b) Show that the function $y(x, t) = x^2$ is *not* a valid solution to Equation (11.6).

(c) What do your answers to Parts (a) and (b) imply about the possible behaviors of a guitar string? *Hint:* You can start the guitar string in any shape you like. The wave equation doesn't tell you what starting shapes the string can or cannot have; it tells you how the string will then behave.

(d) Show that the following function, $y(x, t) = A \sin(3x) \sin(3vt)$, is a valid solution to Equation (11.6).

(e) Show that the following function, $y(x, t) = A \sin(4x) \sin(4vt)$, is a valid solution to Equation (11.6).

(f) Show that the function given by $y(x, t) = A \sin(3x) \sin(4vt)$ is *not* a valid solution to Equation (11.6).

(g) Based on your answers to Parts (d) and (e), write a general function with an arbitrary constant k that solves the wave equation.

(h) Does the function you wrote in Part (g) represent a standing wave, a traveling wave, or neither?

(i) What is the mathematical relationship between the spatial frequency and the (temporal) frequency of the function you wrote in Part (g)?

(j) Explain how your results predict the properties listed in the text for phonons in a crystal lattice, noting in particular the relationship between the spatial wavelength and the temporal frequency.

Chapter Summary

When a very large collection of molecules is held together in a rigid structure, that structure has properties that can be quite different from the properties of individual molecules. While it is not possible to solve Schrödinger's equation for such a large configuration of particles, we can use the principles of quantum mechanics to model those particles and therefore predict the behavior of the solid.

Section 11.1 Crystals

The term "crystal," meaning a regular lattice of molecules, does not only refer to gemstones. Table salt, copper, and many other solids organize themselves into crystalline structures, and much of their behavior is determined by the geometry of those structures.

- Crystals can be broadly divided into four categories based on the primary mechanism binding the atoms together.
  - An "ionic crystal" is held together by attraction between positive and negative ions.
  - A "covalent solid" is held together by covalent bonds.
  - In a "metal" some of the electrons are shared by all of the atoms and can move more or less freely through the lattice as an electron gas. The metal is bound by the attraction between this electron gas and the lattice of positive ions.
  - A "molecular crystal" is a lattice of molecules rather than atoms.
- Many properties of a solid can be roughly predicted based on how the crystal is bonded together. For example, ionic and covalent crystals generally have higher melting points than metals, which in turn tend to have higher melting points than molecular crystals.
- The "unit cell" of a crystal is the smallest group of atoms from which you can make the entire lattice by repeating it.
- Common structures include a "face-centered cubic" lattice (the unit cell is the vertices of a cube plus an atom at the center of each face) and a "body-centered cubic" lattice (the vertices of a cube plus one atom at the center).
- The "cohesive energy" of a solid is the energy required to break it into its constituent atoms. For ionic crystals, one contribution to this energy is the Coulomb potential energy

between ions. That potential energy is proportional to the "Madelung constant," which is associated with the geometric arrangement of ions in the crystal.

Section 11.2 **Band Structure and Conduction**

The energy levels available to an electron in a crystal are organized into "bands." Within a band – a range of allowed energy – the energy levels are effectively continuous rather than discrete. But alternating with these allowed bands are ranges of forbidden energy.

The electrical conductivity of a crystal is determined by where the highest filled energy level falls within that band structure.

- A conductor has a partially filled band of electric states, so the electrons in that band can be easily excited by an externally applied field.
- In an insulator, the highest occupied band is full, so even a fairly strong electric field will barely excite any electrons to higher states.
- In a semiconductor, the highest occupied band is also full, but with a relatively small energy gap between that band and the next available band ($\lesssim 2$ eV).

Heat conduction also involves electrons being excited to higher energy levels, so good electrical conductors are usually good heat conductors and vice versa.

Section 11.3 **Semiconductors and Diodes**

A semiconductor in its pure form ("intrinsic") has a very high resistivity, but you can "dope" it by mixing in small amounts of other metals to make it a better conductor.

- The highest filled energy band in a semiconductor is called the "valence band" and the lowest unfilled band is called the "conduction band."
- You make an "n-type" semiconductor by mixing in atoms that contribute electrons to the conduction band.
- You make a "p-type" semiconductor by mixing in atoms that contribute unfilled states (called "holes") to the valence band.
- By putting an n-type and a p-type semiconductor next to each other you can create a "diode," a circuit element that allows current to flow in one direction much more freely than in the opposite direction.

Much of this section is devoted to the very subtle question of *why* two adjacent doped semiconductors create a one-way circuit element. Critical to this discussion, and also to the next section on transistors, is the "depletion region": a zone around the p-n junction that has no free electrons or holes to carry charge.

Section 11.4 **Transistors**

A transistor is a circuit element with three leads. Small changes in the voltage at one of these leads cause large changes in the resistance between the other two leads.

- In a "field effect transistor," current flows through an n-type or p-type semiconductor that is adjacent to a doped semiconductor of the opposite type. An applied voltage widens or shrinks the depletion region, thus allowing or blocking the current.

- A "logic gate" is a computer element in which the voltage of an output wire depends on some combination of the voltages of input wires. For example, an AND gate has a high output voltage if and only if both of the inputs are high voltage. Because a transistor provides an electronic on/off switch for current, transistors can be combined to make a variety of logic gates.

Section 11.5 Why Do Crystals Have a Band Structure?

The fact that a metal's energy levels come in "bands" explains many important properties of metals, including the specific properties that make semiconductors so useful. Exploring *why* the energy levels are organized into bands takes us back into quantum mechanics.

- One way to understand this phenomenon is to remember the discussion in Chapter 9 of bonding and antibonding states. When two atoms are brought close together, the energy levels available to their electrons split. When many atoms share their electrons, those energy levels form wide bands with a density of states so high that they are effectively continuous.

- An alternative explanation of the same phenomenon comes from analyzing the behavior of the wavefunction of an electron in a periodic potential (such as a crystal lattice). A simple example is the "Kronig–Penney" model, consisting of infinitely many evenly spaced potential barriers.

Section 11.6 Magnetic Materials

Every atom in a solid is a tiny magnet, due to the orbits and the spins of its electrons. If all these small magnetic fields have equal magnitudes, but half of them point up and half point down, the overall solid has no magnetic field at all. But if the up/down balance is biased, the result is magnetism on a macroscopic scale.

- "Diamagnetism" opposes an externally applied magnetic field. A diamagnet in an external field has equal numbers of electrons aligned with or against the field, but those aligned against it produce stronger magnetic fields than the ones aligned with it.

- "Paramagnetism" reinforces an externally applied magnetic field. Paramagnetism occurs in materials in which the atoms or molecules have unpaired electrons and thus intrinsic magnetic moments. The external field causes some of those atoms or molecules to flip into alignment with the field, so that there are more magnetic moments aligned with the field than against it.

- Atoms or molecules that have intrinsic magnetic moments of their own display both diamagnetism and paramagnetism, but the latter is much stronger than the former, so the net effect is to reinforce the external field. Atoms with no intrinsic moments display only diamagnetism, so the net effect is to oppose the external field.

- "Ferromagnetism" reinforces an external field, like paramagnetism but much more strongly. Unlike diamagnets and paramagnets, ferromagnets continue to produce a field even after the external field is removed. That occurs because atoms or molecules keep their neighbors in alignment with them.

Section 11.7 Heat Capacity

Chapter 10 introduced the quantity called heat capacity, roughly the energy required to increase a system's temperature by one degree. This section walks through different models that provide successively more accurate predictions of the dependence of heat capacity on temperature.

- The "law of Dulong and Petit" predicts that the heat capacity per mole should be the same for all solids at all temperatures. This follows from a purely classical model of a solid, and from the principles of statistical mechanics: every atom in a crystal is free to vibrate in three dimensions, and the equipartition theorem (Chapter 10) assigns a thermal energy to each degree of freedom.

- The "Einstein model of heat capacity" considers each atom as a *quantum* harmonic oscillator. At high temperatures the result is not significantly different from the classical version (the correspondence principle at work). But at low temperatures the quantization of energy levels become important, because even the first excited state has too high an energy to be reached. So Einstein's model predicts that the heat capacity per mole should be roughly constant at high temperatures, but should approach zero at low temperatures.

- The "Debye model of heat capacity" treats the atoms, not as individual vibrating entities, but as linked to each other, so that the vibration of each atom affects the vibrations of its neighbors. In the Debye model the degrees of freedom being excited aren't individual oscillations but oscillatory waves, called "phonons." Like Einstein's model, Debye's model predicts constant heat capacity at high temperature and zero heat capacity in the limit $T \to 0$, but the Debye model more accurately describes the exact low-temperature dependence.

- All three models discussed above focus only on the nuclei in the crystal structure. For metals, the heat capacity at very low temperatures also includes the contribution of the electron gas. Adding this contribution to the Debye model creates an extremely accurate prediction of the dependence of heat capacity on temperature.

12

The Atomic Nucleus

A brief history of smallness.

- Seventeenth and eighteenth centuries: Scientists understand that matter is made of particles too small to see, and they devise a host of ingenious experiments to determine the nature and properties of these particles. In hindsight, we would describe much of this work as groping toward a clear distinction between atoms and molecules.

- Early nineteenth century: Avogadro captures that distinction when he describes "compound molecules" made up of "elementary molecules." Subsequent decades see increasingly sophisticated models, with many of the same geometric structures and even the same notation that we use today, to describe how atoms form molecules.

It is an ironic side note that the word "atomic" means "indivisible": the name was applied to the particles that were thought to be fundamental. As our story continues, we discover that atoms are not in fact atomic.

- 1897: J. J. Thomson shows that atoms contain very small negatively charged electrons.

- 1911: Rutherford shows that an atom's positive charge is concentrated in a small region in the center of the atom that he calls the "nucleus."

That takes us as far as Chapter 4. Now we add another crucial moment to the story.

- 1932: Chadwick discovers the neutron. This is the final piece needed to allow our modern picture of the atomic nucleus, a densely packed cluster of positively charged protons and uncharged neutrons.

This chapter is a survey of the structure and behavior of the atomic nucleus. If you don't have time in your course for the whole chapter, it would be perfectly reasonable to get the overview that Section 12.1 provides and stop there. Later sections discuss some of the experimental evidence for nuclear structure, nuclear models, and nuclear reactions. Chapter 13 takes the final step downward, to the particles that make up protons and neutrons.

12.1   What's in a Nucleus?

An atom is a small cluster of protons and neutrons, surrounded at a great distance by electrons. Chapters 7 and 8 talked about those electrons a lot: the energy levels they can occupy, transitions between those levels, and so on. But those discussions treated the nucleus as a positively charged "black box" in the center. What do you see when you peer inside the box?

12.1.1 Discovery Exercise: What's in a Nucleus?

Two protons sit 10^{-15} m away from each other. They are held together by the "strong nuclear force" but repel each other electrically.

1. Find the magnitude of the (positive) electric potential energy of the two protons.
   See Check Yourself #21 at www.cambridge.org/felder-modernphysics/checkyourself

2. For the protons to be bound in the nucleus, they must have a negative potential energy whose magnitude is larger than the electric potential energy you just calculated. To put that number in context, how many times larger is that electric potential energy than the 13.6 eV binding energy of an electron in a hydrogen atom?

12.1.2 Explanation: What's in a Nucleus?

A nucleus is a not-quite-spherical collection of positively charged protons and uncharged neutrons at the center of an atom. Each "nucleon" (proton or neutron) has about 2000 times the mass of an electron. But they're not identical; a neutron is about 0.1% heavier than a proton.

A nucleus is described by three (related) numbers:

Z	Atomic number	The number of protons
N	Neutron number	The number of neutrons
$A = Z + N$	Mass number	The total number of nucleons

Because Z and A are the most commonly discussed quantities, we often write formulas with $A - Z$ instead of explicitly using the letter N.

We can summarize a number of the key facts that describe a nucleus.

- The total charge of a nucleus is eZ, where e is the charge of one proton.
- The total mass of a nucleus is roughly $m_p A$, where m_p is the proton mass. (The actual mass is somewhat smaller than this, because the binding energy – the negative potential energy that holds the nucleus together – reduces the total mass. Section 12.3 will present a more accurate mass estimate called the "semiempirical mass formula.")
- A typical nuclear radius is 1–10 "femtometers" (1 fm $= 10^{-15}$ m), about 10,000 times smaller than the radius of a typical atom.
- The density of a typical nucleus is on the order of 10^{17} kg/m^3. (A quart of nuclear-density milk would weigh more than Mount Everest.) Different nuclei generally have similar densities, so their radii are proportional to $A^{1/3}$ (Problem 8).

Early twentieth-century physicists saw that the charge of any nucleus is an integer (Z) times the charge of a hydrogen nucleus, and that the mass of any nucleus is approximately an integer (A) times the mass of a hydrogen nucleus. This "whole number rule" suggested the existence of protons, indivisible particles that make up every nucleus.

But if protons were the whole story, then A and Z would always be the same as each other. How to explain the fact that, for almost all nuclei, $A > Z$? The obvious answer was that the nucleus contained A protons and $A - Z$ electrons. The electrons would contribute little to the mass but decrease the total charge. But an electron confined to the nucleus would violate the uncertainty principle (Problem 11). Furthermore, it was impossible to reconcile this model with measured values of spin (Question 3). These problems were resolved by Chadwick's discovery of the neutron in 1932, leading to the basic picture of the nucleus that we still use today: Z protons, $A - Z$ neutrons, and no electrons.

We represent a particular nucleus or atom with the notation ${}^A_Z X$, as, for instance,

$${}^{17}_{8} O \quad \text{Oxygen with 8 protons and } 17 - 8 = 9 \text{ neutrons.}$$

(That atom would also generally have 8 electrons, but they won't come up much in this chapter.) Remember that the number of protons defines the element. So sometimes we leave Z out of the notation and write simply ${}^{17}O$, since the O for oxygen implies that $Z = 8$.

Why do we define an element based only on its protons? Because the atomic number determines how an element behaves chemically. Take a hydrogen atom (${}^1H$), add an extra neutron, and you now have a heavier atom called "deuterium" (${}^2H$) that will still bond in the same ways as the original. For instance, two such atoms can bond with each other to form a molecule, or can bond with an oxygen to form a water molecule (called "heavy water").

So atoms with the same atomic number but different mass numbers are not considered different elements, but different "isotopes" of the same element. For instance, ${}^{16}_8 O$ and ${}^{26}_8 O$ are both isotopes of oxygen. The former is a stable isotope. The latter is unstable or "radioactive"; almost as soon as it appears, it decays into a different nucleus. (Isotopes of oxygen range from $A = 11$ to 26, but only 16–18 are stable.)

The word "nuclide" refers to a particular isotope of a particular element: ${}^{16}_8 O$, ${}^{17}_8 O$, and ${}^{12}_6 C$ are all different nuclides.[1]

The "Strong Force" (or "Nuclear Force")

The protons inside a nucleus repel each other with an electric force made extremely powerful by the small distances involved. The nucleus is held together by an even more powerful force, generally called the "strong force" or "nuclear force."

This is where you're probably expecting us to give you the equation for the strong force, equivalent to $F = kq_1q_2/r^2$ for the electrostatic force. No such formula exists, but we can describe this force qualitatively.

By analogy, how would you qualitatively describe the electric force? You would presumably start by discussing what particles are affected: like charges repel, opposite charges attract, and uncharged particles neither exert nor feel the force. You might then show Figure 12.1, the electric potential energy of two protons as a function of the distance between them.

From Figure 12.1, without the benefit of any specific numbers, you can see that the two protons are always driven to increase their mutual distance, and that this force increases the

---

1 Technically, each one of those *in its ground state* is a nuclide. A nucleus in an excited state is considered a separate nuclide.

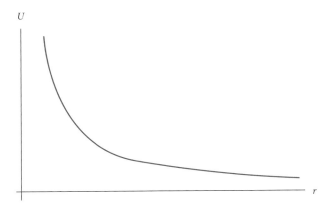

Figure 12.1 **Electrostatic potential energy between two protons.**

closer they draw to each other. We call electricity a "long-range force" because it falls off slowly enough with distance ($F \propto 1/r^2$) that it has significant effects even on macroscopic scales.

To similarly describe the strong force, we begin by saying that it exerts the same force between two protons, two neutrons, or one of each. Electrons neither exert nor feel the strong force.

As Figure 12.2 shows, the strong force between nucleons is a "short-range force" that falls off exponentially with distance, and is essentially negligible beyond about 2 fm. At distances less than 2 fm the attractive strong force between nucleons becomes about 100 times stronger than the electric force between protons at that distance, so it can hold the nucleus together. But at distances less than about half a fm the strong force switches to being repulsive, which explains why the nucleons don't all collapse into a region 1–2 fm wide.

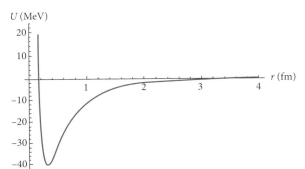

Figure 12.2 **Rough model of the strong force potential energy between two nucleons.**

In Chapter 13 we'll discuss a model in which all forces are caused by the creation and destruction of "force carriers." We're not going to discuss that model in this chapter, but you can explore how it applies to nuclear forces in Problem 12.

Separation Energies

The repulsion between nucleons at distances below ~0.5 fm keeps them at an average separation of about 1.2 fm within a nucleus. That fact, plus the short range (~2 fm) of the attractive nuclear force, means that nuclear forces "saturate," which is a fancy way of saying that each nucleon in a nucleus only feels this force from its nearest neighbors.

So if you add nucleons to a nucleus, the average force felt by each nucleon barely changes. Saturation explains why all nuclei have essentially the same density; the balance of forces that keeps nucleons about 1.2 fm apart from each other doesn't depend on how many nucleons there are, only on how each one interacts with its immediate neighbors.

A nucleon that is surrounded by other nucleons, all about 1.2 fm away, feels almost no net force. But a nucleon at the edge of the nucleus feels a strong attractive force on one side that is not balanced by anything on the other side. The result is that a nucleus can be roughly modeled as a finite square well (Figure 12.3): each nucleon can move freely within the nucleus, but experiences a high potential wall at the surface. The depth of this well is typically tens of MeV.

The energy required to liberate a proton from the nucleus is the gap between the highest filled proton energy state and the top of the well. It's called the "proton separation energy." The corresponding energy gap for neutrons is the "neutron separation energy." Each of those energies is typically around 6–8 MeV. For comparison, the ionization energy required to liberate an electron from an atom is typically tens of eV.

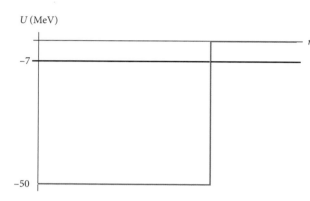

Figure 12.3 The potential well felt by a nucleon in a typical nucleus. The line at $E \approx -7$ MeV shows the highest occupied energy level, which means it takes about 7 MeV to liberate a nucleon from most nuclei. In stable nuclei that line is at approximately the same energy for protons and neutrons.

Binding Energy and Stability

The binding energy B of a nucleus is the difference in energy between the bound state (the nucleus) and the unbound state (all the component nucleons far away from each other). The higher the binding energy, the more stable a nucleus will be. More specifically, nucleons will tend to arrange themselves into whatever combinations have the highest *binding energy per nucleon, B/A*.

Figure 12.4 shows the values of B/A for different mass numbers, and Figure 12.5 shows stable and unstable isotopes by atomic number. Much of this chapter will be built around the information in these two figures: Section 12.3 presents models that can be used to explain and predict binding energy, and subsequent sections discuss the nuclear reactions that result from these energetic differences. For this reason we have posted both of these figures at www.cambridge.org/felder-modernphysics/figures. You may want to bookmark them, print them out, or otherwise keep them handy as you go through the chapter.

To begin your acquaintance with Figure 12.4, note the following:

- As we stressed above, Figure 12.4 is not a graph of total binding energy, but of binding energy *per nucleon*. A high value of B/A tends to make an atom stable. (If you took a collection of atoms with a relatively low B/A-value, and regrouped their components into atoms with a higher B/A-value, net energy would be released.)

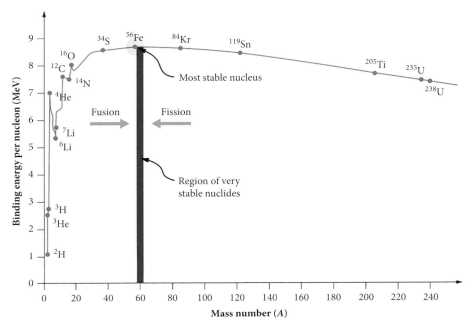

Figure 12.4 Binding energy per nucleon (B/A) as a function of mass number (A).
Source: OpenStax © 2016.

- Starting with carbon, this quantity does not vary much: B/A for any atom with $Z \geq 6$ is roughly 8 MeV. This makes sense when you consider that a typical nucleon only interacts with its immediate neighbors, so adding more nucleons to the whole nucleus does not cause much change to the existing nucleons.

- Below carbon, B/A drops rapidly. This also makes sense when you consider that a nucleon near the *edge* of a nucleus has fewer neighbors, and therefore less binding energy, than an internal nucleon. For a very small nucleus, most of the nucleons are on the edge.

- The quantity B/A generally increases and then decreases (with a few notable exceptions), reaching a maximum around iron. (The most stable nuclide is $^{62}_{28}$Ni, but the processes in stars that generate most of the heavy elements in the universe produce very little $^{62}_{28}$Ni. The most stable nuclide typically found in nature is thus $^{56}_{26}$Fe. For this reason, many sources refer to iron as the most stable element.)

- We will discuss fusion and fission later in this chapter, but the most important insight comes from this drawing. When atoms with $Z < 26$ combine (fusion), increasing their atomic numbers toward iron, they release energy. When atoms with $Z > 26$ break apart (fission), *decreasing* their atomic numbers toward iron, they release energy.

- Finally, note the spikes. Certain nuclei such as oxygen, and especially helium, are more stable than the otherwise smooth curve would predict. Physicists refer to "magic numbers": if Z or N is 2, or 8, or 20, or certain higher numbers, the binding energy is unusually strong. We will see later that the "shell model" of the nucleus explains these magic numbers as filled nuclear shells, just as noble gases can be explained as filled electron shells.

Figure 12.5 shows the combinations of protons and neutrons that lead to stable nuclei (blue), that lead to unstable nuclei (green), or just don't happen (white). For instance, if you look carefully,

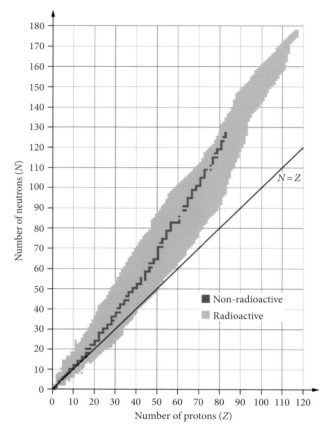

**Figure 12.5 The "curve of stability."
Stable nuclides are shown in blue,
unstable ones in green. No nuclei
can exist in the white region.
Source: OpenStax © 2019.**

you might notice that the longest vertical blue line in the graph occurs at $Z = 50$. This tells us
that tin is the element with the greatest number of stable isotopes, in part because 50 is one of
the "magic numbers" mentioned above. (In fact tin has 10 stable isotopes, including the "doubly
magic" $^{132}_{50}$Sn, as well as 29 unstable isotopes shown in green on the graph.)

For very low atomic masses, we see that the stable nuclei tend to have balanced numbers of
protons and neutrons. (They lie on or near the line $N = Z$.) As the numbers increase we see a
tendency to have more neutrons than protons. That's because the collective electric repulsion
of a large number of protons can make a nucleus unstable, unless balanced by the strong force
attraction of an even larger number of neutrons.

Angular Momentum and Spin

The angular momentum story inside a nucleus is very similar to the story inside an atom.

- There are two sources of angular momentum: the orbital motion of the nucleons inside
  the nucleus, and their intrinsic spins. (Somewhat confusingly, the word "spin" is often
  used to refer to the sum of these two effects inside a nucleus, while the same word is only
  used for intrinsic spin when applied to an electron.)

- Protons and neutrons, like electrons, have intrinsic spin $s = \pm(1/2)\hbar$ along whatever axis
  you measure the spin. Therefore the intrinsic spin of the nucleus is an integer multiple
  of $\hbar$ if A is even, a half-integer multiple if A is odd.

- The half-integer spins of the nucleons means that they are fermions, and therefore subject to the Pauli exclusion principle. Remember, however, that this principle applies only to identical particles. So each neutron claims a quantum state that is unavailable to all the other neutrons in the nucleus, but this has no effect on the protons – and vice versa.

- The total energy of an atom is reduced if the nuclear magnetic moment is parallel to the magnetic moment of the electrons, and increased if they are antiparallel. This very small effect is called "hyperfine splitting." In the presence of an external magnetic field, there is also a nuclear Zeeman effect.

12.1.3 Questions and Problems: What's in a Nucleus?

Conceptual Questions and ConcepTests

1. Briefly define each of the following terms.

   (a) Atomic number

   (b) Mass number

   (c) Femtometer

   (d) Isotope

   (e) Unstable (or radioactive) isotope

   (f) Nuclide

   (g) Saturation

   (h) Proton separation energy

   (i) Binding energy of a nucleus

2. We use the notation $^{17}_{8}O$ for oxygen with 8 protons and 9 neutrons. Write the analogous notation for uranium with 143 neutrons.

3. The nucleus of deuterium, or "heavy hydrogen," has the same charge as a regular hydrogen nucleus ($^{1}_{1}H$) but twice the mass.

   (a) An early model of the nucleus contained only protons and electrons. How many of each would this model expect to find in a deuterium nucleus?

   (b) The measured total spin of a deuterium nucleus is 1. Explain why this cannot be reconciled with the proton–electron model.

   (c) Explain how the right number of protons and neutrons can provide the measured charge, mass, and spin of deuterium.

4. Figure 12.2 shows that the potential energy curve for two nearby nucleons slopes upward for $r \gtrsim (1/2)$ fm, but slopes downward for lower r-values. How would a nucleus behave if that curve did not rise toward ∞ as r approaches zero, but instead continued downward toward $-\infty$?

5. Why do we find so few nuclei of any kind, and especially stable nuclei, below the $N = Z$ line in Figure 12.5?

6. The binding energy per nucleon stays roughly constant at 8 MeV because each nucleon lowers only the energy of its nearest neighbors (see Figure 12.4). Based on that argument, why is B/A so much lower for very low values of A?

Problems

7. Little Elroy Jetson is playing with his new Nuclear Construction for Kids™ set. The starter set comes with three protons and three neutrons. (When answering this problem, keep your eye on Figure 12.4.)

   (a) How much energy will be released if Elroy assembles his particles into three deuterium nuclei, $^{2}_{1}H$?

   (b) How much energy will be released if Elroy instead assembles his particles into one lithium nucleus, $^{6}_{3}Li$?

   (c) If Elroy's kit has thousands of protons and neutrons, and he wants to release as much energy as possible, of what element would you recommend he make the greatest amount?

8. Imagine that every nucleus is a perfect sphere, that the nuclear mass is evenly distributed throughout

that sphere, and that all nuclei regardless of size have exactly the same density. (None of those three assumptions is perfectly true, but they are all reasonable approximations.)

(a) Show that these assumptions lead to the relationship $R = km^{1/3}$, where R is the radius of a nucleus, m is its mass, and k is the same constant for all nuclei.

(b) Explain why this equation implies $R = R_0 A^{1/3}$, not $R = R_0 Z^{1/3}$ or $R = R_0 N^{1/3}$.

(c) What are the units of the constant R_0?

(d) The radius of a uranium nucleus is approximately 5.9 fm. Use this fact to approximate the value of R_0.

(e) Draw a graph of nuclear radius as a function of mass number. The overall shape should be correct, and you should calculate a few values including the lowest $\left({}^{1}_{1}\text{H}\right)$ and the highest $\left({}^{238}_{92}\text{U}\right.$ is a reasonable stopping place, although heavier atoms have been synthesized).

9. Two protons exert an electric repulsion on each other $(U = ke^2/r)$ and a strong-force attraction (Figure 12.2).

(a) Draw a diagram of the total potential energy as a function of distance on the domain $0 < r \le 3\,\text{fm}$. Your graph should be reasonably quantitatively accurate, so you will have to look up the values of the electric constant k and the proton charge e. (Be careful about units!)

(b) Explain why two protons initially at rest at a distance of 4 fm couldn't fuse *classically*.

(c) Now suppose those two protons do fuse because of quantum tunneling. If they did so, how much energy would be released?

(d) If that energy were released as a single photon, where in the electromagnetic spectrum would that photon lie?

10. For two protons separated by about 1 fm, roughly what fraction of the total potential energy is gravitational? *Hint*: You can use Figure 12.2 to estimate the strong-force potential energy.

11. Early twentieth-century physicists imagined a nucleus as a collection of Z protons and $A - Z$ electrons. This "proton–electron model" requires

confining an electron to the width of a nucleus, and that creates problems with the uncertainty principle. As an example, consider ${}^{4}_{2}\text{He}$. The width of this nucleus, and therefore the maximum position uncertainty of any particle confined to this nucleus, is roughly $1\,\text{fm} = 10^{-15}\,\text{m}$.

(a) Calculate the minimum uncertainty in the momentum of such an electron.

(b) Show that non-relativistic approximations aren't adequate for the momentum you just calculated.

(c) So, using relativistic equations, show that an electron with that momentum could not be confined inside a nucleus by Coulomb forces. *Hint*: Think about the conditions required for something to be confined, as opposed to escaping.

12. **Exploration: Nuclear Force as Pion Exchange**

In quantum field theory, forces are viewed as particle exchanges. For example, the electromagnetic force between a proton and an electron comes about because each particle emits photons that the other one absorbs. We believe that the attractive strong force between two nucleons is mediated by a particle called a "pion." You can explain the short range of the nuclear force based on the mass of the pion, using the time–energy uncertainty principle.

(a) A pion's rest energy is about 140 MeV. If we view the pion's brief existence as an energy fluctuation ΔE, what is the shortest time it could possibly exist according to the time–energy uncertainty principle?

(b) Assuming the pion exists for the amount of time you calculated in Part (a), what is the farthest distance it could travel?

(c) It has been speculated that the short-range repulsive force between two nucleons may be mediated by another type of particle. Would you expect that other particle to be heavier or lighter than a pion? Explain.

(d) Why don't arguments of the type we've made in this problem limit the range of the electromagnetic force?

In Chapter 13 we'll see a more fundamental way to describe the strong force, in terms of gluons.

12.2 Experimental Evidence for Nuclear Properties

This section will not catalog all the evidence that determines and verifies the information in Section 12.1. But it will discuss some of the important historical experiments, and give some sense of how we can measure or conclude *anything* about objects that are less than a trillionth of an inch from end to end.

12.2.1 Explanation: Experimental Evidence for Nuclear Properties

Rutherford Scattering

We discussed Ernest Rutherford's discovery of the nucleus in Chapter 4. To quickly recap, Rutherford fired high-energy alpha particles (which we would now call $^4_2$He nuclei) at a thin gold foil.[2] An electron is too light to push an alpha particle around, so any deflection was presumably caused by the positive charge in the gold atoms. In the "plum pudding" model of the atom that was prevalent at the time, an atom's positive charge was thought to be evenly spread throughout the atom. That would have resulted in very small deflections of the alpha particles.

But Rutherford found that a small but non-negligible fraction of the alpha particles rebounded at large angles, suggesting that the mass and positive charge of the atom were concentrated in a tiny region in the center. Of course, we now call that region the nucleus.

If you know the initial kinetic energy of an alpha particle, and how many target nuclei there are per unit volume in the foil, you can calculate the probability that your alpha particle will come within any given distance of a nucleus. (For a thin foil target, the odds of a single alpha particle having close encounters with two nuclei is negligible.) And based on the distance an alpha particle comes to a nucleus, you can predict its angle of deflection. Putting it all together, Rutherford wrote what we now call the "Rutherford scattering formula": the fraction of alpha particles that should be deflected at any given angle, based on his nuclear model of the atom.

In his original 1911 gold-foil experiment, Rutherford's formula matched his data perfectly. In 1919 he repeated the experiment with an aluminum target. The formula still worked for low-energy alpha particles, but for high-energy ones the results differed significantly from his predictions. He correctly reasoned that this was because some of the alpha particles were penetrating *inside the nucleus.*

Active Reading Exercise: A Finite Size Nucleus

The Rutherford scattering formula is based on a point-like nucleus. That model works as long as the alpha particle stays outside the nucleus, because the force from a spherical charge is the same as if it were all at one point. But when the distance of closest approach is smaller than the radius of the nucleus, Rutherford's formula fails.

---

2 Technically his students Hans Geiger and Ernest Marsden fired the alpha particles, but we're pretty sure Rutherford told them to.

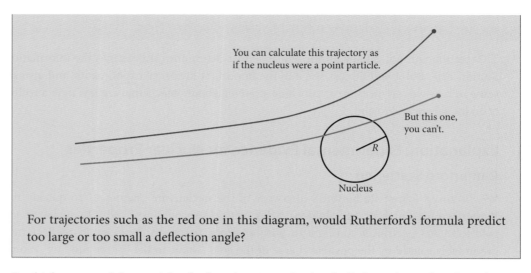

You can calculate this trajectory as
if the nucleus were a point particle.

But this one,
you can't.

R

Nucleus

For trajectories such as the red one in this diagram, would Rutherford's formula predict
too large or too small a deflection angle?

For high-energy alpha particles fired at aluminum, Rutherford's formula predicted too large a
deflection angle. That's because when the alpha particle passes through the nucleus it feels a
smaller net force than it would if the charge were all at the center. That discrepancy meant that
Rutherford's 1919 experiment found fewer scattering events at very high angles than his model
predicted.

Later experiments were able to show deviations from Rutherford's scattering formula for a
variety of target nuclei. Figure 12.6 shows the scattering of alpha particles off silver and lead
nuclei measured by George Farwell and Harvey Wegner in 1954. The horizontal axis is the alpha
particle energy and the vertical axis is proportional to the number of alpha particles deflected
at $60°$. For low energies, Rutherford's formula for a $-k/r$ potential from the nucleus works
perfectly. At high energies, the rate of scattering at this large angle is significantly less than the
formula predicts.

Do you see what this deviation buys us? For any given alpha particle energy we can calculate
the distance of closest approach to the nucleus. For each target nucleus, the alpha particle energy
at which the Rutherford formula fails then tells us the distance at which the alpha particle
is penetrating inside the nucleus. In other words, these measurements give us a value for the
radius of each nucleus. Farwell and Wegner found nuclear radii of roughly 10 fm for a variety
of different elements.[3]

A number of other methods exist for measuring nuclear radii. We describe below a way to
do this using diffraction, and Problem 8 describes a very different technique. These methods all
give answers consistent with $R = R_0 A^{1/3}$ where $R_0 = 1$–5 fm. The wide range in that value is
an indication of the fact that "radius of a nucleus" is not an exactly defined concept.

Electron Scattering

In 1953, Robert Hofstadter performed an experiment very similar to Rutherford's, measuring
scattering from nuclei. But Hofstadter used a beam of electrons, not alpha particles. There are
two main advantages to using electrons. First, electrons don't feel the strong nuclear force, so

---

3 The interpretation of these results is somewhat complicated by the need to make assumptions about the radius of the
   alpha particles.

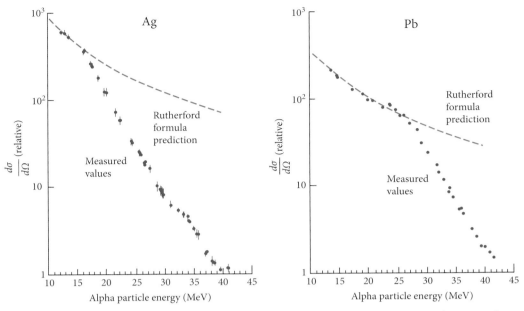

Figure 12.6 The number of alpha particles scattered at 60° by silver and lead nuclei as a function of alpha particle energy. At high energies the observed rate is well below the predictions of Rutherford's formula.

when an electron penetrates a nucleus the resulting interactions are only electromagnetic, and thus much simpler. Second, the protons and neutrons within an alpha particle can be perturbed in ways that result in complicated *internal* dynamics; an electron, being a point particle, avoids such complications.

When Rutherford analyzed alpha particle scattering, he was able to treat the alpha particles classically. But in Hofstadter's experiment, the much lighter and more energetic electrons showed wave behavior. When high-energy electrons are fired at a target, each nucleus acts like a small circular scattering surface, resulting in a pattern essentially identical to single-slit diffraction. For example, the scattering data in Figure 12.7 show the number of electrons scattered (vertical axis) at each angle (horizontal axis). The data show a node, an angle at which very few electrons were scattered because of destructive interference.

To see diffraction, Hofstadter needed electrons with wavelengths comparable to the size of a nucleus. Despite the advantages we listed above of using electrons, their light mass gives them relatively large de Broglie wavelengths, so Hofstadter had to use particle accelerators to get the electrons up to energies of several hundred MeV. (Compare that to the roughly 7 MeV of Rutherford's alpha particles.)

As you may recall – and it's in Appendix E if you don't – diffraction nodes occur at angles given by $\sin \phi = n\lambda/w$. Plugging in the de Broglie wavelength of the electron, you can measure ϕ for the first node ($n = 1$) and solve for the nuclear width w (see Problem 4). Hofstadter won a Nobel prize in 1961 for his electron scattering work.

In addition to giving us another way to measure nuclear width, the details of the electron diffraction patterns tell us that the charge density is spread very evenly throughout the nucleus.

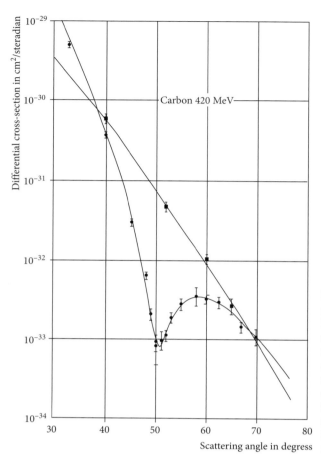

Figure 12.7 Scattering of 420 MeV electrons off ^{12}C nuclei. The horizontal axis is the scattering angle and the vertical axis is proportional to how many electrons were scattered at that angle. The dip in scattering rate at 50° is a diffraction node. (The roughly linear curve comes from inelastic scattering events, which are not relevant for our purposes.) *Source: © The Nobel Foundation 1961.*

This means that the protons themselves are distributed evenly throughout the space. (Other evidence suggests that the same is true of the neutrons.)

Mass Spectrometry

Mass spectrometry is a technique for measuring the masses of charged particles. When that technique is used on ionized atoms, the result is effectively a measurement of nuclear mass since the mass of an atom is almost entirely in the nucleus.

Figure 12.8 shows a schematic for a mass spectrometer. We begin with a stream of singly ionized atoms. Each ion in this stream has a charge of $+e$ and an unknown mass M, which this apparatus will determine.

The beam first passes through a region of crossed electric and magnetic fields, in the cylinder labeled "Source" in the diagram. The only atoms that come through this region undeflected are those for which $eE = Bev$, meaning $v = E/B$. So the cylinder serves as a source of atoms with tightly controlled velocities.

After they leave the source, the electric field is gone but the atoms are still subject to the magnetic field – magnitude B, pointing directly out of the page – causing them to follow a circular path until they reach a photographic plate. Measuring their positions on this plate gives us a measurement of the radius R of their circular path.

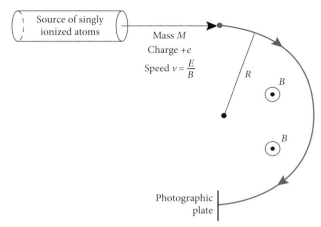

Figure 12.8 Mass spectrometer (Bainbridge design).

Active Reading Exercise: Mass Spectrometry

The centripetal acceleration that keeps the atom in Figure 12.8 moving in a circle is supplied by the magnetic force $\vec{F} = q\vec{v} \times \vec{B}$. Use that fact to calculate the mass M of the atom as a function of e, R, B, and E.

We hope you concluded that $M = eRB^2/E$. With carefully controlled field strengths and a precisely measured radius, that formula can provide an accurate measurement of atomic mass.

When J. J. Thomson conducted the first such tests in 1911, he was surprised to find that neon atoms left images in two different places on his photographic plate (Figure 12.9). Of the

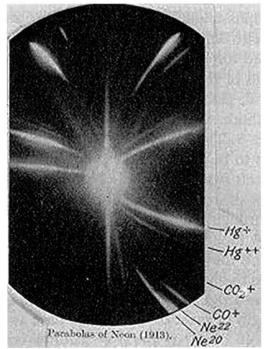

Figure 12.9 Thomson's photographic plate, with distinct lines for ^{20}Ne and ^{22}Ne in the lower right-hand corner. *Source: J.J. Thomson 1913.*

neon atoms, 91% were measured with masses corresponding to $A = 20$, and the other 9% were measured with $A = 22$. He concluded that there were different varieties of the same element, identical chemically but distinguishable physically. Thomson's experiment was the first evidence for different stable isotopes of the same element.

12.2.2 Questions and Problems: Experimental Evidence for Nuclear Properties

Conceptual Questions and ConcepTests

1. When an alpha particle is scattered from a larger nucleus, you can reasonably treat both nuclei as point particles unless the distance of closest approach is smaller than the nuclear radius. Why? *Hint*: Ignore modern physics entirely and treat this as a problem in classical mechanics.

2. Why did Rutherford see deviations from his predictions when he used aluminum, but not with gold?

Problems

3. The Rutherford scattering formula breaks down when an alpha particle actually penetrates inside a nucleus. In this problem, assume that the target nucleus has a radius given by $R = R_0 A^{1/3}$ where $R_0 = 1$ fm, and the incident nucleus (the alpha particle) is a point particle.

   (a) Calculate the kinetic energy an alpha particle would need to have in order to penetrate within the radius of a nucleus of aluminum, $^{27}_{13}\text{Al}$.

   (b) Calculate the kinetic energy an alpha particle would need to have in order to penetrate within the radius of a nucleus of gold, $^{197}_{79}\text{Au}$.

   (c) Briefly explain why Rutherford's scattering formula worked with gold (for all incident particles), and with aluminum for low-energy incident particles, but broke down when high-energy particles were fired at an aluminum target.

4. Figure 12.7 shows the scattering of 420 MeV electrons off ^{12}C nuclei. Each nucleus acts as a "slit" in a diffraction experiment.

   (a) Use the data to estimate the diameter of a carbon nucleus.

   (b) Use your answer to estimate the value of R_0 in the formula $R = R_0 A^{1/3}$.

5. Suppose a 200 MeV electron scatters off a nucleus, which you can treat as a slit 10 fm wide. At what angle would you find the first node of the resulting diffraction pattern?

6. Imagine doing a mass spectrometry experiment like the one in Figure 12.8, but with a *doubly* ionized atom. Calculate the mass of the atom as a function of the proton charge e, the radius R, and the field strengths.

7. Recall that Rutherford knew that any deflection of his alpha particles was caused by their interaction with heavy positive charges, because electrons are too light to cause much deflection in an alpha particle. But the same logic does not apply to Hofstadter's experiment; he was firing electrons. Why weren't they scattered by the atomic electrons before they ever reached the nuclei?

   (a) If you do this calculation classically, you conclude that the incoming electrons *would* in fact be scattered by the electrons in the target. To get a rough approximation of this, use classical equations for conservation of energy and momentum to show that an object of mass m colliding with an initially stationary object of mass m may be deflected by as much as $45°$. *Hint*: The largest angle you can get will occur if the two particles move off at equal and opposite angles from the original path, so you can assume that.

   (b) But Hofstadter's electrons were fast. Show that, for a 500 MeV electron, the relativistic relationship between energy and momentum can reasonably be approximated, *not* by the classical $E = p^2/(2m)$, but by $E = pc$.

   (c) Show that, if the incoming and target electrons both head off at the same speed (which implies opposite angles $\pm\theta$), the deflection angle must be $\theta = 0$: in other words, there is no scattering.

8. **Mirror Nuclei**

"Mirror nuclei" (or "mirror nuclides") are pairs of nuclei with different Z and N but the same total atomic number, such as $_6^{13}C$ and $_7^{13}N$. Because the strong force is independent of charge, two mirror nuclei should have the same strong-force binding energy, and their energy difference should only come from their electrostatic potential energies. The electrostatic potential energy of a uniform sphere of charge q is $(3/5)q^2/(4\pi\epsilon_0 R)$.

(a) Calculate the energy difference ΔE between a nucleus with Z protons and N neutrons and a mirror nucleus with $Z-1$ protons and $N+1$ neutrons. You can assume both nuclei have the same radius R.

(b) Why is it reasonable to assume they both have the same radius?

(c) Solve the equation you found in Part (a) to find R as a function of ΔE.

(d) When $_{16}^{31}S$ decays into $_{15}^{31}P$ it emits 5.4 MeV of radiation, indicating that the sulfur nucleus has 5.4 MeV more energy than the phosphorus. Use that fact to estimate the radius of these nuclei.

(e) Plug your result into the formula $R = R_0 A^{1/3}$ to estimate R_0.

12.3 Nuclear Models

In Chapter 7 we analyzed the behavior of a hydrogen atom (or any other atom with only one electron) by writing down Schrödinger's equation with the appropriate potential and solving it. But our approach to multielectron atoms in Chapter 8 was quite different: we used models, qualitative arguments, plausible extrapolations from the one-electron case, and of course experimental results. You cannot analytically solve Schrödinger's equation for a multielectron atom, so you have to use more indirect methods to predict the system's behavior.

Similarly, we cannot write down and solve a definitive equation that describes an atomic nucleus. Instead we use a variety of models that account for many of the observed properties and interactions of different nuclei.

As we mentioned in Section 12.1, we are going to refer to Figures 12.4 and 12.5 frequently throughout this chapter. You may want to have them handy as you read this section. You can find them at www.cambridge.org/felder-modernphysics/figures

12.3.1 Explanation: Nuclear Models

The data in Figures 12.4 and 12.5 come from measurements, but much can be explained by theoretical models. This section will discuss various models of the nucleus and how they explain different aspects of these data. The rest of the chapter will largely be about the consequences of the differences between nuclei shown in those two graphs.

The Liquid Drop Model

The "liquid drop model" is based on the following similarities between a nucleus and a drop of water.

- The density, measured in particles per unit volume, is roughly constant throughout the interior. The density is also largely independent of the size of the nucleus or the water drop.

- The particles (nucleons or water molecules) are mostly attracted only to their nearest neighbors. That means particles in the interior feel very little net force, while particles at the edge feel a net attraction toward the center.

One of the key insights of the liquid drop model is that the higher potential energy of nucleons near the surface leads to an effective surface tension, which is minimized by minimizing the surface area. That implies that nuclei, like water drops, should be approximately spherical. (Nuclei are actually elliptical, but most have long and short axes within a few percent of each other.)

One of the goals of any nuclear model is to explain Figure 12.4, since those data – the binding energy per nucleon of various nuclides – tell us so much about nuclear behavior. Here are some insights we can glean from the liquid drop model.

- On the far left side of Figure 12.4, we see that nuclides with very low atomic weights have very low B/A-values. This follows from the fact that we noted above, that nuclei near the edge of a nucleus are at a higher potential energy – that is, they are *less bound* – than nucleons buried deeper inside the nucleus. In a nucleus with $A = 7$, a significant fraction of the nucleons are on the edge; for $A = 16$ that fraction is much lower.

- Moving to the right side of Figure 12.4, we see that the binding energy per nucleon starts to decrease as the atomic weight grows to the mid-60s and beyond. As you add more protons, the positive Coulomb potential energy increases. Remember that the electric force is "long-range": every proton in a nucleus repels every other proton, while every nucleon attracts only its nearest neighbors. So, past a certain point, the positive Coulomb energy per nucleon grows faster with Z than the negative nuclear binding energy.

The liquid drop model dominated nuclear physics for several decades, and it is still a very useful starting point. But there are many important phenomena it *can't* explain, such as the "magic numbers." (Why is B/A higher for ^{12}C than for ^{14}N?) And you should keep in mind, if you're picturing a group of spherical protons and neutrons packed rigidly together in a spherical nucleus, that such a picture is useful but limited: the nucleons are in constant motion, and anyway their positions and momenta are always subject to quantum mechanical uncertainty.

The Fermi Gas Model

The liquid drop model is classical; further insight into nuclei requires considering the quantum nature of nucleons.

We mentioned in Section 12.1 that nucleons are approximately in a finite-square-well potential, with a flat potential inside the nucleus and a rapid rise in potential at the boundaries. Since protons and neutrons are fermions, they obey the Pauli exclusion principle. So in the ground state the protons occupy the Z lowest energy levels of the finite square well and the neutrons occupy the N lowest energy levels. (The Pauli exclusion principle only applies to identical fermions, so a proton and a neutron can be in the same energy eigenstate as each other.)

A collection of fermions in a flat potential is called a "Fermi gas," which lends this model its name.

The Fermi gas model explains one of the most important tendencies in Figure 12.5, the curve of stability. In that figure we see that light stable nuclei tend to cluster around $N = Z$, while heavier nuclei require increasing N/Z ratios.

- Imagine a nucleus with significantly more neutrons than protons. (The argument works just as well the other way.) That means that the energy level occupied by the highest-energy neutron is higher than the energy level occupied by any of the protons. Exchanging some of those neutrons for protons would lower the total energy of the nucleus, which explains why $N = Z$ is favorable at the lighter end. See Figure 12.10.

- But for larger nuclei the repulsive Coulomb energy between protons becomes more important. This is because, as we discussed in the liquid drop model above, adding more protons increases the repulsion that every proton feels from every *other* proton. So more neutrons exerting more attractive strong force are required to keep the nucleus stable.

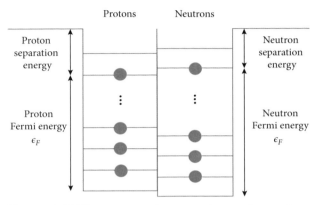

Figure 12.10 The protons and neutrons in a nucleus each occupy a separate square well. The Fermi energy is the gap from the bottom of the well to the highest filled state, and the separation energy is the gap from the highest filled state to the top of the well. The proton well is less deep because of positive Coulomb potential energy, an effect that grows at larger Z. We arbitrarily put the highest filled neutron energy above the highest filled proton energy; either one can be higher.

The highest filled energy level in a Fermi gas in its ground state is called the "Fermi energy" ϵ_F. You can calculate ϵ_F, which tells you the energy gap from the bottom of the well to the highest occupied state (typically a little over 40 MeV). You can also measure the proton or neutron separation energy, which tells you the energy gap from the highest occupied state to the top of the well (typically about 7 MeV). The sum of those two numbers is the depth of the potential well. See Problem 7.

The Shell Model

The main limitation of the Fermi gas model is the assumption that each nucleon feels a flat potential inside the nucleus. That's not a bad approximation, but a more realistic model needs to take into account that each nucleon moves in a roughly spherically symmetric potential created by all the other nucleons.

That potential is *not* the same as the $-k/r$ potential felt by an electron. This potential change is much more gradual within the nucleus, and sudden at the edge. But the potential does increase with r, so there is at least some analogy there. When we solved Schrödinger's equation for the electron, we found that its energy could occupy certain discrete "shells" that explained, for instance, electron emission spectra. The shell model of the nucleus says that the nucleons likewise have discrete energy levels, or shells, available to them.

This model explains the "magic numbers" discussed in Section 12.1. Recall that an atom with a full electron shell is a noble gas and tends to be inert; throw in one extra electron and you get an alkali metal, whose outermost electron is loosely bound and therefore highly prone to ionization. In a very analogous way, a nucleus with a full neutron shell is unusually stable; throw in an extra neutron and you get an isotope which is likely to shed that neutron, spontaneously decaying down to the magic number. That extra stability is reflected in Figure 12.4 as spikes of unusually high B/A for nuclei with full shells.

As in atoms, the lowest shell can hold two identical fermions with opposite spins. So $^{4}_{2}\text{He}$ is "doubly magic": its lowest proton shell is filled, and (separately) its lowest neutron shell is filled. Helium therefore has a very high binding energy per nucleon. The high B/A-value for $^{4}_{2}\text{He}$ is behind the phenomenon of "alpha decay," which we will discuss in Section 12.4.

When you take into account spin interactions between nucleons, pairs of protons and pairs of neutrons tend to bind together into relatively low-energy configurations. (We will not discuss why that occurs, since it involves more complicated quantum calculations than we can get into here.) The fact that nucleons pair up in this way accounts for the observed fact that nuclei with even Z and N tend to be more stable (higher B/A) than those in which either or both of those numbers are odd.

The Collective Model

The "collective model" is a hybrid of the liquid drop and shell models.[4] In the collective model, the completely filled shells are treated as a deformable liquid drop, and the nucleons in the outermost shell are viewed as orbiting about this inner core. Tidal interactions between these outer nucleons and the liquid core lead to accurate predictions of magnetic dipole moments and electric quadrupole moments, neither of which can be explained by any of the other models.

The Semiempirical Mass Formula

Later in this chapter we will look at conversions of one nucleus to another. An essential property of any such reaction is its "Q-value": the difference in energy between the products and the reactants. So it is important to be able to accurately predict the energy – or, equivalently, the mass – of a given nuclide.

Different tin nuclei might have different numbers of neutrons, but any two un-excited $^{132}_{50}\text{Sn}$ nuclei are pretty much identical, and they should have the same mass. Speaking generally, then, there must be a function $M(Z, A)$ that gives the mass of any nucleus in its ground state.

A reasonably accurate formula for that function is the "semiempirical mass formula." This formula was originally formulated by Carl Friedrich von Weizsäcker based on the liquid drop model, but it includes corrections due to quantum effects. In many ways it can be seen as a summary of the effects we've described in several of the nuclear models.

---

4 The collective model was partly developed by Aage Bohr, son of Niels Bohr, and led to their being one of four father–son pairs who have both been awarded Nobel Prizes in physics.

Because protons and neutrons have the same mass m_p to a fair degree of precision, our first stab at such a formula might be $M(Z, A) = m_p A$. Because their masses are not quite identical, a more accurate guess would be $M(Z, A) = m_p Z + m_n(A - Z)$.

But we recall from relativity that the energy inside the nucleus also contributes to its mass. So a complete analysis must take into account all the different forms of energy stored in the nucleus. Below we present a list of factors to build into our model. Note as we go that we assume all nuclei have the same density. That assumption means volume is proportional to A, width goes like $A^{1/3}$, and surface area as $A^{2/3}$.

In addition to the masses of the individual protons and neutrons . . .

- The binding energy due to the strong-force attraction between all the nucleons is a negative contribution to the total energy. If we assume that every nucleon is in the same potential well as every *other* nucleon, this energy should be proportional to A.

- But a nucleon at the outer edge of the nucleus is in a less deep potential well, and contributes less negative energy, than a nucleon in the interior. We build that into our model with a *positive* corrective term proportional to the surface area of the nucleus, which is to say, proportional to $A^{2/3}$.

- The electric repulsion between the protons is a positive contribution to the total energy. In Problem 6 you can show that this term is proportional to $Z(Z - 1)/A^{1/3}$.

- As we noted in our discussion of the Fermi gas model, nuclei tend to have Z close to N – in other words, Z close to $A/2$. We model that tendency with an energy term that is zero if $Z = A/2$, and is increasingly positive with increasing departure from this ideal. This term goes like $(Z - A/2)^2/A$.

- When we discussed the shell model we noted the tendency to have even Z-values and even N-values. So we add a "pairing" term to our model: zero if one of these values is even and the other is odd, negative energy if both are even, and positive energy if both are odd. The dependence of this term on A is difficult to derive and is thus written as A^k, where k is treated as another empirical coefficient to be matched to data. The best current data suggest $k = -1/2$.

Accumulating all these terms gives us a formula with five constants of proportionality:

$$M(Z, A) = m_p Z + m_n(A - Z) - a_1 A + a_2 A^{2/3} + a_3 \frac{Z(Z - 1)}{A^{1/3}}$$

$$+ a_4 \frac{(Z - A/2)^2}{A} + (-1, 0, \text{ or } 1)a_5 A^k. \tag{12.1}$$

The formula is "semiempirical" because the constants are then filled in based on experimental measurements of nuclear masses. This might make you suspicious: after all, with six constants to play with, you could match any six values perfectly, which would not validate any theory. But the semiempirical mass formula accurately predicts the masses of hundreds of stable nuclei and many more unstable nuclei, so the underlying theory is probably reasonably correct.

That said, the formula also has limitations. It does not work well for low A-values, and it does not account for the high binding energy of nuclei with "magic numbers," i.e. full shells of protons and/or neutrons.

12.3.2 Questions and Problems: Nuclear Models

Conceptual Questions and ConcepTests

1. List at least two ways in which a collection of nucleons inside a nucleus is like a collection of molecules in a liquid drop. Then list at least two ways in which it is not.

2. Why is it energetically favorable for a collection of nucleons to arrange themselves into an almost perfectly spherical shape?

3. Why does the shell model (or the Pauli exclusion principle) allow exactly four nucleons in the lowest-energy state?

4. If you add more protons to a nucleus, you increase the total electric repulsion but also increase the total strong-force attraction. You can easily imagine how this might increase or decrease the stability of the nucleus. But adding more neutrons seems like all attraction and no repulsion. So why can't you just throw more and more neutrons into any nucleus, and get more and more stability?

5. The term $a_2 A^{2/3}$ in the semiempirical mass formula is based on the nucleons at the outer edge of the nucleus.

   (a) Why does this term scale like $A^{2/3}$?

   (b) Why is this term positive?

Problems

6. One of the terms in the semiempirical mass formula represents the electric repulsion between the protons.

   (a) Explain why this term has a positive sign.

   (b) Every proton repels every other protons. For Z protons, how many proton pairs are there?

   (c) The potential energy between two protons scales like $1/r$, where r is the distance between them. The protons in a nucleus are distributed evenly (constant density). So the average distance of a proton to its *neighbors* is roughly constant, but the average distance of a proton to *all the other protons in the nucleus* grows with nuclear size. If we say that this average distance grows like r^k for some constant k, argue that this constant must be 1.

   (d) Putting it all together, explain why the proton repulsion term goes like $Z(Z-1)/A^{1/3}$.

7. In this problem you'll use the Fermi gas model to estimate the depth of the potential well felt by a nucleon inside a nucleus. As a rough estimate, we'll assume the potential well has the same energy eigenvalues as a 3D infinite square well. For that case, the Fermi energy of a collection of n identical fermions is[5]

$$\epsilon_F = \frac{3^{2/3} \pi^{4/3} \hbar^2 n^{2/3}}{2m V^{2/3}}.$$

Here m is the particle mass and V is the volume of the well (the nucleus in this case). This formula for the Fermi energy tells you how much energy the most energetic particle in the Fermi gas has, relative to the bottom of the potential well. Calculate the Fermi energy for a nucleus with n protons, n neutrons, and a typical nuclear density of 10^{17} kg/m³. This is a rough estimate, but you should be able to check that your answer matches the order of magnitude in Figure 12.3 on p. 560.

8. A reasonable set of values for the coefficients in the semiempirical mass formula (Equation (12.1)) is $a_1 = 15.8$ MeV, $a_2 = 18.3$ MeV, $a_3 = 0.714$ MeV, $a_4 = 92.8$ MeV, $a_5 = 12$ MeV, $k = -1/2$. Make a table of the predictions of this model as a function of Z and A, including the entire "curve of stability" on the right side of Figure 12.5. Then look up a table of nuclide masses. How well does the formula work out, across a wide range of mass numbers? Did you find any places where it does not work well?

---

5 You calculated this in Problem 32 of Section 10.6.

12.4 Three Types of Nuclear Decay

The phrase "nuclear decay" refers to the spontaneous transformation of a nucleus from one state to another. Of course such transformations go from higher-energy to lower-energy states, releasing energy in the form of radiation. Measuring that radiation gives us important information about nuclear states, just as we saw in earlier chapters that measuring the photons emitted by electron transitions gives us information about the states of atoms and molecules.

12.4.1 Discovery Exercise: Three Types of Nuclear Decay

Table 12.1 lists binding energy per nucleon in MeV for several nuclides.

Consider the following nuclear reaction:

$$^{238}U \rightarrow {}^{234}Th + {}^{4}He.$$

1. Calculate the *total* binding energy (not binding energy per nucleon) before and after the reaction.

   See Check Yourself #22 at www.cambridge.org/felder-modernphysics/checkyourself

2. Would you expect this reaction to occur spontaneously? Why or why not?

3. Answer the same questions for the reaction $^{84}Kr \rightarrow {}^{80}Se + {}^{4}He$.

Table 12.1 Binding energy per nucleon for several nuclides

Nuclide	^{238}U	^{84}Kr	^{234}Th	^{80}Se	^{4}He
B/A (MeV)	7.57	8.72	7.60	8.71	7.07

12.4.2 Explanation: Three Types of Nuclear Decay

As we've noted earlier, a nucleus that can spontaneously decay, meaning turn into a lower-energy combination of nuclei, is called "unstable." It may also be called "radioactive" because the decay process necessarily involves the emission of some kind of particles that carry off the excess energy.

Relativistically, the loss of energy in a nuclear decay can also be described as a loss of mass. The total energy lost is called the "Q-value" of the reaction:

$$Q = (M_{initial} - M_{final})\,c^2.$$

Only reactions with $Q > 0$ happen spontaneously, meaning without any external input of energy.

The simplest form of nuclear decay is when a nucleus simply emits one of its nucleons, but this is rarely energetically favorable (see Question 6). Aside from fission, which we'll discuss in Section 12.5, there are three main types of decay that do commonly occur in nature. These three decay processes were studied and named by Rutherford before they were properly understood. So their names can be interpreted as "This is a reaction that emits a mysterious thing" and "This is a reaction that emits a different mysterious thing," with Greek letters attached to the various mysteries. Today the names survive, although the processes and products are better understood.

- "Alpha decay" occurs when a nucleus emits a $^4_2$He nucleus. We still refer to $^4_2$He nuclei by the name Rutherford gave them, "alpha particles."

- "Beta decay" involves a neutron spontaneously changing into a proton or vice versa. In the process the nucleus emits an electron or positron (referred to in this context as a "beta particle"). We will discuss below the fact that scientists who followed Rutherford realized another particle must be involved as well, leading to the discovery of the neutrino.

- "Gamma decay" occurs when a nucleus drops from an excited state into a lower-energy state, releasing a high-energy photon. Today we still refer to light in that frequency range as "gamma radiation" or "gamma rays."

Alpha Decay

We said above that it is rarely energetically favorable for a nucleus to emit a single nucleon. Because of the unusually high binding energy of $^4_2$He, however, it is often favorable for a heavy nucleus ($Z \gtrsim 82$) to emit a helium nucleus:

$$^A_Z(\text{Parent}) \rightarrow \, ^{A-4}_{Z-2}(\text{Daughter}) + \, ^4_2\text{He}. \tag{12.2}$$

When we say alpha decay is "energetically favorable" for a given nucleus, we are saying that the total mass energy of the original configuration (the parent nucleus) is greater than the total mass energy of the final configuration (the daughter nucleus and the alpha particle, presumed to be very far away from each other). The difference appears as kinetic energy of the alpha particle, which is typically 4–8 MeV.

But in between the state "parent nucleus" and the state "widely separated daughter nucleus and alpha particle," it seems the system needs to pass through the state "daughter nucleus right next to alpha particle." In that state there's an additional Coulomb repulsion energy of about 30 MeV.

These energies are illustrated in Figure 12.11, which shows the potential energy of the alpha particle as a function of distance from the nucleus. The alpha particle has enough energy to *be* outside the nucleus, but not enough to go past the barrier and *get* outside the nucleus.

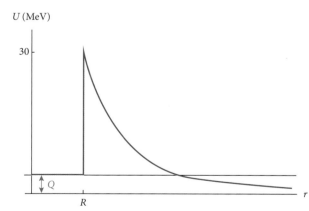

Figure 12.11 The potential energy of an alpha particle as a function of distance from the center of a nucleus of radius R. For $r > R$ the potential energy is the repulsive Coulomb energy. For $r < R$, there is no separate alpha particle and the potential energy reflects the binding energy of the four nucleons inside the nucleus. The difference between the inside-the-nucleus energy and the far-from-the-nucleus energy is the Q-value of the alpha decay.

Alpha decay was one of the first known examples of quantum tunneling. You may recall from Chapter 6 that a quantum mechanical particle can have a non-zero probability of being transmitted to the other side of a potential barrier even if its total energy is less than the height of the barrier.

The maximum height of the Coulomb potential energy for alpha decay is roughly 30 MeV for most nuclei. The higher the Q-value for the decay, the more potential energy the system starts with, so the less of a "jump" the alpha particle has to make to tunnel out. Because the probability of tunneling is exponentially sensitive to the height and width of the barrier, the lifetimes for alpha decays depend very sensitively on Q.

We can illustrate this point by comparing two different isotopes of thorium. ^{218}Th alpha-decays with a Q-value of 9.85 MeV, while the alpha decay of ^{232}Th has $Q \approx 4.01$ MeV. How much does that difference – just over a factor of two – change their decay times? More than you might think. The half-life of the former nuclide is 0.11 μs, while the latter lasts 14 billion years: roughly the current age of the universe.

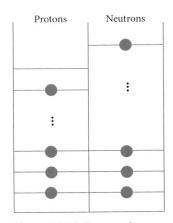

Figure 12.12 For any given mass number A, there is a ratio of neutrons to protons that minimizes the total energy. If a nucleus has an excess of one type of nucleon, that type will fill energy states higher than some of the available states for the other type.

Beta Decay

Within a nucleus, the protons and neutrons each fill up all their available energy states up to the Fermi energy. If one of them has a filled state at a higher energy than an unfilled state in the other one (Figure 12.12), then the nucleus can sometimes shed energy by turning a neutron into a proton or vice versa, a process known as "beta decay."

Active Reading Exercise: Beta Decay

When a neutron decays into a proton it produces other particles as well.

1. What conservation law would be violated if a neutron decayed into a proton and nothing else?
2. What other conservation laws does this process have to obey? See if you can come up with at least three.

If a neutron decayed into a proton and nothing else, that would violate conservation of charge. The beta decay of a neutron into a proton must therefore also produce an electron. This does *not* mean that the electron was hiding inside the neutron all along: as we mentioned in Section 12.1, an electron so confined would violate the uncertainty principle. The neutron decays, creating a proton and an electron that were not previously there. Just as Rutherford used the phrase "alpha particle" for what we would now call a helium nucleus, he detected the emission of "beta rays" that we now identify as electrons.

But neutrons, protons, and electrons are all spin-1/2 particles. If a neutron decayed into a proton and an electron you would start with total spin $\pm 1/2$ and end with total spin -1, 0, or 1 (depending on how the proton and electron spins were aligned). That would violate conservation of angular momentum.

Physicists studying beta decay in the 1920s faced another problem. Conservation of energy suggests that the electron should be emitted with a kinetic energy equal to 0.78 MeV, the difference in mass energy between a neutron and a proton. (Do you see why the proton would carry away negligible kinetic energy? See Problem 19.) But measurements showed that beta decay produced electrons with a range of kinetic energies from 0 up to 0.78 MeV. Some physicists were so vexed by this mismatch that they seriously suggested that energy might not always be conserved!

In 1931 Pauli suggested a better solution to the spin and energy problems. He proposed a third particle emerging from the reaction, extremely difficult to detect, but carrying with it energy and angular momentum (and no charge). Today we call those particles "neutrinos" (meaning "little neutral ones," written ν) and "antineutrinos" (their antimatter counterparts, written $\bar{\nu}$).

There are three different reactions that are classified as beta decay. Each one turns a neutron into a proton or vice versa. Electrons and positrons make up for the charge difference, while neutrinos and antineutrinos balance the scales of spin and energy:

$$\text{Electron emission} \quad n \rightarrow p + e^- + \bar{\nu}$$
$$\text{Electron capture} \quad p + e^- \rightarrow n + \nu$$
$$\text{Positron emission} \quad p \rightarrow n + e^+ + \nu$$

Note that electron capture involves stealing one of the electrons from the inner shells of the atom! The resulting process generally releases more energy than positron emission, so very few atoms spontaneously emit positrons.

If a neutron is not part of a nucleus it will spontaneously beta-decay in about 15 minutes. A proton outside a nucleus will never beta-decay (Question 9).

Gamma Decay and the Mössbauer Effect

Recall that some of the earliest evidence for quantum mechanics was discrete spectral lines. Any given element emits photons with certain specific frequencies, and can only absorb photons with those same frequencies. The Bohr model of the atom associated those values with the energy gaps between excited electron states.

The shell model of the nucleus predicts that nucleons should also occupy discrete energy levels, so we should expect nuclear state changes to also be associated with characteristic photon frequencies. But whereas the energy required to kick an electron into an excited state is typically on the order of eV, the excitation energy of a nucleus is in the range of MeV. So outside certain very specific circumstances – the interiors of stars, artificial reactors, and alpha and beta decays – you generally find nuclei sitting undisturbed in their ground states. When they do change states, the associated photons are in the high-frequency region of the spectrum that, as mentioned above, we still (following Rutherford) call "gamma rays."

In 1958 Rudolf Mössbauer published a technique for accurately measuring nuclear excited states. He received the 1961 Nobel Prize for what we now refer to as the "Mössbauer effect."

The basic idea is illustrated in Figure 12.13. Nucleus A is in an excited state. It drops down to its ground state, emitting a photon which goes on to hit Nucleus B. Since the photon's energy is precisely the energy difference between the ground state and this excited state, the photon can excite Nucleus B into that same excited state.

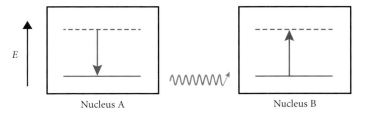

Figure 12.13 A nucleus drops from an excited state to a ground state, emitting a photon. A second nucleus absorbs that photon and goes into the same excited state the first nucleus started in.

But that won't generally happen, and here's why. When Nucleus A drops down, some of its energy goes into recoil of the nucleus itself. So the photon doesn't have enough energy to excite Nucleus B.

Mössbauer's solution was to embed his nuclei in crystals that held them tightly in place, so that all the energy of the nuclear transition went into the gamma rays (Question 10). Those gamma rays were then successfully absorbed by other nuclei. We say that such a system is in "resonance": the frequency of the gamma rays matches the energy required to excite the target nuclei.

Once you have that basic setup, you can tinker with it to learn a great deal about nuclear state transitions.

- You can move the crystals relative to each other, which Doppler-shifts the gamma rays to higher or lower frequencies. When you cross a certain threshold speed, the photons are no longer absorbed by target nuclei. This gives you an experimental measurement of the width (the energy uncertainty) of the excited state.

- You can change the energies of the excited states – for instance, by applying an external magnetic field (the nuclear Zeeman effect). If you change the excitation energy of the emitting nucleus but not of the target nucleus (or vice versa), the photons will no longer be absorbed. You can then move the nuclei relative to each other until the Doppler shifting puts you *back* into resonance. This experiment effectively measures how much you shifted the excitation energies.

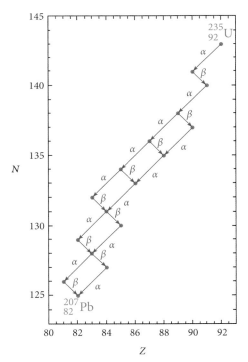

Figure 12.14 A decay chain. ^{235}U alpha-decays into ^{231}Th, which beta-decays into ^{231}Pa. At many points the chain branches, but it always ends at the stable nucleus ^{207}Pb.

Nuclear Decay Chains

Most of the time, when an unstable nucleus decays it produces another unstable nucleus, which in turn decays, until it reaches a stable state. The result is a "decay chain" such as the one shown in Figure 12.14.

Alpha decay reduces the mass number of a nucleus by 4, while beta and gamma decays leave it unchanged, so the quantity A mod 4 doesn't change in nuclear decay. There are thus four separate decay chains. For instance, Figure 12.14 shows the "$4n + 3$ series." Most alpha and beta decays also leave the nucleus in an excited state (because of selection rules), which then drops down in gamma decay. But, because the gamma decay does not change either Z or A, it is generally not listed as part of the chain.

If you look at the curve of stability in Figure 12.5, you can see that alpha decays move nuclei down and to the left along lines of slope 1, while beta decays move nuclei along lines of slope -1. But the curve of stability rises with a slope of roughly $3/2$. So after a few alpha decays, a nucleus typically needs to beta-decay in order to stay on the curve of stability.

Most nuclei in the $4n + 1$ series are not found in nature because the parent nucleus ^{237}Np has a half-life of *only* about three million years, so it all decayed long ago. All that's left are the last two steps in the series: ^{209}Bi (with a half-life of over 10^{19} years) and ^{205}Th (stable). The other three series all end in isotopes of lead.

Rates of Decay, and Some Terminology

If you have a sample of n radioactive nuclei, the number of them that decay per second is proportional to how many you have. So the number of remaining nuclei (or equivalently the mass of your sample) is described by a simple differential equation:

$$\left. \begin{array}{ll} \text{Equation:} & \dfrac{dn}{dt} = -\lambda n \quad \text{where } \lambda \text{ is a } positive \text{ constant} \\ \text{Solution:} & n = n_0 e^{-\lambda t}. \end{array} \right\} \qquad (12.3)$$

Such a decay process can be characterized by a number of related terms:

- The "decay constant" λ is the probability per unit time for a single nucleus to decay. The "mean lifetime" $1/\lambda$ is the time it takes for $1/e$ of the nuclei in your sample to decay.

- The "half-life" $t_{1/2} = (\ln 2)/\lambda$ is the time it takes for half of the nuclei in your sample to decay.

- The "activity" $a = \lambda n$ is the number of decays per unit time. Since a is proportional to n, a also decays exponentially: $a = a_0 e^{-\lambda t}$. The SI unit of activity is the "becquerel" (Bq), equal to one decay per second, but a more convenient and often-used unit is the "curie" (Ci), equal to 3.7×10^{10} decays per second (roughly the activity of a gram of radium).

Of these terms, the most familiar is the half-life, which is a convenient way of describing how long a particular decay typically takes. If you have a gram of a substance, one half-life later you will have half a gram. Wait another half-life, and you will be down to a fourth. The half-life of ^{238}U is about 4.5 billion years; the half-life of ^{14}C is about 5700 years; the half-life of ^{12}O is about 6×10^{-22} seconds.

Radioactive substances like ^{238}U with very long half-lives aren't typically very dangerous because they radiate so slowly. Substances with very short half-lives can be dangerous if they emit a lot of radiation, but the danger is short-lived. What you have to worry about are the substances with intermediate half-lives. Those substances can last long enough to be absorbed or ingested in the body and then emit significant radiation. For example, ^{210}Po alpha-decays with a half-life of about 140 days and is therefore one of the most dangerous radioactive materials known.[6]

---

6 Polonium poisoning has been suspected in a number of assassinations, including that of ex-Palestinian leader Yasser Arafat.

12.4.3 Questions and Problems: Three Types of Nuclear Decay

Conceptual Questions and ConcepTests

1. In order for a given parent nucleus to alpha decay, the daughter nucleus must have ... (Choose one and explain.)

    A. at least a certain minimum binding energy per nucleon.

    B. at most a certain maximum binding energy per nucleon.

    C. exactly the right binding energy per nucleon.

    D. none of the above. The possibility of alpha decay doesn't depend on the product's binding energy per nucleon.

2. $^{215}_{84}$Po can undergo either alpha or beta decay.

    (a) If $^{215}_{84}$Po alpha-decays, what is the daughter nucleus?

    (b) If $^{215}_{84}$Po beta-decays, what is the daughter nucleus? *Hint*: Look at the curve of stability to determine which type of beta decay is likely to be energetically favorable.

3. Equation (12.2) is the generic form of an alpha decay. Use similar notation to represent a beta decay in which a neutron inside the nucleus of a large atom is transformed into a proton (which stays in the nucleus), and an electron and an antineutrino (which fly off).

4. Two substances both decay according to Equation (12.3), but Substance A has a higher value of λ than Substance B. If both substances start off with the same number of atoms, how will their decay processes compare over time?

5. A decay chain begins with ^{238}U. Explain why that chain cannot include ^{205}Tl.

6. Why is it not generally energetically favorable for a nucleus to decay by emitting a single nucleon? *Hint*: What's the binding energy per nucleon of a single nucleon?

7. Figure 12.5 shows that the highest binding energy per nucleon occurs at $A \approx 56$. So a nucleus with $A > 60$ would generally increase its binding energy per nucleon by emitting an alpha particle. Why is

alpha decay only energetically favorable for much heavier nuclei? *Hint*: Consider the binding energy per nucleon of $^{4}_{2}$He.

8. When many identical nuclei decay through electron emission, the emitted antineutrinos emerge with a range of possible kinetic energies. When many identical nuclei decay by capturing an electron from the innermost shell, do the emitted neutrinos ... (Choose one and explain.)

    A. come out with a range of possible kinetic energies?

    B. all come out with nearly the same kinetic energy?

9. Why can't an isolated proton (not in a nucleus) undergo beta decay?

10. When a nucleus drops down from an excited state and emits a photon, conservation of momentum demands that the nucleus recoil; the photon cannot possibly have 100% of the transition energy. How did embedding the nuclei in crystals allow Mössbauer to achieve resonance? (Even though this is a "Question" rather than a "Problem," your answer should include a few equations.)

11. Neutrons are unstable; a neutron spontaneously beta-decays after about 15 minutes. So how can nuclei exist?

Problems

12. An atom of ^{238}U, initially at rest, alpha-decays into an atom of ^{234}Th. The binding energies per nucleon are 7.570 MeV for ^{238}U, 7.597 MeV for ^{234}Th, and 7.074 MeV for an alpha particle. Calculate the kinetic energy you would expect to measure in the emitted alpha particle. *Hint*: The speeds involved do not require relativistic formulas.

13. An excited state of ^{69}As was measured to have a lifetime of 72 fs (meaning 72×10^{-15} s) and it decayed by emitting gamma rays of approximately 8 MeV. Using the time–energy uncertainty principle, estimate the spread in energies of the gamma rays emitted by this transition.

14. While a plant or animal is alive, it keeps the ratio of carbon isotopes in its body at 1 atom of ^{14}C for every 10^{12} atoms of ^{12}C. (This is because animals eat plants, and plants take in carbon from the atmosphere, and the atmospheric ratio is $1 : 10^{12}$.) When the plant or animal dies it stops accumulating carbon, and the ^{14}C in its body decays with a half-life of roughly 5700 years. Measuring the amount of ^{14}C can therefore be used to approximate the age of an organic sample, a process known as "radiocarbon dating." If the carbon in an organic sample is found to have 1 part in 10^{14} of ^{14}C, how long ago did it die?

15. A collection of atoms decaying according to Equation (12.3) has a consistent "half-life": if you wait that long, only half the substance will remain. Does it also have a consistent "third-life" after which a third of the substance will remain? If so, find how the "third-life" is related to the half-life. If not, explain why not.

16. Equation (12.3) presents both the differential equation for radioactive decay, and its solution. Show that the solution does indeed solve the differential equation.

17. A parent nucleus X, at rest, decays into a daughter nucleus Y and an alpha particle α that fly off in opposite directions. The energy Q released in the decay becomes the kinetic energy of the two final particles: $Q = K_\alpha + K_Y$.

    (a) Using conservation of momentum, eliminate K_Y to write an equation relating Q, K_α, M_α, and M_Y. *Hint*: The energies in this problem are so much smaller than the nuclear masses that you can use non-relativistic formulas.

    (b) Approximate the mass of each nucleus (including the alpha particle) as $m_p A$ and solve the equation you derived in Part (a) to get an equation for K_α that depends only on Q and A_X. Simplify as much as possible.

    (c) The binding energy of $^{228}_{90}$Th is 1.743077 GeV, the binding energy of $^{224}_{88}$Ra is 1.720301 GeV, and the binding energy of $^4_2$He is 28.296 MeV. Suppose you witness the alpha decay of thorium to radium, and the alpha particle emerges with only 5.14 MeV. Assuming we can

neglect any difference in the kinetic energy of the Ra nucleus, the excess energy went into an excited state of the resulting $^{224}_{88}$Ra nucleus. When that nucleus subsequently decays down into its ground state, what energy should you see in the resulting gamma ray?

    (d) Why was it reasonable to neglect any change to the kinetic energy of the Ra nucleus in Part (c)?

18. You might imagine that the only nuclei present on Earth today should be those with half-lives that are at least a substantial fraction of the Earth's age, but we find nuclei with half-lives of a fraction of a second. To see how that can be, consider the following example. ^{235}U decays with a half-life of approximately 700 million years into ^{231}Th, which in turn decays into ^{231}Pa with a half-life of one day. Neither ^{235}U nor ^{231}Th is produced or destroyed in significant amounts by any other process. What's the ratio of ^{231}Th to ^{235}U in the Earth today? *Hint*: You can assume that thorium is in a steady state in which the rate of its decay equals the rate of its production.

19. In this problem we will imagine that a neutron – initially at rest – decays into a proton, an electron, and nothing else. (This is how beta decay was understood before Pauli proposed the neutrino.)

    (a) We're going to assume in this problem that the emitted proton is slow (and we can use classical equations), while the emitted electron is fast (and we can use the "ultra-relativistic" approximation $pc \gg mc^2$). Briefly explain why these are reasonable assumptions.

    (b) Based on those assumptions, write the formula for the kinetic energy of the proton as a function of its momentum, and the kinetic energy of the electron as a function of its momentum.

    (c) Write the equation for conservation of energy for this reaction. Remember that $p_{neutron} = 0$. So conservation of momentum dictates that $p_{proton} = p_{electron}$, which you can therefore just call p.

    (d) Solve your equation for p.

    (e) Find the ratio $KE_{proton}/KE_{electron}$. Your answer should be a formula involving the masses of all three particles.

(e) Plugging in numbers for the three masses, show that the Q-value of the reaction should appear almost entirely as the kinetic energy of the emitted electron.

20. When ^{57}Fe drops from its first excited state to its ground state, it emits a 14.4 keV photon. If you put two samples of ^{57}Fe in crystals so that they have negligible recoil, you find that the photons emitted by one sample can cause excitations in the second sample, until you move them toward each other with a relative velocity of $v \geq 5 \times 10^{-5}$ m/s. Find the width ΔE of the first excited state of ^{57}Fe. *Hint*: If your calculator doesn't have enough digits for the calculation, you might need to use computer software or a website.

12.5 Nuclear Fission and Fusion

We began this chapter with a brief history of the discoveries of ever smaller particles. That history, culminating with Chadwick's discovery of the neutron in 1932, put us on the path toward nuclear fission and fusion. Here are some important milestones along *that* journey.

- 1932: Mark Oliphant achieves hydrogen fusion in the lab.
- 1938: Otto Hahn and Fritz Strassmann discover the nuclear fission of heavy elements.
- 1939: Lise Meitner and Otto Frisch provide a theoretical explanation and a name for Hahn's discovery.
- 1939: Hans Bethe's paper "Energy Production in Stars" identifies nuclear fusion as the mechanism underlying sunshine.
- 1945: The first fission-based bomb is tested in the deserts of New Mexico.
- 1952: The first fusion-based bomb is tested in the Marshall Islands.
- 1954: The first fission-based power station to generate electricity for a power grid begins operation in Obninsk in the Soviet Union.
- The future: Fusion-based power plants produce clean, sustainable energy in unlimited quantities.

Hey, if we weren't optimistic, hopeful-about-the-future types, we never would have set out to write a textbook.

12.5.1 Discovery Exercise: Nuclear Fission

Consider the following fission reaction:

$$^{236}_{92}\text{U} \rightarrow {}^{92}_{36}\text{Kr} + {}^{141}_{56}\text{Ba} + ?$$

1. What must be the final product (where we have left a question mark) to keep proton number and neutron number conserved?

2. Use Figure 12.4 to estimate the binding energy per nucleon of each of these three nuclides. Based on those numbers and your answer to Part 1, estimate the energy released by this reaction.

12.5.2   Explanation: Nuclear Fission and Fusion

One of the most important lessons of Figure 12.4 is that the most stable configuration of a collection of nucleons occurs in nuclei with Z of 26–28 (which corresponds to an A of 56–62). So you can release energy by breaking apart larger nuclei, or by combining smaller nuclei. These two processes are nuclear "fission" and "fusion," respectively.

This section will discuss both of those processes, but we'll start with a discussion of nuclear reactions in general, and in particular the measure called the "cross-section" that is used to measure their probabilities.

In a typical nuclear reaction, a high-energy particle strikes a stationary nucleus. The result is a new nucleus (still nearly stationary) and one or more emitted particles:

$$n + {}^{21}_{10}\text{Ne} \rightarrow {}^{18}_{8}\text{O} + {}^{4}_{2}\text{He} \qquad \text{An example of a nuclear reaction.} \qquad (12.4)$$

The time scale for such reactions is typically on the order of 10^{-20} s. Beta decay, the conversion of neutrons into protons and vice versa, takes far longer than that (10^{-10} s or more) so these very fast reactions preserve Z and N independently. For instance, in Equation (12.4) we see 10 protons and 12 neutrons going into the reaction as an isolated neutron incident upon a neon target, and 10 protons and 12 neutrons coming out as a residual oxygen nucleus and an emitted alpha particle.

The Reaction Cross-Section

Imagine firing a high-energy particle at a thin foil target. There are many possible outcomes to such an experiment. The particle might hit a nucleus and cause it to emit a single neutron. Or two neutrons. Or three. Or an alpha particle. Or the particle might not cause any nuclear reactions and simply come out the other side. (That last possibility is often by far the most probable.)

Now suppose that you fire a stream of particles at the target and measure how often each outcome occurs. Let's say you find that 1 out of every 100 incident particles cause a single neutron to be emitted, and 1 out of every 1000 cause an alpha particle to be emitted.

If you were writing up the results of your experiment you could simply report those probabilities, but someone repeating the experiment with a thicker or denser piece of foil would presumably get different numbers. So physicists report reaction probabilities using a measure called a "cross-section."

The term cross-section is defined by a simple fiction. To find the cross-section for the reaction "emit a single neutron," you pretend that every time a particle strikes a nucleus it is guaranteed to produce that reaction. Since that reaction occurred 1/100th of the time, that means that in this pretend scenario the nuclei fill up 1/100th of the area of the foil. If you divide the probability (1/100) by the number of nuclei per unit area in the foil (the actual number, not a pretend number), you get the effective area of each nucleus in this pretend scenario. That effective area per nucleus is written σ, and is called the "reaction cross-section" for the reaction in which a single neutron is emitted:

$$\sigma = \frac{\text{probability of a particular reaction}}{\text{number of nuclei per unit area in the target foil}} \qquad \textit{for a thin target.} \qquad (12.5)$$

The cross-section σ has units of area and is generally measured in "barns" ($1\,\text{b} = 10^{-28}\,\text{m}^2$).

One nice thing about the cross-section is that it gives you a simple picture to go with your probability. Each nucleus takes up an area σ, and the bigger that area is, the more likely the particle is to hit one of them (Figure 12.15). But it's important to remember that σ is not an actual cross-sectional area of the nucleus. Nuclei don't have exactly defined radii, and the cross-section is really just a convenient way of expressing probabilities. In the example we've been describing, for instance, the same nucleus would have 10 times as big a cross-section for emitting a neutron as it would for emitting an alpha particle. The cross-section for any particular reaction also depends on the energy of the incoming particle.

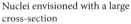
Nuclei envisioned with a large cross-section

The same nuclei envisioned with a smaller cross-section for a different reaction

Figure 12.15 The same set of nuclei can have different cross-sections for different reactions. The cross-section is not the "actual area" of the nucleus (which is not well-defined), but instead a convenient way of expressing reaction probabilities.

Equation (12.5), which we used to define "cross-section," is only valid in the limit of a target thin enough that you can neglect the probability of one nucleus shadowing another. For such a thin target, the reaction probability is simply the cross-section times the number of nuclei per unit area. For a thicker target, calculating probabilities from cross-sections is more complicated. But the same meaning of cross-section applies: if the nuclei were hard spheres such that the reaction would be guaranteed to happen if and only if the incident particle collided with one of them, how big would those spheres have to be to reproduce the measured probabilities?

Fission

Technically alpha decay is a form of fission, but it is usually considered a separate process. So when people talk about nuclear fission they mean a large nucleus splitting into two other relatively large nuclei.

Nonetheless, some of the basic physics of alpha decay applies to fission as well. Fission can only occur spontaneously if the average binding energy per nucleon of the products is greater than that of the parent. And even when that's the case, the fission products must tunnel through an intermediate radius where they are no longer bound by the strong force but are close enough to each other to feel a huge Coulomb repulsion.

For all naturally occurring elements the potential barrier for alpha decay is lower than that for fission, so alpha decay is the primary way in which nuclei spontaneously break into smaller pieces. However, some of those elements will undergo fission if an external particle pushes the nucleus into an excited state, closer to the top of the potential barrier.[7]

Imagine that a particle slams into a heavy nucleus, giving it energy and inducing it to split into two smaller nuclei. Because fission happens much faster than beta decay, the daughter nuclei emerge with the same total neutron-to-proton ratio as the parent. But as you may recall from the curve of stability (Figure 12.5), light nuclei need to have fewer neutrons per proton than heavy nuclei in order to be stable. So the daughter nuclei promptly emit several high-energy neutrons.

---

7 For some artificially produced elements with $Z \gtrsim 100$, spontaneous fission can dominate over alpha decay.

Those neutrons can then slam into neighboring heavy nuclei, inducing *them* to undergo fission, and a chain reaction can occur. Depending on the speed of the chain reaction, the result can either be a nuclear reactor or a nuclear bomb. We will talk about the first possibility.

Nuclear Reactors

An element that can sustain a fission chain reaction is called a "fissile material," the most common example being ^{235}U. (There are other fissile materials, most notably ^{239}Pu.) The products of a fission reaction vary unpredictably: even under identical starting conditions you cannot predict exactly what you will get out. Sometimes there are even three products ("ternary fission"). Far more commonly there are two products ("binary fission"), one with $A \approx 95$ and the other with $A \approx 135$. A typical reaction is shown below:

$$n + {}^{235}_{92}\text{U} \quad \rightarrow \quad {}^{92}_{36}\text{Kr} + {}^{141}_{56}\text{Ba} + 3n. \tag{12.6}$$

Fission of one ^{235}U atom generally releases two or three neutrons, which makes it sound like it would be easy to start a chain reaction. But there are two problems.

The first problem is that the neutrons are produced at very high energies, and the cross-section for neutrons to react with ^{235}U goes down rapidly with energy. The solution is to slow the neutrons down. So the nuclear fuel is surrounded by a "moderator," which can be ordinary water. The neutrons collide with the water molecules and lose energy until they slow down enough to react with a ^{235}U nucleus and cause another fission event. Unfortunately water molecules often absorb the neutrons, turning their ^{1}H nuclei into ^{2}H ("deuterium"). So a better (but more expensive) moderator is "heavy water," in which the hydrogen in the water molecules is already made of deuterium.

The second problem with creating a chain reaction is that ^{235}U, the isotope you want, makes up only 0.7% of naturally occurring uranium. The rest is ^{238}U, which happily absorbs high-energy neutrons but rarely undergoes fission as a result. However, the cross-section for neutron capture by ^{238}U increases with energy (the opposite of ^{235}U). That means moderating the neutrons is once again the answer; if you slow them down before too many of them are absorbed by ^{238}U, enough of them will encounter ^{235}U to sustain a chain reaction.

In practice, to run a reactor with regular water as the moderator you have to first enrich the uranium so that ^{235}U makes up at least 3%–5% of it. That's hard to do because ^{235}U and ^{238}U are chemically identical, so you have to use centrifuges that separate them based on mass. The other option is a heavy-water reactor, which can moderate the neutrons efficiently enough to sustain a chain reaction with natural uranium.

That chain reaction must be precisely controlled, though, so that each fission event generates, on average, exactly one other fission event. If each event generates less than one other event, the reaction will die out. If each one generates more than one more, the rate will grow exponentially until the resulting heat melts the core of the reactor. This fine tuning is achieved with "control rods" made of substances like boron or cadmium that absorb neutrons. Moving the rods in or out of the reactor core can keep the reaction rate stable.

How much energy is released in a single fission reaction? The exact answer depends on the specific reactants and products, but it doesn't vary much. Looking back at Figure 12.4 we see that the binding energy per nucleon is roughly 7.6 MeV for the reactants (uranium, plutonium, or others in that range), and 8.5 MeV for the products. Thus roughly 0.9 MeV is released per nucleon, which comes to about 200 MeV per atom. (That's on the order of 10^8 times greater than the per-atom energy released in chemical reactions, because nuclear binding energies are so much stronger than electron binding energies. Fission of a kilogram of ^{235}U releases approximately three million times as much energy as burning a kilogram of coal.)

The World's First Fission Reactor

In 1972 scientists discovered that two billion years ago a uranium reactor spontaneously formed at a place called Oklo in the modern country of Gabon, in Africa. The Oklo reactor ran for several hundred thousand years in a sandstone deposit suffused with enough water to act as a moderator. Of course this was normal water, not heavy water, and nobody was around at the time to run the uranium through centrifuges. So how was this possible?

Because ^{235}U decays much faster than ^{238}U, the fraction of ^{235}U in natural uranium deposits has been decreasing since the Earth formed. Two billion years ago ^{235}U made up several percent of the uranium, rather than the 0.7% it makes up today, so a chain reaction could occur. We believe the water also served the function of control rods. When the reaction started occurring too quickly the heat evaporated the water, removing the moderator and slowing the reaction.

The Oklo reactor was discovered because scientists noticed that the uranium deposits there currently have a ^{235}U concentration of 0.717%, instead of the usual 0.720% found elsewhere on Earth. That tiny difference was enough to signal that some of the ^{235}U had been used up earlier, and subsequent research confirmed the conclusion by finding the daughter nuclei produced by the natural fission.

Fusion

Nuclear fission as an energy source is tremendously powerful and efficient, but it's also problematic. The products are highly radioactive, leading to a nuclear waste problem that currently has no great solution. Nuclear *fusion* doesn't have this problem; the product is helium. Given the right technology we could literally harness the power that fuels the Sun, producing more than enough clean energy for all the world's needs. However, we have yet to build a fusion reactor that produces more energy than is required to run it. As of the time of this writing there are several reactors being built that may pass that threshold.

We observed in Section 6.5 that fusion in the Sun is an example of quantum mechanical tunneling. As two protons approach each other their potential energy increases because of their mutual Coulomb repulsion. If the two protons get close enough to each other, they fuse into a helium nucleus with a significantly decreased potential energy due to the strong force. But there is an in-between distance where the Coulomb repulsion creates a high potential energy, and the nuclear force is not yet able to fuse them. This creates a potential hill called the "Coulomb barrier" (Figure 12.16). The Sun shines because those protons can tunnel through that barrier and fuse.

As you might expect, the cross-section for that tunneling increases significantly at higher temperatures, since higher kinetic energy effectively lowers the Coulomb barrier. Efficient fusion typically requires temperatures above 10^8 K, as well as extremely high densities. But the core of the Sun is only at about 1.5×10^7 K, which is why stars last so long; on average it takes billions of years for any given proton in the Sun to collide with enough energy to fuse.

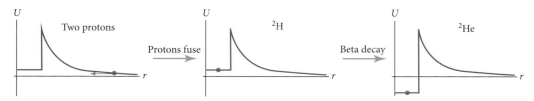

Figure 12.16 Potential energy as a function of separation for two colliding protons. At large distances they repel electrically. If they get within nuclear distances the nuclear force takes over and they can fuse. The resulting nucleus can then beta-decay into deuterium, releasing net energy as radiation.

So to make a fusion reactor, we have to create conditions several times hotter than the core of the Sun. But there's another problem: atoms at these temperatures are ionized, so they tend to repel each other at high speeds into the walls of their container. One option for dealing with this problem is a "tokamak," a reactor that uses strong magnetic fields to confine the ions in a toroidal chamber. The other main method is "inertial confinement," heating and compressing the fuel so quickly that it undergoes fusion before it can fly apart. This can be done, for example, by shining powerful lasers from multiple directions on a small fuel pellet.

Unlike fission, fusion does not proceed via a chain reaction. The hydrogen has to be externally heated to a temperature where it can fuse, and the energy released does not come out in forms that produce additional fusion reactions. The lack of a chain reaction makes fusion harder to sustain than fission, but it also means a fusion reactor would not have a risk of a catastrophic meltdown like a fission reactor. (Of course there are fusion-based bombs, but they require enough externally supplied energy to heat all the hydrogen to fusion temperatures. The only way we can do that is to surround the hydrogen with a fission bomb!)

Most of the energy released in fusion is in the form of high-energy neutrons. That energy can be captured by letting them collide with a target such as lithium that can absorb them and turn their energy into heat. It's important to absorb all the neutrons or they could destroy the reactor and/or turn the surrounding materials radioactive. So, in all fairness, fusion has its own set of challenges.

You probably won't be surprised to hear that we're oversimplifying the fusion process in a few ways. Perhaps most importantly, a helium nucleus isn't made of two protons. So the fusion process takes several distinct steps:

1. When two protons fuse they form ^{2}He, sometimes called a "diproton." This process actually *increases* the energy, so a net input of energy is required to make this happen.

2. Because of that high energy, the diproton is unstable, and tends to recollapse back into two protons. However . . .

3. If the diproton lasts long enough, one of its protons can beta-decay into a neutron. The result is ^{2}H (deuterium), which is stable. The beta decay releases more energy than the initial fusion consumed, so the whole process is exothermic.

4. Once the deuterium forms it can react with another proton to form ^{3}He, which can undergo several possible other fusion reactions (fusing with another ^{3}He or with a ^{4}He, for example, each of which gives products which then decay).

The final products are primarily ^{4}He and a lot of energy, but the road to get there is complicated.

12.5.3 Questions and Problems: Fission and Fusion

Conceptual Questions and ConcepTests

1. The cross-section for a nuclear reaction can depend on which of the following? (Choose all that apply.)

   A. The type of incident particle

   B. The type of target nucleus

   C. The energy of the incoming particle

   D. The thickness of the target

2. Suppose nucleus X with mass number A_X splits into nuclei Y and Z, with mass numbers A_Y and A_Z. What must be true of the binding energies per nucleon of the three nuclei if this reaction occurs spontaneously? (Your answer should be in the form of an equation.)

3. Some of the simplest reactions you can imagine are an atom splitting into two identical atoms ($^A_Z\text{Big} \rightarrow 2^{A/2}_{Z/2}\text{Small}$), and the reverse ($2^A_Z\text{Small} \rightarrow {}^{2A}_{2Z}\text{Big}$).

   (a) How big would A have to be for this fission reaction to give off energy (have a positive Q-value)?

   (b) Even when this fission reaction is energetically favorable (meaning, even for the A-values that you calculated in Part (a)), this type of even splitting generally does not occur. Why not?

   (c) How small would A have to be for this fusion reaction to give off energy?

4. Equation (12.5) gives the relationship between the cross-section of a reaction and the probability of that reaction, but this relationship only holds for a thin target. Explain why the probability of a reaction is *not* simply the cross-section times the number of nuclei per unit area for the case of a thick target.

5. Increasing the velocity of an incoming neutron decreases the cross-section of a ^{235}U reaction, and simultaneously *increases* the cross-section of a ^{238}U reaction.

   (a) Explain what that sentence means in terms of reaction probabilities. Do not use the words "cross-section" or "area." (Don't use "barns"

either. You can use "mellifluous" but we'll be impressed if you make it relevant.)

   (b) Explain how fission reactors use that fact to increase the chance of desirable, rather than undesirable, reactions.

6. In a fission reactor, let k be the average number of neutrons from each fission event that lead to another fission event.

   (a) If $k > 1$, will the reaction rate (number of events per second) increase linearly, quadratically, or exponentially? Briefly explain why.

   (b) If $k < 1$, how will the reaction rate decrease? Briefly explain why.

   (c) Name at least three factors that affect the value of k.

7. A fission reactor surrounds the fissile material with water, and then inserts control rods made of a different substance into the water. Briefly describe the effect that the water has on the neutrons, and then the effect that the control rods have. We're particularly interested here in how these two effects differ.

8. When the first atomic bomb was set off in New Mexico in 1945, some physicists were concerned that the intense heat could trigger hydrogen fusion, which might start a chain reaction and cause the Earth's atmosphere to explode. Why didn't that happen?

For Deeper Discussion

9. Normal fusion processes can't produce elements heavier than nickel ($Z = 28$; A is most often 58–60). But heavier elements such as uranium can form in a supernova, an extremely hot, dense plasma filled with elements up to nickel, plus lots of free neutrons. Explain how such conditions could form these heavy elements. *Hint*: Each step of the process doesn't need to jump to the most stable arrangement possible of the nucleons involved, but each step must result in a higher total binding energy than the previous step. A free neutron has *zero* binding energy.

Problems

10. Consider the spontaneous fission reaction:

$$^{256}_{100}\text{Fm} \ \rightarrow \ ^{125}_{49}\text{In} + ^{131}_{51}\text{Sb}.$$

The binding energies per nucleon of $^{256}_{100}\text{Fm}$ and $^{125}_{49}\text{In}$ are 7.432 MeV and 8.408 MeV, respectively. What is the minimum possible value for the binding energy per nucleon of $^{131}_{51}\text{Sb}$ that would make it possible for this reaction to occur spontaneously? (The actual value is 8.393 MeV, so you can check your answer by making sure it's less than that.)

11. A stream of particles is fired at a rate R (particles per unit time) at a thin target of thickness d and number density n (nuclei per unit volume). If the cross-section for alpha particle emission is σ_α, how many alpha particles will be emitted per unit time?

12. When 100 MeV alpha particles are fired at a target of ^{133}Cs, the cross-section to initiate a nuclear reaction is 2.2 barns. Cesium has a density of 1900 kg/m^3. If the cesium target is 1 mm thick, what's the probability of an alpha particle initiating a nuclear reaction? *Hint:* You can use A to estimate the mass of a cesium atom (or just look it up).

13. We said in the Explanation (Section 12.5.2) that ^{235}U produces about 0.9 MeV per nucleon when it undergoes fission.

(a) How much energy would be released if a kilogram of ^{235}U underwent fission?

(b) The United States consumes about 4 million gigawatt hours of electricity per year, and the density of uranium is about 20,000 kg/m^3. How many cubic meters of ^{235}U would be needed to power the US for a year?

(c) One "D–T fusion event" – the fusion of a deuterium nucleus with a tritium nucleus – releases 17.6 MeV. How many kilograms of deuterium and tritium would be required to power the US for a year?

Chapter Summary

This chapter gives a first overview of the atomic nucleus. There is no appendix of collected equations, but Figure 12.4 (p. 561 or online), the binding energy per nucleon for different atomic weights, is central. Much of the chapter is devoted either to explaining where the numbers in that figure come from, or to exploring the reactions to which those numbers lead.

Section 12.1 What's in a Nucleus?

This section summarizes the properties of a nucleus, and of the strong force that holds its nucleons together.

- We represent a particular nucleus or atom with the notation $^{A}_{Z}\text{X}$, where Z is the number of protons and A is the combined number of protons and neutrons.

- The nucleus is held together by the "strong" (or "strong nuclear") force. At distances between 1/2 and 2 fm, this force attracts protons and neutrons with a force many orders of magnitude stronger than the electric repulsion between the protons. Closer than about 1/2 fm, this force becomes repulsive.

- The "proton separation energy" is the energy required to liberate one proton from the nucleus. The "neutron separation energy" is of course the same thing for one neutron.

- The "binding energy" B of a nucleus is the difference in energy between the bound nucleus, and the component nucleons far away from each other. The higher the binding energy per nucleon (B/A), the more stable a nucleus will be.

- Figure 12.4 shows the values of B/A for different mass numbers. The figure peaks at $A = 62$, so energy is released when lighter elements increase toward that number (fusion), or when heavier elements decrease toward that number (alpha decay or fission). Some elements with so-called "magic numbers" of protons or neutrons have anomalously high binding energies.

- Figure 12.5 shows stable and unstable isotopes by atomic number.

- A nucleon in a nucleus, like an electron in an atom, has angular momentum due to both its orbit and its intrinsic spin. Protons and neutrons have half-integer spins, which means that, as fermions, they are subject to the Pauli exclusion principle.

Section 12.2 Experimental Evidence for Nuclear Properties

This section discusses some of the experiments that led to our current understanding of the nucleus.

- Rutherford scattering, as previously discussed in Section 4.1, contradicted the "plum pudding" model of the atom and implied the existence of a small, dense center of positive charge. Eight years later, Rutherford found that his formula broke down for high-energy incident particles, which actually penetrated into the nuclei of his aluminum foil target, allowing for an initial calculation of the nuclear radius.

- Over four decades after Rutherford's experiment, Robert Hofstadter fired a high-energy beam of electrons at a target. The electron experiment acts as a diffraction apparatus, with the target nuclei serving as the sources of the diffraction (the "slits"). The locations of the resulting nodes and antinodes gives an independent measure of the nuclear radius.

- In mass spectrometry, the trajectory of a charged particle through a magnetic field allows for a calculation of the particle's mass. Spectrometry can be used to measure the masses of different isotopes of the same element.

Section 12.3 Nuclear Models

It is not possible to analyze the nucleus by directly solving Schrödinger's equation for a collection of nucleons, but different models of the nucleus offer different insights into nuclear behavior.

- The "liquid drop model" makes use of similarities between a collection of nucleons forming a nucleus and a collection of molecules forming a drop of water. Inside the nucleus, the density of nucleons is roughly constant and the nucleons feel very little net force. Nucleons at the edge of the nucleus are at a higher potential energy, so they feel an attraction toward the center (which leads to surface tension and a roughly spherical shape).

- The "Fermi gas model" supplements the liquid drop model with quantum mechanical considerations, considering the nucleons as a collection of fermions that are generally in their ground states. This model explains why low-mass nuclei tend to be found fairly close to the $N = Z$ line on the curve of stability, while higher-mass nuclei require a few more neutrons than protons.

- The "shell model" posits discrete allowable energy levels for nucleons, analogous to the levels (or shells) allowed to the electrons in an atom. Just as a full shell of electrons leads to a fairly non-reactive atom (a noble gas), a full shell of protons or neutrons leads to a highly stable nucleus (corresponding to the "magic numbers").

- The "collective model" treats the filled shells of a nucleus as a deformable liquid drop, with the nucleons in the (unfilled) outermost shell orbiting around this inner core. This model accurately predicts the magnetic dipole moment and electric quadrupole moment of a nucleus.

- The "semiempirical mass formula" builds many of these considerations into a mathematical model for the mass of a nucleus as a function of Z and A. (Remember that the mass is not simply the sum of the proton and neutron masses; every change in the binding energy is registered as a mass change.) This formula accurately predicts the masses of hundreds of nuclei.

Section 12.4 Three Types of Nuclear Decay

The "Q-value" of a nuclear reaction is the energy difference between the initial and final states, $E_{\text{initial}} - E_{\text{final}}$. A reaction will spontaneously occur if $Q > 0$.

- Alpha decay involves a nucleus emitting a helium nucleus (also called an "alpha particle"):

$$^A_Z(\text{Parent}) \rightarrow {}^{A-4}_{Z-2}(\text{Daughter}) + {}^4_2\text{He}.$$

- Beta decay involves a neutron changing into a proton, or vice versa. Conservation of charge demands the emission of an electron or positron, or sometimes the capture of an electron from the atom. Conservation of energy and of angular momentum demand the emission of a neutrino or an antineutrino.

- Gamma decay involves a nucleus dropping down from an excited state to a lower state, emitting high-energy photons in the process. The study of gamma decay in the Mössbauer effect led to a greater understanding of nuclear energy levels.

- A high-weight nucleus will often undergo a "chain" of alpha and beta decays until it reaches its final, stable destination.

- A sample of radioactive atoms will decay down exponentially.

Section 12.5 Nuclear Fission and Fusion

Because the binding energy per nucleon (B/A) peaks around iron, heavier and lighter nuclei can reduce their energy by shifting their atomic weights toward roughly 56.

- When a high-energy particle strikes a target, the probability that it will produce any particular reaction is given by a "cross-section" for that reaction. The cross-section can be interpreted as an effective target area of each nucleus. In this (fictional) image, the incident particle will produce that reaction if and only if it strikes that effective target.

- When a neutron with the right energy level slams into the nucleus of a "fissile material" such as ^{235}U, it can cause the large nucleus to split into two smaller nuclei and a few leftover neutrons. This process releases large amounts of energy, since the total binding energy of the products is higher than the binding energy of the original nucleus. The released neutrons can go on to split other nuclei, leading to a chain reaction.

- In a fission reactor, a "moderator" (usually water) lowers the energy of the free neutrons, thus enabling a chain reaction. "Control rods" absorb excess neutrons, keeping the reaction rate stable.

- The Sun is powered by hydrogen nuclei fusing into helium and other heavier elements, thus increasing their total binding energy. Artificial fusion reactors have achieved the same thing, but so far none produces more energy than it consumes.

13

Particle Physics

Following Chadwick's discovery of the neutron in 1932, it might have seemed that all forms of matter could be explained as different combinations of fewer than 100 elements, and those elements in turn could be explained as different combinations of protons, neutrons, and electrons. Add photons to the list, and you pretty much had the universe summed up. This is the kind of model physicists like: complicated behavior arising from a few simple building blocks.

If there was a moment of such confidence, it didn't last long. Carl Anderson identified a positively charged electron in 1932 (dubbed a "positron" by the *Physical Review*), and over the next few decades hundreds of new particles were discovered. When Nobel Laureate Isidor Rabi was informed of a particle essentially identical to the electron but with larger mass (now called a "muon"), he responded: "Who ordered that?" Enrico Fermi supposedly quipped: "If I could remember the names of all these particles, I would have been a botanist."

In the 1960s, Murray Gell-Mann found a way out. He proposed that most of the apparently fundamental particles were made from more basic building blocks that he called "quarks." In the 1970s a small group of fundamental particles, including Gell-Mann's quarks, were formalized into the "standard model of particle physics", which is still used today. That model accounts for all the particles we have ever observed, and for three of the four known fundamental forces of nature.

The first two sections of this chapter survey the overwhelming "zoo" of particles that emerged over decades of experiments, and the standard model that brought order to the chaos. If you want a general overview of particle physics, you could stop after those two sections. The third section fills in some of the experimental evidence that uncovered so many particles. The rest of the chapter explores topics like the strong force, conservation laws and symmetry, and quantum field theory: the theory we use to describe how all these particles interact. Along the way we'll highlight some open questions and areas of active research in particle physics today.

13.1   Forces and Particles

In this section we're going to survey particle physics from the 1930s to the 1960s, a period marked by a dizzying proliferation of seemingly fundamental particles. In the next section we'll see how the "standard model" of the 1960s and 1970s simplified the list down to roughly 20.

13.1.1   Explanation: Forces and Particles

What defines a particular type of particle?

That's one of those questions that you've probably never bothered to ask. If you were sorting a set of blocks into different types, you might look at their shapes or their sizes or their colors. But an electron and a neutrino don't have shapes, sizes, or colors. (Or smells or textures or patterns or specific heats or favorite football teams.) What makes them different particles?

> ### Active Reading Exercise: Different Types of Particles
>
> Consider two point particles. List all the properties you can think of that might distinguish these two particles, such that we would say that they are not the same type of particle. (With a bit of thought, you should be able to think of the three we list below.)

Perhaps the most obvious property on the list is mass. Every electron has the same mass as every *other* electron. A less obvious property is spin. Remember that "fermions" have half-integer spins along any axis of measurement, and obey the Pauli exclusion principle, while "bosons" have whole-integer spins and do not.

The third property we hope you identified is charge, but we want to frame that one as part of a larger category: one of the most important properties of a particle is its *reaction to forces*. An electron can be attracted or repelled by the electric force, but is unaffected by the strong force. A neutron can be attracted by the strong force, but is unaffected by the electric force. A proton feels both.

The discovery of new particles over a 30-year period was largely a discovery of particles that reacted in different ways to different forces. For this reason, before we start cataloging particles, we need to take a look at the modern idea of a force.

What is a Force?

Newton's first law says that an object with no force on it moves at constant velocity. So in classical mechanics we can define a force as anything that causes a particle to deviate from that behavior.

In quantum mechanics a particle doesn't have a well-defined position, velocity, or acceleration, but we can offer an analogous definition. In the absence of a force – or, equivalently, in a constant potential energy field – a particle will have the energy eigenstates and time evolution of a free particle, $\Psi(x, t) = e^{i(kx - \omega t)}$ in 1D. Anything that causes any other behavior is a force.

A force, thus defined, can have many different measurable effects. It can of course have essentially the classical effect of imparting a non-zero average acceleration to a particle. We have also seen examples where a force splits a wavefunction into transmitted and reflected waves, and where a force binds a particle and restricts its energy levels to certain discrete values, among others. But there is another reaction that might not have occured to you:

When one type of particle turns into a different type, that is a reaction to a force.

You may well feel that we've stretched the word "force" past its breaking point. Particle decay is not the same thing as acceleration, and it isn't even a change in the eigenstates that come out of Schrödinger's equation. In fact, particle decay cannot be predicted by Schrödinger's equation at all; it has to be introduced to standard quantum mechanics as a new postulate. So what does it have to do with anything we associate with the word "force"?

The connection becomes clear in quantum field theory, currently our most complete and accurate model of particle behavior. In fact, within the framework of quantum field theory, all forces – all particle interactions – are based on particle transformations! We'll offer a general overview of quantum field theory in Section 13.5, and leave the rest for later courses. But for our purposes here, as we go about discovering and classifying particles, it's enough to know that in the *absence* of any forces, a particle will remain the same particle and evolve as a free particle. Any other behavior is a response to a force.

Current theory recognizes four fundamental forces. The most significant effect of each force is as follows:

- Gravity causes every object in the universe to attract every other object.
- Electromagnetism causes charged particles to attract and repel each other.
- The strong force causes nucleons (protons and neutrons) to attract each other, thus holding nuclei together.
- The weak force is responsible for beta decay, in which one type of nucleus transforms into another while emitting other particles (e.g. electrons and neutrinos).

Properties of the Four Forces

The Earth and the Moon are both full of protons, neutrons, and electrons. But the mutual interaction of the Earth and Moon is all gravitational: the electromagnetic and strong forces are entirely negligible. On the other hand, when we analyze the binding of an oxygen atom to two hydrogen atoms in a water molecule, we look only at the electromagnetic force. And for two protons in a nucleus, the strong force dominates.

Which force dominates in any given situation is primarily determined by the following properties of these forces, which are summarized in Table 13.1:

- *Particles affected*: Gravity affects everything in the universe. (It has to, since we now regard gravity as a force that warps spacetime itself.) The weak force also affects every particle (or at least, all those we know of now). The electromagnetic force, as you know, affects only charged particles. You also know that the strong force affects protons and neutrons but not electrons; below we will generalize this rule to the wider collection of particles that are, and are not, affected.
- *Strength*: Any strength comparison between forces is somewhat arbitrary since the forces affect different particles differently and depend on distance in different ways. Nonetheless, it is possible to associate with each force a dimensionless "coupling constant" related to its strength. (For electromagnetism the coupling constant is the fine structure constant that we introduced in Chapter 7: $\alpha \approx 1/137$.) These coupling constants tell us that when the strong force matters at all, it is indeed the strongest of the forces. After that comes electromagnetism, and then the weak force. Gravity is negligible at the particle level.
- *Range*: Gravity and electromagnetism both have potential functions proportional to $1/r$. (Take the derivative of that, or the gradient in more than one dimension, to get the

familiar $1/r^2$ dependence for the forces.) Over great distances these "long-range forces," although diminished, can still have important effects. The strong and weak forces, the "short-range forces," have potentials roughly proportional to $e^{-r/r_0}/r$: a much more rapid decay. The strong force becomes irrelevant over any distance greater than a few fm, and the weak force on scales only a thousandth of that![1]

- *Time scale*: Remember that an "unstable" particle is one that, given enough time, will decay into other particles – and that this decay must be caused by one of the forces listed above. The exact half-life of a particle decay depends on the details of the particles involved, but in general each force has a range of characteristic time scales for decays associated with that force. That time scale is roughly inversely proportional to the strength of the force. For example, if a particle decays in less than 10^{-22} seconds it's a safe bet that the decay is driven by the strong force, while if it takes about 10^{-17} seconds the decay is an electromagnetic process. The characteristic time scales are pretty much non-overlapping, so they provide crucial information for teasing out what's happening in particle interactions.

Table 13.1 The quantities in this table don't all have precise definitions, but these values give a general idea of the properties of the four forces.

	Particles affected	Relative strength	Range	Characteristic time scale
Strong	Hadrons and gluons	1	10^{-15} m	$< 10^{-22}$ s
Electromagnetic	Charged	10^{-2}	Long-range	$10^{-20} - 10^{-14}$ s
Weak	All	10^{-8}	10^{-18} m	$10^{-13} - 10^{-8}$ s
Gravity	All	10^{-39}	Long-range	n/a

The Particle Zoo

You may recall that we embarked on our discussion of forces to help enumerate particles. One important way of classifying particles is by whether they react to the strong force. (The categories in Figure 13.1 don't include photons, or a few other particles that we'll discuss in Section 13.2.)

- "Leptons" do not participate in strong-force interactions. All leptons have spin $1/2$. We consider these to be fundamental particles, and we have never measured a non-zero radius for them. There are six of them in all: electrons, muons, tauons, and the three types of neutrinos.
- "Hadrons" are those particles that have strong-force interactions. While they were initially thought to be fundamental, we now know that hadrons are made of "quarks," which we'll discuss in Section 13.2. Hadrons typically have a radius of about 1 fm. There are hundreds of known hadrons, and they are categorized into two groups:

---

1 A thousandth of a femtometer is an attometer, or am, in case you were wondering.

- – "Mesons" are hadrons with integer spin. The most common type of meson is a "pion," which comes in three varieties: negative (π^-), positive (π^+), and neutral (π^0).
- – "Baryons" are hadrons with half-integer spin (e.g. 1/2 or 3/2). The best known examples are protons and neutrons, each with spin 1/2.

Remember that spin determines whether a particle is a fermion or a boson. Leptons and baryons are fermions, and obey the Pauli exclusion principle. Mesons are bosons, and do not.

<u>Hadrons</u> feel stong force	
<u>Mesons</u> have integer spin: $\pi^+, \pi^-, \pi^0, \eta, \eta', \eta_c, \eta_b,$ $K^+, K^0_s, K^0_l, D^+, D^0, D^+_s,$ and many dozens more	<u>Baryons</u> have half-integer spin: $p, n, \Lambda^0, \Lambda^+_c, \Lambda^0_b, \Sigma^+\Sigma^0, \Sigma^-,$ $\Xi^0, \Xi^-, \Xi^+_c \Omega^0_c, \Omega^-_b, \Omega^+_{cc}, \Omega^0_{cb},$ and many dozens more

<u>Leptons</u> have spin 1/2 and don't feel strong force: $e, \nu_e, \mu, \nu_\mu, \tau, \nu_\tau$ (This is the complete list.)

Figure 13.1 Particle physics circa 1960. We'll present a more up-to-date view in Section 13.2.

Antimatter

Each particle described above also has an associated antiparticle, with the same mass and types of interactions, but with opposite charge. When a particle encounters its own antiparticle, the two annihilate and their energy is converted to other forms (often photons). For example, a neutron and an antineutron behave nearly identically, but you can distinguish them by putting them both next to a normal neutron and seeing which one explodes.

The antielectron is called a "positron," but most antiparticles have no special name and are just given the prefix "anti." Antiparticles are written with a bar, e.g. $\bar{n}$ for an antineutron. Alternatively, a charged antiparticle may be indicated by specifying its charge, such as p^- for an antiproton.[2]

A few particles such as photons are their own antiparticles. You could equivalently say that photons have no antiparticle since photons do not annihilate when they meet. (What would they produce?) For technical reasons in quantum field theory, we say instead that the photon is its own antiparticle.

We call the particles around us "matter" and their exotic partners that we produce in a lab "antimatter," but the laws of physics are almost exactly the same for both. So there might be parts of the universe made of what we call antimatter, and scientists there would presumably use the opposite names from what we do. We have good reason to believe, however, that all of the galaxies we can see are made of what we call matter.

Conservation Laws and Other Particle Categories

We have said several times that the mid-twentieth century was characterized by the discovery of an onerous bestiary of particles and properties.

---

2 A positron can be written as e^+ but never as p^+ since p refers to a proton.

> *Nowadays, atomic particles appear suddenly, out of the blue, doing somersaults. The*
> *physics of yesteryear was a bit like ballroom dancing to Mozart, while now it's more*
> *like a fairground with halls of mirrors, labyrinths, target-shooting booths and men*
> *hawking phenomena.*
>
> – Ernesto Sabato, "Física escandaloso" in *Uno y el Universo* (1945)

A lot of these discoveries came from conservation laws. In addition to the three classical conservation laws (energy, momentum, and angular momentum), particle physics introduces new quantities that must also be conserved. For instance, we saw in Chapter 12 that conservation of spin led physicists to conclude that the reaction $n \rightarrow p + e$ must produce another particle, which we now identify as an antineutrino. The proliferation of particles went hand-in-hand with the proliferation of conservation laws.

Our goal here is not to enumerate all the properties and particles that emerged during this turbulent period, but to convey a sense of the process. To that end, we offer the example of the kaon. This particle (also called the "K meson") was first identified in the 1940s. Its behavior was so confusing that it, and other particles like it, came to be categorized as "strange" particles.

Experiments with strong-force interactions revealed patterns in the types of allowable decay:

$$p + p \rightarrow p + p + K^0 + \overline{K}^0 \qquad\qquad \text{occurs}$$
$$p + p \rightarrow p + p + K^0 \qquad\qquad \text{never occurs}$$

Why can kaons be produced in particle–antiparticle pairs, but never singly? The obvious inference is some kind of conserved quantity. But a kaon has no charge and no spin,[3] and none of the other then-known conserved quantities worked either, so they invented a new one called "strangeness." If a kaon has a strangeness of 1, and an antikaon −1, then the production of a kaon–antikaon pair is allowed because it conserves strangeness. The production of a single kaon never occurs, because it violates this conservation law.

But other experiments cast doubt on this conclusion:

$$K^0 \rightarrow \text{various mesons and leptons} \qquad\qquad \text{occurs on a time scale} \sim 10^{-10} \text{ s.}$$

Mesons and leptons have no strangeness, so why is that decay allowed? The vital clue is the time scale, a smoking gun that this interaction – unlike the ones we mentioned above – is a reaction to the weak force. The conclusion from all of this is that strangeness is conserved in strong interactions, but not in weak interactions. (It also seems to be conserved in electromagnetic interactions.)

The strong-force conservation of strangeness, together with a few other observations, allows us to narrate the life story of a kaon:

- Kaons are produced via strong interactions. Therefore, conservation of strangeness demands that a kaon is never produced alone, but always in combination with an antikaon.

- However, there are no lighter "strange" particles for a kaon (or antikaon) to decay into. So after 10^{-22} seconds has elapsed . . . nothing happens. The kaon *cannot* experience any strong-force decay.

---

3 Technically we're speaking of a particular kind of kaon represented as K^0; there are other kaons that do have charge. That in itself gives you a good sense of the mess we're talking about.

- But a very long time later, on the order of 10^{-10} seconds, the kaon decays through the weak force, as described above. The antikaon does the same, independently.

You may be able to suggest an alternative story. After they are produced, couldn't the kaon and antikaon annihilate each other in a strong-force interaction, preserving strangeness? The answer is that yes, they could, but that's generally not what happens. They typically fly apart and then decay separately as described above.

Again, for our purposes here, the kaon and the strangeness property are just examples of what early to mid-twentieth-century particle physics was like. After the discovery of strangeness, another set of particles was found with another property that seemed to be conserved in strong but not weak interactions. Continuing the whimsical naming, physicists called that property "charm."

. . . And so it went on. By the 1960s, well over 100 hadrons were known with a variety of masses, spins, conserved or semiconserved quantities, decay times, and other seemingly random collections of properties. It became increasingly difficult to believe that all these particles and all their properties were fundamental building blocks with no deeper explanation.

In the 1960s and 1970s a theory was developed that explained these particles and their properties from a much smaller set of building blocks. That theory, the standard model of particle physics, is the topic of the next section.

13.1.2 Questions and Problems: Forces and Particles

Conceptual Questions and ConcepTests

1. In the interactions between the Earth and the Moon, the only relevant force is gravity.

   (a) Both objects have lots of hadrons, but the strong force is not relevant in their mutual interaction. Why?

   (b) Both objects also have lots of charged particles, but the electromagnetic force is also not relevant in their mutual interaction. Why?

   Based on the 1960s-era physics we described in this section, your answers to Parts (a) and (b) should not be the same. Section 13.2 will present an updated model in which the two explanations are more parallel.

2. In Question 1 you looked at three different forces – gravity, strong, and electromagnetic – and explained why only the first is relevant in interactions between the Earth and Moon.

   (a) Which of these three forces is dominant in interactions between the oxygen atom and the two hydrogen atoms in a water molecule?

Explain, for each of the other two forces, why we can ignore them for this scenario.

   (b) Which of these three forces is dominant in interactions between two protons in a nucleus? Explain, for each of the other two forces, why we can ignore them for this scenario.

3. Two important conserved quantities are baryon number (number of baryons minus number of antibaryons), and lepton number (number of leptons minus number of antileptons). There is no conserved "meson number." Based on those facts, classify each of the following reactions as "could happen" or "could never happen." (Here p is a proton, n is a neutron, and π^0 is a pion.)

   (a) $p + p \rightarrow p + p + \pi^0$

   (b) $p + p \rightarrow p + p + \pi^0 + \pi^0$

   (c) $p + p \rightarrow p + p + n$

4. When a single particle decays, the decay products cannot have more combined mass than the original particle (no matter how much kinetic energy the original particle had). But two particles can collide

and be converted into particles more massive than the combined mass of the originals. Explain why this is possible for the collision but not for the single-particle decay. *Hint*: Choose an appropriate reference frame.

Problems

5. When two protons collide they tend to bounce off each other because of their electromagnetic repulsion. But if they get close enough, the strong force takes over and they fuse together. The exact probability of fusing depends in a complicated way on their energies and involves tunneling probabilities, but as a reasonable model we can say that they will fuse if they have enough energy to come within 1.85×10^{-14} m of each other.

   (a) How much kinetic energy must two protons have in order to collide and fuse?

   (b) In a reference frame in which one proton is at rest, how fast must the other be moving in order for them to fuse?

   (c) The temperature in the center of the Sun is about 15 million kelvins. Use the Maxwell speed distribution (Appendix I) to estimate the probability that a proton collision leads to fusion. You can make the simplifying assumptions that one proton is at rest and that the other proton is moving precisely toward the first one. *Hint*: If you can't get your calculator to evaluate the integral, you can try mathematical software such as wolframalpha.com.

   (d) A proton in the Sun's core collides with other protons approximately 10^8 times per second. Roughly how long will a proton last before it fuses with another proton?

13.2 The Standard Model

The 1950s accumulated an unwieldy list of hadrons with seemingly random masses, charges, and other properties. In 1964 Murray Gell-Mann and George Zweig independently realized that those properties could all be explained if hadrons were composed of smaller particles. These particles are often found in groups of three, so Gell-Mann named them "quarks" after the line "Three quarks for Muster Mark" in James Joyce's *Finnegan's Wake*.

Gell-Mann originally suggested three different types of quarks, but that turned out to be insufficient. Expanding Gell-Mann's original model to six types, physicists were able to explain all the mesons and baryons that had been detected and cataloged. Building on that foundation, they created the "standard model" that is still the basis of particle physics today. In this section we will show how that model uses fewer than 20 fundamental particles to explain literally hundreds of no-longer-quite-fundamental particles, three forces, and the existence of mass itself. We will close by mentioning some components of the known universe that the standard model *cannot* explain.

13.2.1 Discovery Exercise: The Standard Model

🔍 You have been given a construction kit with four kinds of pieces:

- The "down quark" has charge $-1/3$, and the "antidown quark" has charge $+1/3$.
- The "up quark" has charge $+2/3$, and the "antiup quark" has charge $-2/3$.

All the quarks in your set have spin $1/2$, but with a flexible addition rule: two spin-$1/2$ particles can combine to a spin of either 0 or 1, depending on whether they are aligned or anti-aligned.

You have an unlimited supply of all four types. Show how you can combine quarks to make each of the following particles:

1. A proton (charge $+1$, spin 1/2)
2. A neutron (charge 0, spin 1/2)
3. A neutral pion (charge 0, spin 0)
4. A negatively charged pion (charge -1, spin 0)

13.2.2 Explanation: The Standard Model

In Section 13.1 we divided particles into hadrons and leptons, and subdivided hadrons into baryons and mesons. It's a simple scheme that encompasses almost all the particles in the universe. So what's the problem?

A quick glance at Wikipedia's list of particles gives you a good sense of where the trouble lay. The list contains all six leptons: so far, so good. But it also contains about 100 types of baryons, and an even larger number of mesons. Each particle type comes with a daunting list of properties such as isospin, parity, and decay behavior. The list does not read like a particle version of the periodic table, with neat rows and columns designating properties in an orderly way, and perhaps an occasional gap where a new one *ought* be discovered but hasn't yet. It reads more like a grocery list written by a random number generator. Each new particle comes as a surprise, and offers no hint as to what the next one will be.

Starting with the same list of properties, you could easily make up your own particle: this mass, this charge, this spin, this strangeness, and so on. But the odds are very good that the particle you made up would not exist. Why? Why is there a particle with this particular set of properties, but not with that set of properties? Why is this particular property conserved, but not that one? You can see why mid-twentieth-century physicists came to believe that there had to be a simpler underlying structure.

The key to bringing order to the chaos came with the realization that hadrons are made of smaller particles, now called "quarks." You can see the effects of this change in Figure 13.2, which shows all the particles in the current "standard model" of particle physics. No hadrons appear in this figure because they are no longer seen as fundamental particles. Instead, a list of fewer than 20 particles suffices to explain every baryon, meson, or other particle we've ever produced in a lab.

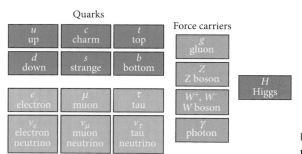

Figure 13.2 The standard model of particle physics. Appendix J reproduces this figure with more details.

This section explains each of the pieces of the standard model.

Matter, aka Quarks and Leptons

Figure 13.3 shows the first part of the standard model, quarks and leptons. In particle physics the word "matter" usually means these particles or anything made out of them.

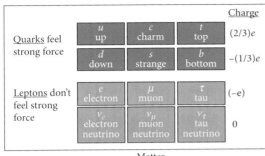

Figure 13.3 The particles that make up matter. Each quark box represents a "flavor" (up, down, etc.). Each column of two quarks and two leptons is called a "family." The particles in each row get more massive as you move to the right.

Compared to the chaos of the particle zoo, Figure 13.3 is not only a short list, but it has some of the same orderly feeling that the periodic table brought to the study of the elements. Each column is sometimes called a "family" of four particles. Each row represents three particles that are almost identical to each other, but with increasing mass as you move from left to right. (A muon has more mass than an electron, and a tauon still more.) Those mass differences make the second and third families unstable; they decay into particles of the first family. It's possible to create an atom with muons instead of electrons, and/or strange and charm quarks instead of up and down. But such an exotic atom doesn't last long, so all the matter we observe in nature is made from the first family: protons and neutrons are made from up and down quarks (*uud* and *udd*, respectively), and electrons are, well, just electrons.

The quark model was able to reveal an underlying order in the chaotic list of hadrons because there are simple rules for how quarks combine. By following those rules you can predict what hadrons will exist, and many of their properties. Those rules are based on a property not shown in Figure 13.3: a property called "color."

Quarks, Color, and How to Make a Hadron

In Section 13.1 we defined a hadron as a particle that participates in strong interactions. We can now give the more fundamental definition: "A hadron is a particle made of quarks." And by the way, quarks have strong interactions.

Figure 13.3 shows the six types (or "flavors") of quarks: up, down, etc. It notes that all quarks have spin 1/2 (which makes them fermions). The quarks in the top row (up, charm, and top) each have charge $+2/3$ (in units of proton charge), and the quarks in the bottom row (down, strange, and bottom) have charge $-1/3$. Their antiparticles, not shown in the figure, have the opposite charges.

But every quark also has another property called its "color." This property (red, blue, green, antired, antiblue, or antigreen) plays a role in the strong force similar to the role played by charge (positive or negative) in the electromagnetic force.

Color in the quark sense has nothing to do with actual color, as in different frequencies of visible light. The metaphor was chosen because our eyes perceive all light as combinations of three primary colors. (Quark color also has nothing to do with the color scheme of Figure 13.3, but we did at least avoid using blue, red, or green in our figures to minimize confusion.)

So every quark is characterized by a flavor (shown in Figure 13.3) and a color (*not* shown in the figure). For instance, there are three different types of up quark: red, blue, and green, each with charge $(2/3)e$. Each one of these has an antiquark that reverses both its color and its charge, such as an antiup quark whose color is antired and whose charge is $-(2/3)e$. So the box labeled "up" in our standard model represents six possible particles, and the box labeled "down" another six, and so on.

Color determines whether the strong force between quarks is attractive or repulsive, analogously to the role charge plays for electric force. Different colors attract (e.g. blue attracts red) and like colors repel (blue repels blue). The repulsion between same-colored quarks is stronger than the attraction between differently colored quarks. For colors interacting with anticolors, opposites attract and other pairs repel (blue attracts antiblue, but repels antired and antigreen). The forces between quarks are literally on the order of *tons*.

Those various combinations of attraction and repulsion between quarks give rise to the following simple rule that explains all hadrons (Question 4):

Quarks combine into color-balanced combinations.

There are three common color-balanced combinations. (See Question 7 for some more exotic varieties.)

- Red–green–blue makes a baryon.
- Antired–antigreen–antiblue makes an antibaryon.
- Any combination of a color and its anticolor makes a meson. (It's often arbitrary which member of a meson–antimeson pair you call the particle and which one the antiparticle.)

A particular type of hadron is defined by its flavor combination. For instance, *uud* (two up quarks and one down) makes a proton. Those three quarks *must* include one red, one blue, and one green, but it doesn't matter which color goes with which flavor. Similarly, the combination $u\bar{d}$ defines a π^+ particle (a type of pion). If the up quark is red, then the antidown quark must be antired – and similarly for blue or green – but the behavior of the particle will be the same in any of these three cases, so all three answers are called π^+ particles.

Active Reading Exercise: Building Hadrons

In the questions below, "quark content" refers to quark flavors, e.g. *uud* for a proton.

1. What's the quark content of the lightest possible baryon with charge -1?
2. What's the quark content of the lightest possible meson with charge -1?

You can find our answers to that exercise at www.cambridge.org/felder-modernphysics/activereadingsolutions

Let's look at a few examples of how this view of hadrons, as color-balanced combinations of quarks, helps explain some facts that were experimentally discovered when every hadron was thought to be a fundamental particle.

- One implication is that all hadrons have integer amounts of charge (Question 3). As an example, a proton is made of *uud*, which adds up to a charge of 1. A neutron is *udd*, which gives it zero charge.

- Spin addition is more complicated because it depends on alignment. If two of three bound quarks have their spins aligned, and the third points the opposite way, the total spin is 1/2, as we see in a proton or a neutron. If all three spins are aligned, the resulting baryon has spin 3/2. When cyclotron experiments in the 1950s revealed something similar to a proton but with spin 3/2, it was thought to be a new type of particle; now this "Δ^+ baryon" is known to be an excited state of a proton.

- Before the standard model, the quantity "baryon number" – defined as 1 for a baryon, -1 for an antibaryon, and 0 for a meson – was known to be conserved. It was also known that there was no analogously conserved "meson number." We can now explain (Question 5) both of those observations by postulating a conserved quark number. (For historical reasons this is still referred to as the "baryon number" and assigned a value of 1/3 for quarks and $-1/3$ for antiquarks, which preserves the original definitions.)

- Before the standard model, some hadrons were known to have quantities like "strangeness" and "charm." There was little rhyme or reason to explain which particles had these properties, but the properties were known to be conserved in strong interactions. We now recognize exactly six such quantities, one for each quark flavor. For instance, strangeness is the number of strange antiquarks minus the number of strange quarks. (This convention, the reverse of what you might expect, keeps the definition consistent with earlier usage.) And there are comparable quantities for down quarks, top quarks, and so on. Each of these six quantities is individually conserved in strong-force interactions, but can change in weak-force interactions. For instance, the decay reaction $K^+ \rightarrow 2\pi^+ + \pi^-$ occurs through a weak-force interaction; it could not occur on a strong-force timescale, as it involves changes in both strangeness and downness.

One property that you might notice missing from that list is mass. Mass *is* important in particle physics, in exactly the ways you would expect. Mass is a property of particle type: that is, every proton has the same mass as every other proton.[4] And of course total energy, including mass energy, is conserved in any interaction. So we would never leave mass out of any list of important particle properties.

But our list above was properties of hadrons that can be determined from the properties of their constituent quarks, and mass does not belong on that list. Quarks are never found alone, so you can't measure their masses in isolation. Furthermore, a hadron is made up of both quarks and "virtual particles," particle–antiparticle pairs that are continually being created and annihilated inside the hadron. We'll discuss virtual particles in Section 13.5, but here it's enough to stress that the mass of a hadron is *not* the sum of the masses of its individual quarks. The total mass is always larger than that – sometimes by a lot.

Even though individual quarks have never been observed, there are many reasons we believe in the quark model. Perhaps the most direct evidence comes from "deep inelastic scattering" experiments. High-energy electrons are fired at nucleons and their scattering is measured at

---

4 Heavier particles can have some quantum uncertainty in their masses, related to their decay lifetime by the time–energy uncertainty principle.

different angles. Just as Rutherford scattering proved that an atom has a small clump of positive charge in its center, these experiments reveal that nucleons have three small clumps of fractional charge inside them, which occasionally give rise to scattering at angles that would be implausible if a proton's charge were uniformly distributed.

There's also indirect evidence for the existence of color. You can have a baryon consisting of three quarks of the same type, such as the Δ^{++} baryon made of *uuu*. From the mass of this particle we can infer that the quarks are all in their ground states, which would be forbidden by the Pauli exclusion principle if the three quarks were indistinguishable.

Leptons

Leptons, the lower group in Figure 13.3, do not feel the strong force. The most familiar example is the electron. Muons and tauons are essentially the same as electrons – same charge, same spin, etc. – except that they are much more massive ($m_\mu \approx 200 m_e$, $m_\tau \approx 3500 m_e$).

Neutrinos are chargeless leptons so they only interact via the weak force (and gravity). As a result, they pass through solid matter with almost no interaction. About 100 billion neutrinos from the Sun pass through each square centimeter of Earth every second, and virtually all of them emerge on the other side of the Earth less than a second later. If a beam of neutrinos encountered a block of lead *a light year thick*, about half of the neutrinos would make it through with no interaction.

We first realized that neutrinos exist because without them some of the reactions we see wouldn't conserve angular momentum; there had to be an as-yet-unknown particle involved. We can now detect neutrinos using multi-ton tanks of liquid. So many neutrinos pass through that occasionally one interacts with a particle in the liquid and produces a burst of radiation that can be captured by detectors around the tank.

Just as baryon number is conserved, there is also a lepton number – positive for leptons, negative for antileptons – that is conserved in all interactions. In fact, with one exception that we'll discuss below, lepton number is generally conserved *within each family* (column in Figure 13.3). For example, in the beta decay $n \rightarrow p + e + \bar{\nu}$ the antineutrino must be an electron neutrino. Otherwise the reaction would produce a positive lepton number in the electron family and a negative one in one of the other families.

The exception to this rule is that neutrinos by themselves oscillate over time between their three different flavors. These neutrino oscillations were discovered when people measured the neutrinos coming from the Sun and found a third as many as our theories predicted. This turned out to be because their detectors were only sensitive to electron neutrinos. The neutrinos produced by the Sun are all electron neutrinos, so this wasn't expected to be a problem. But by the time they reach Earth, they are a roughly equal mix of all three neutrino types. Detectors sensitive to other neutrino types have now confirmed that the total flux of neutrinos from the Sun matches theoretical predictions. You'll investigate neutrino oscillation in Problems 15–18.

Force Carriers

We have not yet gone through the entire standard model shown in Appendix J. So what's left, when we've already said that Figure 13.3 is a complete list of quarks and leptons, and quarks and leptons make up all the matter in the universe?

You may be able to think of one answer to that question – a particle we have discussed many times in this book, but not yet in this section – and that's the photon. The photon, and the other particles grouped with it in Appendix J, enable particles of matter to exert forces on each other.

To see what that means, consider three facts that you already know, but may never have considered all at the same time:

- Classical electromagnetic theory tells us that a charged particle does not directly exert force on another charged particle. Instead, the first particle creates electric and magnetic fields in the space around it, and those fields exert force on the second particle. The electric and magnetic fields mediate the forces between the two particles.

- Classical electromagnetic theory also tells us that light is made of electric and magnetic fields.

- Quantum mechanics tells us that light can also be described as a collection of particles, called photons.

Our modern understanding takes these facts one step farther: all electric and magnetic fields are made of photons. So when one charged particle exerts force on another charged particle, that effect is conveyed from one particle to the other by photons.

We'll talk more about how forces work quantum mechanically in Section 13.5. Our point here is that forces between particles of matter are mediated by particles. These particles are called "force carriers."[5] For the electromagnetic field, the force carriers are photons. The strong force is carried by gluons. The weak force has three force carriers: the positively charged W^+, the negatively charged W^-, and the neutral Z.

Let's briefly survey some properties of these particles in Figure 13.4.

- *Mass*: Photons and gluons have no mass, but the weak-force carriers do. For reasons we'll discuss in Section 13.5, a massive force carrier gives rise to a short-range force. So the masses of the W and Z bosons (each of which is more than 150,000 times the electron mass) explain why the weak force only operates on such small length scales. (If you've been reading carefully, you should now be confused about the strong force. Hold that thought.)

- *Charge*: Photons and gluons have no charge. As we mentioned earlier, charge distinguishes the three different types of weak-force carrier.

- *Spin*: All force carriers are spin-1 particles, and thus bosons.

- *Color*: Photons and weak-force carriers have no color. But gluons do, and that's important, because color determines a particle's participation in the strong force. Remember that photons *carry* the electromagnetic force but, having no charge themselves, are not *affected* by it. Gluons carry the strong force and also exert strong forces on each other, which is one of the reasons the strong force is more complicated to model than electromagnetism.

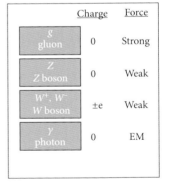

	Charge	Force
g gluon	0	Strong
Z Z boson	0	Weak
W^+, W^- W boson	$\pm e$	Weak
γ photon	0	EM

Force carriers
spin-1 particle

Figure 13.4 Each force is mediated by its own particle(s). So these particles, although *not* constituents of matter like quarks and leptons, are important components of the standard model.

---

5 They're also called "gauge bosons" for reasons we're not going to go into here.

You may have noticed that our discussion of mass seems to contradict itself. If a massive force carrier is what makes a force short-range, and gluons are massless, why is the strong force a short-range force?

The answer is that the strong force between two quarks is in principle a long-range force, but those forces mostly cancel out when color-balanced hadrons interact. The situation is analogous to the forces between polar molecules. Hydrogen molecules are neutral, but if you put two of them near each other they polarize each other and exert an attractive force. While the electric force between charged particles falls off as $1/r^2$, the van der Waals force between two hydrogen molecules falls off as $1/r^6$. In a similar way, two isolated quarks might exert a long-range force on each other, but quarks don't wander the universe in isolation; they exist only in color-balanced combinations such as protons or kaons. These particles are overall color-neutral, but exert the strong force on each other through polarization, and that effect dies down very quickly with distance.

You may also notice that there's no force carrier for gravity in the standard model. Physicists talk about a hypothetical force carrier called a "graviton," but gravity is so weak at the particle level that there is no prospect of detecting gravitons in the foreseeable future.

The Higgs Boson

H
Higgs

spin 0
charge 0

**Figure 13.5
Hi-ho, the
derry-o, the
Higgs stands
alone.**

If you've been following along our standard model in Figure 13.2 or in Appendix J, then you know that the only particle we have not yet discussed gets its own category: see Figure 13.5.

The Higgs boson is the only spin-0 particle in the standard model. All other known spin-0 particles are made from fundamental particles whose non-zero spins cancel.

The Higgs was proposed because of a problem with the theoretical models of the 1950s and 1960s. These models successfully predicted many properties of the observed particles and their interactions, but they also predicted that all of the fundamental particles of nature should be massless. As you might imagine, this caused some discrepancy between theory and experiment. (Easy experimental test: is everything in the room around you moving at the speed of light?)

In the 1960s several groups independently realized that this problem could be overcome if particles didn't have intrinsic masses, but rather *acquired* mass by interacting with a field. This "Higgs field"[6] fills all of our known universe uniformly. By analogy, you might imagine a universe with a uniform magnetic field at every point. But the Higgs field is a scalar field, so unlike a magnetic field it doesn't pick out any special direction in space.

To explain how the Higgs field gives rise to mass, people often use the analogy of pushing an object through water. At its most basic level, mass is the property that makes something difficult to move. A block submerged in water is harder to push than the same block in the open air, so you can imagine that massive objects are difficult to move because they are "submerged" in the Higgs field.

That analogy is helpful as a mental image, but it has its limitations.

---

6 One of the first people to propose this field was British theoretical physicist Peter Higgs.

Active Reading Exercise: The Higgs/Water Analogy

When an object is underwater, it is harder to get it moving, which makes it feel like it's more massive than it would be in air. List at least two ways in which the object's properties underwater are *not* analogous to having more than its usual mass.

You can find our answers to that exercise at www.cambridge.org/felder-modernphysics/activereadingsolutions

So by all means hold onto the Higgs-field-is-kind-of-like-water analogy if it is helpful to you, but keep in mind that no analogy is perfect.

In any case, each particle has its own characteristic strength of interaction with the Higgs field, and that strength of interaction is what we call the particle's mass. Massless particles such as photons and gluons have no interaction with the Higgs field.

The "Higgs boson" is a particle-like excitation of the field. When scientists at the Large Hadron Collider announced that they had found the Higgs boson in 2012, that means they had excited and observed a small, localized quantum ripple in the otherwise uniform Higgs field that surrounds us all. That ripple, the Higgs boson, was the last particle in the standard model to be detected.

Stepping Back: Beyond the Standard Model

When the standard model was developed in the early 1970s, it successfully explained all the particle observations that had been made to date. It also predicted the existence of other particles: in fact, fewer than half of the fundamental particles in the standard model had been detected when the model was developed. Over the next few decades every one of those particles was found, all with properties consistent with the model's prediction. The final confirmation of the standard model came with the discovery of the Higgs boson at the Large Hadron Collider in 2012.

Nonetheless, there is clearly more to fundamental physics than we know so far.

- "Dark matter" makes up about 90% of all the mass in the observable universe. We can detect it through its gravitational effect on ordinary matter, and we know enough about it to say that it is not made of the particles in the standard model.

- Other astronomical observations indicate a field of energy that permeates the observable universe and causes galaxies to accelerate away from each other. We call that field "dark energy," and once again we can rule out its being anything in the standard model.

We'll talk more about dark matter and dark energy in Chapter 14. We will stress then, and we want to stress now, that *they are not the same thing*. They just have similar names. Everyone gets them confused. But now you won't.

Anyway, both of them are forms of energy that are not accounted for by the standard model. Here are a few more hints of physics beyond the standard model.

- In the 1960s several physicists showed that the electromagnetic and weak forces are actually different manifestations of one underlying force, now called the "electroweak force." We're not going to discuss the electroweak force except to note that this theory led to successful predictions of the properties of the W and Z bosons, and thus was part of the development of the standard model.

  The success of electroweak theory has led to a search for a theory that similarly unifies the electroweak and strong forces. Such theories are called "grand unified theories," or GUTs. Several possible versions have been proposed, and they all require the existence of extremely heavy particles not included in the standard model. Unfortunately we are nowhere close to having accelerators capable of producing particles heavy enough to test these theories.

- Gravity presents a unique problem. Our current best understanding of gravity comes from Einstein's general theory of relativity, which views gravity as an interaction between matter and spacetime. Just as Einstein's special theory of relativity introduces new rules for extremely fast objects (rules that approach Newtonian rules at low speed), and quantum mechanics introduces new rules for extremely small objects (rules that approach Newtonian rules at macroscopic sizes), general relativity introduces new rules for extremely massive objects (rules that approach Newtonian rules when gravitational fields are weak).

  Quantum field theory is consistent with special relativity and works for both slow and fast objects, but no one has found a way to unify quantum mechanics and general relativity into a self-consistent theory. The universe as we currently understand it seems to follow one set of laws for very small objects, and a completely different set of laws for very large ones, both of them conveniently approaching Newton in the middle.[7] One of the most promising attempts at a unifying theory of quantum gravity (sometimes called a "theory of everything") is string theory, but its math is so complicated that its equations can only be solved for very specialized conditions, and so far it has not made any testable predictions.

All of these reasons and more suggest that the standard model, successful as it has been, is not the final word.

Last minute addition: Shortly before this book went to press, Fermilab released data on the gyromagnetic ratio (sometimes called the "*g*-factor") of the muon. The gyromagnetic ratio is a dimensionless measure of the muon's magnetic moment. The measured value was $2.00233184122 \pm 0.00000000082$, compared to the standard model's prediction of $2.00233183620 \pm 0.00000000086$. This discrepancy of less than a millionth of a percent is so large, compared to the experimental uncertainty, that it has less than a 1-in-40,000 probability of having been produced by random chance. If confirmed, this measurement will be the first-ever experimental failure of the standard model.

---

7   In principle we could observe both quantum and general relativistic effects at the same time if we had enough mass packed into a small space. The density at which this would occur is called the "Planck density." It's about 10^{96} kg/m^3, or roughly the density of a billion galaxies packed into the size of an atomic nucleus. We're not going to produce that in a lab any time soon. Until we understand how to reconcile quantum mechanics and general relativity, we won't be able to describe the earliest moments of the Big Bang or the center of a black hole, the only two known occurrences of Planck-density matter in nature.

13.2.3 Questions and Problems: The Standard Model

Conceptual Questions and ConcepTests

1. Taking into account both antiparticles and color differences, how many different types of particles does the quark section of Figure 13.3 represent?

2. Each baryon is a combination of three (not necessarily different) quark flavors, such as *uud* for protons. Can any possible combination of three quark flavors be combined into a baryon, or does the color-balancing rule restrict what combinations you can make?

3. Prove, using the rules for the color balancing of quarks, that all hadrons must have integer charge.

4. Any two quarks exert an enormous force on each other, even at large distances. But two baryons (e.g. two neutrons) exert almost no strong force on each other when they are more than a few fm apart. Based on that fact, what can you conclude about the relative strengths of the repulsive force between two like-colored quarks and the attractive force of differently colored quarks?

5. Pre-1960s physicists noted that there is a total baryon number that is conserved. (When you produce a baryon, you must also produce an antibaryon.) They also noted that there is *not* a corresponding conserved meson number. Explain both of these observations based on the conservation of quark number.

6. Why do all baryons have half-integer spins and all mesons have integer spins?

7. In addition to baryons and mesons, the rule that quarks join in color-neutral combinations allows for the existence of "pentaquarks," made of five quarks or antiquarks each. After decades of searches, the first widely accepted detection of a pentaquark happened at LHC in Switzerland in 2015. ("Tetraquarks," made of four quarks each, have also been found.)

    (a) Are pentaquarks (choose one and explain):
        A. fermions?
        B. bosons?
        C. They could be either.

    (b) Give an example of five quark colors that could make up a pentaquark. (Don't worry about the flavors.)

8. Section 13.1 discussed the kaon. Originally thought to be a fundamental particle, the kaon is now known to be a combination of one first-family quark (up or down) with one strange quark. For instance, the K^+ is made of an up quark and a strange antiquark.

    (a) What is the charge of a K^+?

    (b) If you look up the K^+ you will see that it is listed as a spin-0 particle. You may also find a K^{*+} listed as a different particle with all the same properties except a different spin. The K^+ and K^{*+} have the same quark content, so what's the difference that leads them to have different spins, and what is the spin of the K^{*+}? (You don't have to look up any information to answer this question.)

    (c) Could there be a similar particle made from an up quark and a strange quark (without "anti" in front of either)? Why or why not?

    (d) Kaons are produced via strong-force interactions when nuclei are smashed together in accelerators. Explain why this always produces them in kaon–antikaon pairs, rather than individually.

9. The Σ^+ particle is made of two up quarks and a strange quark.

    (a) Is this particle a baryon or a meson?

    (b) What is its total charge?

    (c) If you look up the Σ^+ you will see that it is listed as a spin-1/2 particle. You may also find a Σ^{*+} listed as a different particle with all the same properties except a different spin. The Σ^+ and Σ^{*+} have the same quark content, so what's the difference that leads them to have different spins, and what is the spin of the Σ^{*+}? (You don't have to look up any information up to answer this question.)

(d) Sigma (Σ) particles are produced via strong-force interactions when nuclei are smashed together in accelerators. Explain why this always produces them in sigma–antisigma pairs, rather than individually.

10. In each part below, you are given properties of a hypothetical particle. For each one, either give one possible quark composition of a baryon or meson that would produce these properties, or explain why such a particle would not be possible. (Charges listed are in units of the proton charge. Remember that strangeness is $+1$ for a strange antiquark and -1 for a strange quark.)

(a) Charge $= 1$, spin $= 1/2$, strangeness $= 0$

(b) Charge $= 2$, spin $= 3/2$, strangeness $= 0$

(c) Charge $= 2$, spin $= 1$, strangeness $= 0$

(d) Charge $= 2$, spin $= 1/2$, strangeness $= -1$

(e) Charge $= 4/3$, spin $= 0$, strangeness $= 0$

11. You measure a hadron with a mass of roughly 2520 MeV/c^2, a charge of 2, and a spin of $3/2$.

(a) What are the possible quark makeups of this particle? Explain how you know. You may find it helpful to refer to the quark properties in Appendix J.

(b) What additional property might you measure that would allow you to distinguish between the different possibilities you listed? (Don't worry for the moment about how you would measure it.)

12. Describe an allowed decay process by which a muon can decay into particles entirely in the first "family."

13. In the interaction between the Earth and the Moon, the only relevant force is gravity. Both objects have lots of quarks, but the strong force is not relevant in their mutual interaction. Why? (If you answered Question 1 in Section 13.1, note that your answer here will be different from what it was there.)

For Deeper Discussion

14. In quantum field theory, "fields" and "particles" are ultimately the same thing. For example, a photon is an individual particle but 10^{20} photons moving together looks like a wave in a classical electromagnetic field. In principle we can similarly associate a field with every particle in the standard model. So why do we not observe the "electron field" in the same way we see the "photon field" (aka the "electromagnetic field")?

Problems

The problems in this section explore neutrino oscillations.

At any given moment, a neutrino may be in a superposition of the three neutrino flavors (ν_e, ν_μ, and ν_τ), or it may be in an eigenstate of definite flavor. The neutrino may also be in a superposition of three different masses, or it may be in an eigenstate of definite mass. But the eigenstates of definite flavor are not eigenstates of definite mass, so only one of them can be certain at a time!

In Problems 15–18 you will investigate what this implies for neutrinos and their tendency to oscillate between flavors. We will use a simplified model that has only two flavors: electron neutrino and muon neutrino. The eigenstates of flavor can be expressed as superpositions of the mass eigenstates ψ_1 and ψ_2 as follows:

$$\left.\begin{array}{l} \psi_e = \cos\theta\,\psi_1 + \sin\theta\,\psi_2 \\ \psi_\mu = -\sin\theta\,\psi_1 + \cos\theta\,\psi_2 \end{array}\right\} \quad (13.1)$$

Here ψ_1 represents an eigenstate of mass m_1, and ψ_2 an eigenstate of mass m_2. The "mixing angle" θ is an experimentally determined parameter.

15. Interpreting Equations (13.1)

(a) Briefly describe what these equations would tell us if θ were zero. (It isn't really.)

(b) Based on the rule for the time evolution of a wavefunction, explain why a neutrino in state ψ_1 will *always* have definite mass m_1, but a neutrino in state ψ_e will *not* remain in a state of definite electron-neutrino-hood. (Assume the neutrino in each case has a definite value of momentum.)

(c) Explain why ψ_e *would* be a stationary state if θ were zero.

(d) Explain why ψ_e would also be a stationary state if the two mass states were degenerate ($m_1 = m_2$).

(e) Show that if ψ_1 and ψ_2 are properly normalized, so are ψ_e and ψ_μ.

16. Calculating Neutrino Oscillations Analytically

The neutrinos created in the Sun are electron neutrinos, because of the reactions that create them. (Remember that lepton family number is conserved *except* in neutrino oscillation.) In this problem you'll examine the probability of measuring such a neutrino, at some later time, as a muon neutrino.

(a) Solve Equations (13.1) to express the mass eigenstates ψ_1 and ψ_2 in terms of the flavor eigenstates. (If you're familiar with linear algebra you can recognize this as inverting a 2×2 matrix. If not, just solve two simultaneous equations.)

(b) Write the wavefunction of a solar neutrino at a time t after it was produced, originally in the state ψ_e. Your answer will include the energies E_1 and E_2.

(c) Use your solutions to Parts (a) and (b) to write the neutrino's wavefunction at time t as a superposition of the states ψ_e and ψ_μ. As a check on your answer, it should exactly equal ψ_e at $t = 0$.

(d) If you measure the neutrino's flavor at time t, what is the probability that you will measure it to be a μ neutrino? Simplify your answer as much as possible. You should be able to express it in terms of the energy difference $\Delta E = E_2 - E_1$, without either of the individual energies E_1 or E_2 appearing in the formula. (You can assume here that the eigenfunctions are properly normalized.)

(e) Your answer should still equal zero when $t = 0$, of course. But it should also equal zero when $\theta = 0$ or $\Delta E = 0$, based on the arguments you made in Problem 15. Use all three conditions to check your formula. (You don't need to have done Problem 15 to answer this.)

17. Calculating Neutrino Oscillations Numerically

[*This problem depends on Problem 16.*] Solar neutrinos have total energies far greater than their mass energies. For $mc^2 \ll E$, you can approximate (via a Maclaurin series or binomial expansion) the relativistic energy formula as $E \approx E_0 + m^2c^4/(2E_0)$, where $E_0 = pc$ is *almost* the entire energy of the particle. So $\Delta E \approx \Delta(m^2)c^4/(2E_0)$. (It doesn't matter whether you use E_1 or E_2 for E_0 since the value of E_0 is many orders of magnitude bigger than ΔE.)

(a) Use this approximation to rewrite your answer to Problem 16, the probability of measuring a muon neutrino at time t, in terms of $\Delta(m^2)$.

(b) Use the fact that solar neutrinos travel at almost exactly c to rewrite your formula in terms of the distance L that the neutrino has traveled rather than the time t since it was produced.

(c) Observationally, $\theta \approx 0.58$ rad and $\Delta(m^2)c^4 \approx 7.5 \times 10^{-5}$ eV^2. If a solar neutrino is produced with 400 keV of total energy, how far will it travel before it has the maximum possible chance of being detected as a muon neutrino, and what will that probability be? *You will find it easier to keep everything in units of eV rather than converting to SI units.*

(d) Would it be possible to position a detector at just the right distance from the Sun so that all the solar neutrinos you detected show up as electron neutrinos? Explain.

18. Representing Neutrino Oscillations Graphically

 [*This problem depends on Problem 17.*] A solar neutrino is produced with 400 keV of total energy. On one plot show the probabilities of detecting the neutrino as an electron neutrino, and as a muon neutrino, as a function of distance from the point where it is produced.

13.3   Detecting Particles

The first two sections of this chapter described the discovery of particles (and then more particles and then even more particles), ultimately leading to the standard model of particle physics. This section looks at some of the experiments behind those discoveries.

Just as in Section 12.2, our goal here is not to overwhelm you with technicalities. Rather, we want to convey a sense of how experimentalists – people who cannot manipulate objects smaller than you can, or see changes faster than you can – can detect a particle that's smaller than a fm and lasts for 10^{-23} s, and can even measure its properties in some detail. We will focus on the two technologies that dominated particle detection for much of the twentieth century: first cloud chambers, and later bubble chambers.

13.3.1   Explanation: Detecting Particles

Charles Wilson invented the "cloud chamber" in 1911, and was awarded the Nobel Prize for it in 1927. The cloud chamber was the principal instrument of particle physics for decades.

The technology of a cloud chamber is straightforward. You can find instructions online for making one yourself using dry ice and rubbing alcohol, and you can use it to track cosmic rays. The basic idea is to fill a chamber with supersaturated vapor. (Wilson originally used water vapor, but most later cloud chambers used alcohol.) When a high-energy charged particle passes through the chamber, it ionizes atoms along its path. Those ions attract the polar molecules of the vapor, causing droplets to condense around the particle's track. Those trails of droplets can be photographed.

When a strong magnetic field is applied, the charged particles follow curved paths through the detector. You can't do this part at home because it takes very strong magnets to noticeably deflect particles moving at high speeds. But when Carl Anderson did this in a lab in 1932, he got the picture on the left side of Figure 13.6. Note that the particle's track takes it *through* the lead plate that runs across the chamber.

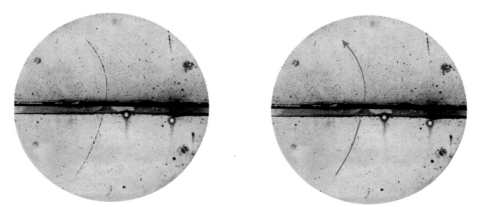

Figure 13.6 On the left is the 1932 cloud chamber photograph with which Carl Anderson discovered the positron. On the right we have added color and an arrow to highlight the positron's path. The thick horizontal line in the middle is a lead plate used to slow the particle down. *Source: American Physical Society © 1933.*

Here are some things we can learn from that figure.

- You know that a charged particle in a uniform magnetic field, moving perpendicular to that field, travels in a circle. You can show in Problem 6 that the radius of that circle is proportional to the momentum of the particle. So by measuring the radius of the circular track in his photograph, Anderson determined the momentum of his particle.

- As the particle moves through the lead plate, it loses energy and slows down. As we just discussed, this causes it to travel along a circle of smaller radius (also known as a larger "curvature" of the path). The photograph shows a higher curvature above the lead plate than below, which is how Anderson determined that his particle had moved upwards. We indicate this direction with the arrow in our enhanced picture on the right. (Fun fact: Millikan opposed the addition of the lead plate to the experimental apparatus, arguing that Anderson should just assume all cosmic rays come from above.)

- Once you know the direction of motion, the direction of the curve tells you the sign of the particle's charge. Anderson's magnetic field was pointing into the figure, and (as you can see) his particle bent to the left, so that means his particle had a positive charge.

- Finally, the particle that made this track must have much less mass than a proton. The reason is that a massive particle, with the momentum indicated by this track, would not have enough kinetic energy to get through the lead plate (Problems 8 and 9).

Putting it all together, Anderson had discovered the positron: a particle with the mass of an electron but a positive charge. Four years later, in the same year he was awarded the Nobel Prize for that discovery, he went on to discover the muon!

In 1952 Donald Glaser invented the "bubble chamber." A bubble chamber (such as the one in Figure 13.7) is similar to a cloud chamber but it uses a liquid medium, usually liquid hydrogen. Just before you want to detect the particles you rapidly expand the chamber. That causes the pressure to drop and puts the liquid into a superheated state, meaning it is hot enough to boil but the liquid is so uniform that bubbles can't form unless there's a "nucleation site" for them to form on. Once again, charged particles passing through the chamber ionize atoms along their paths. In this case, those ions act as sites where bubbles of hydrogen gas can form along the path of the particle. Those bubbles rapidly grow large enough to be visible in a photograph.

As with cloud chambers, bubble chambers are surrounded by large magnets that deflect the paths of charged particles, allowing their properties to be determined.

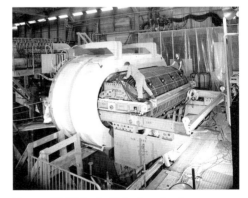

Figure 13.7 The bubble chamber known as "Gargamelle" at CERN. The cylinder on which the people are working is the actual chamber where the liquid is stored. A piston could cause the liquid chamber to expand by about 1%. The large white coils are electromagnets. *Source: CERN © 1977–2022.*

Bubble chambers can be made much larger than cloud chambers and can detect more energetic particles, so they quickly became the standard tool for particle detection, and Glaser won the Nobel Prize in 1960.

More Elusive Particles

Cloud chambers and bubble chambers solve the problem of detecting small particles: the bubble tracks they leave grow to sizes that can easily be photographed. But some particles don't give up their secrets so easily. Here we will consider two difficulties: particles with no charge, and particles with extremely short life spans.

We'll start with electrically neutral particles. Do you see the problem they cause? They do not ionize atoms as they move through a chamber, so they leave no visible tracks.

But remember that these particles come and go as individual steps in larger processes. When charged particles decay into neutral particles, which then decay into other charged particles, the result in a bubble chamber is visible trails with conspicuous gaps. Careful analysis of these gaps (based as always on conservation laws) reveals the properties of the uncharged particles. An example is shown in Figure 13.8. In Problems 11 and 12 you can analyze this image to determine the masses of the neutral particles involved.

Figure 13.8 On the left is a photograph from a bubble chamber, taken on October 17, 1957. On the right we highlight in blue the tracks that we will focus on, and add in red the inferred paths of two particles that are uncharged and therefore left no visible paths. *Source: © The Regents of the University of California, Lawrence Berkeley National Laboratory.*

The other difficulty we wish to consider is life span.

Can you detect the track of a particle that exists for only 10^{-7} seconds? The answer is yes, you can easily do this. Since most particles in high-energy particle experiments move close to the speed of light, such a particle would leave a track clear across your bubble chamber (and be long gone) before it decayed.

In fact, with modern technology and careful photographic techniques, we can detect bubble chamber tracks down to lengths of roughly 10^{-6} m. That should allow us to detect a particle whose lifetime is 10^{-14} s or less. But many particles have much shorter lifetimes than that. That's where relativity steps in to help.

Active Reading Exercise: Relativity and Bubble Chambers

Suppose a charged particle is moving through a bubble chamber at 0.99995 c.

1. How long would it take for that particle to make a 10^{-6} m track, according to the lab frame?

> **2.** How much time would elapse while it made that track, according to its own reference frame?
>
> **3.** If the particle's lifetime were somewhere between your two answers, would it make a track long enough for us to detect or not?

You can find our answers to that exercise at www.cambridge.org/felder-modernphysics/activereadingsolutions

The particles move so fast that they experience extreme time dilation, so they last longer (in the lab frame) and leave longer tracks than you would expect without relativity.

Even with time dilation, however, we can't detect bubble chamber tracks for particles with lifetimes less than about 10^{-16} s. That excludes all particles that decay via strong interactions, so how did we find all those hadrons? To answer that, we have to consider particle collisions in a bit more detail than we have done so far.

When particles collide, the velocities of the outgoing particles are constrained by conservation of energy and momentum, but those conservation laws are not usually enough to uniquely determine the outcome. You can think of this in purely classical terms (Problem 10). A 1D elastic collision between two billiard balls is uniquely determined by the conservation laws. In 2D the conservation laws give you three equations and the outgoing velocities have four components, so there's a distribution of possible outcomes. Which one happens depends on details of the collision, such as the spot on each ball where the collision occurred.

You can imagine a particle experiment as shooting a wide beam of billiard balls, all moving parallel with nearly identical velocities, across an enormous table with countless well-spread-out target balls. As the targets are struck they fly off with final velocities that are more or less randomly distributed among all the possible outcomes consistent with conservation of energy and momentum.

With that in mind, consider a high-energy collision between an electron and a proton. One possible outcome of this collision is for the proton to be replaced with a pion and a neutron:

$$e + p \rightarrow e + \pi^+ + n. \tag{13.2}$$

As expected, when you repeat the experiment many times the three outgoing particles have a wide range of velocities, always consistent with energy and momentum conservation. But the outcomes are not evenly distributed throughout that range. Instead you see a spike, as in Figure 13.9: one particular result that occurs not always but disproportionately often. (We will explain shortly why the "Z parameter" in Figure 13.9 is defined as it is, but our point here is just that it describes the outgoing particles' energies and momenta.)

There is nothing in the mathematics of collisions (classical or quantum mechanical) to suggest that the process described in Equation (13.2) should lead to such a high frequency of one particular outcome. We explain it by positing a different process:

$$e + p \rightarrow e + \Delta^+ \rightarrow e + \pi^+ + n. \tag{13.3}$$

Equation (13.3) introduces a middle step with a new particle, called a "delta baryon." This Δ^+ emerges from the original collision, passes a brief but (we hope) fulfilling 5.6×10^{-24} s, and then decays into the same π^+ and n particles that could have come from the original collision.

Number of events

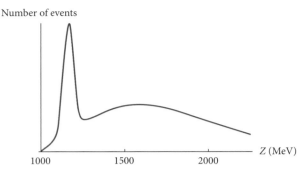

Figure 13.9 Histogram of the number of events observed with different values of the parameter
$$Z = \sqrt{(E_{\pi^+} + E_n)^2 - |\vec{p}_{\pi^+} + \vec{p}_n|^2 c^2}.$$
The spike around 1200 MeV suggests events in which the pion and neutron were produced by a short-lived particle with that mass.

But this time – here comes the key difference – those final particles come from the decay of a single particle, not from a collision, so their final energies are much more consistent.

Hence, Figure 13.9 tells a fairly detailed story. It tells of collisions that often follow the process in Equation (13.2), resulting in Z-values in a smooth and somewhat bell-shaped curve, but that *sometimes* follow instead the process in Equation (13.3), resulting in a much tighter range of results. Analysis of those statistics allows us to calculate the probability that any given collision will lead to the first process or the second. We can also determine the properties of the new particle whose existence we have just inferred, such as its mass (see the following Active Reading Exercise) and its mean lifetime (Question 7).

Active Reading Exercise: Detecting a Short-Lived Particle

A Δ^+ particle at rest in the lab frame decays into a π^+ and n. You capture and measure the energies and momenta of the two outgoing particles. Write the equation for the mass of the Δ^+ particle in terms of your measured quantities. Note that these particles are moving at relativistic speeds, so you need relativistic formulas (Appendix B).

The sum of the outgoing energies equals the Δ^+ energy, and the sum of the outgoing momenta equals the Δ^+ momentum. Using the relativistic equation $E^2 = p^2 c^2 + m^2 c^4$, we conclude that

$$m_{\Delta^+} c^2 = \sqrt{(E_{\pi^+} + E_n)^2 - |\vec{p}_{\pi^+} + \vec{p}_n|^2 c^2}.$$

You can now see why the most convenient independent variable for graphs such as Figure 13.9 is the "Z parameter" $\sqrt{(E_{\pi^+} + E_n)^2 - |\vec{p}_{\pi^+} + \vec{p}_n|^2 c^2}$ for the final particles – in this example the pion and neutron.

Note that if you run this experiment once and get a Z parameter of exactly 1200 MeV, that doesn't mean that a Δ^+ particle was involved; its extremely short life means there's no direct measurement that will confirm its existence. But the large spike at one particular Z-value suggests that *some* of the events do involve a particle with the indicated mass. A great deal of particle physics discoveries involve taking many measurements and concluding statistically that some of them must have involved a particular particle – the Δ^+, the Higgs boson, or whatever else you are looking for – but you can never know for sure about one individual measurement.

Modern Detectors

Cloud chambers were behind many of the discoveries in particles physics from the 1920s into the 1950s, and bubble chambers were the principal tool for detecting particles for several decades after that. But these chambers have a number of limitations and are no longer widely used. For example, a bubble chamber takes a few seconds to reset between uses, whereas a modern detector can take thousands of images per second.

"Scintillation counters" emit flashes of light when struck by high-energy particles. "Silicon detectors" and "multiwire proportional counters" produce currents when hit by particles. A full discussion of modern particle detectors would be long and go out of date quickly, but they are generally based on electronics and produce data rapidly in computer-usable forms.

13.3.2 Questions and Problems: Detecting Particles

Conceptual Questions and ConcepTests

1. When Carl Anderson found a positron in a cloud chamber photograph, he used its motion through a lead plate to show which direction it was moving in the photograph. Why was that information necessary to confirm his discovery?

2. Charged particles move across the page, initially traveling directly upward. A constant magnetic field points directly out of the page.

   (a) Draw the path of a positively charged particle. Make sure to draw arrows to indicate the direction of travel!

   (b) Draw the path of a negatively charged particle, also with arrows.

   (c) Draw the path of a second negatively charged particle with the same mass and charge as the first one, but moving more slowly. How does this path differ from the previous one?

   (d) Briefly explain the significance of this question to the material in this section.

3. If you know a particle's charge, then its curvature in a bubble chamber allows you to find its momentum, but only if you have cameras taking pictures from various angles in 3D. Why would it be impossible to determine this from a single camera image?

4. In the Active Reading Exercise on p. 618 you used time dilation to explain why a particle makes a longer track in a bubble chamber than a non-relativistic calculation would predict: the particle's "decay clock" is running slowly. But that explanation only works from the lab frame. In the particle's frame, the particle's clock is running perfectly normally. (The lab clocks are running slowly, but that doesn't matter, since they don't determine the decay time.) If the particle travels (say) all the way from one end of the bubble chamber to the other without decaying, how do we explain that in the particle's reference frame?

5. Research one type of particle detector in use today, and write a paragraph description of what it does and how. Make sure to include some discussion of how this modern detector is better than a cloud or bubble chamber!

Problems

6. A particle of charge q is moving in a bubble chamber perpendicular to an applied magnetic field B. Express the particle's momentum in terms of B, q, and the radius R of the particle's curved track. (Particles in bubble chambers move at relativistic speeds, but the relativistic calculation in this case gives the same answer as the Newtonian one.)

7. We said that a histogram of the Z parameter of collision products can be used to infer the existence of a short-lived particle. For example, Figure 13.9 suggests the existence of the Δ^+ particle, with a mass of about $1200\,\mathrm{MeV}/c^2$. How can you use the time–energy uncertainty principle to estimate the lifetime of the Δ^+ from that same figure? Explain how to do that, and use the figure to make a numerical estimate.

8. Prove that if two relativistic particles have equal momenta, the one with more mass will always have less kinetic energy.

9. Anderson knew his particle was positively charged because of the direction of its curve. But it still might have just been a proton, right? That possibility was ruled out by the fact that his particle had enough energy to make it through the lead plate. *Hint*: This problem will be easier if you leave the units in MeV and cm rather than converting to SI.

   (a) Based on the curvature of the path, Anderson determined that his particle had a momentum of 63 MeV/c before it hit the lead plate. If the particle had been a proton, what would its kinetic energy have been?

   (b) A proton at those energies passing through lead loses about 680 MeV of kinetic energy per cm. How far would such a proton penetrate a block of lead?

   Anderson's lead block was 6 mm thick. Since the particle got through, you should have concluded (as he did) that it could not have been a proton.

10. A ball with mass m_1 traveling at velocity $v_{1,0}$ slams elastically into a second ball with mass m_2 initially at rest. The speeds are low enough to make Newtonian calculations reasonable.

    (a) If the entire problem takes place in one dimension, calculate the final velocities $v_{1,f}$ and $v_{2,f}$ of the two balls in terms of the given quantities.

    (b) Now suppose the problem takes place in two dimensions, so the balls scatter at angles θ_1 and θ_2. Write the laws of conservation of energy and momentum, and explain mathematically why those equations *cannot* be solved without more information.

11. Figure 13.8 shows a photograph from a bubble chamber. The link below takes you to a larger, printable image of the parts of this photograph that show a pion decaying into a kaon and a lambda, and the lambda then decaying into a proton and a pion:
    www.cambridge.org/felder-modernphysics/figures

In this problem you're going to analyze the second decay reaction, $\Lambda^0 \rightarrow p + \pi^-$, to determine the mass of the lambda particle. The proton has mass 1.7×10^{-27} kg. The pion has mass 2.5×10^{-28} kg, and the same charge as an electron.

(a) The momentum of a particle of charge q is related to the curvature of its track by $p = qRB$ (Problem 6). The field used in this experiment was 1.5 T and the radius of curvature of the pion's track is 0.05 m. Find the magnitude of the pion's momentum.

The proton's track is too straight to measure its curvature, but we can find it from conservation of momentum if we know the direction in which the lambda was moving. We can figure that out since we can assume the lambda particle moved in a straight line from the point where it was produced to where it decayed.

(c) Why can we assume the lambda moved in a straight line?

(d) The pion's initial velocity was at an angle $35°$ to the right of the lambda's path. It looks as though the proton continued directly along the path of the lambda, but careful measurement shows that its initial path deviated to the left by $0.32°$. Find the magnitude of the proton's momentum.

(e) Find the magnitude of the lambda's momentum.

(f) Find the total energy of the lambda particle. Newtonian approximations will not work here; you need the relativistic equations in Appendix B.

(g) Use your results from Parts (e) and (f) to calculate the mass of the lambda. Express your answer in MeV/c^2. (You should find that your answer is reasonably close to the tabulated mass of a Λ^0 particle.)

12. [*This problem depends on Problem 11.*] Perform an analysis similar to Problem 11 to estimate the kaon mass. You'll need to measure the angles of the two pions from the kaon's path. (It may help to print out the online diagram.) You will also need to measure the radius of curvature of the π^+. To do that you

draw a chord connecting the two ends of its arc. If L is the chord length and s is the perpendicular distance from the midpoint of the chord to the arc then $r = L^2/(8s) + s/2$. Once you have calculated that radius, you'll need to rescale it from the size of your printout to the size of the actual image.

13. You perform many particle experiments in which you observe the following interaction: $e + p \rightarrow e + \pi^+ + n$. For simplicity we'll assume you measured all of the initial and final velocities happening in the same direction. Your friend the theorist suggests that the inital collision may be producing a Δ^+, which then decays into the pion and neutron. You measure the pion's energy to be 950 MeV. If your friend is correct, what is the neutron's energy? You'll need to look up some particle data and use relativistic formulas to answer this.

14. A pion π^+ (140 MeV/c^2) collides with a stationary proton p (938 MeV/c^2) and produces a Σ^+ (1189 MeV/c^2) and a K^+ (494 MeV/c^2). The sum of the final masses is 605 MeV/c^2 greater than the sum of the initial masses, so conservation of energy says the pion must have at least 605 MeV of kinetic energy.

    (a) Why must it actually have significantly more kinetic energy than that?

    (b) Prove that, for a given total momentum, the smallest total energy for two particles occurs when they have equal velocities.

    (c) What is the minimum kinetic energy that the pion must have to produce this reaction? (This is called the "threshold energy" for the reaction.)

13.4 Symmetries and Conservation Laws

This section discusses two types of properties that a physical theory can have.

- A *conservation law* describes an unchanging quantity. For instance, in the absence of external forces, $m\vec{v}$ is a conserved quantity in Newtonian mechanics. The same quantity is *not* conserved in relativity. So "conservation of $m\vec{v}$" characterizes some physical theories but not others.

- A *symmetry* indicates a change you can make to a system without altering its resulting behavior. For instance, if you change the time of an event from t to $t + \Delta t$, that event may have different effects on the objects around it. But if you change the time of every event in the universe by the same Δt, the universe unfolds exactly as before; the laws of the universe are the same at $t = 1$ as at $t = 2$. This property, known as "time-translation symmetry," applies to all known physical theories.

In 1915, the German mathematician Emmy Noether proved a remarkable connection between these two types of properties. She showed that, given a certain type of symmetry, it is possible to derive a conserved quantity. We will not prove that theorem here, but we will outline some of the important symmetries in particle physics and the conservation laws they imply.

13.4.1 Discovery Exercise: Symmetries

You are watching a video of a ball falling to the ground and then bouncing back up. But you suspect that the video has been tampered with. For each of the following tamperings, describe how you could detect it – or say that you could not.

1. The video has been rotated by $90°$, so the "down" direction in real life is "right" in the video.

    See Check Yourself #23 at www.cambridge.org/felder-modernphysics/checkyourself

2. The video is being played backward.

3. The video has been speeded up.

4. The video has been left/right reflected.

13.4.2 Explanation: Symmetries and Conservation Laws

If you manufacture 1000 Λ^0 baryons, they will all decay. But their decay times will vary, and the decay *products* will also vary – not because of any difference in initial conditions, but because of quantum mechanical randomness. One might decay into a proton and a π^- pion, another into a neutron and a π^0 pion, and there are several other possibilities, each occurring with predictable probabilities. But you will never see a Λ^0 decay into a proton and a π^0, because that would violate conservation of charge.

We've been discussing conservation laws as restrictions: any event that violates these laws is forbidden. But quantum field theory makes a much stronger statement:

> *Any decay that <u>can</u> happen – that is, any decay that obeys all the conservation laws – <u>will</u> happen.*

That is, the probability of any allowable decay is greater than zero, so it will happen sometimes. If a particular decay never happens, that means there is a conservation law forbidding it.

For instance, a neutron can spontaneously decay into a proton and a lepton–antilepton pair. Because it can, it does: isolated neutrons are unstable.

On the other hand, a proton *cannot* decay into a neutron and other particles, because a proton is lighter than a neutron. In fact, a proton is the lightest baryon you can make, so conservation of energy prevents it from decaying into any other baryon. And conservation of baryon number prevents a proton from decaying into any non-baryon. So protons are stable: no one has ever measured a proton decaying, because there's nothing it can decay into and still follow these conservation laws.

Approximate Conservation Laws

We have seen that some quantities are always conserved: for instance, charge. Other quantities are conserved only in certain types of interactions: for instance, quark flavor is conserved in strong interactions but not in weak ones.

Leptons provide examples of both types of conservation. Total lepton number (number of leptons minus number of antileptons) is always conserved. Lepton number *within a particular family* is conserved in most interactions (when a muon decays it must produce a muon neutrino), but not in neutrino oscillation (that muon neutrino may later turn into an electron neutrino).

Rules that require a particular quantity to be always conserved are referred to as "exact" conservation laws, while conditional rules are called "approximate" conservation laws. Underlying any approximate law is an issue of time scale: the allowed transition is much slower than the forbidden one. So we may see one flavor of quark decay into another, but this will happen in the (relatively long) weak-force time, not in the (relatively short) strong-force time. Similarly, the muon neutrino will probably not oscillate into an electron neutrino until long after the decay process that created it. So every approximate conservation law can be stated as "you will not see such-and-such decay process in less than such-and-such a time period."

Once you have the idea of approximate conservation, you can start to doubt the rules you think you know. Maybe baryon number and lepton number conservation are violated by some unknown, very slow processes. Maybe protons really do decay if you wait long enough!

In fact, the theories that unify strong and electroweak interactions into one force law imply that baryon and lepton number conservation *should* be violated, so if those theories are correct then proton decay should be possible. As of 2021, experimental searches have put a lower limit of over 10^{34} years on spontaneous proton decay. So if baryon number is an approximate conservation law, it's a very good one.

Symmetries

Conservation laws are an important way of characterizing a particular physical theory. Another important property of a physical theory is its "symmetries." Our primary goal in this section is to explain the idea of a symmetry, and discuss some of the most important symmetries in particle physics. But at the end we will circle back to discuss a deep and surprising connection between symmetries and conservation laws.

A symmetry is something that you could change about the universe without any detectable effect. For example, suppose you watch a video of two spaceships interacting. They push off each other, fire thrusters, and then launch projectiles from one ship to the other. You carefully measure the motions in the video and confirm that they all obey Newtonian mechanics. Now someone shows you that same video rotated 90°. The motions are all in different directions in the new video, but they still follow Newtonian mechanics perfectly. We say that Newtonian physics is "symmetric under rotation."

While we believe rotational symmetry is a fundamental property of physics, it is *not* a property of our daily experience. Play a video of you and your friends rotated 90° and the change will be obvious, because you will all be walking on a vertical wall! We say that inside the room, rotational symmetry is "spontaneously broken": one direction (up and down) acts differently from the other two because of the local conditions. But if you zoom the video out far enough, you will see that you are all held to the side of the screen by the Earth to your left. Newtonian physics still works in that video.

Table 13.2 summarizes three of the most important symmetries in physics.

Table 13.2 Continuous symmetries

The following symmetries are exactly obeyed by all known laws of physics.	
Rotational symmetry	The laws of physics are the same in every direction.
Translational symmetry	The laws of physics are the same at every place.
Time-translation symmetry	The laws of physics are the same at every time.

Thinking back to Chapter 1, you may be able to think of another one. We'll remind you of it in Question 10.

C, P, and T Symmetries

Rotation, translation, and time-translation are all "continuous symmetries," meaning they apply to quantities that can vary by arbitrarily small increments. There are also "discrete symmetries." Here we will discuss the three most important ones, which are summarized in Table 13.3.

Table 13.3 Discrete symmetries

The following discrete symmetries are approximately obeyed in particle physics.	
Charge conjugation (C)	The laws of physics are (approximately) the same if you replace every "charge" by its opposite. We are using the word charge here more broadly than we have generally done, as discussed below.
Parity (P)	The laws of physics are (approximately) the same in a mirror image.
Time reversal (T)	The laws of physics are (approximately) the same with time played backwards.

We'll start with the second one. What if – instead of watching the real universe – you watched the universe's reflection in a mirror? If our physical laws obey the symmetry called "parity" (often symbolized by the single letter P) then those laws would function identically in the mirror, so you would have no way of detecting the change.

Second, what if you took a video of two colliding billiard balls, and then played the video backward? The symmetry called "time reversal" (T) says that the motion would still obey the normal laws of collisions, so you would have no way of identifying the reversal. (Our normal experience that forward time is very different from backward time does not result from asymmetries in fundamental laws, but from the increase in entropy. We discussed this in Chapter 10; we'll ask you for a brief example in Question 7.)

Finally, what if you changed all positively charged particles to negative, and vice versa? That's a subtler question than it appears. Consider:

- An electron has charge $-e$ and lepton number 1.
- A positron has charge $+e$ and lepton number -1.
- But *no known particle* has charge $-e$ and lepton number -1.

(Remember that lepton number matters because it is a conserved quantity, so it dictates that some decays are allowed and others are forbidden.) An alternative universe with "symmetry of electric charge" would have to have that third (nonexistent) particle: a particle with all the same properties as a positron, except reversed electric charge. We can therefore see that our universe is *not* symmetric with respect to electric charge. The same example demonstrates that the universe is not symmetric with respect to lepton number either.

So we don't talk about reversing electric charge or lepton number in isolation. "Charge symmetry" (C) says that the universe would be unchanged if you reversed not just the electric charge but all the charges, including baryon number, lepton number, and color (blue ↔ antiblue and so on), of every particle. Under that transformation, the list of particles in the standard model would be unchanged.

For most of the twentieth century it was believed that C, P, and T were exact symmetries of nature. All three of them do indeed work in the vast majority of situations, but exceptions have been found for all three.

Chien-Shiung Wu and Violations of Symmetry

The first violation of one of the above symmetries was found in 1957, when Chien-Shiung Wu showed that neutrinos look different in a mirror image. The Active Reading Exercise below will give you an important insight into her experiment.

Active Reading Exercise: A Right-Handed Clock

The clock in the figure is moving to the right while its hands turn clockwise around the dial. You are watching the clock, and also watching its reflection in a mirror.

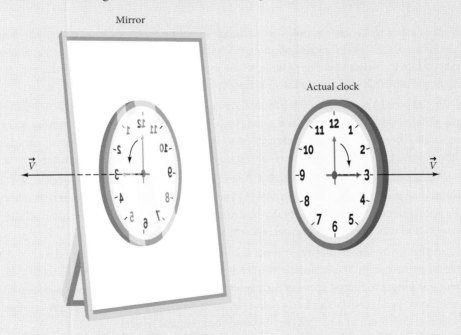

1. What is the direction of the angular momentum vector for the hands of the actual clock?

2. What is the direction of the angular momentum vector for the hands of the reflected clock?

3. A "right-handed clock" has its angular momentum vector parallel to its momentum vector; a "left-handed clock" has the two vectors antiparallel. Classify both the actual clock, and the mirror clock, as right-handed or left-handed.

We hope you determined that the clock is right-handed, and its reflection is left-handed. A similar distinction can be made with particles, with the particle's spin (rather than the clock hands) supplying the angular momentum. None of this implies any violation of P, the parity symmetry. You can imagine a world in which right-handed particles and left-handed particles are randomly distributed, and such a world would follow the same laws in a mirror.

But when Wu measured the beta decay of cobalt-60, she found that the produced antineutrinos tend to come out right-handed. Neutrinos show the opposite behavior, generally coming out left-handed. Countless experiments since have confirmed this bias in the handedness of antineutrinos and neutrinos produced in weak interactions.

Do you see the implications of that result? It simultaneously demonstrates limitations in two different symmetries!

- If you played a video of Wu's cobalt decays in mirror-image, you would see a bunch of left-handed antineutrinos. That would tip you off that the video wasn't following the laws of physics that we see in our non-mirrored universe. Thus P is not an exact symmetry.

- If you replaced Wu's cobalt with "anti-cobalt" (changed the charges of all particles), the decay would produce neutrinos instead of antineutrinos. But you could detect that change by the left-handedness of those resultant particles. Thus C is not an exact symmetry either.

Once it came to be understood that C and P could be violated, it was generally believed the *combination* CP was still a valid symmetry. The charge-switched universe, seen in a mirror, looks exactly like the original universe without the mirror – or at least it does in Wu's experiment. But we now know that CP isn't a perfect symmetry either. And it's a good thing it isn't.

In Chapter 14 we will discuss the creation of matter and antimatter in the Big Bang. Without any details, you can see that in a universe that perfectly followed CP symmetry, matter and antimatter would have been created in equal amounts. Then they would have annihilated each other, leaving only photons. The resulting universe would be bright, but not much fun.

Instead, the evidence shows that CP symmetry was obeyed approximately: for every 10^9 antimatter particles, there were $\left(10^9 + 1\right)$ particles of matter. The particles that remained after the subsequent annihilation, literally one matter particle in a billion, make up all the stars and planets and other matter in our current universe.

So the existence of matter is itself incontrovertible proof of CP violation.[8] CP violation has been found in the standard model both theoretically and experimentally, but it's not nearly strong enough to account for even the one-in-a-billion difference between matter and antimatter produced in the early universe, so the existence of matter is yet another piece of evidence that there must be physics beyond the standard model.

So C is violated, P is violated, and the combination CP is violated. What about T? Does a video played backward obey the same physics as one played forward? Are there any exact, discrete symmetries in nature?

---

8 You could try to get around this by saying the universe had extra matter from the beginning. For reasons we'll discuss in Chapter 14, we are confident that the early universe went through a stage with no matter or antimatter, so some CP-violating process must have caused more matter than antimatter to be made after that.

Yes, there is at least one exact discrete symmetry, and no, it's not T. There is a theorem saying that any quantum field theory must obey the symmetry CPT. If you reverse every type of charge, mirror-image your video, *and* play it backward, it will look indistinguishable from a video you could make of an actual process in the real universe. That fact, combined with the fact that CP is violated, implies that T by itself must be another approximate symmetry.

Noether's Theorem: Symmetries and Conservation Laws

In this section we've talked about two important ideas: conservation laws and symmetries. In 1915 Emmy Noether proved that these two are related.

> *Given any continuous symmetry, Noether's theorem allows you to find a conserved quantity.*

For instance, you can show in Problem 14 that Noether's theorem starts with "Newtonian physics is symmetric under translation," and concludes that "Newtonian physics conserves momentum." Before you do the math, let's consider the result you're going to prove. Imagine a planet orbiting a star. The entire planet-and-star system is symmetric under translation. (Add 5 to both of their x-values and nothing changes.) Correspondingly, the momentum of the entire system is conserved. But the planet by itself, moving through the gravitational field of the star, is *not* symmetric under translation. (Move the planet closer to the star and its behavior will change!) And for a system defined as *only* the planet, momentum is not conserved. So in this simple example, the symmetry and the conservation law do indeed go hand-in-hand.

Noether's theorem is remarkably general. In the form in which she proved it, it applies to every non-quantum theory we have, including Newton's mechanics, Maxwell's electromagnetic fields, and Einstein's relativity (both special and general). Later theorists were able to extend it to quantum mechanics (both "original recipe" and quantum field theory). In short, every physical theory that we use obeys Noether's theorem.[9]

The three most famous applications are the three continuous symmetries described above.

- The fact that the laws of physics are symmetric under translation implies that momentum must be conserved.
- The fact that the laws of physics are symmetric under rotation implies that angular momentum must be conserved.
- The fact that the laws of physics are symmetric under time-translation implies that energy must be conserved.

Noether's theorem was originally formulated in terms of continuous symmetries, but some analogous results have been proven for discrete symmetries as well. For example, each type of particle can be assigned a value of "parity" and the overall parity is conserved in strong and electromagnetic interactions (which obey the P symmetry), but violated in weak interactions (which do not).

---

9 If you've studied mechanics beyond the introductory level you may be familiar with Lagrangians and actions. Noether's theorem is formulated in terms of symmetries of the action, and it applies to all these physical theories because they can all be derived from principles related to the action. The extension of Noether's theorem to quantum field theory is called the "Ward–Takahashi identity," after John Ward and Yasushi Takahashi.

13.4.3 Questions and Problems: Symmetries and Conservation Laws

Conceptual Questions and ConcepTests

1. Your friend says that translation is clearly not a valid symmetry because he can easily tell a video taken in the Sahara Desert from one taken in the middle of the Pacific Ocean. How do you respond?

2. We've been talking about symmetries that apply to the physical laws of the universe, but there are also symmetries that apply in specific situations. For example, an electron in a cubic lattice feels a potential that is periodic.

   (a) Does that periodicity correspond to a continuous symmetry or a discrete symmetry? Explain.

   (b) Give at least one other example of a situation in which a particle experiences a symmetry different from the universal symmetries of nature.

3. Experimental searches for proton decay have been occurring for a few decades. How can a search that has only been going on for that length of time put a lower limit of 10^{34} years on the proton lifetime? (For context, 10^{34} years is over a trillion times the current age of the universe.)

4. Define in your own words the symmetries C, P, and T.

5. The C symmetry is sometimes defined as "change the charge of every particle," where the word "charge" is understood to refer not only to electric charge but also to color charge, baryon number, and so on. In other sources you will see C symmetry defined as "exchange every particle with its antiparticle." Explain why these two definitions are *not* perfectly equivalent.

6. A right-handed particle is defined as one whose spin is parallel to its momentum; a left-handed particle is defined as one whose spin is antiparallel to its momentum.

   (a) No one worries about the question "What if the spin points in some other direction, neither perfectly parallel nor perfectly antiparallel to the momentum?" Why not?

   (b) The names were not chosen arbitrarily. What do these two definitions have to do with "right" and "left"?

7. We described T as an approximate symmetry of nature, but in the macroscopic world it doesn't seem to hold at all. Two spaceships might collide and go flying off as a giant cloud of debris, but you will never see a giant cloud of debris collapse and form into two spaceships, flying backward in opposite directions. Given examples like that one, how can we say T is an approximate symmetry? Equivalently, how could physicists have believed until the late twentieth century that T was an exact symmetry of nature?

8. For each reaction below, indicate whether this reaction will sometimes occur. If it will not, give at least one reason why not. The letters signify a neutron (n), proton (p), electron (e), neutrino (ν), pion (π^0), or neutrino (μ), or their antiparticles.

   (a) $n \rightarrow p + e + \overline{\nu}$

   (b) $n + p \rightarrow n + p + e + \overline{\nu}$

   (c) $n + n \rightarrow p + e + \nu$

   (d) $n + p \rightarrow n + e^+ + \nu$

   (e) $n + p \rightarrow n + n + e^+ + \nu$

   (f) $n + p \rightarrow n + n + e^+ + \overline{\nu}$

9. A new particle called a "moron" (m) is discovered. It undergoes the reaction $m \rightarrow p + p^- + n$ in approximately 10^{-23} s.

   (a) Is the moron a baryon, meson, or lepton? How do you know?

   (b) What force governs the decay of the moron?

   (c) What is the charge of the moron?

   (d) You don't have enough information to determine the precise spin of the moron, but you can narrow it down to two possibilities. What are they?

   (e) What can you say about the quark content of the moron?

(f) The antiparticle of the moron is called the "lesson."[10] Answer all of the above questions for this particle.

10. Translational symmetry says that if you replace the position x for every particle in the universe with position $x + \Delta x$, no experimental results will change. Replacing the velocity v_x of every particle with $v_x + \Delta v_x$ is a symmetry in Newtonian physics, but not in special relativity. What is the relativistic equivalent?

11. Wu's experiment demonstrating parity violation was the first solution to what science writer Martin Gardner called the "Ozma problem." Imagine sending a message to an alien race. You don't know where they are and you can't look at any object in common. Your message is a recording of words (which they can understand) but includes no pictures. How can you tell them which directions we mean by "right" and "left"? It is generally agreed that in a universe with perfect parity symmetry the problem would be unsolvable. Explain how you could solve it in the universe in which we actually live.

For Deeper Discussion

12. In a universe that had the symmetry "rotating every object by $30°$ changes nothing," it would not also be necessary to specify that "rotating every object by $60°$ changes nothing." The former implies the latter, since you can just rotate by $30°$ twice. You might think, by a similar argument, that the rotational symmetry of our universe (you can rotate by any amount in any direction) implies parity (you can reflect through any plane), but it doesn't. Why not?

13. A friend of yours argues that color should be added to the list of discrete symmetries: "Swap every blue particle for antiblue, and red for antired, and green for antigreen, and nothing would change." You disagree: "There's no such thing as a quark with $(2/3)e$ charge and antiblue color. If you swap color you have to also swap electric charge." But your friend persists: "That's just playing with words. Suppose I changed the colors but nothing else. There's no

experiment that would show I had changed anything." Is your friend right? Why or why not?

Problems

14. **Noether's Theorem in Newtonian Physics**

Noether's theorem can be expressed in terms of the "Lagrangian," i.e. the kinetic energy minus the potential energy. The Newtonian Lagrangian in two dimensions can be written

$$L = \frac{1}{2}m\dot{x}^2 + \frac{1}{2}m\dot{y}^2 - U(x, y).$$

The dot indicates a time derivative, so $\dot{x}$ means dx/dt. We are using that notation (rather than just writing v_x) because Cartesian coordinates are just one way to express the Lagrangian, and not always the most convenient. Imagine more generally that you have expressed the Lagrangian in terms of some set of spatial coordinates, and one of those spatial coordinates is called q. Then Noether's theorem states:

If the Lagrangian is invariant under the transformation $q \rightarrow q + \delta q$ (a change of q by an infinitesimal amount), then the quantity $\frac{\partial L}{\partial \dot{q}} \delta q$ is conserved.

In this problem you will show how this very general relationship connects certain specific symmetries with familiar conservation laws.

In general, δq can be a function of the coordinates, but in this problem we're only going to consider constant transformations, so Noether's theorem promises that $\frac{\partial L}{\partial \dot{q}}$ is conserved.

(a) Translational symmetry says that the Lagrangian is invariant under the transformations $x \rightarrow x + \delta x$ and $y \rightarrow y + \delta y$ for constant δx and δy. Show that this symmetry is equivalent to saying that the potential energy is constant everywhere.

(b) Show that, under translational symmetry as defined in Part (a), Noether's theorem becomes conservation of momentum. (This leads to the conclusion that momentum is conserved in a constant potential field – in other words in the absence of external forces – hardly a surprise!)

---

10 Get it?

(c) Now consider rotational symmetry. Use the relationships $x = \rho\cos\phi$ and $y = \rho\sin\phi$ to rewrite the Lagrangian in terms of the polar coordinates ρ and ϕ. (You may be more used to the letters r and θ for polar coordinates. It doesn't matter.) As a check, you may recognize your answer as the kinetic energy terms for motion in the radial and tangential directions.

(d) In terms of the coordinates ρ and ϕ, how would you write a rotation about the origin?

(e) Assuming the Lagrangian is symmetric under such a rotation, find the quantity – a function of some combination of ρ, ϕ, $\dot\rho$, and $\dot\phi$ – that Noether's theorem promises is conserved.

(f) Show that your answer to Part (e) is the particle's angular momentum.

15. [*This problem depends on Problem 14.*] In this problem you're going to re-do Part (e) of Problem 14, leaving the Lagrangian in terms of x and y.

(a) Find the transformation terms δx and δy for a rotation $\delta\phi$ about the origin. You should find that they are *not* constant.

(b) In a case like this where a symmetry applies to two transformations done simultaneously (on x and y), the conserved quantity is the sum $\frac{\partial L}{\partial x}\delta x + \frac{\partial L}{\partial y}\delta y$. Calculate this and show that this can be rewritten to match your answer to Problem 14(e). (Once again, if you find that "the quantity ⟨*something*⟩ times a constant is conserved," you can simplify that to just "⟨*something*⟩ is conserved.")

13.5   Quantum Field Theory

An introductory course in classical mechanics answers three questions:

1. *What are the basic laws?*
    Newton's laws. In particular, the equation $F = ma$ describes how an object will move when subject to forces.

2. *Okay, so what are the forces?*
    Gravity, electric force, tension, spring forces, and many others.

3. *What are the objects, and what are their properties?*
    Rockets, bicycle tires, and frictionless rectangular blocks all feature prominently. When such an object is introduced, its important properties such as mass, coefficient of friction, and moment of inertia must be specified in the problem.

For example, if you want to classically solve for the accelerations of two interacting electrons, you need to know Newton's second law, *and* Coulomb's law for the force, *and* the charge and mass of an electron.

The answers change in modern physics, but those three questions are still useful. This chapter so far has focused on the third question (listing particles), with some space given to the second (listing forces). For the first question – the fundamental laws – the best currently known answer at the particle level is given by "quantum field theory."

13.5.1   Explanation: Quantum Field Theory

The phrase "quantum mechanics" describes the set of ideas that began with Planck's first suggestion of quantized energy levels in 1900, and reached its full development in the 1920s with the work of Bohr, Schrödinger, Heisenberg, and others. Subsequent decades saw the development of a more advanced theory called "quantum field theory" (QFT) that currently forms our most accurate predictive tool for particle behavior, and is therefore one of our best attempts to articulate the fundamental governing laws of the universe.

In this section we will use the phrase "quantum mechanics" exclusively to refer to the earlier, pre-QFT theory. (Some sources use this phrase more broadly to refer to both theories, and may refer to the earlier theory as "Schrödinger mechanics.")

Quantum field theory fills two gaps in quantum mechanics:

1. The great triumph of quantum mechanics was the hydrogen atom. As you saw in Section 7.4, the Schrödinger calculation does not treat an electron as a classical particle, like a planet orbiting a star; rather, the electron's quantum mechanical properties, such as discrete energy levels and uncertain position, are vital clues to its behavior. But – you may not have noticed this at the time – the same calculation treats the *electric field* as classical. When we consider the particles and the fields as interacting quantum entities, we move into the realm of quantum field theory.

2. Much of the current chapter has been devoted to the issues of particle decay and particle creation, neither of which is described by Schrödinger's equation. People do quantum mechanical calculations of processes such as an electron in an atom changing its state and emitting a photon, but these calculations require ad hoc postulates that aren't part of quantum mechanics. Quantum field theory addresses such transformations as integral parts of the theory, and in fact describes all particle interactions – all forces – in terms of particles being created and destroyed.

Quantum field theory is also compatible with special relativity, unlike the original formulation of quantum mechanics. It is *not* compatible with general relativity.

This section is going to be qualitative. We are going to discuss what some QFT calculations do, but we are not going to actually do those calculations, which rely on math beyond the scope of this book.

Path Integrals

Before introducing quantum field theory, we're going to restate the rules of ordinary quantum mechanics in a way developed by Richard Feynman in 1948.

Recall our favorite quantum mechanical experiment, the double slit. A photon is fired from a source at a known position, passes a wall with two narrow slits, and arrives on a second wall at a position that is subsequently measured. How do you calculate the probability that the photon lands at some position x on the back wall (within some tolerance Δx)?

In the traditional formulation, you calculate the wavefunction as it travels through both slits at once, constructively and destructively interferes with itself, and arrives at the back wall. At any given spot on the back wall, the probability density is given by the squared magnitude of the resulting wavefunction. Multiply that density by Δx to get a probability.

Feynman found a mathematically different approach to the same problem. For the particular result whose probability you want to calculate:

1. *List every possible path that the photon could take from its starting point to that particular endpoint.* You might think in this example that there are only two possible paths, one

through each slit. But remember that a wave spreads out in every possible direction from every point. So for any given point on the back wall, there are infinitely many spatial trajectories the photon could have taken to arrive there. And any given spatial trajectory still represents infinitely many paths, since the photon could travel along that trajectory with different or varying energies.

Nonetheless, it is possible to describe the set of all possible paths. Think of it as a list that just happens to be infinitely long.

2. *For each path, perform a certain calculation to find its "probability amplitude."*

The probability amplitude is a complex number associated with a particular path. We will leave the details of this calculation for a later course.

3. *Add up the probability amplitudes of all the paths that lead to the same end result.*

Because the set of all possible paths is continuous, that sum is actually an integral, called a "path integral": another mathematical topic we're going to skip. But remember that destructive interference is a possibility, so a collection of non-zero probability amplitudes could still sum to zero.

The squared magnitude of your final answer is the probability of finding that result in your experiment.

Even without the mathematical details, you can probably tell that actually doing this calculation is almost always impossible! But Feynman proved that if you *could* do this calculation – or when you can reasonably approximate it – the calculated probability will match the traditional Schrödinger calculation.

So what Feynman has given us is an equivalent, but less useful, formulation of quantum mechanics. But Feynman's approach extends to the problems that Schrödinger's equation cannot handle: processes in which particles are created or destroyed.

Feynman Diagrams

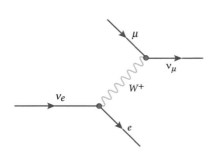

One of Feynman's most important contributions was a visualization tool for particle interactions, now generally referred to as a "Feynman diagram." Figure 13.10 shows an example.

The reaction shown in Figure 13.10 is possible, though its probability is low. But it gives us a simple example of many important features of Feynman diagrams. (The conventions we use here are common, but not universal.)

Figure 13.10 An electron neutrino decays into an electron and a W^+ boson. That boson then collides with a muon and they turn into a muon neutrino.

- A straight line represents a matter particle.
- A wavy line represents a force-carrying particle. This particular wavy line is labeled as a W^+ boson. A wavy line with no label is always assumed to represent a photon.
- The vertices where lines meet represent interactions.
- The horizontal direction represents time, going left to right. In Figure 13.10, the vertex for the ν_e interaction is further left than the vertex for the μ interaction, indicating that it happened earlier. From this we can infer that the ν_e emitted the W^+ and the μ absorbed it.

However, there is no scale. The left-to-right sequence indicates the order of events, but distance indicates nothing about how much time elapsed between those events.

- The vertical direction represents nothing at all. When you construct a Feynman diagram, you just use your two-dimensional space to represent the order of events as clearly as possible.

Hence, Figure 13.10 can be read as follows:

> *You observed an electron neutrino and a muon going in, and an electron and a muon neutrino coming out.*
>
> *One possible path for this result is: the electron neutrino decayed into an electron and a W boson. The muon and the W^+ then decayed into the muon neutrino.*

Note the phrase "one possible path." If we gave you the relevant formulas – which we're not going to do – you could calculate the probability amplitude of this particular reaction, and of every other path that might lead to the same final outcome (an electron and a muon neutrino). You could then sum those probability amplitudes, take the magnitude and square it, to find the probability of that outcome.

But if you saw precisely this outcome in the lab, you would have no way of determining whether this particular path (with the W^+) or some other path was followed. In a very real sense, that question has no meaning. So the boson that you never saw, and that may or may not ever have existed, is referred to as a "virtual particle."

In a moment we will test your ability to interpret a Feynman diagram. But first we have to introduce one more convention: antimatter particles are generally represented with their arrows pointing from right to left. Feynman showed in 1949 that you make all the right predictions if you interpret antimatter as ordinary matter traveling backward in time. (This follows from CPT symmetry, which we discussed in Section 13.4.) But we will always speak of antimatter particles moving *forward* in time, and treat the backward arrows simply as a visual indicator that this particular line represents a particle of antimatter.

Active Reading Exercise: Reading Feynman Diagrams

The figure shows two different Feynman diagrams, representing two different sets of events. The result you would see in the lab – the incoming particles, and the resulting particles – is the same for both diagrams. We therefore refer to these as two different paths for the same reaction.

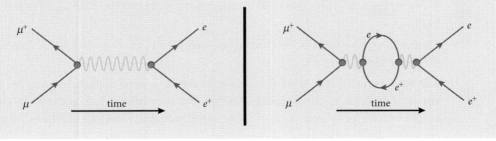

1. What particles go into this reaction? What particles come out? (Remember that time always flows from left to right; the backward arrows simply designate antimatter particles.)

2. Describe the step-by-step process represented by the diagram on the left.

3. Describe the step-by-step process represented by the more complicated diagram on the right.

We hope you saw that . . .

1. A muon and an antimuon go into this reaction; an electron and a positron come out.

2. In the path on the left, the muon and antimuon annihilate, giving off energy in the form of a photon. That photon then pair-produces the electron and positron.

3. The path on the right adds an additional step. Remember *not* to follow the loop in the middle around in a clockwise direction, as the arrows suggest. Rather, move always from left to right. The first photon pair-produces an electron and a positron, which then annihilate each other to form a photon, which then pair-produces the final products.

Feynman developed a set of rules for calculating the probability amplitude associated with each diagram. When a muon and an antimuon meet, one possible outcome is the production of an electron and an antielectron. To calculate the probability of that happening, you would have to add up the probability amplitudes of the two diagrams in the Active Reading Exercise, as well as all the others that contribute to this particular outcome.

Quantum Electrodynamics, or QED

We said that in Newtonian physics it's not enough to know Newton's laws; you have to know force laws in order to know how things will behave. Similarly, it's not enough to know how to calculate probability amplitudes from Feynman diagrams (which we're not going to tell you anyway in this book); you also have to know which diagrams represent allowed processes.

You already know part of the answer: if you compare the set of initial particles to the set of final particles, all relevant conservation laws must be obeyed. But for the processes that go on between the initial and final states, there can be other restrictions. We will focus here on the restrictions in "quantum electrodynamics" (QED), the quantum field theory of electromagnetism:

- Every vertex in an allowable QED diagram must represent the meeting of exactly three paths.

- Charged particles such as electrons must form two of those paths, and a photon must form the third.

- The two charged-particle paths must be for the same type of particle: you can't have a vertex of an electron, a muon, and a photon. But antimatter is not considered a separate set of particles for this purpose, so if one of the particles is an electron, the other particle could be either an electron or a positron.

So now you know the rules for an allowed *vertex*. But one allowable vertex can represent different allowable *reactions*, depending on how you orient it. Figure 13.11 shows the same vertex twice. In the first example – remember that we always read left to right – an electron goes in, and an electron and a photon go out. In other words the electron emits a photon. In the second example an electron and a positron annihilate each other to produce a photon.

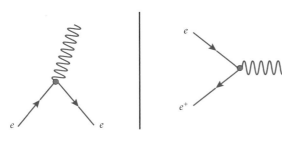

Figure 13.11 The same vertex in two different orientations can represent an electron emitting a photon (left) or an electron and positron annihilating to form a photon (right).

Active Reading Exercise: Creating More Paths

The Active Reading Exercise on p. 635 shows two Feynman diagrams for the process of a muon and an antimuon colliding and producing an electron and a positron. Both diagrams are constructed from allowed QED vertices. Draw two other allowed Feynman diagrams for the same process (meaning the same list of initial and final particles).

You can see a few possible answers to that exercise at www.cambridge.org/felder-modernphysics/activereadingsolutions

The overarching idea is that certain processes – a particle emits and then reabsorbs a photon, a photon spontaneously creates an electron–positron pair which then annihilates, and so on – can be inserted into such a process an arbitrary number of times. Each such insertion creates a new path that contributes to the overall probability of the final result.

So, imagine that you observe a muon–antimuon collision, and you measure an electron and a positron coming out. We have now presented five different Feynman diagrams for that process (counting our answers to the Active Reading Exercise) – you could come up with as many more as you like, once you have the idea – and you might reasonably ask which diagram represents the events that occurred in your lab. How many different photons or particle pairs appeared and then disappeared as in-between steps?

As reasonable as the question may sound, quantum field theory disallows it, in the same way that quantum mechanics says you can't meaningfully ask which slit a photon went through in a double-slit experiment. You measured a muon and an antimuon going in, the universe went into a superposition of all of the processes that could possibly occur from there, and when you measured the outcome it collapsed into an electron and a positron coming out. The probability of that measurement came from the sum of all the probability amplitudes of the diagrams that describe it.

As we have said several times, we're not going to perform the calculations. But it should be clear at this point that for any specific reaction – any set of input and output particles – there are infinitely many Feynman diagrams, representing infinitely many possible processes. How can we possibly sum over all of them? We can tell you two facts that make such calculations possible in QED:

- Every vertex in a QED diagram contributes a number on the order of the square root of the fine structure constant $\alpha \approx 1/137$.

- Part of calculating the probability amplitude for an entire diagram is *multiplying* the numbers associated with all its vertices.

Do you see what those two facts imply? A diagram with 10 vertices has a fairly small probability amplitude; a diagram with 30 vertices has a *very* small probability amplitude. So you can get away with calculating a relatively small set of diagrams to get a very accurate answer for the probability of most events.[11]

By the way, particle physicists don't generally talk about how many vertices are in a diagram but rather how many closed loops. For any given initial and final set of particles, the two are equivalent; to add vertices you have to add closed loops and vice versa.

QED is the most accurately tested physics theory ever developed. You may recall our discussion on p. 350 of the "gyromagnetic" ratio of an electron's magnetic moment to its spin angular momentum, which QED has correctly predicted to more than 10 digits of precision, as of this writing.

Quantum Chromodynamics, or QCD

The quantum field theory that describes the strong force is called quantum chromodynamics (QCD). (The "chromo" comes from the fact that we call strong charges "colors.") The basic rules of QCD are very similar to those of QED: you draw diagrams and sum probability amplitudes.

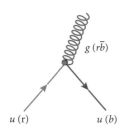

$g\,(r\bar{b})$

$u\,(r)$ $u\,(b)$

Figure 13.12 A QCD vertex in which an up quark emits a gluon and changes from red to blue. (We colored in the up quark in this diagram to illustrate this point, but color is usually left out of Feynman diagrams.)

The basic vertex of a QCD diagram also looks familiar, with two straight lines for one flavor of quark and one wavy line for a gluon. Depending on how you orient it, that vertex can show a quark emitting or absorbing a gluon, or the pair production or pair annihilation of a quark and antiquark.

But there are important differences between these two types of force. A QED vertex represents two particles that carry charge and a photon that conveys the force but has no charge of its own. A QCD vertex represents two quarks that carry color and a gluon that *also* carries color. In fact each gluon has one color and one anticolor, e.g. $g\bar{b}$. That means that a quark can change color at a vertex, consistently with the conservation of color, as illustrated in Figure 13.12.

In QED, a particle with charge can emit or absorb a photon. Similarly, in QCD a particle with color can emit or absorb a gluon. But a gluon is *itself* a particle with color, which means gluons can emit and absorb gluons. In other words you can have a QCD vertex with gluons and no quarks. In principle there should be bound particles consisting of nothing but gluons, but as of this writing no such "glue-balls" have been experimentally detected.

---

11 We're oversimplifying here. As the number of vertices n in a Feynman diagram increases, the probability amplitude of that diagram decreases, as we said. But the total number of diagrams with n vertices increases rapidly with n. Theoretical physicist and mathematician Freeman Dyson showed that the series used to calculate probability amplitudes in QED are *not* convergent, but "asymptotic" series expansions. If you kept adding diagrams indefinitely the sum would eventually stop converging toward the correct answer and start diverging. While this fact is theoretically interesting, it turns out to have no practical impact. For as many diagrams as we can practically calculate, adding more diagrams brings results closer to the experimentally measured values.

Another crucial difference between QED and QCD is the strength of the interactions. You can calculate QED probabilities to high accuracy with a relatively small number of diagrams, because each vertex you add to a diagram reduces the probability amplitude by about $\sqrt{\alpha}$, and α is a very small number. The corresponding number in QCD is of order 1, so the corresponding method doesn't work. (Like a Taylor series, sometimes the expansion in Feynman diagrams doesn't approximate the correct value, no matter how many terms you include.)

Forces in QFT

For Newton a force was a direct, long-distance interaction between two objects. Later we learned to think of forces as mediated by fields: an electron generates an electric field, which then exerts force on another electron. In quantum mechanics those fields are themselves quantized – that is, they are made of discrete particles – so we could say that the first electron emits a sea of photons, and those photons interact with the second electron.

But quantum mechanics does not contain a systematic model for particles emitting and absorbing other particles. Quantum field theory provides just such a model – and indeed, as we have mentioned, such processes underlie all QFT descriptions of forces. To put it another way, it can be shown that in the classical limit, with relatively weak interactions and small uncertainties, the end results of these particle interactions look like classical forces.

Feynman diagrams provide a powerful tool for calculating these forces, and they provide useful intuition for the underlying processes. But, as the following exercise illustrates, you have to be careful about interpreting that intuitive picture in a classical way.

Active Reading Exercise: Force as an Exchange of Particles

The Feynman diagram below contains two vertices, both of which obey the QED rule discussed above. It describes a "scattering" process, meaning that it begins and ends with the same set of particles. The left-to-right timeline tells us that the event "Particle A interacts with photon" precedes Particle B's similar event, so we can read this diagram as saying that Particle A emitted a photon that was subsequently absorbed by Particle B.

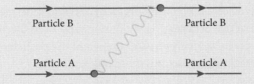

Based on classical conservation of momentum, does this interaction …

    A. accelerate both particles toward each other?

    B. accelerate both particles away from each other?

    C. accelerate one toward the other and the other one away?

Briefly explain your answer.

We hope you concluded that the force is repulsive. When Particle A emits the photon it recoils, and when Particle B absorbs that photon it gets pushed back. Total momentum is conserved because the particles have both been pushed away from each other.

So "away from each other" is the answer we wanted you to get, because it's the answer that makes sense classically. But it's not actually the full story. QFT uses interactions like that one to describe *all* forces, including both attraction and repulsion. How can photon exchange be the mechanism behind attractive forces?

Some sources answer that question with elaborate images such as representing the exchange particle (the photon) as a spring that pulls both sides inward, or even as a boomerang that is thrown backward and loops around. These pictures are fun, but the physics they describe has almost nothing to do with actual QFT calculations.

The real answer is that you can't interpret a Feynman diagram as a set of classical particle trajectories. Each diagram represents one possible set of particle exchanges, and the final probability depends on the mathematical interference between all such sets. The straight lines and curvy lines represent particles, but they do not represent physical journeys followed by those particles. As such they don't have to follow equations of motion, and can even violate equations like the relativistic formula relating energy to momentum.

We have not found any pictures or descriptions that explain these calculations intuitively, but the math works out the way it must. In the case of two opposite charges, the probability amplitudes for repulsion interfere destructively while those for attraction interfere constructively, and thus attraction is the final result.

Virtual Particles

The short-lived particles created and destroyed in Feynman diagrams are called "virtual particles," whereas the ones that we actually detect at the beginning and end of an experiment are called "real" particles.

Physicists often talk as if the distinction between real and virtual particles were a rigid line, but there can be in-between cases. If a positron and electron annihilate in a supernova and emit photons, and a few million years later one of those photons hits Earth's atmosphere and pair-produces an electron and a positron that we detect, it seems a bit perverse to call that photon a virtual particle. We know that a photon traveled from that supernova to Earth. We know when it was generated, how much energy it had, and when it was annihilated. So our indirect measurement is as good as a direct measurement of the photon itself. (Remember that Section 13.3 discussed the practice of detecting, analyzing, and even discovering particles that left no tracks, based on the particles that preceded them and the particles that followed them. No one suggests that the particles thus discovered are not real!)

In practice, the longer a particle lasts the "more real" it is. The time–energy uncertainty principle tells us that a particle that is created and destroyed in the span of 10^{-25} seconds doesn't have a well-defined energy, while one that lasts for a million years does.

So when an experiment involves particles being rapidly created and destroyed, our measurements don't constrain the properties of those particles very well, or even what types of particles are involved, and we call those particles virtual. In the supernova example, the process takes so long that we can say with negligible uncertainty what happened between the initial and final moments, and for all intents and purposes the particle that hit Earth was a real photon.[12]

---

12 Similarly, once you measure a virtual rabbit it becomes a real rabbit, and once it is real it will always be real.

13.5.2 Questions: Quantum Field Theory

Conceptual Questions and ConcepTests

1. Figure 13.10 on p. 634 shows a Feynman diagram that we could describe in the following way: "An electron neutrino and a muon come in; the electron neutrino emits a W^+ and turns into an electron; the muon then absorbs that W^+ and turns into a muon neutrino." Tell a similar story for each of the Feynman diagrams in Figure 13.13.

2. Figure 13.14 shows five different Feynman diagrams, each of which has its own probability amplitude. Which of those probability amplitudes would it make sense to add to each other as part of calculating a probability, and of what event would you be calculating the probability?

3. Draw a Feynman diagram that shows the following five-step process. An electron and a muon come in;

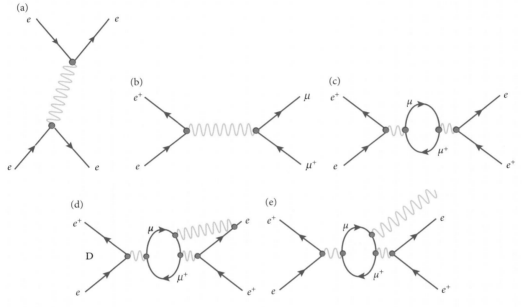

Figure 13.13

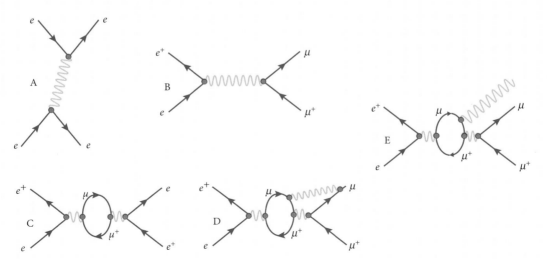

Figure 13.14

the muon emits a photon, which pair-produces an electron–positron pair, which then pair-annihilates to form another photon. That photon is then absorbed by the original electron.

4. Figure 13.11 on p. 637 shows two possible orientations of a QED vertex with different interpretations.

   (a) The figure on the right draws the wavy line perfectly horizontal, but the figure on the left does *not* draw it perfectly vertical. Why didn't we draw that one vertically?

   (b) Draw the same vertex in two other possible orientations by rotating each of those two by 180°, and briefly describe what these two new vertices represent.

5. Suppose you flipped a Feynman diagram upside-down (mirrored, not rotated). How, if at all, would the set of events described in the new diagram differ from the set of events described in the original?

6. Draw four Feynman diagrams for the process "two electrons scatter off each other."

7. The vertex on the right side of Figure 13.11 (p. 637) showing electron–positron annihilation is an allowed vertex in QED, but this process by itself is not physically possible.

   (a) Why not? *Hint*: There is a relatively simple answer once you switch your perspective into the right reference frame.

   (b) Draw a Feynman diagram for an allowed pair-annihilation process. Your initial system will be an electron and a positron, but your final system will not be a single photon.

8. Before the development of quark theory, the strong force between hadrons was described as an exchange of pions. We still view that description as correct, but now we consider it to be a more coarse-grained description of an underlying process involving quarks. Figure 13.15 shows a nuclear process. A proton and a neutron exchange a pion, switching identities in the process. Draw a Feynman diagram that describes this exact same process, but shows the interaction at the level of quarks and gluons instead of hadrons. Show gluons as coiled lines, like so: ⦿⦿⦿⦿⦿⦿⦿⦿.

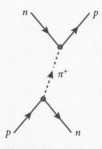

Figure 13.15

For Deeper Discussion

9. In Section 13.3, our explanation of Figure 13.9 hinged on the fact that the process $e + p \rightarrow e + \Delta^+ \rightarrow e + \pi^+ + n$ produced measurably different outcomes from the direct process $e + p \rightarrow e + \pi^+ + n$. But in this section we claimed that, when a set of particles transforms into another set of particles, there is no unique answer to which particles were created in between. How can you resolve that seeming contradiction?

Chapter Summary

The standard model of particle physics comprises a small list of fundamental particles and their properties. Conservation laws describe which particle interactions may possibly occur, and quantum field theory is a mathematical framework for calculating the probabilities of those interactions.

The standard model is summarized in Appendix J.

Section 13.1 Forces and Particles

This section describes the categorization of particles and forces before the introduction of quarks and the standard model.

- One important way to categorize particles is by their reaction to forces. (For instance, the property we call "charge" describes a particle's reaction to the electromagnetic force.) Therefore, before enumerating particles, it's important to understand the quantum mechanical idea of a force.

- A force in quantum mechanics can have many different effects, not just the classical effect of changing a particle's acceleration. In particular, the decay of a particle into a different type of particle is always a response to a force.

- The four fundamental forces – gravity, electromagnetism, strong force, and weak force – cause generally different effects, work on different sets of particles, and operate over different distances in different time frames. Their basic properties are summarized in a table in Appendix J.

- All observed particles of matter are classified as either hadrons (which feel the strong force) or leptons (which don't). Hadrons are further subdivided into mesons (integer spins) and baryons (half-integer spins).

- Every particle has a corresponding antiparticle, with the same properties but opposite charges.

- This section contains a discussion of the kaon, as an example of how particles were discovered during this era, and of the kinds of conserved quantities – in this case "strangeness" – that were invented to explain the patterns of decay.

- The discovery of hundreds of apparently fundamental particles with seemingly random properties prompted theorists to believe there must be an underlying structure.

Section 13.2 The Standard Model

The standard model resolved the chaos of the "particle zoo" into a compact list of fundamental particles by introducing quarks as the building blocks of hadrons. By understanding how this small list of particles interacts and combines, you can predict the much larger list of mesons and baryons and their properties. The standard model is laid out in Appendix J.

- There are six "flavors" of quark: up, down, charm, strange, top, and bottom. Each flavor can come in three "colors": blue, green, or red. Each quark also has an antiquark, with the opposite charge and opposite color (antiblue, antigreen, or antired).

- There are also six types of lepton: electron, muon, and tauon, each with charge $-e$, and a corresponding (chargeless) neutrino for each. Once again, each has an antiparticle.

- Three quarks (one of each color) make a baryon. Two quarks (a color and its anticolor) make a meson. Knowing the quark composition of a hadron allows you to predict its charge, as well as more exotic conserved quantities such as strangeness, but *not* its mass.

- Quarks are never observed individually, but only in the color-balanced combinations that make up hadrons.

- In addition to quarks and leptons, the standard model includes "force-carrying" particles. For instance, the electromagnetic force between any two particles is mediated by photons.

- The final piece of the standard model is the Higgs boson. This particle is responsible for other particles having mass.

The standard model predicted the existence and properties of many particles before they were discovered. The subsequent discoveries of those particles gave scientists great confidence in the model. At the same time, there are known phenomena such as dark matter and dark energy (Chapter 14) that require explanations beyond the standard model.

Section 13.3 Detecting Particles

This section gives a brief overview of the technology that was used through most of the twentieth century to detect particles that in general are too small, too fast, and too short-lived to see with the naked eye.

- A "cloud chamber" is filled with supersaturated vapor. A charged particle passing through the chamber ionizes atoms along its path, leaving behind a trail of droplets that can be photographed. A later technology, the "bubble chamber," replaces the vapor with a liquid medium.

- Electrically neutral particles do not leave trails, but their properties can be inferred from the paths of charged particles that come before and after them.

- Particles that decay in too short a time to leave a visible trail can be inferred based on the particles into which they decay.

Bubble chambers were a significant improvement over cloud chambers. Both technologies have been superseded by modern detectors that detect particles faster and feed their data directly into computers.

Section 13.4 Symmetries and Conservation

One property of any physical theory is its "conservation laws": quantities that remain unchanged under all, or some, circumstances. Another property of any physical theory is its "symmetries": changes you can make to the system without altering the resulting behavior. "Noether's theorem" reveals a surprising, and very general, connection between these two types of properties.

- Some quantities in the standard model are conserved "exactly": i.e. they are conserved under all circumstances. To the best of our current knowledge, exactly conserved quantities include energy, momentum, angular momentum, charge, baryon number (number of baryons minus number of antibaryons), and lepton number (number of leptons minus number of antileptons).

- Some quantities are only conserved "approximately": they are conserved in some situations. For instance, charm (the number of charm quarks minus the number of charm antiquarks) is conserved in interactions involving the strong force, but this conservation can be violated by the (much slower) interactions involving the weak force. The same rule applies to each of the other five quark flavors.

- Any interaction that does *not* violate a conservation law has a non-zero probability of occuring.

- If you watched a video of particle interactions at a $30°$ tilt, those interactions would obey the same laws of physics as the untilted version. We say that the laws of physics are "symmetric under rotation." Other exact, continuous symmetries include symmetry under spatial translation and symmetry under time-translation.

- The discrete symmetry called "C" says that the universe would be unchanged if you reversed the charges of all particles (turning positive into negative, blue into antiblue, and so on). The discrete symmetries called "P" and "T" say, respectively, that the laws of physics look the same in a mirror (parity), and that the laws of physics function the same in backward time as in forward time. All three symmetries work in most, but not all, situations.

- CPT symmetry says that if you switched charge *and* parity *and* time, the resulting universe would be indistinguishable from our own. We believe this to be an exact symmetry.

- Emmy Noether's theorem says that every continuous symmetry of a theory implies an associated conservation law. Later work showed similar results for discrete symmetries.

Section 13.5 Quantum Field Theory

Quantum field theory (QFT) is a generalized version of the quantum mechanics described in earlier chapters. In quantum field theory all forces are described as particle interactions and decays.

- A QFT process can be visually represented in a "Feynman diagram," showing a possible set of particle interactions occurring between the final and initial states.

- A Feynman diagram may contain many intermediate reactions such as photons being emitted and reabsorbed, or matter–antimatter particle pairs being created and destroyed. The "virtual particles" in these in-between steps are not detected, but they contribute to the overall probability of the final outcome.

- A Feynman diagram indicates the order of events (left to right in the diagram in the convention we're using), but it does not indicate the spatial paths taken or the times elapsed between events.

- A Feynman diagram is associated with a complex number called its "probability amplitude." To find the probability of the final outcome, sum the probability amplitudes of all possible paths to that outcome. (These probability amplitudes can interfere constructively or destructively.) The squared amplitude of the final sum is the probability.

- Traditional descriptions of particle interactions, such as "an electron and a proton exerted a mutual attraction on each other" or "an electron and a positron annihilated to produce a muon and antimuon," are explained in quantum field theory as the results of the constructive and destructive interference of infinitely many particle interactions. The resulting predictions, when tested by experiment, are more accurate than any other currently known physical theory.

14
Cosmology

Much of this book has focused down to ever smaller scales, from atoms to nuclei to fundamental particles. At the other extreme is cosmology, the study of the overall structure and history of the universe. This chapter will introduce the Big Bang model of cosmology, what it does and doesn't explain about the history of the universe, and some of the evidence for the model. Then it will explain some problems left unexplained by the Big Bang model, and how some of these problems can be resolved by an event called "inflation" in the first fraction of a second after the Big Bang.

14.1 The History of the Universe

This section discusses what it means to say that "the universe is expanding," and what the Big Bang was. (Neither of those is exactly what most people think.) It then presents a timeline of the post-Big-Bang universe.

If you only want a broad overview of cosmology, you can read this section and then stop. But if you continue, you will see that this section sets up many of the questions and issues that are addressed later in the chapter.

14.1.1 Explanation: The History of the Universe

Before we talk about history, we want to offer some mental images to help you picture what's in the universe, and some of the length scales involved.

As Douglas Adams famously said: "Space is big. Really big. You just won't believe how vastly hugely mind-bogglingly big it is. I mean, you may think it's a long way down the road to the chemist,[1] but that's just peanuts to space."

As a stab at contextualizing the sizes involved, imagine that the Earth were a millimeter across (a typical grain of sand).

- The Sun would be 11 cm across, or a little bit bigger than a baseball.
- The Sun and Earth would be about 12 meters apart, so picture the grain of sand and the baseball at opposite ends of a school bus.

---

1 American readers are advised to substitute "drug store" for "chemist." To keep things fair, British readers can plan on substituting "cricket ball" for "baseball" in a few sentences.

- The outermost planet in the solar system (Neptune) would be about 350 meters from the Sun. Put the Sun on the ground, with the Earth one bus-length higher, and Neptune would be floating a little over the top of the Eiffel Tower.

- With our solar system occupying the Eiffel Tower in Paris, the next solar system over would be more than halfway across the Atlantic, and only two other solar systems would be closer to us than New York. *The galaxy is a far emptier place than most of us imagine!*

- The Milky Way galaxy would reach about halfway from the Earth to the Sun, filled with billions of solar systems that are each an Eiffel Tower wide and spaced out a few thousand kilometers apart. As you picture this, remember that each solar system is a baseball-sized star, often surrounded by sand-grain-sized planets.

- Andromeda, the nearest other large galaxy, would be floating somewhere out past Saturn.

- In this scale model that we're building, the radius of the observable universe would stretch almost from us to the next solar system. If we brought the scale down so the Milky Way and Andromeda were at opposite ends of the Eiffel Tower, the observable universe would encompass most of the Earth.

Now that you have some idea of what the universe looks like, we can turn our attention to how it has changed over time.

The Expanding Universe

Where do we begin our history of the universe? Everyone knows the answer to that question: we start at the Big Bang, the instant when our universe first came into existence from a point with zero volume and infinite density. But as so often happens, what "everyone knows" is not quite right. The Big Bang was not necessarily the instant when our universe came into existence, and the universe at the moment we call the Big Bang did not have zero volume or infinite density.

To say what we do mean by the Big Bang, we need to first ask what it means to say that the universe is expanding. The room that you are in is not getting bigger over time. Also the Earth is not getting bigger, our solar system is not getting bigger, and our galaxy is not getting bigger.

What is happening, however, is that the distance between galaxies is getting bigger. That's not always true for galaxies that are close to each other: in our neighborhood we see some dwarf galaxies orbiting our own Milky Way, and Andromeda is heading toward us on a collision course. But if you look at the big picture, what you see is galaxies rushing away from other galaxies.

Should we therefore say that the universe is "getting bigger"? Maybe. If the universe is finite, then it is indeed getting bigger. If the universe is infinite, then to say it's "getting bigger" is meaningless. Section 14.3 will discuss the ideas of a finite universe and an infinite universe, and why we don't know which one we live in. But either way, we can say with confidence that the number of galaxies per unit volume is going down over time. Expansion really means that the universe is getting *less dense*.

If the universe is getting less dense over time, it must have been more dense in the past.[2] A billion years ago the universe was more dense than today, and two billion years ago it was

---

2 Don't you wish you could come up with brilliant insights like that?

more dense than that. We can project back to a time approximately 13.8 billion years ago, when the density of the universe was approximately 10^{96} kg/m^3 – a number known as the "Planck density."

But our ability to project backward stops at that point. Our known laws of physics, from the biggest scales (general relativity) to the smallest (quantum field theory), do not apply to matter above the Planck density. We cannot say anything scientifically meaningful about such dense stuff. In Chapter 13 we briefly discussed the reasons for this limitation of our physics theories. The short version is that, above the Planck density, general relativity and quantum field theory become contradictory, and we don't know how to reconcile them. The important point for this chapter is that we cannot use known laws of physics to extrapolate backward to times with such high densities.

The Big Bang model begins at the moment when the universe was at the Planck density.

Did the universe come into existence at that moment? Had it existed for infinitely long before that? Did the universe contract down to Planck density, turn around, and start expanding again? Any of those are possible because we don't know the right physics to answer these questions.

So our timeline below begins at a "$t = 0$" moment that is not necessarily the birth of our universe, but the earliest moment we know about. At that moment the universe was unimaginably dense, unimaginably hot, rapidly expanding, and subject to physical laws that we don't understand.

From that moment until this, we can describe three broad trends.

1. The universe expanded, which is to say, it became less dense. Much of this was simply a consequence of inertia: things kept moving apart because they were already moving apart. But we will have more to say about the forces involved, both sustaining and opposing this motion, later in the chapter.

2. The universe cooled down. This is a natural consequence of the first trend: when a gas expands, it cools.

3. Things combined into ever larger structures. Quarks bonded with each other to form protons and neutrons, which later fused into nuclei, which later joined with electrons to form atoms and then molecules, which later formed stars and planets and galaxies. This is a consequence of the second trend: bonds break apart at high temperatures, so cooling allowed the formation of stable structures.

Below we present some of the highlights of that journey.

A Timeline of the Universe

The times and temperatures in the description below should be taken as approximate.

$t = 0$: Planck Density

The universe at this moment was unimaginably dense, hot, and expanding very rapidly. We don't know what caused those conditions, or for that matter how our universe came to exist at all. And we don't have the physics to describe whatever happened before this moment. The Big Bang model describes how it has evolved from that moment until today.

$t = $ *an unknown but very small fraction of a second*: Inflation

Shortly after the Big Bang, the universe underwent a period known as "inflation" in which distances grew exponentially with time. The exact timing and exponential growth rate during inflation are unknown, but, in the simplest versions of the theory, distances grew by a factor of *ten to the power of a million* in about 10^{-35} s.

At this point you may be thinking: "How could that happen?" or "Why would anyone believe such an insane theory?" Those questions will be addressed in Section 14.7, which will also explain how this brief period of hyper-accelerated expansion set the initial conditions for all the events in the rest of this timeline.

$t \ll 1$ s: Matter is Produced

At the end of inflation there was a lot of energy, in an exotic form that we will discuss in Section 14.7, but that energy rapidly decayed into matter and electromagnetic radiation. The matter included electrons, quarks, and neutrinos, and all of their corresponding antiparticles. Dark matter and dark energy were also produced at this time; we'll talk about them in Section 14.5.

$t = 10^{-5}$ s: Quarks Combine into Protons and Neutrons. Temperature: 10^{12} K

For the first hundred-thousandth of a second after they were produced, quarks were too energetic to combine. When the temperature and density dropped sufficiently, quarks bonded with each other to form protons and neutrons; antiquarks combined into antiprotons and antineutrons. There have been essentially no free quarks in the universe since.

$t = 1$ s: Neutrinos Decouple and Antimatter Annihilates. Temperature: 10^{10} K

Two important events happened coincidentally at about 1–3 seconds after the Big Bang: the universe became transparent to neutrinos, and antimatter was removed from the universe.

Neutrinos can pass through almost anything without interacting, but during that first second the density was so high that even the neutrinos were continually colliding with other particles. At $t = 1$ s the density dropped enough that the universe became transparent to neutrinos, and the neutrinos that were present at the time have been streaming freely through space ever since. This "cosmic neutrino background" was first detected in 2015.

Matter–antimatter pair production and pair annihilation were both happening continuously at this time. (At $t \lesssim 1$ s, not in 2015.) But as the temperature dropped, it became rarer for photons to have enough energy to pair-produce, so the annihilations began to outpace the production. This happened first for proton–antiproton and neutron–antineutron pairs, and later for electron–positron pairs.

You can imagine a universe in which matter and antimatter were perfectly balanced. In such a universe all the matter and antimatter would have annihilated each other, leaving only radiation. But our universe was not quite that symmetrical: for reasons we do not currently understand, we had a billion and one particles of matter for every billion particles of antimatter. So after the mutual annihilation we were still left with those one-in-a-billion particles, which became all the matter we see making up the universe today. This process of antimatter annihilation was pretty much complete by about 3 s.

$t = 1$ m: Nucleosynthesis. Temperature: 10^9 K

One minute after the Big Bang marks the beginning of "nucleosynthesis," the fusion of protons and neutrons into light nuclei.

Active Reading Exercise: Nucleosynthesis

Before the universe reached the one-minute mark, the ambient temperature was too hot for protons and neutrons to bond. After the temperature dropped sufficiently, nuclei began to form.

The strong force was just as strong before one minute as it was after. So why was nucleosynthesis impossible until the temperature went down enough?

When two protons and two neutrons are bound together in a nucleus, they are in a minimum of their potential energy function, with less overall energy than they would have if they were far away from each other. (If that weren't true, nuclei couldn't hold together.) But the particles won't stay in that potential minimum if they all have enough kinetic energy to break apart.

Before about one minute, the average thermal speed of protons and neutrons was higher than the escape velocity from a nucleus, so the particles couldn't stick together. When the temperature became low enough, particles started falling into those potential wells without enough energy to get out again. The first such events probably began around 10 seconds after the Big Bang, but the pace picked up dramatically by about 1 minute.

Nucleosynthesis didn't last long. It had slowed dramatically by 3 minutes, and was pretty much entirely over by the 20-minute mark. The reason it stopped is once again that the universe cooled. In order for two protons to fuse they have to overcome their electric repulsion and get close enough for the attractive, short-range nuclear force to dominate. For a few minutes the universe was in just the right temperature range: hot enough for protons to get close to each other, but not too hot to prevent them sticking together when they did.

After Big Bang nucleosynthesis, the nuclei in the universe were by mass about 75% hydrogen (a single proton), and 25% $^4_2$He. There were also trace amounts of $^7_3$Li and other isotopes of hydrogen and helium. (Single neutrons are unstable, so the only remaining neutrons were the ones bound into nuclei.) That mix of elements remained unchanged for millions of years, until fusion started up again in the cores of stars.

After the first 20 minutes, the next few hundred thousand years were comparatively boring. The universe consisted of a plasma of nuclei and electrons, plus neutrinos, radiation, dark matter, and dark energy. Because of the charged electrons and nuclei, the plasma was opaque: photons were constantly being scattered or absorbed and re-emitted with a microscopic mean free path.

The universe was still expanding – that is, both its density and temperature were decreasing – but nothing dramatic changed, until . . .

$t = 370,000$ years: Recombination. Temperature: 4000 K

After almost 400,000 years, the universe cooled enough for electrons and nuclei to combine into atoms, a process misleadingly dubbed "recombination."[3] This marks the transition of the

---

3   So far as we know they had never been combined into atoms before then, but the name was given during a period when cosmologists favored the idea of a cyclic universe, with an infinite cycle of expansion and contraction. They thought that near the end of each contraction atoms would be ripped apart, and then recombine during the next expansion phase. The cyclic universe idea is still studied, but is considered unlikely by most cosmologists.

universe from a "plasma" (a collection of freely moving particles that are *not* combined into atoms) into a "gas" (now they are).

The most important residue from this important transition is blackbody radiation. As you may recall (from Section 3.4 or 10.7, for instance), high-temperature bodies radiate a lot. The plasma that filled the universe before recombination was a near-perfect blackbody, continually radiating energy from every point in every direction. But each photon didn't get far; as soon as it was emitted it was absorbed by one of the charged particles (nuclei or electrons) around it. The gas that filled the universe *after* recombination, by contrast, was made of neutral hydrogen atoms, which didn't emit or absorb much of anything.

So in the brief transitionary period from plasma to gas, a lot of radiation was emitted and then *not* absorbed. That radiation has been freely streaming through the universe ever since. As we will discuss later, detection of that radiation, called the "cosmic microwave background," provided pivotal early evidence for the Big Bang model.

The roughly one hundred million years from recombination to the formation of the first stars is called the "dark ages" because no new light was produced during this time. The background radiation was of course still around, so if you had been present at the start of the dark ages you would have seen a red glow all around you. As the universe expanded, however, that radiation lost energy and changed to infrared light. The rest of the dark ages would have looked quite literally dark.

The universe of the early dark ages had an almost perfectly uniform density. At the time of recombination, the most dense points in space exceeded the least dense by about 1 part in 100,000! However, gravity slowly increased these differences. Any spot slightly more dense than average tended to pull matter toward it. Any spot slightly less dense than average had a weaker pull and tended to lose matter. The result was that matter began to form into clumps, with swaths of empty space in between.

$t = 30$–200 million years: A Star is Born

Nobody knows when the first star in the observable universe formed. We have observed starlight[4] that was emitted when the universe was about 400 million years old, and in the near future we will probably push that record further back in time.

We can't assign a temperature to the end of the dark ages because the universe was no longer homogeneous. We can make a statement like "during recombination, the temperature was 4000 K" because the temperature was 4000 K *everywhere*. But by the time stars and galaxies formed, some regions of the universe were very cold and others were very hot.

The formation of stars and galaxies was the result of the gravitational clumping that went on throughout the dark ages. When gravity made clouds of gas dense enough, some regions of those clouds collapsed inward under their own gravity until the resulting heat and pressure was enough to start nuclear fusion. Thus a star was born. That fusion created an outward pressure that balanced the inward pull of gravity, leaving the star in a state of equilibrium.

The first stars were formed almost entirely out of hydrogen and helium – because, as you may recall, that's pretty much all there was lying around. They were much larger and hotter than our Sun and burned through their hydrogen fuel much more quickly. When too much of

---

4 What we observed was the collective light of the stars making up an early galaxy. We can't resolve individual stars at that distance.

that fuel was used up to maintain equilibrium, the stars collapsed inward, heated up, and then exploded. Astronomers speculate that these early stars may have had lifetimes as short as a few million years.

Fusion in these early stars created heavy elements, and the explosions of these stars sent those elements out into space. When subsequent generations of stars formed, they included some of those heavy elements.

Our own Sun is a third-generation star, which formed about 4.6 billion years ago. Astronomers tend to think of time backward from today, so stars like our Sun are called "Population I." The previous generation is "Population II" and the first stars in the universe made up "Population III." As of this writing no Population III stars have been identified, but we see many Population I and II stars in our own galaxy.

We believe the Milky Way was formed not long after the end of the dark ages and probably had Population III stars. Presumably we don't see them because they burned out long ago.

When a Population I star like our Sun forms out of a collapsing dust cloud, some of the heavier elements get left behind in a disk orbiting the new star. The small fragments of matter collide and merge into a solar system of planets, asteroids, and other bodies. Most stars in our galaxy have planets orbiting them.

The Future

There are two main possibilities for the future of our universe.

If the mutual gravity of the galaxies is strong enough, it may someday stop the universe's expansion. In that case matter will get pulled back together. The burned-out remnants of stars and galaxies will collide and merge, forming ever-larger black holes as the overall density increases. Eventually the average density will reach the Planck density, and, since we don't know physics above the Planck density, we don't know what will happen after that. This collapse scenario is called the "Big Crunch."

If the mutual gravity isn't strong enough, the expansion will continue forever. All stars and galaxies will burn out. Given an infinite amount of time, all macroscopic objects will eventually decay. For quantum mechanical reasons, even black holes can evaporate if given enough time, so eventually the universe will contain nothing but thinly spread out particles and radiation, expanding and cooling for eternity. This scenario is called the "Heat Death" of the universe.

In Section 14.4 we'll discuss what determines which of these two scenarios is going to occur, and how we might someday determine the universe's ultimate fate.

14.1.2 Questions and Problems: The History of the Universe

Conceptual Questions and ConcepTests

1. The word "nucleosynthesis" refers to protons and neutrons fusing into nuclei.

   (a) In the first minute after the Big Bang, the universe was too hot. Why did that prevent nucleosynthesis?

   (b) After about 20 minutes, the universe was too cold. Why did that prevent nucleosynthesis?

   (c) Millions of years later, nucleosynthesis started happening again. Why?

2. Why is the correct temperature for nucleosynthesis many orders of magnitude higher than the correct temperature for recombination?

3. Why did the universe at recombination radiate the photons that we today call the cosmic microwave background (or CMB)? (Choose one.)

A. It was blackbody radiation emitted by the hot plasma just before recombination.

B. It was the result of particles dropping down from more excited to less excited energy levels.

C. It was the result of matter–antimatter annihilation.

D. It was the energy by-product of nuclear fusion.

4. The universe of the early dark ages had an almost uniform density, but not perfectly uniform. Choose one of the following to describe how such a universe should evolve. Then explain *why* it should evolve in that way.

A. The density distribution should stay statistically the same, although the specific regions of higher and lower density will move around.

B. Over time the variations in density will smooth out, leading to a more even density.

C. Over time the variations in density will be exaggerated, leading to a more uneven density.

5. The text describes two possible long-term fates for the universe, the "Big Crunch" and the "Heat Death." Neither one sounds good. What fundamental law of physics, which we discussed in an earlier chapter, precludes the possibility that the universe will indefinitely continue to have stars and planets and wildflowers and roller coasters?

6. The text gives a scale model of the universe starting with the Earth as a grain of sand. Describe a similar model, with examples to illustrate distances, starting with the diameter of the Earth at a nanometer (roughly the size of a molecule).

For Deeper Discussion

7. Positrons (the antimatter counterparts of electrons) were pretty much all annihilated because after about three seconds there weren't any photons around with enough energy to produce new electron–positron pairs. But the electron–positron annihilations produced photons with that much energy, so how could pair production ever stop?

8. If you have studied statistical mechanics (e.g. Chapter 10), you know that particles in a gas have average kinetic energy $(3/2)k_B T$. Setting that equal to 13.6 eV, you can calculate that electrons and protons should fuse into hydrogen atoms below $T \approx 100,000$ K. But recombination didn't happen until the universe dropped to about 4000 K. Why?

Problems

9. You can make an estimate of the density at which quantum gravity effects become important by dimensional analysis. The scale of quantum effects is governed by $\hbar$ and the scale of relativistic effects is governed by c and G.

(a) Combine those constants into a number with units of density and give its value in SI units. The result is the Planck density. Your answer should be within an order of magnitude of the estimate given in this section.

(b) The Earth's mass is about 6×10^{24} kg. To what radius would you have to shrink the Earth in order for it to have the Planck density?

10. The most lightly bound nucleus is ^{2}H, two protons stuck together, with a potential energy per particle of about -1 MeV. *To answer this problem you will need to be familiar with the equipartition theorem (Section 10.5).*

(a) At what temperature would an average proton have too much energy to get bound into an ^{2}H nucleus?

(b) In fact, the temperature at which many nuclei started to form was somewhat lower than what you should have calculated in Part (a). Give at least one plausible reason why.

14.2 How Do We Know All That?

We are all used to the idea that the universe as we know it started billions of years ago in the Big Bang, and has been expanding ever since. But as recently as the early twentieth century most scientists viewed the universe as eternal, and even Einstein considered the Big

Bang model ridiculous when it was first introduced. So what led the scientific community to embrace the idea of the Big Bang and the story of the universe's evolution that we described in Section 14.1?

In this section we'll look at the story behind the story, in two parts. First we'll go over some of the history: the theoretical and experimental evidence that led some scientists to postulate, and others to eventually believe in, the expanding universe and the Big Bang. Then we'll discuss some of the techniques that were (and are) used to measure the vast distances involved in this research.

14.2.1   Discovery Exercise: Distance to a Star

One night you measure a particular star that is exactly 90° away from the Sun in the sky. Six months later, when the Earth has moved to the opposite side of the Sun, you measure that the star is at 89.9999045° away from the Sun (Figure 14.1).

Figure 14.1 From opposite sides of the Sun, the same star is seen at two different angles.

Taking the distance from the Earth to the Sun as 100 million miles, calculate the distance to the star. Express your final answer in light-years. *Hint*: We didn't specify whether we were asking for distance from the Earth the first time, or the second time, or for distance from the Sun. Pick whichever one you want to calculate, and think about why we didn't have to specify which one we meant.

See Check Yourself #24 at www.cambridge.org/felder-modernphysics/checkyourself

14.2.2   Explanation: How Do We Know All That?

Sections 12.2 and 13.3 discussed how we measure objects that are too small, or too short-lived, to see or measure directly. Now consider the opposite question. Thinking back to our scale model where the Earth was a grain of sand: how could people who are roughly the size of an atom, living on a grain of sand, measure the distance to Saturn? How did we postulate and defend ideas such as universal expansion and a Big Bang?

To answer those questions, we're first going to describe a bit of the history that led to the development of the Big Bang model. Then we'll talk about some of the evidence that has accumulated for the model since it was developed. Finally, we'll discuss how it's possible to measure astronomical distances.

Newton, Einstein, Hubble, and Lemaître

We see the stars rise in the east and set in the west every night. By the time of Newton, it was known that this and a few other cyclic stellar motions were really caused by the movement of the Earth. The night sky we see is virtually identical to the one our ancestors saw thousands of years ago, so Newton and most other scientists of his day believed that the stars were eternal and unchanging.

Active Reading Exercise: Static or Expanding?

1. Imagine you look outside for two seconds. During that two seconds you watch a ball in midair. The only force acting on this ball is the Earth's gravitational pull. (The following questions are as trivial as they sound.)

    (a) Is it possible for the ball to be moving down the whole time you are watching it?

    (b) Is it possible for the ball to moving up the whole time you are watching it?

    (c) Is it possible for the ball to be at rest the whole time you are watching it?

2. Now imagine you spend a billion years looking at an entire universe that consists of 100 galaxies floating in space. There's nothing else in this universe. Once again, the only relevant force is gravity: in this case, the gravitational attraction between all the galaxies.

    (a) Is it possible for the galaxies to be moving toward each other the whole time you are watching them?

    (b) Is it possible for the galaxies to be moving away from each other the whole time you are watching them?

    (c) Is it possible for the galaxies to all be at rest the whole time you are watching them?

    (d) Would any of your answers be different if there were infinitely many galaxies spread through an infinite amount of space?

Initially Newton assumed that the stars only filled a finite region. But the clergyman Richard Bentley convinced Newton that such a system would collapse under its own gravity, eventually all falling into one giant mass. We hope the preceding Active Reading Exercise convinced you of the same thing.

So Newton concluded that the stars must go on forever. In that way each star would be pulled equally in all directions and would experience zero net force. Newton's view of an eternal and infinite universe of stars held in Western science until the twentieth century.

In 1916 Einstein published his general theory of relativity. We discuss some of the key ideas of general relativity (GR) in an online section of this book at www.cambridge.org/felder-modernphysics/onlinesections

General relativity is a big and dense topic, and our online section is only a brief introduction. Nonetheless, many of the ideas in this chapter depend on GR. So on several occasions we're going to have to say "The following result can be proven from general relativity, but right now you'll have to take our word for it."

And here comes the first such moment: GR says the universe must either be expanding (objects on average moving apart from each other) or contracting (objects on average moving toward each other).

Even though we're not showing you the math that leads to this conclusion, the physical reason why in GR the universe can't be static is essentially the same problem that Newton encountered with his finite universe: the mutual attraction of all the stars would pull them together. Just like a ball in midair can be moving up or down, but not hovering in one place, so the universe can be expanding (and slowing down) or contracting (and speeding up). The equations of GR do not allow for a static universe, regardless of whether it's finite or infinite.

The conclusion that the universe can't be static seemed to fly in the face of astronomical observations. It was known by the early twentieth century that the stars move, but they seemed to all move in random directions, like leaves on the surface of a pool. There was no evidence of concerted motion toward or away from each other. In 1917 Einstein responded to this problem by adding a term to the equations of general relativity. The new term, called the "cosmological constant," added a repulsive component of gravity in an attempt to model a static universe.

We now know that the universe is expanding. The reason this wasn't apparent in 1916 is that astronomers were mostly looking at stars within our galaxy.[5] But even before the publication of general relativity, some of the measurements had occurred that led to the discovery of this expansion.

- 1915: Vesto Slipher published measurements of the speed of 15 nebulae, fuzzy smudges in the sky of unknown origin, and found that all but two of them were moving away from us at speeds much greater than the measured speeds of any stars. (We'll describe below how he was able to measure these speeds.)

- 1916–17: Einstein published the general theory of relativity and a year later modified it with the "cosmological constant," intended to reconcile the equations with a static universe.

- 1923: Edwin Hubble measured the distance to Andromeda, proving that it was another galaxy outside the Milky Way. It quickly became clear that virtually all of Slipher's nebulae were in fact galaxies.

- 1927: Georges Lemaître published a paper in which he calculated that, in the expanding universe described by general relativity, we should see other galaxies moving away from us at speeds proportional to their distance from us. Einstein met Lemaître later that year and told him that his physics was "abominable."[6]

- 1929: Hubble measured distances to 24 galaxies and showed that their recession speeds (mostly the ones measured by Slipher) were proportional to their distances from us – just as Lemaître's calculations had predicted.

---

5   You might think that gravity would require the stars in the galaxy to be moving away from or toward each other, just like the galaxies. See Question 2.

6   "Vos calculs sont corrects, mais votre physique est abominable."

- 1931: Lemaître published a paper tracing the expansion of the universe backward to an early, hot, dense state, which he called the "Primeval Atom."

Following Hubble's discovery, Einstein retracted the cosmological constant, calling it the biggest blunder of his life. In 1931, after attending a talk by Lemaître, Einstein reportedly said: "This is the most beautiful and satisfactory explanation of creation to which I have ever listened."

The fact that galaxies recede from us at speeds proportional to their distance from us came to be called "Hubble's Law," but in 2018 the International Astronomical Union voted that it should be renamed the "Hubble–Lemaître law."

The Hubble–Lemaître Law

The velocity with which a galaxy is receding from us (v) is proportional to that galaxy's current distance from us (d):

$$v = Hd.$$

The proportionality constant H came to be known as "Hubble's constant." But as we'll see later, this name was misleading because while H is constant across space, it does change over time. It is now generally called the "Hubble parameter."

The modern term for Lemaître's Primeval Atom wasn't coined until 1949 when Fred Hoyle, an outspoken opponent of the theory, derisively referred to the idea of the universe being created in a "Big Bang."

Evidence for the Big Bang Model

Lemaître's Primeval Atom was highly controversial. Hubble's observations showed conclusively that distant galaxies were receding from each other at speeds proportional to their distances. But many scientists were skeptical of the conclusion that the universe at earlier times had been a hot, dense plasma filling all of space.

In 1948 Fred Hoyle, Hermann Bondi, and Thomas Gold published a "steady-state" model that attempted to reconcile the observed expansion with an eternal, unchanging universe. Their idea was that, through an unknown process, matter was constantly created at a rate that would keep the overall density constant as the universe expands. Because the rate of matter creation would be much too low to directly detect, the theory was consistent with all known observations.

Over the next few decades, however, evidence accumulated in favor of the Big Bang model.

In the same year that Hoyle, Bondi, and Gold published their papers, a paper by Ralph Alpher, Hans Bethe, and George Gamow laid out the theory of Big Bang nucleosynthesis[7] (see "$t = 1\,\text{m}$" in Section 14.1). Subsequent calculations of the production of light elements in the early universe made predictions of the relative abundances that would be produced of hydrogen, helium, lithium, and some of their isotopes. Those relative abundances have been confirmed observationally, and there's no explanation for them in the steady-state model.

---

7 The work was Alpher's PhD thesis and Gamow was his advisor. Bethe had nothing to do with it, but Gamow thought it would be funny to add him as a co-author so the author list would sound like "alpha-beta-gamma."

The most important piece of evidence came from the measurement of the cosmic microwave background radiation ("$t = 370{,}000$ yrs" in Section 14.1). The existence of this background radiation was first predicted by Ralph Alpher and Robert Hermann (once again in 1948). They realized that if the early universe had been filled with a hot plasma it must have emitted radiation with a blackbody spectrum, and that radiation should be filling all of space today. As the universe expanded, the wavelengths of that radiation would stretch. Alpher and Hermann calculated that this redshifting would still leave the radiation with a blackbody spectrum, but at a lower temperature. Today it should be composed almost entirely of microwaves.

It took over 15 years after Alpher and Hermann's paper for instruments to be developed that could detect such weak microwave signals. In 1964 Arno Penzias and Robert Wilson set up a microwave detector to measure signals from sources in the Milky Way. They knew almost nothing about cosmology, and they were annoyed to find that their detector registered a microwave signal from every direction in the sky. Eventually they spoke to the cosmology group at Princeton and were informed that they had just discovered the leftover radiation from the early universe.

After the observations of Penzias and Wilson, nearly all cosmologists considered the Big Bang model to have been conclusively verified, and Penzias and Wilson got the Nobel Prize in Physics in 1978 for their discovery.

What makes the CMB so convincing is the fact that it is almost perfectly uniform across the sky. When we measure visible light or radio signals or almost any other kind of signal, we see lots of it when we point our detectors at sources like stars or galaxies, and not much when we point between them at empty space. But the CMB comes at us almost exactly equally from all directions, suggesting that whatever emitted it must have once filled all space uniformly.

A list of the evidence for the Big Bang model could also include the following:

- We have well-tested models of how stars evolve over their lifetimes, and we can use those models to estimate the ages of stars. We expect many stars to live longer than the current age of the universe, but the oldest stars we observe are slightly younger than the age of the universe that we infer from the current rate of expansion.

- Theories of the early universe predict a cosmic neutrino background ("$t = 1$ s" in Section 14.1) emitted long before the cosmic microwave background. That neutrino background was measured in 2015.

- Looking far enough in any direction, we see galaxies as they were billions of years ago. The properties of the galaxies we observe are the same in every direction, but vary with distance from us in ways that are consistent with the evolution of galaxies described by the Big Bang model.

Since 1964 there has been no serious scientific dispute about the accuracy of the Big Bang model. (An exception is Fred Hoyle, the primary proponent of the steady-state model of cosmology, who went to his grave in 2001 still adamantly opposed to the Big Bang model.) But by the 1970s it was becoming clear that there are a number of features of the universe that the Big Bang model could not account for. In Section 14.6 we'll describe those problems, and in Section 14.7 we'll describe how they were solved by rethinking our understanding of the first fraction of a second after the Big Bang.

But here we will take up another question. The original discovery of the universe's expansion required measuring the velocities and distances of other galaxies.

Velocities can be measured very accurately. You may recall from Chapter 4 that elements all emit and absorb light at characteristic frequencies. By looking for the patterns of spectral lines that characterize each element, we can often tell what astronomical objects are made of. You may also recall from Chapter 1 that light from a moving object is Doppler shifted to lower frequencies ("redshifted") when the source is moving away from you. So Slipher and Hubble measured the velocities of galaxies by seeing how much their spectral lines were redshifted compared to stationary sources.

Measuring astronomical distances is much harder. Below we discuss two techniques that can be used to perform those measurements: parallax, and standard candles.

Parallax

If two people observe an object from different locations, they see it at different angles. Based on their distance to each other and the angles at which they see the object, they can calculate how far away the object is. (This is one of the techniques your brain uses to gauge your distance to reasonably nearby objects, based on the different angles to your two eyes.)

But the stars are *really* far away. Even if you take measurements from opposite sides of the Earth, it's virtually impossible to measure the angular difference. We can get a much longer baseline, however, by using the Earth's orbit. Figure 14.2 shows two measurements of a star taken six months apart, with the Earth on opposite sides of the Sun.

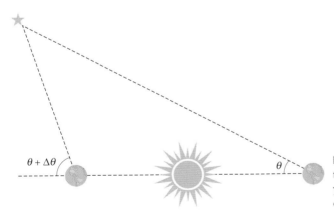

$\theta + \Delta\theta$ θ

Figure 14.2 By measuring the angle to a star at two different times of year, astronomers can calculate the distance to that star.

Even with this baseline of almost 200 million miles, the angular shifts are extremely small. The first successful parallax measurement of a star was in 1838, when Friedrich Bessel measured the star 61 Cygni from each side of the Earth's orbit, and found an angular shift of 0.28 arcseconds, or about 80 millionths of a degree. That measurement indicated a distance of a little more than 10 light years.

Astronomers often measure distances in "parsecs," defined as the distance that would give a **par**allax shift of one arc**sec**ond. A parsec is just over 3 light years, so 61 Cygni is 3.6 parsecs away (1 over 0.28). In this chapter we'll stick with the possibly more familiar unit of light years.

As of this writing, satellite telescopes have used the parallax technique to measure the distances to over a billion stars stretching tens of thousands of light years away: more than a tenth of the way across the Milky Way galaxy. But as impressive as those measurements are, we need a different technique to measure distances beyond our galaxy.

Standard Candles and Cepheid Variable Stars

In 1920 the astronomers Harlow Shapley and Heber Curtis held a public debate on the nature of the cosmos. A key point in the debate centered on the nature of nebulae, glowing blobs of light scattered throughout space. Curtis argued that the nebulae were other galaxies (which he called "island universes") outside the Milky Way, each containing vast numbers of stars. Shapley argued that there was nothing beyond the Milky Way, and that the nebulae are clouds of gas and dust between the stars.

We now know that some of the nebulae are clouds of gas and dust within the Milky Way, and some of them are other galaxies. But determining this requires measuring intergalactic distances, which is far beyond the scope of the parallax technique.

So ... suppose you stand outside on a dark night while 100 people all shine identical flashlights toward you. The nearby flashlights look brighter to you, and the distant ones look dimmer. Specifically, light intensity (power per unit area) falls off as $1/r^2$, so if you are receiving four times as much light from Flashlight B as from Flashlight A, you can infer that Flashlight A is twice as far away as Flashlight B.

Unfortunately, the stars are more like *non*-identical flashlights: some are brighter than others. Likewise, galaxies vary tremendously in brightness. If you don't know the relative brightnesses then you can't figure out the relative distances, and vice versa. In order to calculate distance, you need a "standard candle": a class of objects with equal intrinsic brightness. Remarkably, the universe comes stocked with a few.

In 1908, Henrietta Swan Leavitt analyzed a sample of "Cepheid variable stars" in the Small Magellanic Cloud (SMC). She didn't know what the SMC was (we would now call it a dwarf galaxy), but she reasonably guessed that its size was much smaller than its distance from us. That meant all the Cepheids in her sample were approximately the same distance from us, and their observed brightnesses should be proportional to their intrinsic brightnesses.

But Cepheids have another important property: their luminosity oscillates regularly, with a period that can range from days to months. In a 1912 paper, Leavitt plotted intrinsic brightness vs. period for her Cepheid sample on a log-log plot, and her plot showed a linear relationship (Figure 14.3). So if you measure the periods of two Cepheids, then you know their relative brightnesses: a standard candle! Once you know the ratio of their *intrinsic* brightnesses, the variation in brightness that you actually *measure* tells you their relative distances.

A year after the publication of Leavitt's paper, Enjar Hertzsprung used parallax to measure distances to several Cepheids in the Milky Way, and was thus able to determine the proportionality constant between period and intrinsic brightness. With the scale thus calibrated, the brightnesses of other Cepheids could be used to accurately measure their distances.

In 1924 Hubble wrote to Shapley: "You will be interested to hear that I have found a Cepheid variable in the Andromeda Nebula." Hubble's observation allowed him to prove that Andromeda was far too distant to be in the Milky Way, and the debate about the existence of other galaxies was settled.

Today we measure trillions of galaxies out to distances of billions of light years. We measure Cepheids out to a few tens of millions of light years, and we can use Cepheids to calibrate other

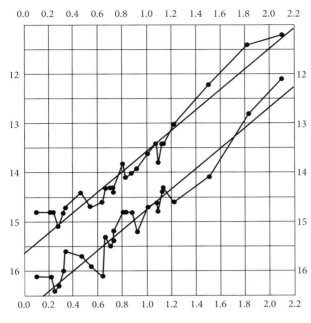

Figure 14.3 Henrietta Leavitt's plot of Cepheid variable stars shows "apparent magnitude" on the vertical axis and the log of the oscillation period on the horizontal axis. The two lines of data are the dimmest and brightest points of the oscillations. Apparent magnitude is proportional to minus the logarithm of the light intensity, so the lower numbers at the top of the graph reflect the brightest stars. *Source: Leavitt and Pickering 1912.*

standard candles that we can observe at even greater distances. This scheme of measurement, using measurements at one scale to calibrate distance measures for farther away objects, is called the "cosmic distance ladder."

14.2.3 Questions and Problems: How Do We Know All That?

Conceptual Questions and ConcepTests

1. What units might be used to measure the Hubble parameter?

2. The universe can't be static because the mutual attraction of all the galaxies would cause them to start falling in toward each other. Our galaxy is neither expanding nor contracting, however. Why doesn't this same argument apply to the stars in our galaxy?

3. The parallax method is used to measure our distance to other stars in our galaxy.

   (a) Why can't we use this method to measure the distance to the other planets in our solar system?

   (b) Why can't we use this method to measure the distance to other galaxies?

4. We define a standard candle as a type of object that always has the same intrinsic brightness, which lets you compare how far away two of them are by seeing how relatively bright they appear. By this definition, Cepheid variable stars are not technically standard candles, because some are more intrinsically bright than others. How are we still able to effectively use them as standard candles?

5. Henrietta Leavitt established the relationship between the periods and brightnesses of Cepheid samples in the SMC.

   (a) What assumption did she have to make about the SMC in order to draw her conclusion?

   (b) Leavitt's discovery was used to establish our distance to other galaxies. Before those measurements could be taken, however, Enjar Hertzsprung needed to measure Cepheids in our own galaxy. Why was that an essential step?

6. You will sometimes hear people associate a particular speed (often the speed of light) with the expansion of the universe. Why is it inherently impossible

to say "Here is the speed at which the universe is expanding"? Failing that, how *can* we assign a numerical value to the rate of expansion?

7. The Active Reading Exercise on p. 655 used a ball, flying through the air, as an analogy for the expanding universe. This is an analogy we will return to, because the ball is being pulled downward by gravity just as the universe's galaxies are being drawn toward each other by gravity.

    (a) Imagine a ball rising upward from the ground. The ball was initially launched, but now there is no force acting on it – not even air resistance – except for the downward gravitational pull. What will happen to that ball in the long term? *Hint*: The answer begins with "It depends." What does it depend on, and how?

    (b) Based on that analogy, what is the ultimate fate of the universe? What does it depend on, and how?

For Deeper Discussion

8. Why are some nearby galaxies moving toward us, but beyond a certain distance *all* galaxies are moving away from us? *Hint*: The answer isn't just because the Milky Way's gravity is strongest nearby.

9. Will Robinson is in a spaceship moving at $0.1c$ relative to Earth. Just as he passes the Earth, he measures the velocities (using redshift) and distances (using Cepheids) to many distant galaxies. Will his measurements obey the Hubble–Lemaître law? (Choose one and explain.)

    A. Yes

    B. No

10. To do a parallax measurement you need to carefully measure the angle to a star two times, spaced months apart from each other. But the Earth is rotating beneath you, so how do you know to such high precision in which direction your telescope was pointing six months earlier? *Hint*: You might think you could measure this by measuring the angle of the star from the Sun. (In fact you might think that because it's how we drew our figures.) But that doesn't generally work since you can't see the Sun at night and you can't see stars during the day. So you need a different standard to make these measurements.

Problems

11. Figure 14.2 on p. 659 shows two measurements of a star taken six months apart.

    (a) Copy this figure and fill in all three angles in the triangle in terms of θ and $\Delta\theta$. (We've already done one, since θ is in the triangle.)

    (b) Let the distance from the Earth to the Sun be D. (Note that D is only half the bottom edge of the triangle.) The law of sines says that in any triangle the ratio of the sine of one of the angles to the length of its opposite side is the same for all three angles. Use the law of sines to write an equation for r, the distance from the star to Earth's left-most position in the diagram. Your answer should be in terms of θ, $\Delta\theta$, and D.

    (c) Take the Earth–Sun distance to be 100 million miles. If you measure the star at $135.000001°$ at one time and $135°$ six months later, how far away is the star?

12. The galaxy UGC 8365 has been measured to be about 18 megaparsecs away. (That's about 59 million light years, but for this problem it will be easier to leave it in megaparsecs.)

    (a) The current value of the Hubble parameter is 73 km/(s Mpc). (That's kilometers per second per megaparsec, and yes those are the units usually used by astronomers for it.) According to the Hubble–Lemaître law, how fast is UGC 8365 moving away from us? (This should be a 20-second calculation.)

    (b) An object's "redshift" z is defined by $1 + z = \lambda_{observed}/\lambda_{emitted}$. What redshift would you predict for UGC 8365 based on its distance?

    (c) An important spectral line in astronomy is the 21 cm wavelength line caused by a hyperfine transition in hydrogen (meaning a transition from a state where the electron and proton spins are aligned to one where they are anti-aligned). By how much would the wavelength of this line shift because of UGC 8365's velocity?

14.3 Infinite Universe, Finite Universe, Observable Universe

No one knows whether the universe is finite or infinite. We can identify properties that distinguish a finite universe from an infinite universe: its energy density, and a more esoteric property called its "curvature." But we can't measure either property accurately enough to answer the question.

In this section we will discuss what our universe might be like if it is finite, and what it might be like if it is infinite. We will also discuss the "observable universe," the space that we can see or measure in any way, which turns out to be a very small percentage of the whole.

14.3.1 Discovery Exercise: An Infinite, Expanding, One-Dimensional Universe

Boris and Natasha both live on a universe that looks suspiciously like an infinitely long ruler, marked off in inches. One day, for no obvious reason, the ruler stretches horizontally: *all horizontal distances on the ruler double, and this expansion takes exactly one second.* (See Figure 14.4.)

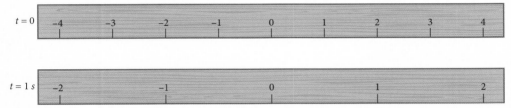

Figure 14.4 Boris and Natasha's universe, before and after its great expansion.

1. Boris is standing on Mark 0 when the great expansion happens. In one second he sees every other mark double its distance from him: for instance, Mark 2 used to be two inches to his right, but now it is four inches to his right. Write down the *average speed* that Boris calculates for Mark 2 during its one-second move.

2. Repeat Part 1 to find the speeds of Marks -2, -1, and 1.

3. The Hubble–Lemaître law in our universe states that the speed with which a galaxy is receding from us is directly proportional to its distance from us. Does Boris calculate the same law in his universe?

   See Check Yourself #25 at www.cambridge.org/felder-modernphysics/checkyourself

4. Natasha is standing on Mark 1. During that same time, what average speeds does she measure for Marks -1, 0, 2, and 3?

5. If Natasha considers herself the center of the universe (which believe us, she does), does she also observe the Hubble–Lemaître law – the speed with which a galaxy is receding *from her* is proportional to its distance *to her* – or does she require a modified law? Show how your answer follows from your calculations.

14.3.2   Explanation: Infinite Universe, Finite Universe, Observable Universe

Is the universe finite or infinite? We don't know, but either answer seems to present a dilemma.

- If the universe is finite, then what's beyond it? If you imagine yourself standing at the edge of the universe and looking out past the boundary, whatever you see there – even if it's just empty space – still counts as part of the universe.
- But if the universe is infinite, how can it expand? You can't get bigger than being infinite.

We're going to address the second question first: how can an infinite universe expand? Then we'll tackle the first: if we define the universe as "everything that exists," how can it be finite? Finally we'll talk about how to tell which kind of universe you're living in – and why we haven't been able to tell so far.

An Infinite Universe

It is one of the more remarkable features of the human mind that even though we cannot picture infinity – we have never seen or experienced infinity – we can still *discuss* infinity and draw logical conclusions about it and do math with it.

To help us think about an infinite universe, we'll start with the simplest case of an infinite *one-dimensional* universe.

Imagine a line that goes on forever in both directions, and imagine dots representing galaxies spaced evenly along this line. We'll call our galaxy Number 0. The ones to the right are Galaxy 1, Galaxy 2, and so on, and to the left are Galaxy −1, Galaxy −2, and so on, going on without end. Just to have a definite image, let's say there's one inch between adjacent dots. (Admittedly that would require some very small galaxies.)

Now imagine that all distances between galaxies double. Standing in Galaxy 0 we now see Galaxy 1 two inches to our right, and Galaxy 2 four inches to our right, and so on, still stretching out to infinity.

Figure 14.5 brings out a number of points about an infinite expanding universe: some fairly obvious, some less so.

- We can say that an infinite universe is "expanding" but we cannot say that it is "getting bigger." If the universe is infinite now, then it has always been infinite; nothing we know could make a finite thing infinite or vice versa. In strict mathematical language, we can even say that the length of this universe after doubling is *the same infinity* as its length before doubling.

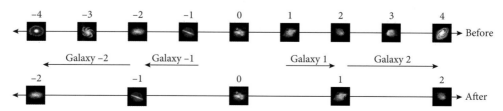

Figure 14.5 An expanding universe in one dimension.

- But if expansion does not imply an increase in size, it does imply a decrease in density. In our example, the density of our 1D universe dropped from 12 galaxies per foot to 6 galaxies per foot.

- Less obviously, this kind of uniform expansion (the distance between any two points doubles) implies the Hubble–Lemaître law. The speed at which a distant galaxy is receding is directly proportional to that galaxy's distance from us. You can see this in the diagram above by looking at how far each of the galaxies moved in the same amount of time; the galaxies farthest from Galaxy 0 moved fastest.

- Perhaps least obvious, but no less important, the expansion has no center. From our perspective, of course, it appears that our galaxy (which we have self-centeredly numbered zero) is the center of the expansion: galaxies far to the right of us are moving quickly to the right, and galaxies far to the left are moving quickly to the left. But Bob, looking outward from Galaxy 3, would see the universe expanding away from him. Alice, on Galaxy −3000, would see the universe expanding away from her. They would probably name their own galaxies "Galaxy 0" and they would observe the Hubble–Lemaître law exactly as we do.

You can extend this picture to two dimensions by picturing a rubber sheet with thumbtacks stuck in it. If you stretch the sheet, all the thumbtacks get farther apart. In three dimensions you can imagine a raisin cake rising in the oven. As the dough stretches, the raisins all get farther apart. In any number of dimensions, a uniformly expanding universe will obey the Hubble–Lemaître law from the point of view of any observer.

A Finite Universe

So that's a quick description of how an infinite universe expands. What about a finite universe? How can that even be possible?

The universe can't end at some point in space. If you think you've found the edge of the universe, whatever lies beyond it, even if it's just empty space, still counts as part of the universe.

But that argument doesn't actually imply that the universe must be infinite. Instead it tells us that if the universe is finite then it must be what we call "periodic." That means that if you travel far enough in any direction you end up back where you started.

A finite, periodic universe is in some ways even harder to picture than an infinite universe. There is once again an analogy that can help, but this analogy can also be misleading in ways we'll explain in a moment.

The analogy is a spherical surface like the surface of a globe. Imagine a two-dimensional being who lives on that surface. We're going to name him Edwin – not after Hubble, but after Edwin Abbott, the author of the charming book *Flatland* that introduced your authors and many other readers to the idea of a two-dimensional world.

If Edwin starts at the North Pole and begins moving in any direction, he will end up at the South Pole. If he keeps going without turning, he will eventually find himself back at the North Pole where he started (Figure 14.6).

The periodic two-dimensional universe that Edwin lives in makes a good analogy for the three-dimensional universe that we *might* live in, in a number of ways.

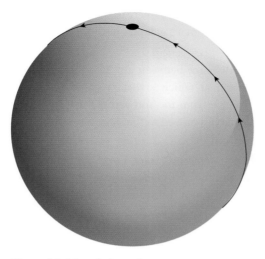

Figure 14.6 In a finite universe you can move without turning and end up back where you started. The surface of a sphere is an analogy for a finite, two-dimensional universe.

- When Edwin walks along the path we just described, he does not see himself as turning, but rather as moving in a straight line within his own world. And still, no matter what direction he starts in, he ends up back where he started.

- Edwin's universe is finite but it has no center and no edges. To put it another way, any point in his universe has as good a claim of center-hood as any other point. (If you're thinking "What about the actual center of the sphere?" be careful: that point doesn't exist. The surface of the sphere is all of space.)

- Now, what happens if Edwin's spherical universe starts to expand, like a perfectly spherical balloon blowing up? Careful measuring will lead him to the Hubble–Lemaître law: every point is moving away from him, at a speed proportional to its distance from him. And once again, Edwin will reach this conclusion no matter where he is standing. No point is special.

As powerful as this analogy is, it can also be misleading. As three-dimensional beings looking down at Edwin's universe, we invariably conclude that he ended up back where he started because he didn't really walk in a straight line. If he could burrow through the center of the globe or fly off the surface of the globe then he would move in a truly straight line, and never come back to his starting place.

That's true, but it's just a limitation of the analogy. There is no inside-the-globe region, or outside-the-globe region. The surface of the globe is the entire universe, and it just has this strange mathematical property that it connects back to itself. Imagining the universe as a globe is a trick for us to picture what that means.

If our 3D universe is finite, then it has the property that a line drawn in any direction from Earth would eventually hit Earth again. That does *not* mean that our universe is a curved surface wrapped around some four-dimensional region. In this model our 3D space is all there is, and it's periodic. Since some curved surfaces are also periodic, they provide a useful mental image if you don't take them too literally.

As a final note, remember that an *infinite* universe expands (becomes less dense) without ever changing its actual volume. When a *finite* universe expands, its size – the total amount of space – does increase over time. If our universe is finite, the distance you would have to travel to get back to where you started is longer now than it was a billion years ago.

The Observable Universe

It seems like it should be easy to determine which kind of universe we live in. Just look far enough in any direction, and if the universe is finite you should see the back of your head.

Looking out as far as we can see in every direction, we don't see the universe repeating itself. From that we conclude that the universe is either infinite, or it's finite but bigger than we can see.

So is the answer to build better telescopes? No. We are already looking as far into space as it is physically possible to see. The reason for that has to do with the speed of light.

When we look at the Sun, we see what it looked like eight minutes ago. As you may recall from Chapter 1, that time gap is a fundamental limitation dictated by relativity: an event on the Sun at noon cannot cause any event on Earth before 12:08, so we cannot *know* about such an event until 12:08. We see Andromeda as it was 2 million years ago, and some further galaxies as they were billions of years ago.

Now imagine a galaxy 100 billion light-years away from us. Do we see the light that left that galaxy 100 billion years ago? No, we can't – because the Big Bang was less than 14 billion years ago. We have no reason to believe there *aren't* galaxies that far away, but if there are, their light won't reach us for billions of years.

All of this leads to a distinction between "the universe" (everything that exists), and "the observable universe" (everything we can see).

Modern telescopes show us galaxies that are so far away that we see them as they were when galaxies in the universe were first forming, just a few hundred million years after the Big Bang. Beyond that in time we see the CMB: you may recall that this light was emitted by the hot glowing gas only a few hundred thousand years after the Big Bang, so the microwaves we detect come from many billions of light-years away. From further still we detect the neutrinos emitted only a second after the Big Bang. We cannot possibly see or detect anything much beyond that – not because there is nothing farther out, but because nothing from farther away can possibly have reached us.

Our observable universe is a sphere with a radius a little over 45 billion light-years, centered on us. Observers in other galaxies would have different observable universes. Ours is the part of the universe that it's possible to see, even in principle, from the Milky Way galaxy.

If you're paying close attention, you might wonder how our observable universe can extend over 45 billion light-years away from us when the Big Bang only happened 14 billion years ago. To answer that, consider a galaxy that formed shortly after the Big Bang. For 14 billion years the light from that galaxy has been moving toward us, while the galaxy itself has been moving away from us. So if the light from the formation of that galaxy is just reaching us now, then the galaxy is now almost 45 billion light-years away from us. (Wait a minute. 14 billion years . . . , 45 billion light-years . . . Does that mean the galaxy is moving faster than the speed of light? In fact, it does: as the Hubble–Lemaître law promises, any galaxy whose distance from us is greater than c/H recedes from us faster than c. Relativity allows such speeds when they are caused by the stretching of space itself, rather than by the movement of objects through space.)

Looking out with our telescopes, we conclude that the universe is at least as big as our observable universe: a sphere with a roughly 45 billion light-year radius. Is the entire universe a trillion light-years across? A trillion trillion light-years across? Infinite? We don't know, because our view is limited.

It may help to consider what all this looks like to Edwin. Just as our 3D observable universe is a sphere centered on us, Edwin's 2D observable universe is a circle centered on him. If that circle only covers a small part of his entire balloon surface, he can't see whether his universe is periodic or not (Figure 14.7).

Figure 14.7 If Edwin's observable universe (the green circle) only covers a small portion of his entire universe, he can't tell whether his universe is finite.

There is a way we could in principle tell whether the universe is finite or infinite, even without being able to see all of it. First we would measure the density ρ, meaning energy per unit volume, of the universe. Then we would compare this to a certain density that we can calculate called the "critical density" ρ_c. According to general relativity, if $\rho > \rho_c$ the universe is finite, and if $\rho \leq \rho_c$ the universe is infinite. This result is usually expressed in terms of the ratio $\Omega = \rho/\rho_c$, called the "density parameter." The universe is finite if $\Omega > 1$.

We're going to make more references to the critical density, so we need to say a few more words about that idea.

- The density of the universe ρ changes over time, as you know. The critical density ρ_c *also* changes over time. The thing that cannot change is their inequality: if $\rho > \rho_c$ was true at the Big Bang then it's still true now, and the universe is finite.

- We haven't said anything about *why* the inequality $\rho > \rho_c$ implies a finite universe, and we're not going to. This is one of those take-our-word-for-this-GR-result moments.

So we've measured the total amount of energy in the observable universe, and divided that by the size of the observable universe, to calculate its density. The result is that our observable universe is so close to critical density – within less than 1% – that we can't tell whether we're above or below it. For reasons we'll explain later in the chapter, we think the universe is probably so close to critical density that we will never be able to measure whether we are above or below it. So once again, we can't tell whether our universe is infinite or finite.

Stepping Back: The Limits of Our Knowledge

The Big Bang model is based on looking at the universe today and extrapolating backward to what it must have been like at earlier times. Section 14.1 introduced the idea that this extrapolation is limited by the fact that we don't have a physical theory to describe matter above the Planck density. So for now our knowledge only goes back to the moment about 13.8 billion years ago when our universe was at the Planck density. The phrase "Big Bang" refers to that moment.

This section introduced a different limitation: because of the finite speed of light and the finite age of the universe, we can only see so far into space. Whatever has happened since the Big Bang in the regions beyond our observable universe can't have had any effects on us yet, so we can't know anything about those regions, including whether they exist at all.

For hundreds of years we have built ever-better telescopes that let us see ever-farther into space. That progression is now over; from now on, better technology will let us see more clearly, but not farther.

On the other hand, even without any technological improvements, our observable universe will be significantly bigger a billion years from now than it is today. If you think like a cosmologist, that isn't too long to wait.

14.3.3 Questions and Problems: Infinite Universe, Finite Universe, Observable Universe

Conceptual Questions and ConcepTests

1. A team of astronomers announces that they have found the location in space where the Big Bang occurred. Which of the following reactions would be most appropriate? (Choose one.)

   A. You don't believe them because the Hubble–Lemaître law says that expansion looks the same no matter where you are, so there's no way to tell where the Big Bang occurred.

   B. You don't believe them because the Big Bang probably occurred outside our observable universe, so we wouldn't be able to see it.

   C. You don't believe them because there is no particular place where the Big Bang occurred.

   D. You believe them.

2. Explain why a finite universe must be periodic.

3. Why is an object 20 billion light-years away within our observable universe, when the universe is less than 14 billion years old?

4. Would it be possible for an infinite one-dimensional universe to expand in a way that would *not* obey the Hubble–Lemaître law? If not, why not? If so, how?

5. Suppose you observe a galaxy a billion light years away, and in the opposite direction you observe what you are sure is the same galaxy, but three billion light years away.

   (a) What is the size of your universe? *Hint*: As with almost any conceptual question about finite universes, it's easier to picture if you go down a dimension and imagine 2D Edwin making this observation on the surface of a sphere.

   (b) Even if you are correct that you are seeing the exactly same galaxy in two directions, one of those images might show the galaxy having more stars than the other. Why? (Assume you can see well enough to accurately count the stars in both images.)

6. Suppose you are in a finite universe and your observable universe is large enough to see yourself in the distance. Which direction do you see yourself in, and why? *Hint*: As with almost any conceptual question about finite universes, it's easier to picture if you go down a dimension and imagine 2D Edwin making this observation on the surface of a sphere.

7. The Hubble–Lemaître law says that a galaxy's recession speed from us is proportional to its distance from us: $v = Hd$. Your friend Joyce says that this law implies that the universe must be finite, because a galaxy farther away from us than c/H would be traveling faster than light. Your friend Rose counters that Joyce must be wrong because she read *in a science textbook* that the universe might be infinite, and science textbooks are never wrong. You don't quite believe Rose's logic, but you suspect something is wrong with Joyce's as well. How can you resolve the argument?

For Deeper Discussion

8. The equations of GR allow the possibility of a universe that is shrinking rather than expanding. Such a universe would still obey the Hubble–Lemaître law, $v = Hd$, so sufficiently distant galaxies would be *approaching* us faster than c. If such a galaxy emitted a light beam, could that galaxy reach us before its light beam did? Why or why not?

Problems

9. Figure 14.5 shows a one-dimensional universe. Imagine that Xavier, standing on a galaxy at position x, is watching Galaxy A at position $x + \Delta x$, and also watching Galaxy B at position $x + k\Delta x$. Suddenly the universe expands, as depicted in the drawing: Xavier's galaxy is now at position $2x$.

   (a) Calculate the change in Galaxy A's position from Xavier.

   (b) Calculate the change in Galaxy B's position from Xavier.

   (c) From Xavier's perspective, does the Hubble–Lemaître law describe the expansion of his universe? Show how your answer follows from your calculations.

10. [*This problem depends on Problem 9.*] Ororo Munroe flies past Xavier, passing him at time $t = 0$ at constant velocity v relative to him. From her perspective, of course, she is not moving. Nonetheless, show that the Hubble–Lemaître law does *not* describe the expansion of the universe from her perspective. *Hint*: You can assume $v \ll c$ and use Newtonian calculations.

11. Figure 14.5 on p. 664 shows a one-dimensional universe expanding. Such a universe obeys the Hubble–Lemaître law.

    (a) Write the Hubble–Lemaître law as a differential equation relating a galaxy's position x to its velocity dx/dt.

    (b) Now, supposing that the Hubble parameter in this one-dimensional universe is constant over time (which it isn't in the real universe), write the function $x(t)$ that solves your differential equation. Your solution should include a constant x_0 for the galaxy's initial position.

    (c) Briefly and qualitatively describe what you would see in such a universe, as you watched a distant galaxy recede.

12. Consider an infinite 2D universe with galaxies laid out in a perfect grid with spacing L, meaning there are galaxies at $(0,0)$, $(0,L)$, $(L,0)$, (L,L), and so on. After a time Δt, all the distances have doubled. (Just replace each L with $2L$.)

    (a) You are sitting at the origin. For the three galaxies that start at $(0,L)$, (L,L), and $(L,2L)$, what are their initial distances from you and what are their average speeds relative to you during the time Δt? Do those three galaxies obey the Hubble–Lemaître law from your perspective?

    (b) Repeat the same calculations from the point of view of the galaxy that starts at (L,L), calculating distance and speed for the other two galaxies and for yours (which starts at the origin). Does that observer measure the Hubble–Lemaître law?

13. **Olbers' Paradox**

    Isaac Newton believed that the universe was an infinite expanse of stars spread out more or less evenly throughout space. A universe like that presents a conceptual problem known as "Olbers' Paradox."[8] To explain the paradox we'll assume that each star is a point-like object, so no star blocks any other star. We're also going to assume that each star emits exactly the same amount of light and that the density of stars is uniform throughout space. None of these assumptions is necessary for the paradox but they make the explanation clearer.

    (a) Consider a star at a distance R from you. If L is the total power (light energy per unit time) emitted by the star and A_E is the combined surface area of your eyes, how much power from the star reaches your eyes? *Hint*: Consider a sphere of radius R drawn around the star and figure out what fraction of that sphere is taken up by your eyes.

    (b) Now suppose the whole sky is filled with identical stars with a uniform density (number of stars per unit volume) ρ. Consider all the stars at distances between R and $R + \Delta R$. If $\Delta R \ll R$, we can approximate the volume of this region as the surface area $4\pi R^2$ times the thickness ΔR. We can also assume the power reaching us from each star in this region is the same, because they are all roughly at distance R. Using those approximations, what is the total amount of power reaching you from all the stars whose distance from you is between R and $R + \Delta R$?

    (c) Now consider a star at a distance $2R$ from you. How much power from that star reaches your eyes?

    (d) How much power reaches your eyes from stars at distances between $2R$ and $2R + \Delta R$?

    (e) Based on what you've found, if you were to include the light from stars at all distances from you, how much power would be reaching your eyes? Briefly explain.

---

8 The paradox is named for Heinrich Olbers, who wrote about it in 1823, but others had written about it at least as far back as the sixteenth century.

You should have just concluded that the night sky should appear infinitely bright. You can get around that to some extent by taking into account that stars have a small width and can in principle block other stars, but even taking that into account, this argument would lead you to conclude that the whole night sky should appear as bright as the surface of the Sun. That puzzle plagued astronomers for centuries until it found a clear resolution in the twentieth century.

(f) Is Olbers' paradox a problem in a finite, periodic universe? Why or why not?

(g) Explain how the ideas in Section 14.3 provide a resolution to Olbers' paradox.

14.4 The Friedmann Equations

There hasn't been much math in this chapter, has there?

Without the math, you can learn a lot of cosmology, but much of it falls into the take-our-word-for-it category. The more you learn the underlying equations, the more sense it all makes.

As we've remarked a few times, a full mathematical treatment of the issues in this chapter requires general relativity. But if we assume the universe is homogeneous and isotropic, the core equations of relativity reduce to two differential equations that describe how the universe expands or contracts. We are not going to talk about the derivation that leads *to* those equations, but we are going to talk about some of the conclusions you can draw *from* them.

14.4.1 Discovery Exercise: The Friedmann Equations

Consider a uniform expanding sphere of matter with radius $r(t)$. Let Galaxy X be a galaxy on the outer edge of this sphere. In this exercise you are going to analyze the dynamics of this sphere using entirely Newtonian physics.

1. Write the net gravitational force on Galaxy X. Your answer will depend on the mass of the galaxy m_X and the mass of the entire sphere M_S.

2. Write the gravitational potential energy of Galaxy X.

3. Write an equation expressing conservation of energy for Galaxy X. Your equation should include r, $\dot{r}$ (which is shorthand for dr/dt), and an arbitrary constant.

   See Check Yourself #26 at www.cambridge.org/felder-modernphysics/checkyourself

4. Rewrite your equation so it depends on ρ, the mass per unit volume, rather than M_S.

5. Show that your equation can be rewritten in the following form with correctly chosen constants k_1 and k_2:

$$\left(\frac{\dot{r}}{r}\right)^2 = k_1 \rho - \frac{k_2}{r^2}.$$

The equation you just derived using Newtonian physics also turns out to be the equation that describes an expanding universe in general relativity. In this section you'll explore some of the implications of this equation.

14.4.2 Explanation: The Friedmann Equations

The central equation of general relativity is the "Einstein equation," which equates a 4×4 matrix describing the energy and momentum of matter to another 4×4 matrix that describes the curvature of spacetime. In 1922 Alexander Friedmann analyzed Einstein's equation under the assumptions that the universe is homogeneous (the same everywhere) and isotropic (no preferred direction) – both reasonable assumptions at very large scales, although not of course at smaller ones. He showed that for that case all of the components of Einstein's equation reduce to two equations, now known as the "Friedmann equations":[9]

$$\left(\frac{\dot{a}}{a}\right)^2 = \frac{8\pi G}{3c^2}\rho - \frac{kc^2}{a^2} \tag{14.1}$$

$$\frac{\ddot{a}}{a} = -\frac{4\pi G}{3c^2}(\rho + 3P) \tag{14.2}$$

- a is a function of time called the "scale factor" that describes the expansion of the universe. Remember that dots represent time derivatives, so $\dot{a}$ means da/dt, and $\ddot{a}$ means d^2a/dt^2.

- k is a constant related to the curvature of space.

- G and c are respectively Newton's gravitational constant and the speed of light.

- ρ and P are respectively the energy density (energy per unit volume), and the pressure, of the fluid filling the universe. They are *not* constants. (If they were, you could just write down the solutions to both equations right now!) But they are related to each other by an "equation of state" – based on what kind of stuff fills the universe – and once you plug that in, you can solve the equations.

We discuss a, k, and the relationship between ρ and P below.

The Scale Factor and the Hubble Parameter

The scale factor $a(t)$ is a unitless measure of distance between objects in an expanding universe. You can set $a = 1$ at any reference time you wish. The time when $a = 0.5$ is then the time when all distances were half as large as they were at your reference time.

Since a is proportional to the distance between objects, $\dot{a}$ is proportional to the relative velocity of those objects. The combination $\dot{a}/a$ is therefore the Hubble parameter H, which we defined as the ratio of a galaxy's recession velocity to its distance from us. So the first Friedmann equation is an equation for the Hubble parameter as a function of the density of matter and the curvature of space.

The statement "$a = 0.5$" doesn't mean anything by itself, since the choice of a scale ($a = 1$) is arbitrary. But the Hubble parameter $H = \dot{a}/a$ is a meaningful quantity: in the universe today it's about 7×10^{-11} yr^{-1}. The quantity $1/H$ is sometimes called the "Hubble time." To within an order of magnitude, the Hubble time is both the current age of the universe and the amount of time we'll have to wait from now before all distances double (see Problem 10).

---

[9] We're going to refer to these equations a lot in the rest of the chapter, both in our explanations and in the problems. You might find it handy to save a copy of them somewhere you can easily refer to.

The Curvature of the Universe

The parameter k describes deviations from Euclidean geometry, which are a central feature of general relativity.

As an example, let's return to Edwin's spherical universe. Edwin's universe doesn't look like the surface of a sphere to him. The paths he walks along are perfectly straight to him, although they look curvy to us. He can never step off his globe, either outside or inside, because those regions literally don't exist; the entire universe in this analogy is the two-dimensional surface of his sphere. His observable universe is a disk around himself. If that observable universe is smaller than the whole sphere, he can never see a light beam loop back to its starting point.

But there is another experiment that Edwin can do to determine the curvature of his universe. If he draws a triangle (all straight lines!), and measures its angles carefully, he will find that they add up to more than $180°$.

For clarity, Figure 14.8 shows a triangle that spans a big chunk of Edwin's universe. If he draws a smaller triangle the angles will still add up to more than $180°$, but not by as much. The smaller Edwin's triangle, the more accurately he will need to measure it to detect the curvature of his universe.

A universe that behaves like the surface of a sphere is called a "closed universe." A closed universe is periodic and finite, and its triangles have angles that add up to more than $180°$. Mathematically, a closed universe corresponds to a positive value of k in the Friedmann equations.

There are two other options. If $k < 0$ we have an "open universe." An open universe is infinite and its triangles have angles that add up to *less* than $180°$. The surface of a sphere perfectly represents the geometry of a closed 2D universe, but there is no 2D surface that exactly reproduces the geometry of an open 2D universe. Nonetheless, you can get a reasonable image by picturing the surface of a saddle (Figure 14.9).

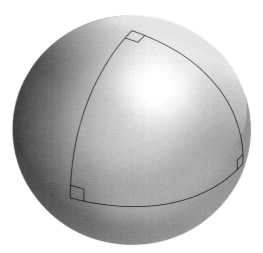

Figure 14.8 In a finite, periodic universe, you can draw three lines that all meet at right angles. From the point of view of a 2D observer in this universe, the lines are all straight.

The third option, a "flat universe," corresponds to $k = 0$ precisely: a plane with perfectly Euclidean geometry. So, to determine which type of universe he inhabits, Edwin might measure the angles of a triangle. If his triangle is big enough, and his measuring devices are accurate enough, he can detect deviations from $180°$ and determine the curvature of his universe.

We can picture all that in two dimensions, at least approximately, by visualizing a sphere, a saddle, and a plane. We cannot visualize a closed or open universe in 3D, but we might live in one anyway. So we've done the experiment. Scientists have looked at triangles formed by traveling light beams that are literally billions of light years across. For the best measurements of which current technology is capable, the angles add up to $180°$.

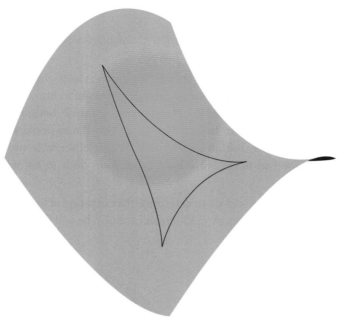

Figure 14.9 In an open universe the angles of a triangle add up to less than $180°$. You can picture this with the analogy of a saddle shape, where the angles are smaller because the lines seem to bend inwards toward each other.

In Section 14.3 we said that by measuring the ratio of actual density to critical density you can tell whether the universe is finite or infinite, but in practice the result is so close to 1 that we can't tell. Now we've seen an independent way to measure whether the universe is infinite or finite: by measuring the geometry of a triangle. Once again, the answer is too close to flatness for us to tell which side we're on. What we can say is that our universe is *remarkably close* to being perfectly flat.

Curvature and the Critical Density

The curvature of the universe is given by the term kc^2/a^2 in the first Friedmann equation. The parameter k is a constant, so the time dependence of that term comes entirely from the a^2 in the denominator.

The value of a can be set to 1 at any time, but the curvature kc^2/a^2 has a measurable value, so any rescaling of a has to be compensated by an appropriate rescaling of k.[10]

A more common way to express the curvature is in terms of density.

Active Reading Exercise: Critical Density

For a universe expanding with Hubble parameter H, the "critical density" ρ_c is the density that universe would have to have in order for it to be flat ($k = 0$).

---

10 It's common in cosmology to set k for a closed or open universe to ± 1, which in principle fixes the moment you call $a = 1$. In our universe, as far as we can measure, $k = 0$, so the choice of when $a = 1$ is still arbitrary.

1. Using the first Friedmann equation, write a formula for the critical density ρ_c as a function of H.

2. In a closed universe ($k > 0$), is the actual density ρ bigger than, equal to, or smaller than ρ_c? How do you know?

You can answer the first question by plugging $k = 0$ into the first Friedmann equation and solving for ρ, which gives

$$\rho_c = \frac{3c^2}{8\pi G}H^2. \tag{14.3}$$

You can answer the second question by solving for ρ without setting k to zero:

$$\rho = \frac{3c^2}{8\pi G}H^2 + \frac{3c^4}{8\pi G}\frac{k}{a^2} = \rho_c + \frac{3c^4}{8\pi G}\frac{k}{a^2}. \tag{14.4}$$

It's not hard to get those equations from the first Friedmann equation, but take a moment to notice what they're telling us.

- Section 14.3 introduced the idea of the critical density ρ_c, and connected it to the question of whether the universe is finite or infinite. Now we can see where the critical density comes from: Equation (14.3) gives us ρ_c for a flat universe. Since ρ_c is a function of H, and H changes over time (as governed by the first Friedmann equation), ρ_c changes too.

- From Equation (14.4) you can see that $k > 0$ implies that $\rho > \rho_c$, and vice versa. So we can connect density to curvature: instead of defining a closed universe by the property $k > 0$ (as we did previously), we can define a closed universe by the equivalent statement $\rho > \rho_c$.

The first Friedmann equation expresses the Hubble parameter as a sum of a density term and a curvature term. When the density term is much bigger than the curvature term, the universe is approximately flat ($\rho \approx \rho_c$). When the curvature term is comparable to or larger than the density term, the curvature is significant. (Perfect flatness means that the density term completely dominates, i.e. $k = 0$.)

Although k is a constant, a and ρ can both change, which can alter the balance in that equation. We will see below that the expansion of the universe can cause the curvature to become less flat – or, under certain unusual circumstances, more flat.

The Equation of State

Mathematically, the Friedmann equations are two coupled ordinary differential equations (ODEs) for three functions of time: a, ρ, and P. So you can't solve them without an additional equation. Physically, the reason you need additional information is because the way in which the universe expands depends on what's in it. So you have to describe some property of the energy that makes up the universe to complete the Friedmann equations.

Typically that extra information is in the form of an "equation of state" that relates the pressure and energy density of the material filling the universe. Two of the most important cases are a universe filled with matter, and a universe filled with electromagnetic radiation. As we said above, we know that our universe has very low curvature, so we'll look at both of these cases with $k = 0$ for simplicity. Later in the section we'll come back and talk about how curvature can change the evolution of the universe.

Matter: Cosmologists use the word "matter" to refer to any form of energy with zero pressure. You might reasonably object that gases do have pressure, and in the first few million years after the Big Bang that pressure was extremely high by human standards. Looking at Equation (14.2), though, the question is how large the pressure is compared to the energy density.[11] As long as the particles are moving much more slowly than c, the energy is dominated by the mass energy (mc^2) and the pressure is negligible compared to the energy density. So the equation of state for matter is $P = 0$. In Problem 14 you can plug this assumption into the Friedmann equations and show that

$$\text{Matter:} \quad P = 0 \quad \rightarrow \quad a \propto t^{2/3}, \rho \propto a^{-3}.$$

Electromagnetic Radiation: You can show from Maxwell's equations that the pressure and energy density of electromagnetic radiation are related by $P = \rho/3$. In Problem 15 you can plug that relation into the Friedmann equations and show that

$$\text{Radiation:} \quad P = \frac{\rho}{3} \quad \rightarrow \quad a \propto t^{1/2}, \rho \propto a^{-4}.$$

Based on those different relationships between ρ and a, you can quickly conclude that the total energies of matter and of radiation will behave differently in an expanding universe.

Active Reading Exercise: Matter, Radiation, and Energy Density

Suppose a region of space expands, and the scale factor $a(t)$ eventually reaches double its original value.

1. Remember that $a(t)$ is a unitless measure of the distance between objects. So when a doubles, what happens to the volume of this region?

2. Matter obeys the relation $\rho \propto a^{-3}$, where ρ is the energy density (measured for instance in J/m³). So when a doubles, the energy density of matter goes down by a factor of 8. Based on that and the volume change you just calculated, what happens to the total energy of the matter in this region?

3. Radiation obeys the relation $\rho \propto a^{-4}$. Based on that and the volume change you just calculated, what happens to the total radiation energy in this region?

In this scenario the volume increases by a factor of 8, the energy density of matter decreases by a factor of 8, and the energy density of radiation decreases by a factor of 16. This brings us to an important conclusion about the total energy in an expanding region:

---

11 You can easily check that force per unit area (pressure) has the same units as energy per unit volume (energy density), so you can meaningfully compare the two.

As the universe increases, the total amount of energy in matter stays the same, but the total amount of energy in radiation decreases.

That's what you would expect for matter: you had mc^2 before and you have mc^2 after, just spread over a wider region. But what happened to the radiation?

One way to understand this result is to consider the radiation as a gas of photons. As the universe expands the number of photons in any region stays constant, just as the total number of matter particles stays constant. But the expansion of space stretches the wavelength of each photon proportionally to a. Since the energy of a photon is inversely proportional to its wavelength, the total energy of the radiation decreases proportionally to a.

The photon argument correctly predicts the result $\rho \propto a^{-4}$, but it still begs an obvious question: where did all that energy go? The answer is analogous to the answer to: "Where does the kinetic energy of a ball go if I throw it up in the air?" As a ball rises it loses kinetic energy, but we know that it could regain that kinetic energy if we let it fall back down again. So we say it has stored that energy as potential energy. That potential energy has no direct effects when the ball is at high altitude; it is simply a measure of how much kinetic energy the ball *could* regain if we let it fall back.

When a radiation-dominated universe expands it loses energy, but if it were to contract again it would gain all that energy back. So we interpret the lost energy as a potential energy associated with the expansion of the universe. Like the ball's potential energy, the universe's potential energy doesn't directly affect anything. In particular, it is not included in the ρ term in the Friedmann equations.

Curvature and the Fate of the Universe

Above we gave you the equations of state for matter and energy, and then we discussed the resulting dependences of ρ on a and t. Those calculations (which we have left for you in Problems 14 and 15) assume $k = 0$, which corresponds to a Euclidean (flat) universe. Even a very small curvature term could have large effects on the future of our universe, however.

Active Reading Exercise: The Fate of the Universe

The Hubble parameter $H = \dot{a}/a$ describes how fast the universe is expanding. Equation (14.1) expresses the Hubble parameter squared as the difference between two terms: one term based on the density ρ (which is itself a function of time), and the other on the curvature k (a constant).

As the universe expands, ρ decreases (which causes the first term to decrease), while a increases (which causes the second term to *also* decrease).

1. In a matter-dominated universe, $\rho \propto a^{-3}$. If a doubles, the ρ-based term in Equation (14.1) will drop by what factor? The k-based term in Equation (14.1) will drop by what factor? Which one is decreasing faster?

2. Assuming $k > 0$ (still for a matter-dominated universe), what does your answer to Part 1 imply about what's going to eventually happen to H? What would that mean for the expansion of the universe?

In a matter-dominated universe, the density term in Equation (14.1) decreases like $1/a^3$, while the curvature term decreases like $1/a^2$. Both terms decrease, but the density term decreases faster. Where does that take us in the long term?

Closed universe: If $k > 0$, the curvature term cannot be bigger than the density term. (Otherwise H would be imaginary.) But as the universe expands, the curvature term eventually catches up and becomes equal to the density term. For one instant $H = 0$, and the expansion of the universe stops. Think of a ball rising upward with less than the escape velocity: the ball moves upward for a while because of inertia, but eventually gravity stops it and for one moment it has zero velocity. Then the Earth's gravity immediately starts pulling it back down. If $k > 0$ in a matter-dominated universe, the galaxies will slow down until they momentarily stop moving away from each other. Then gravity will pull them back toward each other. (Remember that Equation (14.1) is about $\dot{a}^2$; like an equation for $(1/2)mv^2$ for our ball, it looks the same whether the velocity is positive or negative.) The universe contracts again, leading to the "Big Crunch" scenario we described in Section 14.1.

Open or flat universe: If $k < 0$, both terms on the right of Equation (14.1) are positive but decreasing. H will asymptotically approach zero, so the universe will expand more and more slowly without ever stopping, like a ball rising at above the escape velocity. For $k = 0$ (a flat universe) the qualitative picture is the same as for an open universe, although the mathematical form of $a(t)$ at late times is different (Problem 16). Either way, the long-term future looks like the "Heat Death" we described in Section 14.1.

We've been talking about a matter-dominated universe, but all the same conclusions would be true for a radiation-dominated universe. In that case $\rho \propto 1/a^4$, so the curvature term would catch up with the density term even faster than it would for matter.

Table 14.1 summarizes the three possible geometries. The table includes almost everything we've said in this section, but make sure you follow where it all comes from:

Table 14.1 Three possible geometries

	Curvature	Density	Size	Fate (matter or radiation domination)
Closed universe	$k > 0$	$\rho > \rho_c$	finite	will eventually recollapse
Open universe	$k < 0$	$\rho < \rho_c$	infinite	will expand forever
Flat universe	$k = 0$	$\rho = \rho_c$	infinite	will expand forever

- The fact that $k > 0$ implies $\rho > \rho_c$, and vice versa, follows directly from the first Friedmann equation.

- The connection between those statements and the ultimate fate of the universe also follows from the Friedmann equations, but only when you also build in the equation of state for either matter or radiation domination. Section 14.5 will introduce another type of energy that might cause even a closed universe to expand forever.

- The fact that $k > 0$ implies a finite universe, and vice versa, is in the "you have to take our word until you study general relativity" bucket.

A Mixed Universe

We've talked about a universe entirely dominated by matter and a universe entirely dominated by radiation. A universe with some of each (and other types as well, as we shall see) is more complicated. But mathematically, you can treat the different types of energy separately: the energy density of matter scales like a^{-3} and the radiation scales like a^{-4}. The result is that the ratio of matter energy to radiation energy grows as the universe expands.

After all the antimatter annihilated with nearly all the matter roughly one second after the Big Bang, the universe had a billion times more energy in radiation than in matter. The universe expanded as $a \propto t^{1/2}$ for about the next 50,000 years, at which point matter overtook radiation as the dominant energy in the universe.

After that, matter dominated for billions of years. But most of the energy in today's universe is "dark energy," with a very different equation of state. We'll discuss dark energy and its implications for the future of the universe in Section 14.5.

14.4.3 Questions and Problems: The Friedmann Equations

Conceptual Questions and ConcepTests

1. In a universe with matter and radiation, and positive curvature ($k > 0$), which of the following decreases the fastest? (Choose one and explain.)

   A. The energy density of matter

   B. The energy density of radiation

   C. The curvature term in the first Friedmann equation (p. 672)

2. Equation (14.1) is for $\dot{a}^2$; the equation looks the same whether the universe is expanding ($\dot{a} > 0$) or collapsing ($\dot{a} < 0$). (It's like writing that the kinetic energy of a ball is $(1/2)mv^2$, regardless of which way the ball is going.) Make a qualitative argument that if $\dot{a}$ reaches zero (as it does in the $k > 0$ scenario), the universe will then recollapse rather than continuing to expand.

3. Suppose the universe 13 billion years ago was determined to be open ($\Omega < 1$, where $\Omega = \rho/\rho_c$). Which of the following would be true about the universe today? (Choose one and explain.)

   A. It might be open, flat, or closed, depending on how it evolved over that time.

   B. It would definitely still be open, but Ω might not have the same value as it did then.

   C. Ω would be the same today as it was then.

4. Can the curvature term in Equation (14.1) ever be bigger in magnitude than the density term? (Choose one and explain.)

   A. Yes, but only for a closed universe

   B. Yes, but only for an open universe

   C. Yes, regardless of the type of universe

   D. No, regardless of the type of universe

5. We argued that, for matter or radiation domination, an open universe will expand forever and a closed one will eventually recollapse. What will happen to a flat universe ($k = 0$)? (Choose one and explain how you can tell from the first Friedmann equation.)

   A. It will expand forever.

   B. It will eventually recollapse.

   C. It could do either, depending on other factors.

For Deeper Discussion

6. Give an example of a situation where a region of space could be homogeneous but not isotropic.

7. We argued from Equation (14.1) that a matter-dominated universe with $k > 0$ must someday stop expanding and momentarily stop. We then argued on physical grounds that what happens next is that

gravity pulls galaxies toward each other and the universe begins contracting. But how do we know that physical intuition is correct? Equation (14.1) would be perfectly satisfied if a remained constant from that moment on, with $H = 0$ and the two terms on the right equal and opposite. Aside from physical intuition, how can we know that the universe at that moment will contract and not remain static?

Problems

8. Show from the Friedmann equations that the condition $k > 0$ implies $\rho > \rho_c$, and vice versa, where $\rho_c = 3c^2 H^2/(8\pi G)$.

9. In this question you will consider the effect of k on Equation (14.1).

    (a) For any given energy density ρ, the equation implies a maximum allowed value of k. What is it?

    (b) Draw a graph of H^2 as a function of k according to Equation (14.1). Label the values of any points where your graph crosses either axis, and label the regions of your graph that correspond to an open universe, a closed universe, and a flat universe.

10. Consider a flat, matter-dominated universe ($a = a_0 t^{2/3}$). If you start observing at time t_0, how much time will elapse before distances double? Express your answer in terms of the "Hubble time" $1/H_0$, where H_0 is the Hubble parameter at time t_0.

11. With the curvature term negligible, the first Friedmann equation becomes the following equation, where r is the distance from us to any arbitrarily chosen galaxy and κ is a constant:

$$\frac{dr}{dt} = \kappa \sqrt{\rho}\, r.$$

    (a) In this problem you will replace ρ in that equation with $Mc^2/[(4/3)\pi r^3]$. What mass does the constant M represent in that formula? What assumptions does this formula make about the energy that makes up the universe?

    (b) Using the density formula from Part (a), solve the differential equation to find $r(t)$.

    (c) Show that your answer can be written in the form $r = (Qt + C)^{2/3}$ by finding the constant Q in terms of constants given in the problem. Leave C as an arbitrary constant for now.

    (d) Using $t = 0$ as the moment when all distances are zero, solve for the arbitrary constant C.

    (e) Our universe is currently 13.8 billion years old. Under the assumptions in this exercise, how long will it be from now before distant galaxies are twice as far away from us as they are now?

12. We defined the density parameter Ω as ρ (the actual density) divided by ρ_c (the critical density). In this problem you'll consider the behavior of Ω as the universe expands, assuming matter is the only form of energy in the universe (but not necessarily assuming that the density term dominates over the curvature term). For convenience, you can work in units in which $G = c = \hbar = 1$, which makes the Planck density equal to 1. As your initial condition (which you can call $t = 0$), take ρ to be one tenth of the Planck density. Since we are considering matter, the density at later times is $\rho = \rho_0/a^3$. You can also set $a = 1$ initially for simplicity, so now Equation (14.1) and your initial conditions have nothing in them but numbers and the undetermined parameter k.

    (a) Numerically solve Equation (14.1) three times with these initial conditions: once with $k = 0$, once with $k = 0.01$, and once with $k = -0.01$. For each solution, plot Ω as a function of time from $t = 0$ to $t = 1000$.

    (b) One of the three scenarios approaches a vertical asymptote. (The asymptote may not be obvious from your drawing, but it shouldn't contradict your drawing.) Which of the three scenarios does that, and what does this behavior imply about the long-term fate of the universe? *Hint*: You can work your way from Ω to ρ_c to H to $\dot{a}$, and figure out what's happening to the universe based on the behavior of $\dot{a}$.

    (c) What do the other two graphs do, and what do those graphs imply about the fate of the universe?

13. A gas of particles can be considered pressureless ($P \ll \rho$) as long as the average speed of the particles is much less than c. Using the equipartition theorem, estimate the temperature at which a gas of protons stops being effectively pressureless. (By the way, the universe dropped below this temperature about 10^{-5} s after the Big Bang.)

14. Consider a universe that is flat ($k = 0$) and matter-dominated ($P = 0$).

    (a) Algebraically eliminate ρ from the two Friedmann equations to get a single differential equation for $a(t)$.

    (b) Guess a solution of the form $a = a_0 t^p$ with some as-yet-unknown power p. Plug this guess into the differential equation from Part (a) to find p.

    (c) Plug your solution for a into the first Friedmann equation to express $\rho(t)$ in the form $\rho_0 t^n$, where you have specified the power n.

    (d) Use your solutions for $a(t)$ and $\rho(t)$ to express ρ as a function of a.

15. Do all of the steps of Problem 14 for a flat, radiation-dominated universe ($P = \rho/3$). *You do not need to have solved Problem 14 to do this one.*

16. Consider a universe in which the only form of energy is matter, with $\rho \propto 1/a^3$. You're going to solve for the late-time behavior of this universe for open and flat geometries. To keep your answers simple, use $t = 0$ to mean the time when $a = 0$.

    (a) First, assume $k < 0$ (an open universe). At late times, which term on the right side of Equation (14.1) will dominate? Setting the subdominant term to zero, solve for the resulting differential equation to find the late-time behavior of $a(t)$.

    (b) Now do the same for $k = 0$.

    (c) Based on your answers, describe the motion of a distant galaxy at late times in each of these two scenarios.

17. **Exploration: Size of the Observable Universe**

    To calculate the size of the observable universe, you can calculate how far away an object is whose earliest light is just reaching us now. An equivalent and simpler approach is to consider a light beam

emitted from our location at early times, and calculate how far that beam has reached. Suppose at some moment t a light beam is a distance x from us, moving directly away from us. In a time dt, it travels a distance $c\,dt$ further away because of its own motion. It also travels away from us because of the expansion of the universe.

    (a) How far does the light beam move in the time interval dt as a result of the expansion of the universe? Your answer should depend on H, the Hubble parameter.

    (b) Adding up the two motions (the light's proper motion and the expansion), write the total amount dx by which its distance from us increases in the time dt. Divide both sides by dt to get an expression for dx/dt.

    (c) Now let's assume a flat, matter-dominated universe: $a \propto t^{2/3}$. Calculate $H(t)$ and plug it into the equation you just wrote for dx/dt in order to get a differential equation for $x(t)$.

    (d) The generic solution to the linear, first-order differential equation $dx/dt + f(t)x = g(t)$ is given by

    $$x(t) = \frac{1}{I(t)} \int I(t)g(t)dt$$

    where $\quad I(t) = e^{\int f(t)dt}.$

    Solve your equation to find $x(t)$, the distance of the light beam from us as a function of time.

    (e) Your solution should have two terms. Which one will dominate at late times?

    (f) Neglecting the subdominant term, write a simple expression for $x(t)$ valid at late times. You can use this expression to answer the rest of the problem.

    (g) If the universe had been matter-dominated from the Big Bang ($t = 0$) until today ($t = 13.8$ billion years), what would the radius of the observable universe be?

    (h) The actual radius is about 46 billion light years. Why is this not the same as your answer?

18. **Exploration: Deriving the Second Friedmann Equation**

    Both Friedmann equations (Equations (14.1) and (14.2)) can be derived directly from Einstein's equation. But you can also derive the second Friedman equation from the first using thermodynamics.

(a) Multiply both sides of Equation (14.1) by r^2 and then take the derivative with respect to time of both sides. The resulting equation should have $\dot{a}$, $\ddot{a}$, and $\dot{\rho}$.

To get to Equation (14.2) you're going to need to get rid of $\dot{a}$ and $\dot{\rho}$.

(a) Let E be the total energy (not energy density) in a spherical region of radius a. (You can always initially define $a = R$ in the units in which you are measuring R, and by definition a and R will then grow at the same rate.) Write ρ as a function of E (this is trivial) and use that to write $\dot{\rho}$ as a function of $\dot{E}$ (this requires a short

calculation). Plug this in, so your equation for $\ddot{a}$ now contains $\dot{E}$ instead of $\dot{\rho}$.

(b) Thermodynamics tells us that when a fluid with pressure P expands its volume by an amount dV, the fluid loses an amount of energy equal to $dE = -P\,dV$. Use that fact to rewrite your equation for $\ddot{a}$ in terms of $\dot{V}$ instead of $\dot{E}$.

(c) Write $\dot{V}$ (the rate of change of the sphere's volume) in terms of $\dot{a}$ (the rate of change of its radius). Plug the result in to eliminate $\dot{V}$ from your equation for $\ddot{a}$.

(d) Finally, write your equation in terms of ρ but not E, leaving you with an equation for $\ddot{a}$ that only depends on ρ, P, and constants. Do whatever algebra you need to in order to get Equation (14.2).

14.5   Dark Matter and Dark Energy

What's the universe made of? The things we see around us are made of protons and neutrons (which are themselves made of quarks) and electrons. There are also photons (which are technically the only particles we can see). In our daily experience, that's pretty much the full list. We can also add "neutrinos" to the list of abundant particles, although they pass through normal matter with so little interaction that it takes specialized equipment to detect them.

It turns out that those known particles, and every other particle we have ever produced or detected, add up to about 5% of the energy in the universe. The rest is in two forms that we've never directly observed: "dark matter" and "dark energy." In this section we'll discuss how we know they exist, and what we do and don't know about their properties.

14.5.1   Discovery Exercise: Rotation Curves

The Milky Way galaxy is a wide disk of stars orbiting a dense, central bulge. For simplicity we'll assume here that essentially all of the mass M of the galaxy is in the bulge and that the stars outside the bulge are in circular orbits.[12]

With those assumptions, use Newton's second law and Newton's law of gravity to calculate the orbital speed of a star outside the bulge, as a function of its distance from the center of the galaxy. Make a sketch of your calculated function $v(r)$.

You'll see in this section that the observed function $v(r)$ for essentially all galaxies is very different from what you just calculated. That discrepancy between prediction and observation led to the discovery of "dark matter," one of the two largest contributors to the overall energy of the observable universe.

---

12 The second assumption is reasonable. The first one is not – less than half the observed mass in the galaxy is in the central bulge – but it will work to make the qualitative point we want to illustrate here.

14.5.2 Explanation: Dark Matter and Dark Energy

The fraction of each type of energy making up the observable universe is called the universe's "energy budget." Our current best models put it as follows:

- About 68% of the energy is "dark energy," a uniform field filling all space within the observable universe.
- About 27% is "dark matter," an unknown type of particle clustered in and around galaxies.
- The remaining 5% consists of every type of energy we have ever directly observed. That 5% is mostly protons and neutrons.

We know dark matter and dark energy exist because we observe their gravitational effects, but as of this writing we have not yet directly detected either one.

Dark Matter

Dark matter was discovered by watching how stars orbit in galaxies and how galaxies orbit in galaxy clusters.

A large spiral galaxy like our own has a high-density bulge at the center, with a thin outer ring of stars orbiting around that central bulge. To predict the behavior of those outer stars, we note that the centripetal force that holds them in orbit is their gravitational attraction to the central bulge:

$$\frac{m_{star}v^2}{r} = \frac{GM_{galaxy}m_{star}}{r^2}.$$

We thus conclude that, if all the mass of the galaxy were in the central bulge, the orbital speed of any given star would be proportional to $1/\sqrt{r}$, where r is the distance of that star from the center of the galaxy.

The relation $v \propto 1/\sqrt{r}$ works very well for planetary orbits in our solar system because virtually all the mass of our solar system is concentrated in the Sun. In our galaxy, by contrast, less than half of the mass is in the central bulge, so the calculation of $v(r)$ is more complicated. But we can measure the observed density distribution in a galaxy, do the somewhat messier math (still based on setting centripetal force equal to gravitational force), and arrive at a prediction for the rotational behavior of a galaxy. Those calculations predict that speed should fall off with radius outside the central bulge, approaching $v \propto 1/\sqrt{r}$ at large radii.

That prediction turns out to be wrong.

In the 1970s Vera Rubin measured the orbital speeds of stars in many galaxies, and constructed their "rotation curves" (plots of speed vs. radius, such as Figure 14.10). She expected the speeds to fall off with radius as we described above. Instead she found rotation curves that were flat (velocity independent of radius), or in many cases *rose* out to very large radii.

Rubin's conclusion was that every galaxy has 5–10 times more mass than we can see, and that this mass is less concentrated in the central bulge than the things that we do see. This postulated matter is invisible: it doesn't emit, absorb, or reflect light in the visible spectrum, radio spectrum, or any other spectrum that we measure when we monitor the skies.

In fact, the same conclusion had been reached in 1933 by Fritz Zwicky when he measured the rotation curve for galaxies orbiting the central mass of the Coma Cluster, a huge collection of

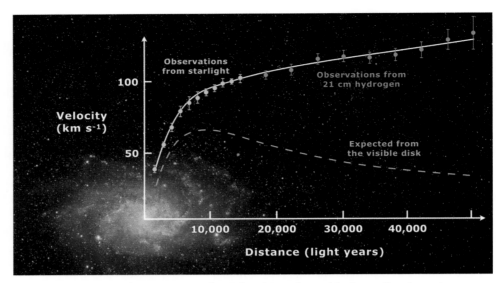

Figure 14.10 Measured rotation curve for Galaxy M33, along with the predicted rotation curve based on the visible matter. *Source: De Leo © 2018.*

galaxies. Zwicky coined the term "dark matter," but it wasn't until Rubin's work in the 1970s that his conclusions could be independently verified. After Rubin's work, the astronomy community came to accept the existence of dark matter.

Given the measured rotation curves, we can infer how the density of dark matter is distributed in and around galaxies. The conclusion is that every galaxy is filled with and surrounded by a dark matter "halo." While the visible matter often lies in a flat disk (as in our own galaxy), the dark matter haloes are nearly spherical (Figure 14.11). The density of dark matter falls off with radius more slowly than the density of visible matter, and the haloes extend several times farther from the center than the visible matter does. (We can measure dark matter beyond the disk by measuring the orbital velocities of objects orbiting the galaxies, such as the dwarf galaxies that orbit the Milky Way.)

The existence of dark matter has now been confirmed by numerous other measurements. The most direct is "gravitational lensing," the bending of light when it passes near a strong source of gravity. Measurements of gravitational lensing by galaxy clusters show an amount of mass consistent with the amount of dark matter measured by rotation curves.

Dark matter has also been indirectly confirmed by simulations of structure formation. If we assume the amount of dark matter indicated by other measurements, our computer simulations reproduce patterns of galaxies consistent with what we see. Without dark matter, or with significantly different amounts of dark matter, the patterns don't match. Those same simulations tell us that dark matter is "cold," meaning the particles are moving at non-relativistic speeds. That implies that dark matter obeys the same $\rho \propto 1/a^3$ rule as ordinary matter.

Dark matter is *not* made up of the particles in the "standard model" of particle theory. We believe that dark matter is made of a type of particle that interacts very weakly with other matter. As of this writing, no non-gravitational interactions with dark matter have yet been observed, but there are numerous searches trying to directly detect dark matter particles.

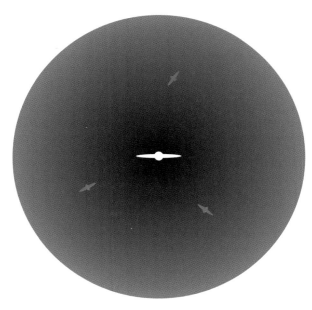

Figure 14.11 The Milky Way galaxy (white) is surrounded by a dark matter spherical halo more than five times wider than the disk of the visible matter. We can measure it far beyond the disk by measuring the rotation of dwarf galaxies (red).

Dark Energy

Dark energy was discovered by measuring how the expansion of the universe changes with time.

In an expanding universe filled with matter, the expansion gradually slows over time, because of the gravitational attraction of galaxies. Because we see distant galaxies as they were in the distant past, we should be able to measure the fact that their universe was expanding faster than our universe is today.

In the 1990s two different groups, one headed by Saul Perlmutter and the other by Brian Schmidt and Adam Riess, set out to make those measurements. In 1998 data were released by both groups. Both groups reached the same conclusion: galaxies are actually speeding up as they move away from us!

Imagine throwing a ball upward. It goes up for a while and then falls back down. If you could throw the ball hard enough (with velocity greater than the escape velocity) it would escape into space and never fall back down. But under no circumstances should the ball continue to accelerate upward after it leaves your hand. Yet that's what the galaxies appear to be doing.

These results suggest that something is exerting a repulsive gravitational force stronger than the combined attractive force of all the galaxies on each other. Not knowing what that something is, Michael Turner gave it the name "dark energy," by analogy with the also-unidentified "dark matter." Both dark matter and dark energy are "dark" in the sense of being invisible, and in the sense of being poorly understood. But the analogy ends there. Dark energy causes repulsive gravity, while dark matter is responsible for most of the attractive gravity in the observable universe.

While dark energy is mysterious and unexpected, it is consistent with general relativity. We saw in Section 14.4 that you can start with a description of the stuff that fills the universe (in the form of a $\rho(a)$ function), and derive from the first Friedmann equation a prediction of how the universe will grow (an $a(t)$ function). For matter, $\rho \propto a^{-3}$ leads to $a \propto t^{2/3}$. For radiation, $\rho \propto a^{-4}$ predicts $a \propto t^{1/2}$. Dark energy obeys the same Friedmann equations, but with a different density relationship that leads to a different growth pattern.

Active Reading Exercise: Dark Energy

Consider a universe dominated by a type of energy density ρ that is constant as the universe expands. (That's constant energy *density*, not total energy.) For a flat universe, the first Friedmann equation then says:

$$\left(\frac{\dot{a}}{a}\right)^2 = \frac{8\pi G}{3c^2}\rho \quad \text{is constant.}$$

You can solve that as is, but the math looks simpler if we use H for $\dot{a}/a$ as usual:

$$\left(\frac{\dot{a}}{a}\right)^2 = H^2 \quad \text{is still constant.}$$

1. Solve this equation for $a(t)$. Your solution will have the constant H in it.

2. Looking at your solution to Part 1, would you see a distant galaxy's motion away from you speeding up or slowing down? How can you tell?

You can harmlessly drop the "squared" from both sides, since nothing in that equation is negative. You can solve the resulting equation by separating variables, or possibly by just looking at it, to arrive at $a = a_0 e^{Ht}$. In short, in a universe dominated by a constant energy density, distances grow exponentially with time.

To find acceleration, you calculate $\ddot{a}$. For matter or radiation domination, $\ddot{a}$ is negative, meaning galaxies slow down as they move away from us. For a universe with constant energy density, a grows exponentially and $\ddot{a} > 0$, so galaxies speed up as they move away from each other.

We've shown that a constant energy density leads to accelerated expansion. You can get accelerated expansion with a changing energy density, as long as it changes slowly enough (Problem 8), but measurements of our universe's acceleration indicate that dark energy does have almost perfectly constant energy density.

We describe the effect of dark energy as "accelerated expansion," but be careful how you interpret that phrase. We measure the expansion rate of the universe by the Hubble parameter, and H is *not* increasing. A billion years from now, galaxies that are, say, five billion light-years away from us will be receding *more slowly* than galaxies that are five billion light-years from us now are receding. But the fact that $\ddot{a} > 0$ means that in a billion years, each particular galaxy will be receding from us faster than it is today.

All of this follows from the remarkable fact that dark energy maintains a constant energy density as the universe expands. What kind of energy would behave that way? Doesn't this violate conservation of energy?

We'll address the second question first. We said in Section 14.4 that the total energy of radiation in any region of the universe decreases as the universe expands, and that the lost energy is stored as a kind of gravitational potential energy. Since the total amount of dark energy in any region *increases* as the universe expands, that means the universe is storing up *negative* potential energy. Recall, though, that this potential energy associated with expansion is not part of the energy density ρ in Friedmann's equations, so it has no measurable effect.

We're still left with the question of what type of energy would behave that way. Cosmologists mostly talk about two possibilities, either one of which might account for dark energy:

- A "scalar field" (like the Higgs field in particle physics) can under certain circumstances have a constant ρ. We'll discuss this possibility further in Section 14.7.

- Another possibility is the cosmological constant that Einstein inserted into the equations of general relativity to try to make the universe stable.

Einstein thought of the cosmological constant as a modification of the law that determines how gravity responds to matter, in the form of a constant term on the left side of the first Friedmann equation.[13] Modern cosmologists think of the cosmological constant as a new form of energy rather than a change to the law of gravity. The effect is the same, but it's generally written now as a constant on the *right* side of the first Friedmann equation.

The difference between a scalar field with constant energy density and a cosmological constant is that a scalar field could eventually decay and lose that energy. If dark energy is truly a cosmological constant, that means there is an energy density intrinsic to space itself. This "vacuum energy" will presumably remain unchanged forever. So the distinction between dark-energy-as-scalar-field and dark-energy-as-cosmological-constant might make no discernible difference now, but might be crucial in determining the ultimate fate of the universe. If dark energy never decays, then the universe will keep expanding forever regardless of whether it is open, closed, or flat (Question 3).

Whatever dark energy is, we can see from the measured rate of accelerated expansion that it makes up about 68% of the energy in the observable universe. Unlike matter (of the visible or dark varieties), dark energy seems to be uniformly spread out throughout space.

The increasing rate of universal expansion is pretty conclusive evidence for dark energy, i.e. some form of constant energy density that makes up most of the energy in the observable universe. But we also have an independent confirmation of the existence of dark energy: the universe's curvature.

Recall that the Friedmann equation provides a connection between the universe's density and its curvature. If the density of the universe is below critical density ($\rho < \rho_c$), then the *curvature* of the universe is negative ($k < 0$, open universe). Based on the low ρ-values measured in the early 1990s, the curvature should be measurably negative. But it isn't; as we mentioned in Section 14.4, all attempts to measure the curvature of the universe have found none.

Dark energy explains that result perfectly. When you add the amount of dark energy inferred from the expansion to the matter (mostly dark matter), you get $\rho = \rho_c$, predicting a flat ($k = 0$) universe to as high an accuracy as we can measure. A lot of the reason why cosmologists have confidence in our measurements of dark matter and dark energy is that we now have three independent, consistent measurements of two quantities: ρ_{DE} and ρ_{matter}.

1. By observing galaxies and clusters, using rotation curves and gravitational lensing, we can add up the average density of matter in the observable universe, including both visible and dark matter. The result is that ρ_{matter} is about $1/3$ of the critical density.

---

13 Einstein added the term to one side of the tensor equation from which the Friedmann equations are derived, but the result is that this constant shows up in the Friedmann equations.

2. By measuring the changes in the expansion rate over time, we get a measure of the difference between the repulsive effect of dark energy and the attractive effect of matter. The result shows that $\rho_{DE} - \rho_{matter}$ is about 1/3 of the critical density.

3. By measuring the curvature of our observable universe, we can determine the total energy density. The result is that $\rho_{DE} + \rho_{matter}$ equals the critical density.

Any two of these measurements are enough to conclude that the density of matter is about 1/3 of the critical density and the density of dark energy is about 2/3 of critical density. The fact that these three all agree adds greatly to our confidence in our cosmological models.

14.5.3 Questions and Problems: Dark Matter and Dark Energy

Conceptual Questions and ConcepTests

1. In your own words, briefly describe how we know that dark matter and dark energy exist. (The two answers are different.)

2. Suppose Galaxy A is currently 8 billion light-years away from us, and Galaxy B is going to be 8 billion light-years away from us in 3 billion years. Which of the following are true about these galaxies in a universe dominated by dark energy (such as ours)? (Choose all that apply.)

   A. In 3 billion years, Galaxy A will be receding from us faster than it is now.

   B. In 3 billion years, Galaxy A will be receding from us more slowly than it is now.

   C. In 3 billion years, Galaxy B will be receding from us faster than Galaxy A is currently receding from us.

   D. In 3 billion years, Galaxy B will be receding from us more slowly than Galaxy A is currently receding from us.

3. Explain why a universe dominated by a cosmological constant will expand forever, regardless of the sign of k. Hint: Review our arguments in Section 14.4 for why the sign of k determines the fate of the universe for matter and for radiation domination.

4. Shortly after antimatter finished annihilating, the energy in radiation was much greater than the energy in matter. At that time was the density of dark energy ... (Choose one and explain.)

   A. much smaller than the density of matter?

   B. comparable to the density of matter?

   C. much larger than the density of matter?

5. We said that we can verify the existence of dark matter by observing the gravitational lensing it produces. Dark energy is twice as dense as dark matter, but we don't observe it causing gravitational lensing. Why not?

For Deeper Discussion

6. One of the strange features of dark energy is that it has *negative* pressure. You might reasonably think that positive pressure would accelerate galaxies away from each other and negative pressure would suck them in, thus increasing the attraction due to gravity. But from the Friedmann equations we know that what's needed for repulsive gravity is a kind of energy that *increases* as the universe expands. Explain why a fluid with negative pressure would have an increasing overall energy in any region of an expanding universe.

Problems

7. What would have to be true of the equation of state of a form of energy ($P(\rho)$) in order for it to cause accelerated expansion ($\ddot{a} > 0$)?

8. In Section 14.4 and this section, we showed that a matter-dominated universe ($\rho \propto a^{-3}$) expands as $a \propto t^{2/3}$, a radiation-dominated universe ($\rho \propto a^{-4}$) expands as $a \propto t^{1/2}$, and a dark-energy-dominated universe with constant energy density expands as $a = a_0 e^{Ht}$.

(a) Show that matter- and radiation-dominated universes have negative acceleration, while a dark-energy-dominated universe has positive acceleration.

(b) Assume $\rho \propto a^x$ for some constant x. Starting from the first Friedmann equation, determine what would have to be true of x in order for the universe to have accelerated expansion. Assume a flat universe for simplicity.

9. Consider a universe that is flat ($k = 0$) and dominated by dark energy (constant ρ, $a = a_0 e^{Ht}$).

(a) Use both Friedmann equations (p. 672) to find the equation of state relating P and ρ for dark energy. Your answer should have nothing but P and ρ in it.

(b) Recalling the definition of pressure as force per unit area exerted by a fluid, what does your result tell you about the force exerted by a region of dark energy on its surroundings?

10. In our current universe, the expansion is accelerating ($\ddot{a}$ is positive), but the Hubble parameter H is decreasing. Your answers to both parts of this problem will come from the definition $H = \dot{a}/a$.

(a) Show that in a decelerating universe (negative $\ddot{a}$), H must always be decreasing.

(b) In an accelerating universe, H will decrease for small $\ddot{a}$ but increase for sufficiently large $\ddot{a}$.

Find the threshold value of $\ddot{a}$, as a function of a and $\dot{a}$.

11. Einstein's cosmological constant effectively added a constant term to the left side of the first Friedmann equation:

$$\left(\frac{\dot{a}}{a}\right)^2 - \Lambda = \frac{8\pi G}{3c^2}\rho - \frac{k}{a^2}.$$

Cosmologists often interpret this new term as due to vacuum energy and take ρ in that equation to mean the density of everything else in the universe. What is the energy density of the vacuum in this interpretation? *Hint*: It depends on Λ, but it can't be Λ because that wouldn't have the right units.

12. Solve the first Friedmann equation to find $a(t)$ for our universe: a flat universe with $1/3$ of its energy currently in the form of matter ($\rho \propto a^{-3}$) and $2/3$ of its energy in the form of dark energy (constant ρ). The current energy density is 10^{-9} J/m^3. You can take $t = 0$ and $a = 1$ initially.

(a) Make a plot of $a(t)$ from today until the time when distances are 10 times what they are today.

(b) How long will it take from today before distances double?

(c) How long will it take from today before distances multiply by 10?

(d) What will the ratio of dark matter to dark energy be at that time?

14.6 Problems with the Big Bang Model

A tremendous amount of evidence supports the Big Bang model as an accurate description of the evolution of the universe from a fraction of a second after the Big Bang until today, almost 14 billion years later. But there are also problems. This section describes some of those problems, and Section 14.7 explains how they can be resolved with an addition to the Big Bang model called "inflation."

14.6.1 Discovery Exercise: Problems with the Big Bang Model

When the universe dropped below the Planck density, as far as we know it could have had any value of curvature. But we can constrain that value based on our current observations.

For simplicity, assume throughout this exercise that the universe has been matter-dominated from the beginning through today.

1. If the universe today is 13.8 billion years old, by what factor has the scale factor increased since it was one second old?

    See Check Yourself #27 at www.cambridge.org/felder-modernphysics/checkyourself

2. We observe that today the density term in the first Friedmann equation is at least 100 times larger than the curvature term. How much larger must the density term have been at $t = 1$ s?

14.6.2   Explanation: Problems with the Big Bang Model

Section 14.2 discussed the initial reluctance of many physicists to accept the Big Bang theory, and the evidence that ultimately won the scientific community over. That evidence is as compelling today as it ever was. It's difficult to imagine any discovery that would seriously suggest, at this point, that the Big Bang model isn't an accurate description of the last 14 billion years or so of our universe's history.

Nonetheless, the theory in its original form did present serious problems. Resolving those problems has led to new ideas about the very early universe.

The Flatness Problem

We observed in Section 14.4 that the universe is, to the best of our ability to measure it, flat. Mathematically that means the ρ-based term in the first Friedmann equation dominates the k-based term, to the point that the latter is insignificant:

$$\frac{8\pi G}{3c^2}\rho \gg \frac{kc^2}{a^2}.$$

Why is that a problem? For most of the universe's history the energy density has been dominated by matter, with $\rho \propto 1/a^3$. Every time a doubles, the k-based term drops by a factor of four, but the ρ-based term drops by a factor of eight. So no matter what the initial ratios are, if you expand the universe enough, the curvature term should eventually come to dominate.

Well, we've expanded quite a bit. Since matter came to dominate the universe at about 50,000 years after the Big Bang, the scale factor has grown by a factor of nearly 3000. Before that, during radiation domination, ρ shrank as $1/a^4$, so density was shrinking relative to curvature even faster.

You can calculate (Problem 8) that in order for the curvature term to be as insignificant as it is today, it had to start out, at the time of Planck density, no more than 10^{-29} times as large as the density term. We can state that another way, remembering that curvature is related to density: the universe's initial density had to equal the critical density to 29 decimal places.

Nothing in our theories says that ρ *shouldn't* have been within 29 decimal places of ρ_c, but nothing says it should either. Imagine you lose a diamond ring and the next day your "friend" shows up with a new diamond that weighs the same as your diamond did, to within one part in 10^{29}. You would probably suspect that this was not a coincidence.

The coincidence looks even more suspicious when you *need* it to come out perfect. If ρ had started out just a tiny bit higher, the universe would have recollapsed so quickly that no structures would ever have formed. A little bit lower, and the universe would have expanded outward so fast that no structures could have formed. Only a density that was almost exactly equal to ρ_c would produce an interesting universe, and for some reason, that's exactly what the initial density was.

The fact that the initial curvature of the universe had to be so precisely fine-tuned in order to produce the universe we see today is called the "flatness problem."

The Homogeneity and Horizon Problems

The world around us is not homogeneous. We're sitting on a dense rock, while hundreds of miles above us is a near-vacuum. We're in a galaxy full of stars, and a few million light years over is another galaxy, while the space in between is mostly empty.

But on the scale of the observable universe those are all local variations. If we zoom out and look at regions more than a few hundred million light years across, we find that they generally look pretty much the same. Large chunks of the observable universe in any direction in the sky look more or less like the large chunks in other directions: same average density, same distribution of galaxy types, and so on.

But remember that gravitational clumping tends to reduce uniformity over time, as high-density regions pull matter away from low-density regions. So for the current-day universe to be generally homogeneous on large scales, the early universe must have been *extremely* homogeneous. That fact is confirmed by measurements of the cosmic microwave background: the brightest directions exceed the dimmest ones by about one part in 100,000.

Going back even further, at the time of Planck density the region that is now our observable universe must have been almost perfectly homogeneous. How did it get to be that way?

You might attempt to explain the near-perfect homogeneity through some process that regulated the density of the early universe, keeping it all uniform. But that doesn't work, and here's why.

The edge of our observable universe is called our "horizon." The horizon is the limit of what we can see, and the farthest region that has had time to affect us in any way. In the standard Big Bang model, no physical process of any kind taking place since the time of Planck density could have created a correlation between conditions here and conditions more than about 45 billion light years away from here.

When we look at a region of space 45 billion light years away in one direction, and another region of space the same distance away in the opposite direction, we are seeing two regions that are outside each other's horizons. So it seems impossible for any early-universe process to have causally set those two regions to have such similar densities; in the Big Bang model as we've described it so far, this uniformity *has* to be a coincidence.

To confound the puzzle further, at earlier times our observable universe represented less of the universe than it does today. We don't just mean that the universe is expanding. We mean that the region that we can observe is growing faster than that expansion, so over time more galaxies come within our horizon (Figure 14.12).

At the time of recombination, our current observable universe contained thousands of regions that had not yet had time to causally affect each other. Nonetheless, we observe that they all had the same density as each other to within one part in 100,000. How did that come about?

The unexplained high level of uniformity of the early universe is called the "homogeneity problem." The related fact that any kind of correlations existed between causally disconnected regions in the early universe is called the "horizon problem."

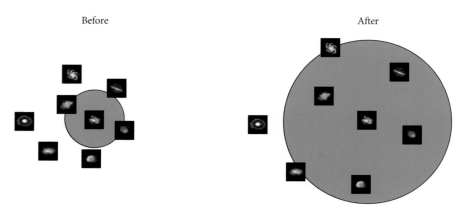

Before   After

Figure 14.12 In a matter-dominated universe, the observable universe (green) grows faster than the expansion of the universe, so more of the universe comes into view as time passes.

The Monopole Problem

The universe around us is made almost entirely of protons, neutrons, electrons, photons, neutrinos, dark matter, and dark energy. Particle accelerators produce many particles that are not on that list: kaons, tauons, Higgs bosons, the whole standard model. Those other particles are unstable, however, and quickly decay into the relatively small number of stable forms of energy that we just listed.

However, a certain class of particle theories called "grand unified theories" (GUTs) predict the existence of an as-yet-unobserved stable particle called a "magnetic monopole."

All magnetic particles we know of are magnetic *dipoles*, meaning they have one north side and one south side, as illustrated in Figure 14.13. A magnetic monopole would have magnetic field lines going out or in like the electric field lines of a charged particle. It would also, according to GUT theories, have a mass of about 10^{16} GeV: 10 orders of magnitude greater than the highest energies we can currently produce in an accelerator.

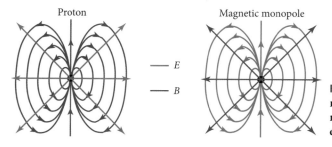

Proton   Magnetic monopole

—— E

—— B

Figure 14.13 A proton is an electric monopole and a magnetic dipole. A magnetic monopole would be an electric dipole.

But if monopoles are too massive to be created in our current labs, they're still well within the energy levels of the early universe. In 1979 a particle physicist named Alan Guth, who was studying GUT theories, calculated how many monopoles should have been created in the dense, hot conditions shortly after the Big Bang. Guth's calculations showed that the number density of monopoles should have ended up about the same as the number density of protons. Because monopoles are stable, that one-to-one ratio would be expected to still hold today.

Since one monopole has more mass than 10^{15} protons, they would completely dominate the current mass and energy of the universe. But there is no evidence that any monopoles exist in the observable universe. If they were produced abundantly in the early universe, where did they go?

The mismatch between the large number of monopoles expected from the early universe and the apparent complete lack of them today is called the "monopole problem." There are high-energy theories that predict the existence of other stable particles that we would expect to be around from the early universe, and they generically go under the label of "relic particles." But the existence of monopoles is the most robust of these theoretical predictions.

Stepping Back: How Bad Were the Problems with the Big Bang Model?

The problems we've described in this section don't suggest that the Big Bang theory is wrong, but they do indicate that the original model is incomplete. That model requires extremely fine-tuned initial conditions, and even then seems to make a wrong prediction about the production of monopoles.

To be fair, you could dismiss the fine-tuning problems (flatness, homogeneity, and horizon) by saying "We don't know physics above the Planck density, so whatever happened before the time of Planck density must have resulted in a nearly homogeneous region with very little (or possibly zero) curvature."

The monopole problem is harder because the monopoles should have been produced *after* the universe dropped below the Planck density. But there is no experimental evidence for GUT theories, so you could conclude that the absence of monopoles in the universe today suggests that those theories are wrong.

There are good theoretical reasons to believe in GUT theories, however, and it was unsatisfying to assume that the unknown physics above the Planck density would somehow magically solve all the fine-tuning problems of the Big Bang model. In 1981 Guth published a paper describing a new theory of the very early universe that solved the problems we've described here, and set the initial conditions required for the subsequent evolution described by the Big Bang model. We'll describe that theory in Section 14.7.

14.6.3 Questions and Problems: Problems with the Big Bang Model

Conceptual Questions and ConcepTests

1. Define each of the following in your own words:
   (a) Flatness problem
   (b) Homogeneity problem
   (c) Horizon problem
   (d) Monopole problem

2. Guth's calculations showed that the lack of magnetic monopoles in the world around us was a mystery that needed to be explained. But the lack of tauons, Higgs bosons, and many other parts of the standard model was *not* considered a mystery that needed to be solved. Why not? (Choose one.)

A. Grand unified theories (GUTs) predict that those particles cannot exist.

B. Those particles, like neutrinos, interact so little with matter that they are almost impossible to detect.

C. Those particles decay almost immediately after being created.

D. Those particles can exist only in conditions of extreme temperature and pressure.

3. We said that at the time of recombination the region that is now our observable universe consisted of thousands of regions that hadn't had time to

influence each other, and yet they were all nearly identical. How do we know that those regions were nearly identical over 13 billion years ago?

4. The air in a room is typically extremely homogeneous, with nearly the same density everywhere. Yet we don't consider that a problem, or a massive coincidence. So why is it surprising that the fluid making up the universe at the time of recombination was very homogeneous?

For Deeper Discussion

5. If magnetic monopoles were created in the early universe, they should still be around today. They should be stable, which means they would never decay. Why?

   (a) Chapter 13 discussed the rule that any decay that can happen, will sometimes happen. So a perfectly stable particle is one that literally cannot possibly decay into anything else. In general, according to quantum field theory, what makes a particular decay reaction illegal?

   (b) What particular property would a magnetic monopole have that would clearly disqualify it from decaying into any currently known particle?

6. If you picture a finite universe such as Edwin's 2D surface-of-a-sphere, you can see that as the universe expands the surface gets flatter. So why is it surprising that we observe a very flat universe today?

7. We said the observable universe grows faster than the expansion of the universe. Would it be possible for it to grow more slowly than the expansion of the universe? Why or why not?

Problems

8. **Does Dark Energy Resolve the Flatness Problem?**

   (If you did the Discovery Exercise for this section, then you've already done the first few parts of this problem, but now you can take that calculation farther.)

   (a) If we assume the universe has always been matter-dominated and is currently 13.8 billion years old, by what factor has the scale factor increased since the universe was one second old?

   (b) We observe that today the density term in the first Friedmann equation is at least 100 times larger than the curvature term. Based on your answer to Part (a), how much larger must the density term have been at $t = 1$ s?

   You should have fiund a very big number. To end up with the universe we see around us today, we would have had to start with a very improbably flat universe. But that calculation was based on a universe made up entirely of matter. Now let's consider a universe that, although still simplified, is a bit closer to reality: a period dominated by matter, followed by a period dominated by dark energy.

   (c) In a universe dominated by dark energy, when a doubles, by what factor does the ratio of density to curvature grow or shrink?

   (d) Given that there is twice as much dark energy as matter in the observable universe, how much has the scale factor increased since dark energy first overtook matter as the dominant form of energy?

   (e) Use your answer to Part (d) to calculate how long ago dark energy overtook matter. Use the value $H = 2.33 \times 10^{-18}$ s^{-1}.

   (f) In Part (b) you calculated the ratio of the Friedmann equation's density term to its curvature term one second after the Big Bang, assuming a matter-dominated universe. Re-do that calculation – still based on the at-least-100-to-1 ratio we see today – but this time assuming a universe that completely switched from matter to dark energy at the moment you calculated in Part (e).

   (g) Choose one of the following, and explain briefly how your answer comes from your calculations above.

   A. The introduction of dark energy makes the flatness problem worse.

   B. The introduction of dark energy doesn't change the flatness problem much.

   C. The introduction of dark energy reduces the flatness problem, but still leaves a sizeable problem that must be resolved.

   D. The introduction of dark energy resolves the flatness problem.

14.7 Inflation and the Very Early Universe

The theory of inflation was first published by Alan Guth in 1981, and is now our mainstream theory for the very early universe, a small fraction of a second after it dropped below Planck density. With a relatively small set of assumptions, this theory explains how the universe came to have the initial conditions for the subsequent evolution described by the Big Bang model, and it explains why we don't observe magnetic monopoles or other heavy, stable relic particles today.

14.7.1 Discovery Exercise: Inflation and the Very Early Universe

We have discussed the remarkable fact that dark energy maintains a constant energy density (measured, for instance, in J/m^3) as the universe expands. If you worked through Problem 8 in Section 14.6, you have seen that the prevalence of dark energy is causing the curvature of the universe to decrease, but not enough to explain the extremely low curvature we measure today.

But what if a type of energy with that same property had been around just after the universe dropped below the Planck density?

1. As you may recall, the first Friedmann equation with a constant energy density and negligible curvature leads to the growth equation $a = a_0 e^{Ht}$, where $H = \sqrt{8\pi G\rho/(3c^2)}$. If the energy density were the Planck density (4.6×10^{113} J/m^3) and the universe expanded in this way for 10^{-35} s, by what factor would the scale factor increase?

   See Check Yourself #28 at www.cambridge.org/felder-modernphysics/checkyourself

2. Suppose the ρ term in Equation (14.1) was 10 times bigger than the curvature term at the start of the 10^{-35} s expansion you just calculated. During that period, ρ stayed constant while a increased. By what factor would the ρ term end up bigger at the end of that expansion?

14.7.2 Explanation: Inflation and the Very Early Universe

The theory of inflation is not a replacement for the Big Bang theory. Inflation takes place *within* the Big Bang theory, filling in one of the first steps after the universe fell below the Planck density. This new step created conditions that resolve the flatness, homogeneity, horizon, and monopole problems discussed in Section 14.6.

The theory of inflation asserts that there was a period of time in the early universe when the dominant form of energy was something called a "scalar field." We said in Section 14.5 that a scalar field under certain circumstances can have an essentially constant energy density as the universe expands and that if it dominated the energy density, that would cause the universe to expand exponentially: $a = a_0 e^{Ht}$ for a constant H.

We will say more about what a scalar field is and why it would have this strange behavior, but first we want to show how such a field would resolve the problems discussed in Section 14.6. As a bonus, we'll see that this theory also explains the origin of all structure in the observable universe.

Because of dark energy, our universe is entering a new period of exponential expansion. The difference between inflation (which took place in the first fraction of a second after the Big Bang) and our current expansion is the exponent, H. Remember that the Friedmann equation says $H^2 \propto \rho$ (as long as the ρ term dominates the k term), so an exponential expansion occurring at close to Planck density would have an exponent almost 10^{70} times larger than the accelerating expansion we're currently experiencing.

Specifically, with the density near the Planck density you get $H = 5.4 \times 10^{43}$ s^{-1}. At that rate . . . here comes the most important sentence in this section . . .

The scale factor during inflation may have increased by a factor of much more than 10^{100} in less than 10^{-35} s!

The number 10^{-35} s is far smaller than even the very small times we have talked about in this book (such as 10^{-20} s for a nuclear reaction). And the number 10^{100} – a googol: what can we compare that number to, to put it in context? The number of seconds since the Big Bang? At 10^{16}, it's not even close. How about the number of elementary particles in the observable universe? That's around 10^{86}. Just multiply that by 10 million, then multiply your answer by 10 million *again*, and you're up to 10^{100}.

So the theory of inflation says that just after the universe dropped below the Planck density it underwent an unimaginable change in scale, in an unimaginably small amount of time. And then inflation stopped, and the universe began the leisurely expansion predicted by the "traditional" Big Bang theory.

Remarkably, as you've seen, such an expansion can be predicted from the Friedmann equations. But even more remarkably, it explains many of the mysteries that the original model could not.

Flatness

The flatness problem is this. During the expansion of either matter or radiation, the density term in the first Friedmann equation shrinks faster than the curvature term. One would expect, given sufficient expansion, that the curvature term would eventually come to dominate that equation. And yet measurements indicate that in our current universe, the curvature term is negligible. That means the universe must have been remarkably flat 14 billion years ago, and nothing in the original Big Bang theory explains why it should have been so.

Now consider inflationary expansion. The k-based term decreases as a increases – so during that brief moment, it decreased by a tremendous amount. But the ρ-based term for a scalar field stays constant. So increasing a makes the k-based term *less* important. If we assume the density and curvature terms were comparable before inflation, 10^{-35} s later the density term would be more than 10^{200} times larger than the curvature term.

We said in Section 14.6 that in the last 14 billion years the ratio of the density term to the curvature term has decreased by a factor of 10^{29}. If that 14 billion years was preceded by a fraction of a second of inflation, the curvature term would still be completely negligible today.

To get a visual image of how inflation solves the flatness problem, return to Edwin living on the surface of a sphere. He doesn't know his universe is closed because his observable universe is so small compared to his entire universe (the sphere) that the biggest triangles he can draw all look like Euclidean triangles. But as his universe expands his observable universe occupies ever more of the sphere, and eventually he'll notice the curvature.

Now imagine that Edwin's universe started with a burst of inflation, and became more than 10^{100} meters across (Figure 14.14). We'll see below that no light or particles survive from the period of inflation, so the farthest objects he can see 14 billion years later are still within an observable universe that's about 45 billion light years in radius. Looking at triangles formed by the light beams he can see, he has no chance of detecting any curvature. Over time his observable universe will grow, but it will remain negligibly small compared to his whole universe.

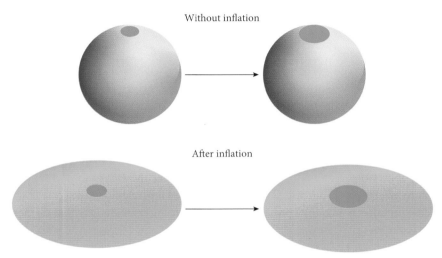

Figure 14.14 In a closed universe without inflation, the observable universe could grow large enough compared to the whole universe for observers to detect curvature. In a post-inflationary universe, the observable universe is such a small fraction of the whole that it would take an unimaginably long time to ever detect curvature. The two lower images show a zoomed-in version of one tiny piece of the post-inflationary universe.

If inflation occurred in the way we think it did, then all the stars will have burned out long before we would ever come close to measuring any deviation from flatness.

Homogeneity and Horizon

The homogeneity problem asks: is there any particular reason why the density of the universe – on very large scales – seems to be uniform throughout the observable universe? The horizon problem in particular asks how there can be a link between the density of different regions that are in our observable universe, but that are not in each other's – or were not in each other's a few billion years ago.

We've been assuming so far that inflation began at the moment when the universe was at Planck density, but there's no particular reason to assume that. It just made our numbers easier. To discuss the homogeneity and horizon problems, it's easier to relax that assumption.

Just to have some definite numbers, suppose inflation began one "Planck time" after the universe was at Planck density: about 5×10^{-44} s. By that time, ordinary causal processes could have created homogeneous regions with widths of about 10^{-35} m. Then inflation happened.

Continuing with the (arbitrary) numbers we used above, we can say that 10^{-35} s later, such a homogeneous region would be about 10^{65} m across. That's almost 40 orders of magnitude larger than the current size of our observable universe. So if inflation preceded power-law expansion, the homogeneity and horizon problems go away.

Fluctuations

Inflation solves the homogeneity problem, but it solves it too well. If the universe expanded exponentially rapidly by a factor of 10^{100} (and in most models of inflation it expanded by much more than that), then all fluctuations would be spread out to distances greater than we can observe, and the whole observable universe should look perfectly homogeneous today. We need *some* variation in density to cause the gravitational clumping that eventually led to stars, galaxies, and so on.

Here's how we can get it. Quantum field theory tells us that all fields experience small, random fluctuations. Normally these fluctuations are too small, and too short-lived, to have any discernible effect on macroscopic scales. But microscopic fluctuations that occurred *during inflation* would immediately be stretched and expanded along with the rest of the scalar field.

The fluctuations near the beginning of inflation were quickly stretched to sizes far larger than the observable universe. In other words, just like pre-inflationary variations, they became undetectable. However, fluctuations that occurred near the *end* of inflation stretched to sizes that were macroscopic, but still small enough to vary noticeably on the scale of our observable universe. In other words they provided the slight variations in density that formed the seeds of later structure.

That sounds reasonable, but did it really happen? Well, if it did, the small fluctuations that resulted should have led to small but detectable variations in the cosmic microwave background. Cosmologists in the 1980s did the calculations to predict the density fluctuations that should have been present right after inflation, and the resulting CMB fluctuations.

Their prediction took the mathematical form of a "power spectrum." Roughly speaking, the CMB power spectrum is a measure of how different the intensity of the CMB is on average between any two directions, as a function of the angle between those two directions.[14]

Figure 14.15 shows the observed power spectrum of the CMB and the predictions of inflation. The match between those predictions and observations is the primary reason most cosmologists now believe in inflation.

Monopoles

In all likelihood, a large number of magnetic monopoles were produced before inflation started. But then the universe rapidly expanded by a factor of more than 10^{100}. By the end of inflation, the density of magnetic monopoles would have been so low that the probability of us ever detecting one in our observable universe is almost exactly zero.

---

14 The power spectrum is analogous to a Fourier transform (Section 6.2), which expresses any function of position by giving the amplitudes of its sinusoidal components as a function of spatial frequency. The difference is that in this case the measured CMB is a function of angle rather than of position, so instead of sine waves we express it as a sum of functions called "spherical harmonics."

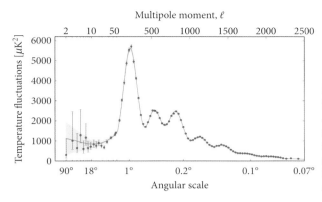

Figure 14.15 The power spectrum of the cosmic microwave background. As you move to the right, you are seeing fluctuations at smaller angular scales. The dots are measured values and the curve is the prediction from inflation. *Source: ESA and the Planck Collaboration.*

So we've solved the monopole problem, right? The problem is that we've once again solved a problem too well. Inflation gets rid of the monopoles, but it also gets rid of the protons, electrons, and pretty much everything else. Remember that the density of the scalar field that drove inflation stays nearly constant as the universe expands, so by the end of inflation the *only* significant form of energy left in the universe is that scalar field.

So how do we end up with a universe filled with matter and radiation?

The answer is that the scalar field, like most forms of energy, is unstable. After a fraction of a second it loses its energy by decaying into particles and radiation. If the energy density at the end of inflation were low enough, this second burst of particle creation would have created protons, electrons, and the other particles that make up the universe today, but would not have produced monopoles.

Scalar Fields

OK, where are we? We have posited a "scalar field" with the (highly unlikely sounding) property that its energy density stays nearly constant as the universe increases. With that equation of state, and with the density constant set close to Planck density, the Friedmann equations predict *tremendous* exponential growth. A brief spurt of such extraordinary growth would leave the observable universe extremely flat, extremely-but-not-entirely homogeneous, and free of monopoles, setting the initial conditions required by the traditional Big Bang model.

We turn our attention now to the question: what is a scalar field, and how could such a field cause inflation to happen?

Mathematically, a "field" is a function of position. So the temperature in the room around you is a scalar field $T(x, y, z)$, and the air velocity in the air around you is a vector field $\vec{v}(x, y, z)$. (Of course fields can change, so a more general description would be a function of x, y, z, and t.)

But in particle physics we often use the word more narrowly, for a mathematical field that is one of the physical building blocks of reality. Neither temperature nor air velocity satisfies that definition, so think instead of an electric or magnetic field. If you ask what those fields are made of, the answer is that they aren't made of anything else; we postulate in our laws of physics that there is a thing called a magnetic field, and we write laws describing how other things like protons and electrons interact with that field.[15]

Particle physics includes many vector fields, but we have only ever observed one physical scalar field: the Higgs field. Most theories of high-energy particle physics, however, predict the existence of other, as-yet-unknown scalar fields that can only be excited at very high energies. These theories associate with any given scalar field $\phi(x, y, z)$ a potential energy function $U(\phi)$. And just as with other physical systems, once you postulate a particular potential energy function, you can use it to predict the behavior of the system.

For example, suppose a particular field has the potential energy function $U = (1/2)k\phi^2$, which looks like the potential energy function for a simple harmonic oscillator. Physically, we're not talking about a mass on a spring: there is no object moving back and forth. But mathematically the systems are perfectly analogous. At any given point in space, the value of the field will oscillate forever between its maximum value ϕ_{max} and its minimum value $-\phi_{max}$.

Unlike a particle's potential energy function, however, $U(\phi)$ specifies the potential energy *density*, or potential energy per unit volume, of the field. To find the total energy you have to integrate ϕ through a finite region of space. But we can avoid calculus if we assume that ϕ is homogeneous throughout the observable universe – which we generally can do, precisely because the process of inflation smoothed out any irregularities. So we will look at the effect of a scalar field that is uniform throughout space but varies with time.

One of Guth's insights was to consider a scalar field trapped at a local minimum of its potential function, but *not* at $U = 0$, as shown in Figure 14.16.

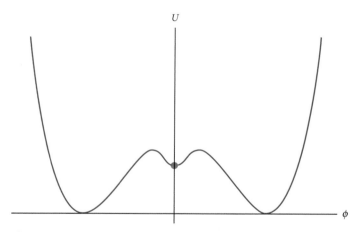

Figure 14.16 A field trapped in a local minimum of its potential energy function will have constant energy density.

As the universe expands, such a field will maintain a constant energy density – because it doesn't have enough energy to get over the hills on either side. And, as we saw above, that constant

---

15 You might object that electric and magnetic fields are "made of" photons, and in a sense you're right. In quantum field theory every fundamental field has an associated particle that is a quantized excitation of the field. So we postulate electric and magnetic fields as fundamental objects in the universe, and then our theories tell us that we can only excite ripples of those fields in quantized jumps, which we call photons. Quantum field theory actually says that every type of particle is a quantized excitation of a fundamental field. So, for example, electrons are excitations of an electron field (completely different from the electric field).

energy density would lead to exponential expansion. Inflation would end when the field found its way out of its potential well by quantum mechanically tunneling to a deeper well; at that point it would find its way down to $U = 0$, which physically corresponds to the scalar field decaying into matter and radiation.

Guth's original model, which depended on tunneling to end inflation, doesn't actually work. The field would tunnel to the minimum at different times in different places, and the near-perfect homogeneity created by inflation would be ruined. Guth acknowledged this problem in his original paper, whose abstract ended with a highly unusual admission: "Unfortunately, the scenario seems to lead to some unacceptable consequences, so modifications must be sought."

In 1982 a solution was found by Andrei Linde, and shortly afterward independently by Andreas Albrecht and Paul Steinhardt. Linde's model, called "new inflation," replaces the local minimum with a very flat potential hill. The energy density is nearly constant as the field slowly moves down the very shallow incline (Figure 14.17), and the field remains homogeneous throughout the process (except for tiny quantum fluctuations).

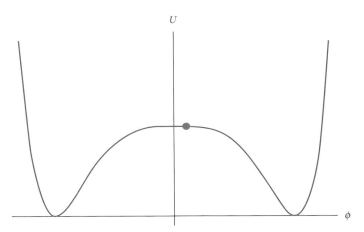

Figure 14.17 In "new inflation" a scalar field has a nearly constant energy density as it moves on a very flat potential. Inflation ends when it rolls to the minimum.

Since new inflation came out, many other models have been proposed. They all have in common that inflation is driven by a scalar field whose energy is mostly, if not entirely, dominated by its potential energy. But we have never observed any of these high-energy scalar fields, so we don't know their potential energy functions. That's why we keep saying things like "If inflation lasted 10^{-35} s . . ." We don't know exactly how long inflation lasted, or the energy density of the universe when it ended, or how much the universe expanded during inflation, because all of those features of inflation depend on the unknown properties of the scalar field that caused it.

The simplest models of inflation lead to far more expansion than is necessary to solve these problems. In the simplest model you can write, during inflation the scale factor would have increased by a factor of over 10 to the power of a million (see Problem 8)!

Stepping Back: The Revised History of the Universe

Our understanding of the universe still begins at the Planck density because we don't have the physics to describe anything before that.

In the original Big Bang model, the part of the universe that grew into our observable universe had to start at the Planck density and almost (but not perfectly) homogeneous, with a density equal to the critical density to at least 29 decimal places, and it had to somehow avoid being completely dominated by magnetic monopoles.

In the inflationary model, the initial conditions just after the Planck density did not have to be homogeneous over any particular length scale, did not have to be geometrically flat, and could contain almost any random assortment of particles and fields. The one initial condition that must have held is that somewhere in that universe there was a region, no matter how small, whose energy was primarily in the form of a potential-energy-dominated scalar field.

As long as that occurred somewhere, that region would have inflated. And since the rest of the universe would not have inflated, a fraction of a second later that inflationary region would have composed essentially 100% of the volume of the universe.

Then that scalar field decayed and inflation ended. From that moment on, the universe expanded as a power law.

- At the end of inflation, the scalar field probably decayed into other unstable fields and particles, which decayed into other unstable fields and particles, until finally the decay chains ended with the small number of low-energy, stable forms of energy that make up the universe today.

- The products of the scalar field decay included radiation, matter, and antimatter. The matter and antimatter annihilated, but because the scalar field decay produced slightly more matter than antimatter, a tiny residue of matter was left.

- From that point on, approximately one second after the end of inflation, the universe evolved in just the way the Big Bang model originally predicted.

There is still a great deal we don't know about cosmology. What happened before Planck density? Why did the decay of the scalar field produce more matter than antimatter? Does the whole universe look like our observable universe, or are there places that look very different? Was the Big Bang a simultaneous event throughout all of existence, or was our Big Bang 14 billion years ago just one of many that happen continually? We discuss some of the work cosmologists are doing on these and other open questions in an online section at www.cambridge.org/felder-modernphysics/onlinesections

14.7.3 Questions and Problems: Inflation and the Very Early Universe

Conceptual Questions and ConcepTests

1. Briefly explain in your own words how inflation solves each of the problems we've discussed: flatness, homogeneity, horizon, and monopole.

2. We said that a universe with constant energy density and zero curvature would expand exponentially, and we then used that result to explain how inflation makes the curvature term negligible. The logic might seem circular. Explain why our conclusions about how much the universe would expand during inflation and how negligible the curvature term would be after inflation would still be valid even if the curvature term started out comparable to the density term, assuming the density ρ remains constant. *You do not need to do any calculations to answer this question.*

3. Today the dominant form of energy in the universe is dark energy, which seems to have a constant

energy density as the universe expands. Mathematically, that means we are entering a second period of exponential expansion. Why did this exponential expansion not empty the observable universe of particles in a fraction of a second in the way early universe inflation did?

4. Guth's original inflationary model (now called "old inflation") suffered from what has come to be known as the "graceful exit problem," which is that the homogeneity created during inflation would have been destroyed at the end of inflation.

    (a) Why would the end of inflation make the universe highly inhomogeneous in this model?

    (b) How does new inflation solve the graceful exit problem?

5. After inflation ends, the scalar field that drove it settles down to its potential minimum at $U = 0$. Why does it settle down instead of oscillating at constant amplitude forever?

For Deeper Discussion

6. A vector field also has a potential energy function, and in principle could also be trapped in a local minimum and cause exponential expansion. But we believe inflation must have been caused by a scalar field. What feature of our universe rules out the possibility that its energy was once dominated by a uniform vector field?

7. In our plot of the scalar field potential energy for new inflation, we showed the scalar field a little to the right of the maximum. Imagine it started exactly at the maximum. Then it would presumably fall to the right in some regions and to the left in others. That would seem to create inhomogeneities in much the same way that tunneling did in Guth's original model. But even if the field starts at the maximum, new inflation is compatible with the observed homogeneity of our observable universe. Why? *Hint*: At any given moment, ϕ is a continuous function of x, y, and z.

Problems

8. **Exploration: Chaotic Inflation**

    "Chaotic inflation" is a model published by Linde in 1983 in which inflation can occur with a scalar

field that isn't in a local minimum or near a local maximum of its potential energy function.

The equation of motion for a homogeneous scalar field in an expanding universe is $\ddot{\phi} + 3H\dot{\phi} + dU/d\phi = 0$, where $U(\phi)$ is the potential energy function. (Recall that U gives potential energy density, not total energy.) For this problem we'll consider the potential energy function $U(\phi) = (1/2)m^2\phi^2$, where m is a constant (non-zero) parameter, so the equation becomes

$$\ddot{\phi} + 3H\dot{\phi} + m^2\phi = 0.$$

You may recognize this as a damped-oscillator equation, where the Hubble parameter provides the damping term. To keep things simple in this problem, we'll work in "Planck units" where $G = c = \hbar = 1$, which also makes the Planck density equal to 1. We'll also assume $k = 0$ (a flat universe). Finally, we'll assume the scalar field energy is dominated by its potential energy: $\rho \approx U(\phi)$.

(a) Use the first Friedmann equation to eliminate H from the equation of motion for ϕ, turning it into a differential equation with nothing but $\phi(t)$ and constants.

(b) Find the "terminal velocity," $\dot{\phi} = v$, at which $\ddot{\phi} = 0$.

(c) As long as ϕ moves at that terminal velocity, $\phi(t) = \phi_0 + vt$. Plug that function into the Friedmann equation and solve for $a(t)$.

(d) Calculate the second derivative $\ddot{a}$. Where $\phi_0 + vt$ appears in your solution, replace it with ϕ to write $\ddot{a}(\phi)$ with no t in it. For what values of ϕ will $\ddot{a}$ be positive?

(e) Assume $\phi_0 = 1/m$, meaning inflation begins with the universe at one-half of the Planck density. If we define the end of inflation as the moment when $\ddot{a}$ stops being positive, by what factor will distances grow during inflation?

(f) From observations of CMB fluctuations, we can calculate that in this model m has to equal 10^{-6} (in the units we're using in this problem). How long does inflation last in this model, and by what factor do distances grow in that time? (To convert your time from Planck units to seconds, divide it by 10^{44}.)

Chapter Summary

This chapter briefly summarizes the origin, history, and ultimate fate of the universe.

Section 14.1 The History of the Universe

The phrase "the universe is expanding" should be properly interpreted as "the average distance between galaxies is increasing." To put it another way, the average *density* of the universe is decreasing.

- Extrapolating current trends backward, we see an increasingly dense universe in the past.
- The limit of our ability to extrapolate back in time is the moment of "Planck density," approximately 10^{96}kg/m^3. We cannot say anything meaningful about prior times, because our current laws of physics do not apply to such high densities. We therefore use the phrase "Big Bang" to refer, not to the actual beginning of the universe (if there was such a thing), but to the moment of Planck density.
- From that moment to this, the universe has been continually expanding, which in turn means that the average temperature has decreased. Decreasing temperature allowed for the formation of ever larger structures: nucleons, then nuclei, then atoms, and eventually stars and galaxies.
- The ultimate fate of our universe will either be a "Heat Death" (energy spread through the cosmos thinly and evenly), or a "Big Crunch" (a recollapse back to the Planck density, beyond which we cannot extrapolate).

Section 14.2 How Do We Know All That?

The expansion of the universe was predicted by theory (a static universe would pull itself inward gravitationally) and confirmed by observation (Hubble used Doppler shifting to measure the recession of distant galaxies). Both theory and observation predict that the expansion should follow the "Hubble–Lemaître law" $v = Hd$: the speed with which a galaxy recedes from us is proportional to its distance to us.

- The most important evidence for the Big Bang came from measurement of the cosmic microwave background (CMB). Calculations based on the Big Bang model predicted microwave energy coming toward us uniformly from every direction, and precisely such energy was detected in 1964.
- Other compelling evidence for the model includes the ages of the oldest observed stars, the cosmic neutrino background, and the evolution of galaxies over time.
- Many of these observations require accurate distance measurements of distant objects. One technique is "parallax": measure the angle from Earth to a distant object, then make the same measurement six months later, and you can infer the distance from how the angle changes. This technique can be used for distances up to tens of thousands of light-years, but that still confines it to objects within our own galaxy.
- More distant objects can be measured by using "standard candles." If all objects of a certain type shine equally brightly, and you can measure the distance to one of them (e.g. by parallax), then you can figure out the distance to all the others by how bright they

appear. Cepheid variable stars provide reliable standard candles because their brightness is correlated with their variation period.

Section 14.3 Infinite Universe, Finite Universe, Observable Universe

We don't know if the entire universe is finite or infinite. We do know that the part of it that we can see is limited by the speed of light.

- To picture an infinite expanding universe, imagine a ruler stretching lengthwise. At any point on the ruler, you would witness the Hubble–Lemaître law: "every other point on this ruler is moving away from me, with a speed proportional to its distance from me." This one-dimensional model can easily be scaled up to two or three dimensions.

- A finite universe is harder to picture. It would be "periodic," meaning that a line extending in any direction would eventually return to its starting point. As an analogy, imagine the surface of a sphere as a two-dimensional universe. As the sphere inflates, once again every point on the sphere measures the Hubble–Lemaître law. Keep in mind, when you work with this analogy, that the space inside the sphere and the space outside the sphere do not exist; the surface of the sphere is the universe.

- Because the universe is only 13.8 billion years old, light from galaxies more than 45 billion light-years away has not had time to reach us. Hence, our "observable universe" – not everything that exists, but everything that we can possibly measure in any way – is a finite sphere centered on us.

Section 14.4 The Friedmann Equations

The Friedmann equations, Equations (14.1) and (14.2), follow from the application of Einstein's theory of general relativity to a uniform and isotropic universe.

- The Friedmann equations are differential equations for the quantity called a, a measurement of the overall scale of the universe. We can choose $a = 1$ at any particular moment. Then $a = 2$ means that intergalactic distances have doubled since that moment.

- The first Friedmann equation includes a constant called k that quantifies the curvature of spacetime. If $k > 0$ then the universe is "closed" and finite (like the sphere in the two-dimensional analogy). If $k < 0$ then then the universe is "open" and infinite (a reasonable 2D analogy is a saddle). To the best of our ability to measure it, k is zero: a "flat" universe.

- A universe dominated by matter or by radiation will eventually recollapse if it is closed and will expand forever if it is flat or open.

- The Friedmann equations can be solved to show that $k > 0$ corresponds to the universe having a density above a certain critical density: $\rho > \rho_c$, or $\Omega = \rho/\rho_c > 1$. Once again, $\rho = \rho_c$ to the best of our ability to measure it.

- To solve the Friedmann equations you need to plug in an "equation of state" describing the nature of the energy that makes up the universe. Plugging in the equation for matter leads to $a \propto t^{2/3}$ and $\rho \propto a^{-3}$. On the other hand, a universe dominated by radiation energy leads to $a \propto t^{1/2}$ and $\rho \propto a^{-4}$.

Section 14.5 Dark Matter and Dark Energy

Much of the energy of the universe is in the form of "dark matter," and even more of it is in the form of "dark energy," leaving relatively little in the form of "everything we see and measure around us."

- Newton's laws predict a relationship between the radius and velocity of an orbiting object. Those laws provide accurate descriptions of the orbits of planets in our solar system, but they fail to correctly model the rotations of galaxies or galaxy clusters. This discrepancy points to the existence of "dark matter": transparent, interacting with normal matter in almost no way but gravitational, but forming far more of the matter of galaxies than the stars and planets we see.

- Gravitational attraction (whether Newtonian or relativistic) predicts that the expansion of the universe should be slowing down, but in fact it's speeding up. This points to the existence of "dark energy," which – unlike either regular matter or dark matter – *repels* instead of attracting.

- The repulsive effect of dark energy is predicted by the Friedmann equations if you assume a type of energy whose density ρ stays constant as the universe expands.

- If dark energy continues to exist forever, then the universe will continue to expand forever, regardless of whether it is closed, flat, or open. If dark energy some day decays to matter and radiation, then the fate of the universe will depend on the sign of k.

Section 14.6 Problems with the Big Bang Model

Although the general history of the universe as described by the Big Bang model has been confirmed enough to ensure its widespread acceptance, the original theory did leave a number of problems to be resolved.

- The "flatness" problem: As the universe expands, its curvature increases. For the modern universe to be as flat as we measure it to be, the early universe must have been almost completely flat. Such a specific match ($\rho = \rho_c$ to within a factor of 10^{-29}) is too perfect to be a coincidence.

- The "homogeneity" and "horizon" problems: Similar to the flatness problem, the near-perfect homogeneity of the early universe demands an explanation. But such an explanation is difficult to craft, in part, because it includes different parts of the early universe that were outside each other's observable universes, and could therefore have had no possible causal connection.

- The "monopole" problem: In the conditions of the very early universe, magnetic monopoles and other "relic particles" should have been created in very large quantities. They should be stable, and should therefore still exist today. But we have never found them.

Section 14.7 Inflation and the Very Early Universe

The theory of inflation posits that shortly after it dropped below the Planck density, the universe experienced a moment of hyper-accelerated expansion: possibly increasing in size by over 10^{100}

in less than 10^{-35} s. The theory explains why such expansion might have occurred, and how such expansion would resolve the problems presented in Section 14.6.

- Inflation caused an extremely small region to expand to much larger than our current observable universe, thus leaving the universe flat and homogeneous to far beyond our ability to measure on the scale of what we can observe. Note that this does not preclude the possibility of inhomogeneities on scales much larger than our observable universe.

- Inflation reduced the density of magnetic monopoles (and all other particles) to almost exactly zero, leaving no particles in our observable universe. The matter that we now see was created *after* inflation, at which point the temperature was too low to create monopoles.

- Inflationary expansion resulted from a "scalar field": a type of energy that keeps a constant density as the universe expands. Recall that dark energy has the same property, and thus accelerates the current expansion of the universe. Inflationary expansion was far more drastic (exponential expansion with a much higher exponent) because of the very high densities involved.

Appendix A
A Chronology of Modern Physics

A more extensive chronology, including references to important papers, is at www.cambridge.org/felder-modernphysics/onlinesections

1900	Max Planck proposes quantization of cavity radiation to resolve the ultraviolet catastrophe.	Section 3.4
1905	Einstein publishes the special theory of relativity.	Chapters 1–2
1905	Einstein uses quantization of light to explain the photoelectric effect.	Section 3.5
1909	G. I. Taylor shows interference from a one-photon-at-a-time double-slit experiment.	Section 3.3
1911	Ernest Rutherford demonstrates the existence of atomic nuclei.	Section 12.2
1913	Niels Bohr proposes that electrons in atoms exist in quantized energy levels.	Section 4.1
1915	Einstein publishes the general theory of relativity, which incorporates gravity.	Online
1915	Emmy Noether proves that every continuous symmetry corresponds to a conservation law.	Section 13.4
1922	Otto Stern and Walther Gerlach demonstrate quantization of angular momentum.	Section 7.5
1924	Louis de Broglie proposes matter waves.	Section 4.2
1925	Wolfgang Pauli proposes the exclusion principle.	Section 8.1
1925	Samuel Goudsmit and George Uhlenbeck propose spin to explain the Stern–Gerlach results.	Section 7.5
1926	Erwin Schrödinger publishes what we now call Schrödinger's equation.	Section 6.6
1926	Max Born proposes the probabilistic interpretation of wavefunctions.	Section 4.3
1927	Werner Heisenberg proposes the uncertainty principle.	Section 4.4
1929	Edwin Hubble measures galaxies receding at rates proportional to their distances.	Section 14.2
1964	John Bell demonstrates that quantum mechanics cannot be reconciled with locality.	Online
1964	Arno Penzias and Robert Wilson measure the cosmic microwave background, confirming the Big Bang model.	Section 14.2
1964	Murray Gell-Mann and George Zweig independently propose the quark model for hadrons, paving the way for the standard model of particle physics.	Section 13.2
1980	Vera Rubin publishes galactic rotation curves, demonstrating the existence of dark matter.	Section 14.5
1981	Alan Guth publishes the theory of inflation.	Section 14.7
2012	The Higgs boson is the last particle in the standard model to be detected.	Section 13.2

Appendix B
Special Relativity Equations

The equations in this appendix are discussed in Chapters 1 and 2.

- The formulas in this appendix assume a reference frame R′ moving with velocity $u\hat{i}$ in Frame R. The event designated $x = y = z = t = 0$ in Frame R is also designated $x' = y' = z' = t' = 0$ in Frame R′.
- Primes denote measurable quantities in Frame R′, *not* derivatives.
- Many formulas below include $\gamma = \dfrac{1}{\sqrt{1 - u^2/c^2}}$. Note that $\gamma > 1$ whenever $0 < u < c$.

The Lorentz Transformations

$$t' = \gamma\left(t - \frac{ux}{c^2}\right) \qquad x' = \gamma(x - ut) \qquad y' = y \qquad z' = z$$

Other Equations

- *Time dilation*: Suppose that in Frame R two events occur at the same place, separated by time Δt (e.g. two ticks of a clock at rest in R). The time elapsed between the events in Frame R′ is $\Delta t' = \gamma\,\Delta t$.
- *Length contraction*: Suppose that in Frame R an object is at rest and has length L in the x direction. In Frame R′ the object has length $L' = L/\gamma$.
- *The relativity of simultaneity*: Frame R′ measures $x = 0, t = 0$ to be simultaneous with $x = a, t = ua/c^2$.
- *The spacetime interval*: The interval $s = \sqrt{c^2\Delta t^2 - \Delta x^2 - \Delta y^2 - \Delta z^2}$ is the same in all reference frames.
- *Velocity transformations*: If $\vec{v}$ is an object's velocity in Frame R, then

$$v'_x = \frac{v_x - u}{1 - v_x u/c^2} \qquad v'_y = \frac{v_y}{\gamma}\frac{1}{1 - v_x u/c^2} \qquad v'_z = \frac{v_z}{\gamma}\frac{1}{1 - v_x u/c^2}.$$

- *Doppler shift*: Light of frequency ν emitted by an object at rest in Frame R is received in Frame R′ with $\nu' = \nu\sqrt{(1 \pm |u|/c)/(1 \mp |u|/c)}$. Use the top signs if the source and receiver are getting closer, and the bottom signs if they are moving apart.
- *Momentum and energy*: In the following equations, $\gamma = 1/\sqrt{1 - v^2/c^2}$, where $\vec{v}$ is an object's velocity in Frame R:

$$\vec{p} = \gamma m\vec{v} \qquad E = \gamma mc^2 = \sqrt{p^2 c^2 + m^2 c^4}.$$

The energy can be broken down into rest energy mc^2 and kinetic energy $\gamma mc^2 - mc^2$.

Appendix C
Quantum Mechanics Equations

Note: Equations specific to the hydrogen atom are discussed in Chapter 7 and summarized in Appendix G.

Schrödinger's Equations (Sections 5.2, 6.6, 7.2)

In one dimension:

$$-\frac{\hbar^2}{2m}\frac{d^2\psi}{dx^2} + U(x)\psi(x) = E\psi(x)$$ The time-independent Schrödinger equation in 1D.

$$-\frac{\hbar^2}{2m}\frac{\partial^2\Psi}{\partial x^2} + U(x)\Psi(x,t) = i\hbar\frac{\partial\Psi}{\partial t}$$ The time-dependent Schrödinger equation in 1D.

In three dimensions:

$$-\frac{\hbar^2}{2m}\nabla^2\psi(x) + U(x)\psi(x) = E\psi(x)$$ The time-independent Schrödinger equation.

$$-\frac{\hbar^2}{2m}\nabla^2\Psi(x,y,z,t) + U(x,y,z)\Psi(x,y,z,t) = i\hbar\frac{\partial\Psi}{\partial t}$$ The time-dependent Schrödinger equation.

Position in One Dimension (Section 4.3)

Probability of finding a particle between $x = a$ and $x = b$: $\displaystyle\int_a^b |\psi(x)|^2\, dx$

Normalization: $\displaystyle\int_{-\infty}^{\infty} |\psi(x)|^2\, dx = 1$

Expectation value: $\displaystyle\langle x\rangle = \int_{-\infty}^{\infty} x\,|\psi(x)|^2\, dx$

Time Evolution (Section 5.6)

If $\Psi(x,0) = \psi_n(x)$ is an energy eigenstate with eigenvalue E_n, then

$$\Psi(x,t) = \psi_n(x)e^{-iE_nt/\hbar}.$$

For a sum or integral over energy eigenstates, apply this rule to each individual eigenstate.

Fourier Transforms and Momentum (Sections 6.2, 6.3)

- The momentum eigenstate is $\psi(x) = Ce^{ikx}$ with eigenvalue $p = \hbar k$.

- $\psi(x) = \dfrac{1}{\sqrt{2\pi}} \displaystyle\int_{-\infty}^{\infty} \hat{\psi}(k)e^{ikx}\,dk$

- $\hat{\psi}(k) = \dfrac{1}{\sqrt{2\pi}} \displaystyle\int_{-\infty}^{\infty} \psi(x)e^{-ikx}\,dx$

- $\displaystyle\int_{a}^{b} \left|\hat{\psi}(k)\right|^2 dk$ is the probability of finding a particle's momentum between $p = \hbar a$ and $p = \hbar b$.

The constants in front of these formulas are chosen so that if the position distribution is properly normalized then the momentum distribution is too, and vice versa.

The Heisenberg Uncertainty Principle (Section 4.4)

- $\Delta x \Delta p \geq \dfrac{\hbar}{2}$

- A Gaussian wavefunction $\psi(x) = Ae^{-k(x-x_0)^2}$ achieves the minimum allowable uncertainty $\Delta x \Delta p = \hbar/2$. Other wavefunctions have higher values of $\Delta x \Delta p$.

- $\Delta E \Delta t \geq \dfrac{\hbar}{2}$

The Infinite Square Well (Sections 5.3, 7.2)

The following equations assume that all spatial coordinates range from 0 to L, and all quantum numbers are positive integers:

$$\text{1D energy eigenstates: } \psi_n(x) = \sqrt{\frac{2}{L}} \sin\left(\frac{\pi n}{L}x\right)$$

$$\text{1D energy eigenvalues: } E_n = \frac{\pi^2 \hbar^2}{2mL^2}n^2$$

$$\text{3D energy eigenstates: } \psi_{abc}(x,y,z) = \sqrt{\frac{8}{L^3}} \sin\left(\frac{a\pi}{L}x\right)\sin\left(\frac{b\pi}{L}y\right)\sin\left(\frac{c\pi}{L}z\right)$$

$$\text{3D energy eigenvalues: } E_{abc} = \frac{\pi^2 \hbar^2}{2mL^2}\left(a^2 + b^2 + c^2\right)$$

The Simple Harmonic Oscillator (Section 5.4)

A simple harmonic oscillator can be defined by the potential energy function $U(x) = (1/2)m\omega^2 x^2$.

- The energy eigenvalues are $E_n = (n + 1/2)\hbar\omega$ for non-negative integers n.
- The corresponding eigenstates are

$$\psi_n(x) = \left(\frac{m\omega}{\pi\hbar}\right)^{1/4} \frac{1}{\sqrt{2^n n!}} H_n\left(\sqrt{\frac{m\omega}{\hbar}}\,x\right) e^{-m\omega x^2/(2\hbar)},$$

where $H_n(x)$ is a "Hermite polynomial," defined as

$$H_n(x) = (-1)^n e^{x^2} \frac{d^n}{dx^n}\left(e^{-x^2}\right).$$

- The first two of these eigenstates are
  - Ground state: $\psi_0 = \left(\frac{m\omega}{\pi\hbar}\right)^{1/4} e^{-m\omega x^2/(2\hbar)}$ with $E_0 = (1/2)\hbar\omega$
  - First excited state: $\psi_1 = \left(\frac{m\omega}{\pi\hbar}\right)^{1/4} \sqrt{\frac{2m\omega}{\hbar}}\, x e^{-m\omega x^2/(2\hbar)}$ with $E_1 = (3/2)\hbar\omega$

Free Particles (Sections 6.2, 6.4)

A free particle can be defined by the potential energy function $U(x) = 0$.

- The energy eigenstate of a free particle is $\psi(x) = Ce^{ikx}$, with eigenvalue $E = \dfrac{k^2\hbar^2}{2m}$.
- With time evolution, $\Psi(x,t) = Ce^{i(kx-\omega t)} = Ce^{i(px-Et)/\hbar}$.
- Watch the signs: e^{3ix} and e^{-3ix} are *different* wavefunctions that correspond to the *same* energy. The former represents a wave moving to the right, the latter to the left.

Appendix D
The Electromagnetic Spectrum

The following graph and table give the ranges for different types of electromagnetic radiation. There are no universally accepted conventions for the exact cutoffs between the different types,[1] but this gives a general idea. The columns of the table are the frequency ν, the wavelength ($\lambda = c/\nu$), the energy of a photon with that frequency ($E = h\nu$), and the temperature of a blackbody whose spectrum would peak at that frequency ($T \approx \frac{h\nu}{2.82 k_B}$). Following the tables we give some notes about the different bands in the spectrum.

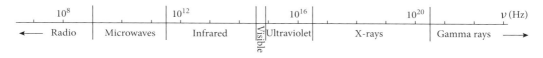

Category	Frequency (Hz)			Wavelength (m)			Photon energy (eV)			Blackbody temp. (K)		
Gamma rays	3×10^{20}	$\rightarrow$			$\leftarrow$	10^{-12}	1.2×10^6	$\rightarrow$		5×10^9	$\rightarrow$	
X-Rays	3×10^{16}	$\leftrightarrow$	3×10^{20}	10^{-12}	$\leftrightarrow$	10^{-8}	1.2×10^2	$\leftrightarrow$	1.2×10^6	5×10^5	$\leftrightarrow$	5×10^9
Ultraviolet	7.7×10^{14}	$\leftrightarrow$	3×10^{16}	10^{-8}	$\leftrightarrow$	3.9×10^{-7}	3	$\leftrightarrow$	120	1.3×10^4	$\leftrightarrow$	5×10^5
Visible	3.8×10^{14}	$\leftrightarrow$	7.7×10^{14}	3.9×10^{-7}	$\leftrightarrow$	7.8×10^{-7}	1.6	$\leftrightarrow$	3	6000	$\leftrightarrow$	13000
Infrared	3×10^{11}	$\leftrightarrow$	3.8×10^{14}	7.8×10^{-7}	$\leftrightarrow$	10^{-3}	0.0012	$\leftrightarrow$	1.6	5	$\leftrightarrow$	6000
Microwave	10^9	$\leftrightarrow$	3×10^{11}	10^{-3}	$\leftrightarrow$	3×10^{-1}	4×10^{-6}	$\leftrightarrow$	1.2×10^{-3}	0.02	$\leftrightarrow$	5
Radio		$\leftarrow$	10^9	0.3	$\rightarrow$			$\leftarrow$	4×10^{-6}		$\leftarrow$	0.02

Visible light		
Color	**Frequency (THz)**	**Wavelength (nm)**
Red	384 $\leftrightarrow$ 482	622 $\leftrightarrow$ 781
Orange	482 $\leftrightarrow$ 503	596 $\leftrightarrow$ 622
Yellow	503 $\leftrightarrow$ 520	577 $\leftrightarrow$ 596
Green	520 $\leftrightarrow$ 610	492 $\leftrightarrow$ 577
Blue	610 $\leftrightarrow$ 659	455 $\leftrightarrow$ 492
Violet	659 $\leftrightarrow$ 769	390 $\leftrightarrow$ 455

See the next page for notes on these tables.

---

1 The numbers we are using are a combination of values from the NASA website www.nasa.gov/directorates/heo/scan/ spectrum/txt_electromagnetic_spectrum.html, the ISO standard "ISO 21348: Definitions of Solar Irradiance Spectral Categories," and the textbook *Optics* by Eugene Hecht (Pearson Education, 2017).

Notes

- Physicists often call radiation from nuclear transitions "gamma rays" and radiation from highly energetic electron transitions "X-rays."[2] Because these are defined in terms of their sources and not just their frequencies, the two ranges can overlap. Hecht's book *Optics*, for example, extends the gamma ray range as low as 2.4×10^{18} Hz while putting the X-ray range up to 5×10^{19} Hz.

- Not only do the color cutoffs differ in different sources, even the named colors that are included differ widely. The once widely taught "ROYGBIV" classification of colors is not commonly used in physics.

- A number of sources include the microwave range as part of "radio," but others consider the radio region to stop where microwaves begin. The ISO standard extends the radio range straight through microwaves, overlapping with infrared.

- Some sources define a minimum frequency for radio waves and/or a maximum one for gamma rays, but there are no commonly accepted terms for electromagnetic radiation beyond those ranges.

- Just as visible light is subdivided into colors, several other bands have important subdivisions. For example, infrared is commonly divided into "near," "middle," and "far" infrared (with "near" being the closest to visible light). There are no universal conventions for the boundaries of these divisions. There are many divisions of radio frequencies (e.g. "very high frequencies" and "ultra high frequencies"), mostly based on their use in communications, and many of these divisions do have precise definitions.

---

2 It is generally agreed that any radiation that turns scientists into rampaging green monsters should be classified as gamma rays.

Appendix E
Interference and Diffraction

Interference and diffraction are mostly discussed in Chapters 3 and 4.

Each section below describes a different geometry. In each case a wave moves along two or more separate paths. Places where the waves from the different paths interfere constructively to produce the maximum possible amplitude are called "antinodes," and places where they interfere destructively to produce no oscillation are called "nodes." In the case of light, these correspond to bright and dark fringes.

Double Slit

A wave of wavelength λ passes through two narrow[1] slits separated by a distance $2d$. The wave is then measured on a back wall a distance L behind the wall with the slits (Figure E.1).

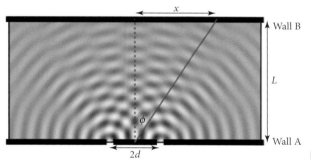

Figure E.1 Double-slit experiment.

The distance from the center of the back wall to the nth antinode is

$$x = \frac{n\lambda}{2}\sqrt{\frac{4\left(d^2 + L^2\right) - n^2\lambda^2}{4d^2 - n^2\lambda^2}}.$$

In this and the approximate formulas below, you can find the nodes by putting in a half-integer value for n.

The far-field approximation: Assuming $L \gg d$ (the back wall is much farther away than the spacing between the slits), the antinodes occur at evenly spaced values of $\sin\phi$ (see Figure E.1):

$$\sin\phi \approx \frac{n\lambda}{2d}.$$

The small-angle approximation: If we additionally assume $d \gg n\lambda$, then $\phi \ll 1$ and we can use $\sin\phi \approx \tan\phi = x/L$. This leads to the antinodes being evenly spaced along the back wall: $x = n\lambda L/2d$. This approximation is never valid if $d \lesssim \lambda$, but if $d \gg \lambda$ then it works for all the peaks up to $d \approx n\lambda$.

---

1 "Narrow" in this context means that the width of each individual slit is much smaller than λ.

Diffraction grating: A diffraction grating is a long line of evenly spaced slits. Assuming the back wall is far enough away that the waves from all slits travel at approximately the same angle to each point, the nodes and antinodes follow the same equations as the far-field and small-angle equations for a double slit. The difference is that the fringes show greater contrast when there are more slits.

Bragg's Law

Bragg's law refers to the scattering of an incoming wave off multiple layers of a crystal (Figure E.2).

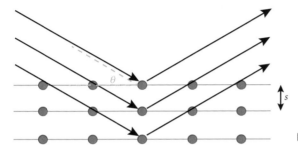

Figure E.2 Waves reflecting off a crystal.

You get a strong reflected beam when the glancing angle θ meets the following condition for integer n:

$$\sin \theta = \frac{n\lambda}{2s}.$$

This looks similar to the far-field formula for a double slit, but here θ is the angle of the incoming beam and s is the spacing between atomic layers.

By convention, Bragg's law is expressed in terms of glancing angle θ rather than angle of incidence ($90° - \theta$).

Single Slit

A diffraction pattern occurs when a wave passes through a single slit whose width w is comparable to the wavelength λ (Figure E.3).

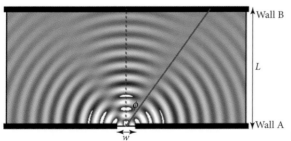

Figure E.3 Single-slit diffraction.

This produces one intense central maximum with smaller maxima on the sides, separated by nodes (dark fringes). In the far-field approximation the nodes occur at the following angles for positive integers n:

$$\sin \phi = \frac{n\lambda}{w}.$$

Appendix F
Properties of Waves

The equations in this appendix are discussed and/or derived in Chapters 3 and 6.

Real-Valued Sinusoidal Functions of One Variable

$$y = A \sin(\omega t) = A \sin(2\pi v t) = A \sin\left(\frac{2\pi}{T} t\right)$$

$$y = A \sin(kx) = A \sin(2\pi f x) = A \sin\left(\frac{2\pi}{\lambda} x\right)$$

Variables (time)	Equations		Equivalent variables (space)
A: amplitude			A: amplitude
T: period			λ: wavelength
ω: angular frequency	$\omega = \dfrac{2\pi}{T}$	$k = \dfrac{2\pi}{\lambda}$	k: wave number
v: frequency	$v = \dfrac{\omega}{2\pi} = \dfrac{1}{T}$	$f = \dfrac{k}{2\pi} = \dfrac{1}{\lambda}$	f: spatial frequency

Complex Exponential Waves

A sinusoidal wave can also be represented by a complex exponential function, related to sines and cosines by Euler's formula: $e^{ix} = \cos x + i \sin x$.

$y = A e^{i\omega t}$ is a complex wave with amplitude A and angular frequency ω.

Standing and Traveling Waves

A standing wave is a wave in space with fixed nodes and antinodes and a time-varying amplitude:

$y = A \sin(\omega t) \sin(kx)$ This is a standing wave with amplitude A, angular frequency ω, and wave number k.

A traveling wave keeps the same amplitude and shape but moves through space:

$y = A \sin(kx - \omega t)$ or $A e^{i(kx - \omega t)}$ This is a wave moving with velocity $v = \omega/k$.

The velocity of a traveling wave can equivalently be written $v = \lambda v$.

The Simple Harmonic Oscillator Equation

Many waves arise as solutions to the following differential equation:

$$\frac{d^2y}{dt^2} = -\omega^2 y \qquad \text{The simple harmonic oscillator equation.}$$

The three general solutions below are equivalent, meaning any one can be rewritten to look like either of the others:

$$y(t) = A\sin(\omega t) + B\cos(\omega t)$$
$$y(t) = C\sin(\omega t + \phi)$$
$$y(t) = De^{i\omega t} + Fe^{-i\omega t}$$

Appendix G
Energy Eigenstates of the Hydrogen Atom

- Section 7.1 discusses the general properties of the hydrogen atom eigenstates and the meaning of the quantum numbers n, l, and m_l.
- Section 7.4 outlines the derivation of the eigenstates.
- Section 7.5 discusses the basic properties of spin.
- This appendix summarizes key points from those sections and gives the full mathematical form of the hydrogen atom eigenstates.

The energy eigenstates of the hydrogen atom are of the form

$$\psi_{nlm_l}(r,\theta,\phi) = R_{nl}(r)Y_l^{m_l}(\theta,\phi).$$

A full specification of the quantum state of a hydrogen atom also includes the spin state (up or down) of the electron.

The Quantum Numbers

The energy eigenstates of the hydrogen atom are states of definite energy, magnitude of angular momentum, and z component of angular momentum. The values of those quantities are labeled by the quantum numbers n, l, and m_l, respectively. These three quantum numbers must be integers, and the spin quantum number m_s must be $1/2$ (spin-up) or $-1/2$ (spin-down). Other restrictions are given below.

Letter	Name	Range	Physical interpretation
n	Principal quantum number	$1 \le n < \infty$	$E_n = -(1/n^2)$ Ry (1 Ry $\approx$ 13.6 eV)
l	Angular momentum quantum number	$0 \le l \le n-1$	$\lvert \vec{L} \rvert = \sqrt{l(l+1)}\,\hbar$
m_l	Magnetic quantum number	$-l \le m_l \le l$	$L_z = m_l\hbar$
m_s	Spin quantum number	$m_s = \pm 1/2$	$S_z = m_s\hbar$

The energy eigenvalues of a hydrogen atom depend only on n (to a very good approximation). The following formulas use m_e and e for the electron mass and charge, respectively:

$$E_n = -\frac{m_e e^4}{32\pi^2\epsilon_0^2\hbar^2 n^2} = -\frac{1}{2n^2}m_e c^2\alpha^2 = -\frac{1}{n^2}1\,\text{Ry} = -\frac{1}{n^2}13.6\,\text{eV} = -\frac{1}{n^2}2.18\times 10^{-18}\,\text{J}.$$

- Note the introduction of the "fine structure constant" $\alpha = \dfrac{1}{4\pi\epsilon_0}\dfrac{e^2}{\hbar c} \approx \dfrac{1}{137}$ (Section 7.7).
- For a "hydrogen-like" atom with Z protons and one electron, the energies are multiplied by Z^2.

Probability Distributions

For any particle with wavefunction ψ in 3D, the probability per unit volume of finding the particle in a given region is $|\psi|^2$. For an electron in a hydrogen atom energy eigenstate ψ_{nlm_l}, the following probability distributions also apply:

- The probability of finding the electron between radii r_1 and r_2 is $\int_{r_1}^{r_2} r^2 R^2 \, dr$.

- The probability of finding the electron in a specified range of angles is $\int_{\theta_1}^{\theta_2} \int_{\phi_1}^{\phi_2} \sin\theta |Y_l^{m_l}|^2 \, d\phi \, d\theta$.

These integrals are normalized so that $\int_0^\infty r^2 R^2 \, dr = \int_0^\pi \int_0^{2\pi} \sin\theta |Y_l^{m_l}|^2 \, d\phi \, d\theta = 1$.

The Functions $R_{nl}(r)$ and $Y_l^{m_l}(\theta, \phi)$: Properties and Examples

Note: If you look up special functions like "spherical harmonics" and "associated Laguerre polynomials," you will find that different sources define them differently. However, all sources agree about the energy eigenstates of the hydrogen atom. The disagreement is just due to arbitrary decisions about which constants to put where. The conventions we use here are designed to properly normalize the probability distributions given above.

The "spherical harmonic" $Y_l^{m_l}(\theta, \phi)$ is a polynomial in $\sin\theta$ and $\cos\theta$ multiplied by $e^{im_l\phi}$. $R_{nl}(r)$ is a polynomial in r times a decaying exponential $e^{-r/(a_0 n)}$, where a_0 is a constant called the "Bohr radius":

$$a_0 = \frac{4\pi \epsilon_0 \hbar^2}{m_e e^2} \approx 5 \times 10^{-11} \text{ m.}$$

The first several radial and angular wavefunctions are listed below.

The First Few Radial Wavefunctions of the Hydrogen Atom $R_{nl}(r)$

$n = 1$	$R_{1,0} = 2a_0^{-3/2} e^{-r/a_0}$	
$n = 2$	$R_{2,0} = \dfrac{1}{\sqrt{2}} a_0^{-3/2} \left(1 - \dfrac{1}{2}\dfrac{r}{a_0}\right) e^{-r/(2a_0)}$	$R_{2,1} = \dfrac{1}{\sqrt{24}} a_0^{-3/2} \dfrac{r}{a_0} e^{-r/(2a_0)}$
$n = 3$	$R_{3,0} = \dfrac{2}{\sqrt{27}} a_0^{-3/2} \left(1 - \dfrac{2}{3}\dfrac{r}{a_0} + \dfrac{2}{27}\left(\dfrac{r}{a_0}\right)^2\right) e^{-r/(3a_0)}$	$R_{3,1} = \dfrac{8}{27\sqrt{6}} a_0^{-3/2} \left(1 - \dfrac{1}{6}\dfrac{r}{a_0}\right) \dfrac{r}{a_0} e^{-r/(3a_0)}$
	$R_{3,2} = \dfrac{4}{81\sqrt{30}} a_0^{-3/2} \left(\dfrac{r}{a_0}\right)^2 e^{-r/(3a_0)}$	

The First Few Spherical Harmonics $Y_l^{m_l}(\theta, \phi)$

$l = 0$	$Y_0^0 = \sqrt{\dfrac{1}{4\pi}}$		
$l = 1$	$Y_1^0 = \sqrt{\dfrac{3}{4\pi}} \cos\theta$	$Y_1^{\pm 1} = \mp\sqrt{\dfrac{3}{8\pi}} \sin\theta \, e^{\pm i\phi}$	
$l = 2$	$Y_2^0 = \sqrt{\dfrac{5}{16\pi}} (3\cos^2\theta - 1)$	$Y_2^{\pm 1} = \mp\sqrt{\dfrac{15}{8\pi}} \sin\theta \cos\theta \, e^{\pm 2i\phi}$	$Y_2^{\pm 2} = \sqrt{\dfrac{15}{32\pi}} \sin^2\theta \, e^{\pm 2i\phi}$

The Full Definition of $R_{nl}(r)$

The radial wavefunction is given by the following:

$$R_{nl}(r) = \sqrt{\left(\frac{2}{na_0}\right)^3 \frac{(n-l-1)!}{2n[(n+l)!]^3}} \left(\frac{2r}{na_0}\right)^l e^{-r/(na_0)} L_{n-l-1}^{2l+1}\left(\frac{2r}{na_0}\right).$$

The L toward the end of that definition is called an "associated Laguerre polynomial" and is defined as follows:

$$L_p^k(x) = (-1)^k \frac{d^k}{dx^k}\left[e^x \frac{d^{p+k}}{dx^{p+k}}\left(e^{-x} x^{p+k}\right)\right].$$

Note: If $k = 0$ then the above formula calls for the "zeroth derivative." The zeroth derivative of a function is defined as the function itself.

Example: Calculating a Radial Wavefunction

Use the formula above to find $R_{2,0}$.

Answer: Plugging in $n = 2, l = 0$ gives

$$R_{2,0}(r) = \sqrt{\left(\frac{1}{a_0}\right)^3 \frac{1!}{4[(2)!]^3}} \left(\frac{r}{a_0}\right)^0 e^{-r/(2a_0)} L_1^1\left(\frac{r}{a_0}\right) = \frac{1}{4\sqrt{2}}\left(\frac{1}{a_0}\right)^{3/2} e^{-r/(2a_0)} L_1^1\left(\frac{r}{a_0}\right).$$

Next we have to find the function $L_1^1(x)$:

$$L_1^1(x) = (-1)^1 \frac{d}{dx}\left[e^x \frac{d^2}{dx^2}\left(e^{-x} x^2\right)\right] = -\frac{d}{dx}\left[e^x \frac{d}{dx}\left(e^{-x} 2x - e^{-x} x^2\right)\right] = -\frac{d}{dx}\left[e^x\left(e^{-x} 2 - 2e^{-x}(2x) + e^{-x} x^2\right)\right]$$

$$= -\frac{d}{dx}\left[2 - 4x + x^2\right] = 4 - 2x.$$

Plugging in $x = r/a_0$ and putting this in our expression for $R_{2,0}$ gives

$$R_{2,0} = \frac{1}{4\sqrt{2}} a_0^{-3/2} e^{-r/(2a_0)}\left(4 - 2\frac{r}{a_0}\right).$$

Take a factor 4 out of the parentheses and you can see that this matches the value of $R_{2,0}$ listed above.

The Full Definition of $Y_l^{m_l}(\theta, \phi)$

Spherical harmonics are products of complex exponentials in ϕ and polynomials in $\cos\theta$:

$$Y_l^{m_l}(\theta, \phi) = \pm\sqrt{\frac{2l+1}{4\pi} \frac{(l-|m_l|)!}{(l+|m_l|)!}} P_l^{m_l}(\cos\theta) e^{im_l\phi} \quad \text{where } l \text{ and } m_l \text{ are integers and } -l \le m_l \le l.$$

Two things about that definition need explaining. The P at the end is an "associated Legendre polynomial," defined by

$$P_l^{m_l}(x) = (1-x^2)^{|m_l|/2} \frac{d^{|m_l|}}{dx^{|m_l|}}\left[\frac{1}{2^l l!} \frac{d^l}{dx^l}\left(x^2 - 1\right)^l\right].$$

The second thing is the $\pm$ at the start of the left-hand side of the formula for the spherical harmonics; this equals $+1$ for all negative m_l and all even values of m_l, but equals -1 for odd positive values. So the only differences between $Y_l^{m_l}$ and and $Y_l^{-m_l}$ are that they have different signs if m_l is odd, and the exponent in $e^{im_l\phi}$ changes sign between them.

Appendix H
The Periodic Table

Chapter 8 discusses how the quantum mechanics of multielectron atoms leads to the structure of the periodic table. This is given on the next page.

Each subshell is labeled with a number for n and a letter for l.
$s\,(l=0),\ p\,(l=1),\ d\,(l=2),\ f\,(l=3)$

Most electron configurations fill in according to the $n+l$ rule. The exceptions are marked below, showing the previous noble gas plus the electrons added beyond it. (Subshells without a superscript have a single electron.)

☐ : Metal   ▨ : Metalloid   ▨ : Non-Metal

Transition metals

	Alkali 1	Alkaline earths 2		3	4	5	6	7	8	9	10	11	12	13	14	15	16	Halogens 17	Noble gases 18
1s	1 **H** Hydrogen																		2 **He** Helium
2s	3 **Li** Lithium	4 **Be** Beryllium	2p											5 **B** Boron	6 **C** Carbon	7 **N** Nitrogen	8 **O** Oxygen	9 **F** Fluorine	10 **Ne** Neon
3s	11 **Na** Sodium	12 **Mg** Magnesium	3p											13 **Al** Aluminum	14 **Si** Silicon	15 **P** Phosphorus	16 **S** Sulfur	17 **Cl** Chlorine	18 **Ar** Argon
4s	19 **K** Potassium	20 **Ca** Calcium	3d	21 **Sc** Scandium	22 **Ti** Titanium	23 **V** Vanadium	24 **Cr** Chromium $[Ar]4s3d^5$	25 **Mn** Manganese	26 **Fe** Iron	27 **Co** Cobalt	28 **Ni** Nickel	29 **Cu** Copper $[Ar]4s3d^{10}$	30 **Zn** Zinc						
4p														31 **Ga** Gallium	32 **Ge** Germanium	33 **As** Arsenic	34 **Se** Selenium	35 **Br** Bromine	36 **Kr** Krypton
5s	37 **Rb** Rubidium	38 **Sr** Strontium	4d	39 **Y** Yttrium	40 **Zr** Zirconium	41 **Nb** Niobium $[Kr]5s4d^4$	42 **Mo** Molybdenum $[Kr]5s4d^5$	43 **Tc** Technetium	44 **Ru** Ruthenium $[Kr]5s4d^7$	45 **Rh** Rhodium $[Kr]5s4d^8$	46 **Pd** Palladium $[Kr]4d^{10}$	47 **Ag** Silver $[Kr]5s4d^{10}$	48 **Cd** Cadmium						
5p														49 **In** Indium	50 **Sn** Tin	51 **Sb** Antimony	52 **Te** Tellurium	53 **I** Iodine	54 **Xe** Xenon
6s	55 **Cs** Cesium	56 **Ba** Barium	5d	71 **Lu** Lutetium	72 **Hf** Hafnium	73 **Ta** Tantalum	74 **W** Tungsten	75 **Re** Rhenium	76 **Os** Osmium	77 **Ir** Iridium	78 **Pt** Platinum $[Xe]6s4f^{14}5d^9$	79 **Au** Gold $[Xe]6s4f^{14}5d^{10}$	80 **Hg** Mercury						
6p														81 **Tl** Thallium	82 **Pb** Lead	83 **Bi** Bismuth	84 **Po** Polonium	85 **At** Astatine	86 **Rn** Radon
7s	87 **Fr** Francium	88 **Ra** Radium	6d	103 **Lr** Lawrencium $[Rn]7s^25f^{14}7p$	104 **Rf** Rutherfordium	105 **Db** Dubnium	106 **Sg** Seaborgium	107 **Bh** Bohrium	108 **Hs** Hassium	109 **Mt** Meitnerium	110 **Ds** Darmstadtium	111 **Rg** Roentgenium	112 **Cn** Copernicium						
7p														113 **Nh** Nihonium	114 **Fl** Flerovium	115 **Mc** Moscovium	116 **Lv** Livermorium	117 **Ts** Tennessine	118 **Og** oganesson

Lanthanides (rare earths) 4f	57 **La** Lanthanum $[Xe]6s^25d$	58 **Ce** Cerium $[Xe]6s^24f5d$	59 **Pr** Praseodymium	60 **Nd** Neodymium	61 **Pm** Promethium	62 **Sm** Samarium	63 **Eu** Europium	64 **Gd** Gadolinium $[Xe]6s^24f^75d$	65 **Tb** Terbium	66 **Dy** Dysprosium	67 **Ho** Holmium	68 **Er** Erbium	69 **Tm** Thulium	70 **Yb** Ytterbium
Actinides 5f	89 **Ac** Actinium $[Rn]7s^26d$	90 **Th** Thorium $[Rn]7s^26d^2$	91 **Pa** Protactinium $[Rn]7s^25f^26d$	92 **U** Uranium $[Rn]7s^25f^36d$	93 **Np** Neptunium $[Rn]7s^25f^46d$	94 **Pu** Plutonium	95 **Am** Americium	96 **Cm** Curium $[Rn]7s^25f^76d$	97 **Bk** Berkelium	98 **Cf** Californium	99 **Es** Einsteinium	100 **Fm** Fermium	101 **Md** Mendelevium	102 **No** Nobelium

Appendix I
Statistical Mechanics Equations

The equations in this appendix are discussed and/or derived in Chapter 10. Blackbody radiation is also discussed in Section 3.4.

Definitions

Name	Symbol	Definition
Binomial coefficient	$\binom{n}{m}$	$\dfrac{n!}{m!\,(n-m)!}$ The number of ways to choose a group of m items from a list of n distinct items
Boltzmann's constant	k_B	1.38×10^{-23} J/K or 8.62×10^{-5} eV/K
Density of states	$g(E)$	The number of states per unit energy
Entropy	S	$k_B \ln \Omega$
Heat	Q	Energy transferred spontaneously from a hot system to a cold system When heat Q flows into a system, $\Delta S \geq Q/T$.
Heat capacity	C	$\dfrac{dE}{dT}$ (taking dE to be the heat input, not work) Heat capacity is written C_P if heat is added at constant pressure, and C_V if it's added at constant volume. In general, $C_P > C_V$.
Ideal gas		A gas of non-interacting, free molecules
Multiplicity	Ω	The number of microstates associated with a macrostate
Temperature	T	$\dfrac{1}{dS/dE}$ (assuming dE is only heat and there are no other changes)
Work	W	Any energy transfer other than heat (e.g. via macroscopic forces)

The Boltzmann Distribution

For a system in equilibrium with a large reservoir at temperature T, the probability of its being in a given microstate with energy E is

$$P = \frac{1}{Z} e^{-E/(k_B T)}.$$

The "partition function" Z is defined to normalize the probabilities:

$$Z = \sum_{\text{all microstates}} e^{-E/(k_B T)} = \sum_{\text{all energies}} \Omega(E) e^{-E/(k_B T)}.$$

The Equipartition Theorem

If the energy of a system in equilibrium with a large reservoir at temperature T depends quadratically on a continuous degree of freedom (e.g. x, y, v_x, ω_x, ...), the average thermal energy associated with that degree of freedom is $(1/2)k_B T$.

- If the degree of freedom is quantized, the equipartition theorem still holds in the limit where the spacing between energy levels is small compared to the system's energy.

- Even when the conditions for the equipartition theorem don't apply, it is usually still true that the thermal energy of a single particle, atom, or molecule is of order $k_B T$, provided the density of states is relatively uniform and the spacing between available energy levels is much smaller than $k_B T$.

- Common examples of the equipartition theorem include:
  - A free, classical particle in three-space has a translational kinetic energy that depends quadratically on three degrees of freedom (v_x, v_y, and v_z), so its average thermal energy is $(3/2)k_B T$.
  - A monatomic atom (He) has no rotational degrees of freedom. A diatomic molecule (H_2) has two, so $\text{KE}_{\text{rot}} = k_B T$. A polyatomic molecule (CO_2) has three, so $\text{KE}_{\text{rot}} = (3/2)k_B T$. (Molecular vibrations are typically frozen out at room temperature.)
  - A one-dimensional simple harmonic oscillator has two quadratic degrees of freedom, $E = (1/2)mv^2 + (1/2)kx^2$, so its average thermal energy is $k_B T$.

Quantum Statistics

For a system of identical particles in equilibrium with a large reservoir at temperature T, the average occupation number $\bar{n}$ of a microstate is given by

$$\text{The Fermi–Dirac distribution (for fermions): } \bar{n} = \frac{1}{e^{(E-\mu)/(k_B T)} + 1}$$

$$\text{The Bose–Einstein distribution (for bosons): } \bar{n} = \frac{1}{e^{(E-\mu)/(k_B T)} - 1}$$

The "chemical potential" μ is defined to normalize the total occupation number $\left(\sum \bar{n} = N \right)$.

- For massless particles such as photons, $\mu = 0$.
- For fermions, $\bar{n}$ is greater than $1/2$ for energies below μ, and less than $1/2$ for higher energies.
- For bosons, μ is always below the ground state energy.

The Blackbody Spectrum

The energy density of radiation inside an enclosed cavity in equilibrium is the integral of the "Planck spectrum," which can be expressed in terms of frequency or wavelength, $\rho = \int_0^\infty u(\nu)d\nu = \int_0^\infty u(\lambda)d\lambda$:

$$u(\nu) = \frac{8\pi h}{c^3} \frac{\nu^3}{e^{h\nu/(k_B T)} - 1}$$

$$u(\lambda) = \frac{8\pi hc}{\lambda^5} \frac{1}{e^{hc/(k_B T\lambda)} - 1}$$

The intensity (energy per time per surface area) emitted by a blackbody is $c/4$ times the energy density of radiation in a cavity:

$$I = \frac{2\pi h}{c^2} \int_0^\infty \frac{\nu^3}{e^{h\nu/(k_B T)} - 1} d\nu = 2\pi hc^2 \int_0^\infty \frac{1}{\lambda^5 \left(e^{hc/(k_B T\lambda)} - 1\right)} d\lambda = \sigma T^4.$$

- Wien's law for peak frequency: $h\nu_{\text{peak}} = 2.82\, k_B T$. (For wavelength, $hc/\lambda_{\text{peak}} = 4.97\, k_B T$.)
- Stefan's law for total intensity: $I = \sigma T^4$, where $\sigma = 5.67 \times 10^{-8}\, \text{W}/(\text{m}^2\text{K}^4)$.

The Maxwell Speed Distribution

The probability that a (non-relativistic) molecule of mass m in an ideal gas at temperature T has speed in the range $v_1 < v < v_2$ is

$$P = 4\pi \left(\frac{m}{2\pi k_B T}\right)^{3/2} \int_{v_1}^{v_2} v^2 e^{-mv^2/(2k_B T)} dv.$$

Appendix J
The Standard Model of Particle Physics

The material in this appendix is discussed in Chapter 13.

Figure J.1 shows the particles that make up the standard model of particle physics. Detailed properties of the particles are given on p. 729.

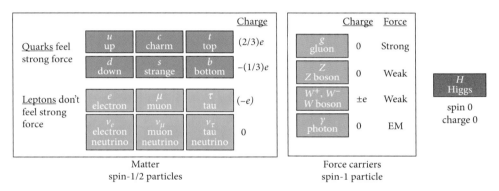

Figure J.1 The standard model.

Notes

- Every quark has a flavor (the six types listed in the table) and a color (not shown in the table). For instance, an up quark can be red, blue, or green.

- Each quark and each lepton has an antiparticle, not shown in the table. For instance, an antiup quark has charge $-(2/3)e$, and its color can be antired, antiblue, or antigreen.

- Each column of quarks and leptons is called a "family." The particles in the left-hand family are the lightest; the right-hand family is the heaviest.

- Hadrons are particles made of quarks. This includes protons (uud) and neutrons (udd). Thus all atoms are made from particles in the first column above.

- A baryon is made of three quarks (red, green, and blue). An antibaryon is made of three antiquarks (antired, antigreen, and antiblue). A meson is made of one quark and one antiquark, of opposite colors.

Forces

	Range	Relative strength	Characteristic time scale	Force carriers
Strong	10^{-15} m	1	$< 10^{-22}$ s	g
Electromagnetism	Long-range	10^{-2}	$10^{-20} - 10^{-14}$ s	γ
Weak	10^{-18} m	10^{-8}	$10^{-13} - 10^{-8}$ s	Z, W^+, W^-
Gravity	Long-range	10^{-39}	n/a	graviton (theorized)

The quantities in this table don't all have precise definitions, but these values give a good general idea of the properties of the four forces. All particles feel gravity and all known particles feel weak interactions. Electromagnetism is felt by charged particles. The strong force is felt by quarks and gluons.

Some Conserved Quantities (Exact and Approximate)

Conserved quantity	Notes
Energy	Implies that a particle can only decay into lighter particles
Momentum	
Angular momentum	Implies that spin is conserved in decays
Charge	
Color	Since gluons carry a color and an anticolor, a quark can change color by absorbing or emitting a gluon.
Baryon number	Grand unified theories (GUTs) imply that baryon number and lepton number aren't conserved, but their sum is.
Lepton number	No experimental violation of baryon or lepton number has been observed.
The quantities below are observed to be conserved in some types of interactions and not in others.	
Quark flavor	Conserved in strong but not weak interactions Includes strangeness, charm, etc.
Lepton number within family	Violated by neutrino oscillations

Particle Properties

	Name	Symbol	Mass (MeV/c^2)	Charge	Spin
Quarks	down	d	4.6	$-1/3$	1/2
	up	u	2.2	2/3	1/2
	strange	s	96	$-1/3$	1/2
	charm	c	1280	2/3	1/2
	bottom	b	4180	$-1/3$	1/2
	top	t	173,000	2/3	1/2
Leptons	electron	e	0.511	-1	1/2
	electron neutrino	ν_e	??	0	1/2
	muon	μ	106	-1	1/2
	muon neutrino	ν_μ	??	0	1/2
	tauon	τ	1777	-1	1/2
	tau neutrino	ν_τ	??	0	1/2
Force carriers	gluon	g	0	0	1
	Z boson	Z	0	0	1
	W boson	W^+, W^-	80.4	± 1	1
	photon	γ	0	0	1
Higgs	Higgs boson	H (or H^0)	125	0	0

- Mass in particle physics is usually given as mass energy in electron volts. Saying the electron's mass is 0.5 MeV/c^2 is equivalent to saying it has a mass of roughly 9×10^{-31} kg.
- The quark masses are approximate since free quarks are never observed. Because of virtual particles, the mass of a hadron is always greater than the sum of the masses of its constituent quarks.
- As of this writing in 2021, the neutrino masses are not known, but they are not zero. The electron neutrino is known to have mass less than 2.2×10^{-6} MeV/c^2, the muon neutrino less than 0.170 MeV/c^2, and the tau neutrino less than 15.5 MeV/c^2.
- Charge is in units of proton charge, or 1.6×10^{-19} C.

Index

Constants

Fundamental Constants

Speed of light	$c = 3.000 \times 10^8$ m/s
Planck's constant	$h = 6.626 \times 10^{-34}$ m^{2}kg/s (or J s)
	$= 4.136 \times 10^{-15}$ eV s
Reduced Planck's constant	$\hbar = h/(2\pi) = 1.055 \times 10^{-34}$ m^{2}kg/s (or J s)
	$= 6.582 \times 10^{-16}$ eV s
Permittivity of free space	$\epsilon_0 = 8.854 \times 10^{-12}$ s^{4}A^2/(m^{3}kg) (or C^2/(J m))
Permeability of free space	$\mu_0 = 1.257 \times 10^{-6}$ kg m/(s^{2}A^2)
Boltzmann's constant	$k_B = 1.381 \times 10^{-23}$ J/K
	$= 8.617 \times 10^{-5}$ eV/K
Electron mass	$m_e = 9.109 \times 10^{-31}$ kg
	rest energy $= 5.110 \times 10^5$ eV
Proton mass	$m_p = 1.673 \times 10^{-27}$ kg
	rest energy $= 9.383 \times 10^8$ eV
Neutron mass	$m_n = 1.675 \times 10^{-27}$ kg
	rest energy $= 9.396 \times 10^8$ eV
Electron charge	$e = -1.602 \times 10^{-19}$ C
Avogadro's number	$N_A = 6.022 \times 10^{23}$

Derived Constants

Coulomb's law constant	$\dfrac{1}{4\pi\epsilon_0} = 8.988 \times 10^9$ J m/C$^2 = 5.610 \times 10^{28}$ eV m/C^2
Compton wavelength of electron	$\dfrac{h}{m_e c} = 2.426 \times 10^{-12}$ m
Bohr radius	$a_0 = \dfrac{4\pi\epsilon_0 \hbar^2}{m_e e^2} = 5.292 \times 10^{-11}$ m